Mammal Community Dynamics
Management and Conservation in the Coniferous Forests
of Western North America

Conservation of mammals in the coniferous forests of western North
America has shifted in recent years from species-based strategies to
community- and ecosystem-based strategies. This paradigm shift has
resulted in an increasing number of ecosystem-level studies and
conservation plans that have yielded information on mammalian
communities and generated interest in their management. This book
focuses on contemporary forest mammals and their associated
ecosystems, providing a synthesis of the published literature on the role
of forest mammals in community structure and function, with emphasis
on their management and conservation. There are three main sections in
the book: the effects of forest management on various taxonomic groups;
community and ecosystem relations; and conservation issues and
strategies. In addition to coverage of some of the charismatic megafauna
such as grizzly bears, gray wolves, mountain lions, elk, and moose, the
book also provides a thorough treatment of small terrestrial mammals,
arboreal rodents, bats, medium-sized carnivores, and ungulates. The
unique blend of theoretical and practical concepts makes this book
suitable for college courses, researchers and their students, as well as for
wildlife and forest managers and land-use policy-makers. Managers,
educators, and research biologists will find it a valuable reference to the
recent literature on a vast array of topics on mammalian ecology.

CYNTHIA J. ZABEL is a Research Wildlife Biologist with the US Forest
Service and adjunct Professor at Humboldt State University.

ROBERT G. ANTHONY is Leader (Wildlife Program), Oregon
Cooperative Fish and Wildlife Research Unit, US Geological Survey, and
Professor of Wildlife Ecology, Oregon State University.

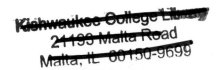

Dedication

To Don and Barbara Zabel for their love, support, and understanding of CJZ's busy career and to Alan Borchardt for his encouragement and patience while CJZ worked long hours into the night for several years to complete this project. Their support made this project possible.

To Reed D. (deceased) and Mary L. Anthony for their love and encouragement throughout childhood and RGA's professional career even when it resulted in living apart for much of our lives. To Libby Bailey, partner in life, for her understanding and patience while RGA worked long hours in order to complete this project. Her encouragement and inspiration made the project that much easier.

Mammal
Community
Dynamics

Management and Conservation in the Coniferous Forests of Western North America

Edited by

CYNTHIA J. ZABEL
USDA Forest Service

ROBERT G. ANTHONY
USDI Geological Survey

PUBLISHED BY THE PRESS SYNDICATE OF THE UNIVERSITY OF CAMBRIDGE
The Pitt Building, Trumpington Street, Cambridge, United Kingdom

CAMBRIDGE UNIVERSITY PRESS
The Edinburgh Building, Cambridge CB2 2RU, UK
40 West 20th Street, New York, NY 10011-4211, USA
477 Williamstown Road, Port Melbourne, VIC 3207, Australia
Ruiz de Alarcón 13, 28014 Madrid, Spain
Dock House, The Waterfront, Cape Town 8001, South Africa

http://www.cambridge.org

First published 2003

Printed in the United Kingdom at the University Press, Cambridge

Typefaces Lexicon No. 2 9/13 pt. and Lexicon No. 1   *System* LʌTₑX 2ₑ   [TB]

A catalog record for this book is available from the British Library

Library of Congress Cataloging in Publication data

Mammal community dynamics : management and conservation in the coniferous
forests of western North America / editors, Cynthia J. Zabel, Robert F. Anthony.
     p.   cm.
Includes bibliographical references and index.
ISBN 0 521 81043 4 (hardback) – ISBN 0 521 00865 4 (paperback)
1. Mammals – Ecology – West (US)   2. Mammals – Ecology – Canada, Western.
3. Mammals – Effect of forest management on – West (US)   4. Mammals –
Effect of forest management on – Canada, Western.   5. Wildlife conservation –
West (US)   6. Wildlife conservation – Canada, Western.   I. Zabel, Cynthia J.,
1952– .   II. Anthony, Robert F., 1944– .
QL715 .M26   2003
599.173′0978 – dc21   2002072898

ISBN 0 521 81043 4 hardback
ISBN 0 521 00865 4 paperback

Contents

Part III
Conservation issues and strategies

Contributors

Cynthia J. Zabel
USDA Forest Service
Redwood Sciences Laboratory
1700 Bayview Drive
Arcata, California 95519
USA

Robert G. Anthony
US Geological Survey
Oregon Cooperative Fish and Wildlife Research Unit
Oregon State University
104 Nash Hall
Corvallis, Oregon 97331-3803
USA

Jack Ward Thomas
Boone and Crockett Professor of Wildlife Conservation
University of Montana
Missoula, Montana 59812
USA

Keith B. Aubry
USDA Forest Service
Pacific Northwest Research Station
3625-93rd Avenue SW
Olympia, Washington 98512
USA

Brian L. Biswell
USDA Forest Service
Pacific Northwest Research Station
3625-93rd Avenue SW
Olympia, Washington 98512
USA

Rudy Boonstra
University of Toronto at Scarborough
Division of Life Sciences
Scarborough
Ontario
Canada M1C 1A4

Stan Boutin
Department of Biological Sciences
University of Alberta
Edmonton
Alberta
Canada T6G 2E9

R. Terry Bowyer
University of Alaska Fairbanks
Institute of Arctic Biology and
Department of Biology and Wildlife
Fairbanks
Alaska 99775
USA

Steven W. Buskirk
Dept. of Zoology and Physiology
University of Wyoming
Laramie, Wyoming 82071
USA

Efren Cázares
Department of Forest Science
Oregon State University
Corvallis
Oregon 97331
USA

Andrew W. Claridge
New South Wales National Parks and Wildlife Service
PO Box 2115
Queanbeya, NSW 2620, Australia

Norris L. Dodd
Arizona Game and Fish Department
PO Box 2326
Pinetop
Arizona 85935-2326
USA

James A. Estes
US Geological Survey
A316 E&MS Building
University of California
Santa Cruz, CA 95064
USA

James G. Hallett
School of Biological Sciences
Washington State University
PO Box 644236
Pullman, Washington 99164-4236
USA

John P. Hayes
Department of Forest Science
Oregon State University
201 Richardson Hall
Corvallis, Oregon 97331
USA

Miles A. Hemstrom
USDA Forest Service
ICBEMP, 4th Floor
PO Box 3623
Portland, Oregon 97208
USA

N. Thompson Hobbs
Natural Resources Ecology Laboratory
Colorado State University
Fort Collins
Colorado 80523
USA

Katherine M. Jacobs
Department of Forest Science
Oregon State University
Corvallis
Oregon 97331
USA

John G. Kie
USDA Forest Service
Pacific Northwest Research Station
1401 Gekeler Lane
La Grande, Oregon 97850
USA

Charles J. Krebs
University of British Columbia
Department of Zoology
6270 University Blvd
Vancouver
British Columbia
Canada V6T 1Z4

Kyran. E. Kunkel
University of Montana
Turner Endangered Species Fund
1123 Research Drive
Bozeman, Montana 59718
USA

Kevin P. Lair
University of Montana
School of Forestry
Wildlife Biology Program
Missoula
Montana 59812
USA

Timothy E. Lawlor
Dept. Biological Sciences
Humboldt State University
Arcata, California 95521
USA

Daniel L. Luoma
Department of Forest Science
Oregon State University
Corvallis, Oregon 97331
USA

Chris C. Maguire
Department of Forest Science
Oregon State University
201 Richardson Hall
Corvallis
Oregon 97331-7501
USA

Bruce G. Marcot
USDA Forest Service
602 SW Main Street
Suite 400
Portland, Oregon 97205
USA

Karl. J. Martin
Bureau of Integrated Science Services
Wisconsin Department of Natural Resources
Rhinelander, Wisconsin 54501
USA

William C. McComb
Department of Natural Resources Conservation
University of Massachusetts
Amherst, Massachusetts 01003
USA

L. Scott Mills
Wildlife Biology Program
School of Forestry
University of Montana
Missoula, Montana 59812
USA

Margaret A. O'Connell
Eastern Washington University
Biology Department and Turnbull Laboratory for
    Ecological Studies
258 Science
Cheney
Washington 99004-2440
USA

Michael M. Pollock
National Oceanic and Atmospheric Administration
Northwest Fisheries Research Center
2725 Montlake Blvd
Seattle
Washington 98112
USA

Michael K. Schwartz
University of Montana
School of Forestry
Wildlife Biology Program
Missoula
Montana 59812
USA

Anthony R.E. Sinclair
University of British Columbia
Centre for Biodiversity Research
6270 University Blvd
Vancouver
British Columbia
Canada V6T 1Z4

Francis J. Singer
US Geological Survey
Midcontinent Ecological Science Center
Natural Resource Ecology Laboratory
Colorado State University
Fort Collins, Colorado 80523-1499
USA

Winston P. Smith
USDA Forest Service
Pacific Northwest Research Station
Forestry Sciences Laboratory
Juneau, Alaska 99801-8545
USA

Kelley M. Stewart
University of Alaska Fairbanks
Institute of Arctic Biology and Department of Biology and Wildlife
Fairbanks
Alaska 99775
USA

David A. Tallmon
University of Montana
Division of Biological Sciences
Missoula
Montana 59812
USA

James M. Trappe
Department of Forest Science
Oregon State University
Corvallis
Oregon 97331
USA

Guiming Wang
Natural Resources Ecology Laboratory
Colorado State University
Fort Collins
Colorado 80523
USA

Jeffrey R. Waters
USDA Forest Service
Redwood Sciences Laboratory
1700 Bayview Drive
Arcata
California 95519
USA

Jerry O. Wolff
University of Memphis
Department of Biology
Memphis, Tennessee 38152
USA

William J. Zielinski
USDA Forest Service
Redwood Sciences Laboratory
1700 Bayview Drive
Arcata
California 95519
USA

Foreword

Until the late 1970s, biologists tended to deal with one species at a time in terms of consideration in forest management. By that time, the consequences and requirements of the plethora of environmental laws passed during that decade (and the regulations issued pursuant to those laws) were coming to bear through land-use planning and court actions.

Managers and biologists were suddenly faced with having to broaden their considerations to include the entire spectrum of wildlife species in both planning and management. One phase in the planning regulations developed by the US Forest Service to be responsive to the cumulative requirements of the National Environmental Policy Act, The Endangered Species Act, and the National Forest Management Act was to play a critical role in changing biologists' focus and activities. Those regulations required that "…viable populations of all native and desirable non-native vertebrates be maintained within the planning area." And, as a result, the world of forest managers and biologists was forever changed.

The first try at summarizing information on the relationship of the entire spectrum of wildlife species to forest habitats was developed in 1976 and published in 1979 (Thomas 1979). It was specific to the Blue Mountains of Oregon and Washington and was based on published information and expert opinion. At the time, the balance was tilted more toward expert opinion relative to hard data.

That effort was quickly emulated for a number of ecosystems in the United States and elsewhere in the world. But as new information and insights evolve from targeted research efforts, it is essential that such new knowledge be packaged in technically appropriate and user-friendly ways so that it may be brought to bear in management and to target new research efforts.

In the preface to the Thomas (1979; p. 6) publication there was a statement that is as true now as it was then and certainly applies to this book. "Perhaps the greatest challenge that faces professionals engaged in forest research and management is the organization of knowledge and insights into forms that can be readily applied ..." Cynthia Zabel and Robert Anthony have done a splendid job of taking on that challenge as it applies to the mammal communities of coniferous forests of western North America. They have recruited the outstanding authorities on the subject to address various critical aspects of how mammal communities in that region are organized, how they respond to various ecosystem components, and how they influence the ecosystems of which they are part.

Biologists working with managers in the coniferous forests of western North America will likely find this book critical to their efforts, as will those who will critique proposed management approaches and judge effects of management activities. Those managers and biologists who ignore this book will likely be found wanting by both peers and in myriad appeals and legal actions that plans and associated documents inevitably face.

Research scientists will find the inclusive state-of-knowledge helpful as they design new research to improve our understanding of the forest ecosystems in question. This effort simultaneously provides a platform for management decisions and a launching pad for the unending search for knowledge as to how forested ecosystems function. And, those that simply seek knowledge and understanding of how mammals live in and influence forested ecosystems will find this book both informative and a pleasure to read and ponder.

JACK WARD THOMAS

Literature cited

Thomas, J.W. 1979. Wildlife Habitats in Managed Forests: the Blue Mountains of Oregon and Washington. USDA Forest Service. Agriculture Handbook No. 533. Pacific Northwest Forest and Range Experiment Station. Portland, Oregon, pp. 104–127.

Acknowledgment

This book is the product of the work, support, and co-operation of many scientists, agencies, universities, and organizations. The project was initiated as a full-day symposium at the 2000 Annual Conference of The Wildlife Society in Nashville, Tennessee. We thank The Wildlife Society for hosting the symposium, which gave us a forum for beginning the process. We are grateful to our many authors for their hard work and dedication to the effort; we thank them for the contributions and for their patience in working with us co-operatively to complete this project. This book would not have been possible without their knowledge, expertise, and great writing skills. We give special thanks to the many reviewers of the draft chapters whose comments and suggestions greatly improved the quality of the book. The reviewers who gave freely of their time are listed. We thank our employing agencies for their support and encouragement while we worked on this project, specifically the USDA Forest Service, Pacific Southwest Research Station, Redwood Sciences Laboratory (CJZ) and USDI Geological Survey, Oregon Cooperative Fish and Wildlife Research Unit, Oregon State University (RGA). We both have benefited and learned from the creativity and hard work of our exceptional graduate students and employees over the years. Colleagues and friends who have inspired us and encouraged us are too numerous to name here. However, we would like to acknowledge James A. Estes (CJZ), Norman S. Smith (RGA), and E. Charles Meslow (RGA) for mentoring us over the years of our careers, teaching us to be critical thinkers, and for their friendship and encouragement. We thank James Estes for his contributions to the Introduction and Jack Ward Thomas for writing the Foreword. We are particularly indebted to Cambridge University Press for agreeing to publish the book and to their highly professional staff, particularly Ellen Carlin, Maria Murphy, and Carol Miller, for their technical advice, marketing skills, and editorial comments.

Reviewers

James K. Agee, University of Washington

Erik Beever, US Geological Survey, Forest and Rangeland Ecosystem Science Center

Paul Beier, Northern Arizona University

Louis Best, Iowa State University

John Bissonette, US Geological Survey, Utah Cooperative Fish & Wildlife Research Unit

Vernon Bleich, California Department of Fish & Game

Mike Bogan, US Geological Survey

Jeff Bowman, Carleton University

Mark Brigham, University of Regina

J. David Brittell, Washington Department of Fish & Wildlife

Fred Bunnell, University of British Columbia

Leslie Carraway, Oregon State University

David L. Certain, The Nature Conservancy of Texas

Wes Colgann III, Louisiana Tech University

James Estes, US Geological Survey

Kerry R. Foresman, University of Montana

Eric Forsman, USDA Forest Service

Russ Graham, Denver Museum of Nature and Science

James Hallett, Washington State University

Mark Harmon, Oregon State University

Greg Hayward, University of Wyoming

Larry Heaney, The Field Museum, Chicago, Illinois

Ed Heske, Illinois Natural History Survey

Thomas R. Horton, State University of New York

Larry Irwin, National Council for Air and Stream Improvement, Inc.

Kurt Jenkins, US Geological Survey, Forest & Rangeland Ecosystem Science Center

Jonathon Jenks, South Dakota State University

Tom Keter, USDA Forest Service

Gary M. Koehler, Washington Department of Fish & Wildlife

Paul R. Krausman, University of Arizona

Thomas H. Kunz, Boston University

Tim Lawlor, Humboldt State University
John F. Lehmkuhl, USDA Forest Service
Susan C. Loeb, USDA Forest Service
Chris C. Maguire, Oregon State University
William McComb, University of Massachusetts
Dale R. McCullough, University of California, Berkeley
Kevin McGarigal, University of Massachusetts
Kevin McKelvey, USDA Forest Service
Steve Mech, Albright College
Charles Meslow (retired), Corvallis, Oregon
Scott Mills, University of Montana
Dennis Murray, University of Idaho
Malcolm North, USDA Forest Service
Thomas O'Neil, consultant, Northwest Habitat Institute
Dave Perault, Lynchburg College
Daniel H. Pletscher, University of Montana
Roger Powell, University of Wyoming
Martin Raphael, USDA Forest Service
Bill Ripple, Oregon State University
Robert Rose, Old Dominion University
John Sawyer, Humboldt State University
Richard Schmitz, Oregon State University
John Stuart, Humboldt State University
Fred Wagner, Utah State University
Richard Waring, Oregon State University
John O. Whitaker Jr. (retired), Indiana State University
Jerry Wolff, University of Memphis

Part I

Management and conservation issues for various taxa

CYNTHIA J. ZABEL, ROBERT G. ANTHONY,
AND JAMES ESTES

1

Introduction and historical perspective

Conservation and management of mammals in western coniferous forest has been one of the important topics in ecology over the
last two decades. The level of attention that such species as the grizzly bear, lynx, gray wolf, elk, and bison have attracted, plus the time
and funds that have been expended on behalf of their management,
exemplifies this interest. Listings of various species as threatened or
endangered under the Endangered Species Act (ESA), particularly grizzly
bears and gray wolves, has stimulated much research, initiated management activities, and prompted many political and legal decisions concerning the management of mammalian species. During this time period, however, a paradigm shift has changed emphasis from single-species
conservation to conservation strategies for communities and ecosystems.
This paradigm shift and the many legal challenges to federal agency management and policy decisions have resulted in the development of several
ecosystem-level management plans, including the Northwest Forest Plan,
the Interior Columbia River Basin Ecosystem Management Project, the
Tongass National Forest Management Plan, the Sierra Nevada Ecosystem
Project, and the Southern California Forest-Urban Interface Plan (Johnson
et al. 1999). As a result of these research and conservation activities, increased information on mammalian communities, as well as interest in
their management, has become available.

The geographic focus of this book is from the east side of the Rocky
Mountains to the Pacific Ocean and from the southern United States to
northern Canada and Alaska. We choose this area primarily because of

the tremendous amount of research that has been conducted there over the last century. However, the autecology and community relations of species should be broadly applicable to other regions and countries. The vegetative communities within western coniferous forests varies from coastal redwoods in the west to juniper woodlands in the east and from mixed-conifer forests on southern mountain tops to isolated patches of spruce-fir forests on the arctic tundra (see Hemstrom 2003).

In the following chapters, authors were asked to review the state of our knowledge on the ecology and community relations of mammals in coniferous forests of western North America, summarize and critique various conservation or management strategies, and suggest future research topics. We hope that the book will provide timely and useful information on the ecology and management of mammals in these forests. We intend the book to be used as a reference for researchers and primary text for graduate courses in wildlife ecology and field biology. The targeted audience includes college students, researchers, academicians, and land managers; however, other interested groups may include policy makers, conservation organizations, and politicians from the West. To initiate this effort, we organized and chaired a full day symposium at The Wildlife Society annual meeting in Nashville, Tennessee in September, 2000 where most of the authors gave presentations on their respective chapters. Since most of the chapters are reviews and syntheses, most of the material has been published, but never integrated into one document. Most of the material in each of the chapters is available in published journals, theses, or government documents (general technical reports), the latter of which are more difficult for most people to access.

The book is organized into three parts. The first part begins with chapters on vegetation types that occur in western coniferous forests, and a chapter on historic changes in mammal distribution and the likely causes of these changes. The following six chapters each focus on a particular taxonomic group of mammals and how they may be affected by land management practices and natural perturbations. They include chapters on bats, terrestrial small mammals, arboreal rodents, small to medium-sized carnivores, large carnivores, and ungulates. The second part contains several chapters on the community relations of specific groups of mammals, their prey, and/or the vegetative community that they affect and vice versa, as well as ecosystem processes in these communities. Topics include the importance of coarse woody debris and micorrhizal fungi to mammals as cover and food, respectively. The influence of large trees and riparian systems on mammal community relations is also covered in this part along with the influence of ungulates on plants, community structure, and ecosystem processes in National Parks. Finally, there is a chapter on the role of lynx-hare cycles in forest community dynamics in boreal forests of Canada. The final part includes chapters that are theoretical by topic, but are related to or provide implications for conservation including habitat fragmentation and connectivity of suitable habitats for mammals. There are also chapters on the role of dispersal in colonization of fragmented landscapes by mammals and the functional diversity of mammals in coniferous forests. Many of the chapters in the book discuss how human intervention and natural disturbance have altered the landscape and how these changes have affected mammal communities. Lastly, we included a synopsis of the book which provides a perspective for future research and conservation.

Historical perspective

This volume focuses on contemporary forest mammals and their associated ecosystems. Forest ecosystems, like nature everywhere, are in the midst of immense change. Modern humans have altered landscapes and elevated rates of extinction to such degrees that neither biodiversity nor the ecosystem services upon which we depend can possibly be sustained (Wilson 1992, Vitousek et al. 1997). Most educated people, including many scientists, see these changes as being unprecedented; the inevitable consequence of technology, human population size, and a global economy. However, many dimensions to nature were changed in dramatic ways long before modern humans arrived on the scene. The New World, in particular, has been vastly altered by human activities since the end of the Pleistocene.

No one major group of organisms has changed more spectacularly during the late Pleistocene/Holocene than the New World mammals, especially the larger species. Two-thirds of the native mammalian megafauna in western North America disappeared abruptly about 13 000 years ago (Martin and Klein 1984). Besides the well-known sabertooth (*Smilodon fatalis*), the extinct species include bears, camels, wolves, horses, ground sloths, mammoths and mastadons, deer, lions, and cheetahs. The cause of this dramatic disappearance of large mammals has been the topic of extensive research and heated debate for more than three decades. The

Pleistocene–Holocene transition was a period of rapid climatic change that some believe must have contributed to the losses, if not caused them outright. This explanation, however, leaves several key facts unresolved. The Pleistocene–Holocene transition was only the last glacial–interglacial transition in a large series of such events that have occurred over the past five million or so years during the current glacial age (Pielou 1991). If climate change was the key driver of mammalian extinctions in the New World, we must reconcile why this didn't happen earlier. Paul Martin (1973) was the first to propose a role by aboriginal people. Modern humans crossed the Bering Land Bridge and invaded the New World at the end of the Pleistocene, bringing with them well developed hunting cultures. As human populations increased and spread across the New World, they encountered prey that were ecologically and behaviorally naive, thus supposedly hunting many of these species to extinction. The growing weight of evidence seems to support this hypothesis, at least in significant part. Similar patterns of megafaunal extinction following human invasions have been documented elsewhere in the world, especially in the Austral-New Zealand region and the Pacific islands (Diamond 1997, Duncan et al. 2002), and recent modeling efforts demonstrate that the observed pattern of large mammal extinctions in the New World was an almost inevitable consequence of human hunting (Alroy 2001). Furthermore, Martin and Szuter (1999) provide intriguing evidence that the distribution and abundance of the surviving large mammals were strongly influenced by pre-Columbian native cultures in western North America.

Regardless of the exact cause, there were pre-historic changes in distribution, abundance, and diversity of large mammals in forest ecosystems of western North America, and these changes almost certainly have had profound influences on associated species and ecosystems. This claim rests on the assumption that some, if not many species of mammals are strong interactors (Berlow et al. 1999, Paine 2000). Janzen and Martin (1982) were the first to propose such effects, arguing that the workings of neotropical forests changed markedly with the extinction of gomphotheres and other large, frugivorous mammals, and that these events and processes had been largely unappreciated or even ignored in a large body of ecological and evolutionary research on neotropical forest ecosystems. The same argument should hold for forest ecosystems in western North America. Terborgh et al. (1999) made this point for the megacarnivores by summarizing evidence that plant–herbivore relationships have been altered fundamentally by the recent extinctions of grizzly bears, wolves, felids, and perhaps other species of megacarnivores from much of their historic range in the New World. Intriguing hints of the breadth and complexity of these effects were provided by a number of recent studies (e.g., McLaren and Peterson 1994, Ripple and Larson 2000, Berger et al. 2001*a,b*, Terborgh et al. 2001). Thus, while this volume focuses primarily on forest ecosystems in today's world and the changes that have occurred in those systems during the past few decades, the deeper dimensions of time provide an essential context for the interpretation of these findings.

Literature cited

Alroy, J. 2001. A multi-species overkill simulation of the end-Pleistocene mass extinction. *Science* **292**:1893–1896.

Berger, J., P.B. Stacey, L. Bellis, and M.P. Johnson. 2001*a*. A mammalian predator–prey imbalance: grizzly bear and wolf extinction affect avian neotropical migrants. *Ecological Applications* **11**:967–980.

Berger, J., J.E. Swenson, and I.L. Persson. 2001*b*. Recolonizing carnivores and naïve prey: conservation lessons from Pleistocene extinctions. *Science* **291**:1036–1039.

Berlow, E.L., C.J. Briggs, M.E. Power, S.A. Navarrete, and B.A. Menge. 1999. Quantifying variation in the strengths of species interactions. *Ecology* **80**:2206–2224.

Diamond, J.M. 1997. *Guns, Germs, and Steel: the Fates of Human Societies*. Norton & Co., New York, New York, USA.

Duncan, R.P., T.M. Blackburn, and T.H. Worthy. 2002. Prehistoric bird extinctions and human hunting. *Proceedings of the Royal Society of London B* **269**:517–521.

Hemstrom, M.A. 2003. Forests and woodlands of western North America. Pages 9–40 *in* C.J. Zabel and R.G. Anthony, editors. *Mammal Community Dynamics. Management and Conservation in the Coniferous Forests of Western North America*. Cambridge University Press, Cambridge, UK.

Janzen, D.H. and P.S. Martin. 1982. Neotropical anachronisms: the fruits the gomphotheres ate. *Science* **215**:19–27.

Johnson, N.K., F.J. Swanson, M. Herring, and S. Greene. 1999. *Bioregional Assessments: Science at the Crossroads of Management and Policy*. Island Press, Washington DC and Covello, California, USA.

Martin, P.S. 1973. The discovery of America. *Science* **179**:969–974.

Martin, P.S. and R.G. Klein, editors. 1984. *Quaternary Extinctions: A Prehistoric Revolution*. University of Arizona Press, Tucson, Arizona, USA.

Martin, P.S. and C.R. Szuter. 1999. War zones and game sinks in Lewis and Clark's west. *Conservation Biology* **13**:36–45.

McLaren, B.E. and R.O. Peterson. 1994. Wolves, moose and tree rings on Isle Royale. *Science* **266**:1555–1558.

Paine, R.T. 2000. Phycology for the mammalogist: marine rocky shores and mammal-dominated communities: how different are the structuring processes? *Journal of Mammalogy* **81**:637–648.

Pielou, E.C. 1991. *After the Ice Age*. University of Chicago Press, Chicago, Illinois, USA.

Ripple, W.J. and E.J. Larsen. 2000. Historic aspen recruitment, elk, and wolves in northern Yellowstone National Park, USA. *Biological Conservation* **95**:361–370.

Terborgh, J., J.A. Estes, P.C. Paquet, K. Ralls, D. Boyd-Heger, B. Miller, and R. Noss. 1999. Role of top carnivores in regulating terrestrial ecosystems. Pages 39–64 *in* M.E. Soulé and J. Terborgh, editors. *Continental Conservation: Design and Management Principles for Long-term, Regional Conservation Networks*. Island Press, Washington DC and Corello, California, USA.

Terborgh, J., L. Lopez, P. Nuñez, M. Rao, G. Shahabuddin, G. Orihuela, M. Riveros, R. Ascanio, G.H. Adler, T.D. Lambert, and L. Balbas. 2001. Ecological meltdown in predator-free forest fragments. *Science* **294**:1923–1925.

Vitousek, P.M., H. A. Mooney, J. Lubchenco, and J. M. Melillo. 1997. Human domination of earth's ecosystems. *Science* **277**:494–499.

Wilson, E.O. 1992. *The Diversity of Life*. Harvard University Press, Cambridge, Massachusetts, USA.

2

Forests and woodlands of western North America

Introduction

Forests and woodlands of western North America vary from massive red-wood stands with lush herb and shrub understories to scattered juniper in near-desert environments; from pine and oak stands characteristic of semi-tropical Mexico to wind-blasted spruce on coastal headlands and isolated tree patches in Arctic tundra. Disturbance (including human activities), stand development, and succession create additional variation in each forest and woodland environment. Wildfire visits some environments infrequently, if ever, while others experience under-burns every 5–20 years. Boreal and moist coastal forests continuously blanket hundreds of thousands or millions of hectares while mountain tops in the Great Basin support a few hectares of trees kilometers from their nearest neighbors. Such tremendous forest variability provides habitat for a large number of mammal species. This chapter describes forests and woodlands, focusing on mature and older conditions, from the western edge of the Great Plains to the Pacific Ocean in Canada and the United States (Fig. 2.1). Discussion is organized using the geographic regions from Bailey (1995, 1998) aggregated to resemble ecological regions described by Barbour and Billings (2000). Readers interested in discussion of the causal environmental factors associated with plant community distribution should consult Bailey (1995, 1998) and Chabot and Mooney (1985).

For the purposes of this discussion, forests are composed of trees with crowns overlapping, generally producing $\geq 60\%$ vertically projected canopy cover (Anderson et al. 1998). Closed forests generally have $\geq 75\%$

canopy cover. Woodlands are composed of open stands of trees with crowns not touching, generally 25% to 60% projected canopy cover (Anderson et al. 1998). General descriptions of forests and woodlands draw from a variety of sources, perhaps the most comprehensive of which is Barbour and Billings (2000). Descriptions of mature and old forest and woodland structure draw from various sources, especially efforts by the USDA Forest Service to characterize old-growth forests (e.g., Jimerson et al. 1991*a*, *b*, Boughton et al. 1992, Green et al. 1992, Hamilton 1993, USDA Forest Service 1993). Old-growth forests, in this context, are defined by a combination of structural and compositional characteristics that relate to ecosystem process and function, including wildlife habitat (Franklin et al. 1981).

Taiga and boreal forest

Tiaga and boreal forests extend across a large area from central Alaska well into British Columbia (Fig. 2.1). The climate is cool and moist with cold, snowy winters and only one month with an average temperature above 10°C. The relatively little precipitation is concentrated in the warm summer months (Bailey 1995, 1998). Winter is the predominant season, with subfreezing temperatures lasting six or seven months. Permafrost prevails under large areas. Soils are often wet, strongly leached, and acidic.

Boreal forests

Boreal forests of white spruce (*Picea glauca*), black spruce (*P. mariana*), paper birch (*Betula papyrifera*), balsam poplar (*Populus balsamifera*), and quaking aspen (*P. tremuloides*) dominate upland and lowland boreal in the far north (Elliot-Fisk 2000). Lichen woodlands, muskeg, and closed-canopy forest often alternate across landscapes depending on topography. Black spruce and white spruce are the dominant conifers. Black spruce becomes more important with increasing latitude. Large areas of boreal forests consist of spruce-feather moss communities (Elliot-Fisk 2000). Black spruce-feather moss (*Hylocomium splendens*) communities typically have a uniform tree stratum of moderate density with almost continuous bryophyte cover beneath. Conversely, the tree stratum of white spruce forests tends to be more irregular and open, with patches of broadleaf shrubs, dwarf shrubs, and patchy bryophytes. Plant species richness is often higher in white spruce forests.

Mature boreal forests are often dense (frequently >500 trees/ha) and composed of relatively small to moderately sized trees (Nienstaedt and

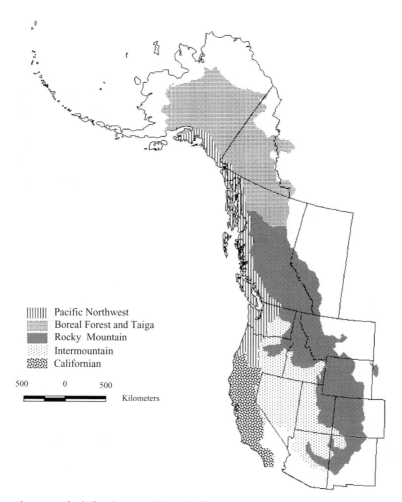

Fig. 2.1. Ecological regions containing significant area of forest or woodland in western North America.

Zasada 1990, Viereck and Johnston 1990, Boughton et al. 1992, Elliot-Fisk 2000). In general, white spruce are larger than black spruce. Mature white spruce average 60–90 cm diameter breast height (dbh – 1.4 m above the ground) and 30+ m tall while black spruce are considerably smaller (13–23 cm dbh and 8–20 m tall) in environments that are more productive. In less productive areas, upper-canopy white spruce may be less than 30 cm dbh and black spruce less than 20 cm dbh. Old-growth forests generally contain relatively larger trees and abundant down wood and snags. Old-growth black spruce forests are particularly dense, often with

$\geq$370 trees/ha over 13 cm dbh and numerous small snags and pieces of down wood.

Taiga

Taiga is typically dominated by black spruce, white spruce, birch (*Betula* spp.), quaking aspen, and the lichen *Stereocaulon paschale* (Elliot-Fisk 2000). Patches of conifers, lichens, shrubs, and herbs extend from the boreal forest to the northern tree limit, becoming open lichen woodlands and scattered trees at the tundra–forest ecotone (Elliot-Fisk 2000). The transition from boreal forests to lichen woodlands is sharp in some areas and gradual in others. While tree density in the taiga is less than in boreal forests, individual trees may be as tall as or taller than their boreal forest counterparts, with branches occurring to ground level (Elliot-Fisk 2000). Large trees may be >22 cm dbh and 10 m tall (Viereck et al. 1992). Some stands contain $\geq$200 trees/ha but tree density is highly variable and stands often patchy (Viereck et al. 1992).

Fire and permafrost largely determine vegetation pattern and succession. Fire is the natural dominant disturbance in most boreal forests (Elliot-Fisk 2000). Succession may generate trends counter to those typically found in forests elsewhere (Elliot-Fisk 2000). Tree survival often becomes more difficult as the canopy closes because the organic mat increases, nutrient availability decreases, summer soil temperature decreases, permafrost rises toward the surface, soil drainage decreases, and frost heave and frost thrust increase. Late-seral conditions vary from forest to muskeg or bog.

Pacific Northwest forests

The Pacific Northwest receives abundant rainfall from maritime polar air masses and experiences a relatively narrow annual temperature range (Bailey 1995, 1998). The average temperature of the warmest month is <22°C, but at least four months have an average temperature of $\geq$10°C. The average temperature of the coldest month is >0°C. Though precipitation occurs throughout the year, it is substantially reduced in summer. Summers are typically cool and often dry. Winters tend to be relatively mild with abundant rain or snow, depending on elevation and latitude. Maritime influence is high, including cool temperatures, high precipitation, and frequent fog. The mountainous portion generally experiences more variable climatic conditions than the lowlands. Considerable snow packs accumulate at mid and upper elevations. The growing season can be short. Sharp precipitation gradients often occur on the leeward side of mountain ranges. The coastal mountains receive 1530–3000 mm or more precipitation annually, with lesser amounts falling farther south (Bailey 1995, 1998). Soils are highly variable, reflecting diverse parent materials and topography (Franklin and Dyrness 1973). Heavy surface organic accumulations are common.

Douglas-fir and western hemlock forests

Douglas-fir (*Pseudotsuga menziesii* var. *menziesii*) and western hemlock (*Tsuga heterophylla*) forests dominate the mountains of the Coast Ranges, Olympic Mountains, and Cascade Mountains north of about 43°N latitude and occur from sea level to >900 m elevation depending on local conditions (Franklin and Dyrness 1973, Franklin and Halpern 2000). Tall, large, Douglas-fir, western hemlock, and western redcedar (*Thuja plicata*) tower above understories ranging from densely shrubby to herbaceous. Western hemlock forests north of about 54° latitude contain mountain hemlock (*Tsuga mertensiana*) and Sitka spruce (*Picea sitchensis*) rather than Douglas-fir (Little 1971, Boughton et al. 1992). Several deciduous tree species, especially red alder (*Alnus rubra*) and bigleaf maple (*Acer macrophyllum*) frequently contribute to diverse canopies, especially in disturbed or riparian areas.

Abundant large trees, snags and down logs provide diverse horizontal and vertical structure in unmanaged mature and old forests. Live trees are often >100 cm dbh and 50 m tall (Hermann and Lavender 1990, Spies and Franklin 1991, USDA Forest Service 1993, Franklin and Halpern 2000). Abundant large live trees, snags, down wood, and multiple canopy layers provide complex habitat structure for mammals in Douglas-fir and western hemlock old-growth forests (Franklin et al. 1981, USDA Forest Service 1993). Mature and old forests on moderately productive sites often include ≥400 trees/ha, ≥50 m²/ha live tree basal area, ≥10 snags/ha at ≥25 cm dbh, and ≥15 down logs/ha with ≥25 cm diameter at the large end. The understory of mature and old forests ranges from rich herbaceous carpets on moist sites to dense shrubs on drier sites. Fire is the principal natural disturbance, with return periods ranging from decades to several centuries (Franklin and Dyrness 1973, Hemstrom and Franklin 1982), though large areas have been logged over the last 100 years. Early seral forests vary considerably in structure, depending on the nature of disturbance events, and are often dominated by Douglas-fir and shrub thickets.

Mixed-conifer montane forests

Mixed-conifer montane forests similar to those found in the northern Sierra Nevada Mountains occur at mid-elevations in the southern Cascade Range and the eastern Siskiyou Mountains (Franklin and Halpern 2000). Several coniferous trees, including Douglas-fir, sugar pine (*Pinus lambertiana*), white fir (*Abies concolor*), ponderosa pine (*Pinus ponderosa* var. *ponderosa*), incense-cedar (*Calocedrus decurrens*), and Shasta red fir (*Abies magnifica* var. *shastensis*), commonly dominate these diverse forests. Forest composition varies considerably as a function of elevation, latitude, and local site conditions. Douglas-fir is characteristic of warmer, drier sites while white fir is common throughout. Understories range from herbaceous on moist sites to shrubby on drier sites. Shasta red fir, often mixed with mountain hemlock, indicates a transition to subalpine environments.

The structure of mature and old forests is similar to that of Pacific silver fir forests to the north and Californian mixed-conifer montane forests to the south. Jimerson et al. (1991*a*) describe Douglas-fir-dominated old-growth forests typical of mixed-conifer montane forests in the southern Cascades and Klamath Mountains. Scattered large, emergent conifers dominate a multiple-layered canopy. Stand basal area averages 62 m²/ha. On moderately productive sites, the majority of stand basal area is in trees that are >76 cm dbh, but most trees are smaller. Large snags (≥51 cm dbh and 5 m tall) and large down wood (≥51 cm diameter at the large end and 3 m long) are abundant.

Fire is the dominant natural disturbance. Many natural stands are composed of multiple-aged cohorts resulting from patchy, mixed-severity wildfires. Logging has altered stand structure in many areas.

Mixed evergreen forests

Mixed evergreen forests dominate highly variable montane environments in the Klamath Mountains (Franklin and Halpern 2000). Precipitation, temperature, and soil gradients are often sharp, reflecting diverse topography, geology, and degree of coastal influence. Forests typically consist of large conifers, especially Douglas-fir, emerging from a diverse canopy of evergreen hardwoods that often includes tanoak (*Lithocarpus densiflorus*), canyon live oak (*Quercus chrysolepis*), Pacific madrone (*Arbutus menzieseii*), and chinquapin (*Chrysolepis chrysopylla*). Understory vegetation can be densely shrubby. Sugar pine and ponderosa pine occur on drier sites. Port Orford cedar (*Chamaecyparis lawsoniana*) and Jeffrey pine (*Pinus jeffreyi*) are often present on soils developed from ultrabasic parent materials. California black oak (*Quercus kellogii*), Oregon white oak (*Q. garryana*), and canyon live oak are common at lower elevations on relatively dry sites. A large number of endemic or relict plant species occur.

Mature and old mixed evergreen forests typically contain multiple canopy layers, including several conifer and at least one hardwood species. Large, live conifers often exceed 80 cm dbh and 60 m in height (USDA Forest Service 1993). Smaller evergreen hardwoods are generally present in an intermediate canopy. Stand density is highly variable, generally 10–30 m²/ha of live tree basal area, and composed of relatively few large trees with or without thickets of smaller trees. Old-growth forests are often composed of scattered large, living trees and two or more canopy layers, scattered large snags, and abundant down wood (USDA Forest Service 1993). Natural stands are often of mixed age and patchy from multiple, mixed-severity wild fires. Much of the forest is young stands or openings from logging over the last 100 years or more.

Mountain hemlock

Mountain hemlock forests occur in cold, moist environments in southeastern Alaska, the British Columbia Coastal Mountains, the Olympic Mountains, the Cascade Mountains, and in the central and northern Sierra Nevada Mountains (Franklin and Halpern 2000). Winter snowfalls of 400–1400 cm are common and snow frequently accumulates to ≥7 m by the end of winter. At the highest elevations, mountain hemlock forests become increasingly mixed with alpine parklands. Pacific silver fir (*Abies amabilis*) is a common associate in maritime environments while subalpine fir (*Abies lasiocarpa*) and Engelmann spruce (*Picea engelmannii*) occur in colder, drier areas. Alaska yellowcedar (*Chamaecyparis nootkatensis*) is a common associate from northern Oregon northward.

Environmental conditions limit tree growth in mountain hemlock forests, but some species are long-lived and attain relatively large size (Means 1990, Diaz et al. 1997, Franklin and Halpern 2000). Large, mature trees often exceed 50 cm dbh and 30 m in height. Stand basal area is generally 30–80 m²/ha, often composed of many relatively small trees. Several canopy layers, abundant large trees, and common large snags characterize old-growth forests (USDA Forest Service 1993, Diaz et al. 1997). Most mature, unmanaged stands contain several large snags per hectare and several tonnes per hectare of large woody debris. Infrequent wildfire was the prevalent disturbance in natural conditions. Root rots form characteristic circular-patchy disturbances in some areas (Hemstrom and Logan 1986). Old-growth mountain hemlock forests at the northern end of their range in south central Alaska generally consist of smaller, more numerous trees (Boughton et al. 1992).

Pacific silver fir

Pacific silver fir forests characterize upper montane forests from the central Oregon Cascades through southern British Columbia and the Olympic Mountains (Franklin and Halpern 2000). Cooler temperatures and deeper, more persistent snow packs distinguish these forests from those of lower elevations. Tree species composition is variable, depending on latitude, elevation, and site conditions. Several conifer species, including Pacific silver fir, western hemlock, Douglas-fir, noble fir (*Abies procera*), western redcedar, and western white pine may be important. Lodgepole pine (*Pinus contorta*), subalpine fir, and Engelmann spruce often occur in subalpine forests at higher elevations or climates that are more continental. Alaska yellowcedar is common at upper elevations, especially to the north. Pacific silver fir and Mountain hemlock dominate cold, snowy subalpine forests in many places. Shasta red fir forms extensive stands at upper elevations in the southern Cascades and Klamath Mountains. Understory vegetation is highly variable, ranging from lush herbaceous cover on relatively moist sites to densely shrubby on drier or less fertile sites.

Large live trees tend to be smaller than those in Douglas-fir-dominated stands at lower elevations, but can be impressive nevertheless. Many stands contain live trees over 60 cm dbh and 50 m tall (Brockway et al. 1983, Hemstrom et al. 1987, Crawford and Oliver 1990, Franklin 1990, Franklin and Halpern 2000). Stand basal area often exceeds 100 m²/ha with more than 500 live trees per hectare. Snags and down wood are often abundant. Old-growth forests typically contain many trees ≥50 cm dbh, multiple canopy layers, and abundant snags and down logs (Hemstrom et al. 1987, USDA Forest Service 1993). Early seral forests vary in structure and are sometimes dominated by forbs or shrubs for several decades following disturbance (Franklin and Halpern 2000). Stand-replacing wildfires at several-century intervals were the prevalent natural disturbance, but extensive areas have recently been logged (Franklin and Dyrness 1973, Hemstrom and Franklin 1982). Snow breakage and wind commonly produce disturbances and regeneration patches of several hectares.

Sitka spruce-western hemlock forests

Sitka spruce-western hemlock forests dominate coastal lowlands in many places, often forming a belt a few kilometers wide along the coast (Franklin and Halpern 2000). Mixed forests of Sitka spruce and lodgepole pine (*Pinus contorta*) occur in the salt-spray zone near the ocean. Common associates include Douglas-fir, western redcedar, and, to the south, grand fir (*Abies grandis*) (Franklin and Halpern 2000). Deciduous tree species, such as red alder and bigleaf maple, frequently occur in disturbed areas, often forming large patches or mixed forests.

Stands often contain abundant large structures. Live trees frequently exceed 100 cm or, in some cases, 200 cm dbh and 50–70 m in height (Hemstrom and Logan 1986, Franklin and Halpern 2000). Stands can be dense in any seral stage. Mature and old forests usually contain multiple canopy layers, numerous large live trees per hectare, $\geq$50–80 m^2/ha of basal area, $\geq$20 large snags/ha, and $\geq$40 tonnes/ha of down wood (Hemstrom and Logan 1986, USDA Forest Service 1993).

Mature and old Sitka spruce forests at the northern end of their range in south central Alaska contain substantially smaller trees. Old-growth forests on productive alluvial sites typically include $\geq$59 trees/ha of $\geq$41 cm dbh, $\geq$7 snags/ha $\geq$41 cm dbh and $\geq$3 m tall, $\geq$5 pieces/ha of down wood $\geq$41 cm diameter at the largest point and 3 m long, and multiple canopy layers (Boughton et al. 1992). Old-growth western hemlock forests on well-drained sites in south central Alaska usually include Sitka spruce or mountain hemlock (Boughton et al. 1992). The canopy is multi-layered. Large trees average 21–36 m in height and 36–89 cm dbh, though the largest individuals may exceed 42 m in height and 140 cm dbh. Down woody debris is abundant, resulting from both wind throw and tree breakage. Large snags are common. Wind throw is the most common natural stand-replacing disturbance, though many areas have been logged.

Californian upland forests and woodlands

The Mediterranean climate of the Californian region has cool, wet winters during which more than 65% of the annual precipitation (averaging 275–900 mm) falls, and warm, dry summers (Bailey 1995, Barbour and Minnich 2000). Upper elevations (>1000 m in the north and 2000 m in the south) are too cool and moist to be classified as Mediterranean (Bailey 1995). The combination of wet winters and dry summers favors the development of lower-elevation forest and woodlands dominated by hard-leafed evergreen trees and shrubs capable of withstanding two to four rainless summer months and severe evapotranspiration demands (Bailey 1998).

Californian forest and woodland vegetation includes: (1) coast redwood forests, (2) northern oak woodlands, (3) southern and coastal oak woodlands, (4) mixed evergreen forests, (5) midmontane forests, (6) upper montane forests, (7) subalpine woodlands, and (8) Sierran east-side and Baja California montane forests (Barbour and Minnich 2000, Franklin and Halpern 2000).

Coast redwood forests

Coast redwood (*Sequoia sempervirens*) forests occupy a narrow strip along the coast in southern Oregon and northern California (Franklin and Halpern 2000). Redwood forests are restricted to a band generally ≤35 km wide that experiences frequent summer fog and abundant winter precipitation. Several other tree species frequently occur, including Douglas-fir, western hemlock, California myrtle (*Umbellularia californica*), and red alder on moist sites; Sitka spruce near the ocean; and Pacific madrone and incense-cedar on drier sites. Tanoak is a late-seral tree species throughout much of the redwood zone.

Massive and long-lived coast redwood characterizes the unique forest structure. The greatest forest biomass known has been measured in redwood forests (Franklin and Halpern 2000). Typical mature trees average >100 cm dbh and 75 m tall (USDA Forest Service 1993, Franklin and Halpern 2000). Some trees, especially redwood, may exceed 300 cm dbh and 100 m in height. Mature stands often consist of widely spaced redwood with a lower canopy of Douglas-fir, western hemlock, and other species. Unmanaged mature and old stands generally consist of ≥69 m²/ha basal area and 37 trees ≥101 cm dbh with abundant large snags and large down wood (Borchert 1992, USDA Forest Service 1993). Age structure analysis of cut-over stands indicates that, although redwood reproduces slowly, its reproductive rate is sufficient to perpetuate populations in the absence of fire and other disturbances (Franklin and Halpern 2000).

Northern oak woodlands

Northern oak woodlands, generally including blue oak (*Q. douglasii*), form a nearly continuous ring between 100 m and 1200 m elevation on foothills around the Central Valley (Barbour and Minnich 2000). Vegetation structure is typically two-layered. The overstory is usually 5–15 m tall with 10%

to 60% canopy closure of various oak species and California foothill pine (*Pinus sabiniana*). California buckeye (*Aesculus californica*), a shorter deciduous tree, often forms a second canopy layer. Mature trees range 10–15 cm dbh in dense stands and 15–30 cm dbh in more open conditions, generally with a combined basal area of 5–9 m²/ha.

Southern and coastal oak woodlands

Southern and coastal oak woodlands dominated by coast live oak (*Quercus agrifolia*) extend from about 30° to 40° N latitude in the California coast range (Barbour and Minnich 2000). The patchy overstory averages 9–22 m tall. Several other tree species may be present, including California walnut (*Juglans californica*) and Engelmann oak (*Q. engelmannii*). Stands range from being open and park-like (10% to 50% canopy cover) to being relatively dense.

Mixed evergreen forests

Mixed evergreen forests characterized by canyon live oak (*Q. chrysolepis*) occur at elevations above oak woodlands (Barbour and Minnich 2000). Common associates of canyon live oak include bigleaf maple, California myrtle, bigcone Douglas-fir (*Pseudotsuga macrocarpa*), and Coulter pine (*Pinus coulteri*). Coniferous overstory trees, when present, are generally scattered and 30–65 m tall. A closed canopy of broadleaf evergreen and deciduous trees, 15–30 m tall, lies beneath.

Douglas-fir characterizes mixed evergreen forests in the north Coast Ranges, western Klamath region, and the northwestern portion of the Sierra Nevada range (Barbour and Minnich 2000). Several other conifers may also be present, including white fir, incense-cedar, sugar pine, ponderosa pine, Pacific yew (*Taxus brevifolia*) and coast redwood. Broadleaf trees are generally present as well, especially Pacific madrone, canyon live oak, tanoak, coast live oak, chinquapin (*Chrysolepis chrysophylla*), interior live oak (*Quercus wislizenii*), and California bay (*Laura nobilis*). The canopy of mature and old forests often consists of a layer of emergent conifers, often >90 cm dbh, and a hardwood understory averaging <40cm dbh (Barbour and Minnich 2000). Mature and old Douglas-fir/tanoak/ Pacific madrone forests in northern California generally contain many conifers ≥76 cm dbh, an average of 64 m²/ha of live tree basal area, ≥6 snags/ha at ≥51 cm dbh, and abundant down wood (Jimerson et al. 1991*b*).

Farther south, mixed evergreen forests characterized by Coulter pine extend from about 38° N latitude in the California Coast Range into the Peninsular Mountains of Baja California (Barbour and Minnich 2000).

Coulter pine is particularly evident in the Transverse and Peninsular Ranges between 1200 m and 1800 m elevation. Several other conifer and hardwood tree species, such as bigcone Douglas-fir, Santa Lucia fir (*Abies bracteata*), ponderosa pine, canyon live oak, California black oak, and coast live oak, are often present, depending on site conditions and geographical location.

Midmontane forests

Midmontane forests are perhaps the most extensive forest type in California, occurring from about 800–2600 m elevation, depending on local conditions (Barbour and Minnich 2000). These forests have often been described as mixed-conifer, usually containing ponderosa pine with several other coniferous and hardwood tree species. Jeffrey pine, rather than ponderosa pine, dominates in the eastern and southern Sierra Nevada mountains and the Transverse and Peninsular ranges. Mixed-conifer forests typically contain four canopy layers at maturity: (1) conifers 30–60 m tall and often >1m dbh and 50% to 80% canopy cover, (2) a patchy subdominant tree layer 5–15 m tall usually consisting of deciduous broadleaf species, (3) a shrub canopy <2 m tall and contributing 10% to 30% canopy cover, and (4) a herbaceous layer with 5% to 10% cover (Barbour and Minnich 2000). Mature and old-ponderosa pine-dominated midmontane forests generally consist of a combination of large conifer ($\geq$76 cm dbh) and large hardwood ($\geq$38 cm dbh) trees averaging about 7 trees/ha total, $\geq$5 large snags/ha, and relatively little down wood (Smith et al. 1991). More mesic sites often support white fir forests that contain larger and more numerous live trees (mostly conifers), abundant large snags, and abundant down wood (Fites et al. 1992).

Upper montane forests

Upper montane forests generally contain lodgepole pine and several other associated species (Barbour and Minnich 2000). These forests occur at relatively high elevations, ranging from about 1800–2400 m in the north to 2200–3000 m in the south. Most of the precipitation falls as snow, generating snow packs of 2.5–4 m that last nearly 200 days/year. Barbour and Minnich (2000) recognize three major phases: (1) a lodgepole pine phase that is moderately dense (55% to 80% cover) and of modest stature (<20 m tall), (2) a red fir phase (*Abies magnifica*) on relatively mesic sites, containing trees 30–45 m tall with 60% average crown cover, and (3) a quaking aspen parkland phase in the Sierra Nevada Mountains where overstory trees may reach 65 cm dbh, 20 m in height, and provide an average of 60% canopy cover in clonal patches of at least several hectares.

Mature and old upper montane lodgepole pine forests on more productive sites develop from dense young stands following stand-replacement disturbances, generally fire (Potter et al. 1992a). Subsequent competition and mortality produce multi-layered, irregularly structured stands containing ≥50 live trees/ha of ≥60 cm dbh and ≥9 snags/ha, and scattered down wood. Large, live trees generally average about 33 m²/ha basal area. Large trees, snags and down wood are less abundant on poorer sites. Red fir forests tend to occur in environments that are more productive and contain more large trees, snags, and logs (Potter et al. 1992b). Large trees generally exceed 33 m in height and 76 cm dbh. Old-growth red fir stands often contain ≥71 large trees/ha, 7 snags/ha of ≥76 cm dbh, and 13 large pieces/ha of down wood.

The mixed subalpine forests and woodlands

The mixed subalpine forests and woodlands occur in more severe environments above upper montane forests (Barbour and Minnich 2000). Most of the precipitation occurs as snow that accumulates in deep and persistent snow packs. Elevations range from 1900 m in the Klamath region to 3500 m in the Transverse and Peninsular Ranges. Some individual trees may reach 25 m height, but the typical canopy is 10–15 m tall in patches of widely spaced individual trees averaging <40% canopy cover. Whitebark pine (*Pinus albicaulis*), limber pine (*P. flexilis*), mountain hemlock, western white pine, and foxtail pine (*P. balfouriana*) may be dominant, depending on conditions and geographical location. Whitebark pine and mountain hemlock are more important in the north, while limber pine dominates in the south.

Potter et al. (1992c) describe the structural characteristics of subalpine white pine, mountain hemlock, and quaking aspen old-growth forests that are representative of the lower-elevation, more productive portions of the subalpine zone. Subalpine white pine stands are often relatively dense with ≥51 trees/ha of ≥76 cm dbh and an average of 3 large snags/ha. Subalpine mountain hemlock forests occurring in the more northerly portions of California also tend to be relatively dense. There are generally ≥55 trees/ha of ≥76 cm dbh, ≥4 snags/ha of ≥76 cm dbh, and ≥9 pieces/ha of down wood of ≥64 cm diameter at the large end. Old-growth quaking aspen stands are usually mixtures of quaking aspen and conifers. Large quaking aspen (46–64+ cm dbh and 19 m tall) are relatively abundant (≥42 trees/ha) and usually accompanied by a few large conifers (≥16 trees/ha at ≥76 cm dbh). Most of the snags are quaking aspen (averaging ≥6 snags/ha of ≥46–64 cm dbh), but there are generally a few large (≥76 cm dbh) conifer snags per hectare. Down wood is often abundant, but mostly from small-sized quaking aspen trees.

East-side Sierran forests

East-side Sierran forests occur on the eastern slopes of Californian mountains, grading into conditions more typical of the Intermountain West. Sharp climatic gradients and discontinuous distribution characterize the subalpine-upper montane and midmontane forest zones (Barbour and Minnich 2000). The subalpine-upper montane zone lies between 2500 m and 2900 m elevation in the northern Sierra Nevada and between 2800 m and 3400 m in the south. Jeffrey pine often dominates both the upper montane and midmontane forests. Red fir is prominent in the subalpine-upper montane zone, often with lodgepole pine. Ponderosa pine replaces Jeffrey pine in the northernmost portions of the area. Midmontane forests often contain open, park-like stands of trees that may reach 40 m in height and 120 cm dbh. The understory typically contains shrubs more characteristic of the Intermountain region, including big sagebrush (*Artemisia tridentata*), Parry's rabbitbrush (*Chrysothamnus parryi*), goldenbush (*Haplopappus bloomeri*), antelope bitterbrush (*Purshia tridentata*), greenleaf manzanita (*Arctostaphylos patula*) and ceanothus (*Ceanothus* spp). Old-growth Jeffrey pine forests are often composed of widely spaced, large pine and, depending on disturbance history, an open understory or relatively dense conifer regeneration (Potter et al. 1992*d*).

Rocky mountain forests and woodlands

The southern Rocky Mountains consist of steep foothills, mountains, and deeply dissected high plateaus (Bailey 1998, 1995). Elevations range from about 1300–3000 m with some mountain peaks reaching >3400 m. The climate varies with elevation. Temperatures average about 4°C at upper elevations and increase to about 13°C in the foothills. Average annual precipitation ranges from approximately 260–890 mm (Bailey 1995). Mountains become more heavily glaciated to the north, where the highest peaks exceed 4000 m. Local relief often varies from 900 m to 2100 m. Annual temperatures average 2°C to 7°C in most of the area, reaching 10°C in the lowest valleys. Climatic conditions are strongly influenced by the prevailing winds and general north–south orientation of the mountain ranges. The base of mountain ranges may receive only 260–510 mm of precipitation per year while precipitation may exceed 1020 mm at higher elevations (Bailey 1995). Much of the precipitation falls as snow in high mountains, though permanent snowfields and glaciers are not common in most areas.

The most important vegetation gradients correspond to changes in elevation and topographic moisture (Peet 2000). Temperature and precipitation often change sharply with elevation. Peet (2000) illustrates this with data from a series of weather stations across an elevation gradient in the central Rocky Mountains, west of Boulder, Colorado. In this case, the mean annual temperature drops from 8.8°C at 1603 m elevation to −3.3°C at 3750 m. Monthly precipitation is highest in May and increases from 395 mm/year at 1603 m to 1050 mm/year at 3750 m elevation.

Ten vegetative types contain significant tree components: (1) riparian and canyon forests, (2) pygmy conifer (pinyon-juniper) woodlands, (3) ponderosa pine woodlands, (4) madrean pine-oak woodlands, (5) Douglas-fir forests, (6) Cascadian forests, (7) montane seral forests, (8) spruce-fir forests, (9) subalpine white pine forests, and (10) tree line vegetation (Peet 2000).

Riparian and canyon forests

Riparian and canyon forests are often dominated by various cottonwoods (*Populus* spp.) and willows (*Salix* spp.) with an increasing coniferous component at upper elevations (Peet 2000). Deciduous species often change with elevation. For example, eastern cottonwood (*P. deltoides*) and Fremont cottonwood (*P. fremontii*) may be replaced by narrowleaf cottonwood (*P. angustifolia*) at mid-elevations and by gray alder (*Alnus incana*), water birch (*Betula occidentalis*), and willows (*Salix* spp.) at higher elevations. Blue spruce, Rocky Mountain Douglas-fir (*Pseudotsuga menziesii* var. *glauca*), white fir, and other conifers are often important at middle elevations, giving way to subalpine fir and Engelmann spruce at upper elevations.

Pinyon-juniper woodlands

Pinyon-juniper woodlands, dominated by widely spaced junipers (*Juniperus* spp.) and pinyon (*Pinus* spp.), are widespread in the southern and central Rocky Mountains and the Intermountain West (Peet 2000). Utah juniper (*J. osteosperma*) dominates at the lowest elevations, with various pinyons assuming dominance at higher elevations. Rocky Mountain juniper (*J. scopulorum*) is more abundant, often with pinyons, at higher elevations and into the adjacent ponderosa pine woodlands. Alligator juniper (*J. deppeana*) and Mexican pinyon (*P. cembroides*) are common in the southern Rocky Mountains. Pinyon-juniper woodlands appear to have expanded significantly in recent times, at least partly as a result of domestic livestock grazing, invasion of exotic plants, and changes in fire regimes, perhaps confounded by climate change (Tausch et al. 1981, Gruell 1999, Tausch 1999).

Trees are generally short (seldom exceeding 7 m in height), with rounded crowns, multiple stems, and a shrub-like or multiple-stemmed appearance (Noble 1990, Ronco 1990, Peet 2000). Individual mature and old trees vary considerably in size, but are often ≥30 cm diameter at the root collar (drc). Relatively open stands with little understory are the rule, though there may be ≥500 small trees/ha. Old-growth woodlands consist of an open canopy, many trees ≥30 cm drc with relatively few snags and little down wood (Mehl 1992).

Ponderosa pine woodlands and forests occur at the lower elevational bounds of closed-canopy forests, often interlaced with pinyon-juniper woodland, from portions of Mexico to the dry interior valleys of southern British Columbia (Steele 1988, Peet 2000). Ponderosa pine forms forests with increasing elevation or site moisture, often becoming successional to Rocky Mountain Douglas-fir, grand fir, blue spruce, subalpine fir, or Engelmann spruce (Steele 1988, Peet 2000). Ponderosa pine woodlands become increasingly sparse at lower elevations, until only a few trees remain in rocky areas.

Two varieties of ponderosa pine correspond to important vegetation and environmental differences (Steele 1988). Pacific ponderosa pine (*Pinus ponderosa* var. *ponderosa*) occurs in the drier portions of a climatic regime characterized by cloudy winters with light precipitation and sunny summers with prolonged drought, extending from California through Oregon, Washington, British Columbia, and western Montana (Steele 1988). Rocky Mountain ponderosa pine (*P. ponderosa* var. *scopulorum*) is a dominant low-elevation forest species in the more continental climate of eastern Montana, Wyoming, North Dakota, South Dakota, Nebraska, Utah, Colorado, New Mexico, and Arizona. Rocky Mountain ponderosa pine becomes restricted to the fringes of the Rocky Mountains east of the continental divide and north of the Colorado–Wyoming border, but remain abundant in outlying ranges to the east and on rocky scarps in the western Great Plains. Rocky Mountain ponderosa pine may be the late-seral conifer species in significant areas at the lower forest edge in South Dakota, Arizona, and New Mexico (Steele 1988). Elsewhere, both varieties may be late-seral species on sites in the transition from forest and woodland to grassland.

Dramatic changes have occurred in the structure of many ponderosa pine forests and woodlands since Euro-American settlement (e.g., Covington and Moore 1994, Hann et al. 1997, Peet 2000). Much of the historical ponderosa pine woodland and forest area was open and park-like prior to Euro-American settlement. Frequent, low-intensity fires killed small trees and maintained open structures. Large, thick-barked trees were fire-resistant and often occurred in clumps or groups with grassy or forb-rich vegetation between clumps. Fire suppression, livestock grazing, timber harvest, and other changes to disturbance regimes frequently allowed the development of dense understories. A great deal of the area previously in open stands of large trees now contains abundant small trees and fine fuels. Fires have become infrequent and intense events that kill even most large trees. Livestock grazing, beginning in the latter half of the nineteenth century in most places, also contributed to altered fire regimes by removing the fine fuels that typically carried low-intensity ground fires.

Structural conditions in existing mature and late-successional ponderosa pine forest vary widely, depending on site environment, disturbance history, fire regime, and other factors (Oliver and Ryker 1990, Green et al. 1992, Peet 2000). Mature trees may be large, ranging about 76–127 cm dbh and 27–40 m tall, and tend to be widely spaced. Stand densities vary but most stands contain relatively dense conifer regeneration resulting from fire suppression. A few stands have relatively simple structures similar to those produced by historical disturbance regimes, with widely spaced large trees, sparse regeneration, and little down wood. Snags and down wood may be common to abundant in denser stands, where fires have been suppressed and dead wood has not been mechanically removed, but relatively uncommon in the more open, park-like stands. Typical multiple-story old-growth forest structure includes more than one canopy layer, abundant trees ≥50 cm dbh, numerous smaller trees, scattered snags ≥30 cm dbh, and at least some large down wood (Green et al. 1992, Mehl 1992, USDA Forest Service 1993).

Madrean pine-oak woodland

Madrean pine-oak woodland reaches its northern limit in the mountains of southern Colorado and northern New Mexico (Peet 2000). Various pine and oak species dominate species-rich woodlands that extend into Mexico. The northern fringes typically contain ponderosa pine with an understory of Gambel oak (*Quercus gambelii*), sumac (*Rhus* spp.), mountain mahogany (*Cercocarpus* spp.), and other shrubs and extend as far north as Colorado Springs, Colorado on the east side of the Rocky Mountains. Extensive shrublands, also dominated by Gambel oak, sumac, and other shrubs, exist in western Colorado and adjacent Utah, where they often merge with ponderosa pine woodlands or are successional to Rocky Mountain ponderosa pine, Rocky Mountain Douglas-fir, or white fir.

Madrean pine-oak woodlands grade from grassland fringes through savanna to closed woodland (Peet 2000). Gottfried et al. (1995) describe woodlands that contain several oak and pine species as relatively small, multiple-stemmed, irregularly formed trees intermingled with shrubs, grasses, forbs, and succulents. Tree densities vary from scattered individuals to several hundred stems per hectare.

Rocky Mountain Douglas-fir forests

Rocky Mountain Douglas-fir forests are found throughout the montane zone of the Rocky Mountains from central British Columbia into Mexico (Peet 2000). They are broadly distributed in the northern portions of this area, extending from lower tree line or ponderosa pine woodland up to the interface with spruce-fir forests of western Montana. Extensive Rocky Mountain Douglas-fir forests also occur on the western-side of the Rocky Mountains and, to a lesser extent, the eastern mountain ranges of the Great Basin (Little 1971). Rocky Mountain Douglas-fir is the characteristic species and is often found with other shade-intolerant conifers, such as lodgepole pine and ponderosa pine. On mesic, fertile slopes, Rocky Mountain Douglas-fir is often associated with or is successional to white fir or blue spruce (Peet 2000). Combinations of blue spruce and Rocky Mountain Douglas-fir often occur in relatively moist canyon bottoms.

Mature and old Rocky Mountain Douglas-fir forests usually consist of widely spaced, moderately sized large trees and variable amounts of snags and down wood. Understories may be filled with dense regeneration where fire has been absent, or relatively open where fire plays a more natural role. Old-growth forests typically include numerous trees ≥40–50 cm dbh, more than one canopy layer, variable numbers of snags, and scattered down wood (Green et al. 1992, Mehl 1992, USDA Forest Service 1993).

Historical disturbance regimes were characterized by infrequent stand-replacement wildfire with intervening under-burns. Fire suppression, timber harvest, and other activities have altered many stands by increasing the numbers of small trees and reducing the numbers of large trees and large snags. Fine fuels may be abundant.

Cascadian forests

Cascadian forests more typical of the Pacific Northwest are found as far east as the western slope of Glacier National Park, Montana (Peet 2000). Moist, cool maritime air masses penetrate inland in this area, bringing abundant rainfall and moderate temperatures. Several species may form dense forests, including western hemlock, western redcedar, Douglas-fir, grand fir, Pacific yew, mountain hemlock, and (at the upper tree line) alpine larch (*Larix lyallii*). At low elevations and on relatively moist sites, grand fir, western hemlock, and western redcedar dominate forests that are similar in structure and composition to those found on the western side of the Cascade Range in Oregon and Washington. These stands can be impressive, with abundant large trees, snags, and down wood. Grand fir is often the late-seral species on moderate sites environmentally between the moist western hemlock-western redcedar forests and relatively dry Douglas-fir forests. Lodgepole pine and, to a lesser degree, western larch (*Larix occidentalis*) may form dense early-seral stands. Western white pine forests, though formerly important, have been substantially reduced by an introduced blister rust (*Cronartium ribicola*).

Several different environments and forest structures exist. Forests characterized by Douglas-fir and grand fir are perhaps most extensive and representative. The structure of unmanaged mature and old western hemlock and western redcedar forests is reminiscent of Cascadian conditions, with abundant large trees, large snags, and down wood. Somewhat drier but more common Douglas-fir and grand fir forests typically contain numerous trees ≥ 51 cm dbh, basal area between 37 and 62 m²/ha, and abundant large snags (Green et al. 1992, Peet 2000). Old-growth forests have multiple canopy layers, abundant trees ≥ 53 cm dbh, abundant snags ≥ 36 cm dbh, and many pieces per hectare of down wood (Green et al. 1992, USDA Forest Service 1993).

Fire is the major natural disturbance, though insects and disease also generate extensive forest mortality (Hann et al. 1997). Mixed-severity fires produced patchy landscapes in most areas, sometimes occurring across hundreds or thousands of hectares. Fire suppression, logging and other activities have changed fire regimes by increasing fine fuels, lengthening the fire-return interval and increasing burn severity.

Seral forests

Seral forests dominated by lodgepole pine and quaking aspen are abundant in the Rocky Mountains. Lodgepole pine dominates many post-fire forests in the central and northern portions of the Rocky Mountains and forms dense stands, frequently seral to subalpine fir and Engelmann spruce, that may be both highly flammable and subject to insect outbreaks (Peet 2000). Mature stands often contain abundant small live trees and down wood. Quaking aspen is the most widely distributed tree species in North America and is the most important deciduous tree in the Rocky Mountains (Peet 2000). It occurs from near the arctic tree line south into Mexico. Quaking aspen generally regenerates from root suckers following fire, often forming extensive clonal patches. Quaking aspen stands tend to occur on fine textured, fertile soils, compared to lodgepole pine, although the ecological relationships between the two species are far from clear (Peet 2000). In addition to lodgepole pine and quaking aspen, several other species may play important seral roles, including western larch, western white pine, Douglas-fir, limber pine, and ponderosa pine.

Mature lodgepole pine stands often contain more than 1000 trees/ha (Lotan and Critchfield 1990, Green et al. 1992, Mehl 1992, USDA Forest Service 1993, Peet 2000). Individual trees tend to be relatively small (less than 30 cm dbh) and short (less than 25 m tall). Depending on site environment, a few to several dozen larger trees ($\geq$33 cm dbh) may be present on a typical hectare. Snags and down wood are often small and abundant. Stand-replacing wildfires, often in conjunction with insect epidemics at century or longer intervals, are the major natural disturbances (Mehl 1992, Schmidt and Amman 1992, Peet 2000).

Mature aspen stands consist of numerous large trees, generally $\geq$18–30 cm dbh and 20–25 m tall, above an understory that usually includes abundant small root-sprouts and often a lush herbaceous layer (Perala 1990, Peet 2000). Old-growth stands are often single-storied, with abundant trees $\geq$36 cm dbh, and relatively few snags and down wood pieces (Mehl 1992).

Engelmann spruce – subalpine fir

Engelmann spruce – subalpine fir forests occur throughout the Rocky Mountain subalpine zone from Mexico to the Yukon (Peet 2000). Subalpine fir and Engelmann spruce are the primary species and are genetically similar to, and may interbreed with, balsam fir (*A. balsamea*) and white spruce in the far north. The structure and composition of spruce-fir forests are remarkably consistent in the Rocky Mountains (Peet 2000). Open forests on slopes often have an understory of low-growing blueberries (*Vaccinium* spp.) and thick moss. Herbs are usually sparse in closed stands on drier sites. Lush herbaceous and shrubby layers may be present on moist sites.

Mature stands often consist of relatively abundant small fir trees and saplings with fewer, larger spruce. Though the relative successional status of the two species is in some doubt, subalpine fir appears to replace Engelmann spruce in the absence of disturbance (Peet 2000). Large trees, especially Engelmann spruce, are generally at least 30–80 cm dbh and 14–30 m tall (Alexander and Shepperd 1990, Alexander et al. 1990, Green et al. 1992, Peet 2000). Mature forests are often dense and irregularly patchy, containing more than 500 trees/ha and 38 m²/ha basal area. Snags tend to be relatively small and abundant. Old-growth forests are multi-layered, with abundant trees ≥40 cm dbh, and several to many snags and pieces of down wood per hectare (Mehl 1992). Infrequent, stand-replacing wildfires, often in association with insect epidemics, often generate large areas of relatively uniform forest (Swetnam and Lynch 1993, Hann et al. 1997), though old forests often contain frequent small openings due to disease, insect mortality, snow breakage, and wind throw.

Subalpine white pine forests

Subalpine white pine forests consisting of short, open-crowned, widely spaced Rocky Mountain bristlecone pine (*P. aristata*), Great Basin bristlecone pine (*P. longaeva*), whitebark pine, or limber pine with intermingled shrubs and grasses grow on exposed ridges and dry slopes in the subalpine zone (Peet 2000). Environments are typically severe with skeletal soils, short growing seasons, cold winters, and desiccating winter winds. Spruce or fir may become established under white pine clumps in less severe areas and grow through the canopy to replace the pines. Fires often burn uphill from adjacent spruce-fir forests, causing mortality in subalpine stands. Recovery in the harsh environment is slow as individual trees and small clumps regenerate in the most favorable micro-sites. Many whitebark pine stands in the central and northern Rockies have suffered a dramatic decline from an introduced fungal rust (*Cronartium ribicola*) (Hann et al. 1997, Peet 2000).

Whitebark pine is the dominant white pine in the central Rockies south through the Wind River Range of Wyoming (Peet 2000). Rocky Mountain bristlecone pine replaces whitebark pine in the southern Rocky Mountains. Great Basin bristlecone pine, a closely related species, usually replaces Rocky Mountain bristlecone pine on high peaks in the Great Basin. Both these latter species may live 2000–5000+ years, though Great Basin bristlecone pine is apparently longer-lived (Peet 2000). Southwestern white pine (*P. strobiformis*) is important in mixed subalpine forests farther south. Limber pine may form subalpine stands, but is more often found on lower-elevation dry ridges.

Most mature and old subalpine whitebark pine forests have experienced considerable recent blister-rust-related mortality. Consequently, snags are often abundant in unmanaged stands. Whitebark pine stands in eastern Montana illustrate typical structural characteristics (Green et al. 1992) with the largest trees averaging about 74 cm dbh. Stands are patchy and variable, consisting of 42–188 trees/ha over 33 cm dbh, basal area ranging 32–62 m²/ha and highly variable amounts of snags and down logs (Green et al. 1992).

Upper tree line vegetation

Upper tree line vegetation is generally wind-beaten, stunted woodland of spruce and fir. Rocky Mountain bristlecone pine, Great Basin bristlecone pine, limber pine, whitebark pine, alpine larch, or mountain hemlock may be locally important (Peet 2000). The elevation of the upper tree line varies significantly with latitude. In New Mexico, the upper tree line generally occurs at about 3800 m (Peet 2000), dropping to about 3000 m in parts of southern Montana, 2600 m in Glacier National Park, Montana, and <1500 m in the Logan Mountains of the Yukon Territory (Peet 2000).

Intermountain West

Pronounced drought and a relatively short period of more moist conditions characterize climatic conditions in the Intermountain West (Bailey 1995, 1998). Most of the precipitation falls in winter, despite a peak in May. The northerly portion of the area is semi-desert, with precipitation distributed through the year and a shorter summer dry period. In general, precipitation averages 130–490 mm and is lowest in the rain shadows of mountain ranges. Winters are often long and cold with frequent sub-freezing conditions. The Intermountain West consists of separate interior basins, only a few of which drain to the ocean. Saline or alkaline soils with shrubby vegetation exist in many lowland areas. Mountains rise sharply from semiarid plains covered with sagebrush (*Artemisia* spp.).

Temperate deserts of the Intermountain West consist mostly of various sagebrush vegetation types (Bailey 1995, West and Young 2000). Much of the native big sagebrush vegetation has been altered by livestock grazing, the invasion of exotic plants, and altered fire regimes. Higher elevations may support woodlands, forests, shrublands, or grasslands, depending on local conditions. Forests, if any, are generally sparse and open.

Pinyon-juniper woodlands

Pinyon-juniper woodlands generally occur above, and inter-fingered with, sagebrush at moderate elevations (Bailey 1995). Pinyon-juniper woodlands found in western Utah and much of Nevada are similar to those in the Rocky Mountain region, but are characterized by single-leaf pinyon (*Pinus monophylla*) rather than Colorado pinyon (*P. edulis*) (Hamilton 1993, Charlet 1996, West and Young 2000). Intermountain pinyon-juniper woodlands can be pure stands of either species or a mixture (Hamilton 1993). Pure juniper is found at elevations below the range of single-leaf pinyon. Single-leaf pinyon may be mixed with juniper, mountain shrub-lands, limber pine, Jeffrey pine, ponderosa pine, or white fir at upper elevations (Hamilton 1993). Pure single-leaf pinyon stands occur at higher elevations across western Nevada and eastern California. Junipers may be found in pure stands in the northern Great Basin and into adjacent Idaho, northern Utah, and western Wyoming. Single-leaf pinyon grows to a rela-tively large size on more moist and productive sites, but junipers are gen-erally small (Hamilton 1993).

Individual trees in pinyon-juniper woodlands are relatively small, gen-erally <50cm drc and 15 m tall (Dealy 1990, Noble 1990, Ronco 1990, Hamilton 1993, West and Young 2000). Stands are often composed of numerous, small, widely spaced trees. Old-growth stands on poor sites contain ≥30 trees/ha of ≥30 cm drc, with more than one tree-size class, and variable amounts of down wood and standing dead trees (USDA Forest Service 1992, Hamilton 1993). Trees are larger on better sites, where there are frequently ≥74 trees/ha of ≥46 cm drc, but other characteristics are similar (USDA Forest Service 1992, Hamilton 1993).

Ponderosa pine forests

Ponderosa pine forests and woodlands often occur at slightly higher elevations and extend to exposed slopes in the montane zone. Pacific ponderosa pine forests in the northern portion of the intermountain region are similar to those in northern Idaho, Montana, and portions of Washington, Oregon, and eastern California. Rocky Mountain ponderosa pine forests and woodlands in the central and southern intermountain region are similar to those in the central and southern Rocky Mountains (Hamilton 1993). Ponderosa pine is absent from, or occurs in isolated patches in, much of the interior intermountain region (Hamilton 1993, Charlet 1996). Where ponderosa pine does occur, it frequently begins as scattered individuals at lower elevations, often mixed with or adjacent to pinyon-juniper woodlands. Stands become denser with increasing elevation and moisture (Mehl 1992, Hamilton 1993, Peet 2000). Where wildfire has been suppressed, stands often consist of emergent large ponderosa pine with a thick understory of ponderosa pine, other conifers, or shrubs. On drier sties and where wildfire plays a relatively natural role, ponderosa pine forests are usually composed of widely spaced, relatively large trees in a park-like setting with few snags and little down wood (Harrington and Sackett 1992, Mehl 1992, Hamilton 1993, Covington and Moore 1994).

Individual mature trees in mature and old stands vary considerably in size, depending on site conditions (Hamilton 1993). Old-growth stands usually contain ≥ 35 trees/ha of ≥ 51 cm dbh on more productive sites and more than one canopy layer where fire has been suppressed. Snags >38 cm dbh are common to abundant, as are large pieces of down wood (Mehl 1992, Hamilton 1993). Stands are open and park-like where fire continues to play a natural role or on drier sites where ponderosa pine is a late-seral species. On these sites, old-growth stands typically include ≥ 17 trees of ≥ 41 cm dbh, ≥ 2 snags/ha at ≥ 36 cm dbh, sparse down wood, and one or more canopy layers (Hamilton 1993).

Rocky Mountain Douglas-fir

Rocky Mountain Douglas-fir forms montane and upper montane forests in portions of the intermountain region (Mehl 1992, Hamliton 1993, Peet 2000), but is absent from substantial areas in the isolated mountain ranges of the interior Great Basin (Charlet 1996). These forests are often associated with ponderosa pine at lower elevations and spruce-fir forests at upper elevations. Douglas-fir can be either an early or a late-seral species, depending on site conditions. Douglas-fir forests may be open and savannah-like on the driest sites (Hamilton 1993). Douglas-fir is often both the early- and late-seral dominant species on low-productivity dry or cold sites. It is also a major seral forest type in more moist environments, occasionally in association with lodgepole pine or aspen (Hamilton 1993, Peet 2000).

Mature forests generally consist of relatively widely spaced trees of ≥51 cm dbh, a few snags/ha of ≥51 cm dbh, and variable amounts of down wood (Hamilton 1993). Old-growth Douglas-fir forests on more productive sites generally include ≥37 trees/ha of ≥61 cm dbh, ≥2 snags/ha of ≥51 cm dbh and 6 m tall, at least a few pieces per hectare of down wood ≥30 cm diameter at the large end and 5 m long, and two or more canopy layers (Hamilton 1993). Trees are smaller, snags fewer, and down wood infrequent on poorer sites.

Subalpine fir and Engelmann spruce

Subalpine fir and Engelmann spruce (spruce-fir) forests are common in subalpine environments in portions of the intermountain region, but missing from substantial areas of the interior Great Basin mountains (Hamilton 1993, Charlet 1996, Peet 2000). Subalpine fir is absent from most of the interior Great Basin in Nevada, except for the northernmost portions, while Engelmann spruce is confined to extreme east-central Nevada (Charlet 1996). In fact, subalpine fir and Engelmann spruce may not co-occur in the basin and range mountains of Nevada (Charlet 1996). In the areas where they do co-occur, subalpine fir and Engelmann spruce form mixed forests or pure stands of either species. Pure stands of Engelmann spruce exist in some broad river valleys at relatively low elevations with cold air accumulations and high water tables. Pure subalpine fir stands are common at high elevations where moisture may be limiting (Hamilton 1993). Both species are susceptible to fire, insects, and disease, but Engelmann spruce typically lives longer. Insect-induced mortality or windthrow may generate abundant dead trees that fuel stand-replacing wildfire (Rebertus et al. 1992, Schmidt and Amman 1992, Hamilton 1993).

Individual large trees in mature and old spruce-fir stands typically exceed 30 cm dbh and 14 m in height, though Engelmann spruce may be considerably larger in moist areas. Old-growth forests on productive sites typically contain ≥60 trees/ha of ≥51 cm dbh, ≥2 snags/ha of ≥30 cm dbh and 4.5 m tall, 1 down log/ha of 30 cm diameter at the large end and

2.4 m long, and more than two canopy layers (Hamilton 1993). Old-growth forests in cold, dry environments have fewer, smaller trees, smaller snags, and often more down wood of smaller size.

Quaking aspen

While it is typically an early seral stage to conifers, quaking aspen forms apparently stable, regenerating stands in some areas of the Intermountain West (Hamilton 1993, Peet 2000). Aspen is an aggressive pioneer species in some areas, especially following wildfire, and may persist for hundreds of years in the absence of fire until it is replaced by shade-tolerant conifers (Hamilton 1993). Aspen stands often contain abundant conifers (commonly >10% of the tree canopy) (Hamilton 1993). In those areas where aspen forms long-term stable forests, conifers are generally less abundant or absent. Mature and old aspen stands frequently contain lush understories. Individual aspen trees seldom survive >200 years, but the underground clone from which they arise may live much longer. Large individual trees in mature and old stands are generally >18 cm dbh and 20 m tall. Stands may contain dense small-tree understories or not, depending on disturbance and site conditions. Old-growth stands on mesic sites typically consist of $\geq$50 trees/ha of $\geq$30 cm dbh, $\geq$5 snags/ha of $\geq$25 cm dbh and 4.5 m tall, and $\geq$25 pieces of down wood/ha $\geq$20 cm diameter at the large end and 3 m long (Hamilton 1993). Live trees are less abundant and often smaller on drier sites.

Subalpine white pine forests

Subalpine white pine forests, sometimes of gnarled Great Basin bristlecone pine over a thousand years old, occur at the upper timber line throughout the intermountain region (Hamilton 1993, Bailey 1995, Peet 2000). Such harsh, high-elevation sites often support nearly pure stands of Great Basin bristlecone pine, occasionally associated with Engelmann spruce or limber pine. Most stands are open and rocky with a few shrubs, grasses, and herbs between widely spaced trees (Hamilton 1993). Wildfire is uncommon due to low fuel amount and continuity, but seems responsible for stand replacement in some areas. Reforestation can be very slow. While Great Basin bristlecone pine may reach relatively large size under favorable conditions, most trees in mature and old stands are small (Hamilton 1993). Old-growth stands characteristically contain $\geq$14 trees/ha of $\geq$25 cm dbh with little standing or down dead wood (Hamilton 1993).

Limber pine is the most widely distributed high-elevation conifer in Nevada and occurs from about 1800–3500 m on dry, high-elevation ridges (Charlet 1996). It may form nearly pure, open stands or may be mixed with several other conifers, including Great Basin bristlecone pine, western white pine, Jeffrey pine, lodgepole pine, Engelmann spruce, subalpine fir, and white fir (Charlet 1996). Stand structure is similar to that of Great Basin bristlecone pine forest, though limber pine may extend to considerably lower elevations.

Summary

Recurring broad-scale themes and high levels of local variability characterize western North American forests. Common broad-scale themes include: (1) a dominance of coniferous vegetation, (2) the nearly universal occurrence of spruce-fir and lodgepole pine forests at upper elevations, (3) the widespread distribution of ponderosa pine and pinyon-juniper forests and woodlands at the transition to dry shrub and grasslands, (4) the wide distribution of quaking aspen as a generally seral species at middle to upper elevations in the interior, (5) the dominance of western hemlock or mountain hemlock in maritime environments along the Pacific coast, and (6) the nearly ubiquitous presence of Douglas-fir at middle and lower elevations. These broad-scale trends apply to much of western North America, but are a thin veneer over highly variable local conditions. Substantial areas of red alder, bigleaf maple, and other deciduous trees occur in riparian or disturbed areas along the coast. Extensive mixed pine-oak forests and oak woodlands are found in the summer-dry Mediterranean climates in central California and parts of Oregon and Washington. The basin and range mountains of the interior west contain island forests of a few species on isolated peaks, including some of the oldest trees on record.

Forests of western North America have experienced considerable change in the last 100 years or more. Most low-elevation forest and woodland areas have undergone alteration from timber harvest or other mechanical treatment and altered fire regimes. In some cases, this has led to a reduction of particular forest habitats (particularly late-successional or old forests with natural structural conditions) and in others to an increase in particular habitats (e.g., the expansion of pinyon-juniper woodlands). Old-growth forests have been substantially reduced in the moist, productive areas near the coast. High-elevation and high-latitude cold forests are perhaps least altered from their historical condition, while low-elevation, park-like, old-growth ponderosa pine forests may be the most altered.

Literature cited

Alexander, R.R. and W.D. Shepperd. 1990. *Picea engelmannii* Parry ex Engelm. Englemann spruce. Pages 187–203 *in* R.M. Burns and B.H. Honkala, technical coordinators. *Silvics of North America*. Volume 1. *Conifers. Agriculture Handbook* **654**. USDA Forest Service, Washington DC, USA.

Alexander, R.R., R.C. Shearer, and W.D. Shepperd. 1990. *Abies lasiocarpa* (Hook.) Nutt. Subalpine fir. Pages 60–70 *in* R.M. Burns and B.H. Honkala, technical coordinators. *Silvics of North America*. Volume 1. *Conifers. Agriculture Handbook* **654**. USDA Forest Service, Washington DC, USA.

Anderson, M., P. Bourgeron, M.T. Bryer, R. Crawford, L. Engelking, D. Faber-Langendoen, M. Gallyoun, K. Goudin, D.H. Grossman, S. Landaal, K.D. Patterson, M. Pyne, M. Reid, L. Sneddon, and A.S. Weakley. 1998. International classification of ecological communities: terrestrial vegetation of the United States. Volume **2**. *The National Vegetation Classification System: List of Types*. The Nature Conservancy, Arlington, Virginia, USA.

Bailey, R.G. 1995. Description of ecoregions of the United States. Miscellaneous Publication No. **1391** (rev.). USDA Forest Service, Washington DC, USA.

Bailey, R.G. 1998. Ecoregions map of North America: explanatory note. Miscellaneous Publication No. **1548**, USDA Forest Service, Washington DC, USA.

Barbour, M.G. and W.D. Billings. 2000. *North American Terrestrial Vegetation*. Second Edition. Cambridge University Press, Cambridge, UK.

Barbour, M.G. and R.A. Minnich. 2000. Californian upland forests and woodlands. Pages 161–203 *in* M.G. Barbour and W.D. Billings, editors. *North American Terrestrial Vegetation*. Cambridge University Press, Cambridge, UK.

Borchert, M. 1992. Interim guidelines defining old growth stands: coast redwood (SAF232) of southern Monterey County California. USDA Forest Service, Pacific Southwest Region, Vallejo, California, USA.

Boughton, J., M. Copenhagen, T. Faris, R. Flynn, L. Fry, P. Huberth, G. Lidholm, R. Loiselle, J. Martin, F. Samson, L. Shea, D. Williamson, S. Borchers, W. Oja, and K. Winterberger. 1992. Definitions for old-growth forest types in southcentral Alaska. R10-TP-**28**. USDA Forest Service, Juneau, Alaska, USA. 30 p.

Brockway, D.G., C. Topik, M.A. Hemstrom and W.H. Emmingham. 1983. Plant association and management guide for the Pacific silver fir zone. Publication R6-Ecol-**130a-1983**, USDA Forest Service, Portland, Oregon, USA.

Chabot, B.F. and H.A. Mooney, editors. 1985. *Physiological Ecology of North American Plant Communities*. Chapman and Hall, New York, New York, USA.

Charlet, D.A. 1996. *Atlas of Nevada Conifers*. University of Nevada Press, Reno, Nevada, USA.

Covington, W.W. and M.M. Moore. 1994. Southwestern ponderosa forest structure changes since Euro-American settlement. *Journal of Forestry* **92**:39–47.

Crawford, P.D. and C.D. Oliver. 1990. *Abies amabilis* Dougl. ex Forbes Pacific silver fir. Pages 17–25 *in* R.M. Burns and B.H. Honkala, technical coordinators. *Silvics of*

North America. Volume 1. *Conifers. Agriculture Handbook* **654**. USDA Forest Service, Washington DC, USA.

Dealy, J.E. 1990. *Juniperus occidentalis* Hook. Western Juniper. Pages 109–115 *in* R.M. Burns and B.H. Honkala, technical coordinators. *Silvics of North America*. Volume 1. *Conifers. Agriculture Handbook* **654**. USDA Forest Service, Washington DC, USA.

Diaz, N.M., C.T. High, T.K. Mellen, D.E. Smith, and C. Topik. 1997. Plant association and management guide for the mountain hemlock zone, Gifford Pinchot and Mt. Hood National Forests. Publication R6-MTH-GP-TP-**08-95**, USDA Forest Service, Portland, Oregon, USA.

Elliot-Fisk, D.L. 2000. The taiga and boreal forest. Pages 41–74 *in* M.G. Barbour and W.D. Billings, editors. *North American Terrestrial Vegetation*. Cambridge University Press, Cambridge, UK.

Fites, J., M. Chappel, B. Corbin, M. Newman, T. Ratcliff, and D. Thomas. 1992. Preliminary ecological old-growth definitions for mixed conifer (SAF TYPE 243) in California. USDA Forest Service, Vallejo, California, USA.

Franklin, J.F. 1990. *Abies procera* Rehd. Noble fir. Pages 80–87 *in* R.M. Burns and B.H. Honkala, technical coordinators. *Silvics of North America*. Volume 1. *Conifers. Agriculture Handbook* **654**. USDA Forest Service, Washington DC, USA.

Franklin, J.F. and C.T. Dyrness. 1973. Natural vegetation of Oregon and Washington. General Technical Report PNW-GTR-**8**. USDA Forest Service, Pacific Northwest Forest and Range Experiment Station, Portland, Oregon, USA.

Franklin, J.F. and C.B. Halpern. 2000. Pacific Northwest forests. Pages 123–160 *in* M.G. Barbour and W.D. Billings, editors. *North American Terrestrial Vegetation*. Cambridge University Press, Cambridge, UK.

Franklin, J.F., K. Cromak, Jr., W. Denison, A. McKee, C. Maser, J. Sedell, F. Swanson, and G. Juday. 1981. Ecological characteristics of old-growth Douglas fir forests. General Technical Report PNW-GTR-**118**. USDA Forest Service, Pacific Northwest Forest and Range Experiment Station, Portland, Oregon, USA.

Gottfried, G.J., P.F. Ffolliott, and L.F. DeBano. 1995. Forests and woodlands of the Sky Islands: stand characteristics and silvicultural prescriptions. Pages 152–164 *in* Biodiversity and management of the Madrean Archipelago: the Sky Islands of southwestern United States and northwestern Mexico. General Technical Report RM-**264**, USDA Forest Service, Rocky Mountain Forest and Range Experiment Station, Ft. Collins, Colorado, USA.

Green, P., J. Joy, D. Sirucek, W. Hann, A. Zack, and B. Naumann. 1992. Old-growth forest types of the Northern Region. R-1 SES 4/92, USDA Forest Service, Northern Region, Missoula, Montana, USA.

Gruell, G.E. 1999. Historical and modern roles of fire in pinyon-juniper. Pages 24–28 *in* S.B. Monson, and R. Stevens, compilers. Proceedings: ecology and management of pinyon juniper communities within the interior west. Publication P-**8**, USDA Forest Service, Rocky Mountain Research Station, Ft. Collins, Colorado, USA.

Hamilton, R.C. 1993. *Characteristics of Old-Growth Forests in the Intermountain Region*. USDA Forest Service, Ogden, Utah, USA.

Hann, W.J., J.L. Jones, M.G. Karl, P.F. Hessburg, R.E. Keane, D.G. Long, J.P. Menakis, C.H. McNicoll, S.G. Leonard, R.A. Gravenmier, and B.G. Smith. 1997. Landscape dynamics of the basin. Pages 337–1055 *in* T.M. Quigley and S.J. Arbelbide, technical editors. An assessment of ecosystem components in the interior Columbia basin and portions of the Klamath and Great Basins: volume 2.

General technical report PNW-GTR-**405**. USDA, Forest Service, Pacific Northwest Research Station, Portland, Oregon, USA.

Harrington, M.G. and S.S. Sackett. 1992. Past and present fire influences on southwestern ponderosa pine old growth. Pages 44–50 *in* M.R. Kaufmann, W.H. Moir, and R.L. Bassett. Old-growth forests in the Southwest and Rocky Mountain Regions. General Technical Report RM-**213**, USDA Forest Service, Rocky Mountain Forest and Range Experiment Station, Ft. Collins, Colorado, USA.

Hemstrom, M.A. and J.F. Franklin. 1982. Fire and other disturbances of the forests in Mount Rainier National Park. *Quaternary Research* **18**:32–51.

Hemstrom, M.A. and S.E. Logan. 1986. Plant association and management guide Siuslaw National Forest. Publication R6-Ecol-**220-1986a**, USDA Forest Service, Portland, Oregon, USA.

Hemstrom, M.A., S.E. Logan, and W. Pavlat. 1987. Plant association and management guide Willamette National Forest. Publication R6-Ecol-**257-b-86**, USDA Forest Service, Portland, Oregon, USA.

Hermann, R.K. and D.P. Lavender. 1990. *Pseudotsuga menziesii* (Mirb.) Franco Douglas-fir. Pages 527–540 *in* R.M. Burns and B.H. Honkala, technical coordinators. *Silvics of North America*. Volume 1. *Conifers. Agriculture Handbook* **654**. USDA Forest Service, Washington DC, USA.

Jimerson, T., B.B. Bingham, D. Solis, and S. Macmeeken. 1991*a*. Ecological definition for old-growth Pacific Douglas-fir (Society of American Foresters' type 229). USDA Forest Service, Vallejo, California, USA.

Jimerson, T., B.B. Bingham, D. Solis, and S. Macmeeken. 1991*b*. Ecological definition for old-growth Douglas-fir/tanoak/madrone (Society of American Foresters' type 234). USDA Forest Service, Vallejo, California, USA.

Little, E.L. 1971. *Atlas of United States Trees*. Volume 1. *Conifers and Important Hardwoods*. Miscellaneous Publication **1146**. USDA, Washington DC, USA.

Lotan, J.E. and W.B. Critchfield. 1990. *Pinus contorta* Dougl. ex. Loud. Lodgepole Pine. Pages 302–315 *in* R.M. Burns and B.H. Honkala, technical coordinators. *Silvics of North America*. Volume 1. *Conifers. Agriculture Handbook* **654**. USDA Forest Service, Washington DC, USA.

Means, J. 1990. *Tsuga mertensiana* (Bong.) Carr. Mountain hemlock. Pages 623–634 *in* R.M. Burns and B.H. Honkala, technical coordinators. *Silvics of North America*. Volume 1. *Conifers. Agriculture Handbook* **654**. USDA Forest Service, Washington DC, USA.

Mehl, M.S. 1992. Old-growth descriptions for the major forest cover types in the Rocky Mountain Region. Pages 106–120 *in* M.R. Kaufmann, W.H. Moir, and R.L. Bassett, technical coordinators. Old-growth forests in the Southwest and Rocky Mountain Regions. General Technical Report RM-**213**, USDA Forest Service, Rocky Mountain Forest and Range Experiment Station, Ft. Collins, Colorado, USA.

Nienstaedt, H. and J.C. Zasada. 1990. *Picea glauca* (Moench) Voss White Spruce. Pages 204–226 *in* R.M. Burns and B.H. Honkala, technical coordinators. *Silvics of North America*. Volume 1. *Conifers. Agriculture Handbook* **654**. USDA Forest Service, Washington DC, USA.

Noble, D.L. 1990. *Juniperus scopulorum* Sarg. Rocky Mountain Juniper. Pages 116–126 *in* R.M. Burns and B.H. Honkala, technical coordinators. *Silvics of North America*.

Volume 1. *Conifers. Agriculture Handbook* **654**. USDA Forest Service, Washington DC, USA.

Oliver, W.W. and R.A. Ryker. 1990. *Pinus ponderosa* Dougl. ex Laws. Ponderosa pine. Pages 413–424 *in* R.M. Burns and B.H. Honkala, technical coordinators. *Silvics of North America*. Volume 1. *Conifers. Agriculture Handbook* **654**. USDA Forest Service, Washington DC, USA.

Peet, R.K. 2000. Forests and meadows of the Rocky Mountains. Pages 75–122 *in* M.G. Barbour and W.D. Billings, editors. *North American Terrestrial Vegetation*. Cambridge University Press, Cambridge, UK.

Perala, D.A. 1990. *Populus tremuloides* Michx. Quaking aspen. Pages 555–569 *in* R.M. Burns and B.H. Honkala, technical coordinators. *Silvics of North America*. Volume 2. *Hardwoods. Agriculture Handbook* **654**. USDA Forest Service, Washington DC, USA.

Potter, D., M. Smith, T. Beck, W. Hance, and S. Robertson. 1992*a*. Ecological characteristics of old growth lodgepole pine in California. USDA Forest Service, Vallejo, California, USA.

Potter, D., M. Smith, T. Beck, W. Hance, and S. Robertson. 1992*b*. Ecological characteristics of old growth red fir in California. USDA Forest Service, Vallejo, California, USA.

Potter, D., M. Smith, T. Beck, W. Hance, and S. Robertson. 1992*c*. Ecological characteristics of old growth in California mixed subalpine forests. USDA Forest Service, Vallejo, California, USA.

Potter, D., M. Smith, T. Beck, W. Hance, and S. Robertson. 1992*d*. Ecological characteristics of old growth Jeffrey pine in California. USDA Forest Service, Pacific Southwest Region, Vallejo, California, USA.

Rebertus, A.J., T.T. Veblen, L.M. Roovers, and J.N. Mast. 1992. Structure and dynamics of old-growth Engelmann spruce-subalpine fir in Colorado. Pages 139–153 *in* M.R. Kaufmann, W.H. Moir, and R.L. Bassett. Old-growth forests in the Southwest and Rocky Mountain Regions. General Technical Report RM-**213**, USDA Forest Service, Rocky Mountain Forest and Range Experiment Station, Ft. Collins, Colorado, USA.

Ronco, F.P. 1990. *Pinus edulis* Engelm. Pinyon. Pages 327–337 *in* R.M. Burns and B.H. Honkala, technical coordinators. *Silvics of North America*. Volume 1. *Conifers. Agriculture Handbook* **654**. USDA Forest Service, Washington DC, USA.

Schmidt, J.M. and G.D. Amman. 1992. *Dendroctonus* beetles and old-growth forests in the Rockies. Pages 51–59 *in* M.R. Kaufmann, W.H. Moir, and R.L. Bassett. Old-growth forests in the Southwest and Rocky Mountain Regions. General Technical Report RM-**213**, USDA Forest Service, Rocky Mountain Forest and Range Experiment Station, Ft. Collins, Colorado, USA.

Smith, S., W. Laudenslayer, J. Trask, and M. Armijo. 1991. Interim guidelines for old growth stands: Pacific ponderosa pine (SAF 245). USDA Forest Service, Vallejo, California, USA.

Spies, T.A. and J.F. Franklin. 1991. The structure of natural young, mature, and old-growth Douglas-fir forests in Oregon and Washington. Pages 91–109 *in* L.F. Ruggiero, K.B. Aubry, A.B. Carey, and M.H. Huff, technical coordinators. Wildlife and vegetation of unmanaged Douglas-fir forests. General Technical Report PNW-GTR-**285**, USDA Forest Service, Pacific Northwest Research Station, Portland, Oregon, USA.

Steele, R. 1988. Ecological relationships of ponderosa pine. Pages 71–71 *in* D.M. Baumgartner and J.E. Lotan, editors. Ponderosa pine: the species and its management. Cooperative Extension Service, Washington State University, Pullman, Washington, USA.

Swetnam, T.W. and A.M. Lynch. 1993. Multicentury, regional-scale patterns of western spruce budworm outbreaks. *Ecological Monographs* **63**:399–424.

Tausch, R.J. 1999. Historic pinyon and juniper woodland development. Pages 12–19 *in* S.B. Monson, S. Richards, R.J. Tausch, R.F. Miller, and C. Goodrich, compilers. Publication RMRS-P-**9**. US Department of Agriculture Forest Service, Rocky Mountain Research Station, Ft. Collins, Colorado, USA.

Tausch, R.J., N.E. West, and A.A. Nabi. 1981. Tree age and dominance patterns in Great Basin pinyon-juniper woodlands. *Journal of Range Management* **34**:259–264.

USDA Forest Service. 1992. Recommended old-growth definitions and descriptions and old-growth allocation procedure. USDA Forest Service, Albuquerque, New Mexico, USA.

USDA Forest Service. 1993. Region 6 interim old growth definition[s] [for the] Douglas-fir series, grand fir/white fir series, interior Douglas-fir series, lodgepole pine series, Pacific silver fir series, ponderosa pine series, Port Orford cedar series, tanoak (redwood) series, western hemlock series. USDA Forest Service, Portland, Oregon, USA.

Viereck, L.A. and W.F. Johnston. 1990. *Picea mariana* (Mill.) B.S.P. Black Spruce. Pages 227–237 *in* R.M. Burns and B.H. Honkala, technical coordinators. *Silvics of North America*. Volume 1. *Conifers. Agriculture Handbook* **654**. USDA Forest Service, Washington DC, USA.

Viereck, L.A., C.T. Dyrness, A.R. Batten, and K.J. Wenzlick. 1992. The Alaska vegetation classification. General Technical Report PNW-GTR-**286**. USDA, Forest Service, Pacific Northwest Research Station, Portland, Oregon, USA.

West, N.E. and J.A. Young. 2000. Intermountain valleys and lower mountain slopes. Pages 255–284 *in* M.G. Barbour and W.D. Billings, editors. *North American Terrestrial Vegetation*. Cambridge University Press, Cambridge, UK.

3

Faunal composition and distribution of mammals in western coniferous forests

Mammals are prominent features of the landscape in coniferous forests of western North America. Large herbivores such as deer, elk, moose, and caribou are particularly noticeable. Smaller, though no less conspicuous, mammals include diurnal ground squirrels, marmots, pikas, and numerous species of chipmunks, and nocturnal bats. Overall, the mammalian faunas of western coniferous forests are rich and varied. Knowledge of their taxonomic makeup and biogeography provides a foundation for understanding current patterns of distribution, habitat use, and management. An examination of the distribution and diversity patterns of coniferous-forest species, together with an assessment of the factors influencing these patterns, forms the basis for this review.

Faunal composition

Nearly half (194) of approximately 400 species of mammals occurring in North America north of Mexico occupy western coniferous forests and associated meadow, brush, or riparian habitats (Table 3.1, Appendix). Eight orders and 27 families of mammals are represented. Per unit area, western North America has a much higher species richness than other temperate or boreal areas of the continent. This difference is readily discernible when comparing western and eastern coniferous forests (Hallett et al. 2003). Chipmunks (*Tamias* spp.) constitute a striking example: 21 species of chipmunks occupy western coniferous forests; there is only one (*Tamias striatus*) in eastern forests. Similar, though less dramatic examples, occur among shrews (*Sorex* spp.), bats (*Myotis* spp.), ground squirrels (*Spermophilus* spp., *Marmota* spp.), and other rodents (e.g., *Peromyscus* spp.).

[41]

The list of species (Table 3.1, Appendix) is somewhat loosely drawn. In the broadest sense, coniferous forests embrace a number of habitats that otherwise are distinctive in their own right, such as streamside habitats and meadows within forests. Also, some species, or populations of certain species, are impermanent residents of forest habitats, using them on a seasonal basis for breeding purposes (some bats), movement corridors (caribou, pronghorns), or for temporary shelter. Ecological relationships of many coniferous-forest mammals are treated comprehensively elsewhere in this volume (Anthony et al. 2003, riparian species; Aubry et al. 2003, arboreal species; Buskirk and Zielinski 2003, small carnivores; Hallett et al. 2003, terrestrial small mammals; Hayes 2003, bats; Kie et al. 2003, ungulates; Kunkel 2003, large carnivores; and Smith et al. 2003, arboreal rodents).

Many species have cosmopolitan distributions in a wide variety of habitats including, but not limited to, coniferous forests (Table 3.1, Appendix). Proportionately more of these taxa are mobile species, such as bats, carnivores, and artiodactyls. Most coniferous-forest species frequent many, if not all, of the six broad habitat types, and relatively few species are confined to only one or two. Still others inhabit coniferous forests only in unusual circumstances. For example, the 13-lined ground squirrel (*Spermophilus tridecemlineatus*), a denizen of short-grass prairies, also occurs in openings among ponderosa pines in southern Colorado (Armstrong 1972).

Less than half (84) of the 194 species are found only within the confines of western coniferous forests; a large portion of these consists of chipmunks. There are even fewer coniferous-forest obligates (perhaps fewer than 25) and, unsurprisingly, these are usually arboreal forms. Still fewer species depend directly upon conifers to support all resource needs. Instead, coniferous forests provide sites for nesting and other forms of shelter, protection, feeding, reproduction, and other activities, and create suitable microclimates for exploitation by mammals (for example, see Anthony et al. 2003, Hallett et al. 2003).

Aside from seed-eating squirrels, there are few species that depend strictly upon conifers for nutrition. Leaves of conifers contain many secondary compounds (e.g., tannins, flavinoids, terpenes), many of which reduce digestibility (Vaughan and Czaplewski 1985). Folivorous mammals are characterized by low rates of metabolism, small litter sizes, and slow rates of maturation of young (Hamilton 1962, McNab 1978, Vaughan and Czaplewski 1985), and they are rare. Exceptions include porcupines (*Erethizon dorsatum*), whose diet consists of bark and growing shoots of firs, redwoods, and pines; Stephen's woodrat (*Neotoma stephensi*), which consumes juniper leaves and twigs; and two species of tree voles (*Arborimus longicaudus* and *A. pomo*), which specialize on needles of Douglas- and grand firs. Of course, many species exploit resources produced by an extensive forest understory of shrubs and other associated flora, including grasses, fungi, and lichens.

Strong habitat associations with mature and old-growth forests are documented for some species (e.g., see Corn et al. 1988, Raphael 1988*b*, Lomolino and Perault 2000), but an obligatory relationship with such forests has not been established for any well-studied species. Consequently, the decline of old-growth forests has not led to the disappearances of any mammal except on local scales. Indeed, the rarest species are not generally associated with late-stage forests. For example, one of the least common mammals in western coniferous forests is the white-footed vole (*Arborimus albipes*), a species found in several seral stages among western forests but with a small geographic range (Verts and Carraway 1995). The rarest species of all is probably the Idaho ground squirrel (*Spermophilus brunneus*), which occupies forest openings within an extremely limited geographic range. With the advent of fire controls, conifers have invaded meadows, leading to small, scattered populations of squirrels totaling no more than several hundred individuals (Gavin et al. 1999).

Distribution patterns

Geographic ranges of mammals in western coniferous forests vary in predictable ways. Species with catholic diets and habitat preferences are widespread throughout many forest habitats at many elevations (e.g., deer mice, *Peromyscus* spp., and coyotes, *Canis latrans*; Fig. 3.1). Species adapted to habitats occurring in an otherwise heterogeneous landscape, such as those at high elevations, have discontinuous geographic ranges (e.g., heather voles, *Phenacomys intermedius*, Fig. 3.2). For a variety of reasons, still others have very limited distributions within forests (Fig. 3.3). Sister species form parapatric or allopatric arrangements in some places (e.g., western red-backed voles, *Clethrionomys californicus*, and southern red-backed voles, *C. gapperi* on opposite sides of the Columbia River; Figs. 3.3, 3.4). Finally, species with boreal origins typically have northern distributions; in southern coniferous forests they often are supplanted by counterparts with austral affinities (e.g., red foxes, *Vulpes vulpes*, and gray foxes, *Urocyon cinereoargenteus*; Figs. 3.5, 3.6).

Fig. 3.1. Geographic distribution of coyotes (*Canis latrans*) in North America. Note the widespread occurrence of this species. (Map taken from Wilson and Ruff (1999), *The Smithsonian Book of North American Mammals*, ©Smithsonian Institution Press, Washington, D.C.).

Factors bearing upon distributions and diversity

Historical influences of mountain-building activities, glacio-pluvial processes, tectonic activities, and climatic changes have produced a highly fragmented landscape in western North America. Indeed, habitat disturbances over both short and long time-scales are characteristic of much of western North America, including coniferous forests. Coastal areas also are dissected by a large number of major river systems that have the potential to form barriers. Vegetation systems are complex and geographically variable. Collectively, these factors have provided circumstances conducive to the creation of habitat diversity and isolation, which, in turn, have led to high species diversity, discontinuous distributions of species, and to species adaptations that provide resiliency to change.

Fig. 3.2. Geographic distribution of western heather voles (*Phenacomys intermedius*). Note the scattered populations of this species in western North America. (Map taken from Wilson and Ruff (1999), *The Smithsonian Book of North American Mammals*, ©Smithsonian Institution Press, Washington, DC.).

In addition, past land connections across the Bering Sea resulted in invasions of species that have enriched the mammalian fauna of western forests. Examples of relatively recent invaders from Eurasia include caribou (=reindeer, *Rangifer tarandus*), elk (=red deer, *Cervus elaphus*), wolverines (*Gulo gulo*), grizzly (=brown) bears (*Ursus arctos*), red foxes, beavers (*Castor canadensis*), and some other rodents. These mammals have replaced much of the megafauna that became extinct in North America at the end of the Pleistocene. Thus, modern coniferous forests of western North America contain a mixture of long-term residents and relatively recent colonizers.

Not all changes have resulted in increased diversity, however. There were widespread extinctions of species in North America at the end of

Fig. 3.3. Geographic distribution of western red-backed voles (*Clethrionomys californicus*). Note the limited range of this species and its juxtaposition with southern red-backed voles (Fig. 3.4). (Map taken from Wilson and Ruff (1999), The *Smithsonian Book of North American Mammals*, ©Smithsonian Institution Press, Washington, DC.).

the Pleistocene (Kurtèn and Anderson 1980). Increased warming since the Pleistocene, including the markedly warmer hypsithermal period several thousand years ago, caused disappearances of many populations of montane mammals in small, isolated patches of coniferous-forest habitats, such as in the Great Basin and American Southwest (Grayson 1993). Consequently, although species richness is relatively high overall in western North America, individual local faunas in isolated places are often impoverished.

Of course, geological and climatic influences are ultimately ecological ones, and many ecological factors, such as competition, predation, and habitat associations, may play prominent roles in determining modern-day species occurrences and community dynamics. Biotic and abiotic

Fig. 3.4. Geographic distribution of southern red-backed voles (*Clethrionomys gapperi*). Note the juxtaposition of this species with western red-backed voles (Fig. 3.3). (Map taken from Wilson and Ruff (1999), *The Smithsonian Book of North American Mammals*, ©Smithsonian Institution Press, Washington, DC.).

factors may act collectively to affect distributions. Together they often confound our ability to pinpoint causes for particular patterns.

Despite disturbances, there have been no modern extinctions of any mammalian species in western coniferous forests. Broad ecological tolerances may have evolved from exposure of these mammals to varying environmental conditions during and succeeding glacial disturbances (Kirkland 1985). Nevertheless, recent disappearances at local levels have caused extensive modification of modern geographic ranges of many species. These changes have resulted largely from human impacts, especially through habitat modifications (e.g., logging) and intense hunting and trapping efforts. Noteworthy examples are elk (*Cervus elaphus*), grizzly bears, gray wolves (*Canis lupus*), and all larger species of forest mustelids (see contributions in Ruggiero et al. 1994).

Fig. 3.5. Geographic distribution of red foxes (*Vulpes vulpes*) in North America. Note the largely northern (boreal) range of this species in relation to gray foxes (Fig. 3.6). (Map taken from Wilson and Ruff (1999), *The Smithsonian Book of North American Mammals*, ©Smithsonian Institution Press, Washington, DC.).

Responses of mammals

Responses of mammals to these effects are not altogether uniform. They vary by species, habitat, and geography within and among species. This variation is not surprising because we do not understand what defines the habitat and distributional limits of many species and pressures from human intervention disproportionately affect particular target species.

Glacial advances and retreats produced profound effects on the distributions of mammals in northwestern North America (Kurtèn and Anderson 1980). As climates changed during and since the Pleistocene, geographic ranges of mammals shifted markedly. Common responses included the southward displacement of species in the late Pleistocene

Fig. 3.6. Geographic distribution of gray foxes (*Urocyon cinereoargenteus*) in North America. Note the largely southern (austral) range of this species in relation to red foxes (Fig. 3.5). (Map taken from Wilson and Ruff (1999), *The Smithsonian Book of North American Mammals*, ©Smithsonian Institution Press, Washington, DC.).

and subsequent northward expansions in the Holocene. However, range shifts of many species occurred at different times and rates, and often in different directions. These dissimilarities resulted in modern community assemblages that do not necessarily resemble past ones (Graham 1986, Graham et al. 1996). Moreover, some species that intermingled in the Pleistocene occur today in different habitats and at distant locations. Collectively, these observations suggest that many species responded in an individualistic way to climatic change.

In contrast, a concerted response to glacial episodes is suggested by the comparative phylogeography of forest species occurring in the northwestern part of the North American continent. Genetic and morphological studies have revealed distinct coastal and continental clades in at least a dozen species or species complexes with widely varying life histories, including mule deer (*Odocoileus hemionus*; Cronin 1992), black bears (*Ursus americanus*; Byun et al. 1997, Wooding and Ward 1999, Stone and Cook 2000), martens (*Martes americana*; Carr and Hicks 1997), montane shrews (*Sorex monticolus*; Demboski and Cook 2001), tree squirrels (*Glaucomys* and *Tamiasciurus* spp.; Arbogast 1999*a*, Arbogast et al. 2001), deer mice (*Peromyscus* spp.; Hogan et al. 1993), and red-backed voles (*Clethrionomys* spp.; Arbogast 1999*b*). These co-distributed clades probably resulted from geographic segregation of coastal and continental components during contraction, fragmentation, and subsequent expansion of boreal forest during glacial cycles of the Pleistocene and their eventual reunion (for reviews, see Demboski et al. 1999, Arbogast and Kenagy 2001). Predictably, species of coniferous-forest mammals with post-Pleistocene origins in North America do not show these patterns (Arbogast 1999*b*). Thus, northwestern boreal-forest mammals constitute a complex set of intra- and inter-species relationships.

Rapidly retreating glaciers and shifts in vegetation also resulted in remnant populations of otherwise northern species that now occupy southern forest refugia. For example, the Mt. Lyell shrew (*Sorex lyelli*) is separated from its northern counterpart, the masked shrew (*S. cinereus*), by several hundred miles (Figs. 3.7, 3.8), and the mostly north-temperate water shrew (*S. palustris*) has several remnant populations in mountains of Arizona and New Mexico. There are other examples: montane shrews (*Sorex monticolus*), pikas (*Ochotona princeps*), Uinta chipmunks (*Tamias umbrinus*), and western heather voles.

Modern distributions of other coniferous-forest species are very different than they were formerly. The white-footed vole, now occupying moist coastal forests of Oregon and northern California, is reported as a Pleistocene fossil from the Snake Range in eastern Nevada (Maser et al. 1981). Mammal faunas found in isolated mountaintop forests of the Great Basin are missing many of the subalpine species that once characterized them, such as martens, northern bog lemmings (*Synaptomys borealis*), and western heather voles. Remaining subalpine species (e.g., pikas) are spotty in occurrence (Grayson 1993, Lawlor 1998). These impoverished faunas are the result of Holocene warming, which evidently caused widespread extirpations of small, fragmented forest populations that became separated from other populations by unsuitable desert terrain.

Rivers constitute formidable barriers to some coniferous-forest species. Sister species occupying opposite sides of the Columbia River include

Fig. 3.7. Geographic distribution of Mt. Lyell shrews (*Sorex lyelli*). This species is a relictual derivative of the masked shrew (*Sorex cinereus*) (Fig. 3.8), which had a more southerly distribution in the late Pleistocene. (Map taken from Wilson and Ruff (1999), *The Smithsonian Book of North American Mammals*, ©Smithsonian Institution Press, Washington, DC.).

ground squirrels (*Spermophilus saturatus* and *S. lateralis*) and red-backed voles (Figs. 3.3, 3.4). One species (the brush rabbit, *Sylvilagus bachmani*) extends northward only to the Columbia River and has no close counterpart on the other side. In coastal Oregon and California, the Rogue, Klamath, and Eel rivers segregate four closely related species of chipmunks (*Tamias townsendii*, *T. siskiyou*, *T. senex*, and *T. ochrogenys*) (Sutton and Nadler 1974, Sutton and Patterson 2000), and sympatric associations of these species with other chipmunk species (e.g., *T. sonomae*) are uncommon. The mechanisms that maintain these distinctions along margins of contact remain obscure. This arrangement is atypical for chipmunks, however. Other species of chipmunks generally do not form sets

Fig. 3.8. Geographic distribution of masked shrews (*Sorex cinereus*) in North America. Note the range of this widespread species in relation to its close relative, the Mt. Lyell shrew (Fig 3.7). (Map taken from Wilson and Ruff (1999), *The Smithsonian Book of North American Mammals*, ©Smithsonian Institution Press, Washington, DC.).

of allopatric sister species; instead, they assort by microhabitat or dietary differences and often co-exist with one another. I have observed as many as four species of *Tamias* at one location in the Sierra Nevada.

Desert valleys act in similar ways. The patchy occurrences of coniferous-forest species on mountaintops in the Great Basin and in the desert Southwest attest to the effects of forest fragmentation on isolation and local extinctions (Brown 1971*b*, 1978, Patterson 1984, Lomolino et al. 1989, Grayson 1993, Patterson 1995, Lawlor 1998). However, recent evidence strongly suggests that arid valleys are variously permeable to movements of species (Davis and Dunford 1987, Davis et al. 1988, Lawlor 1998). In the Great Basin, for example, most species occupying forest fragments are widespread on most mountaintops; only mammals confined to colder subalpine habitats demonstrate disappearances without replacement as post-Pleistocene warming progressed (Lawlor 1998). By contrast, low-elevation woodland species show clear evidence of cross-valley movements and repeated colonizations. For instance, fossil records from Homestead Cave, Utah, reveal that bushy-tailed woodrats (*Neotoma cinerea*) alternately appeared and disappeared several times within the last 10000 years (Grayson and Madsen 2000, Grayson et al. 1996). A demonstration of the close linkage between geographical and elevational distributions of montane mammals and the relative influences of colonization and extinction was provided by Rickart (2001).

Today, fragmentation of coniferous forests occurs at considerably different temporal and spatial scales than that which shaped the landscape of western North America in the Holocene. Human-induced landscape alterations may be qualitatively different and potentially more threatening to species persistence than the natural-disturbance regimes mammals were exposed to in the past (Spies and Turner 1999, Martin and McComb 2003). The maintenance and recovery of populations and communities of forest-dwelling mammals after fire or logging depend upon many factors, including sizes of disturbed and undisturbed areas, the degree of forest fragmentation, the nature of the matrix separating patches of suitable forest, and the amount of forest edge. Few data are available to assess these influences. For example, the impacts of fragmentation on community composition and species abundances during succession are largely unknown (Holt et al. 1995, Schweiger et al. 2000), and only recently has the landscape-level importance of corridors been established for forest-dwelling mammals (Perault and Lomolino 2000).

Interestingly, Lomolino and Perault (2000) showed that characteristics of mammalian communities in second-growth forests of the Olympic Peninsula of Washington do not appear to be converging on those of old-growth forests. Removal of living trees as well as other elements of structural heterogeneity (snags, downed logs) typical of clear-cutting practices in this region may have long-term negative impacts on the restoration of mammalian community composition. Whether the same outcome would prevail in areas where selective cutting methods were used and where downed logs and snags were preserved (e.g., Sierra Nevada forests; Franklin and Fites-Kaufmann 1996) is an intriguing question.

Studies of recent disturbances reveal considerable variation in the impact on individual species. Species that differ with respect to habitat requirements, mobilities, and other natural history traits doubtlessly perceive landscape patterns in different ways. Arboreal species, or species dependent on diets prevalent in particular habitats, are most vulnerable to disturbance. For sciurids, downed logs and vertical heterogeneity are important habitat features (Carey 1991, 1995, Ransome and Sullivan 1997). Of course, clear-cutting causes dramatic declines in abundances of tree squirrels, flying squirrels, and other arboreal mammals. Yet terrestrial species are affected as well. For example, in the Pacific Northwest, populations of vagrant shrews (*Sorex vagrans*), deer mice (*Peromyscus maniculatus*), Oregon voles (*Microtus oregoni*), and Townsend's chipmunks (*Tamias townsendii*) generally increase in response to clear-cutting, whereas Trowbridge's shrews (*Sorex trowbridgii*) and western red-backed voles (*Clethrionomys californicus*) decline (see Corn et al. 1988 and references therein).

Tree voles, which occupy specialized arboreal niches in fir forests, appear to differ geographically in their response to logging of old-growth tree stands. In Oregon, several studies have demonstrated a predilection of *Arborimus longicaudus* for old-growth forests (Aubry et al. 1991, Corn and Bury 1991, Gilbert and Allwine 1991). By contrast, assessments of habitat associations of *Arborimus pomo* in northwestern California using nest surveys and pitfall trapping (Meiselman 1992, Meiselman and Doyle 1996, Raphael 1988a, Thompson and Diller 2002) have proven inconclusive.

Diversity of forest species might be expected to decline with logging. Although species compositions and abundances vary considerably, species numbers apparently do not differ substantially among seral stages (Raphael 1991, Lomolino and Perault 2000).

At Mount St. Helens, re-establishment of viable populations of species has been unpredictably widespread and rapid for an area so recently devastated (MacMahon et al. 1989). One clear message seems to be that habitat suitability, more than any other single factor, constitutes the key element in the successful invasion and maintenance of mammalian populations in western forests.

Competitive interactions are implicated in establishing and maintaining the co-existence of some, and the altitudinal segregation of other, chipmunk species in the Sierra Nevada and southern Rocky Mountains (Brown 1971a, Chappell 1978, Patterson 1984). As well, the juxtaposition of geographic ranges of closely related species, such as those of the Colorado and Uinta chipmunks (*Tamias quadrivittatus* and *T. umbrinus*) in the Rocky Mountains (Armstrong 1972, Fitzgerald et al. 1994) and of red and Douglas squirrels (*Tamiasciurus hudsonicus* and *T. douglasii*) along the Cascade rim (Smith 1981), may be attributable to competition.

Selection of forest habitats by bats is determined by roost availability and foraging requirements (Kunz 1982, Fenton 1990). The extent to which species assemblages of bats are affected by logging activities and resulting habitat fragmentation remains to be fully explored (for several studies, see Barclay and Brigham 1996). Clear-cuts may produce foraging opportunities for species that forage in uncluttered areas, such as silver-haired bats (*Lasionycteris noctivagans*), or those that feed along forest edges. However, many of these same species also require roost sites located in old stands of trees, and others (e.g., many *Myotis* species) seem heavily dependent upon forests for both foraging and resting places. Details regarding the ecology of bats in western coniferous forests are provided by Hayes (2003).

Conservation implications

Historical and contemporary information about compositions and distributions of mammals in western coniferous forests have potential conservation and management implications. That no species has become extinct in the face of extensive modifications of western coniferous forests attests to the resiliency and adaptation of mammals. However, most species frequenting coniferous forests are also found in a variety of other habitats. Those that are not are more prone to disturbances.

Accelerated local extirpations can be expected to lead to increased metapopulation structure in species across the larger landscape, even for widespread and evenly distributed taxa with good dispersal capabilities. Relatively sedentary or habitat-restricted species with already fragmented distributions can be expected to lose still more populations as habitat patches become smaller and more remote from one another. Impoverished communities will result, as demonstrated by documented historic changes in the compositions of mammalian communities occupying mountaintop forests of North American deserts. For example, species occurring in subalpine habitats in these isolated ranges are especially prone to disappearance.

Because they provide supplemental forest habitats for source populations of small mammals as well as dispersal avenues for mobile ones, corridors may ameliorate these effects (Perault and Lomolino 2000). Establishment of adequate connectivity among habitat patches, together with development of appropriately intermingled seral stages and landscape mosaics that mimic historic patterns, can be expected to promote maintenance of populations at both local and large-scale levels.

The discovery of co-distributed coastal and inland clades within species or pairs of sister species in northwestern forests also deserves attention if we desire to preserve their genetic and morphological integrity.

Summary

The dissected character of the western forest landscape created by glaciation, physiographic heterogeneity, and other influences has produced a modern mammalian fauna that has long been exposed to fragmentation processes and is highly diverse. About half of all North American species of mammals occur in western coniferous forests, although fewer than half of those are found wholly within the confines of these forests. Most of the latter are widespread in many habitat types.

Geographic ranges of mammals are, and have been, highly variable in space and time. Responses of species to climatic and other influences on distributions and diversity have been both individualistic and concerted. Fragmentation of formerly widespread coniferous-forest species has produced a patchwork of populations in many parts of the west, especially in southern areas.

Today, the primary influences on diversities and distributions of coniferous-forest mammals stem from habitat disturbance caused by logging and fire. The long-term impacts of these factors remain largely unknown, but recent studies are suggestive. If history provides any lessons, fragmentation will lead to community impoverishment. Habitat suitability will be crucial for the establishment and maintenance of mammal populations, and dispersal avenues containing appropriate habitat will be necessary to support populations, sustain community composition, and preserve species diversity at local and regional levels.

Appendix

Information about distributions and habitats of coniferous-forest mammals (Table 3.1) was obtained from a variety of sources, including the following published compilations for western states and provinces: Alberta (Smith 1993); Arizona (Hoffmeister 1986); British Columbia (Cowan 1973); California (Ingles 1948, 1954); Colorado (Armstrong 1972; Fitzgerald et al. 1994); Idaho (Davis 1939, Larrison and Johnson 1981); Nevada (Hall 1946, 1995); Montana (Foresman 2001); New Mexico (Findley et al. 1975); Oregon (Maser et al. 1981, Maser 1998, Verts and Carraway 1998); Utah (Durrant

1952); Washington (Dalquest 1948); Wyoming (Long 1965). Broader geographic treatments include Banfield (1974), Hall (1981), Ingles (1965), Larrison (1976), and Zeveloff and Collett (1988).

With few exceptions, scientific and vernacular names follow Wilson and Ruff (1999). Some scientific names reflect recent changes and are worth noting. *Lontra* is the newly accepted generic name for otters. *Arborimus* supplants *Phenacomys* and *Corynorhinus* replaces *Plecotus* as generic names of tree and white-footed voles, and big-eared bats, respectively. *Peromyscus keeni* is the current designation for populations of deer mice in the Pacific northwest previously considered populations of *P. maniculatus*, *P. oreas*, or *P. sitkensis*. For references, see Wilson and Reeder (1993) and Wilson and Ruff (1999).

Habitat categories roughly follow classifications used in other mammal surveys (e.g., Fitzgerald et al. 1994). Because geographic ranges of mammals rarely correspond to narrowly defined habitats or vegetation types, the categories used here are considerably more general than those provided by Hemstrom (2003).

Humid coastal forests consist mostly of dense stands of Sitka spruce (*Picea sitchensis*), redwood (*Sequoia sempervirens*), western hemlock (*Tsuga heterophylla*), western redcedar (*Thuja plicata*), grand fir (*Abies grandis*), and Douglas-fir (*Pseudotsuga menziesii*) inhabiting narrow moisture-rich and fog-enshrouded areas in lowlands along the Pacific coast (this category comprises vegetation types included within Pacific northwest forests and Californian upland forests and woodlands as described by Hemstrom 2003). Pinyon-juniper/Steppe forests are relatively open, dry forests occupying high desert valleys and flanks of mountains in the Great Basin, on the Colorado Plateau, and in other areas of southwestern North America. Characteristic species include pinyon pines (*Pinus monophylla, P. edulis*, and *P. quadrifolia*), junipers (*Juniperus* spp.), and a variety of deciduous trees and shrubs (containing elements of Hemstrom's Rocky Mountain forests and woodlands and vegetative types within the Intermountain West). Montane woodland forests are relatively dry open forests, generally at middle elevations, that typically contain ponderosa or Jeffrey pines (*Pinus ponderosa* and *P. jeffreyi*, respectively), Douglas-fir, and true firs (*Abies* spp.). Deciduous trees such as oaks (*Quercus* spp.) and madrone (*Arbutus menziesii*) commonly form a mixed forest with conifers (containing elements of four of the five major vegetation zones described by Hemstrom: Pacific northwest forests, Californian upland forests and Woodland, Rocky Mountain forests and woodlands, and Intermountain West). Subalpine

Table 3.1. Species of mammals and their primary habitat affinities in western coniferous forests. **Boldfaced species entries** identify those that occur wholly or principally in western coniferous forests and associated habitats

Species	Coastal humid forest	Pinyon-juniper/ Steppe forest	Montane woodland forest	Subalpine forest	Northern boreal forest	Forest meadow	Riparian
Didelphimorphia							
Didelphidae							
Didelphis virginiana (Virginia opossum)	X	X	X				X
Insectivora							
Soricidae							
Notiosorex crawfordi (Desert shrew)		X	X				
Sorex arcticus (Arctic shrew)					X	X	X
Sorex arizonae (Arizona shrew)		X	X				X
Sorex bairdii (Baird's shrew)	X		X	X			
Sorex bendirii (Marsh shrew)	X		X				X
Sorex cinereus (Masked shrew)	X	X	X	X	X	X	
Sorex hoyi (Pygmy shrew)				X	X	X	
Sorex lyelli (Mt. Lyell shrew)			X	X			
Sorex merriami (Merriam's shrew)		X	X			X	

Species							
Sorex monticolus (Montane shrew)	X		X	X	X		X
Sorex nanus (Dwarf shrew)	X		X	X	X		X
Sorex ornatus (Ornate shrew)	X			X	X		X
Sorex pacificus (Pacific shrew)	X				X		X
Sorex palustris (Water shrew)	X	X	X	X	X		X
Sorex preblei (Preble's shrew)	X		X	X	X		
Sorex sonomae (Fog shrew)	X			X	X		X
Sorex tenellus (Inyo shrew)	X			X	X		
Sorex trowbridgii (Trowbridge's shrew)	X		X	X	X		X
Sorex vagrans (Vagrant shrew)	X	X	X	X	X		X
Talpidae							
Neurotrichus gibbsii (Shrew mole)	X		X	X	X		X
Scapanus latimanus (Broad-footed mole)	X	X	X	X	X		
Scapanus orarius (Coast mole)	X	X	X	X	X		
Scapanus townsendii (Townsend's mole)	X		X	X	X		

(cont.)

Table 3.1. (cont.)

Species	Coastal humid forest	Pinyon–juniper/ Steppe forest	Montane woodland forest	Subalpine forest	Northern boreal forest	Forest meadow	Riparian
Chiroptera							
Phyllostomidae							
Chaeronycteris mexicana (Mexican long-tongued bat)		X					X
Vespertilionidae							
Antrozous pallidus (Pallid bat)	X	X	X	X		X	X
Corynorhinus townsendii (Townsend's big-eared bat)	X	X	X	X		X	X
Eptesicus fuscus (Big brown bat)	X	X	X	X	X	X	X
Euderma maculatum (Spotted bat)		X	X				X
Idionycteris phyllotis (Allen's big-eared bat)		X	X	X		X	X
Lasionycteris noctivagans (Silver-haired bat)	X	X	X	X	X	X	X
Lasiurus borealis (Red bat)	X	X	X	X		X	X
Lasiurus cinereus (Hoary bat)	X	X	X	X	X	X	X
Myotis auriculus (Southwestern myotis)			X	X			X
Myotis californicus (California myotis)	X	X	X	X		X	X

Myotis ciliolabrum (Western small-footed myotis)						X	X
Myotis evotis (Long-eared myotis)	X		X			X	X
Myotis keenii (Keen's myotis)	X		X				X
Myotis lucifugus (Little brown bat)	X	X	X	X		X	X
Myotis septentrionalis (Northern long-eared myotis)			X				X
Myotis thysanodes (Fringed myotis)	X		X	X		X	X
Myotis velifer Cave myotis			X				X
Myotis volans (Long-legged myotis)	X		X	X	X	X	X
Myotis yumanensis (Yuma myotis)	X		X	X		X	X
Pipistrellus hesperus (Western pipistrelle)	X		X	X		X	X
Molossidae							
Tadarida brasiliensis (Brazilian free-tailed bat)	X		X	X		X	X
Carnivora							
Canidae							
Canis latrans (Coyote)	X		X	X	X	X	
Canis lupus (Gray wolf)	X		X	X	X	X	

(cont.)

Table 3.1. (cont.)

Species	Coastal humid forest	Pinyon-juniper/ Steppe forest	Montane woodland forest	Subalpine forest	Northern boreal forest	Forest meadow	Riparian
Urocyon cinereoargenteus (Gray fox)	X	X	X	X		X	
Vulpes vulpes (Red fox)	X		X	X	X	X	X
Ursidae							
Ursus americanus (Black bear)	X	X	X	X	X	X	X
Ursus arctos (Grizzly bear)	X	X	X	X	X	X	X
Procyonidae							
Bassariscus astutus (Ringtail)	X	X	X	X		X	X
Nasua narica (White-nosed coati)		X	X				X
Procyon lotor (Raccoon)	X	X	X	X		X	X
Mustelidae							
Gulo gulo (Wolverine)	X		X	X	X		
Lontra canadensis (River otter)	X	X	X	X	X	X	X
Martes americana (Marten)	X		X	X	X	X	X
Martes pennanti (Fisher)	X		X	X	X		X

Mustela erminea (Ermine)	X		X	X	X
Mustela frenata (Long-tailed weasel)	X	X	X	X	X
Mustela nivalis (Least weasel)	X		X	X	X
Mustela vison (Mink)	X	X	X	X	X · X
Taxidea taxus (Badger)		X	X	X	X
Mephitidae					
Mephitis mephitis (Striped skunk)	X	X	X	X	X
Mephitis macroura (Hooded skunk)		X			
Spilogale gracilis (Western spotted skunk)	X	X	X	X	X
Felidae					
Felis concolor (Mountain lion)	X	X	X	X	X
Lynx canadensis (Canadian lynx)	X	X	X	X	
Lynx rufus (Bobcat)	X	X	X	X	X
Panthera onca (Jaguar)			X		X · X
Artiodactyla					
Tayassuidae					
Pecari tajacu (Collared peccary)			X		X · X

(cont.)

Table 3.1. (*cont.*)

Species	Coastal humid forest	Pinyon-juniper/ Steppe forest	Montane woodland forest	Subalpine forest	Northern boreal forest	Forest meadow	Riparian
Cervidae							
Alces alces (Moose)	X		X	X	X	X	X
Cervus elaphus (Elk)	X	X	X	X		X	X
Odocoileus hemionus (Mule deer)	X	X	X	X	X	X	
Odocoileus virginianus (White-tailed deer)	X	X	X	X	X	X	X
Rangifer tarandus (Caribou)					X	X	X
Antilocapridae							
Antilocapra americana (Pronghorn)		X	X				
Bovidae							
Bison bison (Bison)	X	X	X	X	X	X	
Oreamnos americanus (Mountain goat)			X	X	X	X	
Ovis canadensis (Bighorn sheep)		X	X	X	X	X	
Ovis dalli (Dall's sheep)					X	X	

Rodentia

Aplodontidae

Aplodontia rufa (Mountain beaver)

Castoridae

Castor canadensis (Beaver)

Sciuridae

Glaucomys sabrinus (Northern flying squirrel)

Marmota caligata (Hoary marmot)

Marmota flaviventris (Yellow-bellied marmot)

Marmota olympus (Olympic marmot)

Marmota vancouverensis (Vancouver marmot)

Sciurus aberti (Abert's squirrel)

Sciurus arizonensis (Arizona gray squirrel)

Sciurus griseus (Western gray squirrel)

Sciurus nayaritensis (Mexican fox squirrel)

Spermophilus armatus (Uinta ground squirrel)

Spermophilus beecheyi (California ground squirrel)

Taxon (common name)							
Rodentia							
Aplodontidae							
Aplodontia rufa (Mountain beaver)	X			X			X
Castoridae							
Castor canadensis (Beaver)	X	X	X	X		X	X
Sciuridae							
Glaucomys sabrinus (Northern flying squirrel)	X		X	X	X		X
Marmota caligata (Hoary marmot)		X		X	X		X
Marmota flaviventris (Yellow-bellied marmot)		X	X	X	X		X
Marmota olympus (Olympic marmot)				X			
Marmota vancouverensis (Vancouver marmot)				X			
Sciurus aberti (Abert's squirrel)				X	X		
Sciurus arizonensis (Arizona gray squirrel)					X		
Sciurus griseus (Western gray squirrel)	X			X	X	X	
Sciurus nayaritensis (Mexican fox squirrel)					X		
Spermophilus armatus (Uinta ground squirrel)			X	X	X	X	X
Spermophilus beecheyi (California ground squirrel)	X		X	X	X	X	X

(cont.)

Table 3.1. (cont.)

Species	Coastal humid forest	Pinyon-juniper/ Steppe forest	Montane woodland forest	Subalpine forest	Northern boreal forest	Forest meadow	Riparian
Spermophilus beldingi (Belding's ground squirrel)		X	X	X		X	
Spermophilus brunneus (Idaho ground squirrel)		X				X	
Spermophilus canus (Merriam's ground squirrel)		X				X	
Spermophilus columbianus (Columbian ground squirrel)			X	X		X	
Spermophilus elegans (Wyoming ground squirrel)		X	X	X		X	
Spermophilus lateralis (Golden-mantled ground squirrel)	X	X	X	X		X	
Spermophilus parryii (Arctic ground squirrel)					X	X	
Spermophilus saturatus (Cascade golden-mantled ground squirrel)			X	X		X	
Spermophilus tridecemlineatus (Thirteen-lined ground squirrel)			X				
Spermophilus variegatus (Rock squirrel)		X	X			X	
Tamias alpinus (Alpine chipmunk)				X			
Tamias amoenus (Yellow-pine chipmunk)		X	X				

Species (common name)						
Tamias caniceps (Gray-footed chipmunk)		X	X	X		
Tamias cinereicollis (Gray-collared chipmunk)		X	X	X		
Tamias dorsalis (Cliff chipmunk)		X	X			
Tamias merriami (Merriam's chipmunk)		X	X			
Tamias minimus (Least chipmunk)		X	X	X	X	X
Tamias obscurus (California chipmunk)		X	X			
Tamias ochrogenys (Yellow-cheeked chipmunk)	X					
Tamias palmeri (Palmer's chipmunk)		X	X	X		
Tamias panamintinus (Panamint chipmunk)		X	X			
Tamias quadrimaculatus (Long-eared chipmunk)		X	X			
Tamias quadrivittatus (Colorado chipmunk)		X	X			
Tamias ruficaudus (Red-tailed chipmunk)			X	X		
Tamias rufus (Hopi chipmunk)		X	X			
Tamias senex (Allen's chipmunk)	X		X	X		
Tamias siskiyou (Siskiyou chipmunk)	X		X	X		

(cont.)

Table 3.1. (cont.)

Species	Coastal humid forest	Pinyon-juniper/ Steppe forest	Montane woodland forest	Subalpine forest	Northern boreal forest	Forest meadow	Riparian
Tamias sononae (Sonoma chipmunk)	X		X				
Tamias speciosus (Lodgepole chipmunk)			X	X			
Tamias townsendii (Townsend's chipmunk)	X		X				
Tamias umbrinus (Uinta chipmunk)		X	X	X			
Tamiasciurus douglasii (Douglas squirrel)	X		X	X			
Tamiasciurus hudsonicus (Red squirrel)	X		X	X	X		
Geomyidae							
Thomomys bottae (Botta's pocket gopher)	X	X	X	X		X	
Thomomys mazama (Western pocket gopher)	X			X		X	
Thomomys monticola (Mountain pocket gopher)			X	X		X	
Thomomys talpoides (Northern pocket gopher)		X	X	X	X	X	
Thomomys umbrinus (Southern pocket gopher)			X	X		X	

Heteromyidae						
Chaetodipus californicus (California pocket mouse)				X		
Chaetodipus hispidus (Hispid pocket mouse)					X	
Dipodomys californicus (California kangaroo rat)			X		X	
Dipodomys heermanni (Heermann's kangaroo rat)			X			
Dipodomys ordii (Ord's kangaroo rat)				X		
Dipodomys panamintinus (Panamint kangaroo rat)				X		
Dipodomys venustus (Narrow-faced kangaroo rat)				X		
Perognathus alticola (White-eared pocket mouse)				X		
Perognathus flavescens (Plains pocket mouse)					X	
Perognathus longimembrus (Little pocket mouse)		X		X	X	
Perognathus parvus (Great Basin pocket mouse)		X		X	X	
Muridae						
Arborimus albipes (White-footed vole)						X
Arborimus longicaudus (Red tree vole)	X					X

(cont.)

Table 3.1. (*cont.*)

Species	Coastal humid forest	Pinyon-juniper/ Steppe forest	Montane woodland forest	Subalpine forest	Northern boreal forest	Forest meadow	Riparian
Arborimus pomo (Sonoma tree vole)	X						
Clethrionomys californicus (Western red-backed vole)	X		X	X			
Clethrionomys gapperi (Southern red-backed vole)	X		X	X	X	X	X
Clethrionomys rutilus (Northern red-backed vole)					X		
Lemmiscus curtatus (Sagebrush vole)		X	X				
Microtus californicus (California vole)	X		X			X	X
Microtus longicaudus (Long-tailed vole)	X	X	X	X	X	X	X
Microtus miurus (Singing vole)				X	X		
Microtus mogollonensis (Mogollon vole)					X		
Microtus montanus (Montane vole)		X	X	X		X	X
Microtus oregoni (Oregon vole)	X		X	X		X	
Microtus pennsylvanicus (Meadow vole)			X	X	X	X	X

Microtus richardsoni (Water vole)	X			X		X	
Microtus townsendii (Townsend's vole)						X	X
Microtus xanthognathus (Taiga vole)					X	X	X
Neotoma albigula (White-throated woodrat)		X	X				
Neotoma cinerea (Bushy-tailed woodrat)	X	X	X	X			
Neotoma fuscipes (Dusky-footed woodrat)	X	X	X	X			
Neotoma mexicana (Mexican woodrat)		X	X				
Neotoma stephensi (Stephen's woodrat)		X	X				
Ondatra zibethicus (Muskrat)	X		X	X			X
Peromyscus boylii (Brush mouse)		X	X	X			
Peromyscus californicus (California mouse)	X		X				
Peromyscus crinitus (Canyon mouse)		X	X				
Peromyscus eremicus (Cactus mouse)		X					
Peromyscus keeni (Northwestern deer mouse)	X		X	X			

(cont.)

Table 3.1. (*cont.*)

Species	Coastal humid forest	Pinyon-juniper/ Steppe forest	Montane woodland forest	Subalpine forest	Northern boreal forest	Forest meadow	Riparian
Peromyscus leucopus (White-footed mouse)		X	X				
Peromyscus maniculatus (Deer mouse)	X	X	X	X	X	X	X
Peromyscus nasutus (Northern rock mouse)		X					
Peromyscus truei (Pinyon mouse)	X	X	X				
Phenacomys intermedius (Western heather vole)					X		
Phenacomys ungava (Eastern heather vole)					X		
Reithrodontomys megalotis (Western harvest mouse)	X	X	X	X		X	X
Synaptomys borealis (Northern bog lemming)					X		X
Dipodidae							
Zapus hudsonius (Meadow jumping mouse)	X				X	X	
Zapus princeps (Western jumping mouse)	X			X	X	X	X
Zapus trinotatus (Pacific jumping mouse)	X		X	X		X	X

Erethizontidae							
Erethizon dorsatum	X	X	X	X	X		
(Porcupine)							
Myocastoridae							
Myocastor coypus	X						X
(Nutria)							
Lagomorpha							
Ochotonidae							
Ochotona collaris	X			X	X		
(Collared pika)							
Ochotona princeps	X		X	X			
(Pika)							
Leporidae							
Brachylagus idahoensis		X	X				
(Pygmy rabbit)							
Lepus americanus	X		X	X	X	X	
(Snowshoe hare)							
Lepus californicus	X	X	X	X		X	
(Black-tailed jackrabbit)							
Lepus townsendii		X	X	X		X	
(White-tailed jackrabbit)							
Sylvilagus audubonii		X	X			X	X
(Desert cottontail)							
Sylvilagus bachmani	X		X	X			X
(Brush rabbit)							
Sylvilagus nuttallii		X	X	X			X
(Mountain cottontail)							

forests embrace forests of red and subalpine firs (*Abies magnifica* and *A. lasiocarpa*, respectively), Engelmann spruce (*Picea engelmannii*), and various pines (e.g., bristlecone (*Pinus aristata*), limber (*P. flexilis*), white (*P. monticola*), whitebark (*P. albicaulis*), lodgepole (*P. contorta*)) extending in elevation from montane woodlands to timberline (comprising components of the four major vegetative zones of Hemstrom cited above for montane woodland forests). Northern boreal forests include the widespread subarctic forests that carpet much of central Canada and Alaska. Communities in these forests, which include the taiga of northernmost areas, are dominated by white and black spruce (*Picea glauca* and *P. mariana*, respectively) and variously contain birch (*Betula papyrifera*), quaking aspen (*Populus tremuloides*), lichens, and feather moss (*Hylocomium splendens*) (equivalent to the taiga and boreal forests of Hemstrom).

Mammals were identified as being associated with meadow or riparian zones only if they predominate in those areas on a broad geographic scale. This approach is somewhat more conservative than that used by Anthony et al. (2003) in their assessment of riparian species in Pacific northwest forests. Meadows, which may be extensive, refer to grasslands interspersed among forests that result from peculiar soil or moisture conditions. Riparian systems comprise aquatic and terrestrial habitats in and adjacent to streams, rivers, lakes, and tidal estuaries. Typically, riparian zones are more highly productive and ecologically rich than surrounding upland areas (Anthony et al. 2003).

Acknowledgments

For improvements to the content of this paper I thank Bob Anthony, Leslie Carraway, Larry Heaney, Barbara Lawlor, Dave Perault, and Cindy Zabel.

Range maps in figures were reproduced from Wilson and Ruff (1999), courtesy of Smithsonian Institution Press, Washington, DC.

Literature cited

Anthony, R.G., M.A. O'Connell, M.M. Pollock, and J.G. Hallett. 2003. Associations of mammals with riparian ecosystems in Pacific Northwest forests. Pages 510–563 *in* C.J. Zabel and R.G. Anthony, editors. *Mammal Community Dynamics. Management and Conservation in the Coniferous Forests of Western North America.* Cambridge University Press, Cambridge, UK.

Arbogast, B.S. 1999a. Mitochondrial DNA phylogeography of the New World flying squirrels (*Glaucomys*): implications for Pleistocene biogeography. *Journal of Mammalogy* **80**:142–155.

Arbogast, B.S. 1999*b*. Comparative phylogeography of North American boreal
    mammals. PhD Dissertation, Wake Forest University, Winston-Salem, North
    Carolina, USA.

Arbogast, B.S., and G.J. Kenagy. 2001. Comparative phylogeography as an integrative
    approach to historical biogeography. *Journal of Biogeography* **28**:819–825.

Arbogast, B.S., R.A. Browne, and P.D. Weigl. 2001. Evolutionary genetics and
    Pleistocene biogeography of North American tree squirrels (*Tamiasciurus*). *Journal
    of Mammalogy* **82**:302–319.

Armstrong, D.M. 1972. Distribution of mammals in Colorado. Volume 3, *Monograph of
    the Museum of Natural History, University of Kansas*, 415 pp.

Aubry, K.B., M.J. Crites, and S.D. West. 1991. Regional patterns of small mammal
    abundance and community composition in Oregon and Washington. Pages
    285–303 *in* L.F. Ruggiero et al., technical coordinators. Wildlife and vegetation
    of unmanaged Douglas-fir forests. USDA Forest Service General Technical
    Report **PNW-285:1–533**, Portland, Oregon, USA.

Aubry, K.B., J.P. Hayes, B.L. Biswell, and B.G. Marcot. 2003. The ecological role of
    tree-dwelling mammals in western coniferous forests. Pages 405–443 in C.J.
    Zabel and R.G. Anthony, editors. *Mammal Community Dynamics. Management and
    Conservation in the Coniferous Forests of Western North America*. Cambridge University
    Press, Cambridge, UK.

Banfield, A.W.F. 1974. *The Mammals of Canada*. University of Toronto Press, Toronto,
    Ontario, Canada.

Barclay, R.M.R., and R.M. Brigham (editors). 1996. Bats and forests symposium,
    October 19–21, 1995, Victoria, British Columbia, Canada. Research Branch,
    British Columbia Ministry of Forests, Working Paper 23/1996:1–292.

Brown, J.H. 1971*a*. Mechanisms of competitive exclusion between two species of
    chipmunks (*Eutamias*). *Ecology* **52**:306–311.

Brown, J.H. 1971*b*. Mammals on mountaintops: non-equilibrium insular biogeography.
    American Naturalist **105**:467–478.

Brown, J.H. 1978. The theory of insular biogeography and the distribution of boreal
    birds and mammals. *The Great Basin Naturalist Memoirs* **2**:209–227.

Buskirk, S.W. and W.J. Zielinksi. 2003. Small and mid-sized cornivores. Pages 207–249
    in C.J. Zabel and R.G. Anthony, editors. *Mammal Community Dynamics.
    Management and Conservation in the Coniferous Forests of Western North America*.
    Cambridge University Press, Cambridge, UK.

Byun, S.A., B.F. Koop, and T.E. Reichmen. 1997. North American black bear mtDNA
    phylogeography: implications for morphology and the Haida Gwaii glacial
    refugium controversy. *Evolution* **51**:1647–1653.

Carey, A.B. 1991. The biology of arboreal rodents in Douglas-fir forests. USDA Forest
    Service General Technical Report **PNW-276:223–250**, Portland, Oregon, USA.

Carey, A.B. 1995. Sciurids in Pacific northwest managed and old-growth forest. *Ecological
    Applications* **5**:648–661.

Carr, S.M., and S.A. Hicks. 1997. Are there two species of marten in North America?
    Genetic and evolutionary relationships within *Martes*. Pages 15–28 *in* G. Proulx,
    H.N. Bryant, and P.M. Woodard, editors. Martes: *Taxonomy, Ecology, Techniques,
    and Management*. Provincial Museum of Alberta, Canada.

Chappell, M.A. 1978. Behavioral factors in the altitudinal zonation of chipmunks
    (*Eutamias*). *Ecology* **59**:565–579.

Corn, P.S., and R.B. Bury. 1991. Small mammal communities in the Oregon Coast Range. Pages 241–254 *in* L.F. Ruggiero et al., technical coordinators. Wildlife and vegetation of unmanaged Douglas-fir forests. USDA Forest Service General Technical Report **PNW-285:1–533**, Portland, Oregon, USA.

Corn, P.S., R.B. Bury, and T. A. Spies. 1988. Douglas-fir forests in the Cascade Mountains of Oregon and Washington: is the abundance of small mammals related to stand age and moisture? Pages 340–352 *in* R. C. Szaro et al., technical coordinators. Management of amphibians, reptiles, and small mammals in North America. USDA Forest Service General Technical Report **RM-166:1–458**, Flagstaff, Arizona, USA.

Cowan, I.M. 1973. The mammals of British Columbia. Fifth edition. Volume 11, *Handbook of the British Columbia Provincial Museum, Victoria*, 414 pp.

Cronin, M.A. 1992. Intraspecific variation in mitochondrial DNA of North American cervids. *Journal of Mammalogy* **73**:70–82.

Dalquest, W.W. 1948. Mammals of Washington. Volume 2, *Monograph of the Museum of Natural History, University of Kansas*, 144 pp.

Davis, R., and C. Dunford. 1987. An example of contemporary colonization of montane islands by small, nonflying mammals in the American Southwest. *American Naturalist* **129**:398–406.

Davis, R., C. Dunford, and M.V. Lomolino. 1988. Montane mammals of the American Southwest: the possible influence of post-Pleistocene colonization. *Journal of Biogeography* **15**:841–848.

Davis, W.B. 1939. *The Recent Mammals of Idaho*. The Caxton Printers, Ltd., Caldwell, Idaho, USA.

Demboski, J.R. and J.A. Cook. 2001. Phylogeography of the dusky shrew, *Sorex monticolus* (Insectivora, Soricidae): insight into deep and shallow history in northwestern North America. *Molecular Ecology* **10**:1227–1240.

Demboski, J.R., K.D. Stone, and J.A. Cook. 1999. Further perspectives on the Haida Gwaii glacial refugium. *Evolution* **53**:2008–2012.

Durrant, S.D. 1952. Mammals of Utah. Taxonomy and distribution. Volume 6, *Monographs of the Museum of Natural History, University of Kansas*, 549 pp.

Fenton, M.B. 1990. The foraging behaviour and ecology of animal-eating bats. *Canadian Journal of Zoology* **68**:411–422.

Findley, J.S., A.H. Harris, D.E. Wilson, and C. Jones. 1975. *Mammals of New Mexico*. University of New Mexico Press, Albuquerque, New Mexico, USA.

Fitzgerald, J.P., C.A. Meaney, and D.M. Armstrong. 1994. *Mammals of Colorado*. Denver Museum of Natural History and University Press of Colorado, Niwot, Colorado, USA.

Foresman, K.R. 2001. *The Wild Mammals of Montana*. Volume 12, Special Publication, American Society of Mammalogists, 278 pp.

Franklin, J.F., and J.A. Fites-Kaufmann. 1996. Assessment of late-successional forests of the Sierra Nevada. Pages 627–656 *in Sierra Nevada Ecosystem Project. Status of the Sierra Nevada*. Volume II, Assessments and scientific basis for management options. Volume 37, Wildland Resources Center Report, Centers for Water and Wildland Resources, University of California, Davis, 1528 pp.

Gavin, T.A., P.W. Sherman, E. Yensen, and B. May. 1999. Population genetic structure of the northern Idaho ground squirrel *Spermophilus brunneus brunneus*. *Journal of Mammalogy* **80**:156–168.

Gilbert, F.F., and R. Allwine. 1991. Small mammal communities in the Oregon Cascade Range. Pages 257–267 *in* L.F. Ruggiero et al., technical coordinators. Wildlife and vegetation of unmanaged Douglas-fir forests. USDA Forest Service General Technical Report **PNW-285:1–533**, Portland, Oregon, USA.

Graham, R.W. 1986. Response of mammalian communities to environmental changes during the Late Quaternary. Pages 300–313 *in* J.R. Diamond and T.J. Case, editors. *Community Ecology*. Harper and Row, New York, New York, USA.

Graham, R.W., E.L. Lundelius, Jr., M.A. Graham, E.K. Schroeder, R.S. Toomey III, E. Anderson, A.D. Barnosky, J.A. Burns, C.S. Churcher, C.K. Grayson, R.D. Guthrie, C.R. Harrington, G.T. Jefferson, L.D. Martin, H.G. McDonald, R.E. Morlan, H.A. Semken, Jr., S.D. Webb, L. Werdelin, and M.C. Wilson. 1996. Spatial response of mammals to late-Quaternary environmental fluctuations. *Science* **272**:1601–1606.

Grayson, D. 1993. *The Deserts Past: A Natural Prehistory of the Great Basin*. Smithsonian Institution Press, Washington, DC, USA.

Grayson, D., and D.B. Madsen. 2000. Biogeographic implications of recent low-elevation recolonization by *Neotoma cinerea* in the Great Basin. *Journal of Mammalogy* **81**:1100–1105.

Grayson, D., S.D. Livingston, E. Rickart, and M.W. Shaver III. 1996. Biogeographic significance of low elevation records for *Neotoma cinerea* from the northern Bonneville Basin, Utah. *The Great Basin Naturalist* **56**:191–196.

Hall, E.R. 1946. *Mammals of Nevada*. University of California Press, Berkeley, California, USA.

Hall, E.R. 1981. *The Mammals of North America*. Two Volumes. John Wiley & Sons, New York, New York, USA.

Hall, E.R. 1995. *Mammals of Nevada*. Second Edition. University of Nevada Press, Reno, Nevada, USA.

Hallett, J.G., M.A. O'Connell, and C.C. Maguire. 2003. Ecological relationships of terrestrial small mammals in western coniferous forests. Pages 120–156 *in* C.J. Zabel and R.G. Anthony, editors. *Mammal Community Dynamics. Management and Conservation in the Coniferous Forests of Western North America*. Cambridge University Press, Cambridge, UK.

Hamilton, W.J., III. 1962. Reproductive adaptations of the red tree mouse. *Journal of Mammalogy* **43**:486–504.

Hayes, J.P. 2003. Habitat ecology and conservation of bats in western coniferous forests. Pages 81–119 *in* C.J. Zabel and R.G. Anthony, editors. *Mammal Community Dynamics. Management and Conservation in the Coniferous Forests of Western North America*. Cambridge University Press, Cambridge, UK.

Hemstrom, M.A. 2003. Forests and woodlands of western North America. Pages 9–40 *in* C.J. Zabel and R.G. Anthony, editors. *Mammal Community Dynamics. Management and Conservation in the Coniferous Forests of Western North America*. Cambridge University Press, Cambridge, UK.

Hoffmeister, D.F. 1986. *Mammals of Arizona*. University of Arizona Press, Tucson, Arizona, USA.

Hogan, K.M., M.C. Hedin, H.S. Koh, S.K. Davis, and I.F. Greenbaum. 1993. Systematic and taxonomic implications of karyotypic, electrophoretic, and mitochondrial-DNA variation in *Peromyscus* from the Pacific Northwest. *Journal of Mammalogy* **74**:819–830.

Holt, R.D., G.R. Robinson, and M.S. Gaines. 1995. Vegetation dynamics in an experimentally fragmented landscape. *Ecology* **76**:1610–1624.

Ingles, L.G. 1948. *Mammals of California*. Stanford University Press, Stanford, California, USA.

Ingles, L.G. 1954. *Mammals of California and Its Coastal Waters*. Stanford University Press, Stanford, California, USA.

Ingles, L.G. 1965. *Mammals of the Pacific States*. California, Oregon, Washington. Stanford University Press, Stanford, California, USA.

Kie, J.G., R.T. Bowyer, and K.M. Stewart. 2003. Ungulates in western coniferous forests: habitat relationships, population dynamics, and ecosystem processes. Pages 296–340 *in* C.J. Zabel and R.G. Anthony, editors. *Mammal Community Dynamics. Management and Conservation in the Coniferous Forests of Western North America*. Cambridge University Press, Cambridge, UK.

Kirkland, G.L., Jr. 1985. Small mammal communities in temperate North American forests. *Australian Mammalogy* **8**:137–144.

Kunkel, K.E. 2003. Ecology, conservation, and restoration of large carnivores in western North America. Pages 250–295 *in* C.J. Zabel and R.G. Anthony, editors. *Mammal Community Dynamics. Management and Conservation in the Coniferous Forests of Western North America*. Cambridge University Press, Cambridge, UK.

Kunz, T.H. 1982. Roosting ecology of bats. Pages 1–56 *in* T. H. Kunz, editor. *Ecology of Bats*. Plenum Press, New York, New York, USA.

Kurtèn, B., and E. Anderson. 1980. *Pleistocene Mammals of North America*. Columbia University Press, New York, New York, USA.

Larrison, E.J. 1976. *Mammals of the Northwest. Washington, Oregon, Idaho, and British Columbia*. Seattle Audubon Society, Seattle, Washington, USA.

Larrison, E.J. and D.R. Johnson. 1981. *Mammals of Idaho*. University of Idaho Press, Pocatello, Idaho, USA.

Lawlor, T.E. 1998. Biogeography of Great Basin mammals: paradigm lost? *Journal of Mammalogy* **79**:1111–1130.

Lomolino, M.V., and D.R. Perault. 2000. Assembly and disassembly of mammal communities in a fragmented temperate rain forest. *Ecology* **81**:1517–1532.

Lomolino, M.V., J.H. Brown, and R. Davis. 1989. Island biogeography of montane forest mammals in the American Southwest. *Ecology* **70**:180–194.

Long, C.A. 1965. Mammals of Wyoming. Volume 14, *Monograph of the Museum of Natural History, University of Kansas*, pp. 493–758.

MacMahon, J.A., R.R. Parmenter, K.A. Johnson, and C.M. Crisafulli. 1989. Small mammal recolonization on the Mount St. Helens volcano: 1980–1987. *American Midland Naturalist* **122**:365–387.

Martin, K.J., and W.C. McComb. 2003. Small mammals in a landscape mosaic: implications for conservation. Pages 567–586 *in* C.J. Zabel and R.G. Anthony, editors. *Mammal Community Dynamics. Management and Conservation in the Coniferous Forests of Western North America*. Cambridge University Press, Cambridge, UK.

Maser, C. 1998. *Mammals of the Pacific Northwest from the Coast to the High Cascades*. Oregon State University Press, Corvallis, Oregon, USA.

Maser, C., B.R. Mate, J.F. Franklin, and C.T. Dryness. 1981. Natural history of Oregon coast mammals. USDA Forest Service General Technical Report **PNW-133:1–496**, Portland, Oregon, USA.

McNab, B.K. 1978. Energetics of arboreal folivores: physiological problems and ecological consequences of feeding on an ubiquitous food supply. Pages 153–162 *in* G.G. Montgomery, editor. *The Ecology of Arboreal Folivores*. Smithsonian Institution Press, Washington, DC, USA.

Meiselman, N. 1992. Nest-site characteristics of red tree voles in Douglas-fir forests of northern California. Master's Thesis, Humboldt State University, Arcata, California, USA.

Meiselman, N. and A.T. Doyle. 1996. Habitat and microhabitat use by the red tree vole *Phenacomys longicaudus*. *American Midland Naturalist* **135**:33–42.

Patterson, B.D. 1984. Mammalian extinction and biogeography in the southern Rocky Mountains. Pages 247–293 *in* M.H. Nitecki, editor. *Extinctions*. University of Chicago Press, Chicago, Illinois, USA.

Patterson, B.D. 1995. Local extinctions and the biogeographic dynamics of boreal mammals in the Southwest. Pages 151–176 *in* C. Istock and R.S. Hoffman, editors. *Storm Over a Mountain: Conservation Biology and the Mount Graham Affair*. University of Arizona Press, Tucson, Arizona, USA.

Perault, D.R. and M.V. Lomolino. 2000. Corridors and mammal community structure across a fragmented, old-growth forest landscape. *Ecological Monographs* **70**:401–422.

Ransome, D.B. and T.P. Sullivan. 1997. Food limitation and habitat preference of *Glaucomys sabrinus* and *Tamiasciurus hudsonicus*. *Journal of Mammalogy* **78**:538–549.

Raphael, M.G. 1988a. Long-term trends in abundance of amphibians, reptiles, and mammals in Douglas-fir forests of northwestern California. Pages 23–31 *in* R.C. Szaro et al., technical coordinators. Management of amphibians, reptiles, and small mammals in North America. USDA Forest Service General Technical Report **RM-166:1–458**, Flagstaff, Arizona, USA.

Raphael, M.G. 1988b. Habitat associations of small mammals in a subalpine forest, Wyoming. Pages 359–367 *in* R.C. Szaro et al., technical coordinators. Management of amphibians, reptiles, and small mammals in North America. USDA Forest Service General Technical Report **RM-166:1–458**, Flagstaff, Arizona, USA.

Raphael, M.G. 1991. Vertebrate species richness within and among seral stages of Douglas-fir/hardwood forest in northwestern California. Pages 415–423 *in* L.F. Ruggiero et al., technical coordinators. Wildlife and vegetation of unmanaged Douglas-fir forests. USDA Forest Service General Technical Report **PNW-285:1–533**, Portland, Oregon, USA.

Rickart, E.A. 2001. Elevational diversity gradients, biogeography and structure of montane mammal communities in the intermountain region of North America. *Global Ecology and Biogeography* **10**:77–100.

Ruggiero, L.F., K.B. Aubry, S.W. Buskirk, L.J. Lyon, and W.J. Zielinski, technical editors. 1994. The scientific basis for conserving forest carnivores. American marten, fisher, lynx, and wolverine in the western United States. USDA Forest Service General Technical Report **RM-254:1–184**, Fort Collins, Colorado, USA.

Schweiger, E.W., J.E. Diffendorfer, R.D. Holt, R. Pierotti, and M.S. Gaines. 2000. The interaction of habitat fragmentation, plant, and small mammal succession in an old field. *Ecological Monographs* **70**:383–400.

Smith, C.C. 1981. The indivisible niche of *Tamiasciurus*:an example of nonpartitioning of resources. *Ecological Monographs* **51**:343–363.

Smith, H.C. 1993. *Alberta Mammals. Atlas and Guide*. Provincial Museum of Alberta, Edmonton, Alberta, Canada.

Smith, W.P., R.G. Anthony, J.R. Waters, N.L. Dodd, and C.J. Zabel. 2003. Ecology and conservation of arboreal rodents of western coniferous forests. Pages 157–206 *in* C.J. Zabel and R.G. Anthony, editors. *Mammal Community Dynamics. Management and Conservation in the Coniferous Forests of Western North America*. Cambridge University Press, Cambridge, UK.

Spies, T.A., and M.G. Turner. 1999. Dynamic forest mosaics. Pages 95–160 *in* M.L. Hunter, Jr., editor. *Maintaining Biodiversity in Forest Ecosystems*. Cambridge University Press, Cambridge, UK.

Stone, K.D., and J.A. Cook. 2000. Phylogeography of black bears (*Ursus americanus*) of the Pacific Northwest. *Canadian Journal of Zoology* **78**:1218–1223.

Sutton, D.A., and C.F. Nadler. 1974. Systematic revision of three Townsend chipmunks (*Eutamias townsendii*). *Southwestern Naturalist* **19**:199–212.

Sutton, D.A., and B.D. Patterson. 2000. Geographic variation of the western chipmunks *Tamias senex* and *T. siskiyou*, with two new subspecies from California. *Journal of Mammalogy* **81**:299–316.

Thompson, J.L., and L.V. Diller. 2002. Relative abundance, nest site characteristics, and nest dynamics of dusky tree voles on managed timberlands in coastal northwest California. *Northwestern Naturalist* **83**:91–100.

Vaughan, T.A., and N.J. Czaplewski. 1985. Reproduction in Stephen's woodrat: the wages of folivory. *Journal of Mammalogy* **66**:429–443.

Verts, B.J., and L.N. Carraway. 1995. Phenacomys albipes. *Mammalian Species* **494**:1–5.

Verts, B.J., and L.N. Carraway. 1998. Land Mammals of Oregon. University of California Press, Berkeley, California, USA.

Wilson, D.E., and D.M. Reeder, editors. 1993. *Mammal Species of the World. A Taxonomic and Geographic Reference*. Smithsonian Institution Press, Washington, DC, USA.

Wilson, D.E., and S. Ruff, editors. 1999. *The Smithsonian Book of North American Mammals*. Smithsonian Institution Press, Washington, DC, USA.

Wooding, S., and R. Ward. 1997. Phylogeography and Pleistocene evolution in the North American black bear. *Molecular Biology and Evolution* **14**:1096–1105.

Zeveloff, S.I., and F.R. Collett. 1988. *Mammals of the Intermountain West*. University of Utah Press, Salt Lake City, Utah, USA.

4
———————

Habitat ecology and conservation of bats in western coniferous forests

Introduction

Until recently, little was known about most aspects of the ecology of bats in forests. A basic understanding of several fundamental elements of the natural history of forest-dwelling bats, such as characteristics of habitat used for foraging and structures used for roosting, was lacking or minimal. In addition, influences of forest structure and management practices on bats were almost entirely unknown. Within the past decade, significant strides have been made in the development of our understanding of the ecology of bats in western coniferous forests and the influences of forest management on bats, but substantial gaps in the knowledge base remain. In this chapter, I review what is known about the habitat ecology of bats in western coniferous forests, evaluate the strengths and weaknesses of the current state of knowledge, suggest directions for future research, and discuss some of the implications for forest management and conservation of bats. I emphasize information based on studies conducted within western coniferous forests whenever possible. Throughout my review, I generally focus on aspects of the ecology of bats in western coniferous forests that are common to several species. Although emphasizing commonalities among species is valuable to help identify unifying principles, it should be recognized that each species has a unique natural history and suite of behavioral characteristics, and that these differences are often important in the conservation of bats.

Forty species of bats occur west of the Mississippi River and north of Mexico. Of these, 19 may be influenced by management of western coniferous forests. The degree of association between bats and coniferous forest habitat varies considerably among these species. Although in reality, association with forested habitat represents a continuum rather than a set of discrete categories, bats can be considered in three broad categories based on their association with western coniferous forested habitat (Table 4.1). The first group includes six species that are primarily associated with forests and regularly occur in western coniferous forests. Although no species of bat is restricted to coniferous forest, some, such as the long-legged myotis (*Myotis volans*) and Keen's myotis (*M. keenii*), are closely associated with coniferous forests throughout much of their range (Warner and Czaplewski 1984, Nagorsen and Brigham 1993). Some in this category, such as the hoary bat (*Lasiurus cinereus*), western red bat (*L. blossevillii*), and northern long-eared bat (*M. septentrionalis*), commonly occur in deciduous forests as well as in coniferous forests (Shump and Shump 1982a, 1982b, Caceres and Barclay 2000). The second group includes 13 species that occur in western coniferous forests but also frequently occur in non-forested habitats, including urban environments, agricultural areas, or arid habitats. The degree of association of these species with western coniferous forests varies considerably among species and with region and forest type. Although some of these species, such as the big brown bat (*Eptesicus fuscus*) and little brown bat (*M. lucifugus*), are abundant in non-forested habitat (Fenton and Barclay 1980, Kurta and Baker 1990), they are often abundant in coniferous forests, and habitat structure and management of western coniferous forests may significantly influence their local or regional abundance. In contrast, the spotted bat (*Euderma maculatum*) occurs most frequently in arid areas and roosts primarily in cliff faces, but forages over meadows and marshes that are frequently in close proximity to woodlands or forests (Leonard and Fenton 1983, Wai-Ping and Fenton 1989, Pierson and Rainey 1998). Management activities in these forests presumably could influence use of foraging habitat by spotted bats. The third category includes 21 species of bats that occur in western North America north of Mexico, but rarely or never occur in western coniferous forests. Many of these species have restricted distributions west of the Mississippi River, with distributions that are centered in the eastern United States (Rafinesque's big-eared bat (*Corynorhinus rafinesquii*), northern yellow bat (*L. intermedius*), Seminole bat (*L. seminolus*), southeastern myotis (*Myotis austroriparius*), gray myotis (*M. grisescens*), Indiana bat (*M. sodalis*), evening bat (*Nycticeius humeralis*), and eastern pipistrelle (*Pipistrellus subflavus*)) or south of the border with Mexico (Mexican long-tongued bat (*Choeronycteris mexicana*), hairy-legged vampire bat (*Diphylla ecaudata*), Underwood's mastiff bat (*Eumops*

Table 4.1. *Species of bats that may be influenced by management of western coniferous forests and their degree of association with forested habitat*

Primarily associated with forested habitat	
Silver-haired bat	*Lasionycteris noctivagans*
Western red bat	*Lasiurus blossevillii*
Hoary bat	*Lasiurus cinereus*
Keen's myotis	*Myotis keenii*
Northern long-eared myotis	*Myotis septentrionalis*
Long-legged myotis	*Myotis volans*
Occur in both forested and non-forested habitats	
Pallid bat	*Antrozous pallidus*
Townsend's big-eared bat	*Corynorhinus townsendii*
Big brown bat	*Eptesicus fuscus*
Spotted bat	*Euderma maculatum*
Allen's big-eared bat	*Idionycteris phyllotis*
Southwestern myotis	*Myotis auriculus*
California myotis	*Myotis californicus*
Western small-footed myotis	*Myotis ciliolabrum*
Long-eared myotis	*Myotis evotis*
Little brown bat	*Myotis lucifugus* [a]
Fringed myotis	*Myotis thysanodes*
Yuma myotis	*Myotis yumanensis*
Big free-tailed bat	*Nyctinomops macrotis*
Rarely or never occur in western coniferous forests	
Mexican long-tongued bat	*Choeronycteris mexicana*
Rafinesque's big-eared bat	*Corynorhinus rafinesquii*
Hairy-legged vampire bat	*Diphylla ecaudata*
Western mastiff bat	*Eumops perotis*
Underwood's mastiff bat	*Eumops underwoodi*
Northern yellow bat	*Lasiurus intermedius*
Seminole bat	*Lasiurus seminolus*
Western yellow bat	*Lasiurus xanthinus*
Southern long-nosed bat	*Leptonycteris curasoae*
Mexican long-nosed bat	*Leptonycteris nivalis*
California leaf-nosed bat	*Macrotus californicus*
Ghost-faced bat	*Mormops megalophylla*
Southeastern myotis	*Myotis austroriparius*
Gray myotis	*Myotis grisescens*
Indiana bat	*Myotis sodalis*
Cave myotis	*Myotis velifer*
Evening bat	*Nycticeius humeralis*
Pocketed free-tailed bat	*Nyctinomops femorosaccus*
Western pipistrelle	*Pipistrellus hesperus*
Eastern pipistrelle	*Pipistrellus subflavus*
Brazilian free-tailed bat	*Tadarida brasiliensis*

[a] In this review, published information on the Arizona myotis (*Myotis occultus*) is included with little brown bats following Valdez et al.'s (1999) assessment that the putative species *M. occultus* is a sub-species of the little brown bat.

underwoodi), western yellow bat (*Lasiurus xanthinus*), southern long-nosed bat (*Leptonycteris curasoae*), Mexican long-nosed bat (*Leptonycteris nivalis*), California leaf-nosed bat (*Macrotus californicus*), ghost-faced bat (*Mormoops megalophylla*), and pocketed free-tailed bat (*Nyctinomops femorosaccus*)). Others occur primarily in arid habitats, and rarely occur in coniferous forests (western mastiff bat (*Eumops perotis*), cave myotis (*Myotis velifer*), western pipistrelle (*Pipistrellus hesperus*), and Brazilian free-tailed bat (*Tadarida brasiliensis*)). Species in this category will not be considered further in this chapter.

Of the 19 species of bats having some level of association with western coniferous forests, 18 are in the family Vespertilionidae and one is in the family Molossidae; roughly half of the species are in the genus *Myotis*. Adults range in size from 4 to 30 g. All are insectivorous (Black 1974, Whitaker et al. 1977, 1981a, 1981b, Barclay 1985, Rolseth et al. 1994, Wilson and Ruff 1999) and most are aerial insectivores, feeding on flying insects. Some species also glean invertebrates from the ground or vegetation. Of the species considered here, gleaning is best documented for the long-eared myotis (*Myotis evotis*; Faure and Barclay 1992), northern long-eared myotis (Faure et al. 1993), southwestern myotis (*M. auriculus*; Fenton and Bell 1979), and pallid bat (*Antrozous pallidus*; Bell 1982), but gleaning also may be used by Allen's big-eared bat (*Idionycteris phyllotis*; Czaplewski 1983), Keen's myotis, fringed myotis (*M. thysanodes*), and other species.

Roosting ecology

Understanding the roosting ecology of bats is fundamental to understanding the ecology of bats in forested ecosystems. Roosts provide bats with sites for resting, protection from weather and predators, rearing young, hibernation, digestion of food, mating, and social interactions (Kunz 1982). For a roost to be suitable for a bat, it must provide a safe refuge, suitable microclimatic conditions, and be accessible to foraging and drinking sites (Kunz 1982). Roosts are hypothesized to be the primary factor limiting distribution and abundance of bats in many areas (Humphrey 1975, Ports and Bradley 1996, West and Swain 1999). Forest management activities can directly influence the distribution, abundance, and survival of bats through influences on the abundance, distribution, and quality of roost sites.

Bats often use different types of roosts for different purposes. For this chapter, I classify roosts into three basic categories. Day roosts are here defined as roosts that are used by bats during daylight hours in the summer months and parts of the spring and fall. In some species, females establish communal day roosts during parturition and lactation. In this chapter I consider these roosts, known as "maternity roosts," along with day roosts used for other purposes. Night roosts are roosts used nocturnally, often for short periods of time (from a few minutes to several hours), during the same time of the year. Winter roosts are roosts used during the winter months and include hibernation sites (hibernacula). In some cases, the same structures are used as day roosts, night roosts, and winter roosts, but most species use different structures seasonally and for different purposes.

With the exception of large communal roosts in caves and man-made structures, until recently descriptions of roosts in forests were largely anecdotal (e.g., Barclay and Cash 1985, Parsons et al. 1986, Clark 1993, Kurta et al. 1993). In the early to mid 1990's, development of light-weight (<0.75 g) radio transmitters spawned a proliferation of studies examining the roosting ecology of small- to medium-sized bats in forests; consideration of these studies reveals a number of basic patterns in use of roosts in western coniferous forests.

Day roosts

Cavities and crevices in trees and snags

Most species of bats that are closely associated with western coniferous forests use crevices or cavities in snags or trees as primary sites for roosting (Table 4.2). Bats frequently roost in cracks or splits in tree trunks or behind loose or sloughing bark (Barclay et al. 1988, Chung-MacCoubrey 1996, Mattson et al. 1996, Vonhof and Barclay 1996, Brigham et al. 1997b, Kalcounis and Brigham 1998, Rabe et al. 1998, Grindal 1999, Barclay and Brigham 2001, Weller and Zabel 2001). Although live trees are sometimes used as roosts by crevice-roosting species of bats, snags are more commonly used (Campbell et al. 1996, Vonhof and Barclay 1996, Brigham et al. 1997b, Betts 1998a, Ormsbee and McComb 1998, Rabe et al. 1998, Waldien et al. 2000, Weller and Zabel 2001). Cavities excavated by woodpeckers are sometimes used as roost sites (Kalcounis and Hecker 1996, Mattson et al. 1996, Vonhof and Barclay 1996, Betts 1998a, Kalcounis and Brigham 1998, Grindal 1999, Bonar 2000). Cavities excavated by woodpeckers provide the dominant roosting habitat for some species of bats in some areas (Kalcounis and Brigham 1998) and may be particularly important as sites for maternity roosts (Mattson et al. 1996). Deep fissures

Table 4.2. *Structures frequently used as day roosts by bats in western coniferous forests*

Species	Crevice or cavity in tree or snag	Foliage	Log	Stump	Cave or mine	Rock crevice	Bridge or building
Primarily associated with forested habitat							
Silver-haired bat	P				S	S	S
Western red bat		P					
Hoary bat		P					
Keen's myotis	P?						
Northern long-eared myotis	P						
Long-legged myotis	P				S	S	
Occur in both forested and non-forested habitats							
Pallid bat	P				S	S	S
Townsend's big-eared bat	S				P	S	S
Big brown bat	P				S	S	S
Spotted bat					S	P	
Southwestern myotis	P						
Allen's big-eared bat						P	
California myotis	P				S	S	S
Western small-footed myotis	S					P	S
Long-eared myotis	P		S	S	S	S	S
Little brown bat	P				S	S	S
Fringed myotis	P				S		S
Yuma myotis	P?				S		S
Big free-tailed bat	S				P		S

Primary roost structures (P) are those most frequently used by the species in or near forested habitat. Secondary roost structures (S) are those that are sometimes used by the species in forested habitats and may be used more frequently in non-forested habitat. "Crevice or cavity in tree or snag" includes hollow trees.

in the bark of trees, such as those often found on the bark of old Douglas-fir (*Pseudotsuga menziesii*) trees, have been postulated to be important roost sites for bats (Perkins and Cross 1988, Christy and West 1993). While fissures in bark of conifers sometimes may be used as roosts, extensive use of these structures has not been documented. Basal hollows in trees are sometimes used for roosting, and this has been observed in western redcedar (*Thuja plicata*; Ormsbee 1996) and extensively documented in redwood (*Sequoia sempervirens*; Rainey et al. 1992, Gellman and Zielinski 1996, Zielinski and Gellman 1999), but likely occurs in other species as well. Stumps also are sometimes used as roost sites (Vonhof and Barclay 1997, Waldien et al. 2000).

The selection of roosts by bats sometimes is constrained by the availability of structures in an area (Campbell et al. 1996). For example, when using cavities excavated by woodpeckers, selection of roost sites is restricted to snags selected by primary excavators (Kalcounis and Hecker 1996, Kalcounis and Brigham 1998). As a consequence, in this case apparent selection by bats for characteristics of a roost (e.g., diameter or species of a snag) may be an artifact of selection by the woodpecker. Similarly, landscape characteristics, such as abundance and characteristics of snags in an area, can affect the types of roosts available and types of roosts used by some species (Waldien et al. 2000). Differences in roost availability may influence the characteristics (e.g., Kalcounis and Hecker 1996) or types (e.g., Brigham 1991, Waldien et al. 2000) of structures used.

Trees and snags used by crevice-roosting bats are typically relatively large in diameter, height, or both (Campbell et al. 1996, Vonhof and Barclay 1996, Brigham et al. 1997b, Betts 1998a, Ormsbee and McComb 1998, Rabe et al. 1998, Waldien et al. 2000, Weller and Zabel 2001). Roost trees and snags in western coniferous forests generally exceed 60 cm in diameter and 18 m in height, but considerable variation exists in size of roosts used (Table 4.3). Trees and snags used as day roosts frequently are taller than the surrounding canopy (Campbell et al. 1996, Vonhof and Barclay 1996, Brigham et al. 1997b, Betts 1998a, Ormsbee and McComb 1998, Waldien et al. 2000, Weller and Zabel 2001), occur in gaps, openings, or areas with low canopy cover (Campbell et al. 1996, Vonhof and Barclay 1996, Brigham et al. 1997b, Waldien et al. 2000), or along forest edges (Grindal 1999, Waldien et al. 2000).

Thermal characteristics of roosts may influence patterns of roost use. Several authors have suggested that bats use tall roost structures in part because the high levels of solar radiation that these structures receive warm them, imparting thermoregulatory benefits to bats that use them (Campbell et al. 1996, Gellman and Zielinski 1996, Vonhof and Barclay 1996, Betts 1998a, Ormsbee and McComb 1998, Weller and Zabel 2001). Roosts located in forest gaps or areas with reduced canopy cover may provide similar thermoregulatory benefits (Betts 1998a, Waldien et al. 2000). Use of warm roosts has been hypothesized to be particularly important for pregnant females because fetal development is related to roost

Table 4.3. *Characteristics of trees and snags used as day roosts by bats in western coniferous forests*

Species	Roost structure used	Diameter (cm)		Height (m)		Stage of decay	No. of roosts	Ref.[h]
		Mean (SD)	Range	Mean (SD)	Range			
Pallid bat	Ponderosa pine snag	69.2 (14.6)[a,b]	31.2–101.6[a]	17.8 (7.9)[a,b]	31.2–101.6[a]	loose bark[a]	3	9
Big brown bat	Ponderosa pine snag	69.2 (14.6)[a,b]	31.2–101.6[a]	17.8 (7.9)[a,b]	31.2–101.6[a]	loose bark[a]	6	9
	Ponderosa pine snag	62.7 (15.2)[b]	–	–	–	–	8	3
	Conifer snag	–	–	–	–	intermediate	2	10
	Aspen tree (living and dead)	35.8 (7.2)	22.8–57.1	25.6 (7.7)	13.6–51.8	trees with cavities	27	6
	Pileated woodpecker cavity	–	–	–	–	–	37[c]	2
Allen's big-eared bat	Ponderosa pine snag	69.2 (14.6)[a,b]	31.2–101.6[a]	17.8 (7.9)[a,b]	31.2–101.6[a]	loose bark[a]	16	9
Silver-haired bat	Conifer snag	–	–	–	–	intermediate	8	10
	Conifer snag	59.8 (14.2)	–	26.7 (7.8)	–	early to intermediate	17	1
	Conifer trees and snags[d]	47	–	–	6.9–15.2	intermediate to late	15	5
	Ponderosa pine trees and snags[e]	39 (12.3)[b]	13–63	14.2 (5.5)[b]	3.7–24.1	intermediate to late	39	7
	Pileated woodpecker cavity	–	–	–	–	–	37[c]	2
Southwestern myotis	Gambel oak cavity (live tree)	–	–	–	–	–	2	9
California myotis	Conifer snag	56 (16.8)	–	27 (7.9)	–	intermediate	19	4
Long-eared myotis	Ponderosa pine snag	69.2 (14.6)[a,b]	31.2–101.6[a]	17.8 (7.9)[a,b]	31.2–101.6[a]	loose bark[a]	24	9
	Conifer snag	–	–	–	–	intermediate	3	10
	Conifer snag	93 (52.3)[b]	–	34 (21.8)[b]	–	intermediate	21	11
	Conifer tree (live)	–	–	–	–	–	3	11
	Gambel oak snag	–	–	–	–	–	2	9
	Gambel oak cavity (live tree)	–	–	–	–	–	2	9
	Hardwood tree (live)	–	–	–	–	–	6	11

Little brown bat[f]	Ponderosa pine snag	69.2 (14.6)[a,b]	31.2–101.6[a]	17.8 (7.9)[a,b]	31.2–101.6[a]	loose bark[a]	21	9
	Douglas-fir snag	—	—	—	—	—	1	9
	Pileated woodpecker cavity	—	—	—	—	—	37[c]	2
Fringed myotis	Ponderosa pine snag	69.2 (14.6)[a,b]	31.2–101.6[a]	17.8 (7.9)[a,b]	31.2–101.6[a]	loose bark[a]	15	9
	Douglas-fir snag	—	—	—	—	—	1	9
	Conifer snag[g]	120.8 (24.9)[b]	58.5–167.0	40.5 (13.6)[b]	15.8–57.5	intermediate	23	12
Long-legged myotis	Ponderosa pine snag	69.2 (14.6)[a,b]	31.2–101.6[a]	17.8 (7.9)[a,b]	31.2–101.6[a]	loose bark[a]	13	9
	Conifer snag	—	—	—	—	intermediate	5	10
	Conifer snag	97 (41.4)[b]	34–172	38 (17.8)[b]	13–72	early to intermediate	36	8
	Live conifer	—	—	—	—	—	4	8

[a] Information for ponderosa pine snags in Rabe et al. (1998) are based on a sample size of 54 for combined information for pallid bats, big brown bats, Allen's big-eared bats, long-eared bats, little brown bats, fringed myotis, and long-legged myotis.

[b] Standard deviation calculated based on information provided for sample size and standard error.

[c] Bonar (2000) did not distinguish between roosts of little brown bats, silver-haired bats, and big brown bats; as a consequence sample size represents the combined number of roosts recorded for these three species combined.

[d] Of 15 roosts examined, 1 was in a live tree, the remainder were in snags; quantitative information presented in Campbell et al. (1996) was not separated by roost type, and thus is combined here.

[e] Of 39 roosts examined, 1 was in an aspen (*Populus tremuloides*); quantitative information presented in Mattson et al. (1996) was not separated by roost type, and thus is combined here.

[f] Information from Rabe et al. (1988) for the Arizona myotis is presented under little brown bat based on Valdez et al.'s (1999) conclusion that the putative species *Myotis occultus* should be considered a subspecies of the little brown bat.

[g] Of 23 roosts examined, 20 were in Douglas-fir snags, 2 were in sugar pine (*Pinus lambertiana*) snags, and 1 was in a ponderosa pine snag; quantitative information presented in Weller and Zabel (2001) was not separated by species of roost tree, and thus is combined here.

[h] References are as follows: 1, Betts 1998a; 2, Bonar 2000; 3, Brigham 1991; 4, Brigham et al. 1997b; 5, Campbell et al. 1996; 6, Kalcounis and Brigham 1998; 7, Mattson et al. 1996; 8, Ormsbee and McComb 1998; 9, Rabe et al. 1998; 10, Vonhof and Barclay 1996; 11, Waldien et al. 2000; 12, Weller and Zabel 2001.

temperature (Racey 1973, Racey and Swift 1981) and low ambient temperatures can reduce reproductive success (Lewis 1993). Fetal and juvenile
growth rates are reduced if pregnant or lactating bats frequently enter torpor (Racey and Swift 1981, Hoying and Kunz 1998). In contrast, males may
incur greater energy savings by roosting in cool sites and entering torpor
to minimize thermoregulatory costs (Tuttle 1976, Williams and Findley
1979, Hamilton and Barclay 1994). Thus, it has been hypothesized that
sites selected by male bats for roosting may be cooler than sites selected by
females (Vonhof and Barclay 1996). Elevational patterns in distributions
of male and female bats (Cryan et al. 2000) and observations that male big
brown bats enter torpor more frequently than do females (Grinevitch et al.
1995) are consistent with this hypothesis, but the hypothesis that selection
of roosts by males and females differs because of use of different strategies for energy conservation has not been adequately evaluated in forested
environments.

Bats can be vulnerable to predation when roosting and when emerging from the roost (see Aubrey et al. 2003); roosting behavior and selection of roost sites should attempt to minimize this risk (Fenton et al.
1994, Kalcounis and Brigham 1994, Fenton 1995). Tall roosts enable bats
to roost far from the ground, potentially providing increased protection
from some types of predation (Vonhof and Barclay 1996, Betts 1998a).
The height at which bats roost and the height of the roost structure are
sometimes positively correlated (Vonhof and Barclay 1996). As predation
on cavity-nesting birds decreases with increasing nest height in some
areas (Nilsson 1984, Rendell and Robertson 1989), bats may also reduce
the risk of predation by roosting high above the ground. Use of sites with
minimal clutter in front of the roost opening may reduce predation by
aerial predators of bats entering or leaving a roost by facilitating direct
access to the roost site by bats (Vonhof and Barclay 1996); this may be especially important for newly volant young (Campbell et al. 1996). However, high levels of clutter in front of a roost also has been postulated to
be advantageous to bats in some situations (Betts 1998a), as maneuverable
bats may be able to better elude capture by aerial predators amongst the
clutter.

Bats also may use tall roosts because they are conspicuous and easily
accessed, reducing energy expenditure while searching for and entering
roosts (Campbell et al. 1996, Vonhof and Barclay 1996, Brigham et al. 1997b,
Betts 1998a, Ormsbee and McComb 1998). Waldien et al. (2000) speculated that the frequent use of gaps and edges for foraging and commuting

by bats (Furlonger et al. 1987, Crome and Richards 1988, Grindal and Brigham 1998) may increase use of roosts in these areas as well.

Several studies have shown that bats prefer particular species of trees as roosts within an area (Barclay et al. 1988, Campbell et al. 1996, Vonhof and Barclay 1996, Brigham et al. 1997*b*, Kalcounis and Brigham 1998, Psyllakis 2001, but see Betts 1998*a*). However, the species of tree used often varies among geographic regions (e.g., big brown bat, Brigham 1991, Vonhof and Barclay 1996, Kalcounis and Brigham 1998, Rabe et al. 1998; silver-haired bat (*Lasionycteris noctivagans*) Campbell et al. 1996, Mattson et al. 1996, Vonhof and Barclay 1996, Barclay et al. 1988, Betts 1998*a*). For most species of bats, tree species appears to be important only as it influences aspects such as characteristics and amount of exfoliating bark, size of roost, thermal characteristics, and presence of woodpecker cavities.

Individual snags provide ephemeral roosting habitat for bats. Longevity of a snag is influenced by a number of factors, including the local disturbance regime, species, cause of mortality, size, morphology, and topographic position (Cline et al. 1980, Morrison and Raphael 1993). In addition, snags are only suitable as roosts for a portion of the time they are standing. Although variation exists among species and locations, bats frequently use snags in intermediate stages of decay (Table 4.3), probably because the presence of exfoliating bark and woodpecker cavities provide opportunities for roosting by bats (Waldien et al. 2000). Bark roosts are particularly ephemeral (Vonhof and Barclay 1996, Barclay and Brigham 2001). The amount and condition of bark on snags can change dramatically over relatively short time periods, and the quality of roost sites beneath bark can change rapidly (Rabe et al. 1998, Barclay and Brigham 2001).

Bats using roost structures that are relatively abundant and ephemeral, such as crevices and cavities in trees and snags, generally exhibit low fidelity to individual roosts (Lewis 1995). Bats roosting in crevices and cavities commonly use multiple roosts and switch roosts every one to several days (Brigham 1991, Chung-MacCoubrey 1996, Crampton and Barclay 1996, 1998, Kalcounis and Hecker 1996, Mattson et al. 1996, Vonhof and Barclay 1996, Brigham et al. 1997*b*, Betts 1998*b*, Kalcounis and Brigham 1998, Ormsbee and McComb 1998, Rabe et al. 1998, Waldien et al. 2000, Weller and Zabel 2001). In some cases, bats from the same colony stay together during movements between roosts (Betts 1998*b*). Bats frequently travel up to several hundred meters between day roosts on consecutive days (Crampton and Barclay 1996, Mattson et al. 1996, Brigham et al.

1997*b*, Betts 1998*b*, Rabe et al. 1998), and may travel up to several kilometers between roosts (Ormsbee 1996). Although individual bats often roost in structures separated by considerable distances over time, bats sometimes exhibit fidelity to general areas rather than specific roosts (Lewis 1995, Kalcounis and Hecker 1996). Lactating females sometimes use a roost for longer periods of time than do non-lactating bats (Kurta et al. 1996, Mattson et al. 1996, Vonhof and Barclay 1996), but even lactating females sometimes switch roosts frequently (Rabe et al. 1998, Waldien et al. 2000). It is possible that although bats regularly switch roosts over short temporal scales, over longer time periods bats demonstrate a high degree of fidelity to a set of roosts. Long-term patterns of use of tree roosts are poorly understood, but the limited information available suggests that crevice- and cavity-roosting bats may infrequently re-use roosts over long time periods (O'Donnell and Sedgeley 1999). Barclay and Brigham (2001) found that California myotis repeatedly used the same trees for roosting over a five-year period, but that the size of colonies using trees decreased substantially over time, suggesting that the quality of the roosts declines rapidly. Long-term patterns of roost use likely vary with species, characteristics of the roost, and geographic area, and additional studies are needed to more fully understand these patterns. A number of hypotheses have been proposed to explain patterns in roost fidelity of bats, including minimization of predation or parasite loads, exploitation of most favorable microclimates, response to changes in structural characteristics of roosts, exploration of alternative roosts to minimize impacts of disturbance or destruction of primary roosts, and minimization of energy costs in relation to shifting forage base (reviewed by Lewis 1995), but the reasons for such frequent roost switching by bats in western coniferous forests remain uncertain.

At larger spatial scales, characteristics of forest stands, such as age and structure, may influence roost site selection by bats. However, stand-scale characteristics generally appear to be less important than characteristics of roost structures and the immediately surrounding microhabitat for many species of bats. For example, Betts (1998*a*) concluded that silver-haired bats will roost in stands of any age as long as sufficient numbers of trees and snags with suitable characteristics are present, and that bats were not selecting stands on the basis of tree species composition. Similarly, Ormsbee and McComb (1998) found that use of roosts by long-legged myotis was not related to stand age. Some studies have found that use of roosts is related to characteristics of the surrounding stand, but this may be related to availability of roost structures rather than stand-level characteristics. For example, all of the roosts of little brown bats and silver-haired bats identified by Crampton and Barclay (1996, 1998) were in old-growth stands, but suitable large snags were not abundant or not present in younger stands in their study area. Despite this, some species of bats may have specific habitat associations at larger spatial scales that influence use of roosts. A paucity of research conducted at large spatial scales currently precludes strong inference at the landscape scale.

Location of roosts relative to other key resources also may influence use by bats. Given the high use of ponds, lakes, and streams by bats for foraging and drinking (Lunde and Harestad 1986, Thomas 1988, Barclay 1991, Brigham et al. 1992, Adams 1997, Grindal et al. 1999), bats may attempt to minimize distance between roosts and water to reduce energy expenditures associated with commuting. However, information concerning spatial relationships between roosts and riparian areas in western coniferous forests is equivocal. Gellman and Zielinski (1996) found that use of redwood hollows increased with decreasing distance to water, Weller and Zabel (2001) found that fringed myotis roosted closer to stream channels than would be expected by chance, and Ormsbee and McComb (1998) reported that long-legged myotis generally roosted outside of riparian areas but closer to water than would be expected by chance. In contrast, other studies found that bats do not preferentially select roosts close to water (Betts 1998a, Waldien et al. 2000), and some have found that use of roosts increases with distance from water (Campbell et al. 1996, Mattson et al. 1996). Brigham (1991) found that big brown bats consistently foraging over water varied locations of roosts considerably, suggesting that distance from water may not be a critical factor in roost site selection. It has been hypothesized that bats may preferentially select upslope sites for roosting because they are warmer than sites in riparian areas (Campbell et al. 1996), and, alternatively, that the relatively high availability of potential roost trees on ridge tops may be responsible for this pattern (Mattson et al. 1996). Spatial relationships between roosts and riparian areas may differ with species, sex, and area, and additional data are needed to fully elucidate patterns.

Few studies have looked at the influence of landscape-level characteristics on use of roosts by bats. Waldien et al. (2000) found that types of roosts used by long-eared myotis varied considerably with landscape condition. In that study, stumps were used most frequently in landscapes subject to extensive clear-cutting, and snags were used predominantly in landscapes dominated by older forests.

Stumps

Spaces between the wood and exfoliating bark of stumps provide roosting opportunities for some species of bats. Although several species infrequently use stumps for roosting (e.g., little brown bat, Kalcounis and Hecker 1996; California myotis (*Myotis californicus*), fringed myotis, and Yuma myotis (*M. yumanensis*), Waldien et al. unpublished data), regular use of stumps as roosts has only been documented for long-eared myotis (Vonhof and Barclay 1997, Waldien et al. 2000). Vonhof and Barclay (1997) only found male and non-reproductive female long-eared myotis roosting in stumps, but Waldien et al. (2000) found reproductive females also extensively roosted in stumps. Stumps used as roost sites by long-eared myotis tend to be relatively tall with deep crevices (Vonhof and Barclay 1997, Waldien et al. 2000). Accessibility of the stump appears to be an important factor in suitability of stumps as roost sites; stumps with vegetation obstructing access (Waldien et al. 2000) and clear-cuts with substantial vegetation (Vonhof and Barclay 1997) are used infrequently for roosting. Characteristics of bark and patterns of bark exfoliation vary substantially among species of trees, and this in turn strongly influences use of stumps as roosts. Vonhof and Barclay (1997) found that stumps of ponderosa pine (*Pinus ponderosa*) and lodgepole pine (*P. contorta*) were preferred in their study area, whereas Waldien et al. (2000) found that stumps of Douglas-fir were selected over western hemlock (*Tsuga heterophylla*) and western redcedar. Stumps only provide suitable roosting habitat for the period between the time that bark has exfoliated adequately to provide crevices and when vegetation in a clear-cut overtops stumps. The interval of time that a stump provides suitable roost sites probably varies regionally and with management history, but may be as short as five years (Waldien et al. 2000). The period of time that stumps provide suitable roosting habitat is further restricted by the fact that crevices in stumps often accumulate moisture following precipitation and thus are suitable for roosting only during dry periods of the year (Waldien et al. 2000).

Foliage

Two species of bats that occur in western coniferous forests, western red bats and hoary bats, roost primarily in the foliage of trees (Shump and Shump 1982*a*, 1982*b*). To date, no intensive studies of roosting habits of these species have been conducted in western coniferous forests. In south-western Oregon, Perkins and Cross (1988) only captured hoary bats in mature and old-growth forests and suggested that the structure of crowns of old trees may be most suitable for roosting of this species. Work on other species of lasiurines in other areas suggest that, like crevice roosting species, lasiurines preferentially select tall, large-diameter trees as roost sites (Menzel et al. 1998, Hutchinson and Lacki 2000). Eastern red bats (*Lasiurus borealis*), historically considered to be conspecific with western red bats (Morales and Bickham 1995), roost primarily in deciduous trees in the eastern United States (Shump and Shump 1982*a*, Menzel et al. 1998, Hutchinson and Lacki 2000) and hoary bats frequently roost in deciduous trees (Shump and Shump 1982*b*), but roosting habits in areas dominated by coniferous forests are unknown. Although lasiurines are abundant in many deciduous forests, they appear to be uncommon in many western coniferous forests, but this may vary geographically. It is possible that the paucity of deciduous trees may result in low abundances of lasiurines in many western coniferous forests, but additional data are needed to examine this hypothesis. However, although many lasiurines are closely associated with deciduous trees, some lasiurines also make extensive use of conifers. In the southeastern United States Seminole bats select pines for roosting, even in landscapes dominated by hardwoods (Menzel et al. 1998).

Other structures

Several species of bats that occur in western coniferous forests sometimes use rock crevices, caves, or man-made structures, such as buildings and bridges, as day roosts (Table 4.2). There has been relatively little research on the influences of forest structure and forest management on use of these structures. It is likely that forest structure and forest management generally influence use of these sites only indirectly, and that primary factors influencing their use are related to characteristics of the roost structure. For species that exhibit flexibility in types of structures used for roosting (e.g., long-eared myotis, little brown bats, and big brown bats), the abundance, distribution, and characteristics of large trees and snags may influence the amount of use of alternative roosting structures. In some cases, changes in forest structure near rock crevices, caves, or man-made structures may influence the quality of these sites for roosting (such as by altering thermal regimes or accessibility of roosts), but there is little information on the way that changes in forest structure influence roost quality in these cases.

Night roosts

Although some bats apparently fly continuously from their departure from the day roost in the evening to their return the following morning (e.g., Wai-Ping and Fenton 1989), a significant portion of the nightly time budget for many bats is spent night roosting (O'Shea and Vaughan 1977, Anthony et al. 1981, Barclay 1982, Hickey and Fenton 1996, Perlmeter 1996). Night roosts are used by bats for resting, energy conservation, mating, social interactions, and consumption and digestion of food (Kunz 1982). In some cases, individual bats exhibit high fidelity to particular night roosts (Lewis 1994).

Information on the use of night roosts in western coniferous forests is limited and largely restricted to man-made structures, especially bridges (Perlmeter 1996, Pierson et al. 1996, Adam and Hayes 2000). The amount of night roosting that occurs at a bridge is strongly influenced by its characteristics. Bats select structures for night roosts that are relatively warm (Barclay 1982). Because concrete bridges maintain a stable thermal environment (Perlmeter 1996, Pierson et al. 1996), they provide ideal night roosts. In the Oregon Coast Range, Adam and Hayes (2000) found that concrete cast-in-place bridges with chambers underneath were used more frequently than other bridge types. Bats were most frequently found in chambers near the ends of the bridge at the junction of the wall of a chamber and the ceiling of the bridge. Concrete flat-bottomed bridges and wooden bridges received little use as night roosts (Adam and Hayes 2000), but use of flat-bottomed concrete bridges can be increased through installation of artificial roosting structures (Arnett and Hayes 2000). Use of concrete bridges as night roosts is positively correlated with size (Adam and Hayes 2000).

Types of natural structures currently used and those used before bridges were prevalent on the landscape are not well known. Extensive use of caves as night roosts has been reported in some areas (e.g., Albright 1959). Large logs (especially those bridging streams), hollow trees, snags, and rock crevices and overhangs also may be used as night roosts in western coniferous forests.

Winter roosts

The overwintering habitat of most species of bats in western coniferous forests is poorly known. Three species (hoary bats, western red bats, and silver-haired bats) are generally considered to be long-distance migrants throughout much of their distributions in western coniferous forests, although details of migration routes and winter habitat are poorly known. Similarly, migratory individuals of these species are thought to remain active throughout the year, but little is known about use of roosts by these species in the winter. Hoary bats are known to overwinter in southern California (Dalquest 1943, Vaughan and Krutzsch 1954, Findley and Jones 1964), and it is possible that Mexico is an important overwintering area for some populations (Findley and Jones 1964). As hoary bats and silver-haired bats have been captured during winter months in more northerly latitudes, some individuals or populations may not migrate, or overlapping north–south shifts may occur in some areas (Schowalter et al. 1978, Nagorsen et al. 1993, Verts and Carraway 1998).

The remaining species are generally thought to be year-round residents, short-distance migrants, or elevational migrants. Throughout most of the region covered by western coniferous forests, winter temperatures drop low enough that bats overwintering need to enter hibernation or torpor for periods of time. Caves and abandoned mines are frequently used as hibernacula by several species of bats, including forest-dwelling species (e.g., Marcot 1984, Perkins et al. 1990, Nagorsen et al. 1993). However, caves with suitable thermal characteristics are absent or rare in many areas; in these areas bats must either migrate to caves or use other types of roosts.

Typical types and locations of sites used as hibernacula or winter roosts are poorly known for most species of bats throughout much of the area covered by western coniferous forests. Bats have been observed hibernating in tree cavities (Fassler 1975) and under the bark of conifers (Nagorsen et al. 1993). However, these observations are anecdotal and the extent of use of trees as hibernacula by bats is unknown. Because of similarities between hollow trees and caves, hollow trees may provide important winter roosts or hibernacula for bats in some areas (Gellman and Zielinski 1996).

In areas where winter temperatures are relatively mild, such as the coastal forests of Oregon or the redwood forests of northern California, some bats exhibit periodic, low levels of activity during the winter (Pearson et al. 1952, Gellman and Zielinski 1996, Hayes 1997). Reasons for activity in winter months are unclear, but in other regions it has been hypothesized that low levels of winter activity may be attributable to bats feeding (Avery 1985, Brigham 1987, Hays et al. 1992), drinking (Speakman and Racey 1989, Hays et al. 1992), or changing hibernacula (Whitaker and Rissler 1992). Types of roosts used by individuals that are active during winter months, and whether characteristics of these roosts are similar to those used as summer day roosts or hibernacula, remain unknown.

Activity areas

Activity areas of bats include sites used for foraging and drinking as well as areas used for commuting among sites used for foraging, roosting, and drinking. Although availability of roosts may be a primary factor limiting bat populations in many areas, characteristics and quality of activity areas may influence the distribution and abundance of bats in some areas. Prey availability may be especially important in some situations (Fenton 1997), particularly in harsh environments at high latitudes or elevation (Humphrey 1975). In other instances, water may be the primary factor limiting bat populations, particularly in arid regions (Szewczak et al. 1998).

Forest and landscape structure may influence the quality of activity areas for bats in a number of ways. Forest structure (the spatial pattern of vegetation in forests; Oliver and Larson 1996) may influence the abundance and distribution of insects on which bats forage and the ease with which bats can fly through stands, echolocate, and capture insects. At the landscape scale, the amount of area comprised of different types of forest stands (landscape composition) may influence the number and species of bats that an area can support, whereas juxtaposition of those habitat types in relation to one another (landscape configuration) determines the amount and characteristics of edge habitat and the level of fragmentation of a site. Both landscape composition and configuration could affect patterns of use, abundance, and population viability of bats.

Development of inexpensive equipment for monitoring the echolocation calls of bats (bat detectors) has spurred a number of studies on spatial and temporal patterns of habitat use by bats, just as the development of lightweight radio transmitters facilitated greater understanding of the roosting ecology of forest-dwelling bats. Use of bat detectors has complemented and expanded studies using radiotelemetry, mist netting, and direct observation to provide a more complete understanding of patterns of habitat use by bats.

Riparian and aquatic habitat

Riparian and aquatic habitats are particularly important activity areas for bats (Lunde and Harestad 1986, Cross 1988, Thomas 1988, Barclay 1991, Brigham et al. 1992, Adams 1997, Grindal et al. 1999, Waldien and Hayes

2001), and water may be especially critical in arid areas (Szewczak et al. 1998). Bats often use aquatic and riparian areas as activity areas more frequently than they use upland areas (Fenton et al. 1983, Krusic et al. 1996, Parker et al. 1996, Walsh and Harris 1996, Grindal et al. 1999, Zimmerman and Glanz 2000, Seidman and Zabel 2001). Aquatic habitat provides two key resources for bats: sources of drinking water and insect prey.

Most species of insectivorous bats appear to require drinking to maintain water balance, with some possible exceptions (e.g., the Brazilian free-tailed bat; Kunz et al. 1995a, 1995b). Most water intake of insectivorous bats comes from water in their insect prey; "typical" diets of insectivorous bats are comprised of approximately 70% water (Kurta et al. 1989a), although this proportion undoubtedly varies among bats depending on the species of prey consumed. Current evidence suggests that roughly one-fifth to one-quarter of the water intake of insectivorous bats must be obtained by drinking (Kurta et al. 1989b, 1990, McLean and Speakman 1999). Water requirements of lactating females exceed those of non-lactating individuals (Kurta et al. 1989b, 1990, McLean and Speakman 1999), leading McLean and Speakman (1999) to speculate that proximity to water may be an important factor in selection of sites for maternity roosts. However, as drinking water appears to be necessary to meet the water needs of non-lactating bats for many species (Kurta et al. 1989a, 1990), access to water may be equally important for lactating and non-lactating bats.

Aquatic areas are favorable foraging sites for bats because they have a relatively high abundance and reliable presence of insect prey (Brigham 1991). As a consequence, a greater amount of foraging activity generally occurs over streams and in riparian areas than in upland areas (Thomas 1988, Krusic et al. 1996, Wilkinson and Barclay 1997, Grindal et al. 1999). When foraging over aquatic areas, bats forage more frequently over still water than over moving water (von Frenckell and Barclay 1987, Krusic et al. 1996, Warren et al. 2000). Bats that forage close to the surface of the water avoid areas with surface clutter (such as rocks) as well as sections of water that generate substantial amounts of noise that may interfere with their echolocation (Mackey and Barclay 1989). Bats that forage higher above the water also avoid areas that generate high levels of surface noise, but are not significantly influenced by surface clutter (Mackey and Barclay 1989).

Water and riparian areas may be focal points of activity for many species of bats in western coniferous forests. Although long-eared myotis obtain much of their diet through gleaning and do not rely on capture of insects in flight, in the western Cascade Mountains of Oregon activity areas of this species were centered on aquatic habitat (Waldien and Hayes 2001). Similarly, in some areas big brown bats use aquatic habitat as focal points for foraging, and females exhibit a high level of fidelity to foraging areas (Wilkinson and Barclay 1997).

Relatively little research has examined the influence of forest-management activities in streamside areas. Hayes and Adam (1996) found that clear-cutting small patches (90 m to 180 m in length) along both sides of the stream significantly influenced use of the area by bats. They reported that activity of *Myotis* species was 4.1 to 7.7 times higher in un-harvested sites than in adjacent logged areas. In contrast, activity of non-*Myotis* species, especially of silver-haired bats, was higher in logged sites.

Influence of clutter, prey availability, and distribution of roosts

Amount of clutter can significantly influence use of habitat by bats. The term "clutter" is derived from radar theory and refers to the number of obstacles a bat must detect and avoid in a given area (Fenton 1990). In cluttered environments bats must discriminate between target and background objects when echolocating and must avoid collisions with obstacles such as branches when flying (Fenton 1990).

The relationship between bats and clutter provides an important foundation for understanding the influence of forest structure and forest management on habitat use by bats. The morphology of bats influences their ability to exploit habitats differing in amount of clutter and structural complexity (Aldridge and Rautenbach 1987, Norberg and Rayner 1987, Crome and Richards 1988, Kalcounis and Brigham 1995). Maneuverable species (generally bats with small bodies and high wing loading) are able to exploit relatively cluttered habitats. Less maneuverable species (generally bats with large bodies and low wing loading) are restricted to more open habitat, such as clear-cuts, gaps in forest stands, forest edges, and the area above the canopy. For example, differences in foraging strategies for hoary bats and silver-haired bats can be understood in terms of the relative ability of the two species to exploit environments differing in amount of clutter (Barclay 1985). Species adapted to cluttered habitat also may use edge and open habitat, but those adapted to open habitat have limited access to more closed habitats (Fenton 1990).

Most species of bats in western coniferous forests generally avoid extensive use of highly cluttered habitat. For example, Townsend's

big-eared bats (*Corynorhinus townsendii*) spend relatively little time foraging in dense forest habitat (Dobkin et al. 1995), and Yuma myotis (*Myotis
yumanensis*; Brigham et al. 1992) and juvenile little brown bats (Adams
1997) forage most frequently in areas with low levels of clutter. Fine-scale
patterns of habitat use relative to amount of clutter have been shown to
vary between sexes or age classes in some studies (Adams 1997) but not
others (Brigham et al. 1992, Kalcounis and Brigham 1995). However, patterns of fine-scale intra-specific partitioning of foraging habitat are complex and appear to vary with population density (Adams 1997), and thus
may not always be evident.

Bats generally use edge habitat for commuting and foraging more frequently than either interior forest habitat or openings such as clear-cuts
or meadows (Furlonger et al. 1987, Clark et al. 1993, Krusic et al. 1996,
Walsh and Harris 1996, Wethington et al. 1996, Grindal and Brigham
1999, Zimmerman and Glanz 2000). Disproportionate use of edge habitat is probably related to avoidance of dense clutter in forest interiors and
avoidance of open areas where abundance of some types of insect prey
are reduced (Hayes and Adam 1996, Grindal and Brigham 1999, Burford
et al. 1999). As bats appear to be more susceptible to predation in open
habitats (Rydell et al. 1996, Duvergé et al. 2000), bats may also avoid open
areas to reduce risk of predation. Grindal and Brigham (1999) found that
biomass of insects along forest-cutblock edges and within the forest did
not differ significantly, although the amount of foraging activity of bats
was significantly higher at the edge than in the forest interior. Similarly,
Brigham et al. (1997*a*) experimentally manipulated the amount of clutter along forest–clear-cut edges by constructing artificial "clutter zones"
along forest–clear-cut edges to evaluate the influence of clutter on use of
habitat by bats. Commuting activity of *Myotis* bats in cluttered and uncluttered habitats did not differ significantly, but foraging activity of *Myotis*
bats was significantly lower in artificial clutter zones despite similar
insect abundance in clutter zones and control sites.

Understanding the influences of prey distribution and availability on
use of habitat by bats is poorly developed. Prey availability probably is
often less important than habitat structure in determining selection of
foraging habitat by aerial insectivorous bats in upland areas (Brigham
et al. 1997*a*, Grindal and Brigham 1999), although this pattern may
not hold for species that glean invertebrates (Grindal and Brigham
1999). However, distribution of moths is strongly influenced by forest-
management practices in eastern Kentucky, and patterns of moth

abundance closely parallel patterns of habitat use by big-eared bats (*Corynorhinus* sp., Burford et al. 1999).

Roost availability can play an important role in determining activity areas used by bats. Bats sometimes forage more frequently in relatively poor quality foraging areas that are in close proximity to roost sites than in high quality foraging areas that are distant from suitable roosting habitat (Geggie and Fenton 1985). Distance from roost to foraging habitat can influence reproductive success in some cases (Tuttle 1976), so it is advantageous for bats to select habitat with high quality roosting and foraging habitat.

Influence of stand age and structure

Amount of bat activity often is related to stand age in western coniferous forests (Thomas 1988, Thomas and West 1991, Erickson 1993, Erickson and West 1996, Humes et al. 1999) and other temperate forests (Crampton and Barclay 1996, 1998, Krusic et al. 1996, Jung et al. 1999). Although the relationships between stand age and bat activity are not simple, some basic patterns emerge from studies conducted in forests differing in age and structure. These findings generally can be understood in terms of the distribution and availability of roosts and the response of bats to clutter.

Bats often use open habitat, including clear-cuts, areas recently cut (within ten years), and meadows, as activity areas more intensively than forested sites (Brigham et al. 1992, Erickson 1993, Erickson and West 1996, Krusic et al. 1996, Grindal and Brigham 1998), although high use of open areas may not occur in all forest types or situations (Lunde and Harestad 1986, Jung et al. 1999). High use of open habitat probably reflects reduced clutter in these areas.

Among forested sites, several studies have documented that bats use old forests more frequently than young forests (Perkins and Cross 1988, Thomas 1988, Thomas and West 1991, Krusic et al. 1996, Crampton and Barclay 1998, Humes et al. 1999, Jung et al. 1999), although Grindal and Brigham (1999) found no significant differences in use among the age classes of forests they examined. High use of old-growth stands by bats has been hypothesized to result from high availability of roosts, especially large diameter snags (Perkins and Cross 1988, Thomas 1988, Thomas and West 1991, Crampton and Barclay 1996, 1998, Humes et al. 1999, Kalcounis et al. 1999). The high amount of activity observed in old-growth stands (Thomas 1988, Erickson and West 1996, Hayes and Gruver 2000) and other forested sites (Hayes 1997, Grindal and Brigham 1999) shortly after dusk and before dawn has been interpreted as the result of bats commuting to and from stands used for roosting (Thomas 1988, Erickson and West 1996, Grindal and Brigham 1999).

The use of upland, intermediate-aged coniferous forests (between 10 and 100 years old) tends to be relatively low (Thomas 1988, Thomas and West 1991, Erickson and West 1996, Krusic et al. 1996, Parker et al. 1996, Crampton and Barclay 1998, Jung et al. 1999), presumably reflecting relatively high clutter and low availability of roost sites in these habitats. However, low use of this age class by bats is probably not a function of age of stands but rather is related to structural characteristics of stands, some of which may be influenced by management activities. For example, Humes et al. (1999) found that use of 50- to 100-year-old Douglas-fir forests by bats varied with stand structure and the management history of the stands; thinned stands ($\bar{x} = 184$ trees/ha, TPH) had significantly more bat activity than unthinned stands ($\bar{x} = 418$ TPH) of the same age.

Bat activity also varies among vertical strata in forests (Bradshaw 1996, Krusic et al. 1996, Jung et al. 1999, Kalcounis et al. 1999, Hayes and Gruver 2000). Difference in use across vertical strata among species is partially a function of differential exploitation of environments by bats based on differences in foraging strategies and flight characteristics (Black 1974, Kalcounis et al. 1999, Hayes and Gruver 2000). In an old-growth forest in western Washington, larger bats used upper vertical strata most frequently, while *Myotis* bats were most active beneath the forest canopy (Hayes and Gruver 2000). In boreal forests of Saskatchewan, vertical patterns of use vary with forest type (Kalcounis et al. 1999). In this study, activity levels varied among vertical strata in aspen stands but not in spruce or jack-pine stands, although the authors noted that small sample size may have obscured some patterns in this study. Vertical partitioning of habitat by bats suggests that structurally complex stands may provide greater foraging opportunities for bats. Young stands with relatively simple forest structure may not provide a diversity of foraging niches for bats, and the structural complexity of old-growth forests may partially account for high levels of activity in old forests (Hayes and Gruver 2000). There are relatively few data concerning use of habitat above the forest canopy in western coniferous forests. Very low levels of activity were found above the canopy of an old-growth Douglas-fir forest (Hayes and Gruver 2000), but substantial activity was recorded above the canopy of some boreal forests (Kalcounis et al. 1999).

Few studies have examined the influence of differences in forest composition on activity levels of bats. In boreal forests of Canada, activity of bats in aspen-white spruce mixedwood forest was higher than that in either aspen-dominated forests or jack-pine forests (Kalcounis et al. 1999). Furthermore, no feeding activity was detected in jack pine, possibly the result of low insect abundance in the structurally simple jack-pine stands (Kalcounis et al. 1999). Differences among forest types were also reported from New Hampshire, where levels of bat activity were higher in older hardwood stands than in older softwood stands (Krusic et al. 1996).

Information gaps

Numerous gaps remain in the understanding of the habitat ecology of bats in western coniferous forests. Additional research on almost any aspect of the ecology of bats will enhance our understanding. Here I focus on areas that I think are most critical to develop a more complete picture of the habitat ecology of bats in forests and to facilitate the development of strategies for their conservation.

A key information gap stems from the lack of a mechanistic understanding of factors responsible for habitat associations of bats in western coniferous forests. Almost all information available on the habitat ecology of bats in western coniferous forests and the influences of forest management on bats is based on observational studies. Although a few field experiments have been conducted (e.g., Mackey and Barclay 1989, Brigham et al. 1997a, Grindal and Brigham 1998), manipulative studies relevant to the ecology of bats in western coniferous forests are rare. It is not possible to infer causality or to fully achieve a mechanistic understanding from observational studies. Results of observational studies have led to a plethora of hypotheses and speculations regarding causal factors responsible for observed patterns of roost site selection and habitat use by bats. These observations and hypotheses can be extremely valuable to help shape conservation strategies and often are essential in developing mechanistic hypotheses. However, the speculative and hypothetical context underlying explanations based on observational work is often lost through time, and hypotheses become dogma. Application of rigorous experimental approaches and use of manipulative studies would be useful to evaluate many of the hypotheses concerning the underlying mechanisms responsible for observed patterns of habitat use by bats. Inference resulting from successful experimental studies would increase our understanding of the habitat ecology of bats and would result in conclusions that can be generalized over a wider range of conditions and situations. Though a significant challenge, conducting such studies should be considered a research priority.

An important limitation underlying current understandings of the ecology of bats in forests and the influences of management activities on bats is the lack of any direct information on population size, viability, reproductive success, and fitness of bats in forests. Long-term assessment of changes in population size by examining changes in population size at roosts (e.g., O'Shea and Vaughan 1999) is rarely possible for forest-dwelling bats because of their dispersed nature and lack of roost site fidelity. Bat detectors can help determine differences in levels of use among areas, but patterns of use do not necessarily reflect differences in abundance (Hayes 2000). Catch-per-unit-effort is sometimes used as an index of abundance for bats (e.g., Perkins and Cross 1988, Adams 1997) and sometimes can provide a valuable first approximation of population differences. However, differences in capture probabilities among sites or sampling period can result in biased estimates and potentially misleading results when using catch-per-unit-effort approaches (Remson and Good 1996, Thompson et al. 1998). More sophisticated approaches to assess abundance or density, such as use of mark-recapture methods, have not been successfully used to estimate the abundance of dispersed populations of bats using multiple roosts because of low recapture probabilities. The development of innovative techniques to assess population trends of bats in western coniferous forests is a critical step to filling this information gap.

Most studies of roosting ecology of bats in western coniferous forests and other forested ecosystems have focused on females. This emphasis is derived from the perspective that maintenance of roosts for females, especially maternity roosts, is key to conserving bats in forests. However, as noted earlier, males and females may adopt different strategies for conservation of energy, and this may result in differences in structures selected for roosting (Tuttle 1976, Williams and Findley 1979, Hamilton and Barclay 1994). Although providing roost sites for females is undoubtedly critical to conservation of bats in forests, maintaining roosts for males is likely also fundamental to maintaining viable populations in many areas. Comparative studies of differences in roost site selection by males and females would be valuable to refine hypotheses concerning strategies for roost site selection and would provide information important to developing approaches for the conservation of bats in forests.

Despite advances in our understanding of the ecology of some species, additional research into several aspects of the roosting ecology of bats in western coniferous forests would be valuable and would help refine conservation strategies for bats. The roosting ecology of several species in this region is poorly known. Very little information is available on roosting ecology of Keen's myotis, Yuma myotis, or foliage-roosting bats in western coniferous forests, and information on several other species is based on limited sample sizes or is restricted in geographic scope. In addition, competition for roosts among species of bats (Reith 1980) or between bats and other taxa (especially birds; Kurta and Foster 1995) may affect local distributions and patterns of activity, but influences of community dynamics and inter-specific interactions on roosting ecology of any species of bat in western coniferous forests are unknown. Furthermore, information on night roosts used by bats is very limited. With few exceptions (e.g., Waldien and Hayes 2001) information on spatial relationships of roosts, foraging areas, and water are poorly understood. Understanding these patterns may be particularly important when managers are interested in maintaining a limited number of potential roost sites while maximizing their conservation benefits. Interactions among multiple spatial scales and their influences on use of roosts and habitat by bats are also poorly understood.

As noted earlier, bats in western coniferous forests generally exhibit minimal fidelity to day-roost sites. However, because of current limitations in longevity of batteries in small radio transmitters, patterns of use of roosts over moderate to long time frames are unknown. As a consequence the number of day roosts required for bats and their spatial relationships to one another and other key habitat resources (such as water and night roosts) are unknown.

Finally, seasonal patterns of movements and overwintering habitat of bats in western coniferous forests are poorly understood. Patterns of long-distance movements of migratory bats and short-distance and elevational movements of resident species are not known with any precision. Hibernacula and overwintering habitat used by most bats in western coniferous forests are unknown. Because of the relatively narrow habitat requirements of hibernating bats coupled with the vulnerability of bats during the winter, protection of winter roosts may be especially important.

Implications for forest management and conservation

Despite gaps in our knowledge, it is possible to evaluate some of the implications of what is known in terms of bat conservation and to suggest management approaches to conserve bat populations. Management of forests for commodity production and conservation of bats often are compatible if appropriate steps are taken. Although numerous management activities in forests, such as use of pesticides, fire control, use of prescribed fire, and recreation management, can influence bats, I emphasize considerations of management activities related to silviculture and timber harvest based on the current state of knowledge.

Management of day roosts should be a cornerstone of strategies to conserve bats in western coniferous forests. For most species, management of day roosts should focus on maintaining tall, large-diameter snags, and trees with structural abnormalities, such as broken tops, dead portions, and lightning scars (often referred to as "defective" or "decadent" trees), through time. Because of their relative rarity in most coniferous forests and their potential importance to bats and other wildlife, special attention should be given to hollow trees. Sustainable management approaches should incorporate both maintenance of existing structures and planning to ensure recruitment of new roosts when existing structures cease to function as roosting habitats. On forest lands intensively managed using traditional silvicultural practices designed to maximize production of wood fiber, large-diameter roost structures will be lost and not replaced over time unless management practices are modified to ensure development and recruitment of these structures (Ohmann et al. 1994). Additional challenges to maintaining these structures result from potential hazards of dead trees to forest workers and safety regulations requiring removal of these structures in some situations (Carey et al. 1999). Approaches to managing snags for bats parallel and complement general strategies for managing dead wood for a diversity of wildlife species and have been described elsewhere (e.g., McComb and Lindenmeyer 1999).

Although the number of day roosts needed to maintain viable populations of bats in western coniferous forests is unclear, three interacting factors suggest that the number of roosts needed is likely to be higher than that required for other species using similar structures. First, several species of bats are dependent on crevices and cavities in trees and snags in western coniferous forests. In addition, individual bats regularly use multiple roosts over short periods of time. Finally, roost quality for bats can change rapidly through time, especially as a result of changes in the amount and characteristics of bark on roost structures. As a consequence, individual snags probably provide habitat for other species of wildlife for a longer period of time than they do for bats. Consequently maintaining relatively high densities of roost structures in an area is probably essential to conserve local bat populations effectively.

Optimal locations on the landscape for maintaining roost structures are not well understood. Because bats in western coniferous forests depend on a source of open water for drinking and many species forage in riparian areas, providing roost structures within 2 or 3 km of open water and riparian habitat will generally be valuable for most species of bats in the region. Resolution of questions concerning the relative quality of roosts located within riparian habitat and those located in upland sites awaits further study. However, maintaining roost structures for bats only in riparian areas or only in upland areas is probably a poor conservation strategy. Bats appear to select roost sites based in part on their thermoregulatory and energetic demands and the thermal characteristics of roosts. Thermoregulatory and energetic demands of bats differ with species, age, sex, reproductive status, season, and ambient conditions, and thermal characteristics of roosts vary with season, ambient conditions, microsite environment, and structural characteristics of the roost. As a consequence, optimal locations for roosts are also likely to vary with individuals, species, characteristics of roosts, and time. Consequently, maintaining potential roosts across the landscape in a variety of topographic settings is an appropriate approach to meet the habitat needs of bats. Strategies that concentrate potential roosts in restricted topographic settings, such as in riparian buffers, may not provide the spectrum of conditions necessary to meet the range of thermal conditions required by bats through time. Furthermore, if trees and snags provide important sites for night roosts and hibernacula, the locations in a landscape that are most appropriate to provide these functions remain uncertain. In the absence of adequate information for winter roosts, maintaining potential roosts in a diversity of landscape positions is a conservative approach.

Where natural roost structures are rare, artificial roost structures, or bat boxes, may be an alternative under some circumstances. Although some studies have examined use of bat boxes in urban settings (e.g., Brittingham and Williams 2000), little research has been published documenting the efficacy of artificial roost structures in forested settings. Experiences of several biologists working in western coniferous forests indicate that artificial roosts are used by bats. Use of innovative designs for artificial roosts, such as the rocket box (Burke 1999), is promising in some situations. However, as characteristics of roosts used by bats probably differ with species, reproductive condition, season, and roost function, it is unlikely that artificial roosts can be designed to meet the diversity of roost requirements of forest-dwelling bats. Thus, while artificial roost structures can be used to supplement roosting opportunities for bats in some situations (e.g., Burke 1999, Arnett and Hayes 2000), I contend that artificial roost structures should only be considered as a temporary supplement for natural roost structures in situations where natural roost structures are rare, and that management and conservation strategies should focus on providing natural roost structures through time.

Influences of management activities on aquatic systems and their adjacent riparian areas should be given special attention because of the importance of sources of open water to bats. Water sites developed and maintained for other purposes, such as for fire control, livestock, or other species of wildlife, can influence distribution, abundance, and patterns of habitat use by bats. Even very small open water sources, such as water troughs, can receive substantial use by bats. Maintaining open water free of obstruction from overtopping vegetation and preventing complete inundation of water by algae or emergent vegetation is a relatively simple conservation measure that would benefit local bat populations, particularly in areas where open water is scarce.

Density management may be a valuable silvicultural tool to improve habitat conditions for bats in young, densely stocked stands in the stem-exclusion stage of stand development (for a description of stand development stages, see Oliver and Larson 1996). Thinning in these stands can be valuable to reduce clutter and increase use by bats (Humes et al. 1999). Thinning also can be used to accelerate development of large-diameter trees and snags (Nyberg et al. 1987, Barbour et al. 1996, Carey and Curtis 1996, Hayes et al. 1997) that provide valuable roosting habitat for bats. In addition, findings that small patch cuts receive high levels of use by bats (Grindal and Brigham 1998) suggest that application of uneven-age approaches to forest management, such as group selection (Nyland 1996), may be beneficial for bats.

Information gathered during recent years on the habitat ecology of bats suggests that conservation of bat populations is possible under a variety of management scenarios. Landscapes with an abundance of high-quality roost sites well distributed across the landscape, adequate foraging habitat, and sources of open water should provide conditions necessary to support healthy populations of bats. A number of approaches can be used to manage forested landscapes to achieve these objectives. Landscapes dominated by old forests will generally provide good habitat for most species of bats occurring in western coniferous forests. Alternatively, in intensively managed landscapes, maintaining a mosaic of forest stands differing in structural characteristics can achieve conservation objectives if sufficient attention is given to tailoring conditions for bats. Maintaining remnant patches of structurally diverse forest with abundant large snags is a useful approach in intensively managed landscapes (Waldien et al. 2000).

Acknowledgments

Ed Arnett, Mike Bogan, Mark Brigham, Steve Cross, Joan Hayes, Tom Kunz, and David Waldien provided helpful suggestions on an earlier draft of this chapter. Preparation of this chapter was supported by the Cooperative Forest Ecosystem Research (CFER) program with funding from the USGS Forest and Rangeland Ecosystem Science Center.

Literature cited

Adam, M.D., and J.P. Hayes. 2000. Use of bridges as night roosts by bats in the Oregon Coast Range. *Journal of Mammalogy* **81**:402–407.

Adams, R.A. 1997. Onset of volancy and foraging patterns in juvenile little brown bats, *Myotis lucifugus. Journal of Mammalogy* **78**:239–246.

Albright, R. 1959. Bat banding at Oregon Caves. *Murrelet* **40**:26–27.

Aldridge, H.D.J.N., and I.L. Rautenbach. 1987. Morphology, echolocation, and resource partitioning in insectivorous bats. *Journal of Animal Ecology* **56**:763–778.

Anthony, E.L.P., M.H. Stack, and T.H. Kunz. 1981. Night roosting and the nocturnal time budget of the little brown bat, *Myotis lucifugus*: effects of reproductive status, prey density, and environmental conditions. *Oecologia* **51**:151–156.

Arnett, E.B., and J.P. Hayes. 2000. Use of boxes installed under flat-bottom bridges as roosts by bats in western Oregon. *Wildlife Society Bulletin* **28**:890–894.

Aubry, K.B., J.P. Hayes, B.L. Biswell, and B.G. Marcot. 2003. The ecological role of tree-dwelling mammals in western coniferous forests. Pages 405–443 *in* C.J. Zabel and R.G. Anthony, editors. *Mammal Community Dynamics. Management and Conservation in the Coniferous Forests of Western North America*. Cambridge University Press, Cambridge, UK.

Avery, M.I. 1985. Winter activity by pipistrelle bats. *Journal of Animal Ecology* **54**:721–738.

Barbour, R.J., S. Johnston, J.P. Hayes, and G.F. Tucker. 1996. Simulated stand characteristics and wood product yields from Douglas-fir plantations managed for ecosystem objectives. *Forest Ecology and Management* **91**:205–219.

Barclay, R.M.R. 1982. Night roosting behaviour of the little brown bat, *Myotis lucifugus*. *Journal of Mammalogy* **63**:464–474.

Barclay, R.M.R. 1985. Long- versus short-range foraging strategies of hoary (*Lasiurus cinereus*) and silver-haired (*Lasionycteris noctivagans*) bats and the consequences for prey selection. *Canadian Journal of Zoology* **63**:2507–2515.

Barclay, R.M.R. 1991. Population structure of temperate zone insectivorous bats in relation to foraging behaviour and energy demand. *Journal of Animal Ecology* **60**:165–178.

Barclay, R.M.R. and R.M. Brigham. 2001. Year-to-year re-use of tree-roosts by California bats (*Myotis californicus*) in southern British Columbia. *American Midland Naturalist* **146**:80–85.

Barclay, R.M.R. and K.J. Cash. 1985. A non-commensal maternity roost of the little brown bat (*Myotis lucifugus*). *Journal of Mammalogy* **66**:782–783.

Barclay, R.M.R., P.A. Faure, and D.R. Farr. 1988. Roosting behavior and roost selection by migrating silver-haired bats (*Lasionycteris noctivagans*). *Journal of Mammalogy* **69**:821–825.

Bell, G.P. 1982. Behavioral and ecological aspects of gleaning by a desert insectivorous bat, *Antrozous pallidus*. *Behavioral Ecology and Sociobiology* **10**:217–223.

Betts, B.J. 1998*a*. Roosts used by maternity colonies of silver-haired bats in northeastern Oregon. *Journal of Mammalogy* **79**:643–650.

Betts, B.J. 1998*b*. Variation in roost fidelity among reproductive female silver-haired bats in northeastern Oregon. *Northwestern Naturalist* **79**:59–63.

Black, H.L. 1974. A north temperate bat community: structure and prey populations. *Journal of Mammalogy* **55**:138–157.

Bonar, R.L. 2000. Availability of pileated woodpecker cavities and use by other species. *Journal of Wildlife Management* **64**:52–59.

Bradshaw, P.A. 1996. The physical nature of vertical forest habitat and its importance in shaping bat species assemblages. Pages 199–212 *in* R.M.R. Barclay and R.M. Brigham, editors. *Bats and Forests Symposium*. British Columbia Ministry of Forests, Victoria, British Columbia, Canada.

Brigham, R.M. 1987. The significance of winter activity by the big brown bat (*Eptesicus fuscus*): the influence of energy reserves. *Canadian Journal of Zoology* **63**:2952–2954.

Brigham, R.M. 1991. Flexibility in foraging and roosting behaviour by the big brown bat (*Eptesicus fuscus*). *Canadian Journal of Zoology* **69**:117–121.

Brigham, R.M., H.D.J.N. Aldridge, and R.L. Mackey. 1992. Variation in habitat use and prey selection by Yuma bats, *Myotis yumanensis*. *Journal of Mammalogy* **73**: 640–645.

Brigham, R.M., S.D. Grindal, M.C. Firman, and J.L. Morissette. 1997*a*. The influence of structural clutter on activity patterns of insectivorous bats. *Canadian Journal of Zoology* **75**:131–136.

Brigham, R.M., M.J. Vonhof, R.M.R. Barclay, and J.C. Gwilliam. 1997b. Roosting behavior and roost-site preferences of forest-dwelling California bats (*Myotis californicus*). *Journal of Mammalogy* **78**:1231–1239.

Brittingham, M.C. and L.M. Williams. 2000. Bat boxes as alternative roosts for displaced bat maternity colonies. *Wildlife Society Bulletin* **28**:197–207.

Burford, L.S., M.J. Lacki, and C.V. Covell, Jr. 1999. Occurrence of moths among habitats in mixed mesophytic forests: implications for management of forest bats. *Forest Science* **45**:323–332.

Burke, H.S., Jr. 1999. Maternity colony formation in *Myotis septentrionalis* using artificial roosts: the rocket box, a habitat enhancement for woodland bats? *Bat Research News* **40**:77–78.

Caceres, M.C. and R.M.R. Barclay. 2000. *Myotis septentrionalis. Mammalian Species* **634**: 1–4.

Campbell, L.A., J.G. Hallett, and M.A. O'Connell. 1996. Conservation of bats in managed forests: use of roosts by *Lasionycteris noctivagans. Journal of Mammalogy* **77**:976–984.

Carey, A.B. and R.O. Curtis. 1996. Conservation of biodiversity: a useful paradigm for ecosystem management. *Wildlife Society Bulletin* **24**:610–620.

Carey, A.B., J.M. Calhoun, B. Dick, D. Jennings, K. O'Halloran, L.S. Young, R.E. Bigley, S. Chan, C.A. Harrington, J.P. Hayes, and J. Marzluff. 1999. Reverse technology transfer: obtaining feedback from managers. *Western Journal of Applied Forestry* **14**:153–163.

Christy, R.E. and S.D. West. 1993. Biology of bats in Douglas-fir forests. USDA Forest Service General Technical Report **PNW-GTR-308**, 28p.

Chung-MacCoubrey, A.L. 1996. Bat species composition and roost use in pinyon-juniper woodlands of New Mexico. Pages 118–123 *in* R.M.R. Barclay and R.M. Brigham, editors. *Bats and Forests Symposium*. British Columbia Ministry of Forests, Victoria, British Columbia, Canada.

Clark, B.S., D.M. Leslie, Jr., and T.S. Carter. 1993. Foraging activity of adult female Ozark big-eared bats (*Plecotus townsendii ingens*) in summer. *Journal of Mammalogy* **74**:422–427.

Clark, M.K. 1993. A communal winter roost of silver-haired bats, *Lasionycteris noctivagans* (Chiroptera: Vespertilionidae). *Brimleyana* **19**:137–139.

Cline, S.P., A.B. Berg, and H.M. Wight. 1980. Snag characteristics and dynamics in Douglas-fir forests, western Oregon. *Journal of Wildlife Management* **44**:773–786.

Crampton, L.H. and R.M.R. Barclay. 1996. Habitat selection by bats in fragmented and unfragmented aspen mixedwood stands of different ages. Pages 238–259 *in* R.M.R. Barclay and R.M. Brigham, editors. *Bats and Forests Symposium*. British Columbia Ministry of Forests, Victoria, British Columbia, Canada.

Crampton, L.H. and R.M.R. Barclay. 1998. Selection of roosting and foraging habitat by bats in different aged aspen mixedwood stands. *Conservation Biology* **12**:1347–1358.

Crome, F.J.H. and G.C. Richards. 1988. Bats and gaps: microchiropteran community structure in a Queensland rain forest. *Ecology* **69**:1960–1969.

Cross, S.P. 1988. Riparian systems and small mammals and bats. Pages 93–112 *in* K.J. Raedeke, editor. *Streamside Management: Riparian Wildlife and Forestry Interactions*. University of Washington Institute of Forest Resources Contribution No. **59**. Seattle, Washington, USA.

Cryan, P.M., M.A. Bogan, and J.S. Altenbach. 2000. Effect of elevation on distribution of female bats in the Black Hills, South Dakota. *Journal of Mammalogy* **81**:719–725.

Czaplewski, N.J. 1983. *Idionycteris phyllotis. Mammalian Species* **208**:1–4.

Dalquest, W.W. 1943. Seasonal distribution of the hoary bat along the Pacific coast. *Murrelet* **24**:20–24.

Dobkin, D.S., R.D. Gettinger, and M.G. Gerdes. 1995. Springtime movements, roost use, and foraging activity of Townsend's big-eared bat (*Plecotus townsendii*) in central Oregon. *Great Basin Naturalist* **55**:315–321.

Duvergé, P.L., G. Jones, J. Rydell, and R.D. Ransome. 2000. Functional significance of emergence timing in bats. *Ecography* **23**:32–40.

Erickson, J.L. 1993. Bat activity in managed forests in the western Cascade Range. Thesis, University of Washington, Seattle, Washington, USA.

Erickson, J.L. and S.D. West. 1996. Managed forests in the western Cascades: the effects of seral stage on bat habitat use patterns. Pages 215–227 *in* R.M.R. Barclay and R.M. Brigham, editors. *Bats and Forests Symposium*. British Columbia Ministry of Forests, Victoria, British Columbia, Canada.

Fassler, D.J. 1975. Red bat hibernating in a woodpecker hole. *American Midland Naturalist* **93**:254.

Faure, P.A. and R.M.R. Barclay. 1992. The sensory basis of prey detection by the long-eared bat, *Myotis evotis*, and the consequences for prey selection. *Animal Behaviour* **44**:31–39.

Faure, P.A., J.H. Fullard, and J.W. Dawson. 1993. The gleaning attacks of the northern long-eared bat, *Myotis septentrionalis*, are relatively inaudible to moths. *Journal of Experimental Biology* **178**:173–189.

Fenton, M.B. 1990. The foraging behaviour and ecology of animal-eating bats. *Canadian Journal of Zoology* **68**:411–422.

Fenton, M.B. 1995. Constraint and flexibility – bats as predators, bats as prey. *Zoological Society of London Symposia* **67**:277–289.

Fenton, M.B. 1997. Science and the conservation of bats. *Journal of Mammalogy* **78**: 1–14.

Fenton, M.B. and R.M.R. Barclay. 1980. *Myotis lucifugus*. *Mammalian Species* **142**:1–8.

Fenton, M.B. and G.P. Bell. 1979. Echolocation and feeding behaviour in four species of *Myotis* (Chiroptera). *Canadian Journal of Zoology* **57**:1271–1277.

Fenton, M.B., H.G. Merriam, and G.L. Holroyd. 1983. Bats of Kootenay, Glacier, and Mount Revelstoke national parks in Canada: identification by echolocation calls, distribution, and biology. *Canadian Journal of Zoology* **61**:2503–2508.

Fenton, M.B., I.L. Rautenbach, S.E. Smith, C.M. Swanepoel, J. Grosell, and J. van Jaarsveld. 1994. Raptors and bats: threats and opportunities. *Animal Behaviour* **48**:9–18.

Findley, J.S. and C. Jones. 1964. Seasonal distribution of the hoary bat. *Journal of Mammalogy* **45**:461–470.

Furlonger, C.L., H.J. Dewar, and M.B. Fenton. 1987. Habitat use by foraging insectivorous bats. *Canadian Journal of Zoology* **65**:284–288.

Geggie, J.F. and M.B. Fenton. 1985. A comparison of foraging by *Eptesicus fuscus* (Chiroptera: Vespertilionidae) in urban and rural environments. *Canadian Journal of Zoology* **63**:263–267.

Gellman, S.T. and W.J. Zielinski. 1996. Use by bats of old-growth redwood hollows on the north coast of California. *Journal of Mammalogy* **77**:255–265.

Grindal, S.D. 1999. Habitat use by bats, *Myotis* spp., in western Newfoundland. *The Canadian Field-Naturalist* **113**:258–263.

Grindal, S.D. and R.M. Brigham. 1998. Short term effects of small-scale habitat disturbance on activity by insectivorous bats. *Journal of Wildlife Management* **62**:996–1003.

Grindal, S.D. and R.M. Brigham. 1999. Impacts of forest harvesting on habitat use by foraging insectivorous bats at different spatial scales. *Ecoscience* **6**: 25–34.

Grindal, S.D., J.L. Morissette, and R.M. Brigham. 1999. Concentration of bat activity in riparian habitats over an elevational gradient. *Canadian Journal of Zoology* **77**:972–977.

Grinevitch, L., S.L. Holroyd, and R.M.R. Barclay. 1995. Sex differences in the use of daily torpor and foraging time by big brown bats (*Eptesicus fuscus*) during the reproductive season. *Journal of Zoology* (London) **235**:301–309.

Hamilton, I.M. and R.M.R. Barclay. 1994. Patterns of daily torpor and day-roost selection by male and female big brown bats (*Eptesicus fuscus*). *Canadian Journal of Zoology* **72**:744–749.

Hayes, J.P. 1997. Temporal variation in activity of bats and the design of echolocation-monitoring studies. *Journal of Mammalogy* **78**:514–524.

Hayes, J.P. 2000. Assumptions and practical consideration in the design and interpretation of echolocation monitoring studies. *Acta Chiropterologica* **2**:225–236.

Hayes, J.P. and M.D. Adam. 1996. The influence of logging riparian areas on habitat utilization by bats in western Oregon. Pages 228–237 *in* R.M.R. Barclay and R.M. Brigham, editors. *Bats and Forests Symposium*. British Columbia Ministry of Forests, Victoria, British Columbia, Canada.

Hayes, J.P. and J.C. Gruver. 2000. Vertical stratification of bat activity in an old-growth forest in western Washington. *Northwest Science* **74**:102–108.

Hayes, J.P, S.S. Chan, W.H. Emmingham, J.C. Tappeiner, L.D. Kellogg, and J.D. Bailey. 1997. Wildlife response to thinning young forests in the Pacific Northwest. *Journal of Forestry* **95**:28–33.

Hays, G.C., J.R. Speakman, and P.I. Webb. 1992. Why do brown long-eared bats (*Plecotus auritus*) fly in winter? *Physiological Zoology* **65**:554–567.

Hickey, M.B. and M.B. Fenton. 1996. Behavioural and thermoregulatory responses of female hoary bats, *Lasiurus cinereus* (Chiroptera: Vespertilionidae), to variations in prey availability. *Ecoscience* **3**:414–422.

Hoying, K.M. and T.H. Kunz. 1998. Variation in size at birth and postnatal growth in the insectivorous bat *Pipistrellus subflavus* (Chiroptera: Vespertilionidae). *Journal of Zoology* (London) **245**:15–27.

Humes, M.L., J.P. Hayes, and M.W. Collopy. 1999. Bat activity in thinned, unthinned, and old-growth forests in western Oregon. *Journal of Wildlife Management* **63**:553–561.

Humphrey, S.R. 1975. Nursery roosts and community diversity of nearctic bats. *Journal of Mammalogy* **56**:321–346.

Hutchinson, J.T. and M.J. Lacki. 2000. Selection of day roosts by red bats in mixed mesophytic forests. *Journal of Wildlife Management* **64**:87–94.

Jung, T.S., I.D. Thompson, R.D. Titman, and A.P. Applejohn. 1999. Habitat selection by forest bats in relation to mixed-wood stand types and structure in central Ontario. *Journal of Wildlife Management* **63**:1306–1319.

Kalcounis, M.C. and R.M. Brigham. 1994. Impact of predation risk on emergence by little brown bats, *Myotis lucifugus* (Chiroptera: Vespertilionidae), from a maternity colony. *Ethology* **98**:201–209.

Kalcounis, M.C. and R.M. Brigham. 1995. Intraspecific variation in wing loading affects habitat use by little brown bats (*Myotis lucifungus*). *Canadian Journal of Zoology* **73**:89–95.

Kalcounis, M.C. and R.M. Brigham. 1998. Secondary use of aspen cavities by tree-roosting big brown bats. *Journal of Wildlife Management* **62**:603–611.

Kalcounis, M.C. and K.R. Hecker. 1996. Intraspecific variation in roost-site selection by little brown bats (*Myotis lucifugus*). Pages 81–90 *in* R. M. R. Barclay and R. M. Brigham, editors. *Bats and Forests Symposium*. British Columbia Ministry of Forests, Victoria, British Columbia, Canada.

Kalcounis, M.C., K.A. Hobson, R.M. Brigham, and K.R. Hecker. 1999. Bat activity in the boreal forest: importance of stand type and vertical strata. *Journal of Mammalogy* **80**:673–682.

Krusic, R.A., M. Yamasaki, C.D. Neefus, and P.J. Pekins. 1996. Bat habitat use in White Mountain National Forest. *Journal of Wildlife Management* **60**:625–631.

Kunz, T.H. 1982. Roosting ecology of bats. Pages 1–55 *in* Kunz, T.H., editor. *Ecology of Bats*. Plenum Press, New York, New York, USA.

Kunz, T.H., O.T. Oftedal, S.K. Robson, M.B. Kretzmann, and C. Kirk. 1995a. Changes in milk composition during lactation in three species of insectivorous bats. *Journal of Comparative Physiology B* **164**:543–551.

Kunz, T.H., J.O. Whitaker, Jr., and M.D. Wadanoli. 1995b. Dietary energetics of the insectivorous Mexican free-tailed bat (*Tadarida brasiliensis*) during pregnancy and lactation. *Oecologia* **101**:407–415.

Kurta, A. and R.H. Baker. 1990. *Eptesicus fuscus*. *Mammalian Species* **356**:1–10.

Kurta, A. and R. Foster. 1995. The brown creeper (Aves: Certhiidae): a competitor of bark-roosting bats? *Bat Research News* **36**:6–7.

Kurta, A., G.P. Bell, K.A. Nagy, and T.H. Kunz. 1989a. Energetics of pregnancy and lactation in free-ranging little brown bats (*Myotis lucifugus*). *Physiological Zoology* **62**:804–818.

Kurta, A., G.P. Bell, K.A. Nagy, and T.H. Kunz. 1989b. Water balance of free-ranging little brown bats (*Myotis lucifugus*) during pregnancy and lactation. *Canadian Journal of Zoology* **67**:2468–2472.

Kurta, A., T.H. Kunz, and K.A. Nagy. 1990. Energetics and water flux of free-ranging big brown bats (*Eptesicus fuscus*) during pregnancy and lactation. *Journal of Mammalogy* **71**:59–65.

Kurta, A., J. Kath, E.L. Smith, R. Foster, M.W. Orick, and R. Ross. 1993. A maternity roost of the endangered Indiana bat (*Myotis sodalis*) in an unshaded, hollow, sycamore tree (*Platanus occidentalis*). *American Midland Naturalist* **130**:405–407.

Kurta, A., K.J. Williams, and R. Mies. 1996. Ecological, behavioural, and thermal observations of a peripheral population of Indiana bats (*Myotis sodalis*). Pages 102–117 *in* R.M.R. Barclay and R.M. Brigham, editors. *Bats and Forests Symposium*. British Columbia Ministry of Forests, Victoria, British Columbia, Canada.

Leonard, M.L. and M.B. Fenton. 1983. Habitat use by spotted bats (*Euderma maculatum*): roosting and foraging behavior. *Canadian Journal of Zoology* **61**:1487–1491.

Lewis, S.E. 1993. Effect of climatic variation on reproduction by pallid bats (*Antrozous pallidus*). *Canadian Journal of Zoology* **71**:1429–1433.

Lewis, S.E. 1994. Night roosting ecology of pallid bats (*Antrozous pallidus*) in Oregon. *American Midland Naturalist* **132**:219–226.

Lewis, S.E. 1995. Roost fidelity of bats: a review. *Journal of Mammalogy* **76**:481–496.

Lunde, R.E. and A.S. Harestad. 1986. Activity of little brown bats in coastal forests. *Northwest Science* **60**:206–209.

Mackey, R.L. and R.M.R. Barclay. 1989. The influence of physical clutter and noise on the activity of bats over water. *Canadian Journal of Zoology* **67**:1167–1170.

Marcot, B.G. 1984. Winter use of some northwestern California caves by western big-eared bats and long-eared myotis. *Murrelet* **65**:46.

Mattson, T.A., S.W. Buskirk, and N.L. Stanton. 1996. Roost sites of the silver-haired bat (*Lasionycteris noctivagans*) in the Black Hills of South Dakota. *Great Basin Naturalist* **50**:247–253.

McComb, W. and D. Lindenmayer. 1999. Dying, dead, and down trees. Pages 335–372 *in* M.L. Hunter, editor. *Maintaining Biodiversity in Forest Ecosystems*. Cambridge University Press, Cambridge, UK.

McLean, J.A. and J.R. Speakman. 1999. Energy budgets of lactating and non-reproductive brown long-eared bats (*Plecotus auritus*) suggest females use compensation in lactation. *Functional Ecology* **13**:360–372.

Menzel, M.A., T.C. Carter, B.R. Chapman, and J. Laerm. 1998. Quantitative comparison of tree roosts used by red bats (*Lasiurus borealis*) and Seminole bats (*L. seminolus*). *Canadian Journal of Zoology* **76**:630–634.

Morales, J.C. and J.W. Bickham. 1995. Molecular systematics of the genus *Lasiurus* (Chiroptera: Vespertilionidae) based on restriction-site maps of the mitochondrial ribosomal genes. *Journal of Mammalogy* **76**:830–749.

Morrison, M.L. and M.G. Raphael. 1993. Modeling the dynamics of snags. *Ecological Applications* **3**:322–330.

Nagorsen, D.W. and R.M. Brigham. 1993. *Bats of British Columbia*. University of British Columbia Press, Vancouver, British Columbia, Canada.

Nagorsen, D.W., A.A. Bryant, D. Kerridge, G. Roberts, A. Roberts, and M.J. Sarell. 1993. Winter bat records for British Columbia. *Northwestern Naturalist* **74**:61–66.

Nilsson, S.G. 1984. The evolution of nest-site selection among hole-nesting birds: the importance of nest predation and competition. *Ornis Scandinavica* **15**:167–175.

Norberg, U.M. and J.M.V. Rayner. 1987. Ecological morphology and flight in bats (Mammalia: Chiroptera): wing adaptations, flight performance, foraging strategy and echolocation. *Philosophical Transactions of the Royal Society of London, B, Biological Sciences* **316**:335–427.

Nyberg, J.B., A.S. Harestad, and F.L. Bunnell. 1987. "Old-growth" by design: managing young forests for old-growth wildlife. *Transactions of the 52nd North American Wildlife and Natural Resource Conference* **52**:70–81.

Nyland, R.D. 1996. *Silviculture Concepts and Applications*. McGraw-Hill, New York, New York, USA.

O'Donnell, C.F.J. and J.A. Sedgeley. 1999. Use of roosts by the long-tailed bat, *Chalinobus tuberculatus*, in temperate rainforest in New Zealand. *Journal of Mammalogy* **80**:913–923.

Ohmann, J.L., W.C. McComb, and A.A. Zumrawi. 1994. Snag abundance for primary cavity-nesting birds on non-federal lands in Oregon and Washington. *Wildlife Society Bulletin* **22**:607–620.

Oliver, C.D. and B.C. Larson. 1996. *Forest Stand Dynamics*. John Wiley and Sons, New York, New York, USA.

Ormsbee, P.C. 1996. Characteristics, use, and distribution of roosts selected by female *Myotis volans* (long-legged myotis) in forested habitat of the central Oregon Cascades. Pages 124–131 *in* R.M.R. Barclay and R.M. Brigham, editors. *Bats and Forests Symposium*. British Columbia Ministry of Forests, Victoria, British Columbia, Canada.

Ormsbee, P.C. and W.C. McComb. 1998. Selection of day roosts by female long-legged myotis in the central Oregon Cascade Range. *Journal of Wildlife Management* **62**:596–603.

O'Shea, T.J. and T.A. Vaughan. 1977. Nocturnal and seasonal activities of the pallid bat, *Antrozous pallidus*. *Journal of Mammalogy* **58**:269–284.

O'Shea, T.J. and T.A. Vaughan. 1999. Populations changes in bats from central Arizona: 1972 and 1997. *Southwestern Naturalist* **44**:495–500.

Parker, D.I., J.A. Cook, and S.W. Lewis. 1996. Effects of timber harvest on bat activity in southeastern Alaska's temperate rainforests. Pages 277–292 *in* R.M.R. Barclay and R. M. Brigham, editors. *Bats and Forests Symposium*. British Columbia Ministry of Forests, Victoria, British Columbia, Canada.

Parsons, H.J., D.A. Smith, and R.F. Whittam. 1986. Maternity colonies of silver-haired bats, *Lasionycteris noctivagans*, in Ontario and Saskatchewan. *Journal of Mammalogy* **67**:598–600.

Pearson, O.P., M.R. Koford, and A.K. Pearson. 1952. Reproduction of the lump-nosed bat (*Corynorhinus rafinesquei*) in California. *Journal of Mammalogy* **33**:273–320.

Perkins, J.M. and S.P. Cross. 1988. Differential use of some coniferous forest habitats by hoary and silver-haired bats in Oregon. *Murrelet* **69**:21–24.

Perkins, J.M., J.M Barss, and J. Peterson. 1990. Winter records of bats in Oregon and Washington. *Northwestern Naturalist* **71**:59–62.

Perlmeter, S.I. 1996. Bats and bridges: patterns of night roost activity in the Willamette National Forest. Pages 132–150 *in* R.M.R. Barclay and R.M. Brigham, editors. *Bats and Forests Symposium*. British Columbia Ministry of Forests, Victoria, British Columbia, Canada.

Pierson, E.D. and W.E. Rainey. 1998. Distribution of the spotted bat, *Euderma maculatum*, in California. *Journal of Mammalogy* **79**:1296–1305.

Pierson, E.D., W.E. Rainey, and R.M. Miller. 1996. Night roost sampling: a window on the forest-bat community in northern California. Pages 151–163 *in* R.M.R. Barclay and R.M. Brigham, editors. *Bats and Forests Symposium*. British Columbia Ministry of Forests, Victoria, British Columbia, Canada.

Ports, M.A. and P.V. Bradley. 1996. Habitat affinities of bats from northeastern Nevada. *Great Basin Naturalist* **56**:48–53.

Psyllakis, J.M. 2001. Bat roosting and foraging ecology in naturally disturbed habitats. M.S. thesis, University of Regina, Regina, Saskatchewan, Canada.

Rabe, M.J., T.E. Morrell, H. Green, J.C. DeVos, Jr., and C.R. Miller. 1998. Characteristics of Ponderosa pine snag roosts used by reproductive bats in northern Arizona. *Journal of Wildlife Management* **62**:612–621.

Racey, P.A. 1973. Environmental factors affecting the length of gestation in heterothermic bats. *Journal of Reproduction and Fertility* **19(Supplement)**:175–189.

Racey, P.A. and S.M. Swift. 1981. Variations in gestation length in a colony of pipistrelle bats (*Pipistrellus pipistrellus*) from year to year. *Journal of Reproduction and Fertility* **61**:123–129.

Rainey, W.E., E.D. Pierson, M. Colberg, and J.H. Barclay. 1992. Bats in hollow redwoods: seasonal use and role in nutrient transfer into old growth communities. *Bat Research News* **33**:71.

Reith, C.C. 1980. Shifts in times of activity by *Lasionycteris noctivagans*. *Journal of Mammalogy* **61**:104–108.

Remsen, J.V., Jr. and D.A. Good. 1996. Misuse of data from mist-net captures to assess relative abundance in bird populations. *Auk* **113**:381–398.

Rendell, W.B. and R.J. Robertson. 1989. Nest-site characteristics, reproductive success and cavity availability for tree swallows breeding in natural cavities. *Condor* **91**:875–885.

Rolseth, S.L., C.E. Koehler, and R.M.R. Barclay. 1994. Differences in the diets of juvenile and adult hoary bats, *Lasiurus cinereus*. *Journal of Mammalogy* **75**:394–398.

Rydell, J., A. Entwistle, and P.A. Racey. 1996. Timing of foraging flights of three species of bats in relation to insect activity and predation risk. *Oikos* **76**:243–252.

Schowalter, D.B., W.J. Dorward, and J.R. Gunson. 1978. Seasonal occurrence of silver-haired bats (*Lasionycteris noctivagans*) in Alberta and British Columbia. *Canadian Field-Naturalist* **92**:288–291.

Seidman, V.M. and C.J. Zabel. 2001. Bat activity along intermittent streams in northwestern California. *Journal of Mammalogy* **82**:738–747.

Shump, K.A., Jr. and A.U. Shump. 1982*a*. *Lasiurus borealis*. *Mammalian Species* **183**:1–6.

Shump, K.A., Jr. and A.U. Shump. 1982*b*. *Lasiurus cinereus*. *Mammalian Species* **185**:1–5.

Speakman, J.R. and P.A. Racey. 1989. Hibernal ecology of the pipistrelle bat: energy expenditure, water requirements and mass loss, implications for survival and the function of winter emergence flights. *Journal of Animal Ecology* **58**: 797–813.

Szewczak, J.M., S.M. Szewczak, M.L. Morrison, and L.S. Hall. 1998. Bats of the White and Inyo Mountains of California-Nevada. *Great Basin Naturalist* **58**:66–75.

Thomas, D.W. 1988. The distribution of bats in different ages of Douglas-fir forests. *Journal of Wildlife Management* **52**:619–628.

Thomas, D.W. and S.D. West. 1991. Forest age associations of bats in the southern Washington Cascade and Oregon Coast Ranges. Pages 295–303 *in* L.F. Ruggiero, K.B. Aubry, A.B. Carey, and M.H. Huff, editors. Wildlife and Vegetation of Unmanaged Douglas-fir Forests. USDA Forest Service, General Technical Report PNW- **285**, Portland, Oregon, USA.

Thompson, W.L., G.C. White, and C. Gowan. 1998. *Monitoring Vertebrate Populations*. Academic Press, Inc., San Diego, California, USA.

Tuttle, M.D. 1976. Population ecology of the gray bat (*Myotis grisescens*): factors influencing growth and survival of newly volant young. *Ecology* **57**:587–595.

Valdez, E.W., R. Choate, M.A. Bogan, and T.L. Yates. 1999. Taxonomic status of *Myotis occultus*. *Journal of Mammalogy* **80**:545–552.

Vaughan, T.A. and P.H. Krutzsch. 1954. Seasonal distribution of the hoary bat in southern California. *Journal of Mammalogy* **35**:431–432.

Verts, B.J. and L.N. Carraway. 1998. *Land Mammals of Oregon*. University of California Press, Berkeley, California, USA. 668pp.

von Frenckell, B. and R.M.R. Barclay. 1987. Bat activity over calm and turbulent water. *Canadian Journal of Zoology* **65**:219–222.

Vonhof, M.J. and R.M.R. Barclay. 1996. Roost-site selection and roosting ecology of forest-dwelling bats in southern British Columbia. *Canadian Journal of Zoology* **74**:1797–1805.

Vonhof, M.J. and R.M.R. Barclay. 1997. Use of tree stumps as roosts by the western long-eared bat. *Journal of Wildlife Management* **61**:674–684.

Wai-Ping, V. and M.B. Fenton. 1989. Ecology of spotted bat (*Euderma maculatum*) roosting and foraging behavior. *Journal of Mammalogy* **70**:617–622.

Waldien, D.L. and J.P. Hayes. 2001. Activity areas of female long-eared myotis in coniferous forests in western Oregon. *Northwest Science* **75**:307–314.

Waldien, D.L.J.P. Hayes, and E.B. Arnett. 2000. Day-roosts of female long-eared myotis in western Oregon. *Journal of Wildlife Management* **64**:785–796.

Walsh, A.L. and S. Harris. 1996. Foraging habitat preferences of vespertilionid bats in Britain. *Journal of Applied Ecology* **33**:508–518.

Warner, R.M. and N.J. Czaplewski. 1984. *Myotis volans. Mammalian Species* **224**:1–4.

Warren, R.D., D.A. Waters, J.D. Altringham, and D.J. Bullock. 2000. The distribution of Daubenton's bats (*Myotis daubentonii*) and pipistrelle bats (*Pipistrellus pipistrellus*) in relation to small-scale variation in riverine habitat. *Biological Conservation* **92**:85–91.

Weller, T.J. and C.J. Zabel. 2001. Characteristics of fringed myotis day roosts in northern California. *Journal of Wildlife Management* **65**:489–497.

West, E.W. and U. Swain. 1999. Surface activity and structure of a hydrothermally-heated maternity colony of the little brown bat, *Myotis lucifugus*, in Alaska. *Canadian Field-Naturalist* **113**:425–429.

Wethington, T.A., D.M. Leslie, Jr., M.S. Gregory, and M.K. Wethington. 1996. Prehibernation habitat use and foraging activity by endangered Ozark big-eared bats (*Plecotus townsendii ingens*). *American Midland Naturalist* **135**:218–230.

Whitaker, J.O. and L.J. Rissler. 1992. Winter activity of bats at a mine entrance in Vermillion county, Indiana. *American Midland Naturalist* **127**:52–59.

Whitaker, J.O., Jr., C. Maser, and L.E. Keller. 1977. Food habits of bats of western Oregon. *Northwest Science* **51**:46–55.

Whitaker, J.O., C. Maser, and S.P. Cross. 1981a. Foods of Oregon silver-haired bats, *Lasionycteris noctivagans. Northwest Science* **55**:75–77.

Whitaker, J.O., C. Maser, and S.P. Cross. 1981b. Food habits of eastern Oregon bats, based on stomach and scat analyses. *Northwest Science* **55**:281–292.

Wilkinson, L.C. and R.M.R. Barclay. 1997. Differences in the foraging behaviour of male and female big brown bats (*Eptesicus fuscus*) during the reproductive period. *Ecoscience* **4**:279–285.

Williams, D.F. and J.S. Findley. 1979. Sexual size dimorphism in vespertilionid bats. *American Midland Naturalist* **102**:113–126.

Wilson, D.E. and S. Ruff. 1999. *The Smithsonian Book of North American Mammals.* Smithsonian Institution Press, Washington DC, USA. 750pp.

Zielinski, W.J. and S.T. Gellman. 1999. Bat use of remnant old-growth redwood stands. *Conservation Biology* **13**:160–167.

Zimmerman, G.S. and W.E. Glanz. 2000. Habitat use by bats in eastern Maine. *Journal of Wildlife Management* **64**:1032–1040.

JAMES G. HALLETT, MARGARET A. O'CONNELL
AND CHRIS C. MAGUIRE

5

Ecological relationships of terrestrial small mammals in western coniferous forests

Small mammals are important components of western forest ecosystems. Their interactions with other organisms and the physical environment are complex. Small mammals are effective predators on seeds, vegetation, and insects, and may influence patterns of forest regeneration (Sullivan 1979, Christy and Mack 1984). Dispersal of seeds, mycorrhizal fungi, and nitrogen-fixing bacteria by small mammals may also affect plant diversity (Maser et al. 1978, Verts and Carraway 1998, Luoma et al. 2003). In addition, small mammals are prey for many carnivorous taxa, and changes in small-mammal abundance may affect the distribution and habitat use of their predators (Carey et al. 1992).

Understanding patterns of abundance and distribution of small mammals and how these species influence forest function and biodiversity are basic ecological problems relevant to both economic and conservation concerns. Exploration of these fundamental ecological relationships is challenging, however, because of the secretive nature of small mammals and their activities at night (e.g., deer mice (*Peromyscus* spp.)), below ground (e.g., gophers (*Thomomys* spp.) and moles (*Scalopus* spp.)), at the soil–vegetation interface (e.g., red-backed voles (*Clethrionomys* spp.) and shrews (*Sorex* spp.)), and in all levels of the canopy (e.g., flying squirrels (*Glaucomys* spp.) and tree squirrels (*Tamiasciurus* spp.)) (Aubry et al. 2003). In this chapter, we emphasize the forest-floor fauna, although not all species discussed confine themselves to the ground stratum (e.g., chipmunks (*Tamias* spp.) and woodrats (*Neotoma* spp.)).

For the forest-floor environment to be suitable small-mammal habitat, it must provide requisites for reproduction and survival including food resources, cover, nest sites, and appropriate microclimatic conditions. These features vary with forest type, seral stage, management history (e.g., harvest activity, site preparation, planting), and soil condition. In addition, the composition of the small-mammal assemblage present in any forest stand is influenced not only by responses of individual species to the forest-floor environment, but also by their interspecific competitors. In this review, we focus on the ecological roles of small mammals in western coniferous forests, describe patterns of species richness and abundance, and evaluate how management activities impact habitat relationships. We begin our discussion with a review of the animal species considered and a highlight of limitations often encountered in their study.

The taxa

Throughout temperate North American forests, small-mammal assemblages are dominated by three families: Soricidae (shrews), Muridae (voles, mice, and woodrats), and Sciuridae (squirrels, chipmunks, and marmots) (Kirkland 1985). In western coniferous forests, species from another five families also occur in some geographic locations and forest types: Talpidae (moles), Geomyidae (gophers), Dipodidae (jumping mice), Aplodontidae (mountain beavers), and Mustelidae (weasels). While our review covers most of the terrestrial small-mammal fauna, highly arboreal rodents (e.g., tree squirrels and tree voles) and small carnivores are covered by other authors in this volume (Aubry et al. 2003, Buskirk and Zielinski 2003, Smith et al. 2003).

As taxonomic distinctions suggest, forest small mammals differ considerably in body size, morphology, physiology, and habits. Shrews are smallest with mean body masses ranging from about 3 g to 12 g. There currently is much interest in the biology of these insectivores (Findley and Yates 1991, Merritt et al. 1994), particularly because of the constraints that high metabolic rates and food requirements place on them (McNab 1991). Across their range, species diversity of shrews generally increases with increasing moisture and reduced temperatures. Cool mesic environments appear to meet the microclimatic requirements of shrews, and they may also ensure a stable invertebrate prey base (Kirkland 1991). The highest abundance and species richness (five species) of shrews are found in moist forests of the Pacific Northwest (e.g., Aubry et al. 1991). Five species of shrews also occur in the drier forests of the Sierra Nevada Mountains in California, but unlike the Pacific Northwest there is limited species overlap along an altitudinal gradient (Williams 1991). Attempts to explain shrew richness center on body size differences among species

(Kirkland 1991, Fox and Kirkland 1992). Small shrews have higher mass-specific metabolic rates and smaller energy stores than large shrews. Additionally, although their total food requirements are lower, the ability of small shrews to survive without food is much less than for large shrews (Hanski 1994). Size also may limit the types of food that shrews exploit. The smaller muscle mass of small shrews confers a bite force that precludes consumption of harder foods (Carraway and Verts 1994). Finally, differences in the abilities of shrews to burrow are related to size (Terry 1981). Lawlor (2003) lists (in Table 3.1 of his Appendix) 19 species of shrews in two genera that occur collectively in western forests. Unfortunately, the ecology of some species is poorly known (e.g., *Sorex preblei, S. nanus*). In our discussion of shrews, we also consider one talpid, the shrew-mole (*Neurotrichus gibbsii*), because it shares many ecological similarities with shrews (Terry 1981).

Terrestrial murid rodents are more diverse than shrews, both taxonomically and ecologically, with representatives from up to seven genera present in western coniferous forest types. Most species have adult body masses between 20 g and 125 g, except for five woodrat species (weights between 200 g and 400 g). *Clethrionomys* (red-backed voles) and *Peromyscus* (deer mice) dominate the murid rodent group throughout western forests. There are three species of western *Clethrionomys*, and all are closely associated with mesic closed-canopy forests; none have overlapping distributions. Although red-backed voles are omnivorous, a large proportion of their diet consists of forest fungi (Hayes et al. 1986). Of nine species of western *Peromyscus*, only the northwestern deer mouse (*P. keeni*) is a forest obligate (Lawlor 2003); the remaining species are habitat generalists and they consume a wide range of food types. Where geographic distributions overlap, up to four *Peromyscus* species may occur in a single stand (Ribble and Stanley 1998). The western heather vole (*Phenacomys intermedius*) and the northern bog lemming (*Synaptomys borealis*) are widely distributed in northern boreal forests, although the distribution of the heather vole extends along the mountains into California and New Mexico. In contrast, the white-footed vole (*Arborimus* (=*Phenacomys*) *albipes*) is restricted to forests in western Oregon and northwestern California and is one of the rarest voles known (Verts and Carraway 1995). Ten species of *Microtus* occupy western forests (Lawlor 2003). These microtines are primarily herbivorous and occur predominantly in early seral stages or montane meadows, where forbs and grasses are abundant. Woodrats (*Neotoma* spp.) utilize a broad array of vegetative types in the west and all forest seral stages. Most species, however, reach their highest densities under narrower habitat conditions (Tevis 1956, Harris et al. 1982, Raphael 1988*b*, Sakai and Noon 1993), and some species overlap. All woodrats are herbivorous.

The three western species of jumping mice (*Zapus* spp.) are found primarily in meadows and adjacent to streams in dense herbaceous vegetation. The large hind feet and long tail of jumping mice facilitate saltatorial locomotion through lush vegetation; the shorter front legs are used to gather food. Jumping mice are omnivorous, feeding opportunistically on a variety of arthropods, fungi, and plant material (Cranford 1978). Prior to entering hibernation for up to nine months, however, jumping mice primarily consume seeds, which provide greater energy.

Of the sciurids, we only consider chipmunks in the genus *Tamias* in this chapter. Within this group, 21 species occur in western forests with the greatest species richness found in the southwestern US (Lawlor 2003). These morphologically similar species are primarily forest occupants, but they vary in their degree of activity on the forest floor and in the tree canopy. They consume a wide variety of seeds, fungi, and insects. Many species cache seeds in underground burrow systems prior to entering torpor during the winter. These food supplies are used periodically throughout the cold months and take the place of fat stores (Verts and Carraway 1998).

One of the most distinctive residents of western forests is the mountain beaver (*Aplodontia rufa*), the most primitive living rodent (Kurtén and Anderson 1980). This herbivorous, fossorial species is limited to forested areas in the far western states and southern British Columbia, primarily from the Coast Range through the western slopes of the Cascade Range; they are absent from interior valleys (Dalquest and Scheffer 1945, Hall 1981, Verts and Carraway 1998). The mountain beaver is found at all elevations below treeline within its range and in forested stands across all seral stages (Carraway and Verts 1993), but it appears to favor mid-seral stages with dense undergrowth (Hacker 1991).

The diverse composition of small-mammal faunas across western North American forests echoes the pronounced differences in species distributions. Current distributions reflect the responses of species to historical changes in geology, hydrology, climate, and vegetation (Lawlor 2003). Kirkland (1985) suggested that fluctuating environments during the four glacial advances of the Pleistocene, as well as more recent disturbance regimes, selected for small-mammal species with broad ecological niches. He further proposed that these generalist species have increased tolerances to anthropogenic disturbances such as forest harvest. Although adaptive trends seem evident for some species, they are not universal as we discuss below.

Assessing populations

Perceptions about the composition of small-mammal communities and species abundances depend on our methods of observation (Kirkland et al. 1998). Because some form of trapping must be employed to survey most small mammals, the choice of trapping method greatly influences the results (e.g., Taylor et al. 1988). As previously noted, species considered in this review vary widely in body size, morphology, and behavior; these characteristics contribute to non-uniform capture probabilities across trap types. Live-traps (e.g., Sherman or Longworth) may not adequately sample shrews because of insufficient trap sensitivity or because shrews avoid these traps (Kirkland and Sheppard 1994). Difficulties inherent in identifying the species of some live shrews and voles may also make live-trapping inappropriate, or require lumping data for two or more species (e.g., Sullivan and Sullivan 1982a).

Large-scale studies conducted in the Pacific Northwest have routinely used removal trapping (e.g., Corn and Bury 1991, Gilbert and Allwine 1991, West 1991, Lehmkuhl et al. 1999). The most common traps employed in these studies are Museum Special snap traps and pitfall traps. In the Oregon Coast Range, McComb et al. (1991) found snap traps to be more effective at sampling certain rodents (e.g., *Peromyscus* and *Tamias*) and pitfall traps for sampling shrews and the Pacific jumping mouse (*Zapus trinotatus*). Similarly, Williams (1991) caught only six shrews in the Sierra Nevada in >20 000 snap trap-nights, but 735 shrews in approximately 25 000 pitfall trap-nights. Because of these differences in trap success, many workers have used both approaches (e.g., Corn and Bury 1991, Gilbert and Allwine 1991, West 1991, Hallett and O'Connell 2000).

Kirkland and Sheppard (1994) recognized the problems associated with comparing capture results from studies that employ different trapping methodologies. As a result, they proposed a standard protocol for sampling shrews using pitfall traps and drift fences, and it was utilized to compare shrew communities in forests of New Mexico and Pennsylvania (Kirkland and Findley 1999). Whether or not this shrew-sampling approach is ultimately adopted across a wide range of studies, the goal to apply similar sampling regimes for small mammals is appropriate. As a review of the literature for this chapter highlighted, one encounters major difficulties when comparing measures of species diversity and abundance among studies that used different trapping approaches.

Additional variation in trapping studies can result from influences of environmental factors such as amount and distribution of rainfall (Maguire 1999) or changes in moonlight through the lunar cycle (Clarke 1983, Kirkland and Sheppard 1994, Zollner and Lima 1999). Unfortunately, few studies have examined the effects of such factors on the behavior or probability of capture of small mammals in western forests. Maguire (1999) found that western red-backed vole (*Clethrionomys californicus*) captures increased in northern California with increasing total weekly rainfall. Individuals of this species appeared to remain underground during the summer dry season and to surface primarily when precipitation was measurable and temperatures were moderate. Similarly, Feldhamer et al. (1993) observed reduced trapping susceptibility of the pygmy shrew (*Sorex hoyi*) in summer in Kentucky. They hypothesized that this shrew may be unable to meet its water requirements above ground under limiting moisture conditions. Studies such as these suggest that forest small mammals confronted with xeric conditions above ground are likely to limit their surface activity and have fewer trap encounters.

Ecological roles of small mammals

Consumers
Diet selection and breadth

Small mammals often are assigned to one of three trophic groups: insectivores (e.g., shrews) that feed largely on insects and other invertebrates, herbivores (e.g., voles) that feed on succulent plant material, and omnivores or omnivore-granivores that feed on a diversity of food types including seeds (e.g., deer mice, chipmunks) (Kirkland 1985). As elsewhere in North America, small mammals in western coniferous forests are predominantly insectivorous and omnivorous, but the emphasized food type may change seasonally. Herbivores typical of grasslands (i.e., *Microtus*) are less abundant in these forest communities, and strict granivores are rare (Fleming 1973).

Trophic generalizations such as those noted above largely reflect the vague understanding of diet selection by small mammals, a subject of great ecological importance. Although the types of foods used by most small-mammal species are known from natural history studies (many reviewed by Verts and Carraway 1998), few quantitative assessments of diet selection and how it varies with food availability or the presence of resource competitors are available. Patterns of co-existence may depend on reduced trophic overlap between (Ryan 1986) and within species (Van Horne 1982), and real and potential food items may determine small-mammal use of and success in a habitat (Gliwicz and Glowacka 2000). Ultimately, small-mammal depredation of seeds, fungi, plants, invertebrates, and bird eggs has strong effects on forest regeneration (e.g., Neal and Borreco 1981, Sullivan et al. 1993), biodiversity (e.g., Maser et al. 1978, McShea 2000), and nutrient cycling (e.g., Sirotnak and Huntly 2000).

Small mammals consume a wide variety of food resources that may vary by geographic region, habitat, and season. For example, in Washington, Gunther et al. (1983) categorized diets of several small-mammal species into five primary groups: fungi and lichens, conifer seeds, herbaceous material, leaves or seeds of grasses, and invertebrates. Invertebrate material made up a large proportion of the diets of two shrews (*S. monticolus* and *S. trowbridgii*) and the shrew-mole (*Neurotrichus gibbsii*), but conifer seeds and fungi or lichens also were important. Deer mice used foods from all categories except grasses. Townsend's chipmunks (*Tamias townsendii*) and southern red-backed voles (*Clethrionomys gapperi*) primarily used fungi and conifer seeds. The relative proportions of food groups varied with season or stand condition for some species. For example, the diet of *N. gibbsii* consisted primarily of invertebrates in burned clear-cuts, but shifted to fungi and conifer seeds in closed-canopy forest. *Sorex trowbridgii* greatly reduced its use of fungi and conifer seeds in autumn when invertebrates were more abundant (Gunther et al. 1983). Similarly, the western jumping mouse (*Zapus princeps*) initially forages on green vegetation after snowmelt, but switches to high-energy seeds to amass sufficient reserves before entering hibernation. Cranford (1978) showed that animals would not attain the necessary hibernation weight during their limited activity period solely on a diet of green vegetation.

Terry (1978) argued that small mammals with high metabolic requirements such as shrews should be food generalists and consume any food that provides sufficient nutrition or energy. She presented results of food preference trials with *Sorex vagrans*, *S. trowbridgii*, and *N. gibbsii* supporting this view. All three species fed on a wide variety of food items including live or dead invertebrates, carrion, and seeds of herbs, shrubs, and trees.

The larger N. *gibbsii*, however, ate foods from the entire range of sizes offered, whereas the two shrew species were unable to handle larger or harder foods (e.g., large beetles and seeds with thick coats).

Further evidence of the generalized diet of shrews comes from a detailed comparative analysis of food habits of five species in Oregon by Whitaker and Maser (1976). They identified 72 food categories, mostly orders and families of insects, ingested by shrews. Individual species consumed 26–47 different food items. Although the shrews shared many food groups, there were proportional differences in major food items used. Whitaker and Maser (1976) related these differences to site-specific availability of food resources and variability in shrew body size. Verts and Carraway (1998), however, re-analyzed Whitaker and Maser's (1976) work by pooling their data into 13 taxonomic groups, and concluded that differences in bite force (i.e., the strength of the masticatory apparatus; Carraway and Verts 1994) could account for differences in the food types selected. They suggested that this was a mechanism for reducing interspecific competition.

The possibility that intraspecific competition also might lead to divergence in food use by demographic groups was explored by Van Horne (1982). She found that seeds and fruits dominated the diet of deer mice, but juveniles fed less on hard-bodied arthropods than adults. When populations were at high densities, arthropod consumption by juveniles was reduced further, supporting the argument that intraspecific competition was underlying food selection. Hanley and Barnard (1999) conducted a 4-year study of diet composition in Sitka mice (*Peromyscus keeni sitkensis*) in Alaska. Although they also found that seeds and fruits were the most important components of the mouse diet, in contrast to Van Horne's (1982) study, no evidence of food niche differentiation between age or sex classes was observed. Apparently, there was little difference in food resources available to mice between the two principal habitats (floodplain vs. upland forest) studied.

The degree to which small mammals exploit a food resource may depend on the presence of non-mammalian taxa that compete for the same resources. For example, in a dry pinyon pine (*Pinus edulis*) forest in northern Arizona, Christensen and Whitham (1993) examined pine cone use by cone-boring insects, birds (jays), and two sciurids (cliff chipmunks (*Tamias dorsalis*) and rock squirrels (*Spermophilus variegatus*)). Harvest rates of all three taxa were negatively correlated. Mammals harvested 2.3 and 2.5 times more cones when insects and birds were excluded, respectively.

Birds increased their percentage of cones harvested five times when mammals were absent. Interestingly, the presence of insects reduced the suitability of a stand for birds and may indirectly have benefited mammals, which could exploit cones that were not attacked by insects.

Population responses to food availability

Although small-mammal populations are assumed to be food limited (Boutin 1990), few studies have examined the population and demographic consequences of changes in food availability on small mammals in western forests. For small mammals, responses to such changes may be pronounced. Tree seeds may be a particularly important food for a number of small mammals in winter and early spring (Hanley and Barnard 1999). Gashwiler (1979) found that overwinter survival and reproduction of deer mice were higher than average immediately following high conifer seed production in Douglas-fir (*Pseudotsuga menziesii*) and western hemlock (*Tsuga heterophylla*) forests in Oregon. In addition, the number of litters per female was higher the year following a good seed crop resulting in higher deer mice populations in autumn. Similar population responses to mast crop fluctuations have been reported for *Peromyscus* species in hardwood forests of the eastern US (Wolff 1996, McShea 2000). Fluctuations in the abundance of invertebrate prey also occur between years as evidenced, for example, by outbreaks of defoliating insects (Torgersen 2001). Shrews may show both numerical (Holling 1959) and functional (Bellocq et al. 1994) responses to changes in insect abundance. In Ontario, for example, the abundance of masked shrews (*Sorex cinereus*) was greater in medium-aged (>40 years) jack pine (*Pinus banksiana*) than in young (20 years) stands, where abundance of preferred prey was lower (Innes et al. 1990). These shrews increased the proportion of lepidopteran larvae in their diet as abundance of this prey type increased (Bellocq et al. 1994). Comparable studies have not been conducted for insectivores in western coniferous forests.

Experimental studies examining the relationship between food abundance and population dynamics of small mammals have provided mixed results. The general approach is to provide supplemental food over an area and then monitor for demographic and density changes. Sullivan et al. (1983), for example, found that population estimates for Townsend's chipmunk increased 40% to 50% over a control population when supplemental food was provided. Juvenile growth and survival also were higher for the experimental population. After food supplementation was stopped, population density and demographic parameters dropped to control levels, indicating that this species is food limited. Similarly, Schweiger and Boutin (1995) reported that persistence of northern red-backed voles over the winter tended to increase with food supplementation. Reproduction commenced two months earlier and resulted in three times as many juveniles on grids with supplemental food than on controls. In addition, immigration onto supplemented grids was up to three times higher than on controls. The higher densities observed in spring after winter food supplementation, however, did not prevent a decline in numbers during the summer and failure to reach a population peak in autumn. Finally, when supplemental seed was provided on forest stands in British Columbia, shrews showed no short-term population response (Sullivan and Sullivan 1982a), even though they will feed on seeds (Terry 1978, Gunther et al. 1983).

Effects on forest regeneration

The foraging activities of small mammals play a relatively minor role in energy turnover of plant materials in forest ecosystems, because only 4% to 13% of plant primary production is available to them (Grodzinski and Wunder 1975). Regardless, the effects that small mammals have on forest regeneration may be substantial. Although predation on seeds and seedlings can hamper attempts to reclaim harvested lands, the dispersal and caching of seeds by small mammals may promote regeneration. In addition, consumption and subsequent dispersal of mycorrhizal fungi by small mammals is also considered important for tree growth (Luoma et al. 2003).

Attempts to regenerate forests after harvest by direct seeding often are unsuccessful because of intense seed predation by *Peromyscus* and *Microtus*, especially when small-mammal densities are high (Sullivan and Sullivan 1982b). In an experimental study in British Columbia, small mammals removed >85% of lodgepole pine (*Pinus contorta*) seeds within three weeks of seeding when densities were greater than five animals per hectare, but only 30% to 40% when there were fewer than three animals per hectare (Sullivan and Sullivan 1982b). Predator satiation could be achieved, however, by seeding with twice as many sunflower seeds as pine seeds. Pine seed survival in the presence of sunflower seeds was 42% to 72% after six weeks versus only 8% to 10% without sunflower seeds. Radvanyi (1970) found that seed predation could also be reduced by seeding immediately before snowfall. White spruce (*Picea glauca*) seed survival in Alberta was 81%

after one year following winter seeding, but only 50% after four months following spring seeding.

In a study designed to determine if small-mammal removal would increase the success of reseeding efforts, animals surrounding the test area rapidly populated the site within 3–12 days after resident animals were removed (Sullivan 1979). Results were comparable between spring when population density was depressed, and autumn when dispersing subadult animals were more abundant. In autumn, 95% of seeds were removed within three days and 93% were depleted within five days in spring. These findings suggest that site-limited population reduction is not an effective method for controlling seed predation.

Tree seeds that successfully germinate to seedlings are also subject to predation by small mammals (Maguire 1989). Seedlings, however, are generally utilized when other foods provide insufficient nutrition or become less available. For example, low protein ferns (common sword-fern (*Polystichum munitum*) and bracken fern (*Pteridium aquilinum*)) normally account for >80% of the mountain beaver diet (Voth 1968, Allen 1969). Lactating females, however, increase their consumption of conifers, especially Douglas-fir, to one-third of the diet for additional needed protein (Voth 1968). Mountain beaver are viewed as significant pests in managed forests of the Pacific Northwest (Crouch 1968, Hooven 1977, Hoyer et al. 1979, Neal and Borreco 1981), particularly in new Douglas-fir plantations (Borrecco and Anderson 1980, Cafferata 1992), because of their consumption of conifer seedlings and their basal/root barking and branch clipping of older trees (Lawrence et al. 1961).

Herbivorous rodents, especially *Microtus* spp., generally feed on seedlings and young saplings under the snow in the winter, when forbs and grasses are less available and are of poorer nutritional quality (Baxter and Hansson 2001). Consumption of bark, vascular tissue, and roots can lead to direct mortality or reduced growth of young trees (Sullivan et al. 2001). The severity of damage to regenerating forests in a given year is variable, but in northern Europe, damage coincides with peaks in small-mammal abundance (Baxter and Hansson 2001). Comparable data are lacking for western coniferous forests of North America. The risk of damage to regenerating forests, however, has led to attempts to prevent seedling predation by providing alternate food resources to small mammals. Sullivan et al. (2001) reduced consumption of lodgepole pine seedlings by *Microtus* by supplemental feeding of alfalfa pellets or bark mulch bound with wax and sunflower oil.

Despite the effectiveness of small mammals as predators of seeds, seedlings, and saplings, their seed-caching behavior (Smith and Reichman 1984) may be essential for the dispersal and success of some plant species when relocation and consumption of the cache by small mammals are incomplete. For example, Vander Wall (1992) placed two arrays of 1064 Jeffrey pine (*Pinus jeffreyi*) seeds in patterns predicted by models of wind dispersal. Yellow-pine chipmunks (*Tamias amoenus*) and three less abundant rodent species removed 95% to 99% of the seeds within two days. Importantly, 35% to 54% of the recovered seeds were buried in shallow surface caches at distances up to 63 m from the source area. Because burial appears to be required for successful germination and not all caches are recovered by rodents (Vander Wall 2000), the dual dispersal system of wind and mammal is an effective strategy for Jeffrey pine. Chambers (2001) reported that seeds of single-leaf pinyon (*Pinus monophylla*) also must be buried by small mammals or birds to successfully germinate. The importance of seed caching by rodents, however, varies with plant species. In western Oregon, West (1968) estimated that 15% of ponderosa pine (*Pinus ponderosa*) and 50% of bitterbrush (*Purshia*) seedlings originated from rodent caches, whereas three other shrub species were not assisted by rodents.

The role of small mammals in dispersing seeds is poorly known for forests more diverse and structurally complex than *Pinus* forests. It also would be valuable to know how vegetational composition of regenerating forest stands is influenced by interspecific relationships in mammalian seed dispersal. For example, several species of larger mammals, including bears (Traveset and Willson 1997) and marten (*Martes americana*) (Hickey et al. 1999), are long-distance seed dispersers. Seeds in the feces of these animals are likely to be unexploited and available for regeneration when small-mammal densities are low, but consumed when populations are high (Bermejo et al. 1998).

Effects on avian species

Small mammals may be important predators on avian nests (Fenske-Crawford and Niemi 1997, Hannon and Cotterill 1998, Blight et al. 1999). Because nest depredation has a major influence on the population dynamics (Ricklefs 1969) and community relationships (Martin 1988a, b) of bird species, many studies have examined how depredation varies with habitat or landscape structure (e.g., Haskell 1995b). Difficulties in finding and monitoring natural nests have led to the use of artificial nests baited with real (e.g., quail) or artificial (e.g., clay) eggs. This approach has its own set of problems in application because of obvious differences between artificial and natural nests (Rangen et al. 2000) and problems in relating results from artificial nests to natural ones (Sieving and Willson 1998). Nonetheless, these studies illustrate the extent to which mammalian predation may be important. Usually predators are determined by sign at the nest (e.g., characteristic tooth marks) or by photographic record (Zegers et al. 2000). Many small-mammal species have been observed preying on artificial nests including southern red-backed voles, deer mice, eastern chipmunks (*Tamias striatus*), red squirrels (*Tamiasciurus hudsonicus*), gray squirrels (*Sciurus carolinensis*), and flying squirrels (*Glaucomys volans*), as have larger species including raccoons (*Procyon lotor*), fishers (*Martes pennanti*), striped skunks (*Mephitis mephitis*), and black bears (*Ursus americanus*) (Fenske-Crawford and Niemi 1997, Zegers et al. 2000).

The relative importance of mammalian versus avian predators (e.g., jays and ravens) may vary with habitat (Sieving and Willson 1998), nest location (e.g., ground or tree nest; Entz 1996), and degree of edge (Fenske-Crawford and Niemi 1997). In southeastern Alaska and adjacent Canada, predation on artificial nests was higher in coniferous than in deciduous forest, which corresponded to the distribution of the principal predator, the red squirrel. In secondary mixed-coniferous forest in northeastern Washington, Entz (1996) compared depredation of artificial nests placed on the ground and in trees. Small mammals, especially southern red-backed voles, accounted for 25% of all predation events, whereas mammals in general accounted for 59% of all predation events and 92% of those on ground nests. Avian predators accounted for 41% of all predation events and 81% of those on tree nests. Small mammals residing within large aspen (*Populus*) woodlots were the primary predators on artificial nests, but crows and jays exploited nests more quickly than mammals in smaller woodlots and at woodlot edges (Hannon and Cotterill 1998).

Haskell (1995a) argued that predation studies using artificial nests with quail eggs might underestimate effects of small mammalian predators such as deer mice because they are unable to handle eggs of that size. Drever et al. (2000) circumvented this problem by using stable isotope methods to examine diets of northwestern deer mice (*Peromyscus keeni*) and Townsend's voles (*Microtus townsendii*) on Triangle Island off the coast of British Columbia. Triangle Island harbors western Canada's largest seabird colony, and eggs are abundant during the avian breeding season. Townsend's voles obtained protein primarily from terrestrial plants and secondarily from terrestrial invertebrates, whereas northwestern deer mice preyed on invertebrates and seabird eggs when they became available. Consumption of egg protein appeared to be a general phenomenon for northwestern deer mice, and Drever et al. (2000) suggested that high densities on the island were likely due to this food source. Similar stable isotope studies could be used to evaluate the importance of bird eggs as a food source for small mammals in other areas with different ranges in egg size and availability.

Prey

Small mammals are important in the diets of many mammalian, avian, and reptilian carnivores in western coniferous forests. The extent to which small mammals are utilized depends on the degree of morphological and behavioral specialization of the predator, but also on the predator's foraging location and season. In addition, for mammalian carnivores, the degree of prey specialization may depend on body size. For example, ermine (*Mustela erminea*) feed primarily on voles, with deer mice and shrews taken to a lesser extent. In contrast, the larger long-tailed weasel (*M. frenata*) may utilize prey up to woodrat size (Rosenzweig 1966). Segregation in prey size may permit co-existence where these predators are sympatric. The larger American marten has an even more generalized diet than ermine or weasels, but it also varies with geographical area and season (Martin 1994). For example, 1014 scat samples of martens collected throughout the year in Oregon included vole-sized prey (62.7% frequency of occurrence), squirrel-sized prey (28.2%), and lagomorphs (2.4%) (Bull 2000). Martens also consume a variety of birds and bird eggs (19.5%), insects (22.4%), and plant materials (13.3%). Martens reduced their use of vole-sized prey from 83.1% to 46.1% and increased their use of squirrel-sized prey from 18.6% to 39.7% in winter (Bull 2000). It is possible that voles are less accessible to martens than larger prey when snow covers the ground (Simms 1979).

There has been little research on the relationship between predators and the dynamics of small-mammal prey populations in western forests. The best documentation comes from studies of specialized predators. For example, Fitzgerald (1977) investigated predation on montane voles (*Microtus montanus*) by two weasel species (*Mustela erminea* and *M. frenata*) in forest meadows in northern California. Voles were the primary food source for both weasels during winter. The proportion of the vole population consumed by *M. erminea* increased to a high of 54% as the vole population declined to its lowest density. Almost all mortality at this time was due to weasel predation. As vole density increased to its peak, however, the proportion of the vole population subject to weasel predation was greatly reduced (6% to 28%). Fitzgerald (1977) argued that predation when vole abundance was lowest was responsible for both the timing and amplitude of the vole cycle.

Considerably more data document the influence of predators on the population dynamics of microtines in Fennoscandia (e.g., Hanski and Henttonen 1996). In particular, evidence has accumulated to link periodic oscillations in vole numbers to predation by weasels (reviewed by Hanski and Henttonen 1996). The greater abundance of more generalized predators (e.g., red foxes (*Vulpes vulpes*) and buzzards (*Buteo* spp.)) in areas of southern Fennoscandia may account for reduced fluctuations in some populations there (Hanski et al. 1991). Because of the similar generalized diets of most predators in western forests, we also may not detect the tight linkages with population dynamics of prey that have been observed elsewhere. Inconsistencies in the estimation of predator densities also make comparative studies difficult (Smallwood and Schonewald 1998).

The demography and behavior of predators may be affected by the distribution and density of small-mammal prey. Owl reproduction may be reduced or fail in years of low prey abundance, suggesting that owls are food limited (Korpimäki 1987, Hamer et al. 2001). For example, boreal owls (*Aegolius funereus*) were found to forage on all small-mammal species <50 g in the northern Rocky Mountains, but southern red-backed voles were the most important prey item in terms of biomass (37%) (Hayward et al. 1993). The owl's foraging habitat was restricted to spruce-fir (*Picea-Abies*) forest (<25% of the forested habitat) where voles were most abundant. When red-backed vole abundance was low, owls included more deer mice, heather voles (*Phenacomys intermedius*), and pocket gophers (*Thomomys talpoides*) in their diet. No owls fledged at times of low vole densities.

In Washington and Oregon, the northern spotted owl (*Strix occidentalis caurina*), an old-forest associate, feeds primarily on small mammals (>90% prey biomass, Forsman et al. 2001). Northern flying squirrels (*Glaucomys sabrinus*) are the largest component of the diet at 45% to 58% of prey biomass depending on geographic region (Forsman et al. 2001, Hamer et al. 2001). Carey et al. (1992) presented some limited evidence suggesting that where predation by spotted owls was high, populations of flying squirrels were depressed in Douglas-fir forests (but see Rosenberg et al. 1994b, Carey 1995). In mixed-coniferous forests in Oregon, where woodrats were available in addition to flying squirrels, owls reduced their home range size from an average of 813 ha to 454 ha in response to the

34% increase in available prey biomass (Carey et al. 1992). In northwestern California and southwestern Oregon, Zabel et al. (1995) also reported smaller home ranges of spotted owls in areas where a greater proportion of the diet consisted of woodrats rather than flying squirrels. Moreover, prey type provided a better predictor of home range size than the proportion of older forest in the owls' home ranges.

How predation on small mammals affects patterns of development and succession for western coniferous forest ecosystems is little understood. In tropical forests of Panama, for example, the absence of top predators and large herbivores resulted in increased seed predation and herbivory by small mammals, which significantly reduced recruitment of some tree species (Asquith et al. 1997). In Virginia, McShea (2000) observed that both population sizes of small mammals (*Peromyscus leucopus*, *Tamias striatus*, and *Sciurus carolinensis*) and rates of predation on artificial bird nests increased the year following large mast crops. However, rates of predation were not correlated with absolute small-mammal densities. Previous studies indicated that medium carnivores accounted for 65% of the predation on artificial nests (Leimgruber et al. 1994). McShea (2000) suggested that increased densities of small mammals maintain populations of medium-sized carnivores during the winter following large mast years, and that these carnivores are responsible for reduced nest success of birds the following spring. Such interactions have not yet been reported for western coniferous forests.

Patterns of species richness

Drawing on trapping studies conducted in different forest types in the western US and Canada, we consider the number of small-mammal species (species richness) that occur in forest ecosystems with respect to differences in forest type and seral stage. Because of differences in sampling design, effort, and methods among studies, we focus our discussion on variation on two groups of small mammals: soricid shrews and murid rodents.

Williams (1991) summarized species richness of *Sorex* assemblages from studies conducted across North America. Based on these data, we calculated an average richness across all assemblages of 2.84 ± 1.19 (SD) species (range: 1–6). Western assemblages have more species than eastern ones on average (3.21 vs. 2.43 species, $P = 0.03$). Historical patterns of diversification of *Sorex* in the west are certainly important underlying causes of continental differences in richness (e.g., Demboski and Cook 2001, Maldonado et al. 2001, Lawlor 2003).

Trapping only a single species of shrew in western forests is unusual, and appears to be restricted to dry forest (e.g., ponderosa pine associations; Rickard 1960) or island situations (e.g., Hanley 1996). Numbers of shrew species may be low, however, and may vary spatially and temporally. Williams (1991) reported altitudinal zonation in distributions of four *Sorex* species on the western slope of the southern Sierra Nevada of California. The species overlapped largely in ecotonal areas between forest types (e.g., ponderosa pine – mixed conifer). When two or more species were present, however, one species was more widely distributed across seral stages of the forest type. Williams (1991) related this pattern to drier conditions in summer or reduced snow pack in winter, an issue we return to later. Kirkland and Findley (1999) assessed shrew richness in different forest types in New Mexico over two years. Dry ponderosa pine-oak (*Quercus*) woodland had the lowest total abundance and highest variability in species richness between years with one or three species present, whereas more mesic spruce-fir (*Picea-Abies*) stands consistently had either two or three species present.

In second-growth, mixed-coniferous forests in northeastern Washington, up to four species of *Sorex* may be present in clear-cuts (5–6 years), regenerating forest (15–20 years), and closed-canopy forest (>60 years) stands (Hallett and O'Connell 1997). These forests are structurally less diverse and drier than unmanaged stands where the highest richness (five species) occurs (Raphael 1988a, Gilbert and Allwine 1991, West 1991). Raphael (1988a) related increased richness to greater maturity of stands and higher moisture in subalpine forests in Wyoming, where there were three and four shrew species in pole and mature lodgepole pine (*Pinus contorta*) stands, respectively, and five species in mature spruce-fir forest. Unmanaged Douglas-fir forests typically have five species of *Sorex* across all seral stages from young to old growth in western Washington and Oregon (Gilbert and Allwine 1991, West 1991). These results taken together suggest that moisture may be the primary determinant of shrew species richness across forest types. The greater abundance of invertebrates in irrigated forests than non-irrigated forests in Pennsylvania lends further support to this hypothesis (McCay and Storm 1997), as does Williams' (1991) finding that some *Sorex* are most abundant near riparian areas.

Murid rodents also show a gradient of increasing richness from drier to more mesic forest types. Working in New Mexico, Kirkland and Findley

(1999) observed six species (three *Peromyscus*, two *Microtus*, and one *Clethri-onomys*), but none of the three forest types they examined had more than three species each. As Lawlor (2003) indicated, many species of mammals are general in their habitat use and occur in forests simply because their resource needs are met (e.g., deer mice), whereas others require specific forest habitats (e.g., red-backed voles; Rosenberg et al. 1994a). Of four *Peromyscus* species at their study site in northern New Mexico, Ribble and Stanley (1998) reported two species to be resident (*P. truei* and *P. boylii*), one that varied in occurrence (*P. maniculatus*), and one that was rare (*P. difficilis*).

In northeastern Washington, average species richness of murids was higher in clear-cuts and declined as the canopy closed during forest regeneration (Hallett and O'Connell 1997). However, during a year of peak abundance for all small mammals, species richness increased in all age classes of the forest. Species typical of clear-cuts or regenerating stands included *Peromyscus maniculatus*, *Microtus montanus*, *M. longicaudus*, *Phenacomys intermedius*, and *Zapus princeps*. These species use a wide variety of shrub, grass, and herbaceous material that is generally absent from closed-canopy stands. The principal rodent species in closed-canopy forests is the southern red-backed vole. Raphael (1988a) observed these same species in Wyoming. He noted little difference in species richness between pole and mature lodgepole pine or mature spruce-fir stands, but mature stands had higher densities of *Clethrionomys* and *Zapus*. For the Oregon Cascades, Gilbert and Allwine (1991) indicated that species richness increased between young (three species) and mature Douglas-fir (five species), but dropped in old growth (four species).

Patterns of abundance

Small-mammal populations vary in abundance both spatially and temporally. An important component of the variance in abundance is certainly variation in the distribution and abundance of resources. However, spatial heterogeneity in density can occur even in habitats that appear to be homogeneous (e.g. Krohne and Burgin 1990), and clarification of the underlying causes will require considerably more research (Bowman et al. 2001). Annual changes in abundance are typical of all small-mammal species, with overwinter mortality resulting in low spring populations, and over-summer reproduction leading to autumn highs (Boutin 1990). Additionally, variation in abundance between years can be significant and in some species may be part of multi-annual population fluctuations.

These annual and multi-annual fluctuations drive patterns of local extinction and colonization.

Despite temporal changes in population numbers, if one sums the abundance of each species across all seral stages in an area, the general pattern observed for temperate forest faunas is the presence of one or two numerically dominant species, usually a *Peromyscus* and/or a *Clethrionomys* (Kirkland 1985). In some western forests, these species are outranked by a species of *Sorex* (e.g., West 1991, Gomez and Anthony 1998, Kirkland and Findley 1999). Numerical dominance frequently means that a species is 10 to >100 times more abundant than most other species. Clearly, dominant species can find and exploit conditions that are highly favorable for survival and reproduction. Additionally, their large numbers ensure that they contribute significantly to forest function and, because they are abundant enough to study, we know the most about them.

Dominance patterns may change with forest age. For example, the red-backed vole is the most abundant rodent in closed-canopy forests in northeastern Washington, but that role is assumed by the deer mouse in clear-cuts (Hallett and O'Connell 1997). Similarly, *Sorex cinereus* is the numerically dominant shrew in closed-canopy stands and *S. vagrans* is dominant in clear-cuts and regenerating stands. Dominant species may also change regionally, as in the southern Washington Cascades where *Sorex trowbridgii* is dominant and *S. vagrans* is much less abundant (West 1991).

As Kirkland (1985) noted, however, it is often the number of rare species that contributes most to species richness; unfortunately, the biology of these species is least understood. In some cases, species are uncommon simply because their specific microhabitat conditions are spatially limited. For example, jumping mice occur in a variety of habitats at low density (Raphael 1988a), but they are locally abundant only where wet areas with lush undergrowth occur (Hayward and Hayward 1995, Gomez and Anthony 1998). Similarly, voles (*Microtus* spp.) are present across all forest ages, but they reach their highest densities in clear-cuts (Morrison and Anthony 1989, Hallett and O'Connell 1997) or meadows where grasses and forbs are abundant (Fitzgerald 1977).

Reasons for the rarity of other species are much less clear. The pygmy shrew (*Sorex hoyi*), for example, is a species with a broad geographic range that is uncommon wherever it occurs. Foresman (1999) collected 116 specimens in >127 000 pitfall trap-nights at five sampling localities in Montana and Idaho over a 13-year period. The pygmy shrew occurred across a range of different forest types, and one specimen was taken in dry sagebrush (*Artemesia tridentata*) steppe. In northeastern Washington, this species showed no abundance differences in any age class of second-growth mixed-coniferous forest, and it varied widely between years in its occurrence across stands (Hallett and O'Connell 1997). This suggests that the pygmy shrew has good colonizing ability, but that local populations may flourish and fade. Although shrews of this size (2–3 g) have smaller food requirements than larger ones, their high mass-specific metabolic rates and reduced body reserves result in hastened starvation when food resources become scarce. Thus, this tiny species probably is affected more quickly by environmental fluctuations than are larger shrews (Hanski 1994).

Evidence that shrews respond to changing environmental conditions comes from observations of increased developmental instability in *Sorex cinereus*, *S. monticolus*, and *S. vagrans* following habitat disturbance in the form of forest harvest (Badyaev and Foresman 2000, Badyaev et al. 2000). Shrews in cutover areas were in poorer physiological condition, had reduced body mass, and had greater asymmetry of the mandible than individuals in undisturbed stands. These measurable changes suggest that environmental stress leads to reduced fitness, and this in turn contributes to population decline. The mechanisms involved in the stress response are largely unknown, but insufficient nutrition for reproductive females may play a role (Badyaev et al. 2000).

Habitat associations

Examination of microhabitat use has been common in small-mammal ecological studies. The objectives of these studies have been varied, but most have focused on the structural or vegetational components of microhabitat associated with use by small mammals (e.g., Belk et al. 1988, Gilbert and Allwine 1991, Hayward and Hayward 1995) and/or on understanding interspecific relationships and niche differentiation (e.g., Dueser and Shugart 1978). Although different spatial scales (e.g., trap station versus trapping grid) may be used, variables describing microhabitat are measured and then related to the presence or absence or number of captures of small mammals. A number of statistical analyses have been used to examine these relationships, ranging from simple to multivariate techniques (reviewed by Morrison et al. 1998). Comparisons among studies are often difficult because of differences in the analyses used and in the choice of habitat features measured.

Significant correlations between habitat variables and population characteristics may suggest potentially important relationships (e.g., Nordyke and Buskirk 1991). For example, the abundance of Townsend's chipmunks was related to the percentage cover of salal (*Gaultheria* spp.) in the Oregon Coast Range (Hayes et al. 1995). When there is a strong correlation between a particular stand feature and mammal abundance, the functional reasons for the association may be difficult to determine. For Townsend's chipmunks, the relationship could be due to salal as a food resource, as cover, or as a surrogate for some other habitat component important to the chipmunk. Hayes et al. (1995) argued that small mammals, such as Townsend's chipmunks, might use forest stands of different ages as long as they provide requisite structural features. This argument is only partly true; the function of the habitat attribute must be provided, and structure per se may support the function in some cases (e.g., protective cover), but not others (e.g., food resources).

Some species have shown consistent relationships with structural microhabitat features. For example, associations of red-backed voles with closed-canopy forests in general, and with litter depth (Rosenberg et al. 1994a) and coarse woody debris in particular, have been observed (e.g., Ramirez and Hornocker 1981, Belk et al. 1988, Tallmon and Mills 1994, Carey and Johnson 1995). Coarse woody debris probably provides several functions including cover from predation, nest sites, runways, and microclimatic conditions favorable to both red-backed voles and organisms they use as food (further discussion in McComb 2003).

For habitat generalists, such as deer mice, attempts to correlate specific components of the vegetation with abundance may fail because requisite resources do not vary with vegetation or the animals are able to substitute resources. For example, Hanley and Barnard (1999) found that population responses of northwestern deer mice in floodplain and upland forests were similar, despite differences in understory vegetation. Morrison and Anthony (1989) examined microhabitat use of seven small-mammal species in young clear-cuts in Oregon. Evaluation of differences in microhabitat by discriminant function analysis led to a model that predicted occurrence of *Peromyscus* with only 40% accuracy. This result was due to generalized habitat use by this species and broad overlap in its use of microhabitats. We note that intersexual competition might underlie broader microhabitat use by a species in some cases, as observed, for example, in montane forests in Utah by Belk et al. (1988) for deer mice.

Understanding habitat associations of small mammals is an important component for moderating the effects of forest management activities. Structural features of the habitat have been the primary focus because they are the easiest to investigate and the easiest to manipulate through forest practices. Success at mitigating the effects of forest harvest will be improved as we learn more about the functional roles of the structures that we measure.

Responses to forest management

Stand structure

Effects of timber harvest on small-mammal populations and communities depend on a variety of factors including the original plant community; the type, size, and timing of harvest or thinning operations; and on-site treatment of slash and snags. The consequences for small-mammal species also depend on the degree to which their life requisites are met by a particular seral stage or forest condition. Many species are likely to benefit from clear-cutting or forest-management practices that set back forest succession, particularly those associated with early-seral communities. Such harvests may mimic natural disturbance and result in a flush of herbaceous vegetation followed later by the development of a shrub layer. Species that can take advantage of these conditions include deer mice, vagrant shrews, mountain beavers, least chipmunks (*Tamias minimus*), yellow-pine chipmunks, northern pocket gophers, heather voles, meadow voles, long-tailed voles, creeping voles, and Pacific jumping mice (Morrison and Anthony 1989, Sullivan et al. 1999). For some small mammals that benefit from timber harvest, abundance does not increase until several years after logging.

Forest-dwelling species and those associated with late-successional forests, however, are often adversely affected by timber harvest that leads to early-seral plant communities. For example, Trowbridge's shrew usually declines in number shortly after clear-cut logging (Tevis 1956, Hooven and Black 1976, Martell 1984) and does not become abundant until several years after cutting (Harris 1968, Simons 1985). Butts and McComb (2000) found this species to be positively associated with cover of woody plants and small-diameter logs, but negatively associated with moss cover, which may not provide adequate protective cover. A litter layer may not develop until several years after harvest; consequently, shrub layers on unscarified clear-cuts may be critical for some shrews (Martell 1983). Southern red-backed voles usually decline following logging (Gashwiler 1967, Corn et al. 1988, Sullivan et al. 1999), fire (Gashwiler 1959, Martell 1984), or herbicide treatment (D'Anieri et al. 1987). Gunther et al. (1983), however, found these voles in clear-cuts shortly after logging in the Cascade Mountains of Washington. These clear-cuts had high levels of coarse woody debris when compared to forested areas. Removal of residual material, however, led to the avoidance of clear-cuts by southern red-backed voles in Alberta (Moses and Boutin 2001). In northeastern Washington, captures of southern red-backed voles were associated with woody debris in all years (Hallett and O'Connell 1997). During a population peak, this vole increased by six times in both clear-cut and regenerating forest stands and captures were positively associated with more classes of woody debris. Despite apparent habitat expansion, reproductive females were more common in closed-canopy stands. Thus, open habitats might provide certain structural features that allow use by dispersing *Clethrionomys gapperi*, but not necessarily the full complement of habitat conditions (e.g., microhabitats supporting hypogeous sporocarps of fungi; Mills 1995) necessary to support sustained populations. Rosenberg et al. (1994a) also found litter depth to be important in the distribution and abundance of western red-backed voles in the Cascade Mountains of Oregon where the species is more abundant in late-successional forests than in managed second-growth forests.

Although timber harvest profoundly affects the composition of small-mammal communities in unmanaged forests of Washington and Oregon, few species appear to be strongly influenced by stand age in Douglas-fir forests west of the Cascade Crest (Aubry et al. 1991, Corn and Bury 1991, Gilbert and Allwine 1991, West 1991, Gomez and Anthony 1998). This suggests that once the forest stand develops the structural features (e.g., litter layer, coarse woody debris) that provide resources required by many small mammals, additional age of the stand does not significantly improve habitat conditions. Young managed forests now far exceed old-growth forests in areal extent, and small-mammal abundance is lower in these forests relative to old-growth (Carey and Johnson 1995). Consequently, there has been growing interest in understanding how the structural features of old-growth forests might be created silviculturally in managed stands. Carey and Johnson (1995) review prescriptions that may maintain the integrity of the forest floor, specifically the coarse woody debris and understory structure on which many small mammals depend. Experimental application of such prescriptions has not yet replicated the community composition of old-growth forests. For example, Wilson and Carey (2000) compared small-mammal communities in stands that had been thinned to encourage development of large trees versus those where live trees, snags, and coarse woody debris from the preceding forest were retained at harvest and subsequently protected. Thinned stands had greater abundance and biomass of small mammals than legacy stands. Community structure differed between treatments, but failed to replicate the species composition of old-growth forests in either case. Current management prescriptions for coarse woody debris may be insufficient to support small-mammal communities that could potentially occur there (Butts and McComb 2000, Maguire 2002).

Landscape change

Most western forests are no longer naturally regenerating; they are largely managed for timber production and they are typically younger and less structurally diverse than the old-growth forests they replaced (Carey and Johnson 1995). Recent timber harvests of second-growth, managed forests have created landscape mosaics populated with stands differing in both age and size. Fragmentation of forested lands can take place over a range of intensities. At one extreme, most of the closed-canopy forest is removed. At the other, only small patches of forest are harvested and most of the landscape remains in closed-canopy forest. The consequences for small mammals, of course, are quite different. Mills (1995) investigated the consequences for western red-backed voles of being restricted to small forest remnants (0.3–3.6 ha) surrounded by clear-cuts. Voles utilized the interiors of these patches, but not the edges. Conditions on a patch edge may differ from the patch interior because of physical (e.g., blowdown, changes in microclimate) or biotic changes (e.g., increase in predators) that take place along the edge. In this case, Mills (1995) found that less food was available near the edge of a patch than in the interior. In contrast, Hayward et al. (1999) examined the responses of southern red-backed voles to small clear-cuts (1–2 ha) and found that voles utilized these areas soon after harvest. Whether this was a function of reduced edge effects or the retention of coarse woody debris on some stands is unclear. Kingston and Morris (2000) found no evidence of edge effects for southern red-backed voles in jack pine-spruce forests in Quebec, but simulations suggested that the habitats they contrasted may have been too similar to see an effect.

In managed forest systems, the degree of fragmentation is usually between the extremes considered above. In northeastern Washington, from

30% to 50% of second-growth closed-canopy forests have typically been removed from watersheds (Hallett and O'Connell 1997). The addition of clear-cuts and young regenerating forests has resulted in species richness of up to 18 species of rodents and insectivores within a region (Hallett and O'Connell 1997). Local richness is usually lower and reflects species-specific responses to microhabitat conditions in a stand, as well as pronounced differences in the relative abundance of species. Where studied, the consequences of forest fragmentation have not appeared to be severe. Based on simulation models of dispersing organisms, Gardner et al. (1991) predict the presence of critical thresholds in habitat connectivity, which when reached will result in rapid decreases in population abundance. Current levels of fragmentation may not have reached such a threshold, and retention of landscape linkages such as riparian areas (Anthony et al. 2003) and forest strips that might act as corridors (Mech and Hallett 2001) might also prevent loss of species. Mills et al. (2003) further consider issues of landscape connectivity.

Summary and future directions

Human activities have greatly altered western coniferous forests, and successful management of these forests for an expanding suite of ecosystem services requires an understanding of the role of small mammals in these systems. Our review of the substantial literature that has developed on the ecological roles of small mammals within western coniferous forests highlights the complex ecological relationships of these mammals with their habitat and points to areas in need of further research.

As consumers, the small mammals of western coniferous forests are predominantly insectivorous and omnivorous, but studies reveal that diets vary with season, intraspecific competition, and competition with mammalian and non-mammalian species. Variability in the availability and use of different foods can have population and demographic consequences (e.g., overwinter survival and reproduction). Variation in diets of small mammals impacts forest dynamics as small mammals can inhibit forest regeneration through predation on seeds and seedlings or, conversely, promote forest regeneration through dispersal of seeds and mycorrhizal fungi. Small-mammal predation on avian nests can be significant, but the effects vary with nest location and habitat. Although the importance of small mammals as consumers to both forest function and biodiversity continues to be documented, we still have limited knowledge about resource use by small mammals and the consequences of changes in resource availability.

Small mammals are important prey for many mammalian, avian, and reptilian carnivores, and changes in small-mammal distribution and density have been shown to impact the reproductive production and habitat use of predators in western coniferous forests. Unfortunately, few studies have attempted to examine the influence of predators on population dynamics of small mammals in western coniferous forests as has been done in forests elsewhere. In addition, we poorly understand how the relative abundances of both small mammals and their predators affect forest function and biodiversity.

Small-mammal assemblages in western forests are typically characterized by a few numerically dominant species and a larger number of less common species. Variation in species richness has been attributed to moisture gradients, forest stand conditions, and changes in small-mammal abundance. Patterns of abundance of the more common small mammals can often be correlated to structural habitat features. However, much less is known about the circumstances that influence these species' habitat choice, including resource availability. In addition, our understanding of factors contributing to patterns of abundance of relatively rare small-mammal species is limited.

The composition of small-mammal communities can be modified by forest-management activities. Thus far, there have been few experimental studies designed to alter forest structural features to achieve a desired outcome in small-mammal abundance and species composition. More experimental research of this nature should be undertaken in a variety of forest types. However, there is also great value in expanding the number of prospective studies that examine structural characteristics over a range of established conditions. Continued research efforts in these areas will increase our understanding of the resources provided by or associated with forest structure. This information will be critical for devising measures to mitigate management practices potentially adverse to forest small mammals.

Finally, influences of the composition and configuration of forested landscapes on small mammals also need to be more fully addressed. High levels of forest fragmentation result in the loss of some species dependent on closed-canopy forest. However, the threshold value of fragmentation that eliminates sufficient habitat or prevents dispersal is unknown. Studies of the genetic structure of small-mammal populations under different landscape configurations and levels of fragmentation will shed greater light on this issue. Such studies could be designed to clarify the spatial scales to which small mammals respond.

As evident from the review provided in this chapter, the published literature leaves little doubt that small mammals perform numerous ecological functions in western forest ecosystems. Also apparent is the limited depth of our knowledge in this area. The challenge is still upon us to more fully expose and understand the complex suite of interactions that small mammals entertain with their forest environment and the species with which they share limited resources.

Acknowledgments

We thank Kerry Foresman, Eric Forsman, Ed Heske, Tim Lawlor, John Lehmkuhl, Tom Manning, Steve Mech, Bob Rose, and Jerry Wolff for their comments on the manuscript.

Literature cited

Allen, L. 1969. Preferential food habits of *Aplodontia rufa*. Thesis. Central Washington State College, Ellensberg, Washington, USA.

Anthony, R.G., M.A. O'Connell, M.M. Pollock, and J.G. Hallett. 2003. Associations of mammals with riparian ecosystems in Pacific Northwest forests. Pages 510–563 *in* C.J. Zabel and R.G. Anthony, editors. *Mammal Community Dynamics. Management and Conservation in the Coniferous Forests of Western North America*. Cambridge University Press, Cambridge, UK.

Asquith, N.M., S.J. Wright, and M.J. Clauss. 1997. Does mammal community composition control recruitment in neotropical forests? Evidence from Panama. *Ecology* **78**:941–946.

Aubry, K.B., M.J. Crites, and S.D. West. 1991. Regional patterns of small mammal abundance and community composition in Oregon and Washington. Pages 285–294 *in* L.F. Ruggiero, K.B. Aubry, A.B. Carey, and M.H. Huff, editors. Wildlife and vegetation of unmanaged Douglas-fir forests. US Forest Service General Technical Report **PNW-GTR-285**, Portland, Oregon, USA.

Aubry, K.B., J.P. Hayes, B.L. Biswell, and B.G. Marcot. 2003. The ecological role of tree-dwelling mammals in western coniferous forests. Pages 405–443 *in* C.J. Zabel and R.G. Anthony, editors. *Mammal Community Dynamics. Management and Conservation in the Coniferous Forests of Western North America*. Cambridge University Press, Cambridge, UK.

Badyaev, A.V. and K.R. Foresman. 2000. Extreme environmental change and evolution: stress-induced morphological variation is strongly concordant with pattern of evolutionary divergence in shrew mandibles. *Proceedings of the Royal Society of London B* **267**:371–377.

Badyaev, A.V., K.R. Foresman, and M.V. Fernandes. 2000. Stress and developmental stability: vegetation removal causes increased fluctuating asymmetry in shrews. *Ecology* **81**:336–345.

Baxter, L. and L. Hansson. 2001. Bark consumption by rodents in the northern and southern hemispheres. *Mammal Review* **31**:47–59.

Belk, M.C., H.D. Smith, and J. Lawson. 1988. Use and partitioning of montane habitat by small mammals. *Journal of Mammalogy* **69**:688–695.

Bellocq, M.I., J.F. Bendell, and D.G.L. Innes. 1994. Diet of *Sorex cinereus*, the masked shrew, in relation to the abundance of Lepidoptera larvae in northern Ontario. *American Midland Naturalist* **132**:68–73.

Bermejo, T., A. Traveset, and M.F. Willson. 1998. Post-dispersal seed predation in the temperate rainforest of southeast Alaska. *Canadian Field-Naturalist* **112**: 510–512.

Blight, L.K., J.L. Ryder, and D.F. Bertram. 1999. Predation on Rhinoceros Auklet eggs by a native population of *Peromyscus*. *Condor* **101**:871–876.

Borrecco, J.E. and R.J. Anderson. 1980. Mountain beaver problems in the forests of California, Oregon and Washington. *Proceedings of the Vertebrate Pest Conference* **9**:135–142.

Boutin, S. 1990. Food supplementation experiments with terrestrial vertebrates: patterns, problems, and the future. *Canadian Journal of Zoology* **68**:203–220.

Bowman, J.C., G.J. Forbes, and T.G. Dilworth. 2001. The spatial component of variation in small-mammal abundance measured at three scales. *Canadian Journal of Zoology* **79**:137–144.

Bull, E.L. 2000. Seasonal and sexual differences in American marten diet in northeastern Oregon. *Northwest Science* **74**:186–191.

Buskirk, S.W. and W.J. Zielinski. 2003. Small and mid-sized carnivores, Pages 207–249 *in* C.J. Zabel and R.G. Anthony, editors. *Mammal Community Dynamics. Management and Conservation in the Coniferous Forests of Western North America.* Cambridge University Press, Cambridge, UK.

Butts, S.R. and W.C. McComb. 2000. Associations of forest-floor vertebrates with coarse woody debris in managed forests of western Oregon. *Journal of Wildlife Management* **64**:95–104.

Cafferata, S.L. 1992. Mountain beaver. Pages 231–251 *in* H.C. Black, editor. Silvicultural approaches to animal damage management in Pacific Northwest forests. US Forest Service General Technical Report **PNW-GTR-287**, Portland, Oregon, USA.

Carey, A.B. 1995. Spotted owl ecology: theory and methodology – a reply to Rosenberg et al. *Ecology* **76**:648–652.

Carey, A.B. and M.L. Johnson. 1995. Small mammals in managed, naturally young, and old-growth forests. *Ecological Applications* **5**:336–352.

Carey, A.B., S.P. Horton, and B.L. Biswell. 1992. Northern spotted owls: influence of prey base and landscape character. *Ecological Monographs* **62**:223–250.

Carraway, L.N. and B.J. Verts. 1993. *Aplodontia rufa*. *Mammalian Species* **431**:1–10.

Carraway, L.N. and B.J. Verts. 1994. Relationship of mandibular morphology to relative bite force in some *Sorex* from western North America. Pages 201–210 *in* J.F. Merritt, G.L. Kirkland, Jr., and R.K. Rose, editors. *Advances in the Biology of Shrews.* Carnegie Museum of Natural History, Special Publication No. **18**, Pittsburgh, Pennsylvania, USA.

Chambers, J.C. 2001. *Pinus monophylla* establishment in an expanding *Pinus-Juniperus* woodland: environmental conditions, facilitation and interacting factors. *Journal of Vegetation Science* **12**:27–40.

Christensen, K.M. and T.G. Whitham. 1993. Impact of insect herbivores on competition between birds and mammals for pinyon pine seeds. *Ecology* **74**:2270–2278.

Christy, E.J. and R.N. Mack. 1984. Variation in demography of *Tsuga heterophylla* across the substratum mosaic. *Journal of Ecology* **72**:75–91.

Clarke, J.A. 1983. Moonlight's influence on predator/prey interactions between short-eared owls (*Asio flammeus*) and deermice (*Peromyscus maniculatus*). *Behavioral Ecology and Sociobiology* **13**:205–209.

Corn, P.S. and R.B. Bury. 1991. Small mammal communities in the Oregon Coast Range. Pages 241–254 *in* L.F. Ruggiero, K.B. Aubry, A.B. Carey, and M.H. Huff, editors. Wildlife and vegetation of unmanaged Douglas-fir forests. US Forest Service General Technical Report **PNW-GTR-285**, Portland, Oregon, USA.

Corn, P.S., R.B. Bury, and T.A. Spies. 1988. Douglas-fir forests of the Cascade Mountains of Oregon and Washington: is the abundance of small mammals related to stand age and moisture? Pages 340–352 *in* R.C. Szaro, K.E. Severson, and D.R. Patton, editors. Management of amphibians, reptiles, and small mammals in North America. US Forest Service General Technical Report **RM-GTR-166**, Portland, Oregon, USA.

Cranford, J.A. 1978. Hibernation in the western jumping mouse (*Zapus princeps*). *Journal of Mammalogy* **59**:496–509.

Crouch, G.L. 1968. Clipping of woody plants by mountain beaver. *Journal of Mammalogy* **49**:151–152.

Dalquest, W.W. and V.B. Scheffer. 1945. The systematic status of the races of the mountain beaver (*Aplodontia rufa*) in Washington. *Murrelet* **26**:34–37.

D'Anieri, P., D.M. Leslie, Jr., and M.L. McCormack, Jr. 1987. Small mammals in glycophosphate treated clearcuts in northern Maine. *Canadian Field-Naturalist* **101**:547–550.

Demboski, J.R. and J.A. Cook. 2001. Phylogeography of the dusky shrew, *Sorex monticolus* (Insectivora, Soricidae): insight into deep and shallow history in northwestern North America. *Molecular Ecology* **10**:1227–1240.

Drever, M.C., L.K. Blight, K.A. Hobson, and D.F. Bertram. 2000. Predation on seabird eggs by Keen's mice (*Peromyscus keeni*): using stable isotopes to decipher the diet of a terrestrial omnivore on a remote offshore island. *Canadian Journal of Zoology* **78**:2010–2018.

Dueser, R.D. and H.H. Shugart, Jr. 1978. Niche pattern in a forest floor small mammal community. *Ecology* **60**:108–118.

Entz, R.D. 1996. Nest predation in riparian areas of northeastern Washington. Thesis. Eastern Washington University, Cheney, Washington, USA.

Feldhamer, G.A., R.S. Klann, A.S. Gerard, and A.C. Driskell. 1993. Habitat partitioning, body size, and timing of parturition in pygmy shrews and associated soricids. *Journal of Mammalogy* **74**:403–411.

Fenske-Crawford, T.J. and G.J. Niemi. 1997. Predation of artificial ground nests at two types of edges in a forest-dominated landscape. *Condor* **99**:14–24.

Findley, J.S. and T.L. Yates, editors. 1991. *Habitats of Shrews (Genus* Sorex*) in Forest Communities of the Western Sierra Nevada, California.* Special Publication of the Museum of Southwestern Biology, University of New Mexico, Albuquerque, New Mexico, USA.

Fitzgerald, B.M. 1977. Weasel predation on a cyclic population of the montane vole (*Microtus montanus*). *Journal of Animal Ecology* **46**:367–397.

Fleming, T.H. 1973. Numbers of mammal species in North and Central American forest communities. *Ecology* **54**:555–563.

Foresman, K.R. 1999. Distribution of the pygmy shrew, *Sorex hoyi*, in Montana and Idaho. *Canadian Field-Naturalist* **113**:681–683.

Forsman, E.D., I.A. Otto, S.G. Sovern, M. Taylor, D.W. Hays, H. Allen, S.L. Roberts, and D.E. Seaman. 2001. Spatial and temporal variation in diets of spotted owls in Washington. *Journal of Raptor Research* **35**:141–150.

Fox, B.J. and G.L. Kirkland, Jr. 1992. An assembly rule for functional groups applied to North American soricid communities. *Journal of Mammalogy* **73**:491–503.

Gardner, R.H., M.G. Turner, R.V. O'Neill, and S. Lavorel. 1991. Simulation of scale-dependent effects of landscape boundaries on species persistence and dispersal. Pages 76–89 *in* M.M. Holland, P.G. Risser, and R.J. Naiman, editors. *Ecotones: the Role of Landscape Boundaries in the Management and Restoration of Changing Environments*. Chapman and Hall, New York, New York, USA.

Gashwiler, J.S. 1959. Small mammal study in west-central Oregon. *Journal of Mammalogy* **40**:128–139.

Gashwiler, J.S. 1967. Conifer seed survival in a western Oregon clearcut. *Ecology* **48**:431–438.

Gashwiler, J.S. 1979. Deer mouse reproduction and its relationship to the tree seed crop. *American Midland Naturalist* **102**:95–104.

Gilbert, F.F. and R. Allwine. 1991. Small mammal communities in the Oregon Cascade Range. Pages 257–267 *in* L.F. Ruggiero, K.B. Aubry, A.B. Carey, and M.H. Huff, editors. Wildlife and vegetation of unmanaged Douglas-fir forests. US Forest Service General Technical Report **PNW-GTR-285**, Portland, Oregon, USA.

Gliwicz, J. and B. Glowacka. 2000. Differential responses of *Clethrionomys* species to forest disturbance in Europe and North America. *Canadian Journal of Zoology* **78**:1340–1348.

Gomez, D.M. and R.G. Anthony. 1998. Small mammal abundance in riparian and upland areas of five seral stages in western Oregon. *Northwest Science* **72**:293–302.

Grodzinski, W. and B.A. Wunder. 1975. Ecological energetics of small mammals. Pages 173–204 *in* F.B. Golley, K. Petrusewicz, and L. Ryszkowski, editors. *Small Mammals: Their Productivity and Population Dynamics*. Cambridge University Press, Cambridge, UK.

Gunther, P.M., B.S. Horn, and G.D. Babb. 1983. Small mammal populations and food selection in relation to timber harvest practices in the western Cascade Mountains. *Northwest Science* **57**:32–44.

Hacker, A.L. 1991. Population attributes and habitat selection of recolonizing mountain beaver. Thesis. Oregon State University, Corvallis, Oregon, USA.

Hall, E.R. 1981. *The Mammals of North America*. Second Edition. John Wiley & Sons, New York, New York, USA.

Hallett, J.G. and M.A. O'Connell. 1997. Habitat occupancy and population patterns of small mammals in managed forests. Pages 2.1–2.16 *in* J.G. Hallett and M.A. O'Connell, editors. *East-side Studies: Research Results*. Volume 3 of Wildlife use of managed forests: a landscape perspective. Final report to the Timber, Fish, and Wildlife Cooperative Monitoring, Evaluation, and Research Committee. **TFW-WL4-98-003**, Washington Department of Natural Resources, Olympia, Washington, USA.

Hallett, J.G. and M.A. O'Connell. 2000. East-side small mammal surveys. Pages 11.1–11.23 *in* M.A. O'Connell, J.G. Hallett, S.D. West, K.A. Kelsey, D.A. Manuwal, and S.F. Pearson, editors. Effectiveness of riparian management zones in providing habitat for wildlife. Final report to Timber, Fish, and Wildlife Program. **TFW-LWAG1-00-001**, Washington Department of Natural Resources, Olympia, Washington, USA.

Hamer, T.E., D.L. Hays, C.M. Senger, and E.D. Forsman. 2001. Diets of northern barred owls and northern spotted owls in an area of sympatry. *Journal of Raptor Research* **35**:221–227.

Hanley, T.A. 1996. Small mammals of even-aged, red alder-conifer forests in southeastern Alaska. *Canadian Field-Naturalist* **110**:626–629.

Hanley, T.A. and J.C. Barnard. 1999. Food resources and diet composition in riparian and upland habitats for Sitka mice, *Peromyscus keeni sitkensis*. *Canadian Field-Naturalist* **113**:401–407.

Hannon, S.J. and S.E. Cotterill. 1998. Nest predation in aspen woodlots in an agricultural area in Alberta: the enemy from within. *Auk* **115**:16–25.

Hanski, I. 1994. Population biological consequences of body size in *Sorex*. Pages 15–26 *in* J.F. Merritt, G.L. Kirkland, Jr., and R.K. Rose, editors. *Advances in the Biology of Shrews*. Carnegie Museum of Natural History, Special Publication No. **18**, Pittsburgh, Pennsylvania, USA.

Hanski, I. and H. Henttonen. 1996. Predation on competing rodent species: a simple explanation of complex patterns. *Journal of Animal Ecology* **65**:220–232.

Hanski, I., L. Hansson, and H. Henttonen. 1991. Specialist predators, generalist predators, and the microtine rodent cycle. *Journal of Animal Ecology* **364**: 232–235.

Harris, A.S. 1968. Small mammals and natural reforestation in southeast Alaska. US Forest Service Research Paper **PNW-75**, Portland, Oregon, USA.

Harris, L.D., C. Maser, and A. McKee. 1982. Patterns of old growth harvest and implications for Cascades wildlife. *Transactions of the North American Natural Resources Conference* **47**:374–392.

Haskell, D.G. 1995a. Forest fragmentation and nest predation: are experiments with Japanese quail eggs misleading? *Auk* **112**:767–770.

Haskell, D.G. 1995b. A reevaluation of the effects of forest fragmentation on rates of bird-nest predation. *Conservation Biology* **9**:1316–1318.

Hayes, J.P., S.P. Cross, and P.W. McIntire. 1986. Seasonal variation in mycophagy by the western red-backed vole, *Clethrionomys californicus*, in southwestern Oregon. *Northwest Science* **60**:150–157.

Hayes, J.P., E.G. Horvath, and P. Hounihan. 1995. Townsend's chipmunk populations in Douglas-fir plantations and mature forests in the Oregon Coast Range. *Canadian Journal of Zoology* **73**:67–73.

Hayward, G.D. and P.H. Hayward. 1995. Relative abundance and habitat associations of small mammals in Chamberlain Basin, central Idaho. *Northwest Science* **69**:114–125.

Hayward, G.D., P.H. Hayward, and E.O. Garton. 1993. Ecology of boreal owls in the northern Rocky Mountains, U.S.A. *Wildlife Monographs* **124**:1–59.

Hayward, G.D., S.H. Henry, and L.F. Ruggiero. 1999. Response of red-backed voles to recent patch cutting in subalpine forest. *Conservation Biology* **13**: 168–176.

Hickey, J.R., R.W. Flynn, S.W. Buskirk, K.G. Gerow, and M.F. Willson. 1999. An evaluation of a mammalian predator, *Martes americana*, as a disperser of seeds. *Oikos* **87**:499–508.

Holling, C.S. 1959. The components of predation as revealed by a study of small-mammal predation of the European pine sawfly. *Canadian Entomologist* **91**:293–320.

Hooven, E.F. 1977. The mountain beaver in Oregon: its life history and control. Research Paper **30**:1–20, Forest Research Laboratory, Oregon State University, Corvallis, Oregon, USA.

Hooven, E.F. and H.C. Black. 1976. Effects of some clear-cutting practices on small-mammal populations in western Oregon. *Northwest Science* **50**: 189–208.

Hoyer, G.E., N. Anderson, and R. Riley. 1979. A case study of six years of mountain beaver damage on Clallam Bay western hemlock plots. DNR Note **28**, Washington Department of Natural Resources, Olympia, Washington, USA.

Innes, D.G.L., J.F. Bendell, B.J. Naylor, and B.A. Smith. 1990. High densities of the masked shrew, *Sorex cinereus*, in jack pine plantations in northern Ontario. *American Midland Naturalist* **124**:330–341.

Kingston, S.R. and D.W. Morris. 2000. Voles looking for an edge: habitat selection across forest ecotones. *Canadian Journal of Zoology* **78**:2174–2183.

Kirkland, G.L., Jr. 1985. Small mammal communities in temperate North American forests. *Australian Mammalogy* **8**:137–144.

Kirkland, G.L., Jr. 1991. Competition and coexistence in shrews (Insectivora: Soricidae). Pages 15–22 *in* J.S. Findley and T.L. Yates, editors. *The Biology of the Soricidae*. Special Publication of the Museum of Southwestern Biology, University of New Mexico, Albuquerque, New Mexico, USA.

Kirkland, G.L., Jr., and J.S. Findley. 1999. A transcontinental comparison of forest small-mammal assemblages: northern New Mexico and southern Pennsylvania compared. *Oikos* **85**:335–342.

Kirkland, G.L., Jr., and P.K. Sheppard. 1994. Proposed standard protocol for sampling small mammal communities. Pages 277–281 *in* J.F. Merritt, G.L. Kirkland, Jr., and R.K. Rose, editors. *Advances in the Biology of Shrews*. Carnegie Museum of Natural History, Special Publication No. **18**, Pittsburgh, Pennsylvania, USA.

Kirkland, G.L., Jr., P.K. Sheppard, M.J. Shaughnessy, Jr., and B.A. Woleslagle. 1998. Factors influencing perceived community structure in nearctic forest small mammals. *Acta Theriologica* **43**:121–135.

Korpimäki, E. 1987. Clutch size, breeding success and brood size experiments in Tengmalm's owl, *Aegolius funereus*: a test of hypothesis. *Ornis Scandinavica* **18**:277–284.

Krohne, D.T. and A.B. Burgin. 1990. The scale of demographic heterogeneity in a population of *Peromyscus leucopus*. *Oecologia* **82**:97–101.

Kurtén, B. and E. Anderson. 1980. *Pleistocene Mammals of North America*. Columbia University Press, New York, New York, USA.

Lawlor, T.E. 2003. Faunal composition and distribution of mammals in western coniferous forests. Pages 41–80 *in* C.J. Zabel and R.G. Anthony, editors. *Mammal Community Dynamics. Management and Conservation in the Coniferous Forests of Western North America*. Cambridge University Press, Cambridge, UK.

Lawrence, W.H., N.B. Kverno, and H.D. Hartwell. 1961. Guide to wildlife feeding injuries on conifers in the Pacific Northwest. Western Forestry and Conservation Association, Portland, Oregon, USA.

Lehmkuhl, J.F., S.D. West, C.L. Chambers, W.C. McComb, D.A. Manuwal, K.B. Aubry, J.L. Erickson, R.A. Gitzen, and M. Leu. 1999. An experiment for assessing vertebrate response to varying levels and patterns of green-tree retention. *Northwest Science* **73**:45–63.

Leimgruber, P., W.J. McShea, and J.H. Rappole. 1994. Predation on artificial nests in large forest blocks. *Journal of Wildlife Management* **58**:254–260.

Luoma, D.L., J.M. Trappe, A.W. Claridge, K.M. Jacobs, and E. Cazares. 2003. Relationships among fungi and small mammals in forested ecosystems. Pages 343–373 *in* C.J. Zabel and R.G. Anthony, editors. *Mammal Community Dynamics. Management and Conservation in the Coniferous Forests of Western North America* Cambridge University Press, Cambridge, UK.

Maguire, C.C. 1989. Small mammal predation on Douglas-fir seedlings in northwestern California. *Wildlife Society Bulletin* **17**:175–178.

Maguire, C.C. 1999. Rainfall, ambient temperature, and *Clethrionomys californicus* capture frequency. *Mammal Review* **29**:135–142.

Maguire, C.C. 2002. Dead wood and the richness of small terrestrial vertebrates in southwestern Oregon. Pages 331–345 *in* W.F. Laudenslayer, Jr., P.J. Shea, B. Valentine, C.P. Weatherspoon, and T.E. Lisle, editors. Proceedings of the Symposium on the Ecology and Management of Dead Wood in Western Forests. US Forest Service General Technical Report **PSW GTR-181**, Berkeley, California, USA.

Maldonado, J.E., C. Vila, and R.K. Wayne. 2001. Tripartite subdivisions in the ornate shrew (*Sorex ornatus*). *Molecular Ecology* **10**:127–147.

Martell, A.M. 1983. Demography of southern red-backed voles and deermice after logging in north-central Ontario. *Canadian Journal of Zoology* **61**:958–969.

Martell, A.M. 1984. Changes in small mammal communities after fire in northcentral Ontario. *Canadian Field-Naturalist* **98**:223–226.

Martin, S. 1994. Feeding ecology of American martens and fishers. Pages 297–315 *in* S.W. Buskirk, A.S. Harestad, M.G. Raphael, and R.A. Powell, editors. *Martens, Sables, and Fishers: Biology and Conservation*. Cornell University Press, Ithaca, New York, New York, USA.

Martin, T.E. 1988*a*. On the advantage of being different: nest predation and the coexistence of bird species. *Proceedings of the National Academy of Sciences U.S.A.* **85**:2196–2199.

Martin, T.E. 1988*b*. Processes organizing open-nesting bird assemblages: competition or nest predation? *Evolutionary Ecology* **2**:37–50.

Maser, C., J.M. Trappe, and R.A. Nussbaum. 1978. Fungal-small mammal interrelationships with emphasis on Oregon coniferous forests. *Ecology* **59**:799–809.

McCay, T.S. and G.L. Storm. 1997. Masked shrew (*Sorex cinereus*) abundance, diet and prey selection in an irrigated forest. *American Midland Naturalist* **138**: 268–275.

McComb, W.C. 2003. Ecology of coarse woody debris and its role as habitat for mammals. Pages 374–404 *in* C.J. Zabel and R.G. Anthony, editors. *Mammalian Community Dynamics in Coniferous Forests of Western North America: Management and Conservation*. Cambridge University Press, Cambridge, UK.

McComb, W.C., R.G. Anthony, and K. McGarigal. 1991. Differential vulnerability of small mammals and amphibians to two trap types and two trap baits in Pacific Northwest forests. *Northwest Science* **65**:109–115.

McNab, B.K. 1991. The energy expenditure of shrews. Pages 35–45 *in* J.S. Findley and T.L. Yates, editors. *The Biology of the Soricidae*. Special Publication of the Museum of Southwestern Biology, University of New Mexico, Albuquerque, New Mexico, USA.

McShea, W.J. 2000. The influence of acorn crops on annual variation in rodent and bird populations. *Ecology* **81**:228–238.

Mech, S.G. and J.G. Hallett. 2001. Evaluating the effectiveness of corridors: a genetic approach. *Conservation Biology* **15**:467–474.

Merritt, J.F., G.L. Kirkland, Jr., and R.K. Rose, editors. 1994. Relationship of mandibular morphology to relative bite force in some *Sorex* from western North America. Carnegie Museum of Natural History, Special Publication No. **18**, Pittsburgh, Pennsylvania, USA.

Mills, L.S. 1995. Edge effects and isolation: red-backed voles in forest remnants. *Conservation Biology* **9**:395–403.

Mills, L.S., M.K. Schwartz, D.A. Tallmon, and K.P. Lair. 2003. Measuring and interpreting connectivity for mammals in coniferous forests. Pages 587–613 *in* C.J. Zabel and R.G. Anthony, editors. *Mammal Community Dynamics. Management and Conservation in the Coniferous Forests of Western North America*. Cambridge University Press, Cambridge, UK.

Morrison, M.L. and R.G. Anthony. 1989. Habitat use by small mammals on early-growth clear-cuttings in western Oregon. *Canadian Journal of Zoology* **67**:805–811.

Morrison, M.L., B.G. Marcot, and R.W. Mannan. 1998. *Wildlife-Habitat Relationships: Concepts and Applications*. University of Wisconsin Press, Madison, Wisconsin, USA.

Moses, R.A. and S. Boutin. 2001. The influence of clear-cut logging and residual leave material on small mammal populations in aspen-dominated boreal mixedwoods. *Canadian Journal of Forest Research* **31**:483–495.

Neal, F.D. and J.E. Borreco. 1981. Distribution and relation of mountain beaver to openings in sapling stands. *Northwest Science* **55**:79–86.

Nordyke, K.A. and S.W. Buskirk. 1991. Southern red-backed vole, *Clethrionomys gapperi*, populations in relation to stand succession and old-growth character in the central Rocky Mountains. *Canadian Field-Naturalist* **105**:330–334.

Radvanyi, A. 1970. Small mammals and regeneration of white spruce forests in western Alberta. *Ecology* **51**:1102–1105.

Ramirez, P., Jr., and M. Hornocker. 1981. Small mammal populations in different-aged clearcuts in northwestern Montana. *Journal of Mammalogy* **62**:400–403.

Rangen, S.A., R.G. Clark, and K.A. Hobson. 2000. Visual and olfactory attributes of artificial nests. *Auk* **117**:136–146.

Raphael, M.G. 1988*a*. Habitat associations of small mammals in a subalpine forest, southeastern Wyoming. Pages 359–367 *in* R.C. Szaro, K.E. Severson, and D.R. Patton, editors. Management of amphibians, reptiles, and small mammals in North America. US Forest Service General Technical Report **RM-166**, Fort Collins, Colorado, USA.

Raphael, M.G. 1988*b*. Long-term trends in abundance of amphibians, reptiles, and mammals in Douglas-fir forest of northwestern California. Pages 23–31 *in*

R.C. Szaro, K.E. Severson, and D.R. Patton, editors. Management of amphibians, reptiles, and small mammals in North America. US Forest Service General Technical Report **RM-166**, Fort Collins, Colorado, USA.

Ribble, D.O. and S. Stanley. 1998. Home ranges and social organization of syntopic *Peromyscus boylii* and *P. truei*. *Journal of Mammalogy* **79**:932–941.

Rickard, W.H. 1960. The distribution of small mammals in relation to the climax vegetation mosaic in eastern Washington and northern Idaho. *Ecology* **41**:99–106.

Ricklefs, R.E. 1969. An analysis of nesting mortality in birds. *Smithsonian Contributions in Zoology* **9**:1–48.

Rosenberg, D.K., K.A. Swindle, and R.G. Anthony. 1994a. Habitat associations of California red-backed voles in young and old-growth forests in western Oregon. *Northwest Science* **68**:266–272.

Rosenberg, D.K., C.J. Zabel, B.R. Noon, and E.C. Meslow. 1994b. Northern spotted owls: influence of prey base – a comment. *Ecology* **75**:1512–1515.

Rosenzweig, M.L. 1966. Community structure in sympatric carnivora. *Journal of Mammalogy* **47**:602–612.

Ryan, J.M. 1986. Dietary overlap in sympatric populations of pygmy shrews, *Sorex hoyi*, and masked shrews, *Sorex cinereus*, in Michigan. *Canadian Field-Naturalist* **100**:225–228.

Sakai, H.F. and B.R. Noon. 1993. Dusky-footed woodrat abundance in different-aged forests in northwestern California. *Journal of Wildlife Management* **61**:343–350.

Schweiger, S. and S. Boutin. 1995. The effects of winter food addition on the population dynamics of *Clethrionomys rutilus*. *Canadian Journal of Zoology* **73**:419–426.

Sieving, K.E. and M.F. Willson. 1998. Nest predation and avian species diversity in northwestern forest understory. *Ecology* **79**:2391–2402.

Simms, D.A. 1979. North American weasels: resource utilization and distribution. *Canadian Journal of Zoology* **57**:504–520.

Simons, L.H. 1985. Small mammal community structure in old growth and logged riparian habitat. Pages 505–506 *in* R.R. Johnson, editor. Riparian ecosystems and their management: reconciling conflicting uses. First North American Riparian Conference. US Forest Service General Technical Report **RM-GTR-120**, Fort Collins, Colorado, USA.

Sirotnak, J.M. and N.J. Huntly. 2000. Direct and indirect effects of herbivores on nitrogen dynamics: voles in riparian areas. *Ecology* **81**:78–87.

Smallwood, K.S. and C. Schonewald. 1998. Study design and interpretation of mammalian carnivore density estimates. *Oecologia* **113**:474–491.

Smith, C.C. and O.J. Reichman. 1984. The evolution of food caching by birds and mammals. *Annual Review of Ecology and Systematics* **15**:329–351.

Smith, W.P., R.G. Anthony, J.R. Waters, N.L. Dodd, and C.J. Zabel. 2003. Ecology and conservation of arboreal rodents of western coniferous forests. Pages 157–206 *in* C.J. Zabel and R.G. Anthony, editors. *Mammal Community Dynamics. Management and Conservation in the Coniferous Forests of Western North America*. Cambridge University Press, Cambridge, UK.

Sullivan, D.S. and T.P. Sullivan. 1982a. Effects of logging practices and Douglas-fir, *Pseudotsuga menziesii*, seeding on shrew, *Sorex* spp., populations in coastal coniferous forest in British Columbia. *Canadian Field-Naturalist* **96**:455–461.

Sullivan, T.P. 1979. Repopulation of clear-cut habitat and conifer seed predation by deer mice. *Journal of Wildlife Management* **43**:861–871.

Sullivan, T.P. and D.S. Sullivan. 1982*b*. The use of alternative foods to reduce lodgepole pine seed predation by small mammals. *Journal of Applied Ecology* **19**:33–45.

Sullivan, T.P., D.S. Sullivan, and C.J. Krebs. 1983. Demographic responses of a chipmunk (*Eutamias townsendii*) population with supplemental food. *Journal of Animal Ecology* **52**:743–755.

Sullivan, T.P., H. Coates, L.A. Jozsa, and P.K. Diggle. 1993. Influence of feeding damage by small mammals on the growth and wood quality in young lodgepole pine. *Canadian Journal of Forest Research* **23**:799–809.

Sullivan, T.P., R.A. Lautenschlager, and R.G. Wagner. 1999. Clearcutting and burning of northern spruce-fir forests: implications for small mammal communities. *Journal of Applied Ecology* **36**:327–344.

Sullivan, T.P., D.S. Sullivan, and E.J. Hogue. 2001. Influence of diversionary foods on vole (*Microtus montanus* and *Microtus longicaudus*) populations and feeding damage to coniferous tree seedlings. *Crop Protection* **20**:103–112.

Tallmon, D.A. and L.S. Mills. 1994. Use of logs within home ranges of California red-backed voles on a remnant of forest. *Journal of Mammalogy* **75**:97–101.

Taylor, C.A., C.J. Ralph, and A.T. Doyle. 1988. Differences in the ability of vegetation models to predict small mammal abundance in different aged Douglas-fir forests. Pages 368–374 *in* R.C. Szaro, K.E. Severson, and D.R. Patton, editors. Management of amphibians, reptiles, and small mammals in North America. US Forest Service General Technical Report **RM-166**, Fort Collins, Colorado, USA.

Terry, C.J. 1978. Food habits of three sympatric species of Insectivora in western Washington. *Canadian Field-Naturalist* **92**:38–44.

Terry, C.J. 1981. Habitat differentiation among three species of *Sorex* and *Neurotrichus gibbsii* in Washington. *American Midland Naturalist* **106**:119–125.

Tevis, L., Jr. 1956. Responses of small mammal populations to logging of Douglas-fir. *Journal of Mammalogy* **37**:189–196.

Torgersen, T.R. 2001. Defoliators in eastern Oregon and Washington. *Northwest Science* **75** [Special Issue]:11–20.

Traveset, A. and M.F. Willson. 1997. Effect of birds and bears on seed germination of fleshy-fruited plants in temperate rainforests of southeast Alaska. *Oikos* **80**:89–95.

Van Horne, B. 1982. Niches of adult and juvenile deer mice (*Peromyscus maniculatus*) in seral stages of coniferous forest. *Ecology* **63**:992–1003.

Vander Wall, S.B. 1992. The role of animals in dispersing a "wind-dispersed" pine. *Ecology* **73**:614–621.

Vander Wall, S.B. 2000. The influence of environmental conditions on cache recovery and cache pilferage by yellow pine chipmunks (*Tamias amoenus*) and deer mice (*Peromyscus maniculatus*). *Behavioral Ecology* **11**:544–549.

Verts, B.J. and L.N. Carraway. 1995. *Phenacomys albipes*. *Mammalian Species* **494**:1–5.

Verts, B.J. and L.N. Carraway. 1998. *Land Mammals of Oregon*. University of California Press, Berkeley, California, USA.

Voth, E.H. 1968. Food habits of the Pacific mountain beaver, *Aplodontia rufa pacifica* Merriam. Dissertation. Oregon State University, Corvallis, Oregon, USA.

West, N.E. 1968. Rodent-influenced establishment of ponderosa pine and bitterbrush seedlings in central Oregon. *Ecology* **49**:1009–1011.

West, S.D. 1991. Small mammal communities in the southern Washington Cascade Range. Pages 269–283 *in* L.F. Ruggiero, K.B. Aubry, A.B. Carey, and M.H. Huff, editors. Wildlife and vegetation of unmanaged Douglas-fir forests. US Forest Service General Technical Report **PNW-GTR-285**, Portland, Oregon, USA.

Whitaker, J.O., Jr., and C. Maser. 1976. Food habits of five western Oregon shrews. *Northwest Science* **50**:102–107.

Williams, D.F. 1991. Habitats of shrews (genus *Sorex*) in forest communities of the western Sierra Nevada, California. Pages 1–91 *in* J.S. Findley and T.L. Yates, editors. *The Biology of the Soricidae*. Special Publication of the Museum of Southwestern Biology, University of New Mexico, Albuquerque, New Mexico, USA.

Wilson, S.M. and A.B. Carey. 2000. Legacy retention versus thinning: influences on small mammals. *Northwest Science* **74**:131–145.

Wolff, J.O. 1996. Population fluctuations of mast-eating rodents are correlated with production of acorns. *Journal of Mammalogy* **77**:850–856.

Zabel, C.J., K. McKelvey, and J.P. Ward, Jr. 1995. Influence of primary prey on home-range size and habitat-use patterns of northern spotted owls (*Strix occidentalis caurina*). *Canadian Journal of Zoology* **73**:433–439.

Zegers, D.A., S. May, and L.J. Goodrich. 2000. Identification of nest predators at farm/forest edge and forest interior sites. *Journal of Field Ornithology* **71**:207–216.

Zollner, P.A. and S.L. Lima. 1999. Illumination and the perception of remote habitat patches by white-footed mice. *Animal Behaviour* **58**:489–500.

WINSTON P. SMITH, ROBERT G. ANTHONY,
JEFFREY R. WATERS, NORRIS L. DODD AND
CYNTHIA J. ZABEL

6
_____

Ecology and conservation of arboreal rodents of western coniferous forests

Introduction

Arboreal rodents were selected as a focal group because of their obvious association with forest canopies and relevance to forest management. The close association of arboreal rodents with trees predisposes them to being impacted by timber harvests (Carey 1989, Aubry et al. 2003, Hallett et al. 2003). Trees provide food, thermal and escape cover, shade, moisture and free water, and cavities that provide nest sites and safe refugia from avian and mammalian predators (Carey 1989, Carey et al. 1999, Aubry 2003). Carey (1989) characterized arboreal rodents according to their degree of dependence on trees for various activities and identified the following stand elements as important to their biology: large live trees, large snags, fallen trees, woody debris, multilayered canopy, overstory and understory diversity, and epiphytes. Additionally, he listed stand stability and landscape contiguity as important attributes with dense underbrush, streamsides, rock, and talus as special features (Carey 1989). More recent studies (e.g., Rosenberg and Anthony 1992, Waters and Zabel 1995, Smith and Nichols 2003) provide new information to examine the conclusions of Carey (1989), which serve as useful hypotheses regarding influences of forest management on arboreal rodent populations in western coniferous forests.

We selected species that are reputed late-seral forest habitat specialists or are known to be important prey for the northern spotted owl (Forsman et al. 2001, Hamer et al. 2001), the American marten (Buskirk and Ruggiero 1994, Ben-David et al. 1997), or the northern goshawk (*Accipiter gentilis*,

Lewis 2001). This chapter will focus on natural history, habitat associations, and conservation of Abert's squirrels (*Sciurus aberti*), Douglas squirrels (*Tamiasciurus douglasii*), red squirrels (*T. hudsonicus*), northern flying squirrels (*Glaucomys sabrinus*), red tree voles (*Arborimus longicaudus*), and Sonoma tree voles (*A. pomo*).

Abert's squirrels

The Abert's squirrel (*Sciurus aberti*), which is also known as the tassel-eared squirrel, is unique among North American arboreal mammals because of its obligate ecological relationship with ponderosa pine (*Pinus ponderosa*) for all aspects of its life history (Keith 1965, Hall 1981, Brown 1984, Snyder 1993, 1998a, States and Wettstein 1998; but see Hutton et al. 2002). There are six geographically isolated subspecies (Hoffmeister and Diersing 1978, States and Wettstein 1998) distributed from central Mexico through Arizona and New Mexico, southeastern Utah and southwestern Colorado, and northward along the Front Range of the Rocky Mountains to extreme southern Wyoming (Nash and Seaman 1977). Although highly dependent on ponderosa pine, the Abert's squirrel is absent from most of the range of this tree species (Lamb et al. 1997). This distributional pattern reflects post-Pleistocene squirrel dispersal northward following the rapid expansion of ponderosa pine (Davis and Brown 1989, Lamb et al. 1997), with squirrels restricted to interior forests that receive relatively limited winter snowfall (Brown 1984) and moderate annual rainfall (<64 cm; McKee 1941, Patton 1975a).

Foraging ecology

The Abert's squirrel is a highly specialized herbivore feeding primarily on ponderosa pine ovulate cone seed, inner bark of terminal shoots, and a diverse assemblage of associated mycorrhizal fungi (Keith 1965, Stephenson 1975, States et al. 1988, States and Gaud 1997, States and Wettstein 1998). It has evolved a unique foraging strategy reflecting its association with ponderosa pine and a highly variable food supply (Snyder 1993, 1998a, States and Wettstein 1998). Abert's squirrels introduced into limited ponderosa pine forests in the Pinaleño Mountains of southeastern Arizona (Brown 1984) were observed feeding in mixed-conifer forests on Douglas-fir (*Pseudotsuga menziesii*), white pine (*Pinus monticola*), and cork-bark fir (*Abies lasiocarpa*) cones, and hypogeous fungi (Hutton et al. 2002). Also, Abert's squirrels were observed taking Engelmann spruce (*Picea engelmanni*) and Douglas-fir cones from middens of Mt. Graham red squirrel

(*T. h. grahamensis*) in mixed-conifer and spruce-fir forests. Abert's squirrels do not cache food (Keith 1965, Patton 1977, Hall 1981), though limited scatter hoarding of fungi and individual pinecones in litter and nests does occur (Stephenson 1975, States et al. 1988, Allred et al. 1994). Of all foods used, pinecone seed has the highest nutritive value (Hall 1981) and caloric content (6.2 kcal/g dry weight; Austin 1990), as well as the highest annual variability in abundance (Larson and Schubert 1970, Hall 1981, States et al. 1988). Pinecone availability strongly influences localized distribution and abundance of squirrels during late summer and fall (Hall 1981, Patton et al. 1985, Dodd et al. 1998, Lema 2001).

Mycelium and hypogeous sporocarps (i.e., truffles) and epigeous fungi are also important sources of food for Abert's squirrels because they are more reliable than pinecone seed and potentially available year-round (States 1985, States et al. 1988, States and Wettstein 1998). Fungi dominate squirrel diets during some seasons (Stephenson 1975, States and Wettstein 1998). Dodd et al. (2003) found evidence of 21 species of fungi in squirrel fecal samples, with highest mean frequency of occurrence in late summer (70.8%, $n = 117$), followed by winter (28.2%, $n = 102$), and spring (9.4%, $n = 163$). Availability of fungi during spring may stimulate onset of breeding (Stephenson 1975) and increase juvenile recruitment and winter survival (Dodd et al. 2003). States and Wettstein (1998) reported squirrel body mass declines that corresponded with diminishing availability of fungi. Caloric content of fungi (4.4 kcal/g dry weight) was lower than pinecone seed, yet mineral and salt concentrations of fungi exceeded those of other foods evaluated (Austin 1990). Fungi are an excellent source of water (Stephenson 1975) and their diversity and abundance facilitates a high foraging efficiency among mycophagous rodents (Smith 1970, States and Wettstein 1998). Many fungi exhibit protein content >20% (some >30%), although some investigators reported that much of the protein of hypogeous fungi is not digestible (Cork and Kenagy 1989, Claridge and Cork 1994, Claridge et al. 1999). Luoma et al. (2003) discuss the nutritional content and value of hypogeous sporocarps to mycophagous mammals.

Selective herbivory by Abert's squirrel appears to be a significant selective force acting on ponderosa pine populations (Snyder 1993, 1998*a*, *b*, States and Wettstein 1998). The ability of this species to utilize inner bark, or phloem, to a significant degree differentiates them from other tree squirrels (States and Wettstein 1998). Squirrels feed on the inner bark of ponderosa pine, despite its low caloric and protein content (Patton 1974, Stephenson 1975). Although use occurs throughout the year, peak use of inner bark occurs during winter when other foods are absent or unavailable due to snow (Stephenson 1975, States et al. 1988). Forage tree selection by squirrels was associated with several factors, including tree diameter and height (Keith 1965, Pederson et al. 1976, Gaud et al. 1993), stand basal area (Ffolliot and Patton 1978), ease of twig peeling (Pederson and Welch 1985), and carbohydrate content (Thomas 1979). However, tree selection is most influenced by genetically determined chemical composition of trees, primarily terpenes in resins (Farentinos et al. 1981, Zang and States 1991, Snyder 1992, 1993, 1994, 1998a, b, Snyder and Linhart 1993). Only moderate to heavy use in consecutive years caused tree damage (Ffolliot and Patton 1978). Squirrel feeding on phloem significantly reduced ponderosa pine reproductive output and growth (Snyder 1993) resulting in a 20% reduction in the developing pinecone crop when mature cone abundance was limited (Allred et al. 1994). Gaud et al. (1993) reported a strong inverse relationship ($r = -0.988$, $P < 0.001$) between cones eaten and terminal shoots clipped by squirrels.

Other seasonally important foods include Gambel oak (*Quercus gambelii*) acorns in fall, staminate cones and apical buds of ponderosa pine in spring, and dwarf mistletoe (*Arceuthobium* spp.) year-round (Stephenson 1975, Hall 1981, States et al. 1988).

Reproductive biology

Onset of breeding in Abert's squirrels coincides with snowmelt and the emergence of staminate cones of ponderosa pine (Stephenson 1975, Brown 1984). Typically, breeding is confined from April to June (Keith 1965, Farentinos 1972a) with a peak in mid-May (Stephenson 1975). However, Pogany and Allred (1995) described Abert's squirrel reproductive strategy as a female-driven system and documented bimodal gestation in winter and spring. Males apparently exhibit consistent prolonged spermatogenesis from October to June, which facilitates capitalizing on unpredictable, short-duration estrus in females. Once females enter estrus, the otherwise largely solitary males become gregarious and triple their home range size while in pursuit of females (Farentinos 1972b, 1979). Lema (2001) noted as many as nine males in a tree with one female during a breeding bout. Males exhibit a social dominance hierarchy during mating (Farentinos 1972b).

Females are receptive to breeding for a limited time (<18 hours) and often are bred by multiple males (Farentinos 1972b, Brown 1984). Gestation was estimated at 40 days (Keith 1965) and 46 days (Farentinos 1972b, Stephenson 1975). Litter size ranged from two to five (placental scars; Keith 1965) and from 2.9 to 3.4 (from nests; Farentinos 1972*b*, Keith 1965, Stephenson 1975). Altricial young remain in natal nests for seven to nine weeks emerging in August (Keith 1965). Young are weaned after 20 days, but remain in family social units associated with natal nests through the fall after which they become solitary (Farentinos 1972*b*).

Population ecology
Movements and home range size

Abert's squirrels are non-territorial (Farentinos 1972*b*, 1979, Patton 1975*a*, Hall 1981, Brown 1984) and exhibit high spatial overlap in home ranges (Farentinos 1972*b*, 1979, Patton 1975*a*, Pederson et al. 1976, Lema 2001). Home range size varies by season, food availability, sex, age, and habitat condition (Keith 1965, Farentinos 1972*a*, 1979, Pederson et al. 1976, Patton et al. 1985). Mean home range during winter in Colorado was larger than in summer for both males (5.3 versus 2.6 ha) and females (4.9 versus 1.3 ha), with more food resources during the summer accounting for the difference (Farentinos 1979). Conversely, Keith (1965) and Lema (2001) reported that the mean home range of squirrels in Arizona was smaller during winter (2.0 and 7.5 ha, respectively) than in other seasons (7.3 and 15.5 ha, respectively). Lema (2001) reported that the mean annual home range of males (26 ha) was larger than that of females (14 ha), but that adults (18 ha) and juveniles (13 ha) had similar home ranges. Farentinos (1979) found that the mean home range of males in the breeding period (20.7 ha) was nearly three times larger than the mean home range during the remainder of the year (7.5 ha), but female home ranges remained relatively stable.

Habitat modification from logging significantly influences the home range size of Abert's squirrels. In Utah, the mean home range increased from 2.5 ha in an unlogged forest to 12.9 ha following harvest (Pederson et al. 1976). Mean home range size for squirrels in northern Arizona increased from 27.2 to 49.8 ha following harvest (Patton et al. 1985). Lema (2001) compared home range sizes at sites of widely divergent habitat condition; mean home range size was similar between dense, multi-aged stands, and open intensively thinned habitats. However, radio-collared squirrels in the open habitat used the edge along adjacent dense habitat to a high degree, particularly during winter (Dodd et al. 1998, Lema 2001).

Information on long-distance movements of Abert's squirrels is limited. Juvenile dispersal occurs during late fall (Farentinos 1972*a*). Dispersal distances of 0.7–1.4 km were recorded for squirrels that left natal nests and established home ranges similar in size to adults (Farentinos 1972*a*).

However, dispersal away from natal nests was not observed among radio-collared juveniles, which had post-natal home ranges that overlapped natal nests (Lema 2001).

Survival

Abert's squirrel survival is influenced by several factors including predation, variable food quality and quantity (Keith 1965, Patton 1974, Hall 1981, States et al. 1988), habitat modification (Farentinos 1972a, Patton 1975a), sport hunting (Patton 1984), and disease (Brown 1984). The amount and duration of winter snow cover is the principal factor affecting survival (Keith 1965, Stephenson 1975, Stephenson and Brown 1980, Hall 1981, Patton 1984). Deep, prolonged snow limits access to fungi and pinecones and relegates squirrels to a diet of pine twig inner bark (States et al. 1988, Austin 1990), which causes squirrels to lose body mass and experience higher mortality rates (Stephenson 1975, Austin 1990). Annual survival rates ranged 0.44–0.78 over eight years, with mortality correlated ($r = 0.85$, $P < 0.01$) with the number of days on which snow depth was ≥ 10 cm (Stephenson and Brown 1980).

Seasonal survival rates (from Jolly–Seber model; Pollock 1982, Pollock et al. 1990) at seven 66-ha sites along a structural habitat gradient in Arizona averaged 0.78 (SE = ±0.04) and ranged from 0.63 (SE = ±0.10) during winter to 0.91 (SE = ±0.11) during late spring–early summer (Dodd et al. 1998). Lema (2001) reported differential Kaplan–Meier survival estimates for a 19-month period in Arizona: adult survival (0.74, $n = 39$) was higher than juvenile survival (0.48, $n = 18$); and adult male survival (0.81, $n = 20$) was higher than adult female survival (0.52, $n = 16$), which was attributed to the cost of parental care of young by females (Lema 2001).

Abert's squirrels are one of 14 key prey species of the northern goshawk (*Accipiter gentilis*) in the Southwest (Reynolds et al. 1992) and are particularly important in winter when other prey hibernate or migrate (Dodd et al. 1998). Lema (2001) documented predation by raptors, presumably goshawks, on ten of 50 radio-collared squirrels. Though collared animals occurred on both open, intensively logged and dense, multi-aged sites, nearly all raptor predation occurred in dense habitats (Lema 2001). Although squirrels in open habitats appeared more vulnerable to predation (Austin 1990), goshawks apparently foraged more frequently in denser forests (Beier and Drennan 1997). Similarly, Sieg (2002) documented 24 incidences of raptor predation on 73 radio-collared squirrels at six sites along a landscape gradient varying in the ratio of optimum to marginal patch area (ROMPA; Lidicker 1988, Krohne 1997). He also noted a positive relationship between ROMPA and the incidence of raptor predation ($r_s = 0.812$, $P = 0.05$).

Habitat relations and forest management

Selected habitats include sites supporting high canopy closure, tree densities, basal area, and interlocking crowns with clumps of large, mature overstory trees (Patton 1975a, Pederson et al. 1976, Patton et al. 1985, Dodd et al. 1998). Among eight study sites in Arizona, juvenile recruitment was correlated ($r_s = 1.00$) with the number of interlocking canopy trees (Dodd et al. 1998) and with the frequency of fungal spores in summer fecal samples ($r_s = 0.943$, $P = 0.005$; Dodd et al. 2003). These were sites that had undergone intensive, even-aged harvests (e.g., shelterwood harvest) and had fewer than three interlocking trees that exhibited minimal or inconsistent recruitment. Tree spacing is important for nest placement (Patton 1975a) and for juveniles and adults traveling from maternal nests, and as cover from aerial and ground predators (Austin 1990). Summer fungal content in the diet was correlated with basal area ($r_s = 0.943$, $P = 0.005$; Dodd et al. 2003), which corroborated the findings of States and Gaud (1997) that hypogeous fungi production is related to ponderosa pine basal area and canopy closure. Juxtaposition of higher quality habitat to marginal habitat was important for squirrels to seasonally exploit food resources associated with poorer quality habitat (Dodd et al. 1998).

Squirrels primarily select for size and chemical cues when feeding on inner bark (Keith, 1965, Patton 1975a, Pederson et al. 1976, Farentinos et al. 1981, Zang and Sates 1991, Snyder 1992). Patton (1975a) found that 75% of nest trees were associated with ≥ 3 trees with interlocking crowns. Trees used by squirrels were significantly larger (diameter) and taller, and were closer to nearest neighbor trees than nearest neighbors were to adjacent trees (Lema 2001).

Abert's squirrel abundance and density estimates vary considerably depending on habitat quality (Patton et al. 1985, Dodd et al. 1998), season (Farentinos 1972a, Dodd et al. 1998), and method of estimation (Brown 1984). Large fluctuations in density among seasons or years are typical for the species (Hall 1981, Brown 1984), though Keith (1965) believed that a general decline in populations occurred due to habitat modification from logging. He found that densities reported earlier (Trowbridge and Lawson 1942) were 10–20 times higher than what he observed for sites in Arizona. Empirically derived estimates of habitat quality based on potential squirrel density varied from 2.45 squirrels/ha in rare "optimum", to 0.35 squirrels/ha in "good", and 0.05 squirrels/ha in "poor" habitats (Patton 1984). Unlogged forests consistently supported densities three to five times higher than adjacent logged areas, regardless of density at any given time (Brown 1984). Similarly, Patton et al. (1985) found a decrease in average abundance when they compared pre-harvest densities (0.30–0.65 squirrels/ha) to post-harvest densities (0.21–0.33) during a 5-year period in northern Arizona. Patton (1984) recommended limiting even-aged shelterwood harvest to small (8 ha) patches, as did Pederson et al. (1987). Ratcliff et al. (1975) found that squirrel abundance was related to basal area ($r = 0.62$, $P < 0.05$) in northern Arizona; adding nest abundance to the model substantially improved the correlation ($r = 0.88$, $P < 0.05$).

At the landscape scale, Dodd (2003) observed marked variation in mean population density and juvenile recruitment over four years along a gradient of ROMPA (Lidicker 1988, Krohne 1997) among nine sites. A threshold response occurred for both parameters between 37% and 42% of optimum patch area, which was similar to values reported for other species (Andren 1994, Bowers and Matter 1997). Squirrel abundance averaged 0.38 squirrels/ha ($n = 20$) above 42% ROMPA, whereas mean density below this value was 0.16 squirrels/ha ($n = 16$). In open marginal habitats, highest mean density occurred at 40–45% ROMPA and highest juvenile recruitment occurred at 35% ROMPA suggesting some positive benefits from habitat mosaic and edge (Patton 1975b). These results underscore the importance of source habitat in maintaining population viability (Andren 1994, Bowers and Matter 1997, Dodd 2003).

Seasonal fluctuations in abundance can be large and typically are attributed to factors that influence annual reproduction, recruitment, and/or juvenile dispersal (Keith 1965, Brown 1984). Dodd et al. (2003) observed uniform mean densities between years (0.33 versus 0.37 squirrels/ha) at eight sites, but recorded large seasonal fluctuations at most sites. At four sites during both years, mean density increased an average of 205% (compared to 17% for the other four sites) from winter (0.23 squirrels/ha) to summer (0.56 squirrels/ha), though juvenile recruitment was insufficient to account for the increases (Dodd et al. 2003). Large fluctuations in squirrel densities from spring to summer were correlated with quadratic mean diameter (diameter of a tree with average basal area) of ponderosa pine (Patton et al. 1985, Dodd et al. 2003), which likely is associated with annual pinecone production and regularity of pinecone crops (Larson and Schubert 1970). Pinecone abundance apparently plays an important role in squirrel population dynamics (Keith 1965, Hall 1981, Patton et al. 1985).

Reproductive performance and juvenile survival vary considerably among habitats and among years within habitats (Brown 1984, Dodd et al. 1998, Dodd 2003). In unlogged forests in Utah, mean recruitment over two years was 1.0 juvenile/female, whereas one site averaged 2.3 juveniles/female after logging (Pederson et al. 1976). During a prolonged drought (60% of normal precipitation), Dodd et al. (1998) reported that the mean recruitment rate at seven sites (over two years) was 0.13 juveniles/female. Following the two driest years (2000 and 2002) of combined winter and spring precipitation on record in Arizona (41% and 28% of normal precipitation, respectively), recruitment averaged 0.06 juveniles/female across nine sites; <3% of adult females showed evidence of successful reproduction (Dodd 2003). In contrast, recruitment at the same sites averaged 0.83 and 1.79 juveniles/female when winter–spring precipitation was 71% and 93% of normal precipitation, respectively. Percent of normal winter–spring (November–May) precipitation accounted for 94% of the variation in recruitment ($r = 0.971$, $p = 0.029$; Dodd 2003). Over the four years, mean density of juveniles recorded in dense (optimum) habitats (0.42/ha) was >2.5 times greater than that observed in open (marginal) habitats (0.16/ha).

Patton et al. (1985) evaluated the effects of timber harvest on Abert's squirrels in northern Arizona and reported higher densities and a greater density response between pre- and post-treatment on unlogged plots. Densities were positively correlated with the number of trees per hectare and showed a positive curvilinear relationship with quadratic mean diameter. Higher tree densities ($\geq$20) in the 30–74 cm diameter at breast height (dbh) range was associated with higher squirrel densities on control plots, likely because of the importance of large trees as food and cover (Patton et al. 1985). Pederson et al. (1987) observed 62% more feeding activity in uncut versus cut forests; uncut forest also had higher (212%) feeding on hypogeous fungi. Reduced fungi production in logged plots was attributed to opening the forest canopy and reducing litter cover. They concluded that lower feeding activity in logged habitat reflected lower squirrel densities and recruitment (Pederson et al. 1987).

At a landscape scale, forest restoration may have significant negative effects on Abert's squirrels, hypogeous fungi, and short-term ecosystem processes. Restoration activities contribute further to cumulative negative impacts on squirrel populations, a consequence of past even-aged management of southwestern forests (Keith 1965, Patton 1984, Dodd and Adams 1989), including reduced stand, patch, and landscape diversity (Patton 1992). Moreover, restoration efforts may result in a loss of remaining limited quality habitats critical for maintaining squirrel populations (Patton 1984, Dodd et al. 1998). The key to maintaining viable Abert's squirrel populations in future managed forests appears to be the retention of sufficient quality habitats (e.g., >35% ROMPA) dispersed among open, thinned stands (marginal habitat) in a landscape mosaic. Application of variable thinning prescriptions and basal area retention in harvest areas will create within- and among-stand structural heterogeneity, with benefits to squirrels, hypogeous fungi, and forest ecosystem processes.

Douglas and red squirrels

The genus *Tamiasciurus* (pine squirrels) is comprised of two species (but see Arbogast et al. 2001), Douglas squirrel and the red squirrel, that are common in late-seral coniferous forests of western North America. They mostly have exclusive ranges, but are sympatric in the Blue Mountains of eastern Oregon and in the Cascade and coastal mountains of north-central Washington and southwestern British Columbia (C. Smith 1968, Steele 1998, 1999, Verts and Carraway 1998). The northern limit of *T. douglasii* corresponds with the northern limit of grand fir. Pacific silver fir with its larger cones replaces grand fir northward along the Pacific coast in the area where red squirrels replace Douglas squirrel (Smith 1981). The red squirrel has a broader geographic distribution, including boreal forests from the northern fringe in Alaska, eastward throughout most of Canada, southward into mixed-conifer forests throughout the Rocky Mountains to southern Arizona and New Mexico, and southward into the Great Lakes states and southern Appalachian Mountains (Flyger and Gates 1982, Steele 1998). Forest composition and food availability west and east of the Cascades may explain some biological variation between the two species (Smith 1970, 1981).

Foraging ecology

Pine squirrels spend most of their time foraging (C. Smith 1968, Ferron et al. 1986) with little difference between sexes (Ferron et al. 1986). Activity patterns vary seasonally and correspond with changes in ambient temperature (Pauls 1978). Pine squirrels have similar food habits (Flyger and Gates 1982), but there are important differences that correspond with their geographical distributions and anatomical differentiation (Smith 1970, 1981). Both species rely on conifer seeds as a staple source of high-energy food (Smith 1970, 1981, Brown 1984), but diet varies with habitat and food availability (Brink and Dean 1966, C. Smith 1968, M. Smith 1968, Sanders 1983, Gurnell 1987, Froehlich 1990, Steele 1998, 1999). Ferron et al. (1986) characterized red squirrels as opportunistic, taking advantage of the continuously changing food resources throughout a growing season. Douglas squirrels adjust their foraging patterns and tree selection in response to changes in cone availability, take more cones from cone-rich trees, but do not preferentially exploit cone-rich patches (Sanders 1983). They are less selective of trees and cones within trees during periods of cone failure (Sanders 1983).

Red squirrels are adapted to feeding on serotinous cones and larger, heavier cones than Douglas squirrels, which are adapted to feed most efficiently on small cones with weak scales and small seeds (Smith 1970, 1981). Both species selectively harvest cones from trees with the highest seed energy/cone, with individual squirrels shifting to tree species with the next highest energy content as cone availability of a preferred tree species diminishes (C. Smith 1968). Selection is based on number of seeds/cone, ratio of seed weight to cone weight, cone hardness, configuration of cones among branches, and distance from middens (C. Smith 1968, 1970, Elliott 1974, 1988, Sanders 1983). Douglas squirrels harvest fewer cones and become more selective as the distance from the cache increases (Sanders 1983). In mixed stands, red squirrels were found to harvest cones first from Pacific silver fir (*Abies amabilis*), Douglas-fir, and western hemlock (*Tsuga heterophylla*) (Steele 1998). Douglas-fir cones were found to be an important source of high-energy seed for Douglas squirrel, but seeds of western redcedar (*Thuja plicata*), red alder (*Alnus rubra*), and maples (*Acer* spp.) were readily consumed (Steele 1999). White fir (*Abies concolor*) cones were selected in the western Sierra Nevada Mountains near the southern limit of its range (Sanders 1983). Lodgepole pine (*Pinus contorta*) is an important source of food for red squirrels because it perennially produces a relatively uniform cone crop with serotinous cones across a large portion of the tree's range. In the Rocky Mountains, blue (*Picea pungens*) and Engelmann spruce cones are selected; blue spruce is the most dependable seed producer in the transition zone of the Colorado East Slope (Finley 1969). Engelmann spruce with lodgepole pine enables red squirrels to occupy 10^3–10^4 km^2 of montane forest where cone supply from other species is erratic or absent (Hatt 1943, Finley 1969). Engelmann spruce,

Douglas-fir, and white fir cones are important foods of the Mt. Graham red squirrel (*T. h. grahamensis*) in southeastern Arizona (Froehlich 1990). White spruce (*Picea glauca*) seeds are essential food in interior Alaska, influencing squirrel geographical distribution and density (Brink and Dean 1966).

Pine squirrels also consume flowers and seeds of angiosperms, fruiting bodies of fungi, and the shoots, buds, and cambium of conifers (C. Smith 1968, Fisch and Dimock 1978, Sanders 1983, Sullivan and Vyse 1987, Sullivan et al. 1996). Cambium is a major dietary component in winter and early spring (Hatt 1943, Steele 1998); terminal buds and shoots of Douglas-fir, spruce, fir, and other conifers become the primary food when cone production is low (M. Smith 1968, Fisch and Dimock 1978, Brown 1984). Squirrels eat both epigeous (mushrooms) and hypogeous (truffles) fungi (McKeever 1964, Flyger and Gates 1982, Sanders 1983, Steele 1998, 1999), especially during summer or other periods of food shortage (C. Smith 1968, M. Smith 1968, Gurnell 1987). Other foods of pine squirrels are leaves, flowers, stalks and fronds of ferns, arthropods, bone, and young birds or small mammals (Steele 1998, 1999). Bones are consumed mostly by pregnant or lactating females or juveniles and may be related to special nutritional needs (Hatt 1943, McKeever 1964, Smith 1981, Sanders 1983).

Pine squirrels cache mushrooms (Froehlich 1990), cones, and seeds near the center of their territories in large ($\leq$10 m diameter, Finley 1969) middens (C. Smith 1968, 1970, 1981). Pine squirrels also stockpile food in secondary caches (M. Smith 1968, Patton and Vahle 1986) that may be critical for overwinter survival (C. Smith 1968) and population growth (Kemp and Keith 1970), especially following a year of low cone production or complete failure (M. Smith 1968, C. Smith 1968, 1981, Finley 1969). Caches are frequently associated with large ($>$50 cm dbh) trees, logs, and dense cover (Patton and Vahle 1986). Squirrels may begin using lodgepole pine cones from the previous season's growth in June (Hatt 1943), but typically begin cutting and caching cones of the current year's production in August (Finley 1969, Froehlich 1990).

Pine squirrels spend considerable time harvesting conifer cones (C. Smith 1968), especially during September to October (Finley 1969, Sanders 1983) and may cut and cache 12000–16000 cones annually (C. Smith 1968). Although squirrels can inflict heavy cone loss to conifers, they do not appear to be a selective agent for conifer characteristics (Sanders 1983; but see Smith 1970 and Elliott 1974). Seeds collected from cone caches have higher germination rates (Wagg 1964) and middens are an important source of conifer seed for tree nurseries (Finley 1969). Still, the prodigious harvesting of cones by squirrels can nearly eliminate a seed crop and impact natural regeneration, especially in ponderosa pine stands (Foiles and Curtis 1965), in low to moderate cone-production years, or following seed tree or shelterwood harvests (Finley 1969, Sanders 1983). Also, damage to managed forest and economic loss from pine squirrels feeding on cambium or conifer buds can be extensive (e.g., 57.5% of young trees, Sullivan and Klenner 1993) and include disruption of regeneration, reduced growth rates, tree mortality, deformed boles, and reduced wood quality (C. Smith 1968, Fisch and Dimock 1978, Flyger and Gates 1982, Sullivan and Vyse 1987, Sullivan et al. 1993, 1996). Several factors influence damage, including availability of cones, proximity of managed stands to mature forests (Fisch and Dimock 1978), spacing of young trees, and adequate cover to reduce risk to predation (Sullivan et al. 1994). Diversionary food was effective in reducing damage during periods of low cone production (Sullivan and Klenner 1993).

Reproductive biology

Pine squirrels are promiscuous breeders and females are spontaneous ovulators (Millar 1970, Koford 1982, Gurnell 1984) that remain in estrus for one day (Millar 1970, Steele 1998, 1999). Timing of ovulation is influenced by photoperiod (Becker 1993) as well as food availability (Millar 1970, Dolbeer 1973, Rusch and Reeder 1978). Breeding mostly occurs among yearlings and adults, but females born early in a year occasionally reproduce later that same year (Koford 1982). Ovulation rates and timing differ between juveniles and adults and among years, a consequence of which is substantial variation in annual breeding period and reproductive rates (Kemp and Keith 1970, Rusch and Reeder 1978). The proportion of reproductive adult and juvenile females varies within and among habitats and years and contributes variability to annual reproductive rates (Kemp and Keith 1970, Dolbeer 1973, Rusch and Reeder 1978, Sullivan 1990). Sullivan (1990) reported variation in proportion of reproductive females between localities representing similar habitat (e.g., 83% versus 45%) that fluctuated annually (e.g., 17% to 83%). Klenner and Krebs (1991), however, reported little variation among juvenile or adult males or females in breeding condition between white spruce and Douglas-fir forests in British Columbia. Similarly, 100% of female red squirrels were in reproductive condition in old-growth and intensively managed second-growth

(20–28 years old) lodgepole pine forests in British Columbia (Ransome and Sullivan 1997).

The breeding season can last for several months, usually begins in early March (Smith 1981, Koford 1982, Steele 1998, 1999), is extremely variable (Millar 1970), and is influenced by food availability (Smith 1981, Sullivan and Sullivan 1982) and weather (Millar 1970, Dolbeer 1973). Pine squirrels usually produce one litter/year in the Rocky Mountain region (Dolbeer 1973) and in the Pacific Northwest (Smith 1981). However, Douglas squirrels have two litters/year where they are sympatric with the red squirrel (Cowan and Guiguet 1960, Smith 1981), or when food is abundant (Koford 1982). Similarly, red squirrels may have two litters/year after a mild winter (Millar 1970) or in more southern latitudes (Froehlich 1990).

Litter size in red squirrels is not determined directly by food availability (Humphries and Boutin 2000), but females apparently adjust timing or abandon reproduction according to availability of food (C. Smith 1968, 1981). Douglas squirrels have larger mean and maximum litter sizes than red squirrels (Smith 1981). Maximum litter size for *T. douglasii* and *T. hudsonicus* was eight and five, respectively in southwestern British Columbia; corresponding mean litter sizes ranged 4.0–5.7 and 2.9–3.8. In other localities throughout its range, mean litter size of red squirrels varied from 3.2 to 5.4 (C. Smith 1968, Kemp and Keith 1970, Millar 1970, Dolbeer 1973, Steele 1998). Within macrohabitats, the mean litter size of red squirrels varied among years or between yearlings and adults within years (Rusch and Reeder 1978). Increases in litter size do not appear to influence maternal survival, but may be associated with reduced offspring survival (Humphries and Boutin 2000). Thus, optimal litter size in pine squirrels may represent a trade-off between offspring number and survival (Humphries and Boutin 2000).

Pine squirrels nest or den in hollow trees, logs, external tree nests, or holes in the ground (Hatt 1943, C. Smith 1968). Red squirrels select tree cavities as nest sites (Froehlich 1990), but use burrows (Steele 1998) or make external nests of twigs, grasses, leaves, moss, and lichens (Hatt 1943, Patton and Vahle 1986). Squirrels typically construct two or more nests near their midden (Hatt 1943, Rothwell 1979, Froehlich 1990). Some nests are used for multiple years (Hatt 1943). External nests are commonly interwoven among branches near the bole of a large tree with dense canopy, typically Douglas-fir or spruce (Rothwell 1979, Brown 1984). Most nests are <8 m (mean = 5.3 m) above ground (Rothwell 1979, Froehlich 1990), but some occur as high as 20 m (Rothwell 1979). Proximate factors associated with nest-site selection in mixed-conifer forests are distance from forest edge, microsite stem density, tree species, bole diameter, structure of tree limbs, and availability of overstory escape routes (Hatt 1943, Rothwell 1979, Patton and Vahle 1986, Steele 1998). Lodgepole pine, especially within interior forests, is highly selected in mixed-conifer forests of the Rocky Mountain region (Hatt 1943, Rothwell 1979). Nest-site selection is critical for thermoregulation and overwinter survival (Hatt 1943, Rothwell 1979), especially for red squirrels because they occupy colder environments than Douglas squirrels (C. Smith 1968, 1981, Brown 1984). Location of nests is also important for efficient use of food resources. Nests typically are within 30 m of primary middens (C. Smith 1968, Rothwell 1979, Patton and Vahle 1986, Froehlich 1990), with the majority ($\approx$90%) within 15 m of a midden (Patton and Vahle 1986, Froehlich 1990).

Population ecology

Movements and home range sizes

Male and female pine squirrels occupy and defend exclusive territories resulting in intense inter-sexual and intra-specific competition (C. Smith 1968, 1981, Kemp and Keith 1970, Gurnell 1984). Where *T. douglasii* and *T. hudsonicus* are sympatric, the two species show congeneric territorial defense (Smith 1981). Pine squirrels have roughly circular territories that center on primary middens (C. Smith 1968, Smith 1970, Patton and Vahle 1986), presumably minimizing time and energy to gather and cache cones and defend middens (C. Smith 1968, Smith 1981, Sanders 1983). Some red squirrels were observed to regularly defend auxiliary caches $\geq$50 m from their primary midden (Patton and Vahle 1986, Froehlich 1990). Although typically solitary (C. Smith 1968), a mother occasionally shares her territory with a juvenile (especially daughters) as a form of extended parental investment (Berteaux and Boutin 2000).

There is essentially no quantitative information about movements or home range size of Douglas squirrels. C. Smith (1968) reported that squirrels of the genus *Tamiasciurus* typically occupy territories that range from 0.2 to 1.2 ha. Territory sizes of red squirrels varied among habitats and locations, ranging from 0.24 ha in white spruce forests of Alberta to 4.24 ha in Englemann spruce forests of Arizona (Froehlich 1990, Steele 1998). Territories may vary in relation to food availability (C. Smith 1968, M. Smith 1968, Klenner 1991). Sullivan (1990) reported a 24 to 50% reduction in territory size of red squirrels following food supplementation in southern British Columbia. M. Smith (1968) reported a mean home range size of 1.6 ha in mature white spruce forests during a winter immediately following a cone failure, which increased to 4.8 ha in the subsequent winter following a second cone failure.

Red squirrels exhibit both natal and breeding dispersal with significant implications for population dynamics (Berteaux and Boutin 2000). Juveniles (especially males) frequently disperse from natal areas, but breeding dispersal and juvenile philopatry are not uncommon (Price et al. 1986, Berteaux and Boutin 2000). Post-breeding females may expand their territories to include more food resources and share a portion or bequeath the entire territory to their young. Prior to dispersal, juveniles undertake extensive ($\leq$900 m) forays, but typically establish territories near ($<$323 m) their mothers (Larsen and Boutin 1994, 1998). Juveniles that settled farther from their mothers had higher predation risks, but secured bigger territories and experienced higher overwinter survival (Larsen and Boutin 1994, 1998). In one study, dispersing females were older, had more juveniles at weaning than resident mothers, and dispersed after breeding in years with higher food availability (Berteaux and Boutin 2000). Survival and future reproduction of dispersing and resident mothers were similar.

Survival

Maximum longevity in the wild is about ten years for red squirrels, but few individuals live for longer than five years (Gurnell 1987). Complete population turnover can occur within a decade (Davis and Sealander 1971). Pine squirrels are prey for many predators (Steele 1998) such as the northern goshawk (*Accipiter gentilis*; Squires and Reynolds 1997, Watson et al. 1998), northern spotted owl (*Strix occidentalis*; Carey 1995, Forsman et al. 2001, Hamer et al. 2001) and American marten (*Martes americana*; Gurnell 1987, Buskirk and Ruggiero 1994). Red squirrels are a key prey species of the northern goshawk in southeast Alaska (Lewis 2001), whereas Douglas squirrels are among the most frequently represented prey in the diet of breeding goshawks in the Washington Cascades (Watson et al. 1998). Observed attempts of predation on Mt. Graham red squirrel occurred by the Mexican spotted owl (*S. o lucida*), bobcat (*Lynx rufus*), and northern goshawk (Schauffert et al. 2002).

Survival rates vary seasonally, annually, among habitats (Table 6.1), among age classes, and between sexes (Kemp and Keith 1970, Rusch and Reeder 1978). Predation was higher for substandard (i.e., physical condition) squirrels (Wirsing et al. 2002). Survival was higher for philopatric than dispersing juveniles, probably because of lower risks of predation

Table 6.1. *Densities (ha) and demography of red squirrels (*Tamiasciurus hudsonicus*) among locations or forest types of western North America*

Forest type/location	Density (ha) Mean	Density (ha) Range	Breeding females (%)	Survival (%)	Reference
White spruce OG interior Alaska		0.21–0.62			M. Smith (1968)
White spruce OG interior Alaska	1.4	1.33–1.52			Searing (1975)
White spruce OG[a] interior Alaska	1.3	1.00–2.08			Wolff and Zasada (1975)
White spruce YG[b] interior Alaska	0.0	0			Wolff and Zasada (1975)
White spruce YG[c] interior Alaska	1.5	0.42–0.71			Wolff and Zasada (1975)
Douglas-fir[d] south-central BC		1.10–3.30			Klenner and Krebs (1991)
White spruce[d] south-central BC		2.11–3.90			Klenner and Krebs (1991)
Lodgepole pine YG[e] south-central BC	1.6	1.56–1.71	100	86	Ransome and Sullivan (1997)
Lodgepole pine MG[f] south-central BC	1.4	1.21–1.59	100	85	Ransome and Sullivan (1997)
Douglas-fir[g] Idaho	2.7	2.15–3.09			Medin (1986)
Mixed-conifer Arizona	0.2				Froehlich (1990)
Engelmann spruce Arizona	0.5				Froehlich (1990)
Jack pine SG[h] Alberta		0.87–2.64		25–63	Rusch and Reeder (1978)
Mixed spruce YG[i] Alberta		1.61–6.85		50–73	Rusch and Reeder (1978)

[a] Computed from pre-harvest data among all stands in Table 1 (Wolff and Zasada 1975).
[b] Young growth stand following clear-cut logging.
[c] Young growth stand following shelterwood logging.
[d] Estimated from Figure 1 of Klenner and Krebs (1991) p. 966; age and history of stands unknown.
[e] 20- to 28-year-old stands.
[f] >120-year-old stands.
[g] Computed from pre-logging data in Table 2 (Medin 1986).
[h] 60-year-old stand.
[i] Age not given, but saplings dominated (600/ha) the understory.
(BC = British Columbia; OG = old-growth, MG = mature growth, SG = second growth, YG = young growth).

(Larsen and Boutin 1994, Berteaux and Boutin 2000). Food (i.e., mostly conifer cones, Smith 1981) has been proposed as a factor limiting red squirrel populations (Sullivan 1990, Klenner and Krebs 1991, Ransome and Sullivan 1997; but see Rusch and Reeder 1978, Koford 1992), possibly through its influence on survival. Mortality apparently varies inversely with fluctuations in cone production (Kemp and Keith 1970, Humphries and Boutin 2000). M. Smith (1968) reported a 67% drop in red squirrel abundance following the second year of cone crop failure in interior Alaska. Several studies demonstrated higher squirrel densities following the experimental addition of food (Sullivan 1990, Klenner and Krebs 1991, Ransome and Sullivan 1997), but supplemental feeding may (Klenner and Krebs 1991) or may not (Sullivan and Klenner 1993, Ransome and Sullivan 1997) improve survival of red squirrels.

Habitat relations and forest management

Red squirrels of western North America select coniferous forests of montane, subalpine, or boreal forest biomes, but also occur in mixed stands with hardwoods, notably *Populus* (C. Smith 1968, Rusch and Reeder 1978, Steele 1998). In southeastern Alaska, they occur in riparian forests that include alder (*Alnus* spp.) or cottonwood (*P. trichocarpa*), but select rainforests of Sitka spruce (*Picea sitchensis*) and western hemlock (Sieving and Willson 1998). Douglas squirrel also occupies several forest types (Steele 1999), but selects dense, mesic coastal forests dominated by western hemlock, Douglas-fir, grand fir (*A. grandis*), Sitka spruce, or western redcedar (Smith 1981). The two squirrel species overlap in a transition zone (mostly >1000 m elevation) between damp coastal forests and interior, drier forests, which are dominated by lodgepole pine and ponderosa pine (Smith 1981).

Important habitat elements for pine squirrels include a diversity of conifers that provide a perennial source of seeds, closed-canopy, multilayered forests with large, old trees (Smith 1981, Vahle and Patton 1983, Carey 1995) and interlocking canopies. These habitats create cool, shaded, moist conditions for preserving food caches, are conducive to the growth of fungi, facilitate efficient foraging, and provide cover from predators (C. Smith 1968, Rusch and Reeder 1978, Vahle and Patton 1983, Brown 1984, Steele 1998, 1999). Large, old trees are better cone producers than younger trees and provide nest sites (C. Smith 1968, Smith 1981, Patton and Vahle 1986, Froehlich 1990). Forest composition is important (Brown 1984); tree species common in pine squirrel habitat include Douglas-fir, western

Table 6.2. *Densities (ha) and demography of Douglas squirrels (* Tamiasciurus douglasii) *among locations and forest types of western North America*

Forest type/location	Density (ha)		Breeding females (%)	Survival (%)	Reference
	Mean	Range			
Douglas-fir OG Washington[a]	0.9	0.40–1.44			Buchanan et al. (1990)
Douglas-fir OG Washington[b]	0.2	0.14–0.27			Buchanan et al. (1990)
Douglas-fir SG[c] Washington	0.2	0.03–0.76			Buchanan et al. (1990)
Western hemlock OG Olympic Peninsula	<0.10				Carey (1995)
Western hemlock SG Olympic Peninsula	<0.10				Carey (1995)
Western hemlock OG[d] North Cascades	0.5				Carey (1995)
Western hemlock SG[e] southwest BC		0.22–0.89	50[f]	60[g]	Sullivan and Sullivan (1982)

[a] Near Columbia Gorge.
[b] 100 km north of Columbia Gorge site.
[c] 42- to 165-year-old stands.
[d] 44- to 67-year-old stands.
[e] 40- to 45-year-old stands.
[f] Average of 1977 (22%; 2/9) and 1978 (63%; 12/19) (Sullivan and Sullivan 1982).
[g] Males.
(BC = British Columbia; OG = old-growth, SG = second growth).

hemlock, Pacific silver fir, grand fir, Engelmann spruce, white spruce, Sitka spruce, subalpine fir *(Abies lasiocarpa)*, and lodgepole pine. Lodgepole pine is a predictable, perennial producer of seeds (Smith 1981, Vahle and Patton 1983). Blue spruce cones are frequently eaten and appear especially valuable for Arizona populations (Brown 1984). Engelmann spruce forests are the primary habitat of the Mt. Graham red squirrel, an endangered subspecies in Arizona (Smith and Mannan 1994). Suitable den or nest sites are likely not limiting (Carey 1991) because pine squirrels use external tree nests or holes in the ground as well as hollow trees and logs (C. Smith 1968).

Both the red squirrel (Table 6.1) and Douglas squirrel (Table 6.2) show considerable geographical and habitat variability in abundance, but some variation is likely due to differences in methods. Food availability and winter severity contributed to seasonal variation in abundance (Kemp and Keith 1970, Rusch and Reeder 1978) that approached a six fold increase

(0.33–2.00 squirrels/ha) between spring and summer (Sullivan and Moses 1986). Pine squirrel populations seem to be limited proximally by food and ultimately through territoriality or spacing behavior (C. Smith 1968, Rusch and Reeder 1978, Sullivan and Sullivan 1982, Buchanan et al. 1990, Klenner 1991, Carey 1995). Territorial behavior was effective in limiting autumn recruitment of juvenile immigrants and breeding adults in spring (Klenner 1991). Seasonal (i.e., spring to fall) variation in Douglas squirrel abundance due to juvenile recruitment can be significant (e.g., 0.22–0.88 squirrels/ha); annual variation in abundance appears to be linked to cone production and contributes to habitat variability (Sullivan and Sullivan 1982). Nearby ($\leq$100 km) populations in similar habitat can differ by an order of magnitude (Buchanan et al. 1990). Douglas squirrel abundance varied from <0.1 to 0.9 squirrels/ha among forest types, with the highest densities occurring in Douglas-fir forests (Buchanan et al. 1990). Winter populations of Douglas squirrels may be three times more abundant in old-growth forests than in younger managed forests (Buchanan et al. 1990), presumably because of a greater, more diverse and predictable food supply (Smith 1981, Buchanan et al. 1990). Old-growth forests of the Washington Cascades have a greater component of western hemlock, which provides a more predictable food source (Buchanan et al. 1990). However, studies conducted elsewhere concluded that the abundance of Douglas squirrel was similar among old-growth, mature, and young forests (Anthony et al. 1987, Carey 1989, 1995).

Red squirrels occur in a wider variety of forest types (Steele 1998, 1999) and show greater variation in population density than Douglas squirrels (Table 6.1). Similar to Douglas squirrels, red squirrels appear to achieve their highest densities in old-growth forests (e.g., Medin 1986). In one study, however, mean red squirrel density was similar between mature (>120 years old) forest and young (20–28 years old) second-growth stands (Table 6.1; Ransome and Sullivan 1997). Near the northern limits of its distribution (64°50′ N), densities (i.e., midden counts) in mature white spruce forest were similar (Searing 1975, Wolff and Zasada 1975). Mt. Graham red squirrels had the lowest mean density in mixed-conifer forests of Arizona (0.2 middens/ha; Froehlich 1990). Mt. Graham red squirrels compete unsuccessfully with the introduced Abert's squirrels, which have increased in abundance at the expense of the endemic subspecies (Brown 1984, Gurnell 1987). Lower densities of Mt. Graham red squirrels in mixed-conifer than in Englemann spruce forest are related to microsite features of primary middens (Froehlich 1990).

There have been no studies of the direct impacts of clear-cut logging on Douglas squirrels. Presumably, the response would be similar to that reported for red squirrels in interior Alaska, where individuals vacated territories after clear-cut logging of mature white spruce forests, and the density of squirrels was reduced from 1.5/ha to 0 after harvest (Wolff and Zasada 1975). Adjacent habitat may provide opportunities for "refugees" to establish new territories (Klenner 1991), but clearly this will depend on the extent to which the landscape has been modified and the amount, quality, and distribution of remaining habitat. The limited information available suggests that converting mature forests to even-aged plantations of coniferous forest will result in lower pine squirrel populations in the short term (Buchanan et al. 1990, Carey 1991, 1995).

Longer-term, cumulative impacts of clear-cut logging will depend on rotation age and the age at which second-growth stands become suitable habitat. In the Oregon Cascade Range, Douglas squirrel populations were well established in young (25–50 years old) second-growth stands, with densities similar to those in mature (130–200 years old) and old-growth (400–450 years old) Douglas-fir forests (Anthony et al. 1987). Red squirrels occupied relatively young (e.g., 17–24 years old) pine stands in southern Canada (Sullivan and Moses 1986, Sullivan et al. 1996). Average densities in lodgepole pine stands were similar between unthinned young stands and mature stands throughout the year (Sullivan and Moses 1986) and were comparable to spring densities (1.1–1.3 /ha) reported between mature pine and mixed-conifer forests (Rusch and Reeder 1978). On some sites, however, densities in unthinned second-growth stands were highly variable (0.13–0.72/ha) and lower than in old-growth stands (0.17–2.15/ha), which were also highly variable. Survival may be lower in young pine stands than in mature forests (Sullivan and Moses 1986).

Structural features of young, managed forests are important characteristics for Douglas squirrels (Carey 1991, 1995). Within stands, factors that likely influence the abundance of cones include density of conifers and the age or size of trees. Although conifers are reproductive at a relatively young age, cone production in young trees generally increases with age (Schopmeyer 1974). Older forests likely support more Douglas squirrels because of the number and diversity of large conifers (Buchanan et al. 1990). However, Waters and Zabel (1998) found that the mean capture rate of Douglas squirrels was significantly greater in mature fir stands (80–100 years old) than in unlogged or shelterwood-logged old-growth fir stands (>250 years old). They speculated that cone production may have been greater in the mature stands than in the old-growth stands because the density of trees of cone-bearing age was greatest in the mature stands.

There has been only one study of Douglas squirrel populations in partially logged forests. Waters and Zabel (1998) compared squirrel abundance in shelterwood-logged, old-growth fir stands (six to seven years post-harvest) to old-growth stands in northeastern California. Despite substantial differences in basal area between logged stands (22.6 m²/ha) and old-growth sites (72.6 m²/ha), Douglas squirrel abundance was similar (Table 6.2). In contrast, red squirrel densities declined by 65% (Wolff and Zasada 1975) and 85% (Medin 1986) when tree densities or the basal area of old coniferous forests was reduced by 83% or 78%, respectively. Animals that remained after logging increased their home range sizes and added adjacent mature forest into their territories (Wolff and Zasada 1975). Similarly, intensive thinning (93% to 95% of stem density) of young (17–27 years old) second-growth stands reduced habitat capability for red squirrels (Sullivan and Moses 1986, Sullivan et al. 1996); mean densities were lower in thinned (0.11 and 0.38/ha) than in unthinned (0.22 and 1.15/ha) second-growth or mature forests (0.58 and 1.11/ha).

Vahle and Patton (1983) found that red squirrel abundance was significantly lower in stands that were selectively logged (≈45% to 60% of basal area) than in mature mixed-conifer forests of Idaho. However, declines were not as great as those observed after higher proportions (≈75% to 80%; Medin 1986) of the overstory trees were removed. Selectively logged stands retained multi-storied and other structural characteristics (e.g., large live trees) typically associated with older forests (Vahle and Patton 1983, Carey 1995, 2000). Low to moderate thinning may improve young pine stands as habitat for red squirrels (Sullivan et al. 1996). Low-intensity thinning of young (17–27 years old) stands resulted in stands with squirrel densities (1.30/ha) that were higher than in heavily thinned stands (0.41/ha) and similar to those in old-growth forests (1.10/ha).

Pre-commercial thinning has often been an effective means of retarding canopy closure in young stands and extending the period in which there is an understory of herbaceous and woody vegetation (Sullivan and Moses 1986, Sullivan et al. 1996). For arboreal rodents, many of these plants (e.g., *Rubus* spp., *Vaccinium* spp.) are important producers of soft mast and seasonally may represent important sources of food (Carey et al. 1999). Pre-commercial thinning releases remaining stems from competition and encourages diameter growth, which ultimately shortens the time interval between harvest and cone production in second-growth stands (Schopmeyer 1974, Johnstone 1981). Commercial thinning further enhances cone production and additionally accelerates the development of second-growth stands toward understory initiation, and ultimately mature forest habitat conditions (Nowacki and Kramer 1998, Carey 2000). In intensively managed landscapes where clear-cut logging is the primary silvicultural prescription, red squirrels may benefit from practices that retain a well-distributed portion of the landscape in older seral stages, include low- to moderate-intensity intermediate stand management (e.g., thinning), and establish a rotation age that retains a relatively small portion of the landscape in young clear-cuts.

Northern flying squirrels

The northern flying squirrel is one of the most common arboreal rodents in the Pacific Northwest. It is nocturnal, typically biphasic, and active year-round (Wells-Gosling and Heaney 1984, Witt 1992), but adjusts its duration and timing of activity according to the onset of darkness and air temperature (Cotton and Parker 2000a). This species occupies conifer and mixed-conifer/hardwood forests throughout much of North America (Wells-Gosling and Heaney 1984, MacDonald and Cook 1996). Several factors apparently influence population densities (Carey et al. 1992; Smith and Nichols 2003); some may not be directly related to habitat (Rosenberg and Anthony 1992). Many facets of their habitat use and behavior appear to be related to reducing predation risks (Pyare et al. 2002), but populations are probably limited by food in many circumstances especially where they are syntopic with Townsend's chipmunk (*Tamias townsendii,* Carey et al. 1999). The northern flying squirrel has been the focus of much research in the Pacific Northwest because it is the primary prey of northern spotted owls throughout much of the owl's range (Forsman et al. 2001, Hamer et al. 2001), and because of its reputed keystone role in fundamental processes of western coniferous forests (Maser et al. 1986, Maser and Maser 1988). The spotted owl (and various mustelids) preys on the squirrel, which eats and disseminates the sporocarps of ectomycorrhizal fungi, which are essential symbionts of dominant tree species (Forsman et al. 1984, Maser et al. 1986, Maser and Maser 1988, Carey 1995, Carey et al. 1999). Thus, this species is an important link in the food chain and forest dynamics of coniferous forests of this region.

Foraging ecology

Although the northern flying squirrel inhabits conifer forests throughout its range, conifer seeds are not a primary food item. Brink and Dean

(1966) found that, unlike red squirrels, captive northern flying squirrels from interior Alaska were unable to maintain body mass on a diet of white spruce cones. Seeds of red fir (*Abies magnifica*) and white fir ranked sixth among nine foods offered to captive northern flying squirrels in a cafeteria-style food preference study (Zabel and Waters 1997). Sporocarps of hypogeous fungi (truffles) and arboreal lichens are the foods most frequently reported in the diets of northern flying squirrels in western North America (Cowan 1936, McKeever 1960, Maser et al. 1985, 1986, Maser and Maser 1988, Hall 1991, Waters and Zabel 1995, Rosentreter et al. 1997, Cazares et al. 1999, Pyare and Longland 2001, 2002, Pyare et al. 2002). Arboreal lichens have been found to be more common in the winter diet than in the summer diet, presumably because truffles are less available during winter (Cowan 1936, McKeever 1960, Mowrey and Zasada 1984, Maser et al. 1985, 1986, Hall 1991, Rosentreter et al. 1997). Spores of truffles in fecal samples collected during winter, however, suggest that flying squirrels cache truffles (Hall 1991, Rosentreter et al. 1997). Mowrey and Zasada (1984) reported that radio-collared flying squirrels ate cached fungi from middens of red squirrels during winter in Alaska. *Bryoria fremontii* is the arboreal lichen that has been reported to occur in the diet of northern flying squirrels most often, and Rosentreter et al. (1997) noted that this species lacks secondary chemical compounds characteristic of other lichens.

Northern flying squirrels are known to eat a variety of other foods, including sporocarps of epigeous fungi (mushrooms) (Laurance and Reynolds 1984, Mowrey and Zasada 1984, Maser et al. 1985, 1986, Hall 1991, Waters and Zabel 1995, Thysell et al. 1997, Cazares et al. 1999, Pyare and Longland 2001, Pyare et al. 2002), invertebrates (Waters and Zabel 1995, Pyare et al. 2002), buds (Connor 1960, Weigl 1978, Mowrey and Zasada 1984), staminate cones of conifers (Weigl 1978, Maser et al. 1985, Waters and Zabel 1995), sap (Schmidt 1931, Foster and Tate 1966), berries (Mowrey and Zasada 1984, Thysell et al. 1997), catkins (Connor 1960, Thysell et al. 1997), seeds (Connor 1960, Ransome and Sullivan 1997, Thysell et al. 1997), conifer seedlings (Thysell et al. 1997), and unidentified plant material (Connor 1960, Waters and Zabel 1995, Rosentreter et al. 1997, Cazares et al. 1999, Pyare et al. 2002). Meat, presumably from nesting birds or small mammals, also was reported in the diet of northern flying squirrels (Connor 1960, McKeever 1960), and the species is known to be readily attracted to traps baited with meat (Jackson 1961). Thysell et al. (1997) concluded that flying squirrels likely eat more non-truffle foods than is indicated by fecal pellet analyses, and that such items may be important nutritional supplements or alternative foods when truffles are scarce.

Reproductive biology

There have been no definitive studies of the reproductive biology of flying squirrels in the Pacific Northwest. They reputedly have a single breeding season during late spring and early summer (Cowan 1936, Wells-Gosling and Heaney 1984, Forsman et al. 1994), although late breeding has been documented in northern California (Raphael 1984). Flying squirrels usually mate during late March–May, have a gestation period of 37–42 days, and young are typically born during late May–June (Booth 1946, Wells-Gosling and Heaney 1984). Females often use tree cavities for maternal dens, but squirrels use a variety of den sites (Cowan 1936, Carey et al. 1997, Bakker and Hastings 2002). Females likely produce only one litter/year, with litter sizes ranging from one to four young (Wells-Gosling and Heaney 1984, Forsman et al. 1994). Forsman et al. (1994) found that breeding occurred from late May to early June on the Olympic Peninsula, Washington, and that most females produced only a single litter in years when they bred. Thus, annual reproductive potential for flying squirrels is low compared to other similar-sized rodents.

Population ecology

Movements and home range size

Movements and home range sizes of northern flying squirrels have been studied with mark-recapture methods on trapping grids and by radio-telemetry techniques. The former method was used to describe movements between successive (24 hour) captures when traps were checked once per day, whereas the latter was used to estimate home range sizes and study movement patterns. There have been at least three studies that described movements between successive captures (Carey et al. 1991, Rosenberg and Anthony 1992, Smith and Nichols 2003), all of which computed the mean maximum distance moved (MMDM) between captures (Wilson and Anderson 1985). Results suggested that northern flying squirrels moved short distances over a 24-hour period, with most movements between successive locations being ≤100 m. Carey et al. (1991) found that flying squirrel movements in the Oregon Coast Range varied only slightly among different stand types with MMDMs of 112, 82, and 93 m in young, mature, and old-growth forests, respectively. Rosenberg and Anthony

(1992) found similar values for MMDM in second-growth (mean = 86.6 m, SE = 6.1 m) and old-growth (mean = 83.5 m, SE = 5.5 m) Douglas-fir stands in the Oregon Cascade Mountains. MMDM ranged from 75 to 142 m in old-growth temperate rainforests of southeastern Alaska, but was similar (mean = 92 m, SE = 9.1 m) between seasons and markedly different habitats (Smith and Nichols 2003). In a radio-telemetry study, Martin and Anthony (1999) found similar distances moved (i.e., successive telemetry locations) over a 24-hour period (range = 64–80 m) between sexes and between second- and old-growth forests in the Oregon Cascades. Thus, daily movements of northern flying squirrels (as indexed by MMDM) appear relatively low, regardless of location or forest type. Movements appear to be influenced mostly by the sex of individuals, or the availability of food resources. Pyare and Longland (2002) noted that movements and habitat use of squirrels in old-growth forests in the Sierra Nevada Mountains were influenced primarily by the distribution of truffles.

There have been two radio-telemetry studies of northern flying squirrel home range size (Witt 1992, Martin and Anthony 1999), and both studies reported home range estimates that averaged less than 5 ha. Witt (1992) estimated home range sizes (minimum convex polygon method) of flying squirrels during winter and spring in old-growth forests of the Oregon Coast Range (mean = 3.7 ha, range = 3.4–4.2 ha). Martin and Anthony (1999) studied the home range sizes (adaptive kernel method) of northern flying squirrels during summer and fall in second- and old-growth forests of the Oregon Cascade Mountains (mean = 4.9 ha, range = 3.6–6.3 ha). In the latter study, home range sizes were similar between years and among stand types, but males (mean = 5.9 ha, SE = 0.75) had larger home ranges than females (mean = 3.9 ha, SE = 0.37). These studies provided minimum estimates, reflecting only a part of the year because the battery life of transmitters was less than six months. Together, however, the two studies provided comparable estimates of home range sizes from two areas during different seasons and substantiated the limited short-term movement patterns of northern flying squirrels in coniferous forests dominated by Douglas-fir.

Survival

Survival rates of northern flying squirrels are not well documented. Villa et al. (1999) reported empirical estimates (proportion of individuals living from one age class to the next) of survival among three age classes in the Oregon Coast Range and Puget Trough of Washington from yearly capture-recapture studies during fall. About 50% of individuals survived from age class I (1–7 months) to age class II (7–21 months) in old-growth and mixed-aged forests in the Oregon Coast Range. Only 7% of individuals captured in young forest stands of the Oregon Coast Range and 30% in young stands in Puget Trough survived from age class I to II. Thirty-three percent of the squirrels survived from age class II to age class III (≥22 months) in the Oregon Coast Range in all stand types, and 22% survived from age class II to III in the Puget Trough. Longevity was ≥7 years in their studies.

Habitat relations and forest management

The northern flying squirrel often is a cavity nester and was thought to be a habitat and diet specialist and thus potentially sensitive to cumulative disturbances (Carey 1989). However, recent studies have shown that this species occurs in young coniferous forests (Aubry et al. 1991, Rosenberg and Anthony 1992, Waters and Zabel 1995) and is common in second-growth stands with residual old-growth components (Rosenberg and Anthony 1992). Northern flying squirrels may exhibit considerable flexibility in their ability to occupy a broad range of environmental conditions (Cotton and Parker 2000b).

The abundance of flying squirrels varies to some extent among forest types and successional stage (Table 6.3). Densities have been estimated by enumeration (minimum number known alive per unit area) or capture-recapture methods using statistical estimators in the program CAPTURE (Otis et al. 1978). Enumeration methods provide a minimum estimate of populations that are biased downward because of the assumption that all individuals were captured. Statistical estimators attempt to account for individuals that are not captured and, therefore, are less biased. Consequently, care must be taken in comparing densities across studies that used different methods to estimate abundance. Overall, northern flying squirrel abundance varied from 0.1 to 4.0 squirrels/ha depending on forest type, seral stage, and management history (Table 6.3). Carey et al. (1992) reported that flying squirrels were more abundant in old-growth forests compared to young forests of the Oregon Coast Range. In contrast, Rosenberg and Anthony (1992) found no significant differences in flying squirrel densities between old-growth and young, managed forests of the Oregon Cascade Mountains. Waters and Zabel (1995) reported greater mean densities in old-growth (>200 years old) fir stands than

Table 6.3. *Abundances of northern flying squirrels* (Glaucomys sabrinus) *among forested habitats of western North America*

Forest type	Age/disturbance history	Season	Density (ha) Mean	Density (ha) Range	Reference
Douglas-fir	Young growth, clear-cut	Spring	1.1	0.7–1.6	Carey et al. (1992)
Douglas-fir	Young growth, clear-cut	Fall	0.5	0.3–0.7	Carey et al. (1992)
Douglas-fir	Young growth, clear-cut	Fall	1.9	1.1–2.5	Rosenberg and Anthony (1992)
Douglas-fir	Old-growth, natural	Spring	1.8	1.1–2.2	Carey et al. (1992)
Douglas-fir	Old-growth, natural	Fall	1.9	1.8–2.2	Carey et al. (1992)
Douglas-fir	Old-growth, natural	Fall	2.3	1.4–3.3	Rosenberg and Anthony (1992)
Douglas-fir	Old-growth, natural	Annual	1.0	0.5–1.8	Witt (1992)
Douglas-fir	Second growth, clear-cut	Annual	0.1	0–0.2	Witt (1992)
W. hemlock/ Sitka spruce	Young growth, clear-cut	Fall	0.2	n/a	Carey et al. (1992)
W. hemlock/ Sitka spruce	Old-growth, natural	Fall	0.5	n/a	Carey et al. (1992)
W. hemlock/ Sitka spruce	Old-growth, natural	Spring	1.8	1.6–2.0	Smith and Nichols (2003)
W. hemlock/ Sitka spruce	Old-growth, natural	Fall	3.2	2.2–4.0	Smith and Nichols (2003)
Mixed conifer	Old-growth, natural	Spring	1.7	0.9–3.2	Carey et al. (1992)
White fir/red fir	Mature, fire replacement	Summer	2.3	2.2–2.4	Waters and Zabel (1995)
White fir/red fir	Old-growth, natural	Summer	3.3	2.8–3.5	Waters and Zabel (1995)
White fir/red fir	Old-growth, shelterwood cut	Summer	0.4	0.2–0.6	Waters and Zabel (1995)

in 75- to 95-year-old fir stands in northeastern California, but the differences were not significant. Studies from the Pacific Northwest and southeastern Alaska suggest a trend toward higher densities in older forests, or managed forests with old-growth components, than in younger, managed stands (Table 6.3). Low densities (0.2–0.5/ha) of the species were reported in western hemlock/Sitka spruce forests of the Olympic Forest in western Washington (Carey et al. 1992). In contrast, Smith and

Nichols (2003) found high densities (1.6–4.0/ha) in western hemlock/Sitka spruce forests of southeastern Alaska. In the Pacific Northwest, densities in Douglas-fir and mixed-conifer forests appeared to be higher than those in hemlock/spruce forests (Table 6.3).

The range of variation in densities among forest types and locations (Table 6.3) raises questions about factors influencing the abundance of flying squirrels at both the micro- and macro-habitat scales. Carey (1995) reported latitudinal variation in abundance, with densities in the southern Coast Ranges and Central Western Cascades of Oregon notably higher than on the Olympic Peninsula or in the North Cascades of Washington. In old-growth forests of the Olympic National Forest of Washington, he found that abundances were associated with the presence of ericaceous shrubs and large snags. Flying squirrels of the Douglas-fir/western hemlock zone in Oregon used habitats with higher decadence and more complex forest canopies (Carey et al. 1999). Large snags are an important feature of northern flying squirrel habitat (Cowan 1936, Carey 1991, 1995, Carey et al. 1997) because flying squirrels often select large, old trees as den sites (Carey et al. 1997, Cotton and Parker 2000, Bakker and Hastings 2002). In southeastern Alaska, flying squirrels used snags more than live trees for denning, and individuals that denned in live trees or snags chose larger diameter trees (Bakker and Hastings 2002). Also, most dens (>72%) were in tree cavities within micro-sites with abundant downed logs. Squirrels selected snags, and larger diameter trees with conks on the bole, broken tops, or visible tree entries. Among live trees, squirrels selected for a higher abundance of hemlock dwarf mistletoe (Bakker and Hastings 2002). It has been suggested that variation in densities between old-growth and young forests of the Pacific Northwest may be related to a limited number of secure cavities for maternal dens (Carey et al. 1997). Witt (1991) reported that supplementary dens increased the carrying capacity of second-growth forests in western Oregon. However, more recent studies suggest that dens are not a dominant factor limiting flying squirrels in second-growth forests of western Oregon (Carey 2002), and that flying squirrels can exhibit considerable plasticity in den selection (Cotton and Parker 2002).

A consistent predictor of flying squirrel abundance or habitat use has been the distribution or abundance of hypogeous fungi. Waters and Zabel (1995) reported a significant correlation between flying squirrel density and truffle frequency in fir forests of northeastern California. Pyare and Longland (2002) concluded that flying squirrels refined their spatial use of old-growth forests according to fine-scale changes in truffle availability. Factors affecting the distribution or abundance of truffles may be important determinants of habitat use and ultimately influence demographics of flying squirrel populations. Although coarse woody debris (CWD) has been reported as an ecological correlate of truffle distribution and abundance, Pyare and Longland (2002) reported that none of the microhabitat variables associated with CWD was important in explaining flying squirrel habitat use in old-growth forests of the Sierra Nevada Mountains. They reported understory cover as the only microhabitat variable measured that was associated with habitat use.

There have been no published studies of northern flying squirrel population responses to recent clear-cut logging; however, flying squirrels were not captured in young (<10 years) clear-cuts in the Oregon Coast Range (Morrison and Anthony 1989). Longer-term and cumulative impacts of clear-cut logging may be inferred from studies that have compared the abundances of this species among different-aged stands. Carey et al. (1992) reported higher densities of flying squirrels in late-seral forests than in younger, managed forests, but younger managed stands appear to sustain northern flying squirrel populations (Rosenberg and Anthony 1992, Carey 1995, 2000, Carey et al. 1999), especially those with legacy components of older forests (Carey 1995). Intensively managed landscapes where clear-cut logging is the primary silvicultural prescription will likely have lower northern flying squirrel populations than largely unmanaged landscapes. Carey (2000) found flying squirrels to be twice as abundant in Douglas-fir forests managed for retention of standing live trees, snags, and fallen trees than in stands intensively managed for timber production. Selective harvests with legacy retention will probably have fewer short-term and cumulative impacts than clear-cut logging on northern flying squirrel populations. Moreover, active stand management, particularly light thinning, likely will further reduce the impacts of timber management on northern flying squirrels and other arboreal rodents (Carey et al. 1999).

The effect of partial overstory removal on flying squirrel populations has been examined in only two studies (Waters and Zabel 1995, Carey 2000). One study examined the effects of shelterwood logging of old-growth fir forests on flying squirrel populations (Waters and Zabel 1995). Flying squirrel densities were consistently higher and averaged an order of magnitude greater in unlogged old-growth fir stands (mean = 2.76/ha, range 2.8–3.5/ha) than in old-growth fir stands that had been logged with a shelterwood harvest system (mean = 0.31/ha, range 0.3–0.6/ha). The body mass of adult males and females, and recapture rates were similar among sites, but recruitment (i.e., % juveniles captured) was higher in unlogged old-growth forests. Lower densities and recruitment in shelter-wood stands suggested that heavy logging and intensive site preparation negatively affected flying squirrel populations (Waters and Zabel 1995).

Tree voles

Two species of tree voles are currently recognized (Wilson and Reeder 1993, Wilson and Ruff 1999): the red tree vole (*Arborimus longicaudus*) and the Sonoma tree vole (*A. pomo*). Tree voles are unique among arvicoline rodents in the degree to which they are arboreal. These voles spend most of their lives in the canopies of trees in mesic coniferous forests of western Oregon and northwestern California. It has been suggested that they are most abundant in old-growth forests (Franklin et al. 1981, Meslow et al. 1981, Carey 1991, Corn and Bury 1991, Ruggiero et al. 1991) and vulnerable to local extirpations due to habitat loss and forest fragmentation (Corn and Bury 1988, Huff et al. 1992). Although natural history information is available on their feeding and breeding behavior, little is known about their population biology. We have found no published documentation of ecological differences between *A. longicaudus* and *A. pomo*. We therefore discuss general ecological characteristics that are assumed to be common to both species.

Confusion exists about the taxonomic status of tree voles. *Phenacomys longicaudus* was originally described by True (1890) from a specimen from Coos County, Oregon. Taylor (1915) described a new subgenus, *Arborimus*, in which he placed *P. longicaudus*. Johnson (1973) elevated *Arborimus* to generic rank for *P. longicaudus* based on a variety of blood protein, reproductive, and morphological characters that differed from *P. intermedius*, the type species for the genus *Phenacomys*. Johnson and Maser (1982) extended the genus *Arborimus* to the white-footed vole (*A. albipes*), but we do not discuss this species in this chapter because it is not considered to be arboreal (Jewett 1920, Maser et al. 1981, Wilson and Ruff 1999, but see Voth et al. 1983). Johnson and George (1991) described a new species, *A. pomo*, for California populations based on differences from *A. longicaudus* in the number of chromosomes, morphology, and results of a laboratory breeding study. Murray (1995) performed two types of molecular analyses and concluded that tree voles from Humboldt, Mendocino, and Sonoma Counties in California were genetically distinct from individuals captured in Del Norte County in northern California and Lincoln and Benton Counties along the central Oregon coast. He suggested that the Klamath River in northern California might be a barrier to gene flow separating *A. long-icaudus* from *A. pomo*. His results also supported separation of *Arborimus* from *Phenacomys* at the generic level. While some mammalogists still recognize the red tree vole in Oregon as *P. longicaudus* (Verts and Carraway 1998), others recognize it as *A. longicaudus* and the Sonoma tree vole in northwestern California as *A. pomo* (Wilson and Reeder 1993, Wilson and Ruff 1999). Research on the genetics of tree voles is ongoing and may lead to future taxonomic revisions (S. Haig, pers. comm.).

Tree voles have a limited geographical distribution. *A. longicaudus* occurs from the Columbia River along the border between Oregon and Washington to northern California. It occurs in the Coast and Klamath Ranges east to the western slope of the Cascade Range. *A. pomo* occurs along the coast and in the Coast Ranges of Humboldt, Mendocino, and Sonoma Counties of California. Zentner (1966) reported tree voles in California as far east as Siskiyou and Trinity Counties, but the eastern extent of the range of *A. pomo* is poorly documented.

Foraging ecology

Tree voles have a unique diet: they forage almost exclusively on conifer needles, primarily those of Douglas-fir. They bite off and discard the lateral resin duct along each edge of the needle and eat the remaining middle portion of the needle (Taylor 1915, Howell 1926, Benson and Borell 1931, Clifton 1960, Maser 1998). Piles of discarded resin ducts are evidence of tree vole presence because no other species is known to forage in this manner on conifer needles. Because the nutritional value is low, tree voles must eat large quantities of conifer needles. Howell (1926) reported that a captive red tree vole ate about 4800 Douglas-fir needles over a two-day period, an average of about 100 needles/per hour. Tree voles typically bite off Douglas-fir twigs and take them back to the nest to eat. Activity outside of the nest occurs primarily at night, but tree voles are thought to feed throughout the day inside their nests (Howell 1926).

Tree voles obtain water from the vegetation they eat and from free water that occurs on conifer needles as a result of dew, rain, and fog condensation. Authors who have kept captive tree voles have commented on the importance of spraying water on Douglas-fir needles that were fed to the voles (Howell 1926, Hamilton 1962).

Reproductive biology

The reproductive potential of tree voles is low compared to that of other arvicoline rodents. Litter size is typically two or three, with litters of one or four occurring less frequently (Hamilton 1962). Gestation is about 27–29 days, and young are weaned at about 30–35 days (Hamilton 1962). Hamilton (1962) speculated that small litters, long gestation period, and slow development of nursing young are reproductive adaptations to energy-conversion difficulties inherent in a diet of conifer needles.

Population ecology

Movements and home range size

We found no published data on movements, dispersal, or home range sizes of these species.

Survival

We found no published data on longevity or survival rates of red tree voles. Red tree voles are prey to numerous avian and mammalian predators. The northern spotted owl preys on tree voles throughout the range of the voles (Barrows 1980, 1987, Forsman et al. 1984, Zabel et al. 1995). Tree voles were the second most common prey species, after northern flying squirrels, of spotted owls in Oregon (Forsman et al. 1984) and the second most common prey species, after dusky-footed woodrats (*Neotoma fuscipes*), of spotted owls in the Coast Range of Mendocino County in northwestern California (Barrows 1980). In two areas along the coast of southwestern Oregon, tree voles were the most common item in spotted owl diets, comprising 49% of the prey captured by two pairs of owls (Forsman et al. 1984:43). Red tree voles were also the most common item (38%) in a sample of 566 prey taken by spotted owls in the foothills of the Coast Range near Corvallis in 1970–1974 (Forsman et al. 1984:43). Other predators reported in the literature include great horned owls (*Bubo virginianus*; Maser 1965), northern saw-whet owls (*Aegolius acadicus*; Forsman and Maser 1970), long-eared owls (*Asio otus*; Reynolds 1970), Stellar's jays (*Cyanocitta stelleri*; Howell 1926) and common ravens (*Corvus corvax*; Thompson and Diller 2002). Published records of carnivore predation are uncommon (Alexander et al. 1995), but red tree voles were a common prey (26%) of ringtails (*Bassariscus astutus*). American martens and long-tailed weasels (*Mustela frenata*) are adept at climbing trees and are likely predators in parts of their ranges where they co-occur with tree voles.

Habitat relations and forest management

Tree voles construct characteristic nests that are primarily found in tree canopies. They usually build their own nests, but often build on top of nests previously built by squirrels, woodrats, and birds (Taylor 1915,

Howell 1926, Maser et al. 1981). Nests are thought to be occupied by individual voles or individual females with young (Howell 1926, Hayes 1996, Thompson and Diller 2002). Nests are typically roughly spherical and comprised of resin ducts, feces, fine twigs ($< \sim 4$ mm), lichen, and needles (Jewett 1920, Howell 1926, Benson and Borell 1931, Maser 1966, Gillesberg and Carey 1991). Nests vary considerably in size, from about 20 cm to 90 cm in diameter, and can be found in all parts of the canopy (Howell 1926, Benson and Borell 1931, Zentner 1966, Gillesberg and Carey 1991). Nests in young trees with small-diameter branches are more frequently positioned against the main bole of the tree, and nests in trees with larger-diameter branches are frequently positioned on branches away from the bole (Benson and Borell 1931, Zentner 1966). Ground nests have also been reported (Howell 1926), but are not thought to be common.

Tree voles are most commonly found in mesic forest habitats dominated by Douglas-fir, and they have been found where Douglas-fir occurs in association with grand fir, Sitka spruce, coast redwood (*Sequoia sempervirens*), and western hemlock. Although they are not considered highly social, tree voles typically have aggregated spatial distributions (Taylor 1915, Benson and Borell 1931, Zentner 1966, Meiselman and Doyle 1996). An association between tree vole distribution and coastal and valley fog has been noted (Hamilton 1962, Meiselman 1992, Maser 1998). Maser (1998) noted that fog along the Columbia River may allow *A. longicaudus* to extend its range farther east than would otherwise be possible.

Quantitative information on tree vole abundance comes from two types of studies: estimates of capture rates in pitfall traps on the ground, and estimates of nest frequency or densities in trees along transects. Published studies on abundance and effects of forest management are limited to retrospective studies that compared tree vole abundance among different age classes of even-age conifer (primarily Douglas-fir) forests. Capture rates of tree voles were highest in old-growth stands and lowest in young stands for each of the pitfall trap studies (Corn and Bury 1986, 1991, Raphael 1988, Gilbert and Allwine 1991, Ralph et al. 1991, Gomez and Anthony 1998, Martin and McComb 2002). Associations between tree vole capture rates and age classes were significant in Corn and Bury (1986, 1991), Gomez and Anthony (1998), and Martin and McComb (2002). Total numbers of tree voles captured in these studies were small, however, ranging from 8 to 42 voles.

Meiselman and Doyle (1996) compared nest frequencies of *A. pomo* in Mendocino County, California among old-growth (> 200 years), mature

(100–200 years), and young (<100 years) age classes. Nest frequency was significantly associated with age class, with most nests found in old-growth stands and the fewest in young stands. Thompson and Diller (2002) compared nest densities of *A. pomo* across six age classes of coastal coniferous forests on commercial timberlands in Humboldt County, California. They found a total of 185 nests and used program DISTANCE (Laake et al. 1996) to calculate nest densities. No nests were found in stands that were 10–19 years old and they concluded that, within their study area, tree voles began re-colonizing Douglas-fir stands about 20 years after clear-cutting. Median nest density varied from 0 to 6.21 active nests/ha, with an increasing trend relative to stand age: 2.1 nests/ha in stands 20–29 years old, 3.8 nests/ha in stands 30–39 years old, 4.6 nests/ha in stands 40–49 year old, 4.9 nests/ha in stands 50–59 years old, and 6.8 nests/ha in stands 60–100 years old. There were significant differences in nest densities among the six age classes, but the differences were not significant if the 10–19 years age class was excluded. Also, nest persistence was similar among age classes. Nests were located higher in larger diameter trees, and more frequently built out on branches away from the bole as stand age increased. Thompson and Diller (2002) hypothesized this may have been related to more developed branch systems that provided good support structure for the nests. They also predicted that inclement weather would affect nests in various age classes differently, and older stands would be more protected. Although results of pitfall studies indicated that tree voles were more abundant in older forests than in younger forests, anecdotal observations of tree voles nesting in young second-growth Douglas-fir are common (Taylor 1915, Jewett 1920, Howell 1926, Benson and Borell 1931, Brown 1964, Maser 1966, Vrieze 1980, Wooster 1994). Results of Thompson and Diller (2002) are consistent with these anecdotal observations and raise questions about the degree to which tree voles are associated with old-growth forests.

Voles need trees for food and as support for their nests, and it is likely that any silvicultural practice that removes significant numbers of trees would have negative effects on tree vole populations. Stand-replacement wildfires must also have negative effects on tree vole populations. Forest fragmentation is expected to influence tree vole distribution because red tree voles are thought to have limited dispersal capabilities. Martin and McComb (2002) found that *A. longicaudus* was associated with unfragmented landscapes in the central Oregon Coast Range, but this is a topic where more information is clearly needed. At least at a within-landscape scale, *A. pomo* does not apparently need continuous forest cover because it has been reported to be common along edges of Douglas-fir stands and in areas where trees are scattered or patchily distributed (Taylor 1915, Howell 1926, Benson and Borrell 1931).

Future research and information needs

Much is known about arboreal rodents in western landscapes in the Pacific Northwest, but many aspects of basic and applied biology remain unknown. The role of predation or competition in population dynamics or geographical variation in abundance as well as population responses to logging, especially reproduction, dispersal, and survival, are areas of information needs. Understanding how species respond to thinning or selective harvests of mature stands and intermediate stand management of second-growth stands following clear-cut logging will be fundamental to sustaining arboreal rodent populations.

Abert's squirrel is an indicator species for the management of the ponderosa pine ecosystem in the southwestern United States. With a strong emphasis on restoration of ponderosa pine forests, a better understanding of the influence of landscape patterns on Abert's squirrel populations is needed. Also, there is limited knowledge of squirrel population dynamics in different habitats, and the role that seasonal food availability plays remains unclear. Preliminary information suggests the existence of thresholds and a non-linear response of populations to diminishing proportions of old forest habitat, yet the influence of patch size and spatial arrangement (e.g., "single-large versus several-small"; Simberloff 1988) remains unknown. There is need to experimentally evaluate an array of forest restoration prescriptions applied at multiple scales to determine which prescription(s) increase(s) Abert's squirrel population densities, recruitment, and survival. Specific information needs for northern flying squirrels include: (1) reproductive biology among forest types; (2) basic biology and habitat associations in coniferous forests of the Rocky Mountain region and southeastern Alaska; (3) influences of forest-management strategies on demography and habitat associations throughout their range; (4) greater emphasis on manipulative, experimental studies; and (5) food preference studies to better understand the importance of different types of fungi and lichens in the nutritional ecology of flying squirrels.

Patterns of tree vole distribution and abundance are poorly documented and information needs are great. Priority research needs include studies to evaluate relationships between forest fragmentation and tree vole distribution, and to evaluate effects of different levels of retention for variable-retention harvesting systems (Franklin et al. 1997) on tree vole populations. Research is also needed on the effects of different spatial patterns of retention. Because tree voles have aggregated spatial distributions, it is possible that aggregated retention patterns would be more conducive to tree vole persistence or re-colonization than would dispersed retention patterns, but this hypothesis needs to be tested.

Summary

An understanding of ecological factors that influence the abundance of arboreal rodents in western coniferous forests is fundamental to establishing science-based policy and identifying effective conservation measures. During the last two decades, considerable research has focused on the natural history and ecology of this group of forest habitat specialists in coastal and interior coniferous forests of the Pacific Northwest and throughout the Rocky Mountains. Similar to many ecological assemblages, arboreal rodents share several ecological attributes, and often affect one another's use of resources and local distribution. This chapter summarizes available ecological information with a goal of gaining additional insights into habitat association patterns and a better understanding of how forest management affects populations of these species. We focused this review on species that have received attention because of their reputed association with late-seral forest habitat and population viability concerns, their importance as prey for other species, and the importance of the functional roles they have in the community dynamics of western coniferous forests.

Forest-management practices that replace mature, multi-layered coniferous forests with young, structurally homogeneous, largely monotypic stands may adversely affect arboreal rodent populations. Also, abundances of arboreal rodents decline with increasing overstory removal. Prescriptions such as selective harvest modify only a portion of the overstory and retain many important structural and biological elements in logged stands. Such managed stands are diverse and structurally heterogeneous forests in contrast to even-aged plantations that are structurally and biologically homogeneous. Furthermore, silvicultural prescriptions that contribute to the development of quality habitat for arboreal rodents will likely provide a wide array of ecological values (e.g., biological diversity) in managed forests. Prescriptions that likely will contribute important habitat elements in managed forests include retaining large snags for cavities, retaining large green trees to ensure production of cones, planting or encouraging mixed species, and using silviculture to create vertical structure and horizontal heterogeneity. Conservation strategies developed within an ecological community framework will likely have higher success than those that emphasize single species.

Acknowledgments

We acknowledge the United States Department of Agriculture, Forest Service, United States Department of Interior, Geological Survey, and Arizona Department of Game and Fish, for supporting the preparation of this manuscript. K. Foresman, E. Forsman, J. Lehmkuhl, T. Hanley, and M. Steele reviewed a draft of this manuscript and offered comments that ultimately improved its quality. We appreciate the editorial review by M. J. Bergener.

Literature cited

Alexander, L.F., B.J. Verts, and T.P. Farrell. 1995. Diet of ringtails (*Bassariscus astutus*) in Oregon. *Northwestern Naturalist* **76**:97–101.

Allred, W.S., W.S. Gaud, and J.S. States. 1994. Effects of herbivory by Abert squirrels (*Sciurus aberti*) on cone crops of ponderosa pine. *Journal of Mammalogy* **75**:700–703.

Andren, H. 1994. Effects of habitat fragmentation on birds and mammals in landscapes with different proportions of suitable habitat: a review. *Oikos* **71**:355–366.

Anthony, R.G., E.D. Forsman, G.A. Green, G. Witmer, and S.K. Nelson. 1987. Small mammal populations in riparian zones of different-aged coniferous forests. *Murrelet* **68**:94–102.

Arbogast, B.R., R.A. Browne, P.D. Weigl. 2001. Evolutionary genetics and Pleistocene biogeography of North American tree squirrels (*Tamiasciurus*). *Journal of Mammalogy* **82**:302–319.

Aubry, K.B., M.J. Crites, and S.D. West. 1991. Regional patterns of small mammal abundance and community composition in Oregon and Washington. Pages 285–294 *in* L.F. Ruggiero, K.B. Aubry, A.B. Carey, and M.H. Huff, technical coordinators. Wildlife and vegetation of unmanaged Douglas-fir forests. USDA Forest Service General Technical Report **PNW-285**, Portland, Oregon, USA.

Aubry, K.B., J.P. Hayes, B.L. Biswell, and B.G. Marcot. 2003. The ecological role of tree-dwelling mammals in western coniferous forests. Pages 405–443 *in* C.J. Zabel and R.G. Anthony, editors. *Mammal Community Dynamics in Western Coniferous Forests: Management and Conservation*. Cambridge University Press, Cambridge, UK.

Austin, W.J. 1990. The foraging ecology of Abert squirrels. Dissertation, Northern Arizona University, Flagstaff, Arizona, USA.

Bakker, V.J. and K. Hastings. 2002. Den trees used by northern flying squirrels (*Glaucomys sabrinus*) in southeastern Alaska. *Canadian Journal of Zoology* **80**:1623–1633.

Barrows, C. 1980. Feeding ecology of the spotted owl in California. *Journal of Raptor Research* 14:73–78.

Barrows, C.W. 1987. Diet shifts in breeding and nonbreeding spotted owls. *Journal of Raptor Research* 21:95–97.

Becker, C.D. 1993. Environmental cues of estrus in the North American red squirrel (*Tamiasciurus hudsonicus* Bangs). *Canadian Journal of Zoology* 71:1326–1333.

Beier, P. and J.E. Drennan. 1997. Forest structure and prey abundance in foraging areas of northern goshawks. *Ecological Applications* 7:564–571.

Ben-David, M., R.W. Flynn, and D.M. Schell. 1997. Annual and seasonal changes in diets of martens: evidence from stable isotope analysis. *Oecologia* 111:280–291.

Benson, S.B. and A.E. Borell. 1931. Notes on the life history of the red tree mouse, *Phenacomys longicaudus*. *Journal of Mammalogy* 12:226–233.

Berteaux, D. and S. Boutin. 2000. Breeding dispersal in female North American red squirrels. *Ecology* 81:1311–1326.

Booth, E.S. 1946. Notes on the life history of the flying squirrel. *Journal of Mammalogy* 27:28–30.

Bowers, M.A. and S.F. Matter. 1997. Landscape ecology of mammals: relationships between density and patch size. *Journal of Mammalogy* 78:999–1013.

Brink, C.H. and F.C. Dean. 1966. Spruce seed as a food of red squirrels and flying squirrels in interior Alaska. *Journal of Wildlife Management* 30:503–511.

Brown, D.E. 1984. *Arizona's Tree Squirrels*. Arizona Game and Fish Department, Phoenix, Arizona, USA.

Brown, L.N. 1964. Breeding records and notes on *Phenacomys silvicola* in Oregon. *Journal of Mammalogy* 45:647–648.

Buchanan, J.B., R.W. Lundquist, and K.B. Aubry. 1990. Winter populations of Douglas squirrels in different-aged Douglas-fir forests. *Journal of Wildlife Management* 54:577–581.

Buskirk, S.W. and L.F. Ruggiero. 1994. American marten. Pages 7–37 *in* L.F. Ruggiero, K.B. Aubry, S.W. Buskirk, L.J. Lyon, and W.J. Zielinski, technical editors. USDA Forest Service General Technical Report **RM-GTR-254**, Fort Collins, Colorado, USA.

Carey, A.B. 1989. Wildlife associated with old-growth forests in the Pacific Northwest. *Natural Areas Journal* 9:151–162.

Carey, A.B. 1991. The biology of arboreal rodents in Douglas-fir forests. USDA Forest Service General Technical Report **PNW-276**, Portland, Oregon, USA.

Carey, A.B. 1995. Sciurids in Pacific Northwest managed and old-growth forests. *Ecological Applications* 5:648–666

Carey, A.B. 2000. Effects of new forest management strategies on squirrel populations. *Ecological Applications* 10:248–257.

Carey, A.B. 2002. Response of northern flying squirrels to supplementary dens. *Wildlife Society Bulletin* 30:547–556.

Carey, A.B., B.L. Biswell, and J.W. Witt. 1991. Methods for measuring populations of arboreal rodents. USDA Forest Service General Technical Report **PNW-273**, Portland, Oregon, USA.

Carey, A.B., S.P. Horton, and B.L. Biswell. 1992. Northern spotted owls: influence of prey base and landscape character. *Ecological Monographs* 62:223–250.

Carey, A.B., T. Wilson, C.C. Maguire, and B.L. Biswell. 1997. Dens of northern flying squirrels in the Pacific Northwest. *Journal of Wildlife Management* 61:684–699.

Carey, A.B., J. Kershner, B. Biswell, and L. Dominguez De Toledo. 1999. Ecological scale and forest development: squirrels, dietary fungi, and vascular plants in managed and unmanaged forests. *Wildlife Monographs* **142**:1–69.

Cazares, E., D.L. Luoma, M.P. Amaranthus, C.L. Chambers, and J.F. Lehmkuhl. 1999. Interaction of fungal sporocarp production with small mammal abundance and diet in Douglas-fir stands of the southern Cascade Range. *Northwest Science* **73**:64–76.

Claridge, A.W. and S.J. Cork. 1994. Nutritional value of hypogeal fungal sporocarps for the long-nosed potoroo (*Potorous tridactylus*), a forest-dwelling mycophagous marsupial. *Australian Journal of Zoology* **42**:701–710.

Claridge, A.W., J.M. Trappe, S.J. Cork, and D.L. Claridge. 1999. Mycophagy by small mammals in the coniferous forests of North America: nutritional value of sporocarps of *Rhizopogon vinicolor*, a common hypogeous fungus. *Journal of Comparative Physiology. B: Biochemical, Systemic, and Environmental Physiology* **169**:172–178.

Clifton, P.L. 1960. Biology and life history of the dusky tree mouse *Phenacomys silvicola* (Howell). Thesis, Walla Walla College, Walla Walla, Washington, USA.

Connor, P.F. 1960. The small mammals of Otsego and Schoharie Counties, New York. New York State Museum and Science Service Bulletin Number **382**, The University of the State of New York, Albany, New York, USA.

Cork, S.J. and G.J. Kenagy. 1989. Nutritional value of hypogeous fungus for a forest-dwelling ground squirrel. *Ecology* **70**:577–586.

Corn, P.S. and R.B. Bury. 1986. Habitat use and terrestrial activity by red tree voles (*Arborimus longicaudus*) in Oregon. *Journal of Mammalogy* **67**:404–406.

Corn, P.S. and R.B. Bury. 1988. Distribution of the voles *Arborimus longicaudus* and *Phenacomys intermedius* in the central Oregon Cascades. *Journal of Mammalogy* **69**:427–429.

Corn, P.S. and R.B. Bury. 1991. Small mammal communities in the Oregon Coast Range. Pages 241–254 *in* L.F. Ruggiero, K.B. Aubry, A.B. Carey, and M.H. Huff, technical coordinators. Wildlife and vegetation of unmanaged Douglas-fir forests. USDA Forest Service General Technical Report **PNW-285**, Portland, Oregon, USA.

Cotton, C.L. and K.L. Parker. 2000*a*. Winter activity patterns of northern flying squirrels in sub-boreal forests. *Canadian Journal of Zoology* **78**:1896–1901.

Cotton, C.L. and K.L. Parker. 2000*b*. Winter habitat and nest trees used by northern flying squirrels in sub-boreal forests. *Journal of Mammalogy* **81**:1071–1086.

Cowan, I.M. 1936. Nesting habits of the flying squirrel *Glaucomys sabrinus*. *Journal of Mammalogy* **17**:58–60.

Cowan, I.M. and C.J. Guiguet. 1960. *The Mammals of British Columbia*. Second edition. Don McDiarmid, Victoria, British Columbia, Canada.

Davis D.W. and J.A. Sealander. 1971. Sex ratio and age structure in two red squirrel populations in northern Saskatchewan. *Canadian Field-Naturalist* **85**:303–308.

Davis, R. and D.E. Brown. 1989. Role of post-Pleistocene dispersal in determining the modern distribution of Abert's squirrel. *Great Basin Naturalist* **49**:425–434.

Dodd, N.L. 2003. Landscape-scale habitat relationships to tassel-eared squirrel population dynamics in north-central Arizona. Arizona Game and Fish Department, Research Branch Technical Guidance Bulletin **6**, Arizona, USA.

Dodd, N.L and S.L. Adams. 1989. Integrating wildlife needs into national forest timber sale planning: a state agency perspective. Pages 131–140 *in* A. Tecle, W.W.

Covington, and R.H. Hamre, technical coordinators. Proceedings of multiresource management of ponderosa pine forests. USDA Forest Service General Technical Report **RM-185**, Fort Collins, Colorado, USA.

Dodd, N.L., S.S. Rosenstock, C.R. Miller, and R.E. Schweinsburg. 1998. Tassel-eared squirrel population dynamics in Arizona: index techniques and relationships to habitat condition. Arizona Game and Fish Department Technical Report **27**, Phoenix, Arizona, USA.

Dodd N.L., J.S. States, S.S. Rosenstock. 2003. Tassel-eared squirrel population, habitat condition, and dietary relationships in northcentral Arizona. *Journal of Wildlife Management* **67**: in press.

Dolbeer, R.A. 1973. Reproduction in the red squirrel (*Tamiasciurus hudsonicus*) in Colorado. *Journal of Mammalogy* **54**:536–540.

Elliott, P.F. 1974. Evolutionary responses of plants to seed-eaters: pine squirrel predation on lodgepole pine. *Evolution* **28**:221–231.

Elliott, P.F. 1988. Foraging behavior of a central-place forager: field tests of theoretical predictions. *American Naturalist* **131**:159–174.

Farentinos, R.C. 1972*a*. Observations on the ecology of the tassel-eared squirrel. *Journal of Wildlife Management* **36**:1234–1239.

Farentinos, R.C. 1972*b*. Social dominance and mating activity in the tassel-eared squirrel (*Sciurus aberti ferrus*). *Animal Behaviour* **20**:316–326.

Farentinos, R.C. 1979. Seasonal changes in home range size of tassel-eared squirrels (*Sciurus aberti*). *Southwestern Naturalist* **24**:49–62.

Farentinos, R.C., P.J. Capretta, R.E. Kepner, and V.M. Littlefield. 1981. Selective herbivory in tassel-eared squirrels: role of monoterpenes in ponderosa pine chosen as feeding trees. *Science* **213**:1273–1275.

Ferron. J., J.P. Ouellet, and Y. Lemay. 1986. Spring and summer time budgets and feeding behaviour of the red squirrel (*Tamiasciurus hudsonicus*). *Canadian Journal of Zoology* **64**:385–391.

Ffolliot, P.F. and D.R. Patton. 1978. Abert squirrel use of ponderosa pine as feed trees. USDA Forest Service Research Note RM-**362**, Fort Collins, Colorado, USA.

Finley, R.B., Jr. 1969. Cone caches and middens of *Tamiasciurus* in the Rocky Mountain Region. *Kansas Museum of Natural History Miscellaneous Publications* **51**:233–273.

Fisch, G.G. and E.J. Dimock, II. 1978. Shoot clipping by Douglas squirrels in regenerating Douglas fir. *Journal of Wildlife Management* **42**:415–418.

Flyger, V. and J.E. Gates. 1982. Pine squirrels: *Tamiasciurus hudsonicus* and *T. douglasii*. Pages 230–238 *in* J.A. Chapman and G.A. Feldhamer, editors. *Wild Mammals of North America*. Johns Hopkins University Press, Baltimore, Maryland, USA.

Foiles, M.W. and J.D. Curtis. 1965. Natural regeneration of ponderosa pine on scarified group cuttings in central Idaho. *Journal of Forestry* **63**:530–535.

Forsman, E. and C. Maser. 1970. Saw-whet owl preys on red tree mice. *The Murrelet* **51**:10.

Forsman, E.D., E.C. Meslow, and H.M. Wight. 1984. Distribution and biology of the spotted owl in Oregon. *Wildlife Monographs* **87**:1–64.

Forsman, E.D., A. Ivy, D.A. Otto, J.C. Lewis, S.G. Sovern, K.J. Maurice, and T. Kaminski. 1994. Reproductive chronology of the northern flying squirrel on the Olympic Peninsula, Washington. *Northwest Science* **68**:273–276.

Forsman, E.D., A. Ivy, D.A. Otto, S.G. Sovern, M. Taylor, D.W. Hays, H. Allen, S.L. Roberts, and D.E. Seaman 2001. Spatial and temporal variation in diets of spotted owls in Washington. *Journal of Raptor Research* **35**:141–150.

Foster, W.L. and J. Tate, Jr. 1966. The activities and coactions of animals at sapsucker trees. *Living Bird* **5**:87–113.

Franklin, J.F., K. Cromack, Jr., W. Denison, A. McKee, C. Maser, J. Sedell, F. Swanson, and G. Juday. 1981. Ecological characteristics of old-growth Douglas-fir forests. USDA Forest Service General Technical Report **PNW-118**. Portland, Oregon, USA.

Franklin, J.F., D.R. Berg, D.A. Thornburgh, and J.C. Tappeiner. 1997. Alternative silvicultural approaches to timber harvesting: variable retention systems. Pages 111–139 *in* K.A. Kohm and J.F. Franklin, editors. *Creating a Forestry for the 21st Century*. Island Press, Washington DC, USA.

Froehlich, G.F. 1990. Habitat use and life history of the Mount Graham red squirrel. Thesis, University of Arizona, Tempe, Arizona, USA.

Gaud, W.S., W.S. Allred, and J.S. States. 1993. Tree selection by tassel-eared squirrels of the ponderosa pine forests of the Colorado Plateau. Pages 56–63 *in* P.G. Rowlands, C.van Piper III, and M.K. Sogge, editors. *Proceedings of the First Biennial Conference on Research in Colorado Plateau National Parks*. NPS/NRNAU/NRTP-**93/10**. Flagstaff, Arizona, USA.

Gilbert, F.F. and R. Allwine. 1991. Small mammal communities in the Oregon Cascade Range. Pages 257–267 *in* L.F. Ruggiero, K.B. Aubry, A.B. Carey, and M.H. Huff, technical coordinators. Wildlife and vegetation of unmanaged Douglas-fir forests. USDA Forest Service General Technical Report **PNW-285**, Portland, Oregon, USA.

Gillesberg, A.M. and A.B. Carey. 1991. Arboreal nests of *Phenacomys longicaudus* in Oregon. *Journal of Mammalogy* **72**:784–787.

Gomez, D.M. and R.G. Anthony. 1998. Small mammal abundance in riparian and upland areas of five seral stages in western Oregon. *Northwest Science* **72**:293–302.

Gurnell, J. 1984. Home range, territoriality, caching behaviour and food supply of the red squirrel (*Tamiasciurus hudsonicus fremonti*) in a subalpine lodgepole pine forest. *Animal Behaviour* **32**:119–131.

Gurnell, J.C. 1987. *The Natural History of Squirrels*. Facts On File Publications, New York, New York, USA.

Hall, D.S. 1991. Diet of the northern flying squirrel at Sagehen Creek, California. *Journal of Mammalogy* **72**:615–617.

Hall, J.G. 1981. A field study of the Kaibab squirrel in the Grand Canyon National Park. *Wildlife Monographs* **75**:1–54.

Hallett, J.G., M.A. O'Connell, and C.C. Maguire. 2003. Ecological relationships of small mammals in western coniferous forests. Pages 120–156 *in* C.J. Zabel and R.G. Anthony, editors. *Mammal Community Dynamics. Management and Conservation in the Coniferous Forests of Western North America*. Cambridge University Press, Cambridge, UK.

Hamer, T.E., D.L. Hays, C.M. Senger, and E.D. Forsman. 2001. Diets of northern barred owls and northern spotted owls in an area of sympatry. *Journal of Raptor Research* **35**:221–227.

Hamilton, W.J., III. 1962. Reproductive adaptations of the red tree mouse. *Journal of Mammalogy* **43**:486–504.

Hatt, R.T. 1943. The pine squirrel in Colorado. *Journal of Mammalogy* **24**:311–345.

Hayes, J.P. 1996. *Arborimus longicaudus. Mammalian Species* **532**:1–5.

Hoffmeister, D. F. and V.E. Diersing. 1978. Review of the tassel-eared squirrels of the subgenus *Otosciurus. Journal of Mammalogy* **59**:402–413.

Howell, A.B. 1926. Voles of the genus Phenacomys. II. Life history of the red tree mouse *Phenacomys longicaudus*. *North American Fauna* **48**:39–65.

Huff, M.H., R.S. Holthausen, and K.B. Aubry. 1992. Habitat management for red tree voles in Douglas-fir forests. Pages 1–16 *in* M.H. Huff, R.S. Holthausen, and K.B. Aubry, technical coordinators. Biology and management of old-growth forests. USDA Forest Service General Technical Report PNW-GTR-**302**, Portland, Oregon, USA.

Humphries, M.M. and S. Boutin. 2000. The determinants of optimal litter size in free ranging red squirrels. *Ecology* **81**:2867–2877.

Hutton, K.A., J.L. Koprowski, V.L. Greer, M.I. Alanen, C.A. Schauffert, and P.J. Young. 2003. Use of mixed conifer and spruce-fir forests by an introduced population of Abert's squirrels (*Sciurus aberti*). *Southwestern Naturalist* **48**(2): in press.

Jackson, H.H.T. 1961. *Mammals of Wisconsin*. University of Wisconsin Press, Madison, Wisconsin, USA.

Jewett, S.G. 1920. Notes on two species of *Phenacomys* in Oregon. *Journal of Mammalogy* **1**:165–168.

Johnson, M.L. 1973. Characters of the heather vole *Phenacomys* and the red tree vole, *Arborimus*. *Journal of Mammalogy* **54**:239–244.

Johnson, M.L. and S.B. George. 1991. Species limits within the *Arborimus longicaudus* species-complex (Mammalia: Rodentia) with a description of a new species from California. *Natural History Museum of Los Angeles County, Contributions in Science* **429**:1–16.

Johnson, M.L. and C. Maser. 1982. Generic relationships of *Phenacomys albipes*. *Northwest Science* **56**:17–19.

Johnstone, D.W. 1981. Precommercial thinning speeds growth and development of lodgepole pine: 25 year results. Northern Forest Research Centre of Canada Information Report NOR-X-**236**. Edmonton, Alberta, Canada.

Keith, J.O. 1965. The Abert squirrel and its dependence on ponderosa pine. *Ecology* **46**:150–163.

Kemp, G.A. and L.B. Keith. 1970. Dynamics and regulation of red squirrel (*Tamiasciurus hudsonicus*) populations. *Ecology* **51**:763–779.

Klenner, W. 1991. Red squirrel population dynamics. II. Settlement patterns and the response to removals. *Journal of Animal Ecology*. **60**:979–993.

Klenner, W. and C.J. Krebs. 1991. Red squirrel population dynamics. I. The effect of supplemental food on demography. *Journal of Animal Ecology* **60**:961–978.

Koford, R.R. 1982. Mating system of a territorial tree squirrel (*Tamiasciurus douglasii*) in California. *Journal of Mammalogy* **63**:274–283.

Krohne, D.T. 1997. Dynamics of metapopulations of small mammals. *Journal of Mammalogy* **78**:1014–1026.

Laake, J.L., S.T. Buckland, D.R. Anderson, and K.P. Burnham. 1996. *DISTANCE 2.2*. Colorado Cooperative Fish and Wildlife Research Unit, Colorado State University, Fort Collins, Colorado, USA.

Lamb, T., T.R. Jones, and P.J. Wettstein. 1997. Evolutionary genetics and phylogeography of tassel-eared squirrels (*Sciurus aberti*). *Journal of Mammalogy* **78**:117–133.

Larsen, K.W. and S. Boutin. 1994. Movements, survival, and settlement of red squirrel (*Tamiasciurus hudsonicus*) offspring. *Ecology* **75**:214–223.

Larsen, K.W. and S. Boutin. 1998. Sex-unbiased philopatry in the North American red squirrel: (*Tamiasciurus hudsonicus*). Pages 21–32 *in* M.A. Steele, J.F. Merritt, and

D.A. Zegers, editors. *Ecology and Evolutionary Biology of Tree Squirrels*. Virginia Museum of Natural History, Special Publication Number **6**, Martinsville, Virginia, USA.

Larson, M.M. and G.H. Schubert. 1970. Cone crops of ponderosa pine in central Arizona, including influence of Abert's squirrels. USDA Forest Service Research Paper RM-**58**, Fort Collins, Colorado, USA.

Laurance, W.F. and T.D. Reynolds. 1984. Winter food preferences of captive-reared northern flying squirrels. *Murrelet* **65**:20–22.

Lema, M.F. 2001. Dynamics of Abert's squirrel populations: home range, seasonal movements, survivorship, habitat use, and sociality. Thesis, Northern Arizona University, Flagstaff, Arizona, USA.

Lewis, S.B. 2001. Breeding season diet of northern goshawks in Southeast Alaska. Thesis, Boise State University, Boise, Idaho, USA.

Lidicker, W.Z., Jr. 1988. Solving the enigma of microtine "cycles". *Journal of Mammalogy* **69**:225–235.

Luoma, D.L., J.M. Trappe, A.W. Claridge, K.M. Jocobs, and E. Cazares. 2003. Relationships among fungi and small mammals in forested ecosystems. Pages 343–373 *in* C.J. Zabel and R.G. Anthony, editors. *Mammal Community Dynamics. Management and Conservation in the Coniferous Forests of Western North America*. Cambridge University Press, Cambridge, UK.

MacDonald, S.O. and J.A. Cook. 1996. The land mammal fauna of Southeast Alaska. *Canadian Field-Naturalist* **110**:571–598.

Martin, K.J. and R.G. Anthony. 1999. Movements of northern flying squirrels in different-aged forest stands of western Oregon. *Journal of Wildlife Management* **63**:291–297.

Martin, K.J. and W.C. McComb. 2002. Small mammal habitat associations at patch and landscape scales in western Oregon. *Forest Science* **48**:255–266.

Maser, C. 1965. Notes on the contents of owl pellets found in Oregon. *The Murrelet* **46**:44.

Maser, C. 1966. Life histories and ecology of *Phenacomys albipes, Phenacomys longicaudus, Phenacomys silvicola*. Thesis, Oregon State University, Corvallis, Oregon, USA.

Maser, C. 1998. *Mammals of the Pacific Northwest from the Coast to the High Cascades*. Oregon State University Press, Corvallis, Oregon, USA.

Maser, C. and Z. Maser. 1988. Interactions among squirrels, mycorrhizal fungi, and coniferous forests in Oregon. *Great Basin Naturalist* **48**:358–369.

Maser, C., B.R. Mate, J.F. Franklin, and C.T. Dyrness. 1981. Natural History of Oregon Coast Mammals. USDA Forest Service and USDI Bureau of Land Management. USDA Forest Service General Technical Report PNW-**133**. Portland, Oregon, USA.

Maser, C., Z. Maser, J.W. Witt, and G. Hunt. 1986. The northern flying squirrel: a mycophagist in southwestern Oregon. *Canadian Journal of Zoology* **64**:2086–2089.

Maser, Z., C. Maser, and J.M. Trappe. 1985. Food habits of the northern flying squirrel (*Glaucomys sabrinus*) in Oregon. *Canadian Journal of Zoology* **63**:1084–1088.

McKee, E. D. 1941. Distribution of tassel-eared squirrels. *Plateau* **14**:12–20.

McKeever, S. 1960. Food of the northern flying squirrel in northeastern California. *Journal of Mammalogy* **41**:270–271.

McKeever, S. 1964. Food habits of the pine squirrel in northeastern California. *Journal of Wildlife Management* **28**:402–404.

Medin, D.E. 1986. The impact of logging on red squirrels in an Idaho conifer forest. *Western Journal of Applied Forestry* **1**:73–76.

Meiselman, N. 1992. Nest-site characteristics of red tree voles in Douglas-fir forests of northern California. Thesis, Humboldt State University, Arcata, California, USA.

Meiselman, N. and A.T. Doyle. 1996. Habitat and microhabitat use by the red tree vole (*Phenacomys longicaudus*). *American Midland Naturalist* **135**:33–42.

Meslow, E.C., C. Maser, and J. Verner. 1981. Old-growth forests as wildlife habitat. *Transactions of the Forty-sixth North American Wildlife and Natural Resources Conference* **46**:329–335.

Millar, J.S. 1970. The breeding season and reproductive cycle of the western red squirrel. *Canadian Journal of Zoology* **48**:471–473.

Morrison, M.L. and R.G. Anthony. 1989. Habitat use by small mammals on early-growth clear-cuttings in western Oregon. *Canadian Journal of Zoology* **67**:805–811.

Mowrey, R.A. and J.C. Zasada. 1984. Den tree use and movements of northern flying squirrels in interior Alaska and implications for forest management. Pages 351–356 *in* W.R. Meehan, T.R. Merrell, Jr., and T.A. Hanley, editors. *Fish and Wildlife Relationships in Old-growth Forests*. Proceedings of a symposium held in Juneau, Alaska, April 12–15, 1982. American Institute of Fishery Research Biologists.

Murray, M.A. 1995. Biochemical systematics of the genus *Arborimus*. Thesis, Humboldt State University, Arcata, California, USA.

Nash, D.J. and R.N. Seaman. 1977. *Sciurus aberti*. *Mammalian Species* **80**:1–5.

Nowacki, G.J. and M.G. Kramer. 1998. The effects of wind disturbance on temperate rain forest structure and dynamics of Southeast Alaska. USDA Forest Service General Technical Report PNW-GTR-**421**. Portland, Oregon, USA.

Otis, D.L., K.P. Burnham, G.C. White, and D.R. Anderson. 1978. Statistical inference from capture data on closed animal populations. *Wildlife Monographs* **62**:1–135.

Patton, D.R. 1974. Estimating food consumption from twigs clipped by the Abert squirrel. USDA Forest Service Research Note RM-**272**, Fort Collins, Colorado, USA.

Patton, D.R. 1975*a*. Abert squirrel cover requirements in southwestern ponderosa pine. USDA Forest Service Research Paper RM-**145**, Fort Collins, Colorado, USA.

Patton, D.R. 1975*b*. A diversity index for quantifying habitat "edge". *Wildlife Society Bulletin* **3**:171–173.

Patton, D.R. 1977. Managing southwestern ponderosa pine for the Abert squirrel. *Journal of Forestry* **75**:264–267.

Patton, D.R. 1984. A model to evaluate Abert squirrel habitat in uneven-aged ponderosa pine. *Wildlife Society Bulletin* **12**:408–413.

Patton, D.R. 1992. *Wildlife Habitat Relationships in Forested Ecosystems*. Timber Press, Inc., Portland, Oregon, USA.

Patton, D.R. and J.R. Vahle. 1986. Cache and nest characteristics of the red squirrel in an Arizona mixed-conifer forest. *Western Journal of Applied Forestry* **1**:48–51.

Patton, D.R., R.L. Wadleigh, and H.G. Hudak. 1985. The effects of timber harvest on the Kaibab squirrel. *Journal of Wildlife Management* **49**:14–19.

Pauls, R.W. 1978. Behavioral strategies relevant to the energy economy of the red squirrel (*Tamiasciurus hudsonicus*). *Canadian Journal of Zoology* **56**:1519–1525.

Pederson, J.C. and B.L. Welch. 1985. Comparison of ponderosa pine as feed and non-feed trees for Abert squirrels. *Journal of Chemical Ecology* **11**:149–157.

Pederson, J.C., R.N. Haysenyager, and A.W. Heggen. 1976. Habitat requirements of the Abert squirrel (*Sciurus aberti navajo*) on the Monticello District, Manti-La Sal

National Forest. Utah State Division of Wildlife Resources Publication **76**–**9**, Salt Lake City, Utah, USA.

Pederson, J.C., R.C. Farentinos, and V.M. Littlefield. 1987. Effects of logging on habitat quality and feeding patterns of Abert squirrels. *Great Basin Naturalist* **47**:252–258.

Pogany, G.C. and W.S. Allred. 1995. Abert's squirrels of the Colorado Plateau: their reproductive cycle. Pages 293–305 *in* van Piper III, C., editor. *Proceedings of the Second Biennial Conference on Research in Colorado Plateau National Parks*. NPS/NRNAU/NRTP-**95**/**11**. Flagstaff, Arizona, USA.

Pollock, K.H. 1982. A capture-recapture design robust to unequal probability of capture. *Journal of Wildlife Management* **46**:752–757.

Pollock, K.H., J.D. Nichols, C. Brownie, and J.E. Hines. 1990. Statistical inference for capture-recapture experiments. *Wildlife Monographs* **107**:1–97.

Price, K., K. Broughton, S. Boutin, and A.R.E. Sinclair. 1986. Territory size and ownership in red squirrels: response to removals. *Canadian Journal of Zoology* **64**:1144–1147.

Pyare, S. and W.S. Longland. 2001. Patterns of ectomycorrhizal-fungi consumption by small mammals in remnant old-growth forests of the Sierra Nevada. *Journal of Mammalogy* **82**:681–689.

Pyare, S. and W.F. Longland. 2002. Interrelationships among northern flying squirrels, truffles, and microhabitat structure in Sierra Nevada old-growth habitat. *Canadian Journal of Forestry Research* **32**: 1016–1024.

Pyare, S., W.P. Smith, J.V. Nichols, and J.A. Cook. 2002. Diets of northern flying squirrels, *Glaucomys sabrinus*, in southeast Alaska. *Canadian Field-Naturalist* **116**:98–103.

Ralph, C.J., P.W.C. Paton, and C.A. Taylor. 1991. Habitat associations of breeding birds and small mammals in Douglas-fir stands in northwestern California and southwestern Oregon. Pages 379–393 *in* L.F. Ruggiero, K.B. Aubry, A.B. Carey, and M.H. Huff, technical coordinators. Wildlife and vegetation of unmanaged Douglas-fir forests. USDA Forest Service General Technical Report **PNW-GTR-285**. Portland, Oregon, USA.

Ransome, D.B. and T.P. Sullivan. 1997. Food limitation and habitat preference of *Glaucomys sabrinus* and *Tamiasciurus hudsonicus*. *Journal of Mammalogy* **78**:538–549.

Raphael, M.G. 1984. Late fall breeding of the northern flying squirrel, *Glaucomys sabrinus*. *Journal of Mammalogy* **65**:138–139.

Raphael, M.G. 1988. Long-term trends in abundance of amphibians, reptiles, and mammals in Douglas-fir forests of northwestern California. Pages 23–31 *in* R.C. Szaro, K.E. Severson, and D.R. Patton, technical coordinators. Management of amphibians, reptiles, and small mammals in North America. Proceedings of the Symposium. USDA Forest Service General Technical Report **RM-166**, Fort Collins, CO, USA.

Ratcliff, T.D., D.R. Patton, and P.F. Ffolliott. 1975. Ponderosa pine basal area and the Kaibab squirrel. *Journal of Forestry* **73**:284–286.

Reynolds, R.T. 1970. Nest observations of the long-eared owl (*Asio otus*) in Benton County, Oregon, with notes on their food habits. *Murrelet* **51**:8–9.

Reynolds, R.T., R.T. Graham, M.H. Reiser, R.L. Bassett, P.L. Kennedy, D.A. Boyce, Jr., G. Goodwin, R. Smith, and E.L. Fisher. 1992. Management recommendations for the northern goshawk in the southwestern United States. USDA Forest Service General Technical Report RM-**217**, Fort Collins, Colorado, USA.

Rosenberg, D.K. and R.G. Anthony. 1992. Characteristics of northern flying squirrel populations in young, second- and old-growth forests in western Oregon. *Canadian Journal of Zoology* **70**:161–166.

Rosentreter, R., G.D. Hayward, and M. Wicklow-Howard. 1997. Northern flying squirrel seasonal food habits in the interior conifer forests of central Idaho, USA. *Northwest Science* **71**:97–102.

Rothwell, R. 1979. Nest sites of red squirrels (*Tamiasciurus hudsonicus*) in the Laramie Range of southeastern Wyoming. *Journal of Mammalogy* **60**:404–405.

Ruggiero, L.F., L.L.C. Jones, and K.B. Aubry. 1991. Plant and animal habitat associations in Douglas-fir forests of the Pacific Northwest: an overview. Pages 447–462 *in* L.F. Ruggiero, K.B. Aubry, A.B. Carey, and M.H. Huff, technical coordinators. Wildlife and vegetation of unmanaged Douglas-fir forests. USDA Forest Service General Technical Report PNW-**285**, Portland, Oregon, USA.

Rusch, D.A. and W.G. Reeder. 1978. Population ecology of Alberta red squirrels. *Ecology* **59**:400–420.

Sanders, S.D. 1983. Foraging by Douglas tree squirrels (*Tamiasciurus douglasii*: Rodentia) for conifer seeds and fungi. Dissertation, University of California, Davis, California, USA.

Schauffert, C.A, J.L. Koprowski, V.L. Greer, M.I. Alanen, K.A. Hutton, and P.J. Young. 2002. Interactions between predators and Mt. Graham red squirrels (*Tamiasciurus hudsonicus grahamensis*). *Southwestern Naturalist* **47**:498–501.

Schmidt, F.J.W. 1931. Mammals of western Clark County, Wisconsin. *Journal of Mammalogy* **12**:99–117.

Schopmeyer, C.S., technical coordinator. 1974. Seeds of woody plants in the United States. *USDA Forest Service Agricultural Handbook* No. **450**. Washington DC, USA.

Searing, G. 1975. Aggressive behavior and population regulation of red squirrels (*Tamiasciurus hudsonicus*) in interior Alaska. Thesis, University of Alaska, Fairbanks, Alaska, USA.

Sieg, M.J. 2002. Landscape composition and Abert squirrel survivorship, predator-based mortality, home-range size and movement. Thesis, Northern Arizona University, Flagstaff, Arizona, USA.

Sieving, K.E. and M.F. Willson. 1998. Nest predation and avian species diversity in northwestern forest understory. *Ecology* **79**:2391–2402.

Simberloff, D. 1988. The contribution of population and community biology to conservation science. *Annual Review of Ecology and Systematics* **19**:473–511.

Smith, A.A. and R.W. Mannan. 1994. Distinguishing characteristics of Mount Graham red squirrel midden sites. *Journal of Wildlife Management* **58**:437–445.

Smith, C.C. 1968. The adaptive nature of social organization in the genus of tree squirrels *Tamiasciurus*. *Ecological Monographs* **38**:31–63.

Smith, C.C. 1970. The coevolution of pine squirrels and conifers. *Ecological Monographs* **40**:349–371.

Smith, C.C. 1981. The indivisible niche of *Tamiasciurus*: an example of nonpartitioning of resources. *Ecological Monographs* **51**:343–363.

Smith, M.C. 1968. Red squirrel responses to spruce cone failure in interior Alaska. *Journal of Wildlife Management* **32**:306–316.

Smith, W.P. and J.V. Nichols. 2003. Demography of the Prince of Wales flying squirrel (*Glaucomys sabrinus griseifrons*): an endemic of southeastern Alaska temperate rainforest. *Journal of Mammalogy* **84**: in press.

Snyder, M.A. 1992. Selective herbivory by Abert's squirrel mediated by chemical variability in ponderosa pine. *Ecology* **73**:1730–1741.

Snyder, M.A. 1993. Interactions between Abert's squirrel and ponderosa pine: the relationship between selective herbivory and host plant fitness. *American Naturalist* **141**:866–879.

Snyder, M.A. 1994. Nest-site selection by Abert's squirrel – chemical characteristics of nest trees. *Journal of Mammalogy* **75**:136–141.

Snyder, M.A. 1998a. Subspecific selectivity by a mammalian herbivore: geographic differentiation of interactions between two taxa of *Sciurus aberti* and *Pinus ponderosa*. *Evolutionary Ecology* **12**:755–765.

Snyder, M.A. 1998b. Abert's squirrel (*Sciurus aberti*) in ponderosa pine (*Pinus ponderosa*) forests: directional selection, diversifying selection. Pages 195–201 *in* M.A. Steele, J.F. Merritt, and D.A. Zegers, editors. *Ecology and Evolutionary Biology of Tree Squirrels*. Virginia Museum of Natural History Special Publication 6, Martinsville, Virginia, USA.

Snyder, M.A. and Y.B. Linhart. 1993. Barking up the right tree. *Natural History* **102**:44–49.

Spies, T.A., J.F. Franklin, and T.B. Thomas. 1988. Coarse woody debris in forests and plantations in coastal Oregon. *Ecology* **69**:1689–1702.

Squires, J.R. and R.T. Reynolds. 1997. Northern goshawk (*Accipiter gentilis*). *The Birds of North America* **298**:1–32.

States, J.S. 1985. Hypogeous mycorrhizal fungi associated with ponderosa pine; scorocarp phenology. Pages 271–272 *in* R. Molina, editor. *Proceedings of the 6th North American Conference on Mycorrhizae*. Forest Research Laboratory, Corvallis, Oregon, USA.

States, J.S. and W.S. Gaud. 1997. Ecology of hypogeous fungi associated with ponderosa pine. I.Patterns of distribution and sporocarp production in some Arizona forests. *Mycologia* **89**:712–721.

States, J.S. and P.J. Wettstein. 1998. Food habits and evolutionary relationships of the tassel-eared squirrel (*Sciurus aberti*). Pages 185–194 *in* M.A. Steele, J.F. Merritt, and D.A. Zegers, editors. *Ecology and Evolutionary Biology of Tree Squirrels*. Virginia Museum of Natural History Special Publication 6, Martinsville, Virginia, USA.

States, J.S., W.S. Gaud, W.S. Allred, and W.J. Austin. 1988. Foraging patterns of tassel-eared squirrels in selected ponderosa pine stands. Pages 425–431 *in* R.C. Szaro, K.E. Severson, and D.R. Patton, technical coordinators. Symposium proceedings on management of amphibians, reptiles and small mammals in North America. USDA Forest Service General Technical Report RM-**166**, Fort Collins, Colorado, USA.

Steele, M.A. 1998. *Tamiasciurus hudsonicus*. *Mammalian Species* **586**:1–9.

Steele, M.A. 1999. *Tamiasciurus douglasii*. *Mammalian Species* **630**:1–8.

Stephenson, R.L. 1975. Reproductive biology and food habits of Abert's squirrels in central Arizona. Thesis, Arizona State University, Tempe, Arizona, USA.

Stephenson, R.L. and D.E. Brown. 1980. Snow cover as a factor influencing mortality of Abert's squirrels. *Journal of Wildlife Management* **44**:951–955.

Sullivan, T.P. 1990. Responses of red squirrel (*Tamiasciurus hudsonicus*) populations to supplemental food. *Journal of Mammalogy* **71**:579–590.

Sullivan, T.P. and W. Klenner. 1993. Influence of diversionary food on red squirrel populations and damage to crop trees in young lodgepole pine forest. *Ecological Applications* **3**:708–718.

Sullivan, T.P. and R.A. Moses. 1986. Red squirrel populations in natural and managed stands of lodgepole pine. *Journal of Wildlife Management* **50**:595–601.

Sullivan, T.P. and D.S. Sullivan. 1982. Population dynamics and regulation of the Douglas squirrel (*Tamiasciurus douglasii*) with supplemental food. *Oecologia* **53**:264–270.

Sullivan, T.P. and A. Vyse. 1987. Impact of red squirrel feeding damage on spaced stands of lodgepole pine in the caribou region of British Columbia. *Canadian Journal of Forestry Research* **17**:666–674.

Sullivan, T.P., H. Coates, L.A. Jozsa, and P.K. Diggle. 1993. Influence of feeding damage by small mammals on tree growth and wood quality in young lodgepole pine. *Canadian Journal of Forest Research* **23**:799–809.

Sullivan, T.P., J.A. Krebs, and P.K. Diggle. 1994. Prediction of stand susceptibility to feeding damage by red squirrels in young lodgepole pine. *Canadian Journal of Forestry Research* **24**:14–20.

Sullivan, T.P., W. Klenner, and P.K. Diggle. 1996. Response of red squirrels and feeding damage to variable stand density in young lodgepole pine forest. *Ecological Applications* **6**:1124–1134.

Taylor, W.P. 1915. Description of a new subgenus (*Arborimus*) of Phenacomys, with a contribution to knowledge of the habits and distribution of *Phenacomys longicaudus* True. *Proceedings of the California Academy of Sciences*, Fourth Series, Volume V, No. **5**:111–161.

Thomas, G.R. 1979. The role of phloem sugars in the selection of ponderosa pine by the Kaibab squirrel. Thesis, San Francisco State University, San Francisco, California, USA.

Thompson, J.L. and L.V. Diller. 2002. Relative abundance, nest site characteristics, and nest dynamics of Sonoma tree voles on managed timberlands in coastal northwest California. *Northwestern Naturalist* **83**:91–100.

Thysell, D.R., L.J. Villa, and A.B. Carey. 1997. Observations of northern flying squirrel feeding behavior: use of non-truffle food items. *Northwestern Naturalist* **78**:87–92.

Trowbridge, A.H. and L.L. Lawson. 1942. Abert Squirrel-Ponderosa Pine Relationships at the Fort Valley Experimental Forest, Flagstaff, Arizona. Arizona Cooperative Wildlife Research Unit, University of Arizona, Tucson, Arizona, USA.

True, F.W. 1890. Description of a new species of mouse, *Phenacomys longicaudus*, from Oregon. *Proceedings of the National Museum*, **13** (Number 826):303–304.

Vahle, J.R. and D.R. Patton. 1983. Red squirrel cover requirements in Arizona mixed conifer forests. *Journal of Forestry* **81**:14–15, 22.

Verts, B.J. and L.N. Carraway. 1998. *Land Mammals of Oregon*. University of California Press, Berkeley, California, USA.

Villa, L.J., A.B. Carey, T.M. Wilson, and K.E. Glos. 1999. Maturation and reproduction of northern flying squirrels in Pacific Northwest forests. USDA Forest Service General Technical Report PNW-GTR-444. Portland, Oregon, USA.

Voth, E.H., C. Maser, and M.L. Johnson. 1983. Food habits of *Arborimus albipes*, the white-footed vole, in Oregon. *Northwest Science* **57**:1–7.

Vrieze, J.M. 1980. Spatial patterning of red tree mouse, *Arborimus longicaudus*, nests. Thesis, Humboldt State University, Arcata, California, USA.

Wagg, J.W.B. 1964. Viability of white spruce seed from squirrel-cut cones. *Forestry Chronicle* **40**:98–110.

Waters, J.R. and C.J. Zabel. 1995. Northern flying squirrel densities in fir forests of northeastern California. *Journal of Wildlife Management* **59**:858–866.

Waters, J.R. and C.J. Zabel. 1998. Abundances of small mammals in fir forests in northeastern California. *Journal of Mammalogy* **79**:244–253.

Watson, J.W., D.W. Hays, S.P. Finn, and P. Meehan-Martin. 1998. Prey of breeding northern goshawks in Washington. *Journal of Raptor Research* **32**:297–305.

Weigl, P.D. 1978. Resource overlap, interspecific interactions, and the distribution of the flying squirrels, *Glaucomys volans* and *G. sabrinus*. *The American Midland Naturalist* **100**:83–96.

Wells-Gosling, N. and L.R. Heaney. 1984. *Glaucomys sabrinus*. *Mammalian Species* **229**:1–8.

Wilson, D.E. and D.M. Reeder, editors. 1993. *Mammal Species of the World: a Taxonomic and Geographic Reference*. Second Edition. Smithsonian Institution Press, Washington DC, USA.

Wilson, D.E. and S. Ruff, editors. 1999. *The Smithsonian Book of North American Mammals*. Smithsonian Institution Press, Washington DC, USA.

Wilson, K.R. and D.R. Anderson. 1985. Evaluation of two density estimators of small mammal population size. *Journal of Mammalogy* **66**:13–21.

Wirsing, A.J., T.D. Steury, and D.L. Murray. 2002. Relationship between body condition and vulnerability to predation in red squirrels and snowshoe hares. *Journal of Mammalogy* **83**:707–715.

Witt, J.W. 1991. Increasing the carrying capacity of second-growth stands for flying squirrels with the use of nest boxes. Page 529 *in* L.F. Ruggiero, K.B. Aubry, A.B. Carey, and M.H. Huff, technical coordinators. Wildlife and vegetation of unmanaged Douglas-fir forests. USDA Forest Service, General Technical Report **PNW-GTR-285**, Portland, Oregon, USA.

Witt, J.W. 1992. Home range and density estimates for the northern flying squirrel (*Glaucomys sabrinus*) in western Oregon. *Journal of Mammalogy* **73**:921–929.

Wolff, J.O. and J.C. Zasada. 1975. Red squirrel response to clear-cut and shelterwood systems in interior Alaska. USDA Forest Service Research Note PNW-**255**. Portland, Oregon, USA.

Wooster, T.W. 1994. *Red Tree Vole (Arborimus longicaudus) Observations in Humboldt, Mendocino, and Sonoma Counties, California 1991–1993*. State of California Department of Fish and Game, Sacramento, California, USA.

Zabel, C.J. and J. R. Waters. 1997. Food preferences of captive northern flying squirrels from the Lassen National Forest in northeastern California. *Northwest Science* **71**:103–107.

Zabel, C.J., K. McKelvey, and J.P. Ward, Jr. 1995. Influence of primary prey on home-range size and habitat-use patterns of northern spotted owls (*Strix occidentalis caurina*). *Canadian Journal of Zoology* **73**:433–439.

Zang, X. and J.S. States. 1991. Selective herbivory of ponderosa pine by Abert squirrels: a re-examination of the role of terpenes. *Biochemical Systematics and Ecology* **19**:111–115.

Zentner, P.L. 1966. The nest of *Phenacomys longicaudus* in northwestern California. Thesis, California State University, Sacramento, California, USA.

7
———

Small and mid-sized carnivores

The small and mid-sized carnivores (Carnivora), or mesocarnivores of western forests comprise 16 species (coyote (*Canis latrans*), red fox (*Vulpes vulpes*), gray fox (*Urocyon cinereoargenteus*), ringtail (*Bassariscus astutus*), raccoon (*Procyon lotor*), marten (*Martes americana*), fisher (*M. pennanti*), ermine (*Mustela erminea*), long-tailed weasel (*M. frenata*), mink (*M. vison*), wolverine (*Gulo gulo*), northern river otter (*Lontra canadensis*), western spotted skunk (*Spilogale gracilis*), striped skunk (*Mephitis mephitis*), Canadian lynx (*Lynx canadensis*), and bobcat (*Lynx rufus*)). The term "forest carnivores" denotes a smaller group of four species – the marten, fisher, lynx, and wolverine – and is only marginally descriptive, inasmuch as it excludes many carnivores that live in forests, and includes the wolverine, which can thrive in the complete absence of trees. The species we consider here represent four (or five (Dragoo and Honeycutt 1997)) taxonomic families and are characterized by adult body weights typically <20 kg. Other mesocarnivores, including the kit fox (*Vulpes macrotis*), swift fox (*V. velox*), least weasel (*Mustela nivalis*), black-footed ferret (*M. nigripes*), and badger (*Taxidea taxus*), occur in the West, occupy habitats near forest edges, and may be conservation concerns. However, they are plains or grassland specialists or, in the case of the least weasel, very poorly known, and cannot be characterized in terms of their needs for forest attributes. So, they are not treated here.

Our understanding of the ecology of carnivores in western coniferous forests varies markedly. American martens have been studied in the West in relation to habitats (at various scales), diets, populations, energetic

[207]

physiology, and reproduction. On some topics, multiple studies have been published. By contrast, ringtails in the north temperate zone are known from only a few studies, and even descriptive habitat use of forests is weakly documented. Perhaps surprisingly, some common species (e.g., raccoon) have received only the most cursory study in relation to habitats in the West. Therefore, we refer to midwestern and eastern studies, recognizing that those results may have limited applicability to habitats in the West.

Perhaps unavoidably, knowledge of habitat requirements of forest obligates is clearer and more specific than for habitat generalists. The latter tend not to receive detailed habitat study because they are not interesting models of habitat selection, and require large sample sizes to even detect selection. Raccoons and ringtails, for example, both rest in cavities in broad-leaved trees and snags, but links between both species and structural or successional stages of upland coniferous forest are unclear, and possibly weak.

Species accounts

We group the mesocarnivores into four habitat strata: habitat generalists, forest specialists, riparian associates, and semi-aquatic species, following the classification of Cooperrider et al. (1999). Habitat generalists occur in a wide range of forested and non-forested habitats, forest specialists are limited to forests and habitats nearby, riparian associates occur primarily in special habitats around water bodies or saturated soils, and semi-aquatic species forage or travel mostly in water, but perform other life functions on land. All species considered here have broad geographical ranges; for example the long-tailed weasel from northern British Columbia to Bolivia. And, all but two species (ringtail and wolverine) have broad distributions or sibling species in eastern North America.

Habitat generalists
Coyote
The coyote is arguably the most cosmopolitan and adaptable mesocarnivore of western North America, having expanded its geographical range and increased its abundance, including in the northwest, over the past 50–100 years (Buskirk et al. 2000*a*). Coyotes are best adapted to non-forested habitats, having been called brush wolves (as contrasted with "timber wolves" (*Canis lupus*)) by early explorers. They have expanded into formerly inaccessible continuous forest as a result of fragmentation,

particularly road building, land conversion, and timber harvesting. Today, coyotes are common or abundant in a wide range of forested ecosystems throughout the West, including the 3030- to 3333-m elevation zone in Colorado (G. Byrne 1998 unpublished, cited in Buskirk et al. 2000a). Coyotes feed on a wide range of items in the West, including mammals, birds, fruits, and invertebrates. Two factors, one abiotic and the other biotic, can be severely limiting to coyotes. Deep, soft snow can limit the distribution of coyotes (Todd et al. 1981), because of their presumably high energetic costs while walking through deep snow, as inferred from foot loadings (Buskirk et al. 2000a). So, areas of the West with deep, soft snow and no avenues for access should have fewer coyotes than elsewhere, although this hypothesis has not been rigorously tested. The presence of sympatric larger-bodied carnivores, particularly wolves, is an important biotic limiting factor to coyotes. Coyotes alter their behaviors, exhibit reduced densities, or become locally extinct in the presence of wolves (reviewed by Buskirk 1999, Buskirk et al. 2000a). Where they are common, on the other hand, coyotes have pivotal community functions because of their predation on and competition with a wide range of smaller carnivores (Sovada et al. 1995, Henke and Bryant 1999). Coyotes are variously credited with suppressing populations of native and non-native mesocarnivores, thereby protecting smaller vertebrate populations (Crooks and Soulé 1999) or threatening native vertebrates (White and Garrott 1997), their perceived value being context- and taxon-specific. But there is increasing recognition that they are important factors in the conservation of mesocarnivores, small birds, and small mammals (Ralls and White 1995, O'Donoghue et al. 1998, Rogers and Caro 1998) by nature of being opportunistic predators, and the largest predators remaining in many areas of the West. Coyotes appear to be benefited by forest-management practices that create openings in previously extensive forests via clearcutting or road building, and may benefit from activities that create trails in deep, soft snow. Foods made available by humans, including refuse, livestock, some crops, and pets, can be important to coyotes. Perhaps most importantly, programs to restore the wolf to its former distribution in the West have important implications for the distribution and abundance of coyotes.

Red fox

With the historical reduction in the distribution of wolves, the red fox now has the widest global distribution of any mammal other than humans and our associated commensal species (Voigt 1987). The Holocene and Recent zoogeography of the red fox in the West is exceedingly complex, involving indigenous and introduced forms (Aubry 1984), some expanding and others shrinking their distributions, presumably in response to human-caused changes. In coniferous forests of the West, the red fox occurs in at least three forms: in Canada and Alaska, native foxes are assigned to three to four subspecies, and occur at various elevations and in various habitats. In the western contiguous United States and southwestern British Columbia, however, foxes include those in primarily lowland habitats, which were introduced from the eastern United States (*V. v. fulva*), and three indigenous subspecies (*V. v. macroura* of the Rocky Mountains, *V. v. cascadensis* of the Cascades, and *V. v. necator* of the Sierra Nevada). The introduced, low-elevation form has expanded its distribution dramatically since the early 1900's (Grinnell et al. 1937) into areas that previously did not have red foxes, whereas the high elevation form has declined in distribution and abundance and represents a conservation concern (Schempf and White 1977, Aubry 1984).

The high-elevation form has a unique evolutionary history; presumptively representing vicariant post-glacial relicts. Its restriction to high elevation, especially alpine habitats (Grinnell et al. 1937, Aubry 1983, Crabtree 1993), seems at odds with the otherwise broad tolerances of the species as suggested by its vast distribution and habitat generalization within the northern part of the New World range. The lowland temperate western red foxes, escaped from fur farms, have expanded their distribution in recent decades into grasslands and coastal marshes (Lewis et al. 1995), as well as interior deserts. Burrows in soil or rock piles are typically used as natal dens. Both introduced and indigenous forms are dietary generalists, feeding on rodents, lagomorphs, insects, fruits, carrion, and birds (Voigt 1987).

The Cascade and Sierra Nevada red fox populations are at the greatest risk of extinction; both appear to be small and in decline (Schempf and White 1977, Aubry 1983). The Sierra Nevada red fox has been considered extremely sensitive to the presence of humans (Grinnell et al. 1937) so that increased recreation within its range could be problematic. The expanding range of the lowland, introduced, red fox poses a risk to the native form via genetic introgression (Aubry 1984, Lewis et al. 1995), which presumably was minimized in the past by densely forested areas that separated them (Aubry 1983). In Yellowstone, the high-elevation red fox found on the Beartooth Plateau has been hypothesized to be negatively affected by coyotes (Buskirk 1999).

Gray fox

The gray fox prefers brushy vegetation in broken terrain and uses woodland habitat more frequently than does the red fox (Samuel and Nelson 1982). Gray foxes are unique among North American canids in their strong tree-climbing ability. In California, riparian areas and old-field habitats were used more frequently than expected (Fuller 1978) and in Utah brushy meadows were favored (Trapp 1978). Natal dens are found in brush piles, rocky outcrops, and hollow trees (Trapp and Hallberg 1975). Lagomorphs, rodents, and small birds are the primary prey, but fruit can dominate the diet in summer and fall (Wilcomb 1948, Trapp 1978). Although trapped for fur in most western states, their fur is not highly prized. They associate with human development to some extent (Riley 1999), but high-density residential development (Harrison 1997) and fragmentation of forested lands (Rosenberg and Raphael 1986) each reduce habitat for gray foxes. Continued regulated trapping should provide data to allow the assessment of status and distribution.

Ermine

The distribution of the ermine overlaps extensively with the larger long-tailed weasel and may be limited by interference competition between the two species (Simms 1979). Ermines occur primarily where voles are common, typically meadows and grasslands in temperate forests. However, this weasel is frequently found in forests and shrub habitats (Hooven and Black 1976, Gilbert and Allwine 1991), unlike in the East (Simms 1979, Verts and Carraway 1998). In the North, this is the only common forest weasel. Grass nests constructed by voles in the winter are usurped by ermines for resting sites (Fitzgerald 1977). Because of its smaller size, the ermine, more than the long-tailed weasel, specializes on voles and mice (Simms 1979), captures them in subnivean spaces (Fitzgerald 1977, Simms 1979) and uses mouse nests for shelter.

Because they associate with meadows in forests, ermines probably are vulnerable to the effects of livestock grazing on vegetation; small mammal prey may be secondarily affected. Encroachment of trees into meadows, due to fire suppression or changes in climate, may also reduce weasel habitat.

Long-tailed weasel

The long-tailed weasel is the largest weasel of the West, and may occur in any habitat, including mature forests, recently logged areas, meadows, and tundra (Simms 1979, Svendsen 1982, Fagerstone 1987). They are common in meadows when vole populations are high and use the nests of their microtine prey during the winter (Fitzgerald 1977). They are less sub-nivean than ermines, presumably being more constrained by their larger bodies (Simms 1979); the northern limit of their distribution may be limited by snow. Long-tailed weasels are primarily carnivorous, consuming a wider range of prey than ermines, but relying predominantly on *Microtus*, *Clethrionomys*, *Peromyscus*, and *Tamias* (Hamilton 1933, Quick 1951). Insofar as they associate with grassland and meadow habitat within forests, long-tailed weasels may be vulnerable to the loss of herbaceous and shrub cover resulting from grazing. Encroachment of trees into meadows, due to fire suppression or changes in climate, may also eliminate some long-tailed weasel habitat.

Wolverine

The wolverine was historically distributed throughout most of Alaska and Canada, southward through the montane West to Utah, Colorado, and California (Banci 1994). Wolverines are now uncommon or rare in the contiguous United States (Wilson 1982, Banci 1994), where the present distribution comprises several peninsular or island extensions of the northern core (Hash 1987, Banci 1994). Wolverines occur in north-central Idaho and northwestern Montana (Banci 1994, Maj and Garton 1994), and may persist in parts of Oregon and Washington (Edelmann and Copeland 1999), but are exceedingly rare in or absent from California and Colorado. Wolverines are generally restricted throughout their temperate western range to boreal forests, tundra, and high-elevation grasslands (Banci 1994). Wolverine habitat, however, is less defined by vegetation types than by where adequate year-round supplies of food occur in extensive wilderness areas. Preferences for particular habitats have been attributed largely to differences in the availability of food (Gardner 1985, Banci 1987), but the availability of coarse woody debris (dead woody material >10 cm diameter: CWD) and rocky areas for resting, rendezvous and denning sites may be critical (Banci 1994). Natal dens tend to occur in rock crevices in talus slopes of high-elevation cirque basins, piles of CWD, and snow tunnels (Hash 1987, Magoun and Copeland 1998), but other protected structures have been reported to be used as well (Grinnell et al. 1937). The diet is composed primarily of the carrion of large mammals, particularly ungulates (Banci 1994). In coastal environments, however, the carrion of marine mammals and salmon is commonly eaten (Banci 1987). In the North, wolverines are closely associated with caribou (*Rangifer tarandus*)

(Banci 1987, 1994) and may follow caribou herds. Small mammals (e.g., ground squirrels (*Spermophilus* spp.), marmots (*Marmota* spp.), snowshoe hares (*Lepus americana*), porcupines (*Erethizon dorsatum*)) and birds (e.g., ptarmigan (*Lagopus* spp.)) are important prey only when the carrion of larger mammals is unavailable (Banci 1987, 1994).

Wolverines are trapped in Canada, Alaska and Montana (Banci 1994) and caution is required to assure that harvests are sustainable. Recent harvest levels in Alaska (10% of fall populations) have exceeded estimates of sustainable levels (Gardner et al. 1993 cited in Banci 1994). The importance of timber harvest to wolverines is not clear; canopy cover and CWD have been reported to be important (Hornocker and Hash 1981, Banci 1994, Copeland 1996). However, most wolverine habitat occurs away from commercial timber harvest so areas of conflict may be quite localized. The greatest threat to populations of wolverines, similar to that for other low-density carnivores, is the fragmentation of their habitat and populations by human structures and activities. Wolverine populations at the southernmost margin of the range may require episodic dispersive inputs to maintain viability, and wilderness habitat corridors may be important for such movements. The frequency and direction of such movements is not known. Wolverines are extremely intolerant of the presence of humans, and the distribution of wolverines in the West coincides largely with the regions of lowest human density (Hornocker and Hash 1981, Banci 1994, Carroll et al. 2001). Perhaps more so than for other species considered here, the mechanism by which human activity excludes wolverines is not known.

Western spotted skunk

The western spotted skunk is parapatric with its eastern counterpart, *S. putorius*, in Colorado and New Mexico. Spotted skunks occur in a variety of habitat types across their range in western North America. In particular regions, however, they are strongly associated with specific cover types and habitat elements. In eastern Oregon, they commonly associate with canyons, cliffs, lava fields, and arid valleys (Bailey 1936), whereas in coastal Oregon they are common in dense forest stands with old-growth elements and dense shrub cover (Carey and Kershner 1997). In California, various subspecies seem to associate with highly varied elements in landscapes from sea level to >2000 m (Orr 1943), but typically with dry rocky uplands characterized by shrubs and little forest cover (Grinnell et al. 1937). Resting occurs in log and tree cavities, rock crevices, wood rat

(*Neotoma* sp.) nests, mountain beaver (*Aplodontia rufa*) tunnels and human structures (Grinnell et al. 1937, Maser et al. 1981, Crooks and Van Vuren 1995). They are excellent climbers. The diet has been described only for the Channel Islands spotted skunk (*S. g. amphiala*), which feeds on deer mice (*Peromyscus maniculatus*), insects, and lizards (Crooks and Van Vuren 1995). The congeneric *S. putorius* feeds primarily on small mammals in winter and arthropods in summer (Crabb 1941). Spotted skunks are trapped for their fur in some western states and provinces, but most animals likely are caught incidental to more valuable furbearers and this does not result in measurable population-level losses (Verts and Carraway 1998). A larger cause of deaths is encounters with humans and their pets around human habitations (Crabb 1948, Verts and Carraway 1998). The incidence of rabies in spotted skunks does not approach that in striped skunks (Krebs et al. 1995*a*), but the perceived association between all skunks and rabies accounts for much unwarranted persecution of spotted skunks.

Striped skunk

Although occurring in forests and forest edges, the striped skunk is mostly a species of open habitats. In coastal regions it can occur in dune, prairie, and meadow communities as well as in early-seral forests (Maser et al. 1981). In interior regions striped skunks are common in valleys with mixtures of pasture, crops, and brush and also in woodlands and open fields broken by wooded ravines and rocky outcrops (Wade-Smith and Verts 1982). They associate closely with human structures in suburban and urban settings (Godin 1982). Typical of lower elevations in California, they seldom occur above 1500 m (Grinnell et al. 1937). They forage on the ground, rarely climb trees (Godin 1982), and may dig shallow burrows for resting sites, but prefer rock crevices, burrows of other species, and holes within or beneath logs (Godin 1982). Striped skunks primarily eat insects and small mammals (Grinnell et al. 1937, Verts 1967). They are still commonly trapped for their skins, but pelt values are low and trapping has not been considered a threat to populations (Godin 1982). Reported rates of occurrence of rabies in striped skunks are among the highest reported for wild or domestic mammals, which contributes to their persecution in many areas.

Bobcat

Within their range, bobcats occur in a wide range of vegetation types, including coniferous forests, marshy hardwood forests, high-elevation shrublands, and deserts. They associate with dense physical structure in the form of shrubs and woody debris of various sizes, and exhibit no particular tie to late-successional stages. Indeed, bobcats in the Southeast have been shown to select for early seres (Miller 1980), and those in Oklahoma heavily used forested sites <10 years after clear-cutting (Rolley and Warde 1985). Bobcats also show some preference for sites with rock ledges and outcrops, in part because of the shelter that such features provide (Zezulak and Schwab 1979). Seasonal shifts in habitat have been described (Rolley 1987); at the northern edge of the range these include winter shifts toward habitats with lower snow depths (McCord 1974), reflecting the heavy foot-loadings for the species (Buskirk et al. 2000a). The importance of trees and particularly CWD to bobcats is apparently low relative to other mesocarnivores. Bobcats in central Idaho were found to prefer open over timbered habitats (Koehler and Hornocker 1991), and no evidence suggests strong ties to late-successional forests in the West.

Foods of bobcats are almost entirely mammals and birds; in the West the most common prey are lagomorphs (Rolley 1987) and rodents (Riley 1999). A general reduction in the distribution of bobcats in the urban East apparently has not been reflected in the West. The principal conservation concerns for bobcats in the West have tended to be related to trapping mortality. Yearly survival rates in heavily trapped populations have been around 0.6 (Rolley 1987), but as low as 0.19 (Fuller et al. 1985). With the decline of the trapping industry for the past 15 years, however, concerns about trapping mortality have dwindled correspondingly. As explained below, bobcats are presumably susceptible to competitive interactions with larger carnivores. No clear approaches to habitat management that would favor bobcats are known for the West. Either the species is insensitive to most common forestry practices, or our knowledge base is too weak to support any salient inferences.

Forest specialists

Marten

The marten is a specialist of coniferous forests of the boreal zone and southern peninsular extensions (Hagmeier 1956, Gibilisco 1994). Martens prefer mesic, conifer-dominated forests with abundant physical structure near the ground and avoid areas lacking overhead cover, particularly in winter (Buskirk and Powell 1994). In some areas of Alaska and northwestern Canada, martens use early seres following burns (Paragi et al. 1996), so that downed tree boles combined with dense herbaceous vegetation can address the need for cover. The avoidance of areas lacking sufficient cover has been called a "psychological need" (Hawley and Newby 1957), and presumably relates to vulnerability to attack by other carnivores and raptors. During the winter, structures near the ground provide access to subnivean spaces for resting and foraging (Buskirk and Ruggiero 1994). Throughout the year martens use cavities in large-diameter live and dead trees as resting sites (Buskirk and Ruggiero 1994) and as natal dens (Ruggiero et al. 1998). In the West, martens are strongly associated with old-growth coniferous forests, but they occur in earlier seres that include remnants of old-growth forest, for example stumps and logs (Baker 1992). Martens generally do not occur in landscapes where more than one-third of the landscape is in openings (Chapin et al. 1998, Hargis et al. 1999, Payer 1999, Potvin et al. 2000). Minta et al. (1999) hypothesized that martens are most selective at the finest scales (scale of foraging and resting sites) and coarsest scales (population selection of landscape types), but less selective at intermediate scales (patch and home range). At regional scales, the amount of habitat and its distance from nearby source populations influence the occurrence of martens on insular mountain ranges in the interior West (Wisz 1999).

Martens eat primarily rodents, lagomorphs, birds, fruit seasonally, and carrion opportunistically (Martin 1994). Voles are the primary prey, particularly in the North (Martin 1994), and *Microtus* spp. and other large-bodied species appear to be preferred because they are consumed in excess of their availability (Weckwerth and Hawley 1962, Buskirk and MacDonald 1984). Larger prey (e.g., lagomorphs, tree squirrels) constitute more of the winter than summer diet (Zielinski et al. 1983, Thompson 1986, 1994). Martens have important ecological relationships with red squirrels (*Tamiasciurus hudsonicus*) and Douglas squirrels (*T. douglasii*) (Buskirk 1984, Corn and Raphael 1992, Sherburne and Bissonette 1993) and in some studies these species compose a significant part of the diet (Martin 1994).

Martens are trapped for their fur in several states and most Canadian provinces (Buskirk and Ruggiero 1994), but are fully protected in a growing number of jurisdictions. Populations are easily over-trapped, which led partly to the reduced distribution and abundances through the early 1900's, including from much of the Coast Ranges of California and Oregon (Zielinski et al. 2001). Even light trapping pressure can inflict high mortality (Schneider 1997); limiting the number of licensed trappers and setting quotas are common practices (Strickland et al. 1982a, b). Timber harvest, especially clear-cutting of mature and old-growth forests, also affects martens negatively (Buskirk and Ruggiero 1994, Thompson and Harestad

1994, Hargis et al. 1999, Payer 1999). Logging results in the direct loss of resting and foraging habitat and increases energetic costs due to avoidance of openings and the maintenance of larger home ranges (Potvin et al. 2000). Partial harvest (thinning) results in stands that are avoided in winter but not summer (Fuller 1999). In Maine, reproduction is higher and mortality from trapping is lower in unharvested forests than in commercial forests (Payer 1999). Trapping mortality also is additive to natural mortality in commercial forests (Payer 1999). Responses of martens to habitat change in general and to landscape fragmentation in particular are not linear (Hargis et al. 1999). Martens exhibit threshold responses to the extent of forest harvesting, and population declines beyond the threshold may be abrupt (Bissonette et al. 1997). Grazing by livestock may affect martens via altering habitat for voles adjacent to resting habitat. Roads are a source of mortality and negatively affect the distribution of marten activity (Robitaille and Aubry 2000).

Fisher

Fishers occur in mesic, conifer-dominated forests of the boreal zone and southern peninsular extensions (Hagmeier 1956, Gibilisco 1994). Primary habitat is dense coniferous forest, usually with a deciduous component and abundant physical structure near the ground (Buskirk and Powell 1994, Powell and Zielinski 1994). Fishers use cavities in large-diameter trees, snags, and logs as daily resting sites (Powell and Zielinski 1994) and as natal dens (Roy 1991, Aubry et al. 1997). Mid-elevation old-growth forests provide needed habitat elements, but fishers travel through, and in some areas occupy, home ranges in regenerating forests that have dense cover and a sufficient number of large structures, especially hardwood boles, for resting (Klug 1996). In more xeric regions, or where logging has removed upland forest cover, riparian zones provide important habitat (Roy 1991, Heinemeyer 1993, Seglund 1995). Fishers avoid regions with deep, soft snow, because of their heavy foot loadings (Krohn et al. 1997). Kelly (1977) and Carroll et al. (1999a) found that habitat selection appeared to be dominated by factors acting at the scale of the home range and above, contrary to the results of Weir and Harestad (1997). In northwestern California, landscape-level indices of canopy closure, size of trees, and percentage conifer interact with regional climatic and geographical variables to explain the occurrence of fishers (Carroll et al. 1999a). Regional habitat of fishers in the Rocky Mountains, modeled using remotely sensed variables, is best described by a combination of precipitation, elevation, "wetness," "greenness," road density, canopy closure, and variation in tree size (Carroll et al. 2001). This confirms the general notion that fishers are strongly associated with productive, low- to mid-elevation forests.

Fishers eat snowshoe hares and porcupines over much of their range (Powell 1993, Martin 1994). Where snowshoe hares and porcupines are uncommon, the diets of fishers are more diverse, including significant quantities of other mammals, reptiles, fruits, insects, and fungi (Zielinski et al. 1999). Two California studies (Grenfell and Fasenfest 1979, Zielinski et al. 1999) showed that fishers eat fruiting bodies of false truffles (*Rhizopogon* spp.).

Fishers are protected from trapping over most of the western states, but are legally trapped throughout Canada and the north-central states (Powell and Zielinski 1994). They are easily over-trapped, even in a short, early trapping season, especially when trapping pressure is heavy (Powell 1979). Fishers are also vulnerable to incidental capture in traps set for other species (Lewis and Zielinski 1996). Where trapping is heavy, limiting the number of licensed trappers and setting quotas based on demographic characteristics of the harvest is necessary (Strickland et al. 1982*a*, Strickland 1994). In the West, fishers avoid recently clear-cut areas and are almost always associated with old stands or old remnant structures in regenerating stands (Buck et al. 1994, Jones and Garton 1994). Their need for continuous forest cover (Rosenberg and Raphael 1986, Carroll et al. 1999*a*) may account for their loss from areas that are intensely managed for timber. Regional persistence seems to reflect the sizes and arrangements of suitable habitat patches; as habitat becomes fragmented, the distribution collapses to include only the largest patches of source habitat (Carroll et al. 2001). All study animals found dead ($n = 7$) in one northwestern California study ($n = 21$) were recovered from clear-cuts, areas lacking overhead cover, or hardwood-dominated stands (Buck et al. 1994), showing that mortality was highest in areas found to be avoided in use-availability analyses. Where sympatric in the West, fishers generally occur at lower elevations than martens (Grinnell et al. 1937, Zielinski et al. 1997, Verts and Carraway 1998), and therefore may encounter roads more frequently.

Canadian lynx

Lynx are now much reduced in distribution and abundance (McKelvey et al. 2000*a*) and contiguous US populations were listed under the

Endangered Species Act in 2000. In the West, lynx are specialists of mesic coniferous forests of the boreal zone and southern peninsular extensions (McKelvey et al. 2000a). Lynx require habitat that will produce sufficient densities of their obligatory prey, snowshoe hares and red squirrels (Aubry et al. 2000). These species, in turn, require high densities of small-diameter woody stems and cone-bearing ages of conifers, respectively. The plant species, successional stages, and moisture regimes that produce these conditions and adequate prey densities are region-specific (Buskirk et al. 2000b). In the North, where populations are strongly cyclic, high densities of small-diameter woody stems tend to occur in early seres, before the closed-canopy stage of stand development. Therefore, in the North early successional stages with dense small woody stems are important to lynx; the role of late successional forests is unclear. However, red squirrels are important prey during periods of hare scarcity (Mowat et al. 2000); therefore, coniferous forest with high densities of red squirrels may be important for the survival of lynx between peaks in hare populations.

In the western contiguous US, snowshoe hare population cycles have small or imperceptible amplitudes (Hodges 2000) and population densities tend to be similar to those during population troughs in the North. Red squirrels tend to be more important prey in the southern part of the distribution, commonly representing 25% to 35% of prey items (Aubry et al. 2000). Here, dense woody stems are absent from some early seres; dry soils during summer can slow regeneration, and dense shrubs are uncommon in early seres, except on moist sites. In such dry landscapes, high-quality lynx habitat may be limited to moist sites: north-facing slopes, riparian zones, and late-successional stands with shaded understories. Cone-bearing ages of conifers important to red squirrels may be more important for lynx in the southern part of the range than in the North, because of the presumably greater importance of red squirrels in the diet.

Although habitat quality for lynx tends to be evaluated in terms of habitat for prey, other factors, including openings in which snowshoe hares can be caught and denning habitat, are important to lynx. Denning sites have tended to be in areas with abundant CWD (Mowat et al. 2000). Management of forests for lynx must be based on site-specific conditions and address all of the life needs of the lynx, including those for reproduction.

Riparian associates

Raccoon

Raccoons occur in a fairly wide range of habitats throughout the West, but to a greater degree than other procyonids, near water. They are uncommon in upland forests and woodlands, and above 2500 m, which effectively excludes them from coniferous forests of the interior West. However, their distribution across habitats, including deserts and interior mountains, and abundance have increased in the last 70 years, abundance by as much as an estimated factor of 20 (Sanderson 1987). Human-caused environmental changes have been credited with contributing to that increase (Gehrt 2003). As a result, raccoons are seldom identified as a conservation concern in forest management. More commonly they are considered pests or, where introduced, threats to indigenous taxa (Hartman and Eastman 1999).

Habitats of raccoons in western forests have been studied little. Riparian forest structure can be important for resting sites, because raccoons rest in tree cavities, and can attain body sizes large enough to require large cavities, and therefore large trees. Over most of the range of raccoons, resting trees have tended to be broad-leaved species. Stuewer (1943) found that resting cavities in Michigan were 3–12 m above ground, with average dimensions 29 cm × 36 cm. Where large trees with hollow cavities are unavailable, raccoons rest in holes in the ground, brush piles, and human structures, but may be more vulnerable to coyotes and other predators when resting near the ground. Raccoons eat an exceptionally wide range of foods (Gehrt 2003), with plant foods tending to predominate at most times. Hard mast is important where available, and corn is important where it is grown. Animal foods tend to be most important in late winter and early spring, and include invertebrates and fish (Tyson 1950), muskrats (Dorney 1954), and crippled waterfowl (Yeager and Elder 1945). In urban and suburban settings, human food sources can be important or dominant (Gehrt 2003). The proximity of foraging areas to resting and denning cover may be crucial, but has not been investigated in the West. The primary mortality causes for raccoons over most of their range are human-related, especially trapping and vehicle collisions (Chamberlain et al. 1999, Gehrt 2003).

Ringtail

The smallest of the north-temperate procyonids, ringtails occur in a wide range of habitats (Orloff 1988). They occur up to 2800 m in elevation (Gehrt

2003), but are most common below 1400 m and in riparian settings (Poglayen-Neuwall and Toweill 1988), especially near rocky outcroppings. The forest and woodland types with which they associate include pinyon pine, juniper, oak, and upland conifers (Orloff 1988, Alexander et al. 1994). The conservation status of ringtails is poorly understood; they are common enough to be trapped for fur in Arizona, New Mexico, and Texas. But, they are fully protected in California, and were thought to be declining in Colorado from 1967 to 1973 (Willey and Richards 1974).

Kinds of resting sites and dens used by ringtails are similar to those for raccoons: rock crevices, cavities in trees or snags, and ground burrows dug by other mammals. Ringtails in Oregon tended to den in Douglas-fir snags 15–30 m tall, but occupied areas that were mostly regenerating (5- to 30-year-old) Douglas-fir (Alexander et al. 1994). So, forestry practices that retain large trees, but include early seres, may be beneficial to this species. Being smaller than raccoons, ringtails should have more potential predators, and physical structure could be correspondingly more important. Diets of ringtails are diverse as for other northern procyonids, but include few aquatic foods and tend to favor small vertebrates: rodents, lizards, and birds. Other foods include insects, fruit, and carrion (Gehrt in press). Lacking more specific information about the habitat needs of this species, it is difficult to identify management approaches that will favor or mitigate impacts on them.

Semi-aquatic species
Mink

The American mink lives in wetlands, along watercourses and lakes, and on coastlines. The permanence of water, proximity to water (Allen 1984) and water quality (Osowski et al. 1995) have been shown to be important predictors of habitat value to minks. Vegetation types with which minks associate are highly varied, ranging from emergent herbs in marshes to stream banks, lakeshores, riparian shrub thickets, and riparian forests. In coastal settings, minks forage primarily in the intertidal zone (Hatler 1976), and short intertidal lengths are preferred (Johnson 1985), as for river otters. Selection for specific timber types is difficult to isolate from other habitat factors, however woody structures, particularly CWD, have appeared important for resting in Idaho (Melquist et al. 1981). Especially in Europe, where the American mink has been studied intensively, resting and denning structures associated with trees have been found to be important (Birks and Linn 1982). In marshes, minks commonly rest and den in abandoned muskrat burrows. Johnson (1985) noted that minks in southeast Alaska were more common than expected on beach segments bordered by clear-cuts, compared to uncut old-growth forest. Clearly, the knowledge base for mink ecology is not sufficient for specific habitat prescriptions for upland coniferous forests. However, the importance of physical structure, including woody vegetation and CWD in riverine settings, seems generally clear.

River otter

River otters occur in marine and freshwater habitats over much of the West. Distribution apparently responds to water quality, and densities tend to be highest in unpolluted coastal waters protected from strong wave action, estuaries, lower stretches of rivers, and coastal marshes (Melquist and Dronkert 1987). After a period of widespread scarcity across the West, natural colonization of otters out of refugia such as Yellowstone National Park has restored parts of the former range (Buskirk 1999). Similarly, widespread re-introductions, not well mapped, have restored parts of the original distribution, including that in southern Colorado and northern Utah.

Diets vary geographically, but are dominated by aquatic prey, particularly fish commonly described as "coarse" or slow-swimming (e.g. sculpins (Cottidae), whitefish (Salmonidae)) and crustaceans. Broad habitat types preferred by river otters vary with season, partly as a function of ice conditions (Reid et al. 1994), which can dramatically affect movements between water and resting sites. Otters in central Idaho preferred valley bottoms to fast-flowing mountain streams, although otters used the latter to access ponds and reservoirs (Melquist and Hornocker 1983). Similarly, otters in arid northeastern Nevada preferred low-gradient streams to high-gradient streams (Bradley 1986). Two factors important to river otters are of particular interest relative to the management of coniferous forests. The first is the overriding importance of woody vegetation, directly and indirectly, to otters, particularly in the interior West. Otters are associated with shrubs and trees directly via log jams and other woody structure used for resting, and indirectly by associating closely with beavers, which require woody vegetation for food, lodges, and dams (Bradley 1986). Otters in Idaho used bank dens and stick lodges of beavers and log jams for resting (Melquist and Hornocker 1983), and those in Nevada were positively associated with the abundance of beavers (Bradley 1986). The second important factor is the proximity of the physical structure where they can rest to water where they can forage. Larsen (1983) found in southeast Alaska that river otters foraging in marine waters preferred to rest in forest <20 m from the shoreline, with a short intertidal length. Melquist and Hornocker (1983) showed that otters in Idaho avoided even high-quality foraging waters if using them required crossing broad open beaches or mudflats to reach resting sites. The proximity of human activities did not prevent habitat use by otters, unless the otters were harassed.

Otters are fully protected over most of the southern part of their western range, but are trapped for fur in the North, and are taken incidentally to beavers wherever the latter are trapped. Forestry management practices that protect streamside woody vegetation, that do not reduce beaver densities, and that retain woody structures near the edges of permanent water should mitigate the effects of timber harvesting on river otters.

Life needs met by forested habitats

Understanding how forests meet specific life needs of animals helps to predict how forest management will affect those animals. For example, knowing that a species avoids predation on its young by giving birth in bole cavities will suggest that reducing the availability of bole cavities may increase neonatal losses to predators. However, it may not; the habitat feature in most cases will not have been shown specifically to limit the distribution or abundance of the species of interest. Often an alternate, unstudied habitat feature can be substituted for one that is removed, with no perceptible loss to the population. This approach – observing some pattern of resource selection and assuming that the resource is limiting – is common, and is one of the few approaches available to resource managers in the face of biological uncertainty (Ruggiero and McKelvey 2000). Among the needs of small and mid-sized carnivores likely to be addressed by forested habitats, the following seem particularly important.

The need for predator escape and avoidance

Most, if not all, small and mid-sized carnivores are vulnerable to predation and aggression by larger-bodied carnivores (Buskirk 1999). One of the most common ways these species avoid predation is by sheltering themselves with a physical structure, which can take the form of burrows, rock crevices, bole cavities, tree branches, log jams, shrubs, and CWD on the forest floor. Of these, some (e.g., burrows, rock crevices) are found in non-forested habitats, but forests are unique among major biomes in the amount of above-ground physical structure they provide. Gray foxes, martens, fishers, river otters, raccoons, ringtails, bobcats, and Canadian lynx are all known to use trees, snags, or CWD to escape predators, and weasels, skunks, and wolverines likely do as well. Martens and fishers are especially averse to traveling in areas without overhead cover (Buskirk and Powell 1994). Rock ledges and outcroppings and boulder fields provide cover for a remarkable range of carnivores, including ringtails, bobcats, and martens. These structures seem to be important mostly for predator avoidance, rather than thermal protection, and are not likely to be affected by forestry practices. Managing for physical structure, especially tree cavities and large-diameter CWD near the ground, is one of the few actions that plausibly could benefit mesocarnivores as a group. Although difficult to generalize, it appears likely that some relationship exists between the size of the animals that need structure and the size of the structures needed. However, species with strong ties to woody structures in some parts of their range can thrive beyond the environmental limits of forests; for example, river otters on the Alaska Peninsula south of the limit of trees.

The need for food

Foods of forest mesocarnivores include a huge range of animals, plants, and fungi: insects, crustaceans, vertebrates from passerine birds to whale carrion, berries, seeds, nuts, and agricultural crops. Although mesocarnivores must respond at some level to site productivity, the tremendous range of foods they eat is produced in such a wide range of environments, and so difficult to measure, that apparent food abundance is not highly predictive of the distribution or abundance of most species. The only possible approach to fulfilling the food needs of mesocarnivores is on a site-specific, species-specific, and time-specific basis, but even this may not be a profitable approach for managers, except via habitat management.

Managing for habitats of the prey of specific mesocarnivores is a common management approach to providing for these species. Red-backed voles (*Clethrionomys* spp.) have well-documented ties to structurally complex mesic coniferous forests, and to large-diameter CWD (Hayes and Cross 1987). Snowshoe hares, by contrast, are associated with moderate to high densities of small-diameter woody stems within reach of the ground or snow surface, and with overhead cover (Mowat et al. 2000). Pine squirrels (*Tamiasciurus* spp.), northern flying squirrels (*Glaucomys sabrinus*), muskrats (*Ondatra zibethicus*), and grouse (Tetraoninae) also have habitat needs that have been documented regionally and should be considered in managing for species that eat them.

The need for access to foraging areas

Access to foraging areas relates to some of the same factors as predator avoidance: physical structure and the juxtaposition of foraging areas with resting or denning sites. For river otters and mink, this translates into ice-free routes from foraging areas to resting sites in winter; lakes with complete ice cover become inaccessible, whereas streams with intermittent open leads or passages beneath the ice may facilitate movements from resting to foraging areas. For martens, the proximity of late successional forests to meadow edges is an important factor in winter foraging (Spencer et al. 1983), as is the subnivean structure mentioned previously (Corn and Raphael 1992). Carnivores that are too large for subnivean foraging (e.g., coyotes, bobcats, and red foxes) and that experience high energetic costs of traveling on the snow surface will be more likely to exhibit seasonal migrations to low-elevation areas or south-facing areas with less snow or crusted snow.

The need for protected thermal environments

By nature of their small body sizes, many small carnivores have inherently high mass-specific metabolic rates, along with limited somatic stores that translate into brief fasting endurance times (e.g., Harlow and Buskirk 1991). Such species are likely to enjoy energetic savings by occupying protected microenvironments that can be warmed by convective heat losses. The thermal energetic basis of micro-habitat selection has been shown most clearly for weasels (Casey and Casey 1979) and the marten (Taylor and Buskirk 1994), but likely applies to small-bodied species generally. A wide range of site types could offer energetic savings, but the kinds of resting sites and dens chosen by diverse species from ringtails to river otters suggest that cavities within or partially enclosed by tree boles are especially important.

Habitat features important to mesocarnivores

Many habitat features are potentially important to small and mid-sized carnivores, but only a few of them tend to be studied: those that are most easily measured, yet have strong predictive value.

Vegetative species dominants

Animals select vegetative species dominants for physical, chemical, or other attributes, not for their taxonomic classification. However, the dominance of some plant species dominants so consistently predicts certain environmental conditions that we can generalize about which animal species will occur with which plant species, with low risk of being wrong. Site moisture is a prime example. Jeffrey pine (*Pinus jeffreyi*), and ponderosa pine (*P. ponderosa*) tend to occur on sites with severe summer droughts and historically frequent fires. With exceptions, such sites tend to have sparse regeneration and little structure near the ground, compared with moist-site species such as Engelmann spruce or subalpine fir. Because martens, fishers, and lynx tend to avoid stands with little understory structure, they tend to avoid dry forest types dominated by Jeffrey or ponderosa pine. Another example relates to the production of hard mast; pinyon pines (e.g. *Pinus edulis*) and whitebark pine (*P. albicaulis*), while ecologically very different, both produce fleshy seeds, in great abundance in some years. These nuts can be keystone resources for communities, but where these species produce few or no seeds, mesocarnivores can be expected to be indifferent to them as a food source.

Successional stage

Successional stage is most commonly mentioned as predictive of habitat value for martens, fishers, and bobcats. The first two species tend to be associated with old-growth stands in the temperate West, although martens have been shown to associate with early post-fire seres in the interior North. Bobcats associate with early seres that produce high densities of lagomorphs in the South. Evidence for lynx varies by region, with evidence suggesting the value of early seres on sites with moist growing seasons. This has not been shown for areas with dry summers in the southern part of the range. Wolverines have not been shown to be associated with specific forest successional stages, but the forests they occupy in the temperate West tend to be subalpine, with long fire-return intervals. Thus, the successional stage of coniferous forests is not highly predictive of habitat value for small and mid-sized carnivores in general, and even for single species may have inconsistent or bi-modal effects.

Snow

Snow attributes have potentially strong explanatory power in the distribution and abundance of western carnivores. Indeed, most habitat attributes dealing with physical structure near the ground must be evaluated in the context of snow depth and characteristics. Thus, CWD close to the ground that provides overhead cover during the snow-off period may, once submerged by snow, serve no function or a different one (e.g., thermal protection). Likewise, snow that impedes movement by all but snow-adapted mammals may, after a brief thaw, develop a crust that allows free movement by many species. Some species, particularly the procyonids, have apparently poor adaptations to snow, and do not occur in the North. Others, like the marten and lynx, are superbly adapted to snow by nature of their light foot loadings, occur only in areas with cold, snowy winters, and likely enjoy competitive advantages over less well-adapted carnivores in extensive areas of soft, deep snow (Buskirk et al. 2000*a*). Snow has been shown to mediate spatial interactions between coyotes and lynx (Murray and Boutin 1991) and between fishers and martens (Krohn et al. 1997), in each case favoring the species with the lightest foot loadings, and presumably greatest ease of travel in deep snow.

Landscape features

Landscape features refer to the amounts and spatial attributes of patch types, including edge densities, road densities, patch connectivity, and amounts of stands in various distance intervals from various structures (e.g., roads and edges). Landscape features are unstudied for most small and mid-sized carnivores. Some evidence suggests that coyotes associate with roads where competition by wolves prevents their use of more remote areas (Thurber et al. 1992), and Buskirk et al. (2000*a*) hypothesized that roads could facilitate movements of coyotes into areas with deep snow in winter. Martens are negatively affected by landscape-scale loss of forest (Bissonette et al. 1997).

Roads

Although recent interest has focused on the effects of roads and vehicles on populations and behaviors of vertebrates (Forman and Alexander 1998), evidence for effects on mid-sized carnivores is scarce and contradictory. A fundamental problem is that the distribution of roads is confounded by other landscape features. For example, roads might be built in productive forests that are chosen for timber harvest, so that isolating the effect of roads from the productivity of the site can be difficult (Cooperrider et al. 1999). Likewise, roads exert influences through various means, including reducing forest cover, improving access for trappers (Hodgman et al.

1994), increasing sedimentation into streams, and providing access for competing carnivores, so that proximal effects may be difficult to isolate. Wolves are important community structuring agents among carnivores, and the distribution of wolves has been shown in several studies to respond to road densities. Wolves tend to be absent where road densities exceed 0.45 km/km^2 (Mladenoff et al. 1995) to 0.6 km/km^2 (Thurber et al. 1994). Bobcats (Lovallo and Anderson 1996, Riley 1999), wolverines (Carroll et al. 2001), lynx (Apps 2000), and martens (Robitaille and Aubry 2000) have been shown to prefer habitats with low densities of roads, avoid crossing roads within their home ranges, or avoid roads in proportion to traffic levels. Lynx in northern Washington, by contrast, were found to use habitats without apparent influence from roads (McKelvey et al. 2000b), and habitats selected by lynx and fishers in the northern Rockies did not respond to road densities (Carroll et al. 2001). Where analyzed, more developed roads (paved, with higher vehicle speeds) tend to have stronger behavioral effects on carnivores.

Non-habitat features

Inter-specific competition

Small and mid-sized carnivores are strongly affected by inter-specific competition and intraguild predation, which are difficult to distinguish from each other in the field. A wide range of species is implicated in these interactions, which are known from various kinds of data: inverse population fluctuations (Latham 1952), aggression directly observed or inferred at necropsy (Schamel and Tracy 1986), and lack of sympatry attributed to competitive exclusion (Krohn et al. 1995). The species involved in these interactions are remarkably diverse and include virtually all those treated here. Coyotes are commonly dominant over smaller canids, bobcats, and lynx (reviewed by Buskirk 1999) and lynx are notorious enemies of red foxes (Stephenson et al. 1991). Fishers are competitively dominant over American martens in settings where snow is shallow or crusted (Krohn et al. 1997). Almost invariably, larger-bodied species dominate interactions with smaller ones, and interactions tend to be stronger between species of similar body form and niche (e.g., coyote and foxes) than between species with highly dissimilar shapes (e.g., otters and badgers). Relative body size also is an important predictor of these interactions. For example, brown bears (*Ursus arctos*, ca. 10^2 kg body weight) compete little with weasels (ca. 10^{-2} kg), but compete fiercely with slightly smaller black bears (*U. americanus*) (Mattson et al. 1992). These interactions are increasingly recognized as fundamental to community structure of carnivores, and are invoked to explain the absence of some species from some sites (e.g., Crooks and Soulé 1999).

Environmental contaminants

Although a number of causes of acute intoxication of wildlife have been reported, chronic exposure to environmental contaminants is an increasingly recognized cause of population-level effects on a number of small and mid-sized carnivores. Forest and wildlife managers will need to be cognizant of the range and potential magnitude of these effects, although they will in few cases have access to site-specific monitoring data. Evidence for chronic intoxication in carnivores has been of several kinds. First, pathological lesions in wild carnivores consistent with those expected from a particular contaminant have shown that disease is occurring among wild animals and can be linked to a specific toxin (Wren 1985). A variant of this line of evidence is finding contaminant levels in wild animals greater than those that have been found to cause symptoms, lesions, or death in captive surrogates (Blus and Henny 1990, Osowski et al. 1995). Second, bioaccumulation of contaminants as they ascend food pyramids, or over the life of an individual animal demonstrates the ecological processes expected to affect top carnivores over their lifetimes (Organ 1989). Third, the absence of carnivores from contaminated areas but not in uncontaminated areas would, with replication, tend to support a contaminant-based explanation for such a distributional anomaly (Szumski 1998). All of these lines of evidence have been documented for North American mesocarnivores, particularly of species involved in aquatic food chains. This leads us to hypothesize that chronic exposure to contaminants is a potentially huge, yet poorly understood, problem for mesocarnivores, particularly those that forage in aquatic systems: minks, river otters, and possibly raccoons.

Changing approaches to and priorities for carnivore conservation

Historically, mesocarnivores were appreciated, if at all, primarily for the economic value of their skins. Indeed, a number of species became so valuable that trapping, weakly regulated, led to major population declines. In California, for example, it was clear by the 1920's that the marten, fisher, and wolverine would require special conservation efforts (Dixon 1924), a concern that recurred in the 1940's (Twining and Hensley 1947). In addition, government agents poisoned and trapped carnivores to protect livestock, and as a general principle of game management. Early conservationists, with little knowledge of the role of carnivores in ecosystems, could predict few tangible consequences of those policies: overabundant ungulates, loss of income from furs, or damage to crops from uncontrolled rodent populations.

The protection of vulnerable fur-bearing species from overexploitation in the mid-1900's was a pivotal conservation measure. This was followed by increased interest in a wide array of species for their intrinsic values and for their contributions to ecosystems. Initially, this increased interest was in individual species at risk, mandated by the Endangered Species Act (ESA). Of the mesocarnivores of western forests, the fisher, wolverine, and lynx have been petitioned for listing under ESA; only the lynx has been listed to date (Federal Register 65(58):16052). A number of other species are protected from trapping, or receive special protection, in individual states or provinces.

The conservation of wild carnivores is shifting away from a focus on individual species and toward the collective role of carnivores in ecosystem functioning (reviewed by Aubry et al. 2003) and monitoring (Noss et al. 1996, Terborgh et al. 1999). Carnivores have two important roles. First, they perform important ecological services, including the cycling of nutrients (Ben-David et al. 1998), transferring energy, completing or interrupting the life cycles of parasites, and structuring the communities of non-prey species, especially each other (Aubry et al. 2003). Several researchers have proposed trophic cascades initiated by carnivores, the effects of which extend to primary producers (McLaren and Peterson 1994, Berger et al. 2001). This "top-down" influence by predators (Matson and Hunter 1992) is supported by anecdotal evidence (predator elimination or introduction: King 1984, McShea et al. 1997; herbivore introduction in the absence of predators, Vitousek 1988) and limited experimental evidence (e.g., Terborgh 1992, Krebs et al. 1995*b*). The extirpation of dominant predators has been described as causing "mesopredator release" (Soulé et al. 1988, Palomares et al. 1995, Crooks and Soulé 1999) wherein the size of the largest predator is reduced by the functional extinction of the previously larger predator, and the newly top predators exert unnaturally high levels of predation on even smaller prey – typically ground-nesting birds and rodents. This form of top-down regulation can feature the same species playing different roles under different circumstances. For example, where large carnivores are absent, coyotes can serve a top-predator role by suppressing the activities of gray foxes and domestic cats (*Felis domesticus*), which, in turn, benefits small bird and mammal populations (Soulé et al. 1988, Crooks and Soulé 1999). The presence of wolves, however, reduces coyote populations, which is expected to have positive effects on other mesocarnivores and on the prey of coyotes (Buskirk 1999). Not all studies support the "top-down" paradigm (e.g., Ryszkowski et al. 1973, Goszczyński 1977, Erlinge et al. 1984, Jedrzejewska and Jedrzejewski 1998) and some (Wright et al. 1994, Polis and Strong 1996) have argued that top-down control exerted by carnivores in nature is uncommon. Regardless of the strength of individual pathways of influence by carnivores on ecosystems, this recent work highlights the diversity and potentially keystone role of carnivores, including mesocarnivores, in ecosystem function, providing further incentives for their conservation and study.

The second important value of carnivores is that they provide information about the integrity of natural systems. Their large home ranges and low population densities make them vulnerable to declines (Weaver et al. 1996, Minta et al. 1999) and, therefore, sensitive indicators of environmental change (Clark et al. 1996, Noss et al. 1996). Species considered here that appear to have the greatest potential as indicators are: (1) those that specialize on late-successional forests (e.g., marten, fisher); (2) those that forage in aquatic environments where contaminants tend to concentrate (e.g., river otter, mink); and (3) those whose space needs are extremely large (e.g., wolverine, lynx).

Considering these important roles of mesocarnivores in natural communities, it is unfortunate that our knowledge of their distributions and abundances in the West is so poor. The declining role of fur-trapping in society means a loss of distributional and demographic information that was generated by fur harvests at little public cost. A number of nonlethal approaches to surveying carnivores (e.g., Zielinski and Kucera 1995) have substituted for this information. Methods that record the tracks of a species on a prepared surface (Zielinski 1995), or that photograph animals remotely (Kucera et al. 1995) have been used to map the distributions of some mesocarnivores (e.g. Kucera et al. 1995, Zielinski et al. 1995) and have been proposed as a means to monitor changes in populations (Zielinski and Stauffer 1996). Similar data have been used to produce regional habitat models (Carroll et al. 1999*b*), which could not have been imagined before remotely sensed satellite images of vegetation,

topography, and climate. Output from these models can be used to compare the habitat suitability of lands with different management histories
(Carroll et al. 1999b). In addition, habitat models developed for separate
species can be compared to understand the habitat attributes that are of
common value, versus the attributes that separate the needs of species
from each other (e.g., Lindenmayer and Lacy 1995, Carroll et al. 1999b,
2001). Even the compilation of carnivore sightings, if screened for reliability, can provide important information about distribution (e.g., Aubry
and Houston 1992) and can be used to model and predict habitat use
(Palma et al. 1999, Carroll et al. 2001). The remote collection of genetic
samples from either hairs or scats (Foran et al. 1997) will likewise provide
understanding of mesocarnivore populations.

These detection methods make it possible to address the distribution
and abundance of mesocarnivores at various spatial scales. Traditionally,
the study of habitat has focused on stand-level attributes, or the aggregation of stand information for analysis at the level of the home range.
The consideration of regional-scale influences on carnivore populations
is relatively new, but such analyses can suggest habitat factors not evident from smaller-scale field studies. For example, correlations between
fisher occurrences and such regional factors as elevation, distance to the
ocean, precipitation, and other geographical variables (Klug 1996, Carroll
1997) may be as strong as, or stronger than, within-patch, home range, or
even landscape-level habitat variables (Carroll et al. 1999a). The relative
contribution of different scales of habitat influence on distribution or
abundance has rarely been evaluated for mesocarnivore species, but is
profoundly important to managers.

Determining the current distribution of mesocarnivores is the first
step toward understanding their status and the first priority for their
conservation. Surveys can collect information simultaneously on most
species of mammals that are attracted to meat baits. Systematic surveys
are also an ideal means to locate patchily distributed populations, even
those that have gone undetected for decades using other methods (e.g., the
Humboldt marten (*Martes americana humboldtensis*) (Zielinski et al. 2001)).
Such large-scale surveys complement, and in some cases substitute for,
data acquired via traditional telemetry studies. The latter method usually
is limited to a relatively small area of a species' range and usually where
the species is common, limiting inferences about areas where it is rare.
We suggest that periodic extensive surveys are the best foundation for
conserving mesocarnivores in western forests. They can describe current

geographical range, be used to develop habitat models at scales that are relevant for decision-making, and provide information on abundance.

The concept of a metapopulation (Levins 1968, Wright 1978, Harrison and Taylor 1997) linked by occasional dispersal may be especially applicable to the conservation of mature-forest specialists in the fragmented habitats of western North America. Unlike birds, forest mammals, particularly the late-successional specialists, are fairly dispersal-sensitive, strongly preferring to travel under dense tree canopies (Buskirk and Powell 1994). The discontinuous patterns of habitat for fishers in the West (Weir and Harestad 1997, Carroll et al. 1999*a*) and for martens on mountain-top islands (Wisz 1999) are examples of habitat discontinuities that could give rise to metapopulations. Alternatively, these spatially structured populations could have the properties of source-sink systems (Pulliam 1996), in which the gradients of dispersal are consistently away from areas that produce a surplus of animals and toward areas that require dispersive inputs for their populations to persist. If clusters of suitable habitat for a species become too small or isolated, the imbalance between immigration and emigration can affect viability (Noon and McKelvey 1996). The decline in distribution of fishers (Powell and Zielinski 1994) and wolverines (Banci 1994) may be due to such regional-level dynamics. Dispersal limitation may affect regional population viability by reducing the connections between subpopulations and lowering the likelihood that suitable habitat patches are colonized and re-colonized over time (Fahrig and Paloheimo 1988). The conservation of forest habitat specialists, such as the fisher, lynx, and marten, as well as other species that require wilderness conditions (e.g., wolverine) will require planning to protect large blocks of suitable habitat and the connections between them (Bissonette and Broekhuizen 1995).

Mesocarnivores in conservation planning

Managing ecosystems requires monitoring the composition, structure, and function of the system. The large area requirements, top trophic status, and low population densities shared by many carnivore species make them ideally suited as indicators of ecosystem integrity and, therefore, as tools for addressing the conservation of ecosystems (Noss et al. 1996, Terborgh et al. 1999). Whether carnivores serve as keystone species in forest ecosystems or not (Mills et al. 1993, Noss et al. 1996), their relatively large area requirements suggest that, if their populations are abundant and well-distributed, the habitat – and perhaps the populations – of many other species will also be protected (Noss and Cooperrider 1996:162). The area requirements and dispersal limitation of some mesocarnivores qualify them as "focal species" for conservation planning (Noss et al. 1996, Lambeck 1997, Carroll et al. 2001).

The utility of carnivores in large-scale conservation planning is exemplified by The Wildlands Project (Foreman et al. 1992, Noss 1992) which has codified its approach to ecosystem protection and planning under the guidance of the "3 Cs": carnivores, cores and connectivity. The size of core areas is determined by the area needed by a subpopulation of focal carnivore species. Protected core areas are the central elements of conservation planning and are similar to the concept of refugia, which have previously been promoted as a means of conserving forest carnivores (de Vos 1951, Buskirk 1994). These core areas are then connected via corridors that are selected to facilitate the movement of the focal carnivores, and other ecosystem components and processes, between core areas. By considering the needs of carnivores, the planner has an objective basis for determining the optimal size of reserve networks (Noss et al. 1996).

Most often it is the large carnivores (wolves, grizzly bears, mountain lions (*Felis concolor*)) that are considered the most useful species for conservation planning. This is primarily due to their large spatial needs, but is probably also influenced by their charismatic appeal to the public. Conservation biologists have recognized that mesocarnivores can have the same role, not only because the home range sizes of some species (i.e., wolverine, lynx) are similar to or greater than those of large carnivores, but because the extirpation of large carnivores in many regions has promoted mesocarnivores to the role of top predators. Including a combination of focal carnivore species, especially mesocarnivores that specialize on habitats of concern, may ultimately be the best approach to large-scale conservation. This is because large carnivores are often generalists whose habitats may not include environments with high value for biodiversity (Noss et al. 1996). Carroll et al. (2001) developed regional habitat models for wolverine, lynx, grizzly bear, and fisher in the Rocky Mountains and found that the occurrence of wolverines and grizzly bears was strongly associated with low levels of roads and human population density, whereas fishers and lynx were less influenced by the density of humans but more affected by the density and extent of low-elevation, productive forests. These results, and those of a similar study that included ten species of carnivores (Carroll et al. 2001), indicate that a comprehensive conservation strategy for carnivores in the

Rocky Mountains should consider the needs of several carnivore species rather than a single presumed umbrella species.

Recent policy developments in the management of federal lands in the United States suggest that regional-scale dynamics are appreciated and considered in land-management planning (e.g., Interior Columbia River Basin Ecosystem Management Project USDA 1996, Northwest Forest Plan, USDA and USDI 1994, Sierra Nevada Framework, USDA 2001). However, many planning processes are still poorly adapted to decision-making across jurisdictional boundaries. Information at the regional scale is the dominant source of knowledge for long-term and large-scale conservation planning (Clark and Minta 1994, Mladenoff et al. 1995, Soulé and Terborgh 1999). If we are to maintain viable populations of species that have large home ranges and are vulnerable to human activities, then the "conservation planner must grabble with the design and management of entire landscapes" (Noss et al. 1996). The conservation of carnivores in landscapes is as much a policy issue as a scientific one (Clark et al. 1996, Primm and Clark 1996). Biological science is fundamental to strategies for conserving mesocarnivores, but political solutions will require integration with the social sciences, education, and law. Therein lies an imposing challenge for coming decades.

Summary

Small and mid-sized carnivores of western forests represent an ecologically diverse and influential guild of forest vertebrates. They comprise over a dozen species with varied life histories and habitat associations, highly species-specific responses to human-caused habitat change, and varied conservation status across western North America. Forested habitats fulfill various of their life needs, including the need for predator escape and avoidance (e.g., marten and river otter), for protected thermal environments (e.g., marten), for specialized foods (e.g., lynx) and for access to specialized foraging areas (e.g., ermines). For several species, the specific life needs fulfilled by forested habitats can only be inferred. Vegetative species dominants are somewhat predictive of the distribution and abundance of mesocarnivores, inasmuch as moist-site species tend to be associated with long fire-return intervals, and seed-producing species (e.g., whitebark pine) attract certain species. Successional stage is highly predictive of the distribution and abundance of some species, particularly martens, fishers, and bobcats, but not at all for others. Several species (e.g., coyote, river otter, ermine) are facultative forest associates and widely distributed and abundant far from trees. The effects of landscape pattern on the distribution and abundance of mid-sized carnivores are just beginning to be understood, with percent of an area in uncut forest, road density, and edge density having been shown important predictors. Inter-specific competition is a powerful influence on the distribution and abundance of mesocarnivores, with size-mediated (top-down) effects known for most species treated here. By the nature of their high trophic positions, carnivores, particularly those that participate in aquatic food webs, are vulnerable to individual- and population-level contaminants, including those introduced or mobilized by human actions. Our values regarding and approaches to monitoring mid-sized carnivores have shifted from managing them for fur and to minimize their depredations to valuing them more broadly. Today, the values attached to them are more intrinsic and esthetic, and related to their contributions to community and ecosystem processes. They are also important tools for monitoring the status of natural systems at various spatial and ecological scales. Large carnivores are more commonly used in conservation plannning, but the habitat specialization and huge spatial requirements of some mesocarnivores may make them more suitable and more constraining conservation planning tools.

Literature cited

Alexander, L.F., B.J. Verts, and T.P. Farrell. 1994. Diet of ringtails (*Bassariscus astutus*) in Oregon. *Northwestern Naturalist* **75**:97–101.

Allen, A.W. 1984. Habitat suitability index models: mink. US Fish and Wildlife Service **FWS/OBS-82/10.61** Revised. 19pp.

Apps, C.D. 2000. Space-use, diet, demographics, and topographic associations of lynx in the southern Canadian Rocky Mountains: a study. Pages 351–371 *in* L.F. Ruggiero, K.B. Aubry, S.W. Buskirk, G.M. Koehler, C.J. Krebs, K.S. McKelvey, and J.R. Squires, editors. *Ecology and Conservation of Lynx in the United States.* University Press of Colorado, Boulder, Colorado, USA.

Aubry, K.B. 1983. The Cascade red fox: distribution, taxonomy, zoogeography and ecology. Dissertation. University of Washington, Seattle, Washington, USA.

Aubry, K.B. 1984. The recent history and present distribution of the red fox in Washington. *Northwest Science* **58**:69–79.

Aubry, K.B. and D.B. Houston. 1992. Distribution and status of the fisher (*Martes pennanti*) in Washington. *Northwestern Naturalist* **73**:69–79.

Aubry, K.B., F.E. Wahl, J. Von Kienast, T.J. Catton, and S.G. Armentrout. 1997. Use role of remote video cameras for the detection of forest carnivores and in radio-telemetry studies of fishers. Pages 350–361 *in* G. Proulx, H.N. Bryant, and

P.M. Woodard, editors. Martes: *Taxonomy, Ecology, Techniques, and Management*. Provincial Museum of Alberta, Edmonton, Alberta, Canada.

Aubry, K.B., G.M. Koehler, and J.R. Squires. 2000. Ecology of Canada lynx in southern boreal forests. Pages 373–396 *in* L.F. Ruggiero, K.B. Aubry, S.W. Buskirk, G.M. Koehler, C.J. Krebs, K.S. McKelvey, and J.R. Squires, editors. *Ecology and Conservation of Lynx in the United States*. University Press of Colorado, Boulder, Colorado, USA.

Aubry, K.B., J.P. Hayes, B.L. Biswell, and B.G. Marcot. 2003. The ecological role of tree-dwelling mammals in western coniferous forests. Pages 405–443 *in* C.J. Zabel and R.G. Anthony, editors. *Mammal Community Dynamics. Management and Conservation in the Coniferous Forests of Western North America*. Cambridge University Press, Cambridge, UK.

Bailey, V. 1936. The mammals and life zones of Oregon. *North American Fauna* **55**: 1–416.

Baker, J.M. 1992. Habitat use and spatial organization of pine marten on southern Vancouver Island, British Columbia. M.S. Thesis. Simon Fraser University, Burnaby, British Columbia, Canada.

Banci, V.A. 1987. Ecology and behavior of wolverine in Yukon. M.S. Thesis. University of British Columbia, Vancouver, British Columbia, Canada.

Banci, V.A. 1994. Wolverine. Pages 99–127 *in* L.F. Ruggiero, K.B. Aubry, S.W. Buskirk, L.J. Lyon, and W.J. Zielinski, editors. The scientific basis for conserving forest carnivores, American marten, fisher, lynx, and wolverine in the western United States. USDA Forest Service General Technical Report **RM-254**, Fort Collins, Colorado, USA.

Ben-David, M., T.A. Hanley, and D.M. Schell. 1998. Fertilization of terrestrial vegetation by spawning Pacific salmon: the role of flooding and predator activity. *Oikos* **83**:47–55.

Berger, J., P.B. Stacey, L. Bellis, and M.P. Johnson. 2001. A mammalian predator-prey imbalance: grizzly bear and wolf extinction affect avian neotropical migrants. *Ecological Applications* **11**:947–960.

Birks, J.D.S. and I.J. Linn. 1982. Studies of home range of the feral mink (*Mustela vison*). *Symposia of the Zoological Society of London* **49**:231–257.

Bissonette, J.A. and S. Broekhuizen. 1995. Martes populations as indicators of habitat spatial patterns: the need for a multiscale approach. Pages 95–121 *in* W.Z. Lidicker, Jr., editor. *Landscape Approaches in Mammalian Ecology and Conservation*. University of Minnesota Press, Minneapolis, Minnesota, USA.

Bissonette, J.A., D.J. Harrison, C.D. Hargis, and T.G. Chapin. 1997. The influence of spatial scale and scale-sensitive properties on habitat selection by American marten. Pages 368–385 *in* J. A. Bissonette, editor. *Wildlife and Landscape Ecology*. Springer-Verlag, New York, NY, USA.

Blus, L.J. and C.J. Henny. 1990. Lead and cadmium concentrations in mink from northern Idaho. *Northwest Science* **64**:219–223.

Bradley, P.V. 1986. Ecology of river otters in Nevada. M.S. Thesis, University of Nevada, Reno, NV, USA.

Buck, S., C. Mullis, and A. Mossman. 1994. Habitat use by fishers in adjoining heavily and lightly harvested forest. Pages 368–376 *in* S.W. Buskirk, A.S. Harestad, M.G. Raphael, and R.A. Powell, editors. *Martens, Sables, and Fishers: Biology and Conservation*. Cornell University Press, Ithaca, New York, USA.

Buskirk, S.W. 1984. Seasonal use of resting sites by marten in southcentral Alaska. *Journal of Wildlife Management* **48**:950–953.

Buskirk, S.W. 1994. The refugium concept and the conservation of forest carnivores. Pages 242–245 *in Proceedings of the XXI International Congress of Game Biologists*, Halifax, Nova Scotia, Canada.

Buskirk, S.W. 1999. Mesocarnivores of Yellowstone. Pages 165–187 *in* T.W. Clark, A.P. Curlee, S.C. Minta, and P.M. Kareiva, editors. *Carnivores in Ecosystems: the Yellowstone Experience*. Yale University Press, New Haven, Connecticut, USA.

Buskirk, S.W. and S.O. MacDonald. 1984. Seasonal food habits of marten in south-central Alaska. *Canadian Journal of Zoology* **62**:944–950.

Buskirk, S.W. and R.A. Powell. 1994. Habitat ecology of fishers and American martens. Pages 283–296 *in* S.W. Buskirk, A.S. Harestad, M.G. Raphael, and R.A. Powell, editors. *Martens, Sables, and Fishers: Biology and Conservation*. Cornell University Press, Ithaca, New York, USA.

Buskirk, S.W. and L.F. Ruggiero. 1994. American marten. Pages 38–73 *in* L.F. Ruggiero, K.B. Aubry, S.W. Buskirk, L.J. Lyon, and W.J. Zielinski, editors. The scientific basis for conserving forest carnivores, American marten, fisher, lynx, and wolverine in the western United Statees. U.S. Forest Service General Technical Report **RM-254**, Fort Collins, Colorado, USA.

Buskirk, S.W., L.F. Ruggiero, and C.J. Krebs. 2000*a*. Habitat fragmentation and interspecific competition: implications for lynx conservation. Pages 83–100 *in* L.F. Ruggiero, K.B. Aubry, S.W. Buskirk, G.M. Koehler, C.J. Krebs, K.S. McKelvey, and J.R. Squires, editors. *Ecology and Conservation of Lynx in the United States*. University Press of Colorado, Boulder, Colorado, USA.

Buskirk, S.W., L.F. Ruggiero, K.B. Aubry, D.E. Pearson, J.R. Squires, and K.S. McKelvey. 2000*b*. Comparative ecology of lynx in North America. Pages 397–417 *in* L.F. Ruggiero, K.B. Aubry, S.W. Buskirk, G.M. Koehler, C.J. Krebs, K.S. McKelvey, and J.R. Squires, editors. *Ecology and Conservation of Lynx in the United States*. University Press of Colorado, Boulder, Colorado, USA.

Carey, A.B. and J.E. Kershner. 1997. *Spilogale gracilis* in upland forests of western Washington and Oregon. *Northwestern Naturalist* 77:29–34.

Carroll, C. 1997. Predicting the distribution of the fisher (*Martes pennanti*) in northwestern California, USA, using survey data and GIS modeling. M.S. Thesis. Oregon State University, Corvallis, Oregon, USA.

Carroll, C., W.J. Zielinski, and R.F. Noss. 1999*a*. Using presence-absence data to build and test spatial habitat models for the fisher in the Klamath Region, U.S.A. *Conservation Biology* 13:1344–1359.

Carroll, C., P.C. Paquet, and R.F. Noss. 1999*b*. *Modeling Carnivore Habitat in the Rocky Mountain Region: a Literature Review and Suggested Strategy*. World Wildlife Fund, Toronto, Ontario, Canada.

Carroll, C., R.F. Noss, and P.C. Paquet. 2001. Carnivores as focal species for conservation planning in the Rocky Mountain region. *Ecological Applications* 11:961–980.

Casey, T.M. and K.K. Casey. 1979. Thermoregulation of arctic weasels. *Physiological Zoology* **52**:153–164.

Chamberlain, M.J., K.M. Hodges, B.D. Leopold, and T.S. Wilson. 1999. Survival and cause-specific mortality of adult raccoons in central Mississippi. *Journal of Wildlife Management* 63:880–888.

Chapin, T.G., D.J. Harrison, and D.D. Katnik. 1998. Influence of landscape pattern on habitat use by American marten in an industrial forest. *Conservation Biology* 12:1327–1337.

Clark, T.W. and S.C. Minta. 1994. *Greater Yellowstone's Future: Prospects for Ecosystem Science, Management, and Policy.* Homestead Press, Moose, Wyoming, USA.

Clark, T.W., A.P. Curlee, and R.P. Reading. 1996. Crafting effective solutions to the large carnivore conservation problem. *Conservation Biology* 10:940–948.

Cooperrider, A., R.F. Noss, H.H. Welsh, Jr., C. Carroll, W. Zielinski, D. Olson, S.K. Nelson, and B.G. Marcot. 1999. Terrestrial fauna of redwood forests. Pages 119–163 *in* R.F. Noss, editor. *The Redwood Forest.* Island Press, Washington DC, USA.

Copeland, J.P. 1996. Biology of the wolverine in central Idaho. M.S. Thesis. University of Idaho, Moscow, Idaho, USA.

Corn, J.G. and M.G. Raphael. 1992. Habitat characteristics at marten subnivean access sites. *Journal of Wildlife Management* 56:442–448.

Crabb, W.D. 1941. Food habits of the prairie spotted skunk in southeastern Iowa. *Journal of Mammalogy* 22:349–364.

Crabb, W.D. 1948. The ecology and management of the prairie spotted skunk in Iowa. *Ecological Monographs* 18:201–232.

Crabtree, R. 1993. Gray ghost of the Beartooth. *Yellowstone Science* 1(spring):13–16.

Crooks, K.R. and M.E. Soulé. 1999. Mesopredator release and avifaunal extinctions in a fragmented system. *Nature* 400:563–566.

Crooks, K.R. and D. Van Vuren. 1995. Resource utilization by 2 insular endemic mammalian carnivores, the island fox and island spotted skunk. *Oecologia* 104:301–307

de Vos, A. 1951. Ecology and management of fisher and marten in Ontario. Ontario Department of Lands and Forests, Technical Bulletin, Wildlife Service 1: 1–90.

Dixon, J. 1924. A closed season needed for fisher, marten and wolverine in California. *California Department of Fish and Game* 11:23–25.

Dorney, R.S. 1954. Ecology of marsh raccoons. *Journal of Wildlife Management* 18: 217–225.

Dragoo, J.W. and R.L. Honeycutt. 1997. Systematics of mustelid-like carnivores. *Journal of Mammalogy* 78:426–443.

Edelmann, F. and J. Copeland. 1999. Wolverine distribution in the northwestern United States and a survey in the Seven Devils Mountains of Idaho. *Northwest Science* 73:295–300.

Erlinge, S., G. Goransson, G. Hogstedt, G. Jansson, O. Liberg, J. Loman, I.N. Nilsson, T. Von Schantz, and M. Sylven. 1984. Can vertebrate predators regulate their prey? *American Naturalist* 123:125–133.

Fagerstone, K.A. 1987. Black-footed ferret, long-tailed weasel, short-tailed weasel, and least weasel. Pages 549–573 *in* M. Novak, J.A. Baker, M.E. Obbard, and B. Malloch, editors. *Wild Furbearer Management and Conservation in North America.* Ontario Ministry of Natural Resources, Toronto, Ontario, Canada.

Fahrig, L. and J. Paloheimo. 1988. Effect of spatial arrangement of habitat patches on local population size. *Ecology* 69:468–475.

Fitzgerald, B.M. 1977. Weasel predation on a cyclic population of the montane vole (*Microtus montanus*) in California. *Journal of Animal Ecology* 46:367–397.

Foran, D.R., S.C. Minta, and K.S. Heinemeyer. 1997. DNA-based analysis of hair to identify species, gender, and individuals for population research and monitoring. *Wildlife Society Bulletin* **25**:840–847.

Foreman, D., J. Davis, D. Johns, R. Noss, and M.E. Soule. 1992. The Wildlands Project mission statement. *Wild Earth* (special issue) **2–3**.

Forman, R.T.T. and L.E. Alexander. 1998. Roads and their major ecological effects. *Annual Review of Ecology and Systematics* **29**:207–231.

Fuller, A.K. 1999. Influence of selection harvesting on American marten and their primary prey in northcentral Maine. M.S. Thesis. University of Maine, Orono, Maine, USA.

Fuller, T.K. 1978. Variable home-range sizes of female gray foxes. *Journal of Mammalogy* **59**:446–449.

Fuller, T.K., W.E. Berg, and D.W. Kuehn. 1985. Survival rates and mortality factors of adult bobcats in north-central Minnesota. *Journal of Wildlife Management* **49**:292–296.

Gardner, C.L. 1985. The ecology of wolverines in southcentral Alaska. M.S. Thesis. University of Alaska, Fairbanks, Alaska, USA.

Gehrt, S.D. 2003. Raccoon and allies. *In* G. Feldhamer, B. Thompson, and J. Chapman, editors. *Wild Mammals of North America*. Second Edition. Johns Hopkins University Press, Baltimore, Maryland, USA: in press.

Gibilisco, C.J. 1994. Distributional dynamics of modern *Martes* in North America. Pages 59–71 *in* S.W. Buskirk, A.S. Harestad, M.G. Raphael, and R.A. Powell, editors. *Martens, Sables, and Fishers: Biology and Conservation*. Cornell University Press, Ithaca, New York, USA.

Gilbert, F.F. and R. Allwine. 1991. Small mammal communities in the Oregon Cascade Range. Pages 269–284 *in* Wildlife and Vegetation of Unmanaged Douglas-fir Forests. USDA Forest Service General Technical Report **PNW-285**, Portland, Oregon, USA.

Godin, A.J. 1982. Striped and hooded skunks. Pages 674–687 *in* J.A. Chapman and G.A. Feldhamer, editors. *Wild Mammals of North America*. Johns Hopkins University Press, Baltimore, Maryland, USA.

Goszczyński, J. 1977. Connections between predatory birds and mammals and their prey. *Acta Theriologica* **22**:399–430.

Grenfell, W.E. and M. Fasenfest. 1979. Winter food habits of fishers, *Martes pennanti*, in northwestern California. *California Fish and Game* **65**:186–189.

Grinnell, J., J.S. Dixon, and J.M. Linsdale. 1937. *Fur-bearing Mammals of California: their Natural History, Systematic Status, and Relations to Man*. Volumes 1, 2. University of California Press, Berkeley, California, USA.

Hagmeier, E.M. 1956. Distribution of marten and fisher in North America. *Canadian Field-Naturalist* **70**:149–168.

Hamilton, W.J., Jr. 1933. The weasels of New York. *American Midland Naturalist* **14**:289–344.

Hargis, C.D., J.A. Bissonette, and D.L. Turner. 1999. Influence of forest fragmentation and landscape pattern on American martens. *Journal of Applied Ecology* **36**:157–172.

Harlow, H.J. and S.W. Buskirk. 1991. Comparative serum and urine chemistry of fasting white-tailed prairie dogs (*Cynomys leucurus*) and American martens (*Martes americana*): representative fat and lean-bodied animals. *Physiological Zoology* **64**:1261–1278.

Harrison, R.L. 1997. A comparison of gray fox ecology between residential and undeveloped rural landscapes. *Journal of Wildlife Management* **61**:112–122.

Harrison, S. and A.D. Taylor. 1997. Empirical evidence for metapopulation dynamics. Pages 27–42 *in* I. Hanski and M. E. Gilpin, editors. *Metapopulation Biology: Ecology, Genetics, and Evolution*. Academic Press, San Diego, California, USA.

Hartman, L.H. and D.S. Eastman. 1999. Distribution of introduced raccoons *Procyon lotor* on the Queen Charlotte Islands: implications for burrow-nesting seabirds. *Biological Conservation* **88**:1–13.

Hash, H.S. 1987. Wolverine. Pages 575–585 *in* M. Novak, J.A. Baker, M.E. Obbard, and B. Malloch, editors. *Wild Furbearer Management and Conservation in North America*. Ontario Ministry of Natural Resources, Ontario, Canada.

Hatler, D.F. 1976. The coastal mink on Vancouver Island, British Columbia. PhD Thesis, University of British Columbia, Vancouver, British Columbia, Canada.

Hawley, V.D. and F.E. Newby. 1957. Marten home ranges and population fluctuations in Montana. *Journal of Mammalogy* **38**:174–184.

Hayes, J.P. and S.P. Cross. 1987. Characterization of logs used by western red-backed voles, *Clethrionomys californicus*, and deer mice, *Peromyscus maniculatus*. *Canadian Field-Naturalist* **101**:543–546.

Heinemeyer, K.S. 1993. Temporal dynamics in the movements, habitat use, activity, and spacing of reintroduced fishers in northwestern Montana. M.S. Thesis. University of Montana, Missoula, Montana, USA.

Henke, S.E. and F.C. Bryant. 1999. Effects of coyote removal on the faunal community in western Texas. *Journal of Wildlife Management* **63**:1066–1081.

Hodges, K.E. 2000. Ecology of snowshoe hares in southern boreal and montane forests. Pages 163–206 *in* L.F. Ruggiero, K.B. Aubry, S.W. Buskirk, G.M. Koehler, C.J. Krebs, K.S. McKelvey, and J.R. Squires, editors. *Ecology and Conservation of Lynx in the United States*. University Press of Colorado, Boulder, Colorado, USA.

Hodgman, T.P., D.J. Harrison, D.D. Katnik, and K.D. Elowe. 1994. Survival in an intensively trapped marten population in Maine. *Journal of Wildlife Management*. **58**:593–600.

Hooven, E.F. and H.C. Black. 1976. Effects of some clear-cutting practices on small-mammal populations in western Oregon. *Northwest Science* **50**:189–208.

Hornocker, M.G. and H.S. Hash. 1981. Ecology of the wolverine in northwestern Montana. *Canadian Journal of Zoology* **59**:1286–1301.

Jedrzejewska, B. and W. Jedrzejewski. 1998. *Predation in Vertebrate Communities: the Białowieża Primeval Forest as a Case Study. Ecological Studies*. Volume 135, Springer-Verlag, New York, New York, USA.

Johnson, C.B. 1985. Use of coastal habitat by mink on Prince of Wales Island, Alaska. M.S. Thesis, University of Alaska, Fairbanks, Alaska, USA.

Jones, J.L. and E.O. Garton. 1994. Selection of successional stages by fishers in northcentral Idaho. Pages 377–388 *in* S.W. Buskirk, A.S. Harestad, M.G. Raphael, and R.A. Powell, editors. *Martens, Sables, and Fishers: Biology and Conservation*. Cornell University Press, Ithaca, New York, USA.

Kelly, G.M. 1977. Fisher (*Martes pennanti*) biology in the White Mountain National Forest and adjacent areas. Dissertation. University of Massachusetts, Amherst, MA, USA.

King, C. 1984. *Immigrant Killers: Introduced Predators and the Conservation of Birds in New Zealand*. Oxford University Press, New York, New York, USA.

Klug, R.R. 1996. Occurrence of Pacific fisher in the redwood zone of northern California and the habitat attributes associated with their detection. M.S. Thesis. Humboldt State University, Arcata, California, USA.

Koehler, G.M. and M.G. Hornocker. 1991. Seasonal resource use among mountain lions, bobcats, and coyotes. *Journal of Mammalogy* 72:391–396.

Krebs, C.J., S. Boutin, R. Boonstra, A.R.E. Sinclair, J.N.M. Smith, M.R.T. Dale, and M.R. Turkington. 1995b. Impact of food and predation on the snowshoe hare cycle. *Science* 269:1112–1115.

Krebs, J.W., M.L. Wilson, and J.E. Childs. 1995a. Rabies – epidemiology, prevention and future research. *Journal of Mammalogy* 76:681–694.

Krohn, W.B., K.D. Elowe, and R.B. Boone. 1995. Relations between fishers, snow, and martens: development and evaluation of two hypotheses. *Forestry Chronicle* 71:97–105.

Krohn, W.B., W.J. Zielinski, and R.B. Boone. 1997. Relations among fishers, snow, and martens in California: results from small-scale spatial comparisons. Pages 211–232 *in* G. Proulx, H.N. Bryant, and P.M. Woodard, editors. Martes: *Taxonomy, Ecology, Techniques, and Management*. Provincial Museum of Alberta, Edmonton, Alberta, Canada.

Kucera, T.E., W.J. Zielinski, and R.M. Barrett. 1995. The current distribution of American marten, *Martes americana,* in California. *California Fish and Game* 81:96–103.

Lambeck, R.J. 1997. Focal species: a multi-species umbrella for nature conservation. *Conservation Biology* 11:849–856.

Larsen, D.N. 1983. Habitats, movements, and foods of river otters in coastal southeastern Alaska. M.S. Thesis, University of Alaska, Fairbanks, Alaska, USA.

Latham, R.M. 1952. The fox as a factor in the control of weasel populations. *Journal of Wildlife Management* 16:516–517.

Levins, R. 1968. *Evolution in Changing Environments*. Princeton University Press, Princeton, New Jersey, USA.

Lewis, J.C. and W.J. Zielinski. 1996. Historical harvest and incidental capture of fishers in California. *Northwest Science* 70:291–297.

Lewis, J.C., R.T. Golightly, and R.M. Jurek. 1995. Introduction of non-native red foxes in California: implications for the Sierra Nevada red fox. *Transactions of the Western Section of the Wildlife Society* 31:29–32.

Lindenmayer, D.B. and R.C. Lacy. 1995. Metapopulation viability of arboreal marsupials in fragmented old-growth forests: comparisons among species. *Ecological Applications* 5:183–199.

Lovallo, M.J. and E.M. Anderson. 1996. Bobcat movements and home ranges relative to roads in Wisconsin. *Wildlife Society Bulletin* 24:71–76.

Magoun, A.J. and J.P. Copeland. 1998. Characteristics of wolverine reproductive den sites. *Journal of Wildlife Management* 62:1313–1320.

Maj, M. and E.O. Garton. 1994. Fisher, lynx, wolverine: summary of distribution information. Pages 169–175 *in* L.F. Ruggiero, K.B. Aubry, S.W. Buskirk, L.J. Lyon, and W.J. Zielinski, editors. The scientific basis for conserving forest carnivores, American marten, fisher, lynx, and wolverine in the western United Statees. USDA Forest Service General Technical Report RM-254, Fort Collins, Colorado, USA.

Martin, S.K. 1994. Feeding ecology of American martens and fishers. Pages 297–315 *in* S.W. Buskirk, A.S. Harestad, M.G. Raphael, and R.A. Powell, editors. *Martens, Sables, and Fishers: Biology and Conservation*. Cornell University Press, Ithaca, New York, USA.

Maser, C., B.R. Mate, J. Franklin, and C.T. Dyfress. 1981. Natural history of Oregon Coast mammals. USDA Forest Service General Technical Report **PNW-133**, Portland, Oregon, USA.

Matson, P.A. and M.D. Hunter. 1992. The relative contributions of top-down and bottom-up forces in population and community ecology. *Ecology* **73**:723.

Mattson, D.J., R.R. Knight, and B.M. Blanchard. 1992. Cannibalism and predation on black bears by grizzly bears in the Yellowstone ecosystem, 1975–1990. *Journal of Mammalogy* **73**:422–425.

McCord, C.M. 1974. Selection of winter habitat by bobcats (*Lynx rufus*) on the Quabbin Reservation, Massachusetts. *Journal of Mammalogy* **55**:428–437.

McKelvey, K.S., K.B. Aubry, and Y.K. Ortega. 2000a. History and distribution of lynx in the contiguous United States. Pages 207–264 *in* L.F. Ruggiero, K.B. Aubry, S.W. Buskirk, G.M. Koehler, C.J. Krebs, K.S. McKelvey, and J.R. Squires, editors. *Ecology and Conservation of Lynx in the United States*. University Press of Colorado, Boulder, Colorado, USA.

McKelvey, K.S., Y.K. Ortega, G.M. Koehler, K.B. Aubry, and J.D. Brittell. 2000b. Canada lynx habitat and topographic use patterns in north-central Washington: a reanalysis. Pages 307–336 *in* L.F. Ruggiero, K.B. Aubry, S.W. Buskirk, G.M. Koehler, C.J. Krebs, K.S. McKelvey, and J.R. Squires, editors. *Ecology and Conservation of Lynx in the United States*. University Press of Colorado, Boulder, Colorado, USA.

McLaren, B.E. and R.O. Peterson. 1994. Wolves, moose and tree rings on Isle Royale. *Science* **266**:1555–1558.

McShea, W.J., J.H. Rappole, and H.B. Underwood. 1997. *The Science of Overabundance: Deer Ecology and Population Management*. Smithsonian Institution Press, Washington DC, USA.

Melquist, W.E. and A.E. Dronkert. 1987. River otter. Pages 627–641 *in* M. Novak, J.A. Baker, M.E. Obbard, and B. Malloch, editors. *Wild Furbearer Management and Conservation in North America*. Ontario Ministry of Natural Resources, Ontario, Canada.

Melquist, W.E. and M.G. Hornocker. 1983. Ecology of river otters in west central Idaho. *Wildlife Monographs* **83**:1–60.

Melquist, W.E., J.S. Whitman, and M.G. Hornocker. 1981. Resource partitioning and coexistence of sympatric mink and river otter populations. Pages 187–220 *in* J.A. Chapman and D. Pursley, editors. *Worldwide Furbearer Conference Proceedings*, Volume 1. Frostburg, Maryland, USA.

Miller, S.D. 1980. The ecology of the bobcat in south Alabama. Dissertation, Auburn University, Auburn, Alabama, USA.

Mills, L.S., M.E. Soulé, and D.F. Doak. 1993. The keystone species concept in ecology and conservation. *BioScience* **43**:219–224.

Minta, S.C., P.M. Kareiva, and A.P. Curlee. 1999. Carnivore research and conservation: learning from history and theory. Pages 323–404 *in* T.W. Clark, S.C. Minta, P.K. Karieva, and A.P. Curlee, editors. *Carnivores in Ecosystems: the Yellowstone Experience*. Yale University Press, New Haven, Connecticut, USA.

Mladenoff, D.J., T.A. Sickley, R.G. Haight, and A.P. Wydeven. 1995. A regional landscape analysis and prediction of favorable gray wolf habitat in the northern Great Lakes region. *Conservation Biology* **9**:279–294.

Mowat, G., K.G. Poole, and M. O'Donoghue. 2000. Ecology of lynx in northern Canada and Alaska. Pages 265–306 *in* L.F. Ruggiero, K.B. Aubry, S.W. Buskirk, G.M. Koehler, C.J. Krebs, K.S. McKelvey, and J.R. Squires, editors. *Ecology and Conservation of Lynx in the United States*. University Press of Colorado, Boulder, Colorado, USA.

Murray, D.L. and S. Boutin. 1991. The influence of snow on lynx and coyote movements: does morphology affect behavior? *Oecologia* **88**:463–469.

Noon, B.R. and K.S. McKelvey. 1996. Management of the spotted owl: a case history in conservation biology. *Annual Review of Ecology and Systematics* **27**:135–162.

Noss, R.F. 1992. The Wildland Project: land conservation strategy. *Wild Earth* (special issue) **1**:10–25.

Noss, R.F. and A.Y. Cooperrider. 1996. *Saving Nature's Legacy: Protecting and Restoring Biodiversity*. Island Press, Washington DC, USA.

Noss, R.F., H.B. Quigley, M.G. Hornocker, T. Merrill, and P.C. Paquet. 1996. Conservation biology and carnivore conservation in the Rocky Mountains. *Conservation Biology* **10**:949–963.

O'Donoghue, M., S. Boutin, C.J. Krebs, G. Zuleta, D.L. Murray, and E.J. Hofer. 1998. Functional responses of coyotes and lynx to the snowshoe hare cycle. *Ecology* **79**:1193–1208.

Organ, J.F. 1989. Mercury and PCB residues in Massachusetts river otters: comparisons on a watershed basis. Dissertation, University of Massachusetts, Amherst, Massachusetts, USA.

Orloff, S. 1988. Present distribution of ringtails in California. *California Fish and Game* **74**:196–202.

Orr, R.T. 1943. Altitudinal record for the spotted skunk in California. *Journal of Mammalogy* **24**:270.

Osowski, S.L., L.W. Brewer, O.E. Baker, and G.P. Cobb. 1995. The decline of mink in Georgia, North Carolina, and South Carolina: the role of contaminants. *Archives of Environmental Contaminants and Toxicology* **29**:418–423.

Palma, L., P. Beja, and M. Rodrigues. 1999. The use of sighting data to analyse Iberian lynx habitat and distribution. *Journal of Applied Ecology* **36**:812–824.

Palomares, F., P. Gaona, P. Ferreras, and M. Delibes. 1995. Positive effects on game species of top predators by controlling smaller predator populations: an example with lynx, mongooses, and rabbits. *Conservation Biology* **9**:294–304.

Paragi, T.F., W.N. Johnson, D.D. Katnik, and A.J. Magoun. 1996. Marten selection of postfire seres in the Alaskan taiga. *Canadian Journal of Zoology* **74**:2226–2237.

Payer, D.C. 1999. Influence of timber harvesting and trapping on habitat selection and demographic characteristics of American marten. Dissertation. University of Maine, Orono, Maine, USA.

Poglayen-Neuwall, I. and D.E. Toweill. 1988. *Bassariscus astutus. Mammalian Species* **327**:1–8.

Polis, G.A. and D.R. Strong. 1996. Food web complexity and community dynamics. *American Naturalist* **147**:813–846.

Potvin, F., L. Belanger, and K. Lowell. 2000. Marten habitat selection in a clear-cut boreal landscape. *Conservation Biology* **14**:844–857.

Powell, R.A. 1979. Fishers, population models, and trapping. *Wildlife Society Bulletin* 7:149–154.

Powell, R.A. 1993. *The Fisher: Life History, Ecology and Behavior.* 2nd edition. University of Minnesota Press, Minneapolis, Minnesota, USA.

Powell, R.A. and W.J. Zielinski. 1994. Fisher. Pages 38–73 *in* L.F. Ruggiero, K.B. Aubry, S.W. Buskirk, L.J. Lyon, and W.J. Zielinski, editors. The Scientific Basis for Conserving Forest Carnivores: American marten, fisher, lynx, and wolverine in the western United Statees. USDA Forest Service General Technical Report **RM-254**, Fort Collins, Colorado, USA.

Primm, S.A. and T.W. Clark 1996. Making sense of the policy process for carnivore conservation. *Conservation Biology* **10**:1036–1045.

Pulliam, H.R. 1996. Sources and sinks: empirical evidence and population consequences. Pages 45–69 *in* O.E. Rhodes, R.K. Chesser, and M.H. Smith, editors. *Population Biology in Ecological Space and Time.* University of Chicago Press, Chicago, Illinois, USA.

Quick, H.F. 1951. Notes on the ecology of weasels in Gunnison County, Colorado. *Journal of Mammalogy* **32**:281–290.

Ralls, K. and P.J. White. 1995. Predation of San Joaquin kit foxes by larger canids. *Journal of Mammalogy* **76**:723–729.

Reid, D.G., T.E. Code, A.C.H. Reid, and S.M. Herrero. 1994. Spacing, movements, and habitat selection of the river otter in boreal Alberta. *Canadian Journal of Zoology* **72**:1314–1324.

Riley, S.P.D. 1999. Spatial organization, food habits and disease ecology of bobcats (*Lynx rufus*) and gray foxes (*Urocyon cinereoargenteus*) in national park areas in urban and rural Marin County, California. Dissertation, University of California, Davis, California, USA.

Robitaille, J. and K. Aubry. 2000. Occurrence and activity of American martens, *Martes americana*, in relation to roads and other routes. *Acta Theriologica* **45**:137–143.

Rogers, C.M. and M.J. Caro. 1998. Song sparrows, top carnivores, and nest predation: a test of the mesopredator release hypothesis. *Oecologia* **116**:227–233.

Rolley, R.E. 1987. Bobcat. Pages 671–681 *in* M. Novak, J.A. Baker, M.E. Obbard, and B. Malloch, editors. *Wild Furbearer Management and Conservation in North America.* Ontario Ministry of Natural Resouces, Ontario, Canada.

Rolley, R.E. and W.D. Warde. 1985. Bobcat habitat use in southeastern Oklahoma. *Journal of Wildife Management* **49**:913–920.

Rosenberg, K.V. and R.G. Raphael. 1986. Effects of forest fragmentation on vertebrates in Douglas-fir forests. Pages 263–272 *in* J. Verner, M.L. Morrison, C.J. Ralph, editors. *Wildlife 2000: Modeling Habitat Relationships of Terrestrial Vertebrates.* University of Wisconsin Press, Madison, Wisconsin, USA.

Roy, K.D. 1991. Ecology of reintroduced fishers in the Cabinet Mountains of northwest Montana. M.S. Thesis. University of Montana, Missoula, Montana, USA.

Ruggiero, L.F. and K.S. McKelvey. 2000. Toward a defensible lynx conservation strategy: a framework for planning in the face of uncertainty. Pages 5–19 *in* L.F. Ruggiero, K.B. Aubry, S.W. Buskirk, G.M. Koehler, C.J. Krebs, K.S. McKelvey, and J.R. Squires, editors. 2000. *Ecology and Conservation of Lynx in the United States.* University Press of Colorado, Boulder, Colorado, USA.

Ruggiero, L.F., D.E. Pearson, and S.E. Henry. 1998. Characteristics of American marten den sites in Wyoming. *Journal of Wildlife Management* **62**:663–673.

Ryszkowski, L., J. Gozczyński, and J. Truszkowski. 1973. Trophic relationships of the common vole in cultivated fields. *Acta Theriologica* **18**:125–165.

Samuel, D.E. and B.B. Nelson. 1982. Foxes: *Vulpes vulpes* and allies. Pages 475–490 *in* J.A. Chapman, and G.A. Feldhamer, editors. *Wild Mammals of North America: Biology, Management, and Economics*. Johns Hopkins University Press, Baltimore, Maryland, USA.

Sanderson, G.C. 1987. Raccoon. Pages 487–499 *in* M. Novak, J.A. Baker, M.E. Obbard, and B. Malloch, editors. *Wild Furbearer Management and Conservation in North America*. Ontario Ministry of Natural Resources, Ontario, Canada.

Schamel, D. and D.M. Tracy. 1986. Encounters between arctic foxes, *Alopex lagopus*, and red foxes, *Vulpes vulpes*. *Canadian Field-Naturalist* **100**:562–563.

Schempf, P.F. and M. White. 1977. Status of six furbearer populations in the mountains of northern California. Unpublished report to USDA Forest Service, Pacific Southwest Region, San Francisco, California, USA.

Schneider, R. 1997. Simulated spatial dynamics of martens in response to habitat succession in the Western Newfoundland Model Forest. Pages 419–436 *in* G. Proulx, H. N. Bryant, and P. M. Woodard, editors. Martes: *Taxonomy, Ecology, Techniques and Management*. Provincial Museum of Alberta, Edmonton, Alberta, Canada.

Seglund, A.E. 1995. The use of resting sites by the Pacific fisher. M.S. Thesis. Humboldt State University, Arcata, CA, USA.

Sherburne, S.S. and J.A. Bissonette. 1993. Squirrel middens influence marten (*Martes americana*) use of subnivean access points. *American Midland Naturalist* **129**:204–207.

Simms, D.A. 1979. North American weasels: resource utilization and distribution. *Canadian Journal of Zoology* **57**:504–520.

Soulé, M.E. and J. Terborgh. 1999. *Continental Conservation: Scientific Foundations of Regional Reserve Networks*. Island Press, Washington DC, USA.

Soulé, M.E., D.T. Bolger, A.C. Alberts, J. Wright, M. Sorice, and S. Hill. 1988. Reconstructed dynamics of rapid extinctions of chaparral-requiring birds in urban habitat islands. *Conservation Biology* **2**:75–92.

Sovada, M.A., A.B. Sargeant, and J.W. Grier. 1995. Differential effects of coyotes and red foxes on duck nest success. *Journal of Wildlife Management* **59**:1–9.

Spencer, W.D., R.H. Barrett, and W.J. Zielinski. 1983. Marten habitat preferences in the northern Sierra Nevada. *Journal of Wildlife Management* **47**:1181–1186.

Stephenson, R.O., D.V. Grangaard, and J. Burch. 1991. Lynx, *Felis lynx*, predation on red foxes, *Vulpes vulpes*, caribou, *Rangifer tarandus*, and Dall sheep, *Ovis dalli*, in Alaska. *Canadian Field-Naturalist* **105**:255–262.

Strickland, M.A. 1994. Harvest management of fishers and American martens. Pages 149–164 *in* S.W. Buskirk, A.S. Harestad, M.G. Raphael, and R.A. Powell, editors. *Martens, Sables, and Fishers: Biology and Conservation*. Cornell University Press, Ithaca, New York, USA.

Strickland, M.A., C.W. Douglas, M. Novak, and N.P. Hunziger. 1982*a*. Fisher. Pages 586–598 *in* J. A. Chapman, and G. A. Feldhamer, editors. *Wild Mammals of North America: Biology, Management, and Economics*. Johns Hopkins University Press, Baltimore, MD, USA.

Strickland, M.A., C.W. Douglas, M. Novak, and N.P. Hunziger. 1982*b*. Marten. Pages 599–612 *in* J. A. Chapman, and G. A. Feldhamer, editors. *Wild Mammals of North*

America: Biology, Management, and Eeconomics. Johns Hopkins University Press, Baltimore, Maryland, USA.

Stuewer, F.W. 1943. Raccoons: their habits and management in Michigan. *Ecological Monographs* **13**:203–257.

Svendsen, G.E. 1982. Weasels. Pages 613–628 *in* J.A. Chapman, and G.A. Feldhamer, editors. *Wild Mammals of North America: Biology, Management, and Economics*. Johns Hopkins University Press, Baltimore, Maryland, USA.

Szumski, M.J. 1998. The effects of mining-related metals contamination on piscivorous mammals along the upper Clark Fork River, Montana. Dissertation, University of Wyoming, Laramie, Wyoming, USA.

Taylor, S.L. and S.W. Buskirk. 1994. Forest microenvironments and resting energetics of the American marten *Martes americana*. *Ecography* **17**:249–256.

Terborgh, J. 1992. Maintenance of diversity in tropical forests. *Biotropica* **24**:283–292.

Terborgh, J., J.A. Estes, P. Paquet, K. Ralls, D. Boyd-Heger, B.J. Miller, and R.F. Noss. 1999. The role of top carnivores in regulating terrestrial ecosystems. Pages 39–64 *in* M. E. Soulé, and J. Terborgh, editors. *Continental Conservation: Scientific Foundations of Regional Reserve Networks*. Island Press, Washington DC, USA.

Thompson, I.D. 1986. Diet choice, hunting behavior, activity patterns, and ecological energetics of marten in natural and logged areas. Dissertation. Queen's University, Kingston, Ontario, Canada.

Thompson, I.D. 1994. Marten populations in uncut and logged boreal forests in Ontario. *Journal of Wildlife Management* **58**:272–279.

Thompson, I.D. and A.S. Harestad. 1994. Effects of logging on American martens, and models for habitat management. Pages 355–367 *in* S.W. Buskirk, A.S. Harestad, M.G. Raphael, and R.A. Powell, editors. *Martens, Sables, and Fishers: Biology and Conservation*. Cornell University Press, Ithaca, New York, USA.

Thurber, J.M., R.O. Peterson, J.D. Woolington, and J.A. Vucetich. 1992. Coyote coexistence with wolves on the Kenai Peninsula, Alaska. *Canadian Journal of Zoology* **70**:2494–2498.

Thurber, J.M., R.O. Peterson, T.D. Drummer, and S.A. Thomasma. 1994. Gray wolf response to refuge boundaries and roads in Alaska. *Wildlife Society Bulletin* **22**:61–68.

Todd, A.W., L.B. Keith, and C.A. Fischer. 1981. Population ecology of coyotes during a fluctuation of snowshoe hares. *Journal of Wildlife Management* **45**:629–640.

Trapp, G.R. 1978. Comparative behavioral ecology of the ringtail and gray fox in southwest Utah. *Carnivore* **1**:3–32.

Trapp, G.R. and D.L. Hallberg. 1975. Ecology of the gray fox (*Urocyon cinereoargenteus*): a review. Pages 164–178 *in* M.W. Fox, editor. *The Wild Canids*. Van Reinhold Co., New York, New York, USA.

Twining, H. and A. Hensley. 1947. The status of pine martens in California. *California Fish and Game* **33**:133–137.

Tyson, E.L. 1950. Summer food habits of the raccoon in southwest Washington. *Journal of Mammalogy* **31**:448–449.

USDA Forest Service and USDI Bureau of Land Management. 1994. Final supplemental environmental impact statement on management of habitat for late-successional and old-growth forest related species within the range of the Northern Spotted Owl. USDA Forest Service, Pacific Northwest Region, Portland, Oregon, USA.

USDA Forest Service. 1996. Status of the Interior Columbia Basin: summary of scientific findings. US Forest Service General Technical Report **PNW-GTR-385**.

USDA Forest Service. 2001. Sierra Nevada Forest plan amendment. Final environmental impact statement. Pacific Southwest Region, Vallejo, CA, USA.

Verts, B.J. 1967. *The Biology of the Striped Skunk*. University of Illinois Press, Urbana, Illinois, USA.

Verts, B.J. and L.N. Carraway. 1998. *Land Mammals of Oregon*. University of California Press, Berkeley, California, USA.

Vitousek, P. 1988. Diversity and biological invasions of oceanic islands. Pages 181–189 *in* E.O. Wilson, and F.M. Peter, editors. *Biodiversity*. National Academy Press, Washington DC, USA.

Voigt, D.R. 1987. Red fox. Pages 379–392 *in* M. Novak, J.A. Baker, M.E. Obbard, and B. Malloch, editors. *Wild Furbearer Management and Conservation in North America*. Ontario Ministry of Natural Resources, Ontario, Canada.

Wade-Smith, J. and B.J. Verts. 1982. *Mephitis mephitis. Mammalian Species* **173**:1–7.

Weaver, J.L., P.C. Paquet, and L.F. Ruggiero. 1996. Resilience and conservation of large carnivores in the Rocky Mountains. *Conservation Biology* **10**:964–976.

Weckwerth, R.P. and V.D. Hawley. 1962. Marten food habits and population fluctuations in Montana. *Journal of Wildlife Management* **26**:55–74.

Weir, R.D. and A.S. Harestad. 1997. Landscape-level selectivity by fishers in south-central British Columbia. Pages 252–264 *in* G. Proulx, H.N. Bryant, and P.M. Woodard, editors. Martes: *Taxonomy, Ecology, Techniques and Management*. Provincial Museum of Alberta, Edmonton, Alberta, Canada.

White, P.J., and R.A. Garrott. 1997. Factors regulating kit fox populations. *Canadian Journal of Zoology* **75**:1982–1988.

Wilcomb, M.J. 1948. Fox populations and food habits in relation to game birds in the Willamette Valley, Oregon. M.S. Thesis. Oregon State College, Corvallis, Oregon, USA.

Willey, R.B. and R.E. Richards. 1974. The ringtail (*Bassariscus astutus*): vocal repertoire and Colorado distribution. *Journal of the Colorado-Wyoming Academy of Science* **7(5)**:58.

Wilson, D.E. 1982. Wolverine. Pages 644–652 *in* J.A. Chapman and G.A. Feldhamer, editors. *Wild Mammals of North America*. Johns Hopkins University Press, Baltimore, Maryland, USA.

Wisz, M.S. 1999. Islands in the Big Sky: equilibrium biogeography of isolated mountain ranges in Montana. M.S. Thesis, University of Colorado, Boulder, Colorado, USA.

Wren, C.D. 1985. Probable case of mercury poisoning in a wild otter, *Lutra canadensis*, in northwestern Ontario. *Canadian Field-Naturalist* **99**:112–114.

Wright, S. 1978. *Evolution and the Genetics of Populations*. Volume 4. University of Chicago Press, Chicago, Illinois, USA.

Wright, S.J., M.E. Gompper, and B. DeLeon. 1994. Are large predators keystone species in neotropical forests? The evidence from Barro Colorado Island. *Oikos* **71**:279–294.

Yeager, L.E. and W.H. Elder. 1945. Pre- and post-hunting season foods of raccoons on an Illinois goose refuge. *Journal of Wildlife Management* **9**:48–56.

Zezulak, D.S. and R.G. Schwab. 1979. A comparison of density, home range, and habitat utilization of bobcat populations at Lava Beds and Joshua Tree national monuments, California. *Bobcat Research Conference, National Wildlife Federation Scientific and Technical Series* **6**:74–79.

Zielinski, W.J. 1995. Track plates. Pages 67–86 *in* W.J. Zielinski, and T.E. Kucera, editors. American marten, fisher, lynx, and wolverine: survey methods for their detection. U.S. Forest Service, General Technical Report **PSW-157**, Albany, California, USA.

Zielinski, W.J., and T.E. Kucera. 1995. American marten, fisher, lynx, and wolverine: survey methods for their detection. U.S. Forest Service General Technical Report **PSW-157** Albany, California, USA.

Zielinski, W.J. and H.B. Stauffer. 1996. Monitoring *Martes* populations in California: survey design and power analysis. *Ecological Applications* **6**:1254–1267.

Zielinski, W.J., W.D. Spencer, and R. D. Barrett. 1983. Relationship between food habits and activity patterns of pine martens. *Journal of Mammalogy* **64**:387–396.

Zielinski, W.J., T.E. Kucera, and R. H. Barrett. 1995. The current distribution of fisher, *Martes pennanti*, in California. *California Fish and Game* **81**:104–112.

Zielinski, W.J., R.L. Truex, C.V. Ogan, and K. Busse. 1997. Detection surveys for fishers and American Martens in California, 1989–1994: Summary and interpretations. Pages 372–394 *in* G. Proulx, H.N. Bryant, and P.M. Woodward, editors. Martes: *Taxonomy, Ecology, Techniques and Management*. Provinicial Museum of Alberta Edmonton, Canada.

Zielinski, W.J., N.P. Duncan, E.C. Farmer, R.L. Truex, A.P. Clevenger, and R.H. Barrett. 1999. Diet of fishers (*Martes pennanti*) at the southernmost extent of their range. *Journal of Mammalogy* **80**:961–971.

Zielinski, W.J., K.M. Slauson, C.R. Carroll, C.J. Kent, and D.G. Kudrna. 2001. Status of American martens in coastal forests of the Pacific states. *Journal of Mammalogy* **82**:478–490.

8
———————

Ecology, conservation, and restoration of large carnivores in western North America

Introduction

Large carnivores operate over large spatial scales and affect and are affected by multiple ecosystem processes (e.g., predation, migration, climate, fire, etc.). Thus carnivore ecologists must deal with expansive areas and multiple scales and disciplines. For this reason, studies of top carnivores have played a significant role in fostering ecosystem approaches among managers and researchers (Minta et al. 1999). Specifically, wolves (*Canis lupus*), cougars (*Felis concolor*), and bears (*Ursus* spp.) have become important symbols for conservation and ecosystem management. Recent research has examined multiple large carnivores (Kunkel et al. 1999, Carroll et al. 2001) and multiple ecosystems in regional conservation networks (Soule and Terborgh 1999). Conservation perspectives resulting from such work can help build strategies to protect appreciable amounts of native biological diversity (Noss and Cooperrider 1994, Paquet and Hackman 1995, Soule and Terborgh 1999). Conservation strategies that focus on charismatic animals such as large carnivores have additional advantages because such a focus motivates organizations and the general public.

Even though large carnivores often are grouped for management and share many traits, some important differences affect their resiliency (Weaver et al. 1996). The most obvious split occurs between the obligatory predators (cougars, wolves, and jaguars (*Panthera onca*)) and the omnivorous facultative predators (bears). Cougars, wolves, and jaguars have higher dispersal capabilities and higher reproductive rates than bears. Wolves are social; cougars, jaguars, and bears are solitary. Cougars, being solitary, are less plastic in predatory behavior than wolves, rely on smaller prey, and are less competitive in multi-carnivore environments

(Kunkel et al. 1999). As stalking predators, cougars are more habitat-specific than coursing wolves. Brown bears (*U. arctos*) pose the most direct perceived conflict with humans, and wolves the greatest perceived conflict with livestock, so both are less socially acceptable and experience higher human-caused mortality. As a result of these varying levels of resiliency, management is easiest for cougars, more difficult for wolves, and most difficult for brown bears.

This chapter reviews and synthesizes information on large carnivores in North America as it applies to their management and their roles in biodiversity conservation in coniferous forests of North America. Treatment of basic ecology will be brief and will focus on aspects related most to management; others (Carbyn et al. 1995, Clark et al. 1999, Soule and Terborgh 1999, Demaris and Krausman 2000, Gittleman et al. 2001) have provided recent reviews of carnivore ecology. Management implications will be indicated throughout and in separate management sections. Black bears (*U. americanus*) are treated more lightly than the other species due to their more secure conservation status. Jaguars also are treated lightly due to their limited range in southwestern coniferous forests, lack of study, and current absence. This does not, however, correlate to a reduced need to consider these species in management.

Current status and ecological roles

Wolves

Historically, wolves were distributed throughout most of North America in all habitats that supported ungulates. By 1930, after decades of persecution, wolves were eliminated from the western U.S. and had declined greatly in western Canada. Wolves started a remarkable comeback in the northwestern U.S. in the 1980's. More than 400 wolves now occupy the northern Rockies as a result of natural re-colonization and re-introductions (Bangs et al. 1998). Populations are centered in the recovery areas of northwest Montana, central Idaho, and the Yellowstone ecosystem of Montana, Idaho, and Wyoming. Re-introductions of Mexican wolves (*Canis lupus baileyi*) were initiated in 1998 in the southwestern U.S., and approximately 30 wolves occupied the area along the New Mexico and Arizona border in 2001 (W. Brown, U.S. Fish and Wildlife Service, unpublished data).

By 1996, eight wolf packs had re-colonized northwest Montana via dispersal from Canada (Boyd and Pletscher 1999). Density of wolves in northwestern Montana is approximately 10 wolves per 1000 km². Annual survival rate of wolves there is about 0.80 and their finite annual rate of increase is about 1.2 (Pletscher et al. 1997). The majority of mortality there, as elsewhere, is illegal and legal kill by humans. Annual finite rates of increase for wolves in North America range from 0.4 to 2.5 (Fritts and Mech 1981, Keith 1983, Fuller 1989).

Wolves released in Yellowstone Park and Idaho have prospered remarkably. Those re-introduced to Yellowstone increased from 31 wolves soft-released (held in pens for acclimation for two months prior to release) in 1996 to 177 in 18 packs in 2000 (Smith et al. 2001). The population in central Idaho increased from 35 hard-released (released immediately to the wild) in 1996 to 191 in nine packs in 2000 (C. Mack, Nez Perce Tribe, unpublished data). At least four wolves have been confirmed to have dispersed from one recovery area to another (Yellowstone to Idaho, northwest Montana to Idaho, and Idaho to northwest Montana), and dispersing wolves have survived to reproduce outside recovery areas.

Impacts of predation

In western North America, wolves prey primarily on elk (*Cervus elaphus*), deer (*Odocoileus* spp.), moose (*Alces alces*), and caribou (*Rangifer tarandus*). Wolves are opportunistic predators and variation in prey preferences have been reported (Huggard 1993, Weaver 1994, Kunkel 1997, Bergerud and Elliot 1998, Smith et al. 2001). Kill rates vary greatly from 2.0 to 7.2 kg per wolf per day (Ballard and Gipson 2000). Wolves generally kill animals that are vulnerable because of age, condition, or habitat and weather circumstances (Mech 1996). Wolf population density varies directly and widely with prey density (Fuller 1989).

The impact of wolf predation on ungulate populations has been much studied and debated. Impacts reported vary from slight to regulating (i.e., density-dependent; Van Ballenberghe and Ballard 1994). Potential reasons for this variation include variation in local conditions (habitat, weather, prey and predator densities, and behavior, etc.; Messier 1995) and the inherent difficulty of field studies of large carnivores that influences data collection and interpretation. There is good evidence for limiting (one factor that far outweighs others in impeding the rate of increase; Leopold 1933:39) but not regulating (density-dependent factors that keep a prey population in equilibrium) impacts of wolf predation (Gasaway et al. 1992, Messier 1994, Van Ballenberghe and Ballard 1994, Boertje et al. 1996,

National Research Council 1997, Kunkel and Pletscher 1999). In northern latitudes with simpler predator–prey systems, the impacts of predation may be regulating, especially when wolves and another predator (bears) prey on one or two prey species (Messier 1994). Where wolves and deer co-exist in the northern U.S. and Canada and have been well studied, their populations have been unstable for the duration they have been examined (the last 20–40 years; Potvin et al. 1988, Fuller 1990, Hatter and Janz 1994). Given the large-scale, density-independent influences such as weather and loose regulatory feedback inherent to these northern systems, this instability is not surprising (Botkin 1990, Bergerud and Elliot 1998).

Human harvests of prey in Alaska and Canada were significantly lower where wolves were not hunted or controlled (Gasaway et al. 1992). Also, hunter success for deer and elk declined as wolves re-colonized northwest Montana and was lower than in areas wolves had not re-colonized (Kunkel and Pletscher 1999). Predator numbers declined as a result of predator-induced declines in prey and this resulted in a rebound in prey in northwest Montana. Mech and Nelson (2000) found no impact of wolf predation on harvest of white-tailed deer (*O. virginianus*) bucks in "good deer habitat" in northern Minnesota, but did find an impact in poor habitat. Additionally, doe harvest had to be eliminated. Despite increasing evidence of wolf predation limiting prey populations, the National Research Council (1997) in a summary of research on effects of wolf control on ungulate populations concluded that there was little evidence to indicate the long-term effectiveness of wolf control for increasing human harvest of prey.

Wolf re-colonization of the western U.S. will likely result in declines of local cervid populations, especially where multiple predators are present (Crete 1999, Kunkel and Pletscher 1999). Managers should expect cervid populations to remain low for extended periods where wolves, bears, cougars, and humans vie for the same prey (Gasaway et al. 1992, National Research Council 1997). Lower cervid densities may result in lower predator densities and thus slow wolf and brown bear recovery (McLellan and Hovey 1995, Boertje et al. 1996, Mladenoff et al. 1997). Depending on objectives, managers should be prepared to reduce hunting pressure on cervids to prevent potentially long-term low densities of prey in such areas (Fuller 1990, Gasaway et al. 1992, Boertje et al. 1996). Management of wolves through harvest may also be an option once wolves are delisted.

Cougars

Cougars presently occupy almost all of their historic range in western North America. The cougar's solitary nature, use of remote and rugged landscapes, relatively uncommon predation on livestock, and relatively high reproductive rate helped it escape the regional extinctions that befell other large carnivores. The recovery of cougars in the West occurred in the equivalent of only three cougar lifetimes (Logan and Sweanor 2000). Cougars are currently expanding into western portions of the Great Plains.

Cougar populations were thought to be regulated through socially controlled land tenure (Hornocker 1970, Seidensticker et al. 1973). Recent research from California, however, indicates that cougars, like other carnivores, were limited in abundance primarily by the supply of food rather than land tenure (Pierce et al. 2000). Density of cougars ranges from 5.8 to 47 cougars per 1000 km^2. Annual survival rates for females average 0.80 (Logan and Sweanor 2000) with humans being the major cause of death for cougars in protected and unprotected populations. Finite annual rates of increase varied from 1.18 to 1.32 for a protected population in New Mexico (Logan et al. 1996).

Impacts of predation

Under some circumstances cougars only minimally affect prey populations. Cougars have had little direct effect on the size of elk and deer populations in the Yellowstone ecosystem (Murphy et al. 1999). Similarly, cougar populations in Idaho, Arizona, and Utah do not prevent elk or mule deer (*O. hemionus*) from increasing (Hornocker 1970, Shaw 1980, Lindzey et al. 1994). Logan et al. (1996) concluded that habitat quality and quantity, not cougars, were the ultimate limiting factors for mule deer in southern New Mexico.

Under other conditions cougars may significantly reduce prey numbers. In northwestern Montana, cougars in combination with wolves and bears limited (as defined above) white-tailed deer and elk populations (Kunkel and Pletscher 1999). In California, cougar predation caused precipitous declines in small bighorn sheep (*Ovis canadensis*) populations where few alternate prey were available (Wehausen 1996, Hayes et al. 2000). Cougars also are limiting recovery of state-endangered desert sheep (*Ovis canadensis mexicanus*) in New Mexico (Fisher et al. 1999). There is no evidence that indiscriminate control of cougars alters these trends (Evans 1983, Hurley and Unsworth 1999; but see Ernest et al. 2002); more targeted control of individual cougars appears most effective for sheep populations (Ross et al. 1997, Wright et al. 2000).

Brown bears

Brown bears currently are found in <50% of their former range in North America and <2% of their former range in the lower 48 United States. Five subpopulations exist in the contiguous United States: (1) Yellowstone ecosystem, (2) northern continental divide ecosystem in northwestern Montana, (3) Cabinet/Yaak ecosystem in northwestern Montana, (4) Selkirk ecosystem in northern Idaho, and (5) North Cascades ecosystem in northern Washington. These areas are dominated by parks and designated wilderness areas. In contrast, only about 12% of the bears are confined to protected areas in British Columbia (McLellan and Hovey 2001).

Densities of brown bear populations range from 3.9 (arctic Alaska) to 551 (southern coastal Alaska) bears per 1000 km^2 (Miller et al. 1997). Brown bears display some of the lowest reproductive rates and rates of increase among terrestrial mammals due to late sexual maturity and protracted reproductive cycles (Jonkel 1987, Eberhardt et al. 1994, Craighead et al. 1995, Hovey and McLellan 1996, Pease and Mattson 1999).

Impacts of predation

Impacts of bear predation on ungulates is mostly on neonates (Schwartz and Franzmann 1991, Kunkel and Mech 1994, Smith and Anderson 1996, National Research Council 1997, Ballard and Van Ballenberghe 1998, Bergerud and Elliot 1998). In combination with other predators, bears may limit some ungulate populations (Messier 1994, Bergerud and Elliot 1998, Kunkel and Pletscher 1999). Messier (1994) suggested that the presence of a bear species may be necessary for wolves to regulate moose at low density. Under such circumstances, reductions in bear densities may or may not benefit ungulate populations (National Research Council 1997, Ballard and Van Ballenberghe 1998).

Predation by bears on salmon (*Onchorynchus* spp.) can be intense. On Chichagof Island in southeast Alaska, over 50% of a sample of 1100 salmon carcasses showed signs of bear predation (Willson et al. 1998). On a Moresby Island stream in western British Columbia, black bears captured over 4200 salmon during the 45-day spawning period in 1993, about 74% of the salmon entering the stream (Reimchen 2000). On large Alaskan rivers with large salmon runs bears take as little as 2.5% of the run but on smaller streams they take up to 85% (Reimchen 2000). Intensive levels of predation may be responsible for the evolutionary selection of some features in salmon including body size and reproductive strategies (Willson et al. 1998, Reimchen 2000).

Black bears

The black bear is the most successful of the world's eight bear species at co-existing with humans. Black bear status in North America varies from pest to threatened (Pelton 2000). The range of the species in the western U.S. is largely associated with public lands in forested mountain terrain. Densities of black bears range from 90 (Alaska) to 1300 (Washington) per 1000 km^2 in western North America (Kolenosky and Strathearn 1987, Miller et al. 1997). Densities and demographic rates are highest in diverse early-successional forests with rich soils and in areas with relatively long foraging seasons (Schwartz and Franzmann 1991). Reproductive rates of black bears are low. If a female bear lives to age 15, she will generally produce a maximum of six litters during her lifetime (Kolenosky and Strathearn 1987). Most mortality is caused by humans and includes hunting, poaching, depredation control, and vehicle collisions. Annual survival rates of adult females average 0.87 (Pelton 2000).

Jaguars

Jaguars were probably eliminated from the northern portion of their range in southern New Mexico and Arizona early in the twentieth century (Valdez 2000). The northern portion of the range of jaguars has receded southward about 1000 km and has been reduced in area by nearly 70% (Swank and Teer 1989). Jaguar elimination resulted from the same predator control programs that reduced the other large carnivores in the West. Limited sightings of probable dispersers from the closest population in Mexico (approximately 200 km south of the US–Mexico border) have been made recently in both New Mexico and Arizona. Jaguars were classed as endangered in the U.S. in 1997, but there is presently no recovery plan in place. Densities of jaguars in tropical forests range from 30 to 70 per 1000 km^2 and likely 13 to 19 per 1000 km^2 in northwestern Mexico (Lopez Gonzales and Brown 2003).

In the tropics, jaguars are usually associated with closed canopy forest and permanent water below 1200 m (Quigley and Crawshaw 1992). Jaguars used riparian forests more than expected and open forests less than expected in the Pantanal region of South America (Crawshaw and Quigley

1991). Jaguars occupy montane oak (*Quercus* spp.), oak-pine (*Pinus* spp.) forests, riparian forests, and mesquite thickets at the northwestern limit of their range (Brown 1983). Jaguars appear particularly adapted for preying on large slow mammals such as peccaries (*Tayassu* spp.), while cougars, being smaller and more agile than jaguars, are more adapted to prey on deer (*Odocoileus* spp.; Aranda 1994). This difference likely reduced competition between the two species (Aranda and Sanchez-Cordero 1996). Like other large carnivores, jaguars are opportunistic and have been recorded to prey on over 85 species (Seymour 1989). Even though jaguars presently occur primarily in the tropics, they originated in the holarctic (Kurten and Anderson 1980) and they no doubt would do well in temperate regions with ample prey (Valdez 2000).

Intraguild dynamics among carnivores

Only recently has comprehensive work examined interactions (predation, competition, kleptoparasitism) among large carnivores in ecosystems (Kunkel 1997, Murphy et al. 1998, Kunkel et al. 1999, Smith et al. 1999, 2001). Wolves and cougars both selected the white-tailed deer for prey in northwestern Montana and they selected similar classes of prey (Kunkel et al. 1999). Wolves kleptoparasitized cougar kills and killed cougars, and wolves and cougars were largely responsible for the white-tailed deer and elk population declines in northwest Montana that resulted in starvation in some cougars. These results suggested exploitation and interference competition between the two species. Exploitation competition rather than interference competition likely was responsible for the population decline in cougars because more cougars starved than were killed by wolves. Such interactions have only been examined in a few other studies worldwide. Iriarte et al. (1990) speculated that prey selection by cougars in the Americas resulted from competitive evolution with sympatric jaguars. Wolf kleptoparasitism of cougar kills may force cougars to increase their kill rate as the wolf population expands (Murphy et al. 1998).

Wolves and bears frequently interact at kill sites, with varying outcomes depending on number of wolves involved. One in three cougar-killed ungulates were scavenged by brown bears in the Yellowstone ecosystem during 1990–1995 and at one in eight carcasses bears displaced cougars (Murphy et al. 1998). These displacement rates were approximately twice as high as rates in the Glacier ecosystem in Montana (Murphy et al. 1998). Habitat and prey size may significantly affect these relationships (Creel 2001).

Umbrella, flagship, top-down, and keystone roles

Because regional land management is highly inefficient if done on a species by species basis (Noss et al. 1997, Simberloff 1998), conservation biologists have attempted to identify and use one or a few species as surrogates for an array of others. Such surrogates have been called indicator, umbrella, flagship, or keystone species depending on their perceived ecosystem roles and utility in addressing conservation problems (Caro and O'Doherty 1999). Definitions of surrogates have varied, and Caro (2000) urged that conservationists should define the goals of conservation projects clearly when using these terms. Power et al. (1996) defined the keystone species as one whose impacts on the community or ecosystem are large relative to the species abundance. Caro and O'Doherty (1999) argued, however, that keystones are not used as a shortcut to describe patterns or processes and have never been successfully used as surrogates, though they may be useful in choosing them (but see Simberloff 1998, Kotliar 2000). The keystone concept may be a useful tool for communicating ecological importance to the public and offsetting unfavorable public opinion of some species (Kotliar 2000). Indicator species are used as surrogates for ecosystem health or areas of high species richness (Landres et al. 1988). Umbrella species differ from indicator species in that they are used to specify the size and type of habitat to be protected rather than its location (Berger 1997). Flagship species are charismatic species used to raise awareness, build public support, or attract funding for a conservation cause (Caro and O'Doherty 1999).

Trophic cascades, top-down effects, and keystones

Removal of top carnivores can lead to a cascade of community alterations including relaxation of predation as a selective force, the irruption of herbivore populations, the spread of disease, and diminished biodiversity (Kay 1994, McShae and Rappole 1997, Wilson and Childs 1997, Berger 1998, Terborgh et al. 1999, 2001). Relaxation of predation was shown in Alaska where field experiments demonstrated that moose are sensitive to vocalizations of ravens (*Corvus corax*) and may rely on their cues to avoid wolf predation (Berger 1999). A similar relationship was absent in areas of Alaska and Wyoming where wolves and bears have been extirpated for 50–70 years. Evidence for diminished biodiversity related to large carnivore absence was provided by Berger et al. (2001a) in a comparison of densities and diversities of riparian birds in areas of high moose density (lacking large carnivores) to areas of lower moose density (human-harvested populations). Willow communities were more altered, and densities and diversity of birds were lower in the areas of higher moose density.

Removal of top predators may result in superabundant populations of herbivores and medium-sized predators. This in turn may result in reproductive failure and local extinction of plants, birds, reptile, amphibians, and rodents (Crooks and Soule 1999, Henke and Bryant 1999, Terborgh et al. 2001). Coyotes expanded their range and densities following the extirpation of wolves in the western U.S. (Johnson et al. 1996), and this may have resulted in decreased densities of red foxes (*Vulpes vulpes*; Peterson 1995) and swift foxes (*Vulpes velox*; Kitchen et al. 1999). Seven years after arrival to Isle Royale, wolves completely eliminated coyotes (Krefting 1969). On the Kenai Peninsula, Alaska (Thurber et al. 1992), northwest Montana (Arjo and Pletscher 1999), and Manitoba (Paquet 1991), however, extensive overlap of wolf and coyote home ranges occurred with little reduction in coyote density. As evidenced in large carnivore interrelationships, habitat and prey size may significantly affect large and smaller carnivore relationships (Creel 2001). Smaller carnivores may be more vulnerable in more open habitats, and wolf and coyote co-existence may be more likely where prey size is larger and consumption by wolves is relatively less, thereby providing food for coyotes (Peterson 1995).

There is little direct evidence indicating that systems are regulated by growth and biomass of plants (bottom-up) or indicating top-down control in terrestrial systems (Gasaway et al. 1992, Crete and Manseau 1996, National Research Council 1997, Soule and Terborgh 1999, Kunkel and Pletscher 1999). Evidence for top-down control is increasing, however (Schmitz et al. 2000, Estes et al. 2001, Halaj and Wise 2001, Miller et al. 2001, Terborgh et al. 2001). On Isle Royale growth rates of balsam fir (*Abies balsamea*) were regulated by moose density, which in turn was reduced by wolf predation (McLaren and Peterson 1994). Ripple and Larsen (2000) provided evidence of a significant decline in aspen overstory recruitment after 1920 in Yellowstone and hypothesized that elimination of wolves and the resultant increase in elk numbers was largely responsible for this decline. Wolf restoration may change this trend by the resulting decrease in elk numbers and subsequent increase in aspen recruitment. An increase in aspen may also result from elk avoiding browsing in aspen stands as an antipredator response to the presence of wolves (Ripple et al. 2001). Boyce and Anderson (1999) believe, however, that fluctuations in vegetation and overwinter mortality will continue to cause much greater annual variation (including major perturbations) of ungulate populations in the Yellowstone ecosystem than wolves. Documenting a fortuitous natural experiment of large predator exclusion on newly created islands in Venezuela, Terborgh et al. (2001) showed that the absence of predators consistently freed certain consumers to increase many times above "normal." This unleashed a trophic cascade whose effects included severely depressed recruitment of canopy trees. Hyperabundant consumers threaten to reduce much plant and animal diversity in these species-rich forests. More opportunistic predators like wolves likely will have greater top-down effects than the more specialized cougars (McCann et al. 1998, Miller et al. 2001).

The top-down or keystone roles of less predatory carnivores (bears) are even less clear, but potentially significant in some systems. The millions of anadramous fish (primarily salmon) that spawn in freshwater streams along the Pacific coast provide a rich food resource that directly affects the biology of terrestrial consumers including bears and indirectly affects the entire food web that ties the water and land together (Willson et al. 1998). The high energy value of this food greatly affects bear reproductive success (Hilderbrand et al. 1999). Bears commonly carry these salmon back to streambanks and tens of meters inshore (Willson et al. 1998). Bears may carry up to 6.7 kg per ha of phosphorus into the terrestrial nutrient cycle, a level similar to the commercial application rate often used for forestry (Willson et al. 1998). This movement of carcasses is also a major source of nitrogen for riparian vegetation (Bilby et al. 1996, Ben-David et al. 1998).

The potential top-down or keystone role of large carnivores is still under debate (Polis and Strong 1996) and largely depends upon the definition of keystone roles (Power et al. 1996). The great complexity of trophic interactions makes the assessment of top-down versus bottom-up control very difficult, as the two processes may not be mutually exclusive and most likely act in concert. Using modeling, Powell (2001) predicted that predators and prey each control systems, but the control acts on different scales with variation in productivity of food causing more variation in herbivore population sizes than variation in predation rates. The best experiment completed examining top-down and bottom-up control in mammalian predator–prey systems indicated that control was both from the top-down and bottom-up in a lynx–hare (*Lynx lynx* and *Lepus americanus*, respectively) system in the boreal forest (Krebs et al. 1995). Both types of control should be anticipated and probably vary depending on spatial and temporal scales and the complexity of the predator–prey systems.

A shift apparently occurs from bottom-up dominance (caribou and moose) in unproductive northern tundra ecosystems to top-down (wolf and bear regulated) dominance in lower-latitude, more productive boreal ecosystems (Messier 1995, Crete and Manseau 1996). The theory of food chain dynamics predicts that trophic levels are added sequentially as primary productivity increases (Fretwell 1987). At low primary productivity, herbivores should have a strong impact on plant biomass, but predators would be absent or unable to regulate their prey (due to the migratory behavior of caribou at high latitudes resulting from low habitat productivity). With increased primary productivity, predators would be able to hold herbivores in check, and herbivores then will have only a small impact on the plant community (e.g., moose at mid-latitudes; Crete 1999). Where wolves are absent cervid biomass is five times greater. The Isle Royale example (McLaren and Peterson 1994) fits this model. The moose-wolf and caribou–wolf systems examined by Messier (1995) fit this theory in a broad sense but departed from it when: (1) habitat quality was high enough for moose to escape regulation by wolves (predator satiation) unless another predator such as bears was also present, and (2) caribou-plant interactions are affected by multi-year time lag effects that produce recurring fluctuations in caribou numbers. The pattern of increasing cervid density with increasing biomass productivity (along the latitudinal gradient) in the absence of wolves and bears predicts that cervid abundance will significantly decrease in the western U.S. with wolf re-colonization (Crete 1999, Oksanen et al. 2001). The equilibrium biomass (<100 kg km^{-2}) of cervids in this region, however, will not be as low as in the moose range of the mid-latitudes because equilibrium density is higher in multi-species assemblages.

Umbrellas and flagships
Because large carnivores have such large home ranges (e.g., 100–2000 km^2) and their habitat encompasses those of many other species, they have been used as umbrella species. It is debatable, however, whether this means large carnivores serve as umbrellas. Noss et al. (1996) and Caro and O'Doherty (1999) could find no definitive published studies documenting the level of protection afforded to other species by a conservation plan

focused on large carnivores. In one of the few preliminary tests of the umbrella concept, Berger (1997) concluded that, despite their large home range, black rhinos (*Diceros bicornis*) did not serve well as an umbrella for species at the same trophic level. While there is great urgency for finding management paradigms to conserve biodiversity, the complexity of ecosystems and our lack of knowledge of them should temper our rush into specific, potentially expensive paradigms (Andelman and Fagan 2000).

In an analysis of the California coastal sage scrub, the Columbia Plateau, and all the U.S. counties, Andelman and Fagan (2000) found that surrogate species did not perform substantially better than randomly selected sets of a comparable number of species. They also found little evidence to support the claim that umbrellas, flagship, or biodiversity indicator schemes (including using large carnivores) have special biological utility as conservation surrogates for protecting regional biota. Extensive reliance on surrogate species may be a poor allocation of scarce conservation resources and even the most carefully selected surrogate might prove inadequate and inefficient (Andelman and Fagan 2000). These authors urged caution in adopting umbrella or flagships until their usefulness as predictors of biological diversity and its persistence has been more fully investigated. One start at such an assessment provided evidence that the red wolves (*C. rufus*) in the southeast served as a successful flagship (Phillips 1990).

Because large carnivores are habitat generalists, protecting their habitat may not necessarily protect the habitat of some specialists. Protection of large areas though will reduce this shortcoming. Large carnivores primarily need sufficient prey and relatively low levels of human-caused mortality, criteria that may not necessarily meet the needs of many other species. The diverse and extensive habitat needs of brown bears, however, may make them a potentially better umbrella species than the other large carnivores. The best surrogate species are those that can be easily monitored (Caro and O'Doherty 1999): bears do not fit this criteria. According to the criteria of Caro and O'Doherty (1999), wolves are the only large carnivore that may serve as a surrogate species and then only as a flagship. In developing a comprehensive conservation strategy for carnivores in the Rocky Mountains, Carroll et al. (2001) concluded that the plan must consider several species rather than a single umbrella species. Even so, they also concluded that the viability of individual species serves as a biological "bottom-line" that allows evaluation of the effectiveness of a conservation strategy in a way not possible with composite indicators of ecosystem function.

Recent expansions by wolf populations provide further evidence of their tenuous role as umbrellas. Restoration of wolves in forested regions of the Great Lakes may not necessarily be a sign that the ecosystem there has been restored to some previous level of ecosystem function (Mladenoff et al. 1997). In fact, wolves may do well because the ecosystem is altered. High deer populations support large numbers of wolves, but they can also negatively affect other important aspects of forest biodiversity (DeCalesta 1994). Wolf recovery in the Great Lakes region and potentially the Yellowstone area has resulted not from restoration of the original ecosystems but from more tolerant human attitudes in combination with human-caused landscape and prey changes (high deer and elk numbers). Management of these systems may require the reduction of prey densities to reduce impacts of high prey numbers on ecosystems (Kay 1994), which would ultimately reduce carrying capacity for wolves. Alternately, wolves may be used as the management tool to reduce those prey densities and associated impacts. See Singer et al. (2003) for more information on this topic.

In some situations, restoration of a large carnivore may have negative consequences for the ecosystem. The expansion of cougars has the potential to extirpate small populations of native vertebrates (e.g., porcupines (*Erethizon dorsatum*) and desert sheep in the Great Basin; Berger and Wehausen 1991, Sweitzer at al. 1997). We must recognize that recent ecosystem changes have altered the dynamics of interacting native species in ways that threaten patterns of biodiversity (Berger and Wehausen 1991, Sweitzer at al. 1997).

Landscape management and large carnivore conservation

Forest, land, and human management

Few areas in western North America combine high biological productivity and low human impact (Carroll et al. 2001). Thus zones of human–carnivore conflict are often in areas of highly productive habitat that have above-average human use, are spatial buffers between large core habitat areas and zones of high human use, or are likely to experience increased human use in the future (Boyd 1997, Mace and Waller 1998, Merrill et al. 1999). Outside of Alaska and northwestern Canada, opportunities for creating single reserves large enough to sustain populations of large carnivores are very limited. Even in Alaska and northwestern Canada, careful management will be required to ensure the long-term persistence of large mammals. Even so, most of western North America still has relatively few people and could accommodate reserves buffered from intensive land use and interconnected by networks covering large areas (Noss 1992).

Wolves

Even though wolves are habitat generalists and ungulate densities explain more than 70% of the variation in wolf densities (Fuller 1989), some natural and anthropogenic landscape variables appear important to wolves (Singleton 1995, Boyd 1997, Kunkel and Pletscher 2000, 2001). In the Great Lakes region, wolves avoid agricultural areas and deciduous forests and favor forests with a conifer component (Mladenoff et al. 1995). Centers of wolf territories are most likely to occur in areas with road densities below 0.23 km per km^2 and nearly all wolves occur where road densities are below 0.45 km per km^2. No wolf territory was bisected by a major highway or where human population densities were >1.5 persons per km^2. Mladenoff et al. (1995) found road density to be the best predictor of wolf habitat in the Great Lakes region. Their data suggested that wolves selected areas to avoid contact with and consequent potential mortality from humans. Wolves in the Rocky Mountains, however, selected for areas with roads, probably because roads coincided with valley bottoms preferred by prey (Boyd 1997, Kunkel and Pletscher 2000), and 75% of human-caused wolf mortality was within 250 m of a road. Similarly, Trans Canada Highway 1 accounted for more than 90% of wolf mortalities in the Bow Valley, British Columbia (Paquet P., World Wildlife Fund, personal communication).

Some models have been developed to predict areas that wolves will re-colonize (Singleton 1995, Boyd 1997, Kunkel 1997, Mladenoff et al. 1999) or where potential conflicts with prey (U.S. Fish and Wildlife Service 1994, Kunkel and Pletscher 1999, 2000, 2001) and livestock (Fritts et al. 1992, U.S. Fish and Wildlife Service 1994, Mech et al. 2000) will be the highest. These models are currently being refined for various regions in the Rocky Mountains. They may help managers to delineate priority areas for managing and monitoring wolves and their prey, and predict and remedy conflicts with landowners and hunters. Only two restrictions on land use have been used to promote wolf recovery in the western U.S. A 1.6-km radius area around dens is sometimes protected from intensive human use between 15 March and 1 July, and USDA Wildlife Services cannot use nonselective predator control in areas occupied by endangered wolves.

Cougars

Topographic relief or abundant vegetative cover are important habitat components for cougars (Seidensticker et al. 1973, Logan and Irwin 1985, Logan et al. 1986, Van Dyke et al. 1986, Laing and Lindzey 1991, Williams et al. 1995). Cover improves success in ambush hunting and provides protection from enemies (Kunkel 1997, Kunkel et al. 1999). Cougar dens are usually located in rock outcrops, in dense shrubfields, or under downed conifers (Murphy et al. 1999). Travel corridors used by cougars in southern California are typically drainage washes or ridges with abundant native woody vegetation that provide security from human disturbance (Beier 1995).

Logging, burning, or grazing may reduce the cover needed by cougars (Logan and Irwin 1985, Van Dyke et al. 1986, Laing and Lindzey 1991). Cougar track density decreased by 61% in timber harvest areas in southern California from 1986 to 1992 (Smallwood 1994). Logging would likely have negligible effects if logged areas were small relative to area requirements of cougars. Like other obligate carnivores, cougar habitat use and density will be affected by landscape alterations as they affect prey spatial temporal distribution, use of hiding cover, and density (Kie et al. 2003). Human activity may reduce habitat quality. Cougars in Utah shifted to nocturnal activity patterns in the presence of human disturbance and crossed less-traveled roads more than higher-use roads (Van Dyke et al. 1986).

Bears

Bear-management strategies presented here apply primarily to brown bears. Because black bears are generally less sensitive to human disturbance than are brown bears, management for brown bears generally benefits and is adequate for black bears. Although brown bears are flexible in the habitats they use (Waller and Mace 1997) protection of certain habitats in forested mountains of the western interior is important. Bears generally select riparian areas and avalanche chutes in spring (Mace et al. 1999, McLellan and Hovey 2001). Bears track plant phenologies by moving up elevation in avalanche chutes during the summer to find the berries (*Vaccinium* spp. and *Shepherdia* spp.) that dominate their diet (McLellan and Hovey 1995). Ensuring wild or prescribed fire is important as berries are found most in open timber and open timber burns 50–70 years old at high elevations, and fire promotes regeneration of important whitebark pine seeds (*Pinus albicaulis*). Timber harvests must be carefully planned as large regenerating timber harvest blocks are rarely used in any season by bears (McLellan and Hovey 2001). Alpine insect aggregations are used by brown bears in summer and fall (Mattson et al. 1991*a*, *b*) and human disturbance to bears in these areas should be minimized.

Human presence in brown bear habitats often leads to bear–human conflicts, often with fatal consequences for bears. Human-defense kills of brown bears occur more frequently in areas of higher human populations (Mattson et al. 1996*a*). Human-habituated brown bears may use native and non-native foods near human developments and are killed more often than non-habituated bears (Mattson et al. 1996*a*). Where available, fish are critical in the summer and fall (Mattson and Reinhart 1995, Hilderbrand et al. 1999). Intense sport fishing and development along rivers and streams can exclude bears from important food sources. When this exclusion is combined with human-induced mortality, mortality of bears can exceed sustainable levels (Mattson and Reinhardt 1995, Schwartz and Arthur 1997). Limiting adult annual female mortality to <10% is key to brown bear conservation in the small, threatened populations of the lower 48 states (Wielgus et al. 1994, Mattson et al. 1996*b*, Mace and Waller 1998, McLellan et al. 1999).

Road construction and timber harvest should avoid riparian areas and avalanche chutes (Mace et al. 1999). Secure cover should be maintained near these chutes; clear-cuts and heavy thinning adjacent to avalanche chutes should be avoided. To maintain existing habitat quality, Craighead et al. (1995) recommended improving sanitation in areas in and surrounding recovery areas and establishing maximum road densities of 1 km per 6.4 km^2 ($\frac{1}{4}-\frac{1}{3}$ the density of current standards). Areas with road densities of >6 km per km^2 were not used by brown bears in western Montana (Mace et al. 1996). Multiple-use lands remote from human population centers may be critical for bears and must be managed for low-density and educated human use (McLellan et al. 1999). Hunters must be educated in identification of brown versus black bears and must handle ungulate carcasses in ways to avoid attracting bears. Craighead and Craighead (1991) recommended that the vegetation of the entire northern Rockies bioregion be satellite-mapped using a regionally consistent, botanically detailed hierarchy of vegetation and landform classifications. Thereby, habitat quality and quantity of bear foods could be defined from ecosystem to ecosystem and carrying capacities could be estimated. This work has recently been completed for the Selway-Bitterroot ecosystem (Merrill et al. 1999, U.S. Fish and Wildlife Service 2000).

Accumulation of energy reserves is crucial for successful reproduction of bears (Hilderbrand et al. 1999). Litter sizes and population densities are linked to dietary meat content (Miller et al. 1997, Hilderbrand et al.1999) indicating the importance of fish and ungulate populations to bears including Yellowstone Lake cutthroat trout (*Oncorhynchus clarki*) and bison (*Bison bison*) and other ungulate carcasses in Yellowstone National Park. Bears with access to abundant meat sources had dietary meat contents generally >70% (Jacoby et al. 1999). Protection and restoration of anadromous fish runs coincident with the maintenance of safe and productive foraging areas for bears may require careful management. Such management is critical in maintaining the historic linkage between these terrestrial and aquatic systems. Restoration of salmon in the Columbian River system including central Idaho, for example, may be important for long-term viability of bears in this system (Craighead et al. 1995, Hilderbrand et al. 1999).

Craighead et al. (1995) outlined goals to achieve population recovery for brown bears in the lower 48 states. They argued that more extensive areas than outlined in the recovery plan (U.S. Fish and Wildlife Service 1993*a*) need to be ascribed with greater connectivity to reduce bear mortality and increase population potential. They recommended following the population viability analysis of Shaffer (1983, 1992) to guide recovery. Most significantly, and most controversially (Schullery 1997), Craighead et al. (1995) recommended developing "ecocenters" for bears of strategically placed food concentration centers. Based on data from bears using dumps prior to closure, Craighead et al. (1995) predicted that ecocenter networks would increase mean rates of natality and survival and carrying capacity, buffer seasonal variation in bear foods, serve to concentrate bears reducing movement into areas where risk of mortality may be high, and increase bear count reliability. They argued that current recovery areas are not large enough to support bears over the long-term and only with husbandry and extra inputs into the system can bears persist.

Brown bear population trends in the Yellowstone ecosystem have been debated (Eberhardt et al. 1994) with Pease and Mattson (1999) concluding the population has changed little since 1975. The population probably increased in whitebark pine crop mast years and declined in years when this crop failed. Mattson and Reid (1991) and Pease and Mattson (1999) cautioned that a conservative approach be maintained because the long-term habitat condition trend resulting from increasing numbers of humans in the area and the potential for global warming is likely downward. They argued that making decisions that have long-term consequences based on short-term trends would be an error, and thus it is premature to remove brown bears from the threatened list. This recommendation is presently under considerable debate (MacCracken and O'Laughlin 1998).

Livestock

Wolves

Control of wolves preying on livestock is one of the greatest management concerns for the species (Mech 1995). Concurrent with the increase in wolves and their range in Minnesota, the number of wolves killed by USDA Wildlife Services for depredation control has increased dramatically, from six in 1979 to 216 in 1997 (Mech et al. 2000). Even with >2000 wolves, the total percentage of farms in wolf range in Minnesota that suffer verified wolf predations is only about 1% per year. Wolf depredations on livestock are relatively low compared to other causes of livestock mortality, but are inordinately controversial (Bangs et al. 1998). Between 1987 and 2000, confirmed minimum livestock losses in northwest Montana totaled 82 cattle, 68 sheep and seven dogs (E. Bangs, USFWS, unpublished data). As a result, 41 wolves were killed and 32 were translocated. On average, <6% of the wolf population is annually affected by agency wolf-control actions (Bangs et al. 1998). Minimum confirmed livestock losses have annually averaged about 3.6 cattle, 27.8 sheep, and 3.8 dogs in the Yellowstone area and 9.2 cattle, 29.4 sheep, and 1.8 dogs in central Idaho. Since 1995, USFWS and the Wildlife Services have killed 18 wolves in central Idaho and 26 in the Yellowstone area because of conflicts with livestock. Since 1987, a private compensation fund administered by Defenders of Wildlife has paid livestock producers about $155 000 for confirmed or highly probable wolf-caused losses in Montana, Idaho, and Wyoming. This compares to an estimated $45 000 000 in annual losses to all causes for livestock producers in Montana alone. While losses to wolf depredation are insignificant to overall losses, losses to individual operators can be significant. It is likely that some form of compensation for losses (private or public) will always be required to ensure persistence of large carnivores in the West, and even then illegal killings of carnivores and low livestock producer support for restoration will remain (Bangs et al. 1998).

To determine best methods for managing depredations, researchers have examined the characteristics of farms experiencing depredations. Farms with chronic losses in Minnesota were larger, had more cattle, and had herds farther from human dwellings than farms with no losses (Mech et al. 2000). Forested public lands intermixed with private farm/ranch lands have experienced the greatest losses in the western recovery areas. The best management prescription for depredations in most cases appears to be lethal control. An analysis for Montana concluded that livestock losses and control costs could be significantly reduced by killing rather than relocating depredating wolves (Bangs et al. 1998). Non-lethal control options may be valuable in certain circumstances especially where wolf populations remain low. Non-lethal tools that will be successful will likely vary by circumstance (Mech 1995, Knowlton et al. 1999).

As wolves adapt to travel through relatively settled and open areas, opportunities for conflicts will increase as wolf populations increase. One partial solution to this is zoning to separate wolf habitat from wolf-free areas (Mech 1995). Zoning at large scales (among states; U.S. Fish and Wildlife Service 1993b) simplifies management. The primary disadvantage of large-scale zoning is that wolves would not be allowed to live in some areas where they could persist. Smaller-scale zoning would be more complex, but would allow wolves to live in many more enclaves. Even though dispersing wolves would likely face higher probabilities of mortality in the small-scale zoning management paradigm, there would be enough populations of wolves in such a meta-population for persistence to be likely (Mech 1995, Haight et al. 1998). Biologically, wolves could occupy parts of almost all regions of the western U.S. For this to occur, however, there must be acceptance by the public to control problem wolves (Mech 1995).

Cougars

In northwestern North America, rates of cougar predation on livestock are generally low. In Montana from 1984 to 1993, only 8.2 predation incidents occurred annually (Montana Fish, Wildlife and Parks 1996). Claims for compensation in Alberta for cougar kills averaged 4.4 per year from 1974 to 1987. For every cougar claim, there were five wolf, 13 bear, and 42 coyote claims over a similar period (Pall et al. 1988). Selective removal of offending individuals is usually a more effective response than other management actions, especially translocation (Ruth et al. 1998). The complete elimination of cougars from problem regions in New Mexico has been attempted three times – twice to protect domestic sheep and once for wild sheep. None of these efforts resulted in a reduction in predation (Evans 1983).

Brown bears

Brown bear depredation on livestock is highly variable among years and areas. Livestock losses from brown bears in the Yellowstone ecosystem averaged 35 cattle and 29 sheep per year (Gunther et al. 1998). Losses averaged eight cattle and 17 sheep per year in the northern continental divide ecosystem (U.S. Fish and Wildlife Service 2000). Management of brown bears that prey on livestock must be more conservative than that for wolves or cougars because of the bear's relatively low reproductive rates. Bears must be provided more leeway than cougars or wolves before direct management actions are taken. Like wolves and cougars, bears that are relocated experience low survival and high return rates to capture sites (Blanchard and Knight 1995). Relocation has been most successful for subadult females. Similar to their wolf program, Defenders of Wildlife pays compensation to ranchers suffering losses of livestock to brown bears. More recently, this program and innovative initiatives by other conservation organizations have expanded to start purchasing grazing permits on public lands and retiring them where chronic conflicts with large carnivores occur. These initiatives hold great potential for reducing many livestock/wildlife conflicts on public lands.

Jaguars

Significant habitat loss and consequent loss of prey base is increasingly forcing jaguars to co-exist with humans and livestock in fragmented areas. The most urgent problem facing jaguar populations is indiscriminate killing of jaguars where conflicts with humans occur. Most research examining livestock predation by jaguars comes from Central and South America where livestock management is often less controlled and where non-lethal methods to manage depredations are less available and known than in North America. As a result, depredation rates there are relatively higher than rates for other large carnivores in North America. Cattle constituted a major part of the jaguar diet in studies conducted on ranches in seasonally flooded savannah woodland in the Venezuelan llanos (Hoogesteijn et al. 1993). Jaguar-caused mortality to calves ranged from 6% on a well-managed ranch to 31% on a smaller ranch in a more agriculturally developed region. Research conducted in Belize, however, indicated that healthy adult male jaguars can range close to livestock without causing problems (Rabinowitz 1986). Formation of protected areas within a network of ranches and ranches with easements may reduce mortality to jaguars but also allow ranching to remain viable (Lopez Gonzalez and Brown 2003).

Landscape effects on carnivore hunting success

Structure and pattern of landscapes can affect the hunting success of carnivores, or alternately, the vulnerability of their prey and thus management of these landscapes can impact carnivores and their prey. Bergerud (1988) postulated that part of the decline of woodland caribou in British Columbia was due to forest harvest practices that concentrated caribou in small patches that were easily accessible and searched by wolves. Prior to the arrival of Europeans, lightning-caused and Indian-caused fires produced more open habitats in many portions of the Rockies (Barrett and Arno 1982). Control of fire in the Rocky Mountains has advanced forest succession which has resulted in an increase in stalking cover for predators (Barrett and Arno 1982). This has potentially altered predator–prey dynamics in certain situations in favor of wolves and cougars. Similar human-caused shifts in balance (disequilibriums) have been hypothesized for declines in bighorn sheep (Berger and Wehausen 1991), moose, and caribou (Bergerud 1981, 1988, but see Kunkel and Pletscher 2000).

Human harvest of carnivores

Due to their threatened and endangered status, wolves and brown bears are not harvested outside of Alaska and Canada. In Alaska, current regulations require wolves and bears that are harvested to be inspected and sealed. The wolf is the only big game animal in Canada that is hunted year-round, has no bag limits in many areas, and does not require a hunting license (Hayes and Gunson 1995). Harvests for wolves are liberal because of concerns for impacts of wolves on big game and because estimates of harvest rates necessary to reduce wolf population densities are >40% (Keith 1983).

Because of the low reproductive rates of brown bears, an overall female mortality of <1–16% (depending on other demographic parameters) was necessary to sustain populations (Eberhardt 1990, McLellan et al. 1999). Sustainable harvest rates range from 2% (Yukon) to 6% (Montana) per year (Miller et al. 1997). Bears in poor habitat can only support the most limited adult female mortality rates; so, harvest rates must be very conservative. Harvest rates of >30% of adult males resulted in a decline of a small brown bear population in Alberta (Wielgus and Bunnell 1994) apparently because of the immigration of subadult male brown bears after mortality of adult males and the resulting infanticide (Wielgus and Bunnell 1995, Swenson et al. 1997). The effects of harvest of males on cub survival remain controversial, but Swenson et al. (1997) concluded that harvesting one adult male brown bear corresponded to harvesting 0.5–1.0 adult females.

Legal harvest is usually the greatest source of mortality for cougars (Murphy 1983, Logan et al. 1986, Anderson et al. 1992, Ross and Jalkotzy 1992). In unhunted cougar populations or where cougar depredation on livestock is substantial, control actions may be the greatest source of human-caused mortality (Cunningham et al. 1995). The strong dispersal capability of cougars leading to immigration may help ameliorate the effects of mortality where cougar habitat is contiguous and exceeds 2200 km² (Beier 1993) and where travel corridors allow free exchange of dispersers among subpopulations (Lindzey et al. 1992, Logan et al. 1996, Murphy et al. 1999). Research and management population data for cougars are often inadequate for managing harvests and thus harvest should be conservative. Harvest of male cougars should not exceed 8%, and hunting of females should be restricted (Logan et al. 1996).

Social systems of all the large carnivores must be taken into consideration in harvest programs, because harvest may disrupt the social system of a species and result in counterintuitive or greater than expected effects (Swenson et al. 1997). There may be significant differences between the behavior of populations of exploited and of non-exploited carnivores and this may impact population trajectories (Seidensticker et al. 1973, Hornocker and Bailey 1986, Kitchen et al. 2000a, b, Wright et al. 2000; but see Meier et al. 1995).

Restoration and conservation over large landscapes

Restoration priorities and techniques

Given the tenuous status of brown bears and wolves in the lower 48 United States and the mandate of the U.S. Federal Endangered Species Act to restore species over a significant portion of their range, restoration is critical for conservation of these species. We should work to restore large carnivores because of the key role they play in ecosystems and should strive to conserve the suite of adaptations of large carnivores to the environmental conditions and prey assemblages in which they live (Wikramanayake et al. 1998). Suitable habitats for wolves and brown bears exist in many places in the West, but, because of high mortality encountered by wolves and bears moving to these areas, re-introductions may be necessary (but see Boyd and Pletscher 1999). Because re-introductions are expensive, often fail, and are very high profile, managers should follow re-introduction guidelines of the International Union for the Conservation of Nature (IUCN; IUCN 1998) and others (Reading and Clark 1996) when considering a re-introduction. The biological feasibility of establishment of the species must be assessed in the area being considered for a re-introduction, and local public support must be established. The monitoring and management programs to be put into place after re-establishment are also critical to success.

Re-introductions of wolves from Canada have apparently been successful in the short term in Yellowstone and central Idaho. Success in the mid-term also seems likely as wolves may reach criteria for downlisting by 2002 (E. Bangs, USFWS, personal communication). Captive-reared Mexican wolves (*C. l. baileyi*) also were re-introduced into eastern Arizona in 1998 in an attempt to establish a wild population of >100 wolves (Parsons 1998). The population was classified as experimental non-essential, and by 2000 54 wolves had been re-introduced, and 30 in five packs were free-ranging (W. Brown, USFWS, unpublished data). Only two pairs have failed to reproduce. To date, five depredations on cattle have occurred and five wolves have been illegally killed.

One of the best areas, both biologically and socially, to restore wolves in the western U.S. is the southern Rockies. A population and habitat viability analysis recently completed for wolf re-introduction into the southern Rockies concluded that biologically the area could support over 2000 wolves (Phillips et al. 2000). Wolf restoration in the southern Rockies could result in a population connected from Alaska to Mexico and meet endangered species recovery requirements of restoring the species to a significant portion of their historic range.

Establishment of a brown bear population has never been attempted via re-introduction, although augmentations have been conducted (Servheen et al. 1995). Subadult female brown bears were successfully translocated from British Columbia to the Cabinet–Yaak ecosystem in northwest Montana to augment that population (Servheen et al. 1995). The re-introduction into the Bitterroot ecosystem in central Idaho of an experimental population of brown bears has been approved by the US Fish and Wildlife Service (2000) and translocation from British Columbia was scheduled to begin in 2002. Assuming a 4% growth rate, recovery of bears in this ecosystem (280 bears) might then occur within

50 years. A Citizen Management Committee would be responsible for management of bears. Integrating stakeholders closely in the development of management objectives and strategies and even in the design and implementation of research increases the positive investment of people most likely to affect bear survival and increases the likelihood that their concerns will be addressed in a constructive preventative manner (Gregory and Keeney 1994, Wondolleck et al. 1994, Mattson et al. 1996a, b). Impacts of re-introduced bears on ungulates, livestock, humans, and land use in the Bitterroot ecosystem were predicted to be minimal (U.S. Fish and Wildlife Service 2000). The southern Rockies (San Juan ecosystem) has been recommended for further evaluation as another brown bear re-introduction site (U.S. Fish and Wildlife Service 1993a, Craighead et al. 1995), and the southwest also deserves consideration.

The high rate of deforestation, settlement, and conversion to livestock ranching are the major threats to jaguar populations in many regions (Valdez 2000). As a result, jaguars have probably lost significant elements of their genetic diversity. Re-introduction in the southwest U.S. should be considered because establishment of a population by natural re-colonization is unlikely, and because a population in the southwest may be necessary to ensure the persistence of jaguars in the northern portion of their range especially as habitat pressures mount in northwestern Mexico.

Connectivity

Wolf re-colonization in northwestern Montana and southeastern British Columbia occurred through natural dispersal (Boyd and Pletscher 1999). Average dispersal distance of female wolves in northwest Montana was 78 km, which fell within the range of mean dispersal distance of females wolves from various studies in North America (65–144 km; Boyd and Pletscher 1999). Annual mean survival rate of dispersing wolves (64%) was lower than for resident wolves (88%; Pletscher et al. 1997). Colonizing wolves moved over large-scale landscapes rather than defined corridors, and the majority of colonizations occurred outside protected areas but originated from them (Boyd and Pletscher 1999). This phenomenon demonstrates the importance of refuges for population resistance and resilience in the face of management mistakes, fragmentation, or natural stochastic events (McCullough 1996).

In areas of low human population density such as southern New Mexico and Montana, cougars are able to disperse across large areas (females averaged 13.1 km and males 116.1 km in New Mexico; Montana Fish, Wildlife and Parks 1996, Sweanor et al. 2000). In areas of high human density, such as southern California, dispersal success has been poor, and subpopulations have become isolated with little chance of rescue (Beier 1993, 1995). Even so, cougars were able to use low-quality corridors for dispersal in these areas (Beier 1995). In areas of extremely fragmented habitat and high human density, corridors should be created along natural travel routes that contain ample woody cover. They should include underpasses with roadside fencing, lack artificial lighting, and have less than one dwelling unit per 16 ha (Beier 1995).

Simulation modeling indicated that a habitat area of 1000–2200 km² is needed to support a cougar population without immigration (15–20 cougars) in southern California with >98% probability of persistence for 100 years (Beier 1993). With the immigration of one female and three males per decade, areas of 600–1600 km² are needed. Such areas are a minimum and do not ensure long-term (centuries) persistence. From his modeling, Beier (1993) concluded that natural catastrophes of moderate severity do not appear important to cougar persistence. These models should be interpreted with caution as analytic models and simulation models incorporating density independence produced much larger minimum areas necessary for cougars.

Brown bear populations required protection of areas of 4000 and 50 000 km² in size to have a 50% and 90% chance, respectively of surviving (Woodroffe and Ginsberg 1998). Juvenile male brown bears dispersed 45–105 km from maternal home ranges through relatively "friendly" habitats in the Yellowstone ecosystem (Blanchard and Knight 1991). Populations isolated by these or greater distances in less friendly habitat would probably not benefit from corridors as traditionally conceived (Mattson et al. 1996a). Rather, movement would depend upon the establishment and survival of adult females in the intervening habitat, that would function as a sequence of demographic stepping stones (linkage zones). Connectivity then depends on creating habitats in these areas where females can survive (Mattson et al. 1996a). The distance between each of the three large brown bear recovery areas in the northern Rockies (Yellowstone, Selway–Bitterroot, and northern continental divide, each >20 000 km²) is <300 km. Minimally disturbed habitat sufficient to support small populations of large carnivores currently exists in intermediate locations between these cores and could be conserved through the establishment of conservation areas and enhanced through the modification of current barriers and prohibition of new barriers (Craighead et al. 1999). No single size, configuration, or suite of attributes exists for designing protected areas for large carnivores, but Mattson et al. (1996a) described a framework and conceptual model (largely incorporating the role of humans) for addressing these issues for brown bears. Carroll et al. (2001) also identified habitats in an area southeast of Wells–Gray Provincial Park in British Columbia and north-central Idaho that are high-quality for carnivores and are unprotected, and thus are priority conservation areas in the Rocky Mountain region. Soule and Terborgh (1999:65–209) offer an excellent synthesis of conservation design including core protected areas.

Little is known about the habitat needs of jaguars in Arizona and New Mexico. More than 3200 km² of protected habitat was estimated to be required to support a minimum population of 50 jaguars in the Pantanal region of South America (Quigley and Crawshaw 1992). A protected area of 6600 km² in eastern Sonora would support an estimated 60–100 jaguars (Lopez Gonzalez and Brown 2003).

Source-sink dynamics

Humans are usually the single greatest cause of large carnivore mortality, and most of this occurs when carnivores stray beyond reserve boundaries (Pletscher et al. 1997, Woodroffe and Ginsberg 1998, McLellan et al. 1999). Border areas around reserves are often population sinks and will be more significant in small reserves where perimeter:area ratios are high and among species that range widely (Woodroffe and Ginsberg 1998). Reserve size was a better predictor of large carnivore disappearance than was population size, and thus stochastic processes were less important than human mortality. Management should focus on maximizing reserve size and reducing persecution in reserve buffer zones (Woodroffe and Ginsberg 1998). Bears that may spend only a short time outside of protected areas due to the attractiveness of resources there may be quite vulnerable to mortality (Mattson et al. 1996a, Samson and Huot 1998).

Although wolves have successfully re-colonized many places outside protected areas in Montana (Boyd and Pletscher 1999), some areas appear to be sinks. Wolves have repeatedly established themselves along the east front of the Rockies in Montana and southern Alberta but have been killed or removed due to livestock predation. In areas like these with open landscapes and high densities of livestock, wolves may never be able to sustain populations over the long-term. In this landscape, cougars and bears have had greater success because they generally have fewer conflicts with livestock, are less visible, and use these areas only seasonally (bears).

Research needs

Large carnivores have been relatively well studied, with wolves among the most studied mammals in the world. The studies of brown bears in the Yellowstone ecosystem and wolves on Isle Royale and northern Minnesota have been some of the longest and most intensive research work done on populations of mammals. Murphy et al. (1999), however, were unaware of a single study of cougar population dynamics that spanned even one full interval of major fluctuation in primary prey. We know much of what we need to manage and restore large carnivores in western North America. Following the declining-population paradigm of Caughley (1994:236), we have deduced why populations of large carnivores have declined and are working to remove the agents of decline. Making this point, Mattson et al. (1996*b*) reported that since the early 1970's, more than 85% of all weaned and older radio-collared brown bears in the Rocky Mountains died because they were killed by people. Further, successful re-introductions of wolves have shown that, for that large carnivore, the cause of decline was successfully deduced and removed. Research now should be focused at continued monitoring of these re-introductions following the scientific method (Caughley 1994). What is primarily lacking for long-term conservation of carnivores is the will to make the sacrifices necessary to live with them. Co-operation and dialogue among all groups including non-governmental organizations and local citizens is essential to moving forward. Research into how best to attain this co-operation is important (Mattson et al. 1996*b*, Ehrlich 2002). Mattson et al. (1996*b*:1221–1223) and Mattson and Craighead (1994:121–125) recommended several new approaches to research, management, and policy measures for moving forward. Craighead et al. (1995) could find no major conservation problem that had been initially recognized and then solved by government agencies without pressure from a critical public. Some of the most progressive research and management of large carnivores is being done by private organizations (e.g., Craighead et al. 1995, Logan et al. 1996, Murphy et al. 1999, Soule and Terborgh 1999:15, Phillips et al. 2000, Berger et al. 2001*b*, Carroll et al. 2001, Sanderson et al. 2002). As stressed by Soule and Terborgh (1999:15), "conservation on the ground must replace the repetitive cycle of conferences, reports, recommendations of governments and ineffective treaties."

Despite all the work that has been done on large carnivores, we have almost no evidence upon which to assess the surrogate value of large carnivores. Restoration of large carnivores is relatively expensive and, in some cases because of this and other reasons, it might not be a conservation priority. We must assess how efficient restoration of large carnivores is for conservation of biodiversity. More research in this area, especially experimental, is needed. Natural expansion and re-introduction of populations of large carnivores into areas where they have been absent provide great opportunities to do this research (Huggard 1993, Kunkel 1997, Phillips et al. 2000, U.S. Fish and Wildlife Service 2000, Berger et al. 2001, Smith et al. 2001). Re-introductions should be designed as experiments to test these hypotheses, and adaptive management principles should be followed. Data on the demographics and behavior of prey and other carnivores, vegetation trends, and species richness prior to, during, and after re-colononization or re-introduction need to be obtained. More research and meta-analysis of data where large carnivores currently exist should be conducted to assess their umbrella, flagship, and keystone roles.

Further research is needed to assess the impacts of human harvest (sport and control harvest) on large carnivore populations, population structure, prey populations, and the impact of this on ecosystems. Sustainable rates of removals for depredating jaguars need to be examined. Further understanding of compensatory versus additive impacts of predation and how competition among carnivores affects this is needed, especially in southern multiple-predator, multiple-prey systems. More innovative work is needed on non-lethal control methods to reduce depredation on livestock. Innovative ways that allow local people, livestock and large carnivores to co-exist are needed, especially for jaguars in northwest Mexico.

Biological requirements to maintain connectivity (linkages) among isolated populations of large carnivores, especially brown bears, are little known. Analyses of how much degradation is too much and how to monitor for degradation must be completed before degradation proceeds (Doak 1995). The role of cover in mitigating impacts of human development on bear occupancy and movement are poorly documented. There is no research concerning the minimum required size of linkage zones or at what level they become ineffective for brown bears (Servheen et al. 2001). Further, classified and validated maps of brown bear habitat are generally non-existent. Despite these shortcomings, many of the basic requirements for connectivity are known, and the larger problem of politically and socially acceptable ways to mitigate for loss of habitat remains.

More research on population monitoring using non-invasive approaches is necessary. These techniques hold great potential for large carnivores, which are notoriously difficult or expensive to monitor (Beier and Cunningham 1996, Kohn and Wayne 1997, Miller et al. 1997, Becker et al. 1998, Woods et al. 1999, Mills et al. 2000). Whitebark-pine nuts are an important food for brown bears in the southern portion of their range, but their availability has decreased markedly in recent years due to whitebark-pine blister rust (*Cronartium ribicola*). Management solutions for blister rust need to be found (Servheen 1998) because brown bear mortality in the Yellowstone area is determined by whitebark-pine seed crop size (Pease and Mattson 1999). Determining jaguar distribution and ecology in northern Mexico should be a research priority. Long-term co-ordinated research and management between Mexico and the U.S. will be required to conserve jaguars in the region. Planning across the complete biological range of jaguars (and all large carnivores) so that all conservation efforts can be placed in the context of the species' biology is important (Sanderson et al. 2002).

Summary

Because of conflicts with humans, large carnivores have been extensively persecuted over the last century, and their populations and range have declined markedly. Recently, however, natural re-colonization and re-introductions of wolves have increased their populations significantly in the northern Rockies. Large carnivores, especially where they occur together, can at times limit and potentially regulate prey populations. However, there is little evidence that control efforts applied toward large carnivores are effective at significantly increasing prey densities over the long term.

At some scales and in some ecosystems, large carnivores likely have top-down regulatory impacts: empirical evidence for this remains slight but is increasing. Largely because research has been inadequate to date, little evidence exists for the umbrella roles of large carnivores. Theoretically, large carnivores serve in some capacity as surrogate species for conservation, especially flagships, and we should exploit this for the carnivores' own sake and larger conservation goals.

Wolves have the potential to occupy many areas, provided that protected source populations exist nearby and wolves are actively managed to reduce conflicts with livestock to obtain a modicum of local support. Because of their greater use of more remote landscapes, cougars are easier to manage and maintain in most western landscapes than are wolves. Reducing human–brown bear conflicts is essential to brown bear persistence. In many cases this simply means reducing numbers of humans in bear habitats through management prescriptions such as road restrictions. Humans must be educated on how to minimize conflicts with bears in bear habitat and be willing to practise this. Larger tracts of habitat with little or no human impacts need to be established and maintained in the lower 48 states of the U.S. and connectivity among these habitats is essential, as is connectivity to more pristine Canadian habitats. Restoration of bears to former habitats is also important to ensure the long-term persistence in the lower 48 states. Building local support and involvement for this restoration is essential for its success.

Large carnivores have and will continue to serve as an important catalyst for conservation. They play important roles in ecosystems, and restoration should be pursued in as many areas as possible. Human encroachment and human-caused mortality are the most significant problems in large carnivore conservation. We must work aggressively to counter these impacts. Conserving animals that are capable of killing us and that need large, wild spaces requires "great commitment on the part of biologists, activists, land managers, and political leaders, and also requires much tolerance from the people who live, work, and play in carnivore habitat" (Noss 1996). We should recognize that because large carnivores do, in some cases, serve as effective flagships and thus attract greater support and attention, funds that are available for wolves, bears, jaguars, and cougars might not be available for work on other less "charismatic" conservation priorities. Therefore, we should take advantage of this and work to ensure that, while restoring large carnivores, we serve as large and significant a conservation goal as possible.

There is much reason for optimism. Dramatic changes in public attitudes toward carnivores have occurred in just a few decades (Kellert et al. 1996). New partnerships among diverse interests are being formed to conserve and restore wildlands and large carnivores (Rasker and Hackman 1996, e. g., citizens brown bear initiative: U.S. Fish and Wildlife Service 2000). Ultimately, the survival of large carnivores and the wildlands they occupy might simply depend on how much we can resolve conflicts among ourselves and adopt more tolerant and less acquisitive lifestyles (McDougal et al. 1988, Daly and Cobb 1989, Mattson et al. 1996*b*). Contrary to conventional wisdom about the cost of wilderness protection to local resource-based economies, recent work indicates that the protection of wilderness habitat that sustains large carnivores does not have a detrimental effect on local economies (Rasker and Hackman 1996). Economic growth is apparently stimulated by environmental amenities (Rasker and Hackman 1996). Much of the world looks to North America for leadership in conservation. Even so, North America needs to "look around the world and learn from those countries where an ecosystem approach has been in use for a long time; where local human populations are recognized as part of the ecosystem; where controlled use is an option once basic ecological values are assured; and where co-operation across cultural, ideological, and political boundaries is a reality" (Weber and Rabinowitz 1996). Some of the poorest countries are making the greatest contributions to conserving large carnivores. Rwanda and Botswana have placed >10% of their land in protected areas while the US has only 4% protected (World Resources Institute 1994, Weber and Rabinowitz 1996).

Biologists have an obligation and responsibility to help protect and preserve the species we are working on by going beyond data collection and serving the larger cause of finding solutions at local levels, as well as national and international levels (Schaller 1996). We must "fight for what remains and restore what has been squandered ... because to save carnivores and their environment is as important to their future as it is to ours" (Schaller 1996).

Acknowledgments
Much of the thought in this chapter results from the privilege of working with great carnivore biologists over the years and the critical thinking I was exposed to while putting together a symposium on the role of top predators in ecological communities for the Society for Conservation Biology meeting in 2000. I thank R.A. Powell, C.E. Meslow, D. Brittel, M. Pokorny, and J.C. Truett for especially helpful reviews.

Literature cited

Andelman, S.J. and W.F. Fagan. 2000. Umbrellas and flagships: efficient conservation surrogates or expensive mistakes? *Proceedings of the National Academy of Sciences* **97**:5954–5959.

Anderson, A.E., D.C. Bowden, and D.M. Kattner. 1992. The cougar of the Umcompahgre Plateau, Colorado. Technical Publication no. 40. Colorado Division of Wildlife, Denver, Colorado, USA.

Aranda, M. 1994. Importancia de los peccaries (*Tayussa* spp.) en la alimentacion del jaguar (*Panthera onca*). *Acta Zoologica Mexicana* **68**:45–52.

Aranda, M. and V. Sanchez-Cordero. 1996. Prey spectra of the jaguar (*Panthera onca*) and puma (*Puma concolor*) in the tropical forests of Mexico. *Studies of the Tropical Fauna and Environment* **31**:65–67.

Arjo, W.M. and D.H. Pletscher. 1999. Behavioral responses of coyotes to wolf recolonization in northwestern Montana. *Canadian Journal of Zoology* **77**:1919–1927.

Ballard, W. and P. Gipson. 2000. Wolf. Pages 321–346 *in* S. Demaris and P.R. Krausman editors. *Ecology and Management of Large Mammals in North America*. Prentice Hall, Upper Saddle River, New Jersey, USA.

Ballard, W.B. and V. Van Ballenberghe. 1998. Moose-predator relationships: research and management needs. *Alces* **34**:91–105.

Bangs, E.E., S.H. Fritts, J.A. Fontaine, D.W. Smith, K.M. Murphy, C.M. Mack, and C.C. Niemeyer. 1998. Status of gray wolf restoration in Montana, Idaho, and Wyoming. *Wildlife Society Bulletin* **26**:785–798.

Barrett, S.W. and S.F. Arno. 1982. Indian fires as an ecological influence in the northern Rockies. *Journal of Forestry* **80**:647–651.

Becker, E.F., M.A. Spindler, and T.O. Osborne. 1998. A population estimator based on network sampling of tracks in the snow. *Journal of Wildlife Management* **62**:968–977.

Beier, P. 1993. Determining minimum habitat areas and habitat corridors for cougars. *Conservation Biology* **7**:94–108.

Beier, P. 1995. Dispersal of juvenile cougars in fragmented habitat. *Journal of Wildlife Management* **59**:228–237.

Beier, P. and S.C. Cunningham. 1996. Power of track surveys to detect changes in cougar populations. *Wildlife Society Bulletin* **24**:540–546.

Ben-David M., T.A. Hanley, and D.M. Schell. 1998. Fertilization of terrestrial vegetation by spawning Pacific salmon: the role of flooding and predator activity. *Oikos* **83**:164–173.

Berger, J. 1997. Population constraints associated with the use of black rhinos as an umbrella species for desert herbivores. *Conservation Biology* **11**:69–78.

Berger, J. 1998. Some consequences of the loss and restoration of large mammalian carnivores on prey. Pages 80–100 *in* T.M. Caro, editor. *Behavioral Ecology and Conservation Biology*. Oxford University Press, New York, New York, USA.

Berger, J. 1999. Anthropogenic extinction of top carnivores and interspecific animal behavior: implications of the rapid decoupling of a web involving wolves, bears, moose, and ravens. *Proceedings of the Royal Society of London B* **266**:2261–2267.

Berger, J. and J.D. Wehausen. 1991. Consequences of a mammalian predator-prey disequilibrium in the Great Basin Desert. *Conservation Biology* **5**:244–248.

Berger, J., P.B. Stacey, L. Bellis, and M.P. Johnson. 2001a. A mammalian predator–prey imbalance: grizzly bear and wolf extinction affect avian neotropical migrants. *Ecological Applications* **11**:947–960.

Berger, J., J.E. Swenson, and I. Persson. 2001b. Recolonizing carnivores and naive prey: conservation lessons from Pleistocene extinctions. *Science* **291**:1036–1039.

Bergerud, A.T. 1981. The decline of moose in Ontario – a different view. *Alces* **17**:30–43.

Bergerud, A.T. 1988. Caribou, wolves, and man. *Trends in Ecology and Evolution* **3**:68–72.

Bergerud, A.T. and J.P. Elliot. 1998. Wolf predation in a multiple-ungulate system in northern British Columbia. *Canadian Journal of Zoology* **76**:1551–1569.

Bilby, R.E., B.R. Fransen, and P.A. Bisson. 1996. Incorporation of nitrogen and carbon from spawning coho salmon into the trophic system of small streams: evidence from stable isotopes. *Canadian Journal of Fish and Aquatic Science* **53**:164–173.

Blanchard, B.M. and R.R. Knight. 1991. Movements of Yellowstone grizzly bears. *Biological Conservation* **58**:41–67.

Blanchard, B.M. and R.R. Knight. 1995. Biological consequences of relocating grizzly bears in the Yellowstone ecosystem. *Journal of Wildlife Management* **59**:560–565.

Boertje, R.D., P. Valkenburg, and M.E. McNay. 1996. Increases in moose, caribou, and wolves following wolf control in Alaska. *Journal of Wildlife Management* **60**:474–489.

Botkin, D.B. 1990. *Discordant Harmonies*. Oxford University Press. New York, New York, USA.

Boyce, M.S. and E.M Anderson. 1999. Evaluating the role of carnivores in the Greater Yellowstone Ecosystem. Pages 265–283 *in* T.W. Clark, A.P. Curlee, S.C. Minta, and P.M. Kareiva, editors. *Carnivores in Ecosystems: the Yellowstone Experience*. Yale University Press, New Haven, Connecticut, USA.

Boyd, D.K. 1997. Dispersal, genetic relationships, and landscape use by colonizing wolves in the central Rocky Mountains. Dissertation, University of Montana, Missoula, Montana, USA.

Boyd, D.K. and D.H. Pletscher. 1999. Characteristics of dispersal in a colonizing wolf population in the central Rocky Mountains. *Journal of Wildlife Management* **63**:1094–1108.

Brown, D.E. 1983. On the status of the jaguar in the southwest. *Southwestern Naturalist* **28**:459–460.

Carbyn, L.N., S.H. Fritts, and D.R. Seip. 1995. *Ecology and Conservation of Wolves in a Changing World*. Canadian Circumpolar Institute, Occasional Publication **35**, Edmonton, Alberta, Canada.

Caro, T.M. 2000. Focal species. *Conservation Biology* **14**:1569–1570.

Caro, T.M. and G. O'Doherty. 1999. On the use of surrogate species in conservation biology. *Conservation Biology* **13**:805–814.

Carroll, C., R.F. Noss, and P.C. Paquet. 2001. Carnivores as focal species for conservation planning in the Rocky Mountain region. *Ecological Applications* **11**:961–980.

Caughley, G. 1994. Directions in conservation biology. *Journal of Animal Ecology* **63**:215–244.

Clark, T.W., A.P. Curlee, S.C. Minta, and P.M. Kareiva. 1999. *Carnivores in Ecosystems: the Yellowstone Experience*. Yale University Press, New Haven, Connecticut, USA.

Craighead, F.L., M.E. Gilpin, and E.R. Vyse. 1999. Genetic considerations for carnivore conservation in the Greater Yellowstone Ecosystem. Pages 285–321 *in* T.W. Clark, A.P. Curlee, S.C. Minta, and P.M. Kareiva, editors. *Carnivores in Ecosystems: the Yellowstone Experience*. Yale University Press, New Haven, Connecticut, USA.

Craighead, J.J. and D.J. Craighead. 1991. New system-techniques for ecosystem management and an application to the Yellowstone ecosystem. *Western Wildlands* **17**:30–39.

Craighead, J.J., J.S. Sumner, and J.A. Mitchell. 1995. *The Grizzly Bears of Yellowstone: their Ecology in the Yellowstone Ecosystem, 1959–1992.* Island Press, Washington, DC, USA.

Crawshaw, P.G. and H.B. Quigley. 1991. Jaguar spacing, activity, and habitat use in a seasonally flooded environment in Brazil. *Journal of Zoology, London* **223**:357–370.

Creel, S. 2001. Four factors modifying the effect of competition on carnivore population dynamics as illustrated by African wild dogs. *Conservation Biology* **15**:271–274.

Crete, M. 1999. The distribution of deer biomass in North America supports the hypothesis of exploitation ecosystems. *Ecology Letters* **2**:223–227.

Crete, M. and M. Manseau. 1996. Natural regulation of cervidae along a 100 km latitudinal gradient: change in trophic dominance. *Evolutionary Ecology* **10**:51–62.

Crooks, K.R. and M.E. Soule. 1999. Mesopredator release and avifaunal extinctions in a fragmented system. *Nature* **400**:563–566.

Cunningham, S.C., L.A. Haynes, C. Gustavson, and D.D. Hayward. 1995. Evaluation of the interaction between mountain lions and cattle in the Aravaipa-Klondyke area of southeast Arizona. Arizona Game and Fish Department Technical Report **no. 17**, Phoenix, Arizona, USA.

Daly H.B. and J.B. Cobb, Jr. 1989. *For the Common Good: Redirecting the Economy Toward Community, the Environment and a Sustainable Future.* Beacon Press, Boston, Massachusetts, USA.

DeCalesta, D.S. 1994. Impact of deer on interior forest songbirds in northwestern Pennsylvania. *Journal of Wildlife Management* **58**:711–718.

Demaris, S. and P.R. Krausman 2000. *Ecology and Management of Large Mammals in North America.* Prentice Hall, Upper Saddle River, New Jersey, USA.

Doak, D.F. 1995. Source-sink models and the problem of habitat degradation: general models and application to the Yellowstone grizzly. *Conservation Biology* **9**:1370–1379.

Eberhardt, L.L. 1990. Survival rates required to sustain bear populations. *Journal of Wildlife Management* **54**:587–590.

Eberhardt, L.L., B.M. Blanchard, and R.R. Knight. 1994. Population trend of Yellowstone grizzly bears as estimated from reproductive and survival rates. *Canadian Journal of Zoology* **72**:360–363.

Ehrlich, P.R. 2002. Human nature, nature conservation, and environmental ethics. *Bioscience* **52**:31–43.

Ernest, H.B., E.S. Rubin, and W.M. Boyce. 2002. Fecal DNA analysis and risk assessment of mountain lion predation of bighorn sheep. *Journal of Wildlife Management* **66**:75–85.

Estes, J., K. Crooks, and R. Holt. 2001. Predation and diversity. Pages 857–878 *in* S. Levin, editor. *Encyclopedia of Biodiversity.* Academic Press, San Diego, California, USA.

Evans, W. 1983. *The Cougar in New Mexico: Biology, Status, Depredation of Livestock, and Management Recommendations.* New Mexico Department of Fish and Game, Santa Fe, New Mexico, USA.

Fisher, A., E. Rominger, P. Miller, and O. Byers. 1999. Population and habitat viability assessment workshop for the desert bighorn sheep of New Mexico (*Ovis canadensis*): Final report. IUCN/SSC Conservation Breeding Specialist Group: Apple Valley, Minnesota, USA.

Fretwell, S.D. 1987. Food chain dynamics: the central theory of ecology. *Oikos* **50**:291–301.

Fritts, S.H. and L.D. Mech. 1981. Dynamics, movements, and feeding ecology of a newly protected wolf population in northwestern Minnesota. *Wildlife Monograph* **80**.

Fritts, S.H., W.J. Paul, L.D. Mech, and D.P. Scott. 1992. Trends and management of wolf-livestock conflicts in Minnesota. United States Fish and Wildlife Service Resource Publication No. **181**. Twin Cities, Minnesota, USA.

Fuller, T.K. 1989. Population dynamics of wolves in north-central Minnesota. *Wildlife Monograph* **105**.

Fuller, T.K. 1990. Dynamics of a declining white-tailed deer population in north-central Minnesota. *Wildlife Monograph* **110**.

Gasaway, W.C., R.D. Boertje, D.V. Grandgaard, D.G. Kellyhouse, R.O. Stephenson, and D.G. Larsen. 1992. The role of predation in limiting moose at low densities in Alaska and Yukon and implications for conservation. *Wildlife Monograph* **120**.

Gittleman, J.L., S.M. Funk, D.W. MacDonald, and R.K. Wayne. 2001. *Carnivore Conservation*. Cambridge University Press, Cambridge, UK.

Gregory, R. and R.L. Keeney. 1994. Creating policy alternatives using stakeholder values. *Management Science* **40**:1035–1048.

Gunther, K.A., M. Bruscino, S. Cain, T. Chu, K. Frey, and R.R. Knight. 1998. Grizzly bear–human conflicts, confrontations, and management actions in the Yellowstone ecosystem 1997. Interagency Grizzly Bear Committee, Yellowstone Ecosystem Subcommittee Report. Yellowstone National Park, Wyoming, USA.

Haight, R.G., D.J. Mladenoff, and A.P. Wydeven. 1998. Modeling disjunct gray wolf populations in semi-wild landscapes. *Conservation Biology* **12**:879–888.

Halaj, J. and D.H. Wise. 2001. Terrestrial trophic cascades: how much do they trickle? *American Naturalist* **157**:262–281.

Hatter, I. and D.W. Janz. 1994. The apparent demographic changes in black-tailed deer associated with wolf control in northern Vancouver Island, Canada. *Canadian Journal of Zoology* **72**:878–884.

Hayes, C.L., E.S. Rubin, M.C. Jorgensen, R.A. Botta, and W.M. Boyce. 2000. Mountain lion predation on bighorn sheep in the Peninsular Ranges, California. *Journal of Wildlife Management* **64**:954–959.

Hayes, R.D. and J.R. Gunson. 1995. Status and management of wolves in Canada. Pages 21–34 *in* L.N. Carbyn, S.H. Fritts and D.R. Seip, editors. *Ecology and Conservation of Wolves in a Changing World*. Canadian Circumpolar Institute, Occasional Publication **35**. Edmonton, Alberta, Canada.

Henke, S.E. and F.C. Bryant. 1999. Effects of coyote removal on the faunal community in western Texas. *Journal of Wildlife Management* **63**:1066–1081.

Hilderbrand, G.V., C.C. Schwartz, C.T. Robbins, M.E. Jacoby, T.A. Hanley, S.M. Arthur, and C. Servheen. 1999. Importance of meat, particularly salmon, to body size, population productivity, and conservation of North American brown bears. *Canadian Journal of Zoology* **77**:132–138.

Hoogesteijn, R., A. Hoogesteijn, and E. Mondolfi. 1993. Jaguar predation vs. conservation: cattle mortality by felines on 3 ranches in Venezuelan llanos. Pages 391–407 *in* N. Dunston and M.L. Gormans, editors. Mammals as predators. Proceeding of the Symposium of the Zoological Society of London **65**. Clarendon Oxford, UK.

Hornocker, M.G. 1970. An analysis of mountain lion predation on mule deer and elk in the Idaho primitive area. *Wildlife Monograph* **21**.

Hornocker, M.G. and T.N. Bailey. 1986. Natural regulation in 3 species of felids. Pages 211–220 *in* S.D. Miller and D.D. Everett, editors. *Cats of the World*. The National Wildlife Federation Institute for Wildlife Research, Washington DC, USA.

Hovey, F.W. and B.N. McLellan. 1996. Estimating population growth of grizzly bears from the Flathead River drainage using computer simulations of reproductive and survival rates. *Canadian Journal of Zoology* **74**:1409–1416.

Huggard, D.J. 1993. Prey selectivity of wolves in Banff National Park. I. Prey species. *Canadian Journal of Zoology* **71**:130–139.

Hurley, M. and J.W. Unsworth. 1999. Southeast Idaho mule deer ecology. Job progress report **W-160-R-25**. Idaho Department of Fish and Game, Boise, Idaho, USA.

IUCN/SSC Re-introduction Specialist Group. 1998. *IUCN Guidelines for Re-introductions*. IUCN, Gland, Switzerland and Cambridge, UK.

Iriarte, J.A., W.L. Franklin, W.E. Johnson, and K.H. Redford. 1990. Biogeographic variation of food habits and body size of the American puma. *Oecologia* **85**:185–190.

Jacoby, M.E., G.V. Hilderbrand, C.S. Servheen, S.M. Arthur, T.A. Hanley, C.T. Robbins, and R. Michener. 1999. Trophic relations of brown and black bears in several western North American ecosystems. *Journal of Wildlife Management* **63**:921–929.

Johnson, W.E., T.K. Fuller, and W.L. Franklin. 1996. Sympatry in canids: a review and assessment. Pages 1289–218 *in* J.L. Gittleman, editor. *Carnivore Behavior, Ecology, and Evolution*. Volume II. Comstock Publishing Associates, Ithaca, New York, USA.

Jonkel, C.J. 1987. Brown bear. Pages 457–473 *in* M. Novak, J.A. Baker, M.E. Obbard, and B. Malloch, editors. *Wild Furbearer Management and Conservation in North America*. Ontario Trapper's Association, North Bay, Ontario, Canada.

Kay, C. 1994. The impact of native ungulates on riparian communities in the Intermountain West. *Natural Resource and Environmental Issues* **1**:23–44.

Keith, L.B. 1983. Population dynamics of wolves. Pages 66–77 *in* L.N. Carbyn, editor. *Wolves in Canada and Alaska: their Status Biology and Management*. Canadian Wildlife Service Report Series No. **45**. Edmonton, Alberta, Canada.

Kellert, S.R., M. Black, C.R. Rush, and A.J. Bath. 1996. Human culture and large carnivore conservation in North America. *Conservation Biology* **10**:977–990.

Kie, J.G., R.T. Bowyer, and K.M. Stewart. 2003. Ungulates in western coniferous forests: habitat relationships, population dynamics, and ecosystem processes. Pages 296–340 *in* C.J. Zabel and R.G. Anthony, editors. *Mammal Community Dynamics. Management and Conservation in the Coniferous Forests of Western North America*. Cambridge University Press, Cambridge, UK.

Kitchen, A.M., E.M. Gese, and E.R. Schauster. 1999. Resource partitioning between coyotes and swift foxes: space, time, and diet. *Canadian Journal of Zoology* **77**:1645–1656.

Kitchen, A.M., E.M. Gese, and E.R. Schauster. 2000*a*. Long-term spatial stability of coyote (*Canis latrans*) home ranges in southeastern Colorado. *Canadian Journal of Zoology* **78**:458–464.

Kitchen, A.M., E.M. Gese, and E.R. Schauster. 2000*b*. Changes in coyote activity patterns due to reduced exposure to human persecution. *Canadian Journal of Zoology* **78**:853–857.

Knowlton, F.F., E.M. Gese, and M.M. Jaegar. 1999. Coyote depredation control: an interface between biology and management. *Journal of Range Management* **52**:398–412.

Kohn, M.H. and R.K. Wayne. 1997. Facts from feces revisited. *Trends in Ecology and Evolution* **12**:223–227.

Kolenosky, G.B. and S.M. Strathearn. 1987. Black bear. Pages 443–454 *in* M. Novak, J.A. Baker, M.E. Obbard, and B. Malloch, editors. *Wild Furbearer Management and Conservation in North America.* Ontario Trapper's Association, North Bay, Ontario, Canada.

Kotliar, N.B. 2000. Application of the new keystone-species concept to prairie dogs: how well does it work? *Conservation Biology* **14**:1715–1721.

Krebs, C.J., S. Boutin, A.R.E. Sinclair, J.N.M. Smith, M.R.T. Dale, K. Martin, and R. Turkington. 1995. Impact of food and predation on the snowshoe hare cycle. *Science* **269**:1112–1118.

Krefting, L.W. 1969. The rise and fall of the coyote on Isle Royale, Michigan. *Naturalist* **20**:24–31.

Kunkel, K.E. 1997. Predation by wolves and other large carnivores in northwestern Montana and southeastern British Columbia. Dissertation, University of Montana, Missoula, Montana, USA.

Kunkel, K.E. and L.D. Mech. 1994. Wolf and bear predation on white-tailed deer fawns in northeastern Minnesota. *Canadian Journal of Zoology* **72**:1557–1565.

Kunkel, K.E. and D.H. Pletscher. 1999. Species-specific population dynamics of cervids in a multipredator ecosystem. *Journal of Wildlife Management* **63**:1082–1093.

Kunkel, K.E. and D.H. Pletscher. 2000. Habitat factors affecting vulnerability of moose to predation by wolves in southeastern British Columbia. *Canadian Journal of Zoology* **78**:150–157.

Kunkel, K.E. and D.H. Pletscher. 2001. Winter hunting patterns of wolves in and near Glacier National Park, Montana. *Journal of Wildlife Management* **65**:520–530.

Kunkel, K.E., T.K. Ruth, D.H. Pletscher, and M.G. Hornocker. 1999. Winter prey selection by wolves and cougars in and near Glacier National Park, Montana. *Journal of Wildlife Management* **63**:901–910.

Kurten, B. and E. Anderson. 1980. *Pleistocene Mammals of North America.* Columbia University Press, New York, New York, USA.

Laing, S.P. and F.G. Lindzey. 1991. Cougar habitat selection in southern Utah. Pages 27–37 *in* C.L. Braun, editor. *Mountain Lion – Human Interaction Symposium.* Colorado Division of Wildlife, Denver, Colorado, USA.

Landres, P.B., J. Verner, and J.W. Thomas. 1988. Ecological uses of vertebrate indicator species: a critique. *Conservation Biology* **2**:316–327.

Leopold, A. 1933. *Game Management.* Charles Scribner's Sons, New York, New York, USA.

Lindzey, F.G., W.D. Van Sickle, S.P. Laing, and C.S. Mecham. 1992. Cougar population response to manipulation in southern Utah. *Wildlife Society Bulletin* **20**:224–227.

Lindzey, F.G., W.D. Van Sickle, B.B. Ackerman, D. Barnhurst, T.P. Hemker, and S.P. Laing. 1994. Cougar population dynamics in southern Utah. *Journal of Wildlife Management* **58**:619–624.

Logan, K.A. and L.L. Irwin. 1985. Mountain lion habitats in the Big Horn Mountains, Wyoming. *Wildlife Society Bulletin* **13**:257–262.

Logan, K.A. and L.L. Sweanor. 2000. Puma. Pages 347–377 *in* S. Demaris and P.R. Krausman, editors. *Ecology and Management of Large Mammals in North America.* Prentice Hall, Upper Saddle River, New Jersey, USA.

Logan, K.A., L.L. Irwin, and R. Skinner. 1986. Characteristics of a hunted mountain lion population in Wyoming. *Journal of Wildlife Management* **50**:648–654.

Logan, K.A., L.L. Sweanor, T.K. Ruth, and M.G. Hornocker. 1996. Cougars of the San Andreas Mountains, New Mexico. Final Report, Federal Aid in Wildlife Restoration Project **W-128-R.** New Mexico Department of Game and Fish, Sante Fe, New Mexico, USA.

Lopez Gonzalez, C.A. and D.E. Brown. 2003. Distribucion y estatus del jaguar (*Panthera onca*) en el Noroeste de Mexico. *In* R.A. Medellin, C. Chetkiewicz, A. Rabinowitz, K.H. Redford, J.G. Robinson, E. Sanderson, and A. Taber, editors. *Jaguars en el Nuevo Milenio*. Fondo de Cultura Economica, Universidad Nacional Autonoma de Mexico Wildlife Conservation Society. In press.

MacCracken, J.G. and J. O'Laughlin. 1998. Recovery policy on grizzly bear: an analysis of two positions. *Wildlife Society Bulletin* **26**:899–907.

Mace, R.D. and J.S. Waller. 1998. Demography and population trend of grizzly bears in the Swan Mountains, Montana. *Conservation Biology* **12**:1005–1016.

Mace, R.D., J.S. Waller, T. Manley, L.J. Lyon, and H. Zuuring. 1996. Relationships among grizzly bears, roads and habitat in the Swan Mountains, Montana. *Journal of Applied Ecology* **33**:1395–1404.

Mace, R.D., J.S. Waller, T.L. Manley, K. Ake, and W.T. Wittenger. 1999. Landscape evaluation of grizzly bear habitat in western Montana. *Conservation Biology* **13**:367–377.

Mattson, D.J. and J.J. Craighead. 1994. The Yellowstone grizzly bear recovery program: uncertain information, uncertain policy. Pages 101–130 *in* T.W. Clark, R. Reading, and A. Clarke, editors. *Endangered Species Recovery: Finding the Lessons, Improving the Process.* Island Press, Washington, DC, USA.

Mattson, D.J. and M.W. Reid. 1991. Conservation of the Yellowstone grizzly bear. *Conservation Biology* **5**:364–372.

Mattson, D.J. and D.P. Reinhardt. 1995. Influence of cutthroat trout (*Oncorhynchus clarki*) in behavior and reproduction of Yellowstone grizzly bears (*Ursus arctos*), 1975–1989. *Canadian Journal of Zoology* **73**:2072–2079.

Mattson, D.J., B.M. Blanchard, and R.R. Knight. 1991*a*. Food habits of Yellowstone grizzly bears. *Canadian Journal of Zoology* **69**:2430–2435.

Mattson, D.J., C.M. Gillin, S.A. Benson, and R.R. Knight. 1991*b*. Bear feeding activity at alpine insect aggregation sites in the Yellowstone ecosystem. *Canadian Journal of Zoology* **69**:1619–1629.

Mattson, D.J., S. Herrero, R.G. Wright, and C.M. Pease. 1996*a*. Designing and managing protected areas for grizzly bears: how much is enough? Pages 133–164 *in* R.G. Wright, editor. *National Parks and Protected Areas: their Role in Environmental Protection.* Blackwell Science, Cambridge, Massachusetts, USA.

Mattson, D.J., S. Herrero, R.G. Wright, and C.M. Pease. 1996*b*. Science and management of Rocky Mountain grizzly bears. *Conservation Biology* **10**:1013–1025.

McCann, K., A. Hastings, and G.R. Huxel. 1998. Weak trophic interactions and the balance of nature. *Nature* **395**:794–798.

McCullough, D.R. 1996. Spatially structured populations and harvest theory. *Journal of Wildlife Management* **60**:1–9.

McDougal, M.S., M.W. Reisman, and A.R. Willard. 1988. The world community: a planetary social process. *University of California Davis Law Review* **21**:807–972.

McLaren, B.E. and R.O. Peterson. 1994. Wolves, moose, and tree rings on Isle Royale. *Science* **266**:1555–1558.

McLellan, B.N. and F.W. Hovey. 1995. The diet of grizzly bears in the Flathead River drainage in southeastern British Columbia. *Canadian Journal of Zoology* **73**:704–712.

McLellan, B.N. and F.W. Hovey. 2001. Habitats selected by grizzly bears in a multiple use landscape. *Journal of Wildlife Management* **65**:92–99.

McLellan, B.N., F.W. Hovey, R.D. Mace, J.G. Woods, D.W. Carney, M.L. Gibeau, W.L. Wakkinen, and W.F. Kasworm. 1999. Rates and causes of grizzly bear mortality in the interior mountains of British Columbia, Alberta, Montana, Washington, and Idaho. *Journal of Wildlife Management* **63**:911–920.

McShae, W.J. and J.H. Rappole. 1997. Herbivores and the ecology of understory birds. Pages 298–307 *in* W.J. McShae, H.B. Underwood, and J.H. Rappole, editors. *The Science of Overabundance*. Smithsonian Institution Press, Washington, DC, USA.

Mech, L.D. 1995. The challenge and opportunity of recovering wolf populations. *Conservation Biology* **9**:270–278.

Mech, L.D. 1996. A new era for carnivore conservation. *Wildlife Society Bulletin* **24**:397–401.

Mech, L.D. and M.E. Nelson. 2000. Do wolves affect white-tailed buck harvest in northeastern Minnesota? *Journal of Wildlife Management* **64**:129–136.

Mech, L.D., E.K. Harper, T.J. Meier, and W.J. Paul. 2000. Assessing factors that may predispose Minnesota farms to wolf depredations on cattle. *Wildlife Society Bulletin* **28**:623–629.

Meier, T.J., J.W. Burch, L.D. Mech, and L.G. Adams. 1995. Pack structure and genetic relatedness among wolf packs in a naturally regulated population. Pages 293–302 *in* L.N. Carbyn, S.H. Fritts, and D. Seip, editors. *Ecology and Conservation of Wolves in a Changing World*. Canadian Circumpolar Institute, Occasional Publication **35**. Edmonton, Alberta, Canada.

Merrill, T., D.J. Mattson, R.G. Wright, and H.B. Quigley. 1999. Defining landscapes suitable for restoration of grizzly bears *Ursus arctos* in Idaho. *Biological Conservation* **87**:231–248.

Messier, F. 1994. Ungulate population models with predation: a case study with the North American moose. *Ecology* **75**:478–488.

Messier, F. 1995. Trophic interactions in two northern wolf-ungulate systems. *Wildlife Research* **22**:131–146.

Miller, B., B. Dugleby, D. Foreman, C. Martinez del Rio, R. Noss, M. Phillips, R. Reading, M.E. Soule, J. Terborgh, and L. Willcox. 2001. The importance of large carnivores to healthy ecosystems. *Endangered Species Update* **18**:202–211.

Miller, S.D., G.C. White, R.A. Sellars, H.V. Reynolds, J.W. Schoen, K. Titus, V.G. Barnes, R.B. Smith, R.R. Nelson, W.B. Ballard, and C.C. Schwartz. 1997. Brown and black bear density estimation in Alaska using radiotelemetry and replicated mark-resight techniques. *Wildlife Monograph* **133**.

Mills, L.S., J.J. Citta, K.P. Lair, M.K. Schwartz, and D.A. Tallmon. 2000. Estimating animal abundance using noninvasive DNA sampling: promise and pitfalls. *Ecological Applications* **10**:283–294.

Minta, S.C., P.M. Kareiva, and A.P. Curlee. 1999. Carnivore research and conservation: learning from history and theory. Pages 323–404 *in* T. Clark, A.P. Curlee, S.P. Minta, and P.M. Karieva, editors. *Carnivores in Ecosystems: the Yellowstone experience*. Yale University Press, New Haven, Connecticut, USA.

Mladenoff, D.J., T.A. Sickley, R.G. Haight, and A.P. Wydeven. 1995. A regional landscape analysis and prediction of favorable gray wolf habitat in the northern Great Lakes region. *Conservation Biology* **9**:279–294.

Mladenoff, D.J., R.G. Haight, T.A. Sickley, and A.P. Wydeven. 1997. Causes and implications of species recovery in altered ecosystems: a spatial landscape projection of wolf population recovery. *BioScience* **47**:21–31.

Mladenoff, D.J., T.A. Sickley, and A.P. Wydeven. 1999. Predicting gray wolf landscape recolonization: logistic regression models vs. new field data. *Ecological Applications* **9**:37–44.

Montana Fish, Wildlife and Parks. 1996. Final environmental impact statement: Management of mountain lions in Montana. Montana Fish, Wildlife and Parks, Helena, Montana, USA.

Murphy, K.M. 1983. Relationships between a mountain lion population and hunting pressure in western Montana. Thesis. University of Montana, Missoula, Montana, USA.

Murphy, K.M., G.S. Felzien, M.G. Hornocker, and T.K. Ruth. 1998. Encounter competition between bears and cougars: some ecological implications. *Ursus* **10**:55–60.

Murphy, K.M., P.I. Ross, and M.G. Hornocker. 1999. The ecology of anthropogenic influences on cougars. Pages 77–101 *in* T.W. Clark, A.P. Curlee, S.C. Minta, and P.M. Kareiva, editors. *Carnivores in Ecosystems: the Yellowstone Experience*. Yale University Press, New Haven, Connecticut, USA.

National Research Council. 1997. *Wolves, Bears, and their Prey in Alaska: Biological and Social Challenges of Wildlife Management*. National Academy Press, Washington DC, USA.

Noss, R.F. 1992. The wildlands project: land conservation strategy. *Wild Earth* (special issue):10–25.

Noss, R.F. 1996. Conservation or conveniences? *Conservation Biology* **10**:921–922.

Noss, R.F. and A. Cooperrider. 1994. *Saving Nature's Legacy: Protecting and Restoring Biodiversity*. Defenders of Wildlife and Island Press, Washington DC, USA.

Noss, R.F., H.B. Quigley, M.G. Hornocker, T. Merrill, and P.C. Paquet. 1996. Conservation biology and carnivore conservation in the Rocky Mountains. *Conservation Biology* **10**:949–963.

Noss, R.F., M.A. O'Connell, and D.D. Murphy. 1997. *The Science of Conservation Planning: Habitat Conservation under the Endangered Species Act*. Island Press, Washington DC, USA.

Oksanen, T., L. Oksanen, M. Schneider, and M. Aunapuu. 2001. Regulation, cycles and stability in northern carnivore-herbivore systems: back to first principles. *Oikos* **94**:101–117.

Pall, O., M. Jalkotzy, and I. Ross. 1988. *The Cougar in Alberta*. Report to Alberta Forestry, Lands, and Wildlife. Associated Resources Consultants, Calgary Alberta, Canada.

Paquet, P.C. 1991. Winter spatial relationships of wolves and coyotes in Riding Mountain National Park, Manitoba. *Journal of Mammalogy* **72**:397–401.

Paquet, P. and A. Hackman. 1995. *Large Carnivore Conservation in the Rocky Mountains: a Long-Term Strategy for Maintaining Free-Ranging and Self-Sustaining Populations of Carnivores*. World Wildlife Fund Canada, Toronto, Ontario, Canada.

Parsons, D.R. 1998. "Green fire" returns to the southwest: reintroduction of the Mexican wolf. *Wildlife Society Bulletin* **26**:799–807.

Pease, C.M. and D.J. Mattson. 1999. Demography of the Yellowstone grizzly bears. *Ecology* **80**:957–975.

Pelton, M.R. 2000. Black bear. Pages 389–408 *in* S. Demaris and P.R. Krausman editors. *Ecology and Management of Large Mammals in North America*. Prentice Hall, Upper Saddle River, New Jersey, USA.

Peterson, R.O. 1995. Wolves as interspecific competitors in canid ecology. Pages 231–243 *in* L.N. Carbyn, S.H. Fritts, and D. Seip, editors. *Ecology and Conservation of Wolves in a Changing World*. Canadian Circumpolar Institute, Occasional Publication **35**, Edmonton, Alberta, Canada.

Phillips, M.K. 1990. Measures of the value and success of a reintroduction project. *Endangered Species Update* **8**:24–26.

Phillips, M., N. Fascione, P. Miller, and O. Byers. 2000. *Wolves in the Southern Rockies. A Population and Habitat Viability Assessment: Final Report*. IUCN/SSC Conservation Breeding Specialist Group, Apple Valley, Minnesota, USA.

Pierce, B.M., V.C. Bleich, and R.T. Bowyer. 2000. Social organization of mountain lions: does a land-tenure system regulate population size. *Ecology* **81**:1533–1543.

Pletscher, D.H., R.R. Ream, D.K. Boyd, M.W. Fairchild, and K.E. Kunkel. 1997. Dynamics of a recolonizing wolf population. *Journal of Wildlife Management* **61**:459–465.

Polis, G.A. and D.R. Strong. 1996. Food web complexity and community dynamics. *American Naturalist* **147**:813–846.

Potvin, F., H. Jolicoeur, L. Breton, and R. Lemieux. 1988. Wolf diet and prey selectivity during two periods for deer in Quebec: decline versus expansion. *Canadian Journal of Zoology* **66**:1274–1279.

Powell, R.A. 2001. Who limits whom: predators or prey. *Endangered Species Update* **18**:98–102.

Power, M.E., D. Tilman, J.A. Estes, B.A. Menge, W.J. Bond, L.S. Mills, J.C. Castilla, J. Lubchenko, and R.T. Paine. 1996. Challenges in the quest for keystones. *Bioscience* **46**:609–620.

Quigley, H.B. and P.G. Crawshaw, Jr. 1992. A conservation plan for *Panthera onca* in the Pantanal region in Brazil. *Biological Conservation* **61**:149–157.

Rabinowitz, A.R. 1986. Jaguar predation on domestic livestock in Belize. *Wildlife Society Bulletin* **14**:170–174.

Rasker, R. and A. Hackman. 1996. Economic development and the conservation of large carnivores. *Conservation Biology* **10**:991–1002.

Reading, R.P. and T.W. Clark. 1996. Carnivore reintroductions: an interdisciplinary examination. Pages 296–336 *in* J.L. Gittleman, editor. *Carnivore Behavior, Ecology, Evolution*. Volume II. Cornell University Press, Ithaca, New York, New York, USA.

Reimchen, T.E. 2000. Some ecological and evolutionary aspects of bear-salmon interactions in coastal British Columbia. *Canadian Journal of Zoology* **78**:448–457.

Ripple, W.J. and E.J. Larsen. 2000. Historic aspen recruitment, elk, and wolves in northern Yellowstone National Park, USA. *Biological Conservation* **95**:361–370.

Ripple, W.J., E.J. Larsen, R.A. Renkin, and D.W. Smith. 2001. Trophic cascades among wolves, elk and aspen on Yellowstone Park's northern range. *Biological Conservation* **102**:227–234.

Ross, P.I. and M.G. Jalkotzy. 1992. Characteristics of a hunted population of cougars in southwestern Alberta. *Journal of Wildlife Management* **56**:417–426.

Ross, P.I., M.G. Jalkotzy, and M. Festa-Bianchet. 1997. Cougar predation on bighorn sheep in southwestern Alberta during winter. *Canadian Journal of Zoology* **74**:771–775.

Ruth, T.K., K.A. Logan, L.L. Sweanor, M.G. Hornocker, and L.G. Temple. 1998. Evaluating cougar translocation in New Mexico. *Journal of Wildlife Management* **62**:1264–1275.

Samson, C. and J. Huot. 1998. Movements of female black bears in relation to landscape vegetation type in southern Quebec. *Journal of Wildlife Management* **62**:718–727.

Sanderson, E.W., K.H. Redford, C.B. Chetkiewicz, R.A. Medelein, A.R. Rabinowitz, J.G. Robinson, and A.B. Taber. 2002. Planning to save a species: the jaguar as a model. *Conservation Biology* **16**:58–72.

Schaller, G.B. 1996. Introduction: carnivores and conservation biology. Pages 1–10 *in* J.L. Gittleman, editor. *Carnivore Behavior, Ecology, Evolution*. Volume II. Cornell University Press, Ithaca, New York, New York, USA.

Schmitz, O.J., P.A. Hamback, and A.P. Beckerman. 2000. Trophic cascades in terrestrial systems: a review of the effects of carnivore removals on plants. *American Naturalist* **155**:347–358.

Schullery, P. 1997. Review of the grizzly bears of Yellowstone. *Journal of Wildlife Management* **61**:1450–1452.

Schwartz, C.C. and S.A. Arthur. 1997. Cumulative effects model verification, sustained yield estimation, and population viability management of the Kenai Peninsula. Alaska brown bear. Federal Aid in Wildlife Restoration Study **4.27**, Alaska Department of Fish and Game, Juneau, Alaska, USA.

Schwartz, C.C. and A.W. Franzmann. 1991. Interrelationship of black bears to moose and forest succession in the northern coniferous forest. *Wildlife Monograph* **113**.

Seidensticker, J.C. IV, M.G. Hornocker, W.V. Wiles, and J.P. Messick. 1973. Mountain lion social organization in the Idaho primitive area. *Wildlife Monograph* **35**.

Servheen, C. 1998. The grizzly bear recovery program: current status and future considerations. *Ursus* **10**:591–596.

Servheen, C., W. Kasworm, and T. Thier. 1995. Transplanting grizzly bears as a management tool: results from the Cabinet Mountains Montana. *Biological Conservation* **71**:261–268.

Servheen, C., J.S. Waller, and P. Sandstrom. 2001. *Identification and Management of Linkage Zones for Grizzly Bears Between the Large Blocks of Public Land in the Northern Rocky Mountains*. U.S. Fish and Wildlife Service Grizzly Bear Recovery Office, Missoula, Montana, USA.

Seymour, K.L. 1989. *Panthera onca. Mammalian Species* **340**:1–9.

Shaffer, M.L. 1983. Determining minimum population sizes for the grizzly bear. *International Conference on Bear Research and Management* **5**:133–139.

Shaffer, M. 1992. *Keeping the Grizzly Bear in the American West: an Alternative Recovery Plan*. The Wilderness Society, Washington, DC, USA.

Shaw, H.G. 1980. Ecology of the mountain lion in Arizona. Final Report P-R Project **W-78-R**. Arizona Game and Fish Department, Phoenix, Arizona, USA.

Simberloff, D. 1998. Flagships, umbrellas, and keystones: is single species management passe in the landscape era? *Biological Conservation* **83**:247–257.

Singer, F.J., G. Wang, and N.T. Hobbs. 2003. The role of ungulates and large predators on plant communities and ecosystem processes in national parks. Pages 444–486 *in* C.J. Zabel and R.G. Anthony, editors. *Mammal Community*

Dynamics. Management and Conservation in the Coniferous Forests of Western North America. Cambridge University Press, Cambridge, UK.

Singleton, P.H. 1995. Winter habitat selection by wolves in the North Fork of the Flathead River, Montana and British Columbia. Thesis. University of Montana, Missoula, Montana, USA.

Smallwood, K.S. 1994. Trends in California mountain lion populations. *Southwestern Naturalist* **39**:67–72.

Smith, B.L. and S.H. Anderson. 1996. Patterns of neonatal mortality of elk in northwest Wyoming. *Canadian Journal of Zoology* **74**:1229–1237.

Smith, D.W., W.G. Brewster, and E.E. Bangs. 1999. Wolves in the greater Yellowstone ecosystem: restoration of a top carnivore in a complex management environment. Pages 103–125 *in* T.W. Clark, R.R. Reading, and A. Clarke, editors. *Carnivores in Ecosystems: the Yellowstone experience*. Yale University Press, New Haven, Connecticut, USA.

Smith, D.W., K.M. Murphy, and D.S. Guernsey. 2001. *Yellowstone Wolf Project: Annual Report, 1999*. National Park Service, Yellowstone Center for Resources, Yellowstone National Park, Wyoming, USA.

Soule, M.E. and J. Terborgh. 1999. *Continental Conservation Scientific Foundations of Regional Reserve Networks*. Island Press, Washington DC, USA.

Swank, W.G. and J.G. Teer. 1989. Status of the jaguar – 1987. *Oryx* **23**:13–21.

Sweanor, L.L., K.A. Logan, and M.G. Hornocker. 2000. Cougar dispersal patterns, metapopulation dynamics, and conservation. *Conservation Biology* **14**:798–808.

Sweitzer, R.A., S.H. Jenkins, and J. Berger. 1997. Near-extinction of porcupines by mountain lions and consequences of ecosystem change in the Great Basin Desert. *Conservation Biology* **11**:1407–1417.

Swenson, J.E., F. Sandegren, A. Soderberg, A. Bjarvall, R. Franzen, and P. Wabakken. 1997. Infanticide caused by hunting of male bears. *Nature* **386**:450–451.

Terborgh, J., J.A. Estes, P. Paquet, K. Ralls, D. Boyd-Heger, B.J. Miller, and R.F. Noss. 1999. The role of top carnivores in regulating terrestrial ecosystems. Pages 39–64 *in* M.E. Soule and J. Terborgh, editors. *Continental Conservation Scientific Foundations of Regional Reserve Networks*. Island Press, Washington, DC, USA.

Terborgh, J., L. Lopez, P. Nunez, M. Rao, G. Shahabuddin, G. Orihuela, M. Riveros, R. Ascanio, G.H. Adler, T.D. Lambert, and L. Balbas. 2001. Ecological meltdown in predator-free forest fragments. *Science* **294**:1923–1925.

Thurber, J.M., R.O. Peterson, J.D. Woolington, and J.A. Vucetich. 1992. Coyote coexistence with wolves on the Kenai Peninsula, Alaska. *Canadian Journal of Zoology* **70**:2492–2498.

U.S. Fish and Wildlife Service. 1993a. Grizzly bear recovery plan. Final report. Missoula, Montana, USA.

U.S. Fish and Wildlife Service. 1993b. Recovery plan for the eastern timber wolf. USFWS, Twin Cities, Minnesota, USA.

U.S. Fish and Wildlife Service. 1994. The reintroduction of gray wolves to Yellowstone National Park and central Idaho. Final Environmental Impact Statement, U.S. Fish and Wildlife Service, Denver, Colorado, USA.

U.S. Fish and Wildlife Service. 2000. Grizzly bear recovery in the Bitterroot ecosystem. Final Environmental Impact Statement, U.S. Fish and Wildlife Service, Missoula, Montana, USA.

Valdez, R. 2000. Jaguar. Pages 378–388 *in* S. Demaris and P.R. Krausman editors. *Ecology and Management of Large Mammals in North America*. Prentice Hall, Upper Saddle River, New Jersey, USA.

Van Ballenberghe, V. and W.B. Ballard. 1994. Limitation and regulation of moose populations: the role of predation. *Canadian Journal of Zoology* **72**: 2071–2077.

Van Dyke, F.G., R.H. Brocke, H.G. Shaw, B.B. Ackerman, T.P. Hemker, and F.G. Lindzey. 1986. Reactions of mountain lions to logging and human activity. *Journal of Wildlife Management* **50**:102–109.

Waller, J.S. and R.D. Mace. 1997. Grizzly bear habitat selection in the Swan Mountains, Montana. *Journal of Wildlife Management* **61**:1032–1039.

Weaver, J.L. 1994. Ecology of wolf predation amidst high ungulate diversity in Jasper National Park, Alberta. Dissertation. University of Montana, Missoula, Montana, USA.

Weaver, J.L., P.C. Paquet, and L.F. Ruggerio. 1996. Resilience and conservation of large carnivores in the Rocky Mountains. *Conservation Biology* **10**:964–976.

Weber, W. and A. Rabinowitz. 1996. A global perspective on large carnivore conservation. *Conservation Biology* **10**:1046–1054.

Wehausen, J.D. 1996. Effect of mountain lion predation on bighorn sheep in the Sierra Nevada and Granite Mountains of California. *Wildlife Society Bulletin* **24**:471–479.

Wielgus, R.B. and F.L. Bunnell. 1994. Dynamics of a small, hunted brown bear *Ursus arctos* population in southwestern Alberta, Canada. *Biological Conservation* **67**:161–166.

Wielgus, R.B. and F.L. Bunnell. 1995. Tests of hypotheses for sexual segregation in grizzly bears. *Journal of Wildlife Management* **59**:552–560.

Wielgus, R.B., F.L. Bunnell, W.L. Wakkinen and P.E. Zager. 1994. Population dynamics of Selkirk mountain grizzly bears. *Journal of Wildlife Management* **58**:266–272.

Wikramanayake, E.D., E. Dinerstein, J.G. Robinson, U. Karanth, A. Rabinowitz, D. Olson, T. Mathew, P. Hedao, M. Conner, G. Hemley, and D. Bolze. 1998. An ecology-based method for defining priorities for large mammal conservation: the tiger as case study. *Conservation Biology* **12**:865–878.

Williams, J.S., J.J. McCarthy, and H.D. Picton. 1995. Cougar habitat use and food habits on the Rocky Mountain Front. *Intermountain Journal of Sciences* **1**:16–28.

Willson, M.F., S.M. Gende, and B.H. Marston. 1998. Fishes and the forest expanding perspectives on fish-wildlife interactions. *Bioscience* **48**:455–462.

Wilson, M.L. and J.E. Childs. 1997. Vertebrate abundance and epidemiology of zoonotic diseases. Pages 224–248 *in* W.J. McShae, H.B. Underwood, and J. H. Rappole, editors. *The Science of Overabundance*. Smithsonian Institution Press, Washington, DC, USA.

Wondolleck, J.M., S.L. Yaffee, and J.E. Crowfoot. 1994. Conflict management perspective: applying the principle of alternative dispute resolution. Pages 305–326 *in* T.W. Clark, R. Reading, and A. Clarke, editors. *Endangered Species Recovery: Finding the Lessons, Improving the Process*. Island Press, Washington DC, USA.

Woodroffe, R. and J.R. Ginsberg. 1998. Edge effects and the extinction of populations inside protected areas. *Science* **280**:2126–2128.

Woods, J.G., D. Paetkau, D. Lewis, B.N. McLellan, M. Proctor, and C. Strobeck. 1999. Genetic tagging of free-ranging black and brown bears. *Wildlife Society Bulletin* **27**:616–627.

World Resources Institute. 1994. *World Resources: 1994–1995*. Oxford University Press, New York, New York, USA.

Wright, A.L., H. Quigley, and K. Kunkel. 2000. Cougars and desert bighorns in the Fra Cristobal Range: scale, geography and seasonality. Proceedings of the 6th Mountain Lion Workshop, San Antonio, Texas, USA, 12–14 December.

9

Ungulates in western coniferous forests: habitat relationships, population dynamics, and ecosystem processes

Introduction

Wild ungulates play important roles in coniferous forest throughout western North America. Their biology is well known compared with that of other species of wildlife. They have sufficiently large home ranges to integrate spatial patterns across landscapes. Finally, they are often migratory (Wallmo 1981, Nicholson et al. 1997). Their life-history characteristics require consideration of entire landscapes rather than isolated patches of habitat for purposes of conservation and management (Hanley 1996, Kie et al. 2002). Ungulates require temporally and spatially diverse habitat components such as food and cover. These mammals can have significant effects on vegetation composition and basic ecosystem processes such as nutrient cycling, thereby acting as keystone species (Molvar et al. 1993, Wallis de Vries 1995, Hanley 1996, Hobbs 1996, Nicholson et al. 1997, Simberloff 1998, Kie et al. 2002).

Ungulates have economic value to society as well. Most species provide recreational hunting opportunities and also can have non-consumptive, aesthetic values (Loomis et al. 1989). Conversely, ungulates can cause damage to gardens and other landscaping (Conover 1997), to agricultural crops, (Austin and Urness 1993), and to new tree seedlings (Bandy and Taber 1974). Damage from deer–vehicle collisions also can be substantial (Romin and Bissonette 1996).

Four species of ungulates, all members of the family Cervidae, commonly occur in coniferous forests in western North America: elk (*Cervus elaphus*), moose (*Alces alces*), mule and black-tailed deer (*Odocoileus hemionus*), and white-tailed deer (*Odocoileus virginianus*). North American elk, or wapiti, are conspecific with European red deer, and are widespread throughout western North America. Rocky Mountain elk (*C. elaphus nelsoni*) occur in forests from southern Arizona and New Mexico northward to British Columbia and Alberta. Roosevelt elk (*C. elaphus roosevelti*) are associated with coastal forests of redwood (*Sequoia sempervirens*), Douglas-fir (*Pseudotsuga douglasii*), and Sitka spruce (*Picea sitchensis*) in northern California, Oregon, Washington, and British Columbia, and have been translocated to Afognak, Raspberry, and Kodiak Islands in Alaska (Bryant and Maser 1982, Wisdom and Cook 2000).

Moose, conspecific with European elk, range from Utah and Colorado to Alaska. Shira's moose (*A. alces shirasi*) occurs in Utah, Colorado, Wyoming, Idaho, Montana, Washington, and southern British Columbia (Franzmann 2000). Northwestern moose (*A. alces andersoni*) occur primarily in British Columbia, Alberta, and other Canadian provinces. Alaskan moose (*A. alces gigas*) occur in boreal forests throughout Alaska and in the western Yukon Territory (Franzmann 2000).

Mule deer of various subspecies are nearly ubiquitous in western North America and often are associated with forested habitats (Kie and Czech 2000). Columbian black-tailed deer (*O. hemionus columbianus*) occur in coastal forests in northern California, Oregon, Washington, and southern British Columbia. This subspecies also has been transplanted to the island of Kauai in Hawaii. Sitka black-tailed deer (*O. hemionus sitkensis*) occur in spruce forests in northern British Columbia and southeastern Alaska (Kie and Czech 2000).

White-tailed deer inhabit coniferous forests in the northwestern United States and western Canada (Baker 1984). Occurrence of this species overlaps the range of mule deer in the west, and they are usually associated with riparian habitats with abundant woody vegetation (Baker 1984). Coues white-tailed deer (*O. virginianus couesi*) occur in ponderosa pine and pinyon-juniper forests in southern Arizona and southwestern New Mexico. Small populations of Columbian white-tailed deer (*O. virginianus leucurus*) are found in western Oregon and Washington.

In addition to these ungulate species commonly found in coniferous forests, populations of two other species can occur locally in the west. Woodland caribou (*Rangifer tarandus caribou*) occur in portions of northeastern Washington, northern Idaho, northwestern Montana, eastern British Columbia, and western Alberta (Miller 1982). The wood bison (*Bison bison athabasca*) occurs in the Rocky Mountain States, Alberta, and other

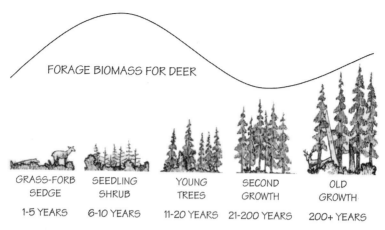

FORAGE BIOMASS FOR DEER

| GRASS-FORB SEDGE | SEEDLING SHRUB | YOUNG TREES | SECOND GROWTH | OLD GROWTH |
| 1-5 YEARS | 6-10 YEARS | 11-20 YEARS | 21-200 YEARS | 200+ YEARS |

Fig. 9.1. Generalized pattern of deer forage supplies during secondary succession in coniferous forest habitat. Duration of stages represents averages for western hemlock-Sitka spruce forests of the northern Pacific coast (after Wallmo and Schoen 1981, Kie and Czech 2000).

Canadian provinces. Its status as a distinct subspecies, however, is in question (Shaw and Meagher 2000).

Habitat relationships

Ungulates and habitat seral stages

Many ungulates in western North America thrive in forests at early successional stages. For example, wildfire, prescribed burning, and clear-cut logging have resulted in increased numbers of mule and black-tailed deer in many areas (Wallmo 1978). In spruce forests of the Pacific northwest, forage for black-tailed deer is most abundant between about five and ten years following disturbance such as logging (Fig. 9.1). As forest stands start to mature and trees shade out desirable forage species, habitat value for deer declines (Wallmo and Schoen 1981, Kie and Czech 2000). In old-growth forests, openings created when large trees fall allow sunlight to reach the forest floor and forage plants for deer to become re-established (Fig. 9.1).

In coniferous forests where snow accumulation is heavy, however, such as in northern British Columbia and southeastern Alaska, new forage created by disturbance is largely unavailable to deer during winter (Wallmo and Schoen 1981). In such situations, resident deer are dependent on old-growth forests where canopy cover of trees is sufficiently dense to intercept snow and provide access to understory forage supplies, as well as a source of arboreal lichens falling from the canopy. In these areas, black-tailed deer are climax-associated or old-forest species rather than successional species (Hanley 1984).

Moose also exhibit variable response to changes in seral stages. Large numbers of yearling moose responded to recently burned areas in Minnesota (Peek 1974), but not in interior Alaska (Gasaway et al. 1989). A lower population density of moose in Alaska prior to burning may have precluded a similar response. Areas in Alaska without a substantial understory of willow (*Salix* spp.) may not be attractive to moose following fire (Weixelman et al. 1998). Finally, woodland caribou often use climax-stage forested habitats (Apps et al. 2001). In conclusion, broad species-wide generalizations about the relationship between seral stages of habitats and ungulates are often only weakly supportable if at all.

Ungulates and food

Ruminants often are characterized by relations among body size, digestive morphology and physiology, and types of forage consumed (Fig. 9.2).

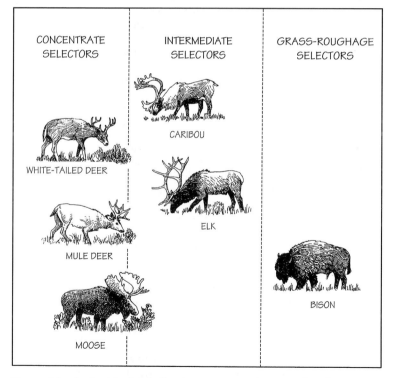

Fig. 9.2. Morphophysiological feeding types among ruminant ungulates from western coniferous forests (adapted from Hoffman 1985).

Typically, larger-bodied ruminants have proportionally larger rumens, lower feeding frequencies, longer forage retention times, and can subsist on forages of lower quality than can smaller-bodied ruminants (Hoffman 1985, Putman 1988). In addition, they have lower metabolic rates than smaller ruminants (Kleiber 1961). As such, these bulk-feeders have the luxury of ingesting fairly coarse forages of low digestibility such as dried grasses, and slowing down the rate of food passage to allow microbial digestion to break down the cell-wall constituents. Small-bodied ruminants such as mule and black-tailed deer must feed on forages in which the nutrients are more concentrated and more digestible. Even large-bodied ruminants such as moose that are adapted to feeding on highly concentrated forages exhibit digestive morphology similar to their smaller-bodied relatives (Hoffman 1985, Putman 1988). Such differences have strong allometric underpinnings (Illius and Gordon 1999).

Mule and black-tailed deer traditionally have been thought of as browsers, particularly during winter, relying primarily on twigs and other vegetative parts of woody plants. Their requirement for forages with concentrated nutrients precludes them from making heavy use of forages of low digestibility such as dried grasses, and may at times result in their preferential use of woody plant material. Nonetheless, the concept of deer as browsers by choice is perhaps one of the oldest and most persistent myths in deer ecology and management (Gill 1976). Given access to seasonally abundant, nutritious, herbaceous plants of high digestibility, deer will tend to select those species in preference to browse species of lower digestibility (Kie and Czech 2000).

Diets of Sitka black-tailed deer on Admirality Island, Alaska, ranged from 57% to 79% non-browse items such as forbs, ferns, grasses, sedges (*Carex nudata*), lichens, algae, and mosses depending on season, although during a winter with deep snow browse consumption peaked at 87% (Hanley et al. 1989). In summer, although shrubs were available to deer, concentrations of digestion-inhibiting compounds were sufficiently high to discourage consumption (Hanley et al. 1989). Diets of Columbian black-tailed deer on oak woodland-annual grass ranges in northern California included as much as 62% newly germinated annual grasses during winter when those species were green and highly digestible (Taber and Dasmann 1958). Other seasonally important forages for mule and black-tailed deer include acorns (Beale and Darby 1991), mistletoe (Urness 1969), lichens (Hanley et al. 1989), mushrooms (Beale and Darby 1991), and succulents (Krausman et al. 1997). Many of those forages are often under-represented in diet studies because of their high digestibility or deficiencies in sampling and analytical methodology (Beale and Darby 1991).

Ungulates and cover

Cover is a loosely defined term, but, in its broadest sense, refers to vegetative, topographic, or other types of environmental structure that enhances reproduction or survival of animals (Bailey 1984). Functions of cover can include concealment and escape from predators, refuge from human disturbances, and shelter from thermal extremes (Patton 1992). Cover also may influence the availability of forage through subtle behavioral mechanisms associated with avoiding predators (Kie 1999, Kie and Bowyer 1999). Of particular interest in forested habitats, however, is the role overstory tree canopies play in intercepting snowfall, and ameliorating thermal extremes during winter.

Even small-bodied ungulates such as mule deer exist across a wide range of temperature gradients averaging from below $-15°$ C during winter in the Rocky Mountains to above $30°$ C in the summer in the southern end of its range (Wallmo 1981). Thermoregulation is accomplished by shivering, changing posture, erecting hair, and by making use of environmental temperature variations afforded by different habitats and topographic features (Mackie et al. 1982). Moreover, cervids are adapted to cold winter temperatures because of the thickness of their winter coats.

The lower boundary of the thermal neutral zone for black-tailed deer in winter pelage was measured at $-10°$ C (Parker 1988). Below that temperature, deer have to expend energy, for example by shivering, to maintain a constant body temperature. Larger-bodied ungulates such as elk and moose are adapted to even colder winter temperatures and more northern environments. The lower boundary of the thermal neutral zone for moose lies somewhere below $-30°$ C (Renecker and Hudson 1986).

The management of forested habitats to provide overstory tree canopy as thermal cover is a long-established paradigm in the northwestern U.S. (Black et al. 1976). Recent research, however, reported that elk in northeastern Oregon lost less body mass during winter when tree canopies were removed (Cook et al. 1998). The ability of elk to absorb solar radiation in opened stands of trees on sunny days may have reduced the need for metabolic thermoregulation and the accompanying expenditure of energy.

Canopy cover in forested habitats does serve other important roles in some instances, however. Previously discussed is the value of old-growth forests in intercepting snow cover and providing access to forage for black-tailed deer during winter in southeast Alaska (Hanley 1984). In addition, snow depth has a great influence on the movements of mule and black-tailed deer (Mackie et al. 1982). Snow depths of 25–30 cm may impede movements of mule deer in Colorado, and depths > 50 cm may completely prevent their use of areas (Loveless 1967). During the heaviest snow accumulations in the Bridger Mountains, Montana, mule deer were restricted to only 20% to 50% of their winter range (Mackie et al. 1982). Forested habitats, therefore, may provide refugia for ungulates during periods of heavy snowfall and extremely low temperatures.

Finally, cover may be very important in forested habitats as concealment from predators and human disturbances. Elk in northeastern Oregon will concentrate in areas away from heavily traveled roads (Rowland et al. 2000); but, in contrast, mule deer in this region attempt to distance themselves from elk and move closer to roads with heavy traffic, where they use vegetative and topographic cover (Wisdom 1998). In such instances, the vertical structure of cover in forested habitats is more important than the amount of overstory canopy.

Ungulates and water

Water usually occurs in sufficient abundance in western coniferous forests to have little effect on the distribution and abundance of ungulates. Furthermore, ungulates in arid and semi-arid environments are adapted to a scarcity of free water. For example, mule deer in Arizona typically visited sources of water once a day and consumed five to six liters of water per visit during the hot summer months, while visitation rates and the amount of water consumed per visit declined during cooler seasons (Hazam and Krausman 1988). Mule deer also obtain water from succulent plant material, dew on the surface of plants, and from metabolic processes. Feeding at night in hot, arid environments not only provides relief from thermal stress but may also be timed to take advantage of diel cycles in plant water content (Taylor 1969). Whether mule deer require free surface water has been debated (Rosenstock et al. 1999, Severson and Medina 1983). Nonetheless, when access to free water is severely restricted in penned white-tailed deer, they reduce their consumption of forage (Lautier et al. 1988). Therefore, although deer in the wild may exist for some periods of time without access to standing water, this poses marginal survival conditions (Severson and Medina 1983).

The abundance and spacing of water sources influence the distribution of mule deer in arid environments. In Arizona and New Mexico, mule deer are usually found within 2.4 km of free water (Wood et al. 1970). Mule deer in northern California averaged 1.19–1.55 km away from water sources, with a mean greatest distance of 2.46 km (Boroski and Mossman 1996). Female mule deer drink more water than males during late summer (Hazam and Krausman 1988). Females are often found closer to sources of water than males, presumably because of demands of lactation (Bowyer 1984, Boroski and Mossman 1996). In many instances, however, females remain close to water sources year-round (Fox and Krausman 1994).

Ungulates and landscapes

Many ungulates are considered well-adapted to habitat edges (Leopold 1933, Clark and Gilbert 1982, Hanley 1983, Kremsater and Bunnell 1992). For example, in California where food and cover occurred in small patches, mule deer did not have to travel far to meet their daily requirements and home ranges were correspondingly small (Leopold et al. 1951). Life-history characteristics in deer, such as the size of the home range, may be related to landscape pattern in more complex ways than simply as a function of the amount of habitat edge.

Size of home ranges in mule and black-tailed deer is correlated with heterogeneity of habitats at a broad spatial scale in California (Kie et al. 2002). The amount of habitat edge measured within 2000 m of the center of the home ranges for 80 female deer (an area larger than most of those home ranges) accounted for 27% of the variability in the size of the home range. Large amounts of edge resulted in small home-range sizes (Fig. 9.3). Spatial heterogeneity, however, measured at the same spatial scale (as indicated by the abundance of different habitat types, the distance between habitat patches of similar type, the shape of the patches, and the amount of structural contrast between patches) accounted for 57% of the variation in home-range size (Kie et al. 2002).

When spatial heterogeneity was measured at smaller scales more closely corresponding to the size of the home ranges, less of the variation in home-range size could be explained (Fig. 9.3). This result has important implications in resource selection studies (Kie et al. 2002). If ungulates perceive potential habitats at scales greater than those they eventually choose as a home range, analysis of habitat selection within the home range may yield biased results. The home range already includes landscape attributes that an individual has selected, whereas other avoided

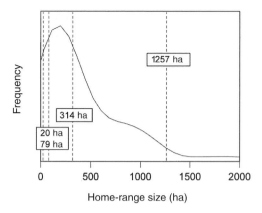

Fig. 9.3. Smoothed frequency distribution of home-range sizes for 80 female mule deer in California (after Kie et al. 2002). Landscape metrics representing spatial heterogeneity were measured at four scales: 20, 79, 314, and 1257 ha (within 250, 500, 1000, and 2000 m from the center of each home range). Home-range size was most closely correlated with landscape metrics measured at the largest spatial scale (1257 ha) even though the majority of the home ranges were smaller in size.

features lie outside the home range. Basing habitat availability on habitats contained only within the home range is circular logic because some degree of selection already has occurred.

Sexual segregation in ungulates

The concept that the sexes use space differently is not new. Charles Darwin (1871), citing previous work by C. Boner, hypothesized the following about red deer (*Cervus elaphus*): "Whilst the horns are covered in velvet, which lasts with red deer for about twelve weeks, they are extremely sensitive to a blow; so that in Germany the stags at this time somewhat change their habits, and avoiding dense forests, frequent young woods and low thickets." Main et al. (1996) summarized previous literature and classified concepts about sexual segregation into three groups: body-size, reproductive-strategy, and social-factor hypotheses.

One body-size hypothesis posits that smaller-bodied females more efficiently use habitats with closely cropped forages, thereby excluding larger males through inter-sexual competition. Sexual segregation between male and female red deer on the island of Rhum in Scotland increased when the population density of females increased, lending support to this hypothesis (Clutton-Brock et al. 1987a). Recent manipulative experiments, however, have cast doubt on that model (Conradt et al. 1999). A newer body-size hypothesis has been proposed whereby male deer may perform better on areas with lower-quality forage than would be ideal for females because of larger rumen sizes (Barboza and Bowyer 2000, 2001). This allows males to increase their intake of less digestible forages and slow passage rates to extract more nutrients. Faced with highly digestible forages, males would actually do less well faced with the possibility of bloat. Conversely, lactating females have the ability to increase post-ruminal components of the digestive system necessary to extract the nutrients from a high-quality diet required for lactation (Fig. 9.4; Barboza and Bowyer 2000). This body-size hypothesis may explain those instances of sexual segregation where males are found in areas characterized by forage that would appear to be less than optimum for females.

The reproductive-strategy hypothesis suggests that the need to bear and raise young poses constraints on habitat use by adult females, thereby limiting their use of space. For example, adult female white-tailed deer in Texas occurred more often in dense chaparral habitats when they had young than did adult males. Males did not have the same anti-predator constraints and were free to use more open savannah habitats where preferred forbs were abundant (Kie and Bowyer 1999). This hypothesis has been suggested for a variety of other ungulates including moose (Miquelle et al. 1992) and bighorn sheep (*Ovis canadensis*; Bleich et al. 1997).

Social factors also have been advanced as mechanisms to explain sexual segregation, whereby behavioral differences might lead to separation of the sexes. Main et al. (1996) rejected many of these hypotheses, but recently several authors have reinvoked such hypotheses to explain sexual segregation (Conradt 1998a, 1998b, Mysterud 2000, Ruckstuhl and Neuhaus 2000). These studies posit behavioral mechanisms such as activity time-budgets that may be associated with segregation, but do not demonstrate actual spatial separation at specific spatial and temporal scales (Bowyer et al. 1996).

Managing forested habitats for ungulates
Habitat seral-stage matters

Some species of forest-dwelling ungulates such as woodland caribou require mid- to late-seral habitats. Other species, such as mule deer, can take advantage of early-seral communities as foraging habitats but black-tailed deer in southeastern Alaska may require climax forests in some instances (Hanley 1984). In addition, even though some ungulates can take

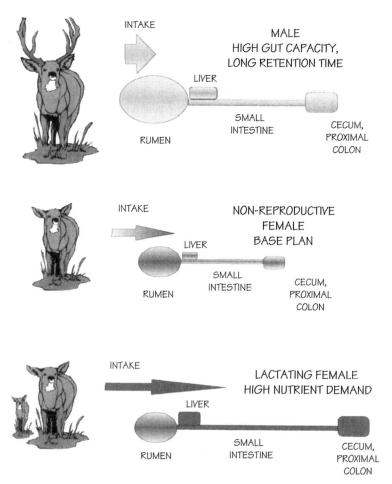

Fig. 9.4. Model of intake and digestive function in non-reproducing females compared with large males and lactating females, which provides a nutritional basis for sexual segregation in polygynous, dimorphic cervids. Width of arrows reflects amount of food intake, length of arrows indicates rate of digesta passage, and shading indicates density of nutrients in food. Diagrams of the digestive tract are shaded to reflect potential changes in fibrosity of food for males, and increases in post-ruminal size (especially the length of the small intestine) and function of lactating females (Barboza and Bowyer 2000). Barboza and Bowyer (2000) provide a complete description of the model and its components.

advantage of early-successional stage habitats following logging, forest practices that speed up and truncate succession (planting new trees, use of herbicides) may severely limit the time period during which those habitats are available (Leopold 1978). Furthermore, where forests have been harvested heavily in the past, habitats at early-successional stages may now be maturing and losing value, with little opportunities for additional logging of harvestable age classes in the coming decades (Jenkins and Starkey 1996).

Large forest clear-cuts likely will not be a silvicultural option on public lands in the future because of adverse public opinion. In addition, prescribed fire at a large enough spatial scale to have a significant effect on deer populations is expensive and may see only limited use. Moreover, let-burn policies are hampered by social and political factors (Kie and Czech 2000). In reality, the extent of early- and mid-successional vegetation seen in the early twentieth century as a result of wildfire, logging, and livestock grazing is unlikely to re-occur in the foreseeable future (Kie and Czech 2000). Consequently, wildlife managers may need to develop new approaches for maximizing habitat benefits for forest-dwelling ungulates in mid- and late-successional habitats (Kie and Czech 2000).

The value of browse as forage may be limited

Concentrate feeders such as mule and black-tailed deer may benefit from abundant new growth on palatable woody plant species. Conversely, they often rely heavily on forbs, new growth of grasses, and a variety of other forages that have high concentrations of nutrients. These forages become even more important in mid- and late-successional stage habitats.

Cover is important but reasons underlying its need may differ

Ungulates are well adapted to winter weather in most instances. Overstory canopy cover of trees, thought to provide thermal cover for elk, is unnecessary for that purpose in many instances (Cook et al. 1998). Hiding and concealment cover may be critical, however, and vertical structure may be more important than canopy cover in this respect. Canopy cover of trees may be important in intercepting snow and providing access to forages during winter in some instances (Hanley 1984).

Water sources may affect the distribution of ungulates

Ungulates can exist in some arid and semi-arid habitats with little or no free water. The abundance and spacing of water sources, however, will influence the distribution of ungulates (Boroski and Mossman 1996).

Landscape structure matters

Diverse mixes of plant communities are best. In general, landscapes rich with different vegetation types, irregularly shaped patches, and a minimum of structural contrast between patches are beneficial for deer (Kie et al. 2002), and maybe for other forest-dwelling ungulates as well.

Male and female ungulates use space differently

The need to bear and raise offspring may often prevent females from making use of newly available supplies of forage (Bowyer et al. 2001). For example, the crushing of willows in interior Alaska resulted in a much greater response by adult male moose than by adult females (Bowyer et al. 2001). In some instances, it may be appropriate to treat males and females as different species for purposes of habitat management (Kie and Bowyer 1999).

Direct manipulation of habitats is expensive and often yields limited benefits for short periods

Seeding, fertilization, and prescribed burning have all been used as tools to improve habitat for ungulates. Such activities are expensive, difficult to accomplish on a scale large enough to substantially increase deer numbers, and may conflict with other land-management goals (Kie and Czech 2000).

Co-ordination of management objectives for ungulates with other resource uses can provide great benefits

In most instances, efforts directed at co-ordinating the management of habitats for ungulates with other land uses have far greater potential benefit than direct improvement of habitat. Other resource uses that can be modified to incorporate habitat management goals for ungulates include timber harvesting, revegetation following wildfire, livestock grazing, development of home sites, and disturbances associated with roads, off-road vehicles, and snowmobiles (Kie and Czech 2000).

Population dynamics

Factors underpinning population dynamics of ungulates have been a topic of considerable debate (Fowler 1981, McCullough 1990, Mackie et al. 1990). Nonetheless, few life-history characteristics of ungulates are free from density-dependent effects (McCullough 1999), and managing either the harvest of, or habitat for, ungulates requires an understanding of density-dependent mechanisms. Ungulates have small litter sizes but long life spans in comparison with small-bodied mammals. Substantial maternal investment in young (especially during the neonate's first year of life) is ubiquitous among ungulates (Clutton-Brock 1991, Rachlow and

Bowyer 1994). Consequently, density dependence is an expected outcome from these and other life-history characteristics of ungulates (Stearns 1977, Stubbs 1977, Fowler 1981, Goodman 1981). Density dependence plays a crucial role in understanding the dynamics of animal populations and is embodied in most models of population growth among ungulates, yet controversy still surrounds this topic.

Detecting and interpreting density dependence

Detecting density dependence is neither simple nor straightforward (Slade 1977, Gaston and Lawton 1987, Pollard et al. 1987). Nevertheless, numerous studies of ungulates have demonstrated the fundamental role that process plays in the dynamics and regulation of populations (Klein 1968, Kie and White, 1985, Skogland 1985, Clutton-Brock et al. 1987b, Boyce 1989, Bartmann et al. 1992, Sand 1996, Singer et al. 1997). Notably, those studies supporting density-dependent mechanisms typically were conducted under circumstances where confounding variables such as predation and severe weather were unimportant or controlled (McCullough 1979, Bowyer et al. 1999). Likewise, research substantiating the importance of density-dependent mechanisms in ungulate populations typically has come from long-term studies in which populations varied markedly in size with respect to carrying capacity (K) of the environment, and appropriate vital rates for populations were collected (McCullough 1979, 1990). Herein we define K to mean the number of animals at or near equilibrium with their food supply. Although many definitions of K exist (McCullough 1979), we constrain our use to the traditional meaning of this term.

What factors might lead to confusion over whether density dependence was affecting the dynamics of an ungulate population? Some obvious problems include examining a population over too small a range of densities (often over too short a time span) to bring about changes in fecundity, recruitment, or survivorship. Also, attempting to compare population density among years when the ability of the environment to support those animals (K) was fluctuating substantially will lead to uncertainty about whether density dependence was operating. We will return to the difficulty of assessing K later. We do not believe that all criticisms of density dependence can be countered by those two potentialities. We note, however, that empirical approaches have some limitations. Careful examination and interpretation of data presented to refute a well-established ecological principle, such as density dependence, are required.

Table 9.1. *Population size, carrying capacity (K), and overwinter mortality in five hypothetical populations of deer inhabiting ranges of varying quality*

Population	K (number of deer)	Population size	Number dying	Overwinter mortality (%)
A	307	321	30	9.4
B	512	507	50	10.0
C	114	124	12	9.7
D	357	362	36	10.0
E	200	214	22	10.3

Note that all populations are at or near K.

We argue that some approaches for assessing density dependence are flawed. Comparing densities among populations to evaluate parameters such as recruitment or mortality is of questionable value (Weixelman et al. 1998, Bowyer et al. 1999). For example, if five populations of deer occurred in habitats that exhibited substantially different carrying capacities (K), and all five populations were at or near K, similar population characteristics such as recruitment or survival would be expected for those populations (Table 9.1). A graph of recruitment rate against population density, however, would lead to the conclusion that density dependence was not operating (a random pattern or a linear fit with a slope near zero), when that mechanism was regulating all five populations (Fig. 9.5). Clearly, habitat quality can confound comparisons of density dependence among populations of ungulates.

Similarly, evaluating density dependence from historical changes in population size can be misleading. Environments and subsequent responses of animals can be quite variable over many years. Assuming that current populations can achieve levels attained long ago involves assuming a similar ability of the habitat to support ungulates (and that the initial population estimates were reliable), which may be incorrect. Moreover, population density can be a misleading indicator of habitat quality (Van Horne 1983). The foregoing examples simply illustrated how data related to density dependence might be viewed improperly and misinterpreted. We suggest that the problem is far more pernicious, and stems from how the basic biology of ungulates and their interactions with their environment are perceived.

We caution that the inverse relation between recruitment rate (young/adult female) and population size need not be linear to infer

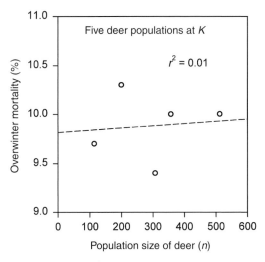

Fig. 9.5. Relationship between overwinter mortality and population size for five hypothetical populations of deer inhabiting ranges of varying quality; all populations were at or near carrying capacity (K). Population size or density should not be used to infer density dependence or independence among different populations. The standard interpretation of this relationship is that the populations are exhibiting density independence (no correlation between population size and overwinter mortality), when all are being regulated by habitat quality (e.g., density dependence) – data from Table 9.1.

density dependence; a non-linear pattern for ungulates may be common, particularly at population densities between maximum sustained yield (MSY) and K (Fowler 1981, Kie and White 1985, McCullough 1999, Person et al. 2001). Moreover, where the difference between MSY and K is small relative to the total size of the population, density dependence may be especially difficult to detect because the onset of obvious density-dependent changes is rapid, and restricted to a narrow range of densities near K.

Another potentially confusing outcome from density dependence relates to the rate of recruitment of young into the population, and the effect of that process on the age structure of the population. A high rate of recruitment (young/adult female) is not necessarily indicative of a productive, increasing population, and a low recruitment rate may not be reflective of a declining population (McCullough 1979). The interaction between population size and recruitment rate (young/adult female × number of adult females) determines the number of young successfully added to the population and, consequently, the allowable harvest (Fig. 9.6). Thus, a high recruitment rate but low population size, and a low

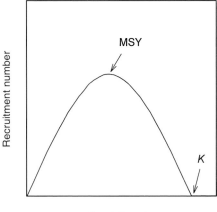

Fig. 9.6. Parabolic relationship between recruitment number (i.e., number of young successfully added to the population) and population size of adult or other reproductive females, showing populations sizes that yield maximum sustained yield (MSY) and carrying capacity (*K*; adapted from McCullough 1979). The parabola need not be symmetrical to infer density dependence.

recruitment rate but a high population size both yield a low number of recruits and, concomitantly, little population change. A population at moderate size and with a modest recruitment rate, however, would produce the greatest number of recruits and maximal amount of population increase (the relation between recruitment number and population size is parabolic, Fig. 9.6).

Further, a population exhibiting a high rate of recruitment might be overharvested and declining, whereas one with a low recruitment rate might be relatively stable near *K*. Likewise, a population with a wide base to an age-class pyramid (numerous recruits and fewer adults, thought to indicate a stationary population) might be either overharvested and declining markedly or increasing rapidly. A population with a small base to a similar age pyramid (few recruits and more adults, thought to identify a declining population) may represent a relatively stationary population at or near *K* (McCullough 1979). Caughley (1974) warned long ago that the age structure of a population should not be used to infer changes in population size without an independent estimate of size for that population. He also wondered why age structure would be useful if an independent assessment of population size was available. We note that the failure to comprehend these inherent relationships among population parameters can lead to misinterpretations as to whether density dependence is operating.

The foregoing descriptions of relations among population size, recruitment rate and number, age structure, and population change may seem simplistic and unnecessary given that those theoretical predictions have been understood for nearly a century, and substantial empirical support exists for them. Unaccountably, this approach for evaluating the population ecology of ungulates is not as widely accepted or practised by managers as might be expected. For instance, some authors have suggested that northern populations of ungulates may not exhibit density dependence (Bergerud et al. 1983, Gasaway et al. 1983, Bergerud 1992, Boertje et al. 1996). Population models that do not include density dependence, however, have ignored much of the biology of ungulates. This approach may risk either a theoretical error or a management catastrophe. Why should this be the case, and what has underpinned this controversy?

Population density, severe weather, management options, and controversies

One problem in understanding population dynamics of ungulates is the potential for an interaction between animal density and severe weather (Bowyer et al. 1986, Sæther 1997, Bowyer et al. 2000). We argue that even where populations of ungulates are reduced to low numbers because of severe weather, there will be effects from density-dependent mechanisms. Individuals in populations near K tend to be in poor physical condition because of intense intra-specific competition (McCullough 1979). Such individuals typically are on a low nutritional plane and often exhibit low rates of reproduction and survivorship, and exacerbated wear on teeth (Skogland 1984, Bowyer et al. 1999). Populations at high densities with individuals in poor physical condition also may be more susceptible to diseases and parasites than those at lower densities and on higher nutritional planes (Eve and Kellogg 1977, Sams et al. 1996). Conversely, animals in populations that are at or below MSY characteristically are in good physical condition resulting from limited intra-specific competition and display high rates of productivity and survival (Fig. 9.7).

Animals in poor physical condition are more likely to be helped or hindered by a variable climate than are animals on a high nutritional plane, which are better buffered against climatic extremes (Fig. 9.8, Bowyer et al. 2000). Consequently, correlations between climatic variables and measures of physical condition, fecundity, or survivorship of animals are

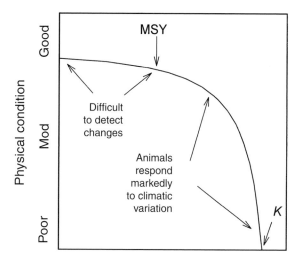

Fig. 9.7. Relationship between physical condition of individuals and population size. Note that changes in physical condition and subsequent effects on reproduction and survivorship may be difficult to detect below maximum sustained yield (MSY), whereas changes in population size between MSY and carrying capacity (K) can be marked.

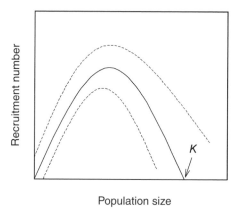

Fig. 9.8. Relationship between increasing variation in recruitment of young and population size of adult females (adapted from McCullough 1979, Bowyer et al. 2000). Variation in recruitment increases from maximum sustained yield (peak of parabola) and carrying capacity (K) because a variable environment has a disproportional ability to help or hinder ungulates in poor physical condition (Fig. 9.7).

expected to be stronger for populations at or near K compared with animals in better physical condition from populations well below K (Figs. 9.7 and 9.8). Hence, a population at K could exhibit high correlations between population parameters and climatic variables. Such an outcome might be interpreted as density-independent limitation when density-related effects are regulating that population. Cause and effect cannot be inferred from correlation (Bowyer et al. 1988). Likewise, teasing apart effects of density from climate in evaluating population parameters may be problematic. Partial correlations for weather-related variables often are robust because they possess a greater range of values than does population density, which often fluctuates about some long-term K.

How then can density-dependent and density-independent effects be discriminated? Assuming a variable climate, a reduction in variance of climate-related effects on population parameters with a substantial lowering of population density with respect to K would be evidence of density dependence (Fig. 9.9); a high and equal variance across a wide range of population densities would be support for density independence. This latter outcome means that climatic effects are so severe they have overwhelmed the ability even of animals in good physical condition to

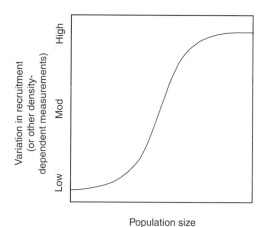

Fig. 9.9. Predicted variation in recruitment or other density-dependent (DD) measurements with increasing population size under conditions of density dependence. The increased variability is caused by interactions between changing physical condition of individuals with increasing population size (Figs. 9.8 and 9.9) and variable environmental conditions. A population limited by density-independent factors such as severe weather should exhibit uniformly high variability across a wide range of population sizes.

compensate by using body reserves. We caution, however, that animals on a sufficiently high nutritional plane may not respond as expected to severe winter conditions with reduced reproduction (Bowyer et al. 1998), or may rebound quickly from overwinter mortality. Ungulates possess an impressive array of adaptation for coping with winter (Telfer and Kelsall 1984).

Long-term data sets necessary to examine factors underpinning population regulation are seldom available (McCullough 1990). Yet, understanding causes of mortality and recruitment are essential for managing populations of ungulates. For instance, low productivity could be dealt with effectively by increasing harvest if a population were experiencing strong density dependence (assuming that was a primary management objective), but such a solution could send populations limited by severe weather or predation spiraling toward a low-density equilibrium or even extirpation. We further caution that too heavy a harvest, even where strong density dependence occurred, could lower recruitment number and population size, driving the population below MSY. This outcome could potentially cause managers to conclude that there was not a density-dependent response; for example, that population size and number of recruits declined rather than increased with increasing harvest (Bowyer et al. 1999). Likewise, too light a harvest relative to K, or harvesting a population that had overshot K, would lead to a similar conclusion – no density-dependent response when that phenomenon was regulating the population. Too often management agencies reduce harvest to compensate for low recruitment without knowing where the population is with respect to K. Where such a population is approaching K, that management results in a further depression in recruitment in the following year as the population nears K (Fig. 9.6). Clearly, change in population size is a necessary metric for assessing density dependence, but should not be used as the sole criterion.

Other pitfalls exist in managing populations of ungulates. If a population exhibits density independence and is reduced by climatic extremes to some low number from which it typically recovers under more hospitable conditions, then that surplus constitutes the allowable harvest (assuming compensatory mortality) – the surplus often determines the harvest under conditions of density independence (Leopold 1933). Conversely, under density dependence, the harvest determines the surplus. For example, recruitment number would increase as the population was harvested back from K toward MSY (Fig. 9.6). Moreover, the sex ratio of the harvest would be relatively unimportant in the dynamics of a population experiencing strong density independence. This occurs because severe density-independent events are thought to kill without regard to sex, age, or physical condition (animals that constituted the surplus would die from causes other than hunting anyway). Under density-dependent circumstances, a far different situation occurs.

Recruitment rate of young for populations of ungulates exhibiting strong density dependence is related primarily to the density of adult females rather than adult males (McCullough 1979, 1984, Bowyer et al. 1999). That outcome occurs because the genders of polygynous ungulates sexually segregate for much of the year (Bowyer 1984, Bleich et al. 1997, Kie and Bowyer 1999, Barboza and Bowyer 2000, Bowyer et al. 2001). Hence, females compete more intensely with other females and young for resources than do adult males because of the spatial separation of the sexes outside rut. That difference in life-history characteristics between the sexes has profound consequences for management of ungulates, especially the sex ratio of the harvest. Populations at or near K will not respond in a density-dependent fashion from a harvest of only males (McCullough 1979, 1984). The population cannot be reduced from K toward MSY with a male-only harvest, and males become a progressively smaller portion of the total population from such a harvest regime (McCullough 1984). Consequently, obtaining the expected increase in recruitment number as a population is reduced from K toward MSY requires a harvest of adult females (McCullough 2001). At MSY, a wider variety of sex ratios can be harvested than at very low or high population densities (McCullough 1979). Harvesting females at densities below a population size that will yield MSY results in a lower recruitment number. Males should dominate in such harvests until the population exceeds the size that will produce the maximal number of recruits (MSY, Fig. 9.6). Thus, the sex ratio of the harvest for populations exhibiting density dependence should vary depending upon where the population is with respect to MSY – a daunting concept to explain to some hunters. As a further confounding factor, management agencies often judge the success of a hunting season by the total harvest (where harvest is the primary management objective). In contrast, hunters may rely more on the number of animals seen in the field or the effort necessary to kill an animal as a measure of satisfaction (Gross 1972, Lautenschlager and Bowyer 1985). For density-dependent populations, those outcomes are inversely related from K to MSY – another potential source of contention in managing ungulates.

Table 9.2. *Life-history characteristics of ungulates that reflect the relative differences in a population at maximum sustained yield (MSY) and at carrying capacity (K)*

Life-history characteristic	Population size at or below MSY	Population size at or near K
Physical condition of adult females	better	poorer
Pregnancy rate of adult females	higher	lower
Pause in annual production by adult females	less likely	more likely
Yearlings pregnant[a]	usually	seldom
Corpora lutea counts of adult females[a]	higher	lower
Litter size[a]	higher	lower
Age at first reproduction for females	younger	older
Weight of neonates	heavier	lighter
Survivorship of young[b]	higher	lower
Age at extensive tooth wear	older	younger

[a] Some species of ungulates may show limited variability in particular characteristics.
[b] In the absence of efficient predators.

Judging the relation of population size to K

One method to help calibrate where a population is in relation to K is to evaluate measures of animal condition and reproduction with respect to population size (Table 9.2). We admonish that such an approach has limitations, but may offer the only data readily available to help determine harvest or to decide whether to manipulate habitat. Measures of animal condition tend to lag behind the ability of the habitat to support them. Thus, a population irruption that resulted in an overshoot of K (Leopold 1943, Klein 1968, Caughley 1970) might be detected far too late to implement meaningful management. Such an overshoot holds the potential to reduce K, further confounding interpretation of animal condition. We discuss the potential for low to moderate densities of herbivores to have a positive effect on their forage later in this chapter. Moreover, an overharvested population may exhibit limited variability in some parameters (for example, litter size may become fixed at some maximum) as the population is driven from MSY toward low density or extirpation. This outcome also makes estimating where the population is with respect to K difficult. Forage-based measures of K exist (Hobbs et al. 1982, Stewart et al. 2000) as do indices of habitat quality based on foraging intensity (Riney 1982). Nonetheless, some metrics require a substantial knowledge of the habitat and nutritional requirements of animals to be of value. Others, such as indices of overgrazing and hedging of trees and shrubs, often lag behind population density of ungulates (Caughley 1977).

K should not be considered a seasonal phenomenon. Most northern ungulates cannot obtain a maintenance diet during winter because of the low quality of forage during that season (Mautz 1978). Body reserves accumulated during spring and summer are essential for reproduction and may help buffer the animal against extreme conditions in winter (Schwartz and Hundertmark 1993, Bowyer et al. 2000). Variation in corpora lutea counts with population density (Teer et al. 1965) would not occur if forage in spring and summer were not in short supply. The world is not completely green (Slobodkin et al. 1967); not all plants are suitable forage for herbivores (Bryant and Chapin 1986). Similarly, inadequate winter forage may exhaust body reserves too rapidly to allow even animals in good condition to survive severe winter weather (Mautz 1978). Thus, animals may balance the need for resources acquired during the growing season with the availability of winter forage. Consequently, K might be set by a variety of conditions related to the quality of summer and winter ranges (Bowyer et al. 1986, Schwartz and Hundertmark 1993). K cannot be determined from either summer or winter alone in a seasonal environment.

Habitat manipulation and population density

Effective manipulation of habitat requires knowledge about the size of the ungulate population with respect to K. For instance, a population held well below K by predation is not limited by available forage, and manipulation of habitat to enhance animal numbers will not likely be successful (Fig. 9.10). A time lag may exist from when the manipulation occurs and the maximal production of forage (Weixelman et al. 1998). But, unless further management is implemented to eliminate other limiting or regulating factors (such as predation or harvest), manipulations of habitat likely will fail to produce the expected results (Fig. 9.10). Conversely, appropriate manipulation of habitat to enhance animal numbers for populations near K holds great promise for increasing population size and allowable harvest (Fig. 9.10). Clearly, the lack of consideration of density-dependent processes can lead to management failures and undermine public confidence in future manipulations of habitat.

Predation and population density

Determining where an ungulate population is in relation to K (Table 9.2) is also key to understanding when management of predators (Gasaway

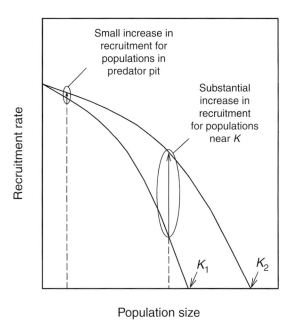

Fig. 9.10. Variation in recruitment rate (young/adult female) with increasing population size relative to changes in carrying capacity (K) caused by habitat manipulation. Note that a substantial improvement in recruitment rate occurs as habitat manipulation such as fire or logging increases carrying capacity from K_1 to K_2. A much smaller change in recruitment would result for populations held at low densities by predation (in a predator pit) because those individuals are not limited nutritionally.

et al. 1983, 1992, Hatter and Jans 1994, Van Ballenberghe and Ballard 1994, Boertje et al. 1996) is likely to affect population dynamics of their primary prey. This perspective is necessary because whether mortality of prey is compensatory or additive changes with population density (Fig. 9.11). Females attempt to produce more young than the environment can support at densities between MSY and K, an antithetical result for those who believe in behavioral regulation of ungulate populations or in group selection (McCullough 1979). Many young produced at high population densities with respect to K are born to mothers with few body reserves and, consequently, neonates exhibit low body mass and survivorship (Schwartz and Hundertmark 1993, Keech et al. 2000). In theory, the loss of another young or adult from the population allows survival of an individual that would not otherwise have been recruited into the population. Nonetheless, many of those young born at high population density are predisposed

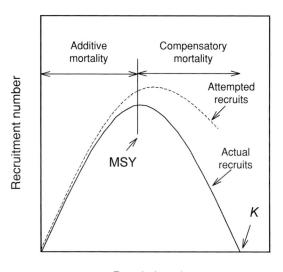

Fig. 9.11. Changes in recruitment number and attempts to recruit young with increasing population size of adult females. Note that individual females attempted to reproduce at a higher level than can be supported by the environment from densities ranging from maximum sustained yield (MSY) to carrying capacity (K), but that attempts to recruit young parallel the recruitment number below MSY because females are in good physical condition (adapted from McCullough 1979). Consequently, mortality tends to become increasingly compensatory from MSY to K (open area) but is largely additive (horizontal lines) below MSY.

to mortality from a variety of sources including predation, and would have perished anyway had they not been killed by a predator – mortality was compensatory.

As populations are reduced from K toward MSY, the proportion of mortality that is compensatory declines with the decrease in "excess" young that females attempt to foster. Females are in sufficiently good physical condition at low densities that many reproductive attempts are successful (Fig. 9.11). Indeed, for populations below MSY, most females conceive and have the necessary body reserves to successfully provision young. Thus, any decimating factor (Leopold 1933), including predation, becomes additive because those neonates would have been recruited had they not been killed. Whether one species of predator or another kills those young is inconsequential – mortality is additive. Hence, a suite of mortality factors near K, where mortality is mostly compensatory, might have a limited effect on the dynamics of an ungulate population. That same combination of mortality sources at or below MSY, where mortality is primarily additive, may have a profound effect on the dynamics of the population and unexpectedly send it declining rapidly toward low density.

Interpreting effects of predation and whether predator control is warranted to meet particular management objectives also requires information about the relationship of the population to K (Table 9.2). For example, a heavy loss of young to predation in an ungulate population near K is not a cause for concern. Those neonates would not likely have been recruited in the absence of predation. Conversely, that same outcome for a population at or below MSY would indicate that predators might be regulating the population, and predator control may be needed to increase the population (assuming that was the management objective). Additive and compensatory effects of predation and harvest on dynamics of ungulate populations are complex, and cannot be understood without considering density-dependent mechanisms (Fig. 9.11).

Ecosystem processes

Ungulates are more than merely products of ecosystems. These large herbivores may serve as regulators of ecosystem processes at several scales of time and space (McNaughton 1985, McNaughton et al. 1988, Hobbs 1996, Kie and Lehmkuhl 2001, Singer et al. 2003). For example, research conducted in coniferous forests in Olympic National Park in western Washington indicated that herbivory, primarily by elk, reduced plant standing crop, increased species richness of forbs, and helped determine the distribution of several shrub species (Woodward et al. 1994, Schreiner et al. 1996). Although the presence of a large number of ungulates in a single herd at a specific location may have profound effects on vegetation, herbivory by large herbivores more strongly affects ecosystems through modification of basic processes, including rates of nutrient turnover, competitive interactions among plant species, and rates and trajectories of successional pathways (Hobbs 1996, Pastor et al. 1997, Augustine and McNaughton 1998). Thus, grazing and browsing by ungulates act primarily as chronic rather than as episodic disturbances, although irruptions and crashes of some populations have occurred episodically (Klein 1968). Population densities of ungulates are also important to consider when examining effects of herbivory (Bowyer et al. 1997). Densities of herbivores and intensity of their foraging may determine whether herbivory increases nutrient cycling and plant productivity (Molvar et al. 1993,

Kielland et al. 1997) or negatively affects plant communities by driving changes in successional pathways (Pastor et al. 1997).

Effects of ungulates on nutrient cycling

Ungulates can increase rates of nitrogen mineralization through consumption of palatable plant species and subsequent fertilization via urine and feces (McNaughton 1992, Seagle et al. 1992, Molvar et al. 1993, Pastor and Cohen 1997). Ungulates increase nitrogen (N) cycling by adding readily available N for microbes, positively affecting conditions for N mineralization, and by changing litter quality (Hobbs 1996, Kielland et al. 1997). Deposition of urine and feces by ungulates affects nitrogen cycling by offering an accelerated alternative to decomposition as a pathway for nitrogen turnover, because nutrients from urine and feces are in forms that are readily accessible by plants (Ruess and McNaughton 1987, 1988, Ruess et al. 1989, Pastor et al. 1993, Frank et al. 1994, Hobbs 1996). Indeed, plants with an evolutionary history of grazing show elevated responses in growth to urea and ammonia, relative to other forms of inorganic N (Ruess 1984, Ruess and McNaughton 1987). Moose browsing on diamondleaf willow (*S. pulchra*) in Alaska increased the rate of nitrogen turnover (Molvar et al. 1993) and litter decomposition (Kielland et al. 1997). Moreover, moose browsing on paper birch (*Betula resinifera*) resulted in more rapid processing of litter by stream insects (Irons et al. 1991).

Nitrogen in feces is positively correlated with forage quality, and fecal nitrogen for cervids is known to increase with increasing forage quality (Leslie and Starkey 1985, Hodgman and Bowyer 1986, Bowyer et al. 1997). High-quality forage promotes fecal decomposition and nitrogen mineralization by microbes, which reduces ammonia losses via volatilization (Ruess et al. 1989). Thus, selection for high-quality forage also has effects beyond the level of herbivore nutrition (Ruess and McNaughton 1987, Ruess et al. 1989, Turner 1989, Frank and McNaughton 1993, Hobbs 1996, Frank and Evans 1997). Grazing by large herds of migratory ungulates such as elk and bison increased the rates of N mineralization in soils in Yellowstone National Park (Frank et al. 1994, Frank and Evans 1997). Moreover, Day and Detling (1990) reported that bison fed selectively on plants that had previously been fertilized by urine.

Ungulate movements within home ranges and following migration tend to concentrate nutrient deposition, because ungulates do not use their environment uniformly (Hilder and Mottershead 1963, McNaughton 1983, 1985, Senft et al. 1987, Coughenour 1991, Ward and

Saltz 1994, Hobbs 1996) and feces tend to be deposited in areas of greater use (Etchberger et al. 1988). Consequently, N consumed over large areas often becomes concentrated spatially and amplifies nutrient returns in areas selected by ungulates (Ruess and McNaughton 1987, Ruess et al. 1989, Turner 1989, Frank and McNaughton 1993, Hobbs 1996, Frank and Evans 1997). For instance, migrations of elk during winter resulted in net movement of N from summer to winter ranges (Frank and McNaughton 1993, Frank et al. 1994, Hobbs 1996).

In forested systems, where foraging is concentrated on woody plants, selective browsing may reduce nutrient cycling (Bryant et al. 1991), particularly when ungulates are at high population densities (Bowyer et al. 1997). Forage selection and litter decomposition are determined by nutrients, structural carbohydrates, lignin, and secondary metabolites (Meetemeyer 1978, Swift et al. 1979, Melillo et al. 1982, Flanagan and Van Cleve 1983, Pastor et al. 1984, Bryant and Chapin 1986, Pastor et al. 1988, Bryant et al. 1991, Pastor et al. 1993). Plant species with high levels of secondary metabolites are not only poor-quality food for herbivores, but also produce low-quality litter. This is because the same chemical properties, such as secondary metabolites, lignin, waxes and cutins, and low nitrogen concentrations, that reduce their quality as food also reduce their value for soil microbes (Flanagan and Van Cleve 1983, Moore 1984, Pastor et al. 1984, 1988, Bryant et al. 1991, Pastor et al. 1993). Soil fertility also declines as the biomass of unpalatable species sequesters increasing amounts of nutrients and as decomposition rates decline because of low litter quality (Bryant et al. 1991). Moreover, chemical defenses of many woody species are correlated directly with their ability to tolerate nutrient stress, and slow-growing species such as white spruce (*Picea glauca*) can persist in nutrient-deficient soils (Bryant et al. 1983, Coley et al. 1985). Those shifts in the dominance of woody species from those that are palatable to herbivores to species that are more chemically defended and less palatable have resulted from high population densities of moose (Pastor et al. 1988).

Important components of determining effects of herbivory on ecosystems are nutrient supplies within the system (Bryant et al. 1991) and population densities of herbivores that feed upon them (Bowyer et al. 1997). Moose populations in interior Alaska are often held at low population densities by predation (Gasaway et al. 1983, 1992), which may prevent heavy browsing and weaken selection toward the dominance of unpalatable species (Molvar et al. 1993). Molvar et al. (1993) suggested that moose act as keystone herbivores that mediate rates of nutrient cycling in northern ecosystems. This observation is consistent with studies of moose herbivory in northern ecosystems, where browsing by moose at low population density increases rates of N turnover (Molvar et al. 1993, Kielland et al. 1997), plant productivity (Molvar et al. 1993), and litter decomposition (Kielland et al. 1997). Conversely, moose at high population densities influenced successional processes, shifting communities from palatable hardwoods to unpalatable conifers with high levels of secondary defense compounds. This ultimately led to slower cycling of nutrients in the systems (Bryant et al. 1991, McInnes et al. 1992, Pastor et al. 1997).

Effects of ungulates on plant productivity

The herbivore-optimization model (Fig. 9.12) predicts that one effect of moderate levels of herbivory on plant production is the enhancement of net primary production of forage plants over ungrazed plants (McNaughton 1979, 1983, 1985, 1986, McNaughton et al. 1988, Georgiadis et al. 1989, Hik and Jefferies 1990, Frank and McNaughton 1992, 1993). Where grazing and browsing by ungulates act to increase the cycling of a nutrient in limited supply, the result may be an overall increase in

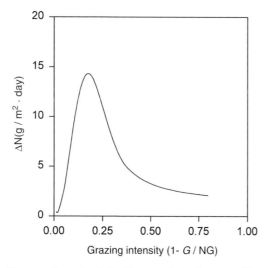

Fig. 9.12. Example of a herbivore optimization curve, illustrating the relationship between increase in grazing intensity and change in above-ground productivity in grasslands, where G = biomass in grazed areas, NG = biomass in ungrazed areas, and ΔN = increase in productivity measured in grams per square meter per day (after McNaughton 1979).

annual net primary productivity (ANPP; McNaughton 1976, 1979, 1983, 1992, Seagle et al. 1992, Frank and Evans 1997, de Mazancourt et al. 1998). Milchunas and Lauenroth (1993) reported that 17% of studies they reviewed showed elevated ANPP in areas grazed compared to areas where grazing was excluded. Williams and Haynes (1995) also noted increased rates of herbage production and increased soil nitrate, phosphate, and organic carbon with inputs of feces compared to control plots.

Adaptation of plants resulting from an evolutionary history of herbivory that provides capacity for re-growth following defoliation involves many factors. These include low stature, deciduous leaves, rhizomatous growth, linear leaf elongation from intercalary meristems, high shoot density, below-ground nutrient reserves, and rapid rates of transpiration and photosynthesis (Coughenour 1985, Hobbs 1996). Moreover, browsing in autumn and winter released stems from apical dominance, resulting in re-growth of larger stems with larger leaves during the following spring and summer (Bergstrom and Danell 1987, Molvar et al. 1993, Bowyer and Bowyer 1997). Likewise, vegetative reproduction by adventitious growth appeared to be enhanced by browsing (Grime 1977). Moose browsing on diamondleaf willow in interior Alaska caused significant increases in the growth of willow stems and leaves (Molvar et al. 1993). Moreover, moose exhibited preference for re-growth of willow stems that had been browsed the previous winter (Bowyer and Bowyer 1997). Compensatory growth in plants is also dependent upon favorable environmental conditions, such as moisture and nutrient availability. Thus, soil characteristics are also important in allowing for compensatory growth following defoliation (Mack and Thompson 1982).

Effects of ungulates on succession and plant species diversity

Browsing by high densities of ungulates can result in decreases in the diversity of plant species (Olff and Ritchie 1998, Riggs et al. 2000). Conversely, low to moderate levels of herbivory by ungulates can enhance species diversity in plant communities by directly reducing the abundance of preferred forages and indirectly influencing competitive interactions among plants (Hobbs 1996, Augustine and McNaughton 1998). Selective browsing can reduce plant species diversity by changing the composition of communities dominated by woody plant species, often increasing the abundance of unpalatable species (Bryant et al. 1991).

In general, plant species richness decreases with even moderate levels of herbivory in nutrient-poor systems and increases in nutrient-rich systems (Proulx and Mazumder 1998). High population densities of large herbivores can contribute to the dominance of conifers and grazing-resistant or unpalatable species (Bryant et al. 1991, Bowyer et al. 1997, Kielland et al. 1997). For example, high population densities of moose on Isle Royale have existed for long periods of time. Selective feeding on deciduous species of willow, aspen (*Populus* sp.), birch (*Betula* sp.) and other palatable hardwoods has resulted in forest composition that has become dominated by white spruce (Pastor et al. 1997, Jordan et al. 2000). Similarly, intense browsing by white-tailed deer in deciduous forests of the eastern U.S. shifted the herbaceous layer toward grasses, unpalatable herbaceous plants, and browse-tolerant woody plants (Tilghman 1989, McShea and Rappole 2000). Browsing by elk and mule deer in the Blue Mountains of Eastern Oregon and Washington resulted in the suppression of palatable understory shrubs when compared with exclosures (Riggs et al. 2000).

Effects of ungulates on other wildlife

Successional changes across landscapes and changes in plant species composition resulting from ungulate herbivory modify habitat for other species of wildlife. Dense populations of white-tailed deer in the northeastern U.S. altered forest stand development by reducing or eliminating young tree seedlings, shrubs, and herbaceous plants (Tilghman 1989, deCalesta 1994, McShea and Rappole 2000). Reduction in the height of woody vegetation with increased density of white-tailed deer reduced the abundance and species richness of birds that traditionally nest at intermediate levels of the canopy (deCalesta 1994). Moreover, deer densities >7.9 deer/km^2 had a significant negative effect on bird populations (de Calesta 1994, McShea and Rappole 2000).

How successional changes in the northwestern U.S. affect songbirds or small mammals has yet to be determined. Effects from loss of habitat with reduction in deciduous species of woody vegetation, however, likely reduce or remove habitat components necessary to other species of wildlife. Many neotropical migrants are relatively well represented in grand fir (*Abies grandis*) communities in the Blue Mountains of the Pacific Northwest. Although some species nest in forest canopies, many use shrubs and saplings either in burns, clear-cuts or under mid- to late-successional canopy in coniferous forest (Riggs et al. 2000). Some shrub-nesting species are relatively sparse or scarce in the Blue Mountains, and are likely linked to herbivory by ungulates as has been reported for deciduous forests in the eastern U.S. (deCalesta 1994, McShea and Rappole 2000, Riggs et al. 2000).

Research needs

Use of forested habitats by ungulates has been studied extensively in the past. Additional research is needed, however, to understand how landscape patterns and characteristics affect behavior, life-history strategies, and other physiological and ecological traits of ungulates. Also, research on the influences of large-scale fires on habitat quality and long-term trends in ungulate populations is needed. Studies also will be needed to determine how post-fire management and rehabilitation practices affect ungulates (Mackie et al. in press).

Determining where a population of ungulates is with respect to K still remains a major challenge. That knowledge is essential for the wise management of populations and in determining when predator control may be effective. Interactions with harvest and severe weather may further complicate matters. Sorting among those factors and how they relate to the size of ungulate populations in relation to K is a fundamental management need. Moreover, there is a need to develop spatially explicit models for ungulate populations (Bowyer et al. in press).

Understanding how ungulates interact with their environment and their role in trophic cascades is an important research need. Evidence exists that ungulate foraging and their deposition of urine and feces may have either positive or negative effects on nutrient cycling and other ecosystem processes. Knowledge of how the population density of ungulates influences those processes is essential. We cannot manage all components of diverse ecosystems, and defining the role of keystone species, such as ungulates, is critical to the future management of coniferous forests (Bower et al. in press).

Summary

Wild ungulates play important roles in coniferous forests throughout western North America. Not only do they respond to habitat changes at the landscape scale, but they also have important effects on basic ecosystem processes, thereby acting as keystone species. Furthermore, the dynamics of ungulate populations are complex. A complete understanding of how ungulate numbers are regulated over time requires consideration of interactions among climate, predation, and density dependence. The failure to integrate all of these concepts into a cohesive management framework is likely to lead to an incomplete understanding of their role in forested ecosystems.

Acknowledgments

The need to distribute writing tasks among three authors resulted in the galling necessity to divide this chapter into three parts (Caughley 1976), corresponding to habitat relationships (J.G.K.), population dynamics (R.T.B.), and ecosystem processes (K.M.S.). As a result, this chapter was a true joint effort. Susan Lindstedt provided artwork for Figs. 9.1 and 9.2. We thank Kurt Jenkins, Michael Wisdom, and Paul Krausman for valuable comments on an earlier draft of this manuscript. The errors that remain, however, are ours.

Literature cited

Apps, C.D., B.N. McLellan, T.A. Kinley, and J.P. Flaa. 2001. Scale-dependent habitat selection by mountain caribou, Columbia Mountains, British Columbia. *Journal of Wildlife Management* **65**:65–77.

Augustine, D.J. and S.J. McNaughton. 1998. Ungulate effects on the functional species composition of plant communities: herbivore selectivity and plant tolerance. *Journal of Wildlife Management* **62**:1165–1183.

Austin, D.D. and P.J. Urness. 1993. Evaluating production losses due to depredating big game. *Wildlife Society Bulletin* **21**:397–401.

Bailey, J.A. 1984. *Principals of Wildlife Management*. John Wiley and Sons, New York, New York, USA.

Baker, R.H. 1984. Origin, classification, and distribution. Pages 1–18 *in* L. K. Halls, editor. *White-tailed Deer: Ecology and Management*. Stackpole Books, Harrisburg, Pennsylvania, USA.

Bandy, P.J. and R.D. Taber. 1974. Forest and wildlife management: conflict and coordination. Pages 21–26 *in* H.C. Black, editor. *Wildlife and Forest Management in the Pacific Northwest*. Oregon State University, Corvallis, Oregon, USA.

Barboza, P.S. and R.T. Bowyer. 2000. Sexual segregation in dimorphic deer: a new gastrocentric hypothesis. *Journal of Mammalogy* **81**:473–489.

Barboza, P.S. and R.T. Bowyer. 2001. Seasonality of sexual segregation in dimorphic deer: extending the gastrocentric model. *Alces* **37**:275–292.

Bartmann, R.M., G.C. White, and L.H. Carpenter. 1992. Compensatory mortality in a Colorado mule deer population. *Wildlife Monographs* **121**:1–39.

Beale, D.M. and N.W. Darby. 1991. Diet composition of mule deer in mountain brush habitat of southwestern Utah. Utah Division of Wildlife Resources, Publication 91–14:1–70.

Bergerud, A.T. 1992. Rareness as an antipredator strategy to reduce predation risk for moose and caribou. Pages 1008–1020 *in* D.R. McCullough and R.H. Barrett, editors. *Wildlife 2001: Populations*. Elsevier, New York, New York, USA.

Bergerud, A.T., W. Wyett, and B. Snider. 1983. The role of wolf predation in limiting a moose population. *Journal of Wildlife Management* **47**:977–988.

Bergstrom, R. and K. Danell. 1987. Effects of simulated winter browsing by moose on morphology and biomass of two birch species. *Journal of Ecology* **75**: 533–544.

Black, H., Jr., R.J. Scherzinger, and J.W. Thomas. 1976. Relationships of Rocky Mountain elk and Rocky Mountain mule deer habitat to timber management in the Blue Mountains of Oregon and Washington. Pages 11–13 *in* S.R. Heib, editor. *Proceedings of the Elk-Logging-Roads Symposium*. Forest, Wildlife, and Range Experiment Station, University of Idaho, Moscow, Idaho, USA.

Bleich, V.C., R.T. Bowyer, and J.D. Wehausen. 1997. Sexual segregation in mountain sheep: resources or predation? *Wildlife Monographs* **134**:1–50.

Boertje, R.D., P. Valkenburg, and M.E. McNay. 1996. Increases in moose, caribou, and wolves following wolf control in Alaska. *Journal of Wildlife Management* **60**: 474–489.

Boroski, B.B. and A.S. Mossman. 1996. Distribution of mule deer in relation to water sources in northern California. *Journal of Wildlife Management* **60**:770–776.

Bowyer, J.W. and R.T. Bowyer. 1997. Effects of previous browsing on the selection of willow stems by Alaskan moose. *Alces* **33**:11–18.

Bowyer, R.T. 1984. Sexual segregation in southern mule deer. *Journal of Mammalogy* **65**:410–417.

Bowyer, R.T., M.E. Shea, and S.A. McKenna. 1986. The role of winter severity and population density in regulating northern populations of deer. Pages 193–204 *in* J.A. Bissonette, editor. *Is Good Forestry Good Wildlife Management?* Maine Agricultural Experiment Station miscellaneous publication No. 689.

Bowyer, R.T., S.C. Amstrup, J.G. Stahmann, P. Reynolds, and F. Burris. 1988. Multiple regression methods for modeling caribou populations. *Proceedings of the North American Caribou Workshop* **3**:89–118.

Bowyer, R.T., J.G. Kie, and V. Van Ballenberghe. 1996. Sexual segregation in black-tailed deer: effects of scale. *Journal of Wildlife Management* **60**:10–17.

Bowyer, R.T., V. Van Ballenberghe, and J.G. Kie. 1997. The role of moose in landscape processes: effects of biogeography, population dynamics, and predation. Pages 265–287 *in* J.A. Bissonette, editor. *Wildlife and Landscape Ecology: Effects and Patterns of Scale*. Springer-Verlag, New York, New York, USA.

Bowyer, R.T., V. Van Ballenberghe, and J.G. Kie. 1998. Timing and synchrony of parturition in Alaskan moose: long-term versus proximal effects of climate. *Journal of Mammalogy* **79**:1332–1334.

Bowyer, R.T., M.C. Nicholson, E.M. Molvar, and J.B. Faro. 1999. Moose on Kalgin Island: are density-dependent processes related to harvest? *Alces* **35**:73–89.

Bowyer, R.T., D.M. Leslie, Jr., and R.L. Rachlow. 2000. Dall's and Stone's sheep. Pages 491–516 *in* S. Demarais, and P.R. Krausman, editors. *Ecology and Management of Large Mammals in North America*. Prentice Hall, Upper Saddle River, New Jersey, USA.

Bowyer, R.T., B.M. Pierce, L.K. Duffy, and D.A. Haggstrom. 2001. Sexual segregation in moose: effects of habitat manipulation. *Alces* **37**:109–122.

Bowyer, R.T., V. Van Ballenberghe, and J.G. Kie. in press. Moose. Pages 000–000 *in* G. Feldhamer, B. Thompson, and J. Chapman, editors. *Wild Mammals of North America*. Johns Hopkins University Press, Baltimore, Maryland, USA.

Boyce, M.S. 1989. *The Jackson Elk Herd: Intensive Wildlife Management in North America*. Cambridge University Press, New York, New York, USA.

Bryant, J.P., and F.S. Chapin III. 1986. Browsing-woody plant interactions during boreal forest plant succession. Pages 313–325 *in* K. Van Cleve, F.S. Chapin III, P.W. Flanagan, L.A. Viereck, and C.T. Dyrness, editors. *Forest Ecosystems in the Alaskan Taiga*. Springer-Verlag, New York, New York, USA.

Bryant, J.P., F.S. Chapin III, and D.R. Klein. 1983. Carbon/nutrient balance of boreal plants in relation to vertebrate herbivory. *Oikos* **40**:357–368.

Bryant, J.P., F.D. Provenza, J. Pastor, P.B. Reichardt, T.P. Clausen, and J.T. du Toit. 1991. Interactions between woody plants and browsing mammals mediated by secondary metabolites. *Annual Review of Ecology and Systematics* **22**:431–446.

Bryant, L.D. and C. Maser. 1982. Classification and distribution. Pages 1–59 *in* J.W. Thomas and D.E. Toweill, editors. *Elk of North America – Ecology and Management*. Stackpole Books, Harrisburg, Pennsylvania, USA.

Caughley, G. 1970. Eruption of ungulate populations, with emphasis on Himalayan thar in New Zealand. *Ecology* **51**:54–72.

Caughley, G. 1974. Interpretation of age ratios. *Journal of Wildlife Management* **38**:557–562.

Caughley, G. 1976. Wildlife management and the dynamics of ungulate populations. Pages 183–246 *in* T.H. Coaker, editor. *Applied Biology*. Volume 1. Academic Press, London, UK.

Caughley, G. 1977. *Analysis of Vertebrate Populations*. John Wiley and Sons, New York, New York, USA.

Clark, T.P. and F.F. Gilbert. 1982. Ecotones as a measure of deer habitat quality in central Ontario. *Journal of Applied Ecology* **19**:751–758.

Clutton-Brock, T.H. 1991. *The Evolution of Parental Care*. Princeton University Press, Princeton, New Jersey, USA.

Clutton-Brock, T.H., G.R. Iason, and F.E. Guiness. 1987*a*. Sexual segregation and density-related changes in habitat use in male and female red deer (*Cervus elaphus*). *Journal of Zoology (London)* **211**:275–289.

Clutton-Brock, T.H., S.D. Albon, and F.E. Guiness. 1987*b*. Interactions between population density and maternal characteristics affecting fecundity and juvenile survival in red deer. *Journal of Animal Ecology* **56**:857–871.

Coley, P.D., J.P. Bryant, and F.S. Chapin, III. 1985. Resource availability and plant antiherbivore defense. *Science* **230**:895–899.

Conover, M.R. 1997. Wildlife management by metropolitan residents in the United States: practices, perceptions, costs, and values. *Wildlife Society Bulletin* **25**:306–311.

Conradt, L. 1998*a*. Measuring the degree of sexual segregation in group-living mammals. *Journal of Animal Ecology* **67**:217–226.

Conradt, L. 1998*b*. Could asynchrony in activity between sexes cause intersexual social segregation in ruminants? *Proceedings of the Royal Society of London (B)* **265**:1359–1363.

Conradt, L., T.H. Clutton-Brock, and D. Thomson. 1999. Habitat segregation in ungulates: are males forced into suboptimal foraging habitats through indirect competition by females? *Oecologia* **119**:367–377.

Cook, J.G., L.L. Irwin, L.D. Bryant, R.A. Riggs, and J.W. Thomas. 1998. Relations of forest cover and condition of elk: a test of the thermal cover hypothesis in summer and winter. *Wildlife Monographs* **141**:1–61.

Coughenour, M.B. 1985. A mechanistic simulation analysis of water use, leaf angles, and grazing in East African graminoids. *Ecological Modeling* **26**:101–134.

Coughenour, M.B. 1991. Spatial components of plant-herbivore interactions in pastoral, ranching, and native ungulate ecosystems. *Journal of Range Management* **44**:530–542.

Darwin, C. 1871. *The Descent of Man and Selection in Relation to Sex*. The Modern Library, New York, New York, USA.

Day, T.A. and J.K. Detling. 1990. Grassland patch dynamics and herbivore grazing preference following urine deposition. *Ecology* **71**:180–188.

de Mazancourt, C., M. Loreau, and L. Abbadie. 1998. Grazing optimization and nutrient cycling: when do herbivores enhance plant production? *Ecology* **79**:2242–2252.

deCalesta, D.S. 1994. Effects of white-tailed deer on songbirds within managed forests in Pennsylvania. *Journal of Wildlife Management* **58**:711–718.

Etchberger, R.C., R. Mazaika, and R.T. Bowyer. 1988. White-tailed deer, *Odocoileus virginianus*, fecal groups relative to vegetation biomass and quality in Maine. *Canadian Field-Naturalist* **102**:671–674.

Eve, J.H. and F.E. Kellogg. 1977. Management implications of abosomal parasites in southeastern white-tailed deer. *Journal of Wildlife Management* **41**:169–177.

Flanagan, P.W. and K. Van Cleve. 1983. Nutrient cycling in relation to decomposition of organic matter quality in taiga ecosystems. *Canadian Journal of Forest Restoration* **13**:795–817.

Fowler, C.W. 1981. Comparative population dynamics in large mammals. Pages 437–455 *in* C.W. Fowler and T.D. Smith, editors. *Dynamics of Large Mammal Populations*. John Wiley and Sons, New York, New York, USA.

Fox, K.B. and P.R. Krausman. 1994. Fawning habitat of desert mule deer. *Southwestern Naturalist* **39**:269–275.

Frank, D.A. and R.D. Evans. 1997. Effects of native grazers on grassland N cycling in Yellowstone National Park. *Ecology* **78**:2238–2248.

Frank, D.A. and S.J. McNaughton. 1992. The ecology of plants, large mammalian herbivores, and drought in Yellowstone National Park. *Ecology* **73**:2043–2058.

Frank, D.A. and S.J. McNaughton. 1993. Evidence for the promotion of aboveground grassland production in Yellowstone National Park. *Oecologia* **96**:157–161.

Frank, D.A., R.S. Inouye, N. Huntly, G.W. Minshall, and J.E. Anderson. 1994. The biogeochemistry of a north-temperate grassland with native ungulates: nitrogen dynamics in Yellowstone National Park. *Biogeochemistry* **26**:163–188.

Franzmann, A.W. 2000. Moose. Pages 578–600 *in* S. Demarais, and P.R. Krausman, editors. *Ecology and Management of Large Mammals in North America*. Prentice Hall, Upper Saddle River, New Jersey, USA.

Gasaway, W.C., R.O. Stephenson, J.L. Davis, P.E.K. Shepard, and O.E. Burris. 1983. Interrelationships of wolves, prey, and man in interior Alaska. *Wildlife Monographs* **84**:1–50.

Gasaway, W.C., S.D. Dubois, R.D. Boertje, D.J. Reed, and D.T. Simpson. 1989. Response of radio-collared moose to a large burn in central Alaska. *Canadian Journal of Zoology* **67**:325–329.

Gasaway, W.C., R.D. Boertje, D.V. Grangard, D.G. Kelleyhouse, R.O. Stephenson, and D.G. Larsen. 1992. The role of predation in limiting moose at low densities in Alaska and Yukon and implications for conservation. *Wildlife Monographs* **120**:1–59.

Gaston, J.J. and J.H. Lawton. 1987. A test of statistical techniques of detecting density dependence in sequential censuses of animal populations. *Oecologia* **74**:404–410.

Georgiadis, N.J., R.W. Ruess, S.J. McNaughton, and D. Western. 1989. Ecological conditions that determine when grazing stimulates grass production. *Oecologia* **81**:316–322.

Gill, R.B. 1976. Mule deer management myths and the mule deer population decline. Pages 99–106 *in* G.W. Workman and J.B. Low, editors. *Mule Deer Decline in the West*. College of Natural Resources, Utah State University, Logan, Utah, USA.

Goodman, D. 1981. Life history analysis of large mammals. Pages 415–436 *in* C.W. Fowler and T.D. Smith, editors. *Dynamics of Large Mammal Populations*. John Wiley and Sons, New York, New York, USA.

Grime, J.P. 1977. Evidence for the existence of three primary strategies in plants and its relevance to ecological and evolutionary theory. *American Naturalist* **111**:1169–194.

Gross, J.E. 1972. Criteria for big game planning: performance measures vs. intuition. *Transactions of the North American Wildlife and Natural Resources Conference* **37**: 246–259.

Hanley, T.A. 1983. Black-tailed deer, elk, and forest edge in a western Cascades watershed. *Journal of Wildlife Management* **47**:237–242.

Hanley, T.A. 1984. Relationships between black-tailed deer and their habitat. USDA Forest Service General Technical Report **PNW-GTR-168**. US Forest Service, Pacific Northwest Research Station, Portland, Oregon, USA.

Hanley, T.A. 1996. Potential role of deer (Cervidae) as ecological indicators of forest management. *Forest Ecology and Management* **88**:199–204.

Hanley, T.A., C.T. Robbins, and D.E. Spalinger. 1989. Forest habitats and the nutritional ecology of Sitka black-tailed deer: a research synthesis with implications for forest management. USDA Forest Service General Technical Report **PNW-GTR-230**. US Forest Service, Pacific Northwest Research Station, Portland, Oregon, USA.

Hatter, I.W. and D.W. Jans. 1994. Apparent demographic changes in black-tailed deer associated with wolf control on northern Vancouver Island. *Canadian Journal of Zoology* **72**:878–884.

Hazam, J.E. and P.R. Krausman. 1988. Measuring water consumption of desert mule deer. *Journal of Wildlife Management* **52**:528–534.

Hik, D.S. and R.L. Jefferies. 1990. Increases in the net above-ground primary production of a salt-marsh forage grass: a test of the predictions of the herbivore-optimization model. *Journal of Ecology* **78**:180–195.

Hilder, E.J., and B.E. Mottershead. 1963. The redistribution of plant nutrients through free-grazing sheep. *Australian Journal of Agricultural Science* **26**:88–89.

Hobbs, N.T. 1996. Modification of ecosystems by ungulates. *Journal of Wildlife Management* **60**:695–713.

Hobbs, N.T., D.L. Baker, J.E. Ellis, and D.M. Swift. 1982. Energy- and nitrogen-based estimates of elk winter-range carrying capacity. *Journal of Wildlife Management* **46**:12–21.

Hodgman, T.P. and R.T. Bowyer. 1986. Fecal crude protein relative to browsing intensity by white-tailed deer on wintering areas in Maine. *Acta Theriologica* **31**:347–353.

Hoffman, R.R. 1985. Digestive physiology of the deer: their morphophysiological specialisation and adaptation. Pages 393–407 *in* P.F. Fennessy and K.R. Drew. *Biology of Deer Production*. Bulletin 22, The Royal Society of New Zealand, Wellington, New Zealand.

Illius, A.W. and I.J. Gordon. 1999. The physiological ecology of mammalian herbivory. Pages 71–96 *in* H.G. Jung and G.C. Fahey, editors. *Nutritional Ecology of Herbivores*. American Society of Animal Science, Savoy, Illinois, USA.

Irons, J.G., J.P. Bryant, and M.W. Oswood. 1991. Effects of moose browsing on decomposition rates of birch leaf litter in a subarctic stream. *Canadian Journal of Fish and Acquatic Sciences* **48**:442–444.

Jenkins, K. and E. Starkey. 1996. Simulating secondary succession of elk forage values in a managed forest landscape, western Washington. *Environmental Management* **20**:715–724.

Jordan, P.A., B.E. McLaren, and S.M. Sell. 2000. A summary of research on moose and related ecological topics at Isle Royale, USA. *Alces* **36**:233–267.

Keech, M.A., R.T. Bowyer, J.M. Ver Hoef, R.D. Boertje, B.W. Dale, and T.R. Stephenson. 2000. Life-history consequences of maternal condition in Alaskan moose. *Journal of Wildlife Management* **64**:450–462.

Kie, J.G. 1999. Optimal foraging in a risky environment: life-history strategies for ungulates. *Journal of Mammalogy* **80**:1114–1129.

Kie, J.G. and R.T. Bowyer. 1999. Sexual segregation in white-tailed deer: density dependent changes in use of space, habitat selection, and dietary niche. *Journal of Mammalogy* **80**:1004–1020.

Kie, J.G. and B. Czech. 2000. Mule and black-tailed deer. Pages 629–657 *in* S. Demarais and P.R. Krausman, editors. *Ecology and Management of Large Mammals in North America*. Prentice Hall, Upper Saddle River, New Jersey, USA.

Kie, J.G. and J.F. Lehmkuhl. 2001. Herbivory by wild and domestic ungulates in the Intermountain West. *Northwest Science* **75** (special issue):55–61.

Kie, J.G. and M. White. 1985. Population dynamics of white-tailed deer (*Odocoileus virginianus*) on the Welder Wildlife Refuge, Texas. *Southwestern Naturalist* **30**:105–118.

Kie, J.G., R.T. Bowyer, M.C. Nicholson, B.B. Boroski, and E.R. Loft. 2002. Landscape heterogeneity at differing scales: effects on spatial distribution of mule deer. *Ecology* **83**:530–544.

Kielland, K., J.P. Bryant, and R.W. Ruess. 1997. Moose herbivory and carbon turnover of early successional stands in interior Alaska. *Oikos* **80**:25–30.

Kleiber, M. 1961. *The Fire of Life*. Wiley, New York, New York, USA.

Klein, D.R. 1968. The introduction, increase, and crash of reindeer on St. Matthew Island. *Journal of Wildlife Management* **32**:350–367.

Krausman, P.R., A.J. Kuenzi, R.C. Etchberger, K.R. Rautenstrauch, L.L. Ordway, and J.J. Hervert. 1997. Diets of desert mule deer. *Journal of Range Management* **50**:513–522.

Kremsater, L.L. and F.F. Bunnell. 1992. Testing responses to forest edges: the example black-tailed deer. *Canadian Journal of Zoology* **70**:2426–2435.

Lautenschlager, R.A. and R.T. Bowyer. 1985. Wildlife management by referendum: when professionals fail to communicate. *Wildlife Society Bulletin* **13**:564–570.

Lautier, J.K., T.V. Dailey, and R.D. Brown. 1988. Effect of water restriction on feed intake of white-tailed deer. *Journal of Wildlife Management* **52**:602–606.

Leopold, A. 1933. *Game Management*. Charles Scribner's Sons, New York, New York, USA.

Leopold, A. 1943. Deer irruptions. *Transactions of the Wisconsin Academy of Sciences, Arts, and Letters* **35**:351–366.

Leopold, A.S. 1978. Wildlife and forest practice. Pages 108–120 *in* H.P. Brokaw, editor. *Wildlife and America*. Council on Environmental Quality, Washington DC, USA.

Leopold, A.S., T. Riney, R. McCain, and L. Tevis, Jr. 1951. The Jawbone deer herd. *California Department of Fish and Game, Game Bulletin* **4**:1–139.

Leslie, D.M. Jr. and E.S. Starkey. 1985. Fecal indices to dietary quality of cervids in old-growth forests. *Journal of Wildlife Management* **49**:142–146.

Loomis, J., M. Creel, and J. Cooper. 1989. Economic benefits of deer in California: hunting and viewing values. Institute of Ecology Report 32, University of California, Davis, California USA.

Loveless, C.M. 1967. Ecological characteristics of a mule deer winter range. Colorado Game, Fish, and Parks Department Technical Bulletin Number 20.

Mack, R.N. and J.N. Thompson. 1982. Evolution in steppe with a few large, hoofed mammals. *American Naturalist* **119**:757–773.

Mackie, R.J., K.L. Hamlin, and D.F. Pac. 1982. Mule deer. Pages 862–877 *in* J.A. Chapman and G.A. Feldhamer, editors. *Wild Mammals of North America: Biology, Management, Economics*. Johns Hopkins University Press, Baltimore, Maryland, USA.

Mackie, R.J., K.L. Hamlin, D.E. Pac., G.L. Dusek, and A.K. Wood. 1990. Compensation in free-ranging deer populations. *Transactions of the North American Wildlife and Natural Resources Conference* **55**:518–526.

Mackie, R.J., J.G. Kie, D.F. Pac, and K.L. Hamlin. in press. Mule deer. Pages 000–000 *in* G. Feldhamer, B. Thompson, and J. Chapman, editors. *Wild Mammals of North America*. Johns Hopkins University Press, Baltimore, Maryland, USA.

Main, M.B., F.W. Weckerly, and V.C. Bleich. 1996. Sexual segregation in ungulates: new directions for research. *Journal of Mammalogy* **77**:449–461.

Mautz, W.W. 1978. Sledding on a bushy hill side: the fat cycle in deer. *Wildlife Society Bulletin* **6**:88–90.

McCullough, D.R. 1979. *The George Reserve Deer Herd: Population Ecology of a K-Selected Species*. University of Michigan Press, Ann Arbor, Michigan, USA.

McCullough, D.R. 1984. Lessons from the George Reserve, Michigan. Pages 211–242 *in* L.K. Halls, editor. *White-Tailed Deer: Ecology and Management*. Stackpole Books, Harrisburg, Pennsylvania, USA.

McCullough, D.R. 1990. Detecting density dependence: filtering the baby from the bathwater. *Transactions of the North American Wildlife and Natural Resources Conference* **55**:534–543.

McCullough, D.R. 1999. Density dependence and life-history strategies of ungulates. *Journal of Mammalogy* **80**:1130–1146.

McCullough, D.R. 2001. Male harvest in relation to female removals in a black-tailed deer population. *Journal of Wildlife Management* **65**:46–58.

McInnes, P.F., R.J. Naimen, J. Pastor, and Y. Cohen. 1992. Effects of moose browsing on vegetation and litter-fall of the boreal forest, Isle Royale, Michigan, USA. *Ecology* **73**:2059–2075.

McNaughton, S.J. 1976. Serengeti migratory wildebeest: facilitation of energy flow by grazing. *Science* **191**:92–94.

McNaughton, S.J. 1979. Grazing as an optimization process: grass-ungulate relationships in the Serengeti. *American Naturalist* **113**:691–703.

McNaughton, S.J. 1983. Serengeti grassland ecology: the role of composite environmental factors and contingency in community organization. *Ecological Monographs* **53**:291–320.

McNaughton, S.J. 1985. Ecology of a grazing ecosystem: the Serengeti. *Ecological Monographs* **55**:259–294.

McNaughton, S.J. 1986. Grazing lawns: on domesticated and wild grazers. *American Naturalist* **128**:937–939.

McNaughton, S.J. 1992. The propagation of disturbance in savannas through food webs. *Journal of Vegetation Science* **3**:301–314.

McNaughton, S.J., R.W. Ruess, and S.W. Seagle. 1988. Large mammals and process dynamics in African ecosystems. *Bioscience* **38**:794–800.

McShea, W.J. and J.H. Rappole. 2000. Managing the abundance and diversity of breeding bird populations through manipulation of deer populations. *Conservation Biology* **14**:1161–1170.

Meetemeyer, V. 1978. Macroclimate and lignin control of litter decomposition. *Ecology* **59**:465–472.

Melillo, J.M., J.D. Aber, and J.F. Muratore. 1982. Nitrogen and lignin control of hardwood leaf litter decomposition dynamics. *Ecology* **63**:621–626.

Milchunas, D.G. and W.K. Lauenroth. 1993. Quantitative effects of grazing on vegetation and soils over a global range of environments. *Ecological Monographs* **63**:327–366.

Miller, F.L. 1982. Caribou. Pages 923–959 *in* J.A. Chapman and G.A. Feldhammer, editors. *Wild Mammals of North America: Biology, Management, and Economics*. Johns Hopkins University Press, Baltimore, Maryland, USA.

Miquelle, D.G., J.M. Peek, and V. Van Ballenberghe. 1992. Sexual segregation in Alaskan moose. *Wildlife Monographs* **122**:1–57.

Molvar, E.M., R.T. Bowyer, and V. Van Ballenberghe. 1993. Moose herbivory, browse quality, and nutrient cycling in an Alaskan treeline community. *Oecologia* **94**:472–479.

Moore, T.R. 1984. Litter decomposition in a subarctic spruce-lichen woodland, eastern Canada. *Ecology* **65**:299–308.

Mysterud, A. 2000. The relationship between ecological segregation and sexual body size dimorphism in large herbivores. *Oecologia* **124**:40–54.

Nicholson, M.C., R.T. Bowyer, and J.G. Kie. 1997. Habitat selection and survival of mule deer: tradeoffs associated with migration. *Journal of Mammalogy* **78**:483–504.

Olff, H. and M.E. Ritchie. 1998. Effects of herbivores on grassland plant diversity. *Trends in Ecology and Evolution* **13**:261–265.

Parker, K.L. 1988. Effects of heat, cold, and rain on coastal black-tailed deer. *Canadian Journal of Zoology* **66**:2475–2483.

Pastor, J., and Y. Cohen. 1997. Herbivores, the functional diversity of plant species, and the cycling of nutrients in ecosystems. *Theoretical Population Biology* **51**:165–179.

Pastor, J., J.D. Aber, C.A. McClaugherty, and J.M. Melillo. 1984. Aboveground production and N and P cycling along a nitrogen mineralization gradient on Blackhawk Island, Wisconsin. *Ecology* **65**:256–268.

Pastor, J., R.J. Naimen, B. Dewey, and P. McInnes. 1988. Moose, microbes, and the boreal forest. *BioScience* **38**:770–777.

Pastor, J., B. Dewey, R.J. Naimen, P.F. McInnes, and Y. Cohen. 1993. Moose browsing and soil fertility of Isle Royale National Park. *Ecology* **74**:467–480.

Pastor, J.R., R. Moen, and Y. Cohen. 1997. Spatial heterogeneities, carrying capacity, and feedbacks in animal-landscape interactions. *Journal of Mammalogy* **78**: 1040–1052.

Patton, D.R. 1992. *Wildlife Habitat Relationships in Forested Ecosystems.* Timber Press, Portland, Oregon, USA.

Peek, J.M. 1974. Initial response of moose to a forest fire in northeastern Minnesota. *American Midland Naturalist* **91**:435–438.

Person, D.K., R.T. Bowyer, and V. Van Ballenberghe. 2001. Density dependence and functional responses of wolves: effects on predator-prey ratios. *Alces* **37**: 253–273.

Pollard, E., K.H. Lakhani, and P. Rothery. 1987. The detection of density-dependence from a series of annual censuses. *Ecology* **68**:2046–2055.

Proulx, M. and A. Mazumder. 1998. Reversal of grazing impact on plant species richness in nutrient-poor vs. nutrient-rich ecosystems. *Ecology* **79**:2581–2592.

Putman, R. 1988. *The Natural History of Deer.* Comstock Publishing Associates, Cornell University Press, Ithaca, New York, USA.

Rachlow, J.L. and R.T. Bowyer. 1994. Variability in maternal behavior by Dall's sheep: environmental tracking or adaptive strategy? *Journal of Mammalogy* **75**:328–337.

Renecker, L.A. and R.J. Hudson. 1986. Seasonal energy expenditures and thermoregulatory responses of moose. *Canadian Journal of Zoology* **64**:322–327.

Riggs, R.A., A.R. Tiedemann, J.G. Cook, T.M. Ballard, P.J. Edgerton, M. Vavra, W.C. Krueger, F.C. Hall, L.D. Bryant, L.L. Irwin, and T. Delcurto. 2000. Modification of mixed-conifer forests by ruminant herbivores in the Blue Mountains Ecological Province. USDA Forest Service Research Paper **PNW-RP-527**. US Forest Service, Pacific Northwest Research Station, Portland, Oregon, USA.

Riney, T. 1982. *Study and Management of Large Mammals.* John Wiley and Sons, New York, New York, USA.

Romin, L.A. and J.A. Bissonette. 1996. Deer-vehicle collisions: status of state monitoring activities and mitigation efforts. *Wildlife Society Bulletin* **24**:127–132.

Rosenstock, S.S., W.B. Ballard, and J.C. DeVos, Jr. 1999. Viewpoint: benefits and impacts of wildlife water developments. *Journal of Range Management* **52**:302–311.

Rowland, M.M., M.J. Wisdom, B.K. Johnson, and J.G. Kie. 2000. Elk distribution and modeling in relation to roads. *Journal of Wildlife Management* **64**:672–684.

Ruckstuhl, K.E. and P. Neuhaus. 2000. Causes of sexual segregation in ungulates: a new approach. *Behaviour* **137**:361–377.

Ruess, R.W. 1984. Nutrient movement and grazing: experimental effects of clipping and nitrogen source on nutrient uptake in *Kyllinga nervosa. Oikos* **43**:183–188.

Ruess, R.W. and S.J. McNaughton. 1987. Grazing and the dynamics of nutrient and energy regulated microbial processes in the Serengeti grasslands. *Oikos* **49**:101–110.

Ruess, R.W. and S.J. McNaughton. 1988. Ammonia volatilization and the effects of large grazing mammals on nutrient loss from East African grasslands. *Oecologia* **77**:550–556.

Ruess, R.W., D.H. Hik, and R.L. Jefferies. 1989. The role of lesser snow geese as nitrogen processors in a sub-arctic salt marsh. *Oecologia* **79**:23–29.

Sæther, B.E. 1997. Environmental stochasticity and population dynamics of large herbivores: a search for mechanisms. *Trends in Ecology and Evolution* **12**:143–149.

Sams, M.G., R.L. Lockmiller, C.W. Qualls, Jr., D.M. Leslie, Jr., and M.E. Payton. 1996. Physiological correlates of neonatal mortality in an overpopulated herd of white-tailed deer. *Journal of Mammalogy* **77**:179–190.

Sand, H. 1996. Life history patterns of female moose (*Alces alces*): the relationship between age, body size, fecundity, and environmental conditions. *Oecologia* **106**:210–220.

Schreiner, E.G., K.A. Krueger, P.J. Happe, and D.B. Houston. 1996. Understory patch dynamics and ungulate herbivory in old-growth forests of Olympic National Park, Washington. *Canadian Journal of Forest Research* **26**:255–265.

Schwartz, C.C. and K.J. Hundertmark. 1993. Reproductive characteristics of Alaskan moose. *Journal of Wildlife Management* **57**:454–468.

Seagle, S.W., S.J. McNaughton, and R.W. Ruess. 1992. Simulated effects of grazing on soil nitrogen and mineralization in contrasting Serengeti grasslands. *Ecology* **73**:1105–1123.

Senft, R.L., M.B. Coughenour, D.W. Bailey, L.R. Rittenhouse, and O.E. Sala. 1987. Large herbivore foraging and ecological hierarchies. *Bioscience* **37**:789–799.

Severson, K.E. and A.L. Medina. 1983. Deer and elk habitat management in the southwest. *Journal of Range Management Monograph Number 2*.

Shaw, J.H. and M. Meagher. 2000. Bison. Pages 447–466 *in* S. Demarais and P.R. Krausman, editors. *Ecology and Management of Large Mammals in North America*. Prentice Hall, Upper Saddle River, New Jersey, USA.

Simberloff, D. 1998. Flagships, umbrellas, and keystones: is single species management passé in the landscape era? *Biological Conservation* **83**:247–257.

Singer, F.J., A. Harting, K.K. Symonds, and M.B. Coughenour. 1997. Density dependence, compensation, and environmental effects on elk calf mortality in Yellowstone National Park. *Journal of Wildlife Management* **61**:12–25.

Singer, F.J., G. Wang, and N.T. Hobbs. 2003. The role of ungulates and large predators on plant communities and ecosystem processes in national parks. Pages 444–486 *in* C. J. Zabel and R. G. Anthony, editors. *Mammal Community Dynamics. Management and Conservation in the Coniferous Forests of Western North America*. Cambridge University Press, Cambridge, UK.

Skogland, T. 1984. Toothwear by food limitation and its life history consequences in wild reindeer. *Oikos* **51**:238–242.

Skogland, T. 1985. The effects of density-dependent resource limitations on the demography of wild reindeer. *Journal of Animal Ecology* **54**:359–374.

Slade, N.A. 1977. Statistical detection of density dependence from a series of sequential censuses. *Ecology* **58**:1094–1102.

Slobodkin, L.B., F.E. Smith, and N.G. Hariston. 1967. Regulation in terrestrial ecosystems, and the implied balance of nature. *American Naturalist* **101**:109–124.

Stearns, S.C. 1977. The evolution of life history traits. A critique of the theory and a review of the data. *Annual Review of Ecology and Systematics* **8**:145–171.

Stewart, K.M., T.E. Fulbright, and D.L. Drawe. 2000. White-tailed deer use of clearings relative to forage availability. *Journal of Wildlife Management* **64**:733–741.

Stubbs, M. 1977. Density dependence in life-cycles of animals and its importance in *K* and *r*-selected strategies. *Journal of Animal Ecology* **46**:677–688.

Swift, M.J., O.W. Heal, and J.M. Anderson. 1979. *Decomposition in terrestrial ecosystems.* Blackwell Press, Oxford, UK.

Taber, R.D. and R.F. Dasmann. 1958. The black-tailed deer of the chaparral: its life history and management in the North Coast Range of California. *California Department of Fish and Game, Bulletin* **8**:1–163.

Taylor, C.R. 1969. The eland and the oryx. *Scientific American* **220**:88–95.

Teer, J.G., J.W. Thomas, and E.A. Walker. 1965. Ecology and management of the white-tailed deer in the Llano Basin of Texas. *Wildlife Monographs* **15**:1–62.

Telfer, E.S. and J.P. Kelsall. 1984. Adaptations of some large North American mammals for survival in snow. *Ecology* **65**:1828–1834.

Tilghman, N.G. 1989. Impacts of white-tailed deer on forest regeneration in northwestern Pennsylvania. *Journal of Wildlife Management* **53**:524–532.

Turner, M.G. 1989. Landscape ecology: the effect of pattern on process. *Annual Review of Ecology and Systematics* **20**:171–197.

Urness, P.J. 1969. Nutritional analyses and in vitro digestibility of mistletoes browsed by deer in Arizona. *Journal of Wildlife Management* **33**:499–505.

Van Ballenberghe, V. and W.B. Ballard. 1994. Limitation and regulation of moose populations: the role of predation. *Canadian Journal of Zoology* **72**: 2071–2077.

Van Horne, B. 1983. Density as a misleading indicator of habitat quality. *Journal of Wildlife Management* **47**:893–901.

Wallis de Vries, M.F. 1995. Large herbivores and the design of large-scale nature reserves in western Europe. *Conservation Biology* **9**:25–33.

Wallmo, O.C. 1978. Mule and black-tailed deer. Pages 31–41 *in* J.L. Schmidt and D.L. Gilbert. editors. *Big Game of North America: Ecology and Management.* Stackpole Books, Harrisburg, Pennsylvania, USA.

Wallmo, O.C. 1981. Mule and black-tailed deer distribution and habitats. Pages 1–15 *in* O.C. Wallmo, editor. *Mule and Black-Tailed Deer of North America.* University of Nebraska Press, Lincoln, Nebraska, USA.

Wallmo, O.C. and J.W. Schoen. 1981. Coniferous forest habitats. Part 2. Forest management for deer. Pages 434–448 *in* O. C. Wallmo, editor. *Mule and Black-Tailed Deer of North America.* University of Nebraska Press, Lincoln, Nebraska, USA.

Ward, D. and D. Saltz. 1994. Foraging at different spatial scales: dorcas gazelles foraging for lilies in the Negev Desert. *Ecology* **75**:45–58.

Weixelman, D.A., R.T. Bowyer, and V. Van Ballenberghe. 1998. Diet selection by Alaskan moose during winter: effects of fire and forest succession. *Alces* **34**:213–238.

Williams, P. H. and R. J. Haynes. 1995. Effect of sheep, deer, and cattle dung on herbage production and soil nutrient content. *Grass and Forage Science* **50**: 263–271.

Wisdom, M.J. 1998. Assessing life-stage importance and resource selection for conservation of selected vertebrates. PhD dissertation, University of Idaho, Moscow, Idaho, USA.

Wisdom, M.J. and J.G. Cook. 2000. North American elk. Pages 694–735 *in* S. Demarais, and P.R. Krausman, editors. *Ecology and Management of Large Mammals in North America.* Prentice Hall, Upper Saddle River, New Jersey, USA.

Wood, J.E., T.S. Bickle, W. Evans, J.C. Germany, and V. W. Howard, Jr. 1970. The Fort Stanton mule deer herd. *New Mexico State University, Agricultural Experiment Station Bulletin* **567**:1–32.

Woodward, A., E.G. Schreiner, D.B. Houston, and B.B. Moorhead. 1994. Ungulate-forest relationships on the Olympic Peninsula: retrospective exclosure studies. *Northwest Science* **68**:97–110.

Part II

Community and ecosystem relations

DANIEL L. LUOMA, JAMES M. TRAPPE,
ANDREW W. CLARIDGE, KATHERINE M. JACOBS
AND EFREN CÁZARES

10
_____

Relationships among fungi and small mammals in forested ecosystems

Introduction

Our approach

Here, we will present information about relationships between small mammals and an important food source, fruitbodies of (predominantly) ectomycorrhizal fungi. After providing some background on the function and diversity of the fungi involved, we will examine historical interest in mycophagy and current questions. The main focus will be on mycophagy (fungi consumption) and potential effects of disturbance on the inter-relationships among trees, truffles, and mammals. We have not limited our discussion to western North America because much relevant research has occurred in Australia.

Mycorrhizae

Different plants form different types of mycorrhizae with different fungi. The Pinaceae are primarily ectomycorrhizal, the Cupressaceae primarily vesicular-arbuscular (VA) mycorrhizal, as are most herbaceous plants. Some genera or families, such as the Salicaceae, can regularly form both ecto- and VA mycorrhizae. The Ericales mostly have their own distinctive mycorrhizae, as do the orchids. The fungi that form these different mycorrhiza types have different dispersal strategies. Though general categories of mycorrhizae are usually defined in morphological terms with little regard to ecology (Smith and Read 1997) mycorrhizal associations can also be categorized ecologically – such as by the degree of dependence of tree species on mycorrhizae for growth and reproduction. This approach lends itself to inclusion of the spore dispersal mechanisms of the mycobiont in a broader ecological context (Trappe and Luoma 1992).

We will focus on ectomycorrhizal fungi (EMF) because those species seem to be the most important with regard to small mammal mycophagy. Ectomycorrhizal fungi form symbiotic relationships with trees and other vegetation. Trees supply carbon from photosynthesis to the fungi, in turn, EMF absorb water, minerals, and nutrients from the soil and transfer them to tree roots (Smith and Read 1997). Mycorrhizae are essential for the survival and growth of most coniferous forest trees and other shrubs and herbaceous vegetation (Smith and Read 1997).

Douglas-fir (*Pseudotsuga menziesii*) forests have tremendous EMF diversity. For example, across its range, Douglas-fir can form mycorrhizal associations with an estimated 2000 species of fungi (Trappe 1977), yet little data exist on the diversity, abundance, synecology, or autecology of these fungi. Ectomycorrhizal fungus species vary in their fruiting season and abundance (Fogel and Hunt 1979, Fogel 1981, Hunt and Trappe 1987, Luoma 1988, 1991, Luoma et al. 1991, Amaranthus et al. 1994, Luoma et al. 1997, North et al. 1997, States and Gaud 1997, Colgan et al. 1999) and in nutritional value (Fogel and Trappe 1978) as further discussed in a subsequent section of this paper.

Sporocarp production

Factors that influence the fruiting of EMF include rainfall, temperature, and various other abiotic and biotic factors (Fogel 1981, Villeneuve et al. 1991). EMF fruiting is often non-uniform, varying from a few, scattered fruit bodies to concentrated clusters of numerous fruit bodies (Fogel 1976, North et al. 1997, States and Gaud 1997, Waters et al. 1997). The presence and abundance of EMF species may change during forest development (Trappe 1977, Mehus 1986, Termorshuizen 1991). Difficulties with sporocarp sampling methodologies and spatial and temporal variability (Luoma 1991) have hampered efforts to integrate sporocarp biomass data with other forest parameters such as wildlife feeding habits and populations (Cázares et al. 1999).

Fleshy sporocarps that form underground may be broadly referred to as truffles for convenience, though some are Ascomycetes (true truffles), others Basidiomycetes (false truffles), and a few sporocarpic Zygomycetes.

Historical precedents

"... for where the fungus is plentiful there the rats are also plentiful, or it may be the other way round."

H. E. PARKS, 1919.

Thus was one of the most crucial questions currently facing wildlife managers succinctly stated by one of the first naturalists in western North America to document the importance of truffles in the diet of certain small mammals. Harold Parks was an indomitable truffle collector and general observer of nature. During his excursions about the San Francisco Bay area, he noted the strong association between woodrats (*Neotoma* spp.) and these underground fruiting bodies of ectomycorrhizal fungi.

Observations of mammal mycophagy have been recorded since at least the 1800's. Reess and Fisch (1887) addressed the dissemination of spores of *Elaphomyces* (stag truffle) by animals (both wild and experimental) and concluded that spores pass through the animals unchanged. Some of the earlier writings on animal mycophagy included an occasional sighting of mushroom consumption but it was not until Cooke (1890) issued a call for a systematic assessment of this behavior that in-depth observations were initiated.

Hastings and Mottram (1916) took up the call and instigated field studies in Great Britain. Based on their observations, they speculated that succulent mushrooms became very important in the diet of rodents during the late fall and, further, that only in the case of buried sporocarps (truffles and false truffles) "do rodents appear to assist materially the fungus in the distribution of the spores."

Buller (1919, 1922) was inspired by the Hastings and Mottram work and published reports on the mycophagy of the red squirrel (*Tamiasciurus hudsonicus*) in North America. Buller provided reports of bulk storage of mushrooms and of fruitbodies being hung individually in the forks of tree branches. Both methods of caching were common in various parts of the squirrel's range.

Mycophagy of the red squirrel in the extreme western part of its range was investigated by Hardy (1949). He noted that this was near the limit of the dry winter conditions necessary for prolonged preservation of fungi. One cache of fungi was composed of 59 specimens in an excellent state of preservation and occupied a volume of about 4 l. The truffle genus *Hymenogaster* comprised about half of the cache and the mushroom genus *Russula* was the next most common item. He also found lichen in the cache and concluded that lichens formed a portion of the squirrel's diet (Hardy 1949).

Fogel and Trappe (1978) reviewed papers from the intervening years and documented a general trend of mycophagy in the diets of many small mammal species. They also posed several questions relevant to the life histories and ecosystematic roles of fungi and animals. Some of those we will address here, such as: what food values do mycophagists derive from fungi and what role does mycophagy play in fungus dispersal? Their questions led to much valuable research that provides the foundation for our section on current issues and research. Others of their questions remain in need of investigation.

Web-of-life relationships

Moving beyond simple observations of mycophagy, many important aspects of the interdependencies among fungi, mycophagists, and forest trees have been explored by researchers during the last 20 years. Truffle fungi are primarily dispersed by small mammals that eat the sporocarps and subsequently disperse spore-packed fecal pellets (Fogel and Trappe 1978, Kotter and Farentinos 1984a, Lamont et al. 1985, Maser and Maser 1988a, Claridge et al. 1992). Spores of a few truffle species, particularly in the genus *Elaphomyces*, are also disseminated by air. The edible outer layer of *Elephomyces* encloses a powdery spore mass that may be discarded while a small mammal is perched above the ground, resulting in the release of spores into the air-stream (Ingold 1973, Trappe and Maser 1977).

Spores can germinate to form new fungal mycelia or fuse with existing fungi, thus colonizing new areas or increasing the genetic diversity of existing fungus populations (Fogel and Trappe 1978, Miller et al. 1994). Forest dwelling small mammal species that depend upon fruiting bodies of EMF contain a diverse array of truffle genera in their fecal material (Maser et al. 1978a, 1985, Colgan et al. 1999, Carey et al. 2001).

As truffles mature, they produce strong, chemically complex odors that attract many small mammals (Trappe and Maser 1977, Donaldson and Stoddart 1994). The scent a truffle exudes may contain chemical compounds similar to certain animal hormones. Human odor trials suggest that males and females may respond to these odors differently (Marin and McDaniel 1987). Responding to these olfactory cues, small mammals are extremely adept at uncovering mature sporocarps (Pyare and Longland 2001a). With consumption of the truffle, many fungal spores are ingested; these spores remain viable after passage through the animal's digestive tract (Trappe and Maser 1976, Kotter and Farentinos 1984b). Some studies suggest that the spores of some truffle species actually require passage through an animal's digestive tract before they will germinate (Lamont et al. 1985, Claridge et al. 1992). Claridge et al. (1992) found that spores obtained directly from sporocarps of the Australian truffle, *Mesophellia pachythrix*, applied to eucalyptus trees didn't form any ectomycorrhizae whereas *M. pachythrix* spores that came from fecal pellets did. However, they could not determine whether it was passage through the gut or some other factor that allowed the spores to germinate in natural forest soil conditions.

As prey for raptors (e.g., goshawks) and mammalian carnivores (e.g., martens and fishers) small mammals form important links in the trophic structure of forest ecosystems (Fogel and Trappe 1978, McIntire 1984, Hayes et al. 1986, Carey 1991). The potential for indirect consumption of truffles by predators of small mammals has been recognized, but there is also evidence that fishers consume truffles directly (Grenfell and Fasenfest 1979, Zielinski et al. 1999). A wide variety of animals and trophic relationships, then, are instrumental in distributing mycorrhizal fungi to new tree roots. The animals at the same time depend on the trees for cover and reproductive sites (Aubry et al. 2003). Disruption of any part of this inter-dependent web of organisms will inevitably affect the others. Improved understanding of these relationships can lead to improved approaches to management of forest ecosystems (Aramanthus and Luoma 1997, Lawrance 1997, Colgan et al. 1999, Carey 2000, Wilson and Carey 2000, Carey 2001, Carey and Harrington 2001, Carey and Wilson 2001).

Mycophagy

Methods

Animal mycophagy studies are mainly based on stomach content or fecal pellet analyses (Tevis 1953, Fogel and Trappe 1978, Maser et al. 1978*a*, Maser and Maser 1988*a*, Carey 1995, Waters and Zabel 1995, Currah et al. 2000). These analyses can provide an accurate record of an animal's recent meals. Fecal samples provide a non-lethal method useful for long-term and integrated studies of diet habits. An effective and widely used method for analyzing fungi in diets was developed by McIntire and Carey (1989). This method utilizes a total of 75 fields-of-view/fecal sample being examined. For each field of view, at 400 × magnification, the presence of each fungal genus and other dietary items is recorded. Colgan et al. (1997) introduced a modification of the method that is useful when a large number of samples needs to be processed and pooling of samples is acceptable under the objectives of the study. When a more detailed assessment of non-fungal components of the diet is needed or if the animal species of interest is thought to be a facultative mycophagist, analysis of gut contents may be desirable (Currah et al. 2000). Working in western Oregon conifer forests, Carey et al. (1999) determined that the number of animals necessary to sample in order to record all fungal taxa recently consumed by a small mammal population in an area was ≥ 7.

Identification of the spores found during mycophagy studies has often been problematic due to the lack of comprehensive resources. Castellano et al. (1989) developed a key to spores for truffles of northern temperate forests. That work has now been extensively revised and released in CD-ROM format (Jacobs et al. 2003). In addition to an interactive key, investigators now have access to high-quality illustrations that depict spores and sporocarps of all 98 genera of sequestrate (truffle-like) fungi known from Northern Hemisphere temperate forests.

Case studies
Flying squirrels
Over most of its range, the threatened northern spotted owl feeds primarily on flying squirrels (Forsman et al. 1984, Thomas et al. 1990, Carey 1991) with the exception of the Klamath Province in northern California and southwest Oregon (Zabel et al. 1995). Northern flying squirrels (*Glaucomys sabrinus*), in turn, require truffles as their primary food source (Maser et al. 1985, Hall 1991, Carey 1995, Waters and Zabel 1995, Colgan et al. 1997, Zabel and Waters 1997, Pyare and Longland 2001*b*). Spring-captured northern flying squirrels from the southern Coast Range of Oregon ate a wider diversity of food items than fall-captured squirrels, though the diet in each season was dominated by fungi (Carey et al. 1999).

Data on stomach contents of northern flying squirrels led McKeever (1960) to conclude that when the snow cover was deep, lichens (also fungal) were the principal food of flying squirrels. With a decrease in snow cover in the spring, the squirrels consumed some truffles. In summer, their entire diet consisted of fungi. In the fall, lichens appeared again but fungi constituted over half the diet. Despite the availability of various seed crops in the three forest types of McKeever's study (*Pinus ponderosa*, *P. contorta*, and mixed *Abies*), no seeds were found in the stomachs. Rosentreter et al. (1997) found a similar seasonal pattern in northern flying squirrel food habits in central Idaho. In contrast, Currah et al. (2000) working in the boreal forest of northeast Alberta, found that flying squirrels consumed substantial amounts of mushrooms and no lichens during the winter. They attributed this result to the ability of flying squirrels to raid the caches of red squirrels.

In one study, the dietary composition of *G. sabrinus* tended to parallel the seasonal availability of sporocarps, suggesting that, in general, it did not prefer particular truffle species under those field conditions (Maser et al. 1986). The notable exception was the consumption of *Rhizopogon* sporocarps, which didn't change with seasonal abundance (Maser et al. 1986). This may have been an artifact of the sample technique or an actual disproportional consumption of *Rhizopogon* by *G. sabrinus*. Subsequent food trial studies under laboratory conditions showed that *G. sabrinus* does have a preference for consuming certain species of truffles over others (Zabel and Waters 1997). Truffles of *Gautieria monticola* and *Alpova trappei* and the lichen *Bryoria fremontii* were the top ranked food items in a comparison of sporocarps, lichen, and seeds. *Rhizopogon* truffles were not included in the experiment, however (Zabel and Waters 1997). Flying squirrels were found to consume significantly more *Gautieria* spores than chipmunks or voles in western Oregon and Washington, though the diets of all three groups were dominated by *Rhizopogon* spores (Jacobs 2002). Since fungal spores may be retained in the gut of flying squirrels for up to 11 days, wide dispersal of spores is possible (Pyare and Longland 2001*b*).

Knowledge of squirrels' food habits provided insight on a formerly puzzling aspect of northern flying squirrel biology. Bobcats (*Lynx rufus*) and coyotes (*Canis latrans*) prey effectively on flying squirrels, yet to do so they must capture the squirrels on the ground. Biologists had wondered what drew "arboreal" squirrels away from the relative safety of the tree crowns. In this case, of course, they are on the ground to dig out their primary food, truffles (Wells-Gosling and Heaney 1984, Maser et al. 1985).

G. sabrinus utilizes a wide range of forest habitats and has a home range of 3–6 ha (Witt 1992, Martin and Anthony 1999). Thus, management practices that cause local reductions in fungal diversity and abundance may not affect this species as much as species with smaller home ranges. In one study, thinning treatments applied to young stands (35–45 years old) showed no strong effect on *G. sabrinus* density but *G. sabrinus* density was highly positively correlated with truffle biomass and frequency (Gomez et al., unpublished data, in Smith et al. 2003).

Voles

Many small mammal mycophagists eat foods other than fungi, but some have evolved to specialize on hypogeous sporocarps. An excellent example is the western red-backed vole, *Clethrionomys californicus*. Its diet consists largely of truffles supplemented by lichens (Ure and Maser 1982, Hayes et al. 1986, Maser and Maser 1988b, Thompson 1996, Cázares et al. 1999). The skull and tooth structure of the coastal subspecies (*C. californicus californicus*) are fragile (Grayson et al. 1990) potentially having evolved with a highly specialized diet of soft, hypogeous sporocarps. Its habitat, the coast and Coast Ranges of northwestern California and western Oregon, lies in a relatively rare climate type that permits year-round fruiting of truffles. In the north Coast Range of Oregon, a mean of 85% and up to 98% of the stomach contents of *C. californicus californicus* were truffles (Ure and Maser 1982). Nineteen truffle genera have been identified in the fecal pellets of *C. californicus* in southern Oregon (Hayes et al. 1986). Due to its reliance on a fungal diet, this vole is an important disperser of truffle spores in the forest (Thompson 1996).

Clethrionomys californicus is strongly affected by clear-cutting and forest fragmentation. A strong negative edge effect has been shown both for *C. californicus* population numbers and truffle production, suggesting that truffle distribution in forest remnants may be one of the factors limiting *C. californicus* populations (Mills 1995). *C. californicus* is rarely found in clear-cuts, and intensively logged areas will be nearly devoid of *C. californicus* until the canopy begins to close (Gashwiler 1970, Ure and Maser 1982). As a result, this animal infrequently disperses spores into severely disturbed areas.

In southern New England, Getz (1968) demonstrated that the southern red-backed vole (*C. gapperi*) has high kidney requirements for water and suggested that habitats with "sufficient water or succulent food items" are a necessity for its survival. Truffles, which are high in water content, could potentially fulfill this role.

Although both *C. californicus* and *C. gapperi* are mycophagous, their diets differ according to their habitat. While *C. gapperi* captured in the lowlands of Washington ate as much fungi as *C. californicus* in early fall, *C. gapperi* found in higher elevations in Washington had a much higher incidence of conifer seeds in their stomachs (Ure and Maser 1982). These observations led Ure and Maser (1982) to conclude that mycophagy for these voles is closely related to habitat and not a feature specific to the species. That conclusion was further supported by Maser and Maser (1988b) wherein the stomach contents of *C. californicus* from western Oregon and *C. gapperi* from various areas across North America were examined. They also found that habitat highly influenced the mycophagy of *C. gapperi*, suggesting a facultative aspect to mycophagy in this species. From the Rocky Mountains west, 23 different fungal genera were observed in the diets of *C. gapperi*, but only seven genera were recorded from animals further east. In comparison, *C. californicus* captured in Oregon consumed 28 fungal genera (Maser and Maser 1988*b*).

Chipmunks

In contrast to the western red-backed vole and the northern flying squirrel, most small mammals feed not only on truffles (when available) but on a wide variety of other foods (Fogel and Trappe 1978, States and Wettstein 1998, Currah et al. 2000). Townsend's chipmunk, *Tamias townsendi*, eats conifer seeds and has often been regarded as a hindrance to reforestation. However, it avidly eats truffles as well (Tevis 1952, 1956, Maser et al. 1978*a*, Carey et al. 1999). The Siskiyou chipmunk (*Tamias siskiyou*) is also highly mycophagous. Animals trapped in Jackson County, Oregon had 16 different genera of truffles in their stomach contents, with 96% to 99% of the stomachs examined containing truffle spores (McIntire 1984).

During the early years following clear-cutting, chipmunks transfer mycorrhizal fungal spores from intact forests into timber harvest units. As the vegetation recovers from disturbance, truffles, fruits, seeds, and other foods become available. In times of shortage of other foods, truffles can be critical to chipmunks. As noted by Tevis (1952) only "individuals living where hypogeous fungi flourished became heavy and fat" before hibernation. The chipmunks' propensity for fungi and their movements between forests and timber harvest units make them an important vector of truffle spores (Tevis 1952, Trappe and Maser 1976, Maser et al. 1978*a*, 1978*b*, Maser and Maser 1988*a*, Rosenberg 1990).

Northern and southern hemisphere parallels

The trees, fungi, and mammals of the northern and southern hemisphere have evolved in striking parallelism since their separation by continental drift. This has become evident from research over the past 25 years, especially in North America and Australia. Australia has a higher diversity of species of truffles than is known from anywhere else. Claridge et al. (2000) found over 250 species in an area of southeastern Australia only about 300 km in diameter, more species than are known from all of Europe. The Australian climate, characterized by warm spells and drying winds intervening in the cool, wet times of year when mushrooms fruit, seems to have provided the selection pressure for fungi to evolve to a fruiting habit below ground (Thiers 1984). In the moist coolness of the soil the fungi can mature their spores regardless of the weather above ground. The success of the hypogeous strategy, however, requires an alternative to the mushroom's discharging of spores using moving air as the agent of dispersal. Mycophagy is one such alternative in Australia, as it is in North America.

In both the northern and southern hemispheres, mammals have evolved or adapted to using hypogeous fungi as an important food source. Some North American rodents such as the northern flying squirrel depend on hypogeous fungi as their major food, whereas others such as deer mice (*Peromyscus* spp.) may eat the fungi opportunistically as a lesser part of their overall diet. In Australia, marsupials vary similarly. The diet of the long-footed potoroo (*Potorous longipes*) is about 90% hypogeous fungi, whereas hypogeous fungi may be only seasonally important in the diet of bandicoots (*Perameles* spp.). Potoroos, bandicoots, bettongs (*Bettongia* spp.), and native Australian rodents such as bush rats (*Rattus fuscipes*) and smoky mice (*Pseudomys fumeus*) are mycophagists analogous to the forest rodents of North America.

Large mammals such as deer (*Odocoileus* spp.), elk (*Cervus elaphus*), mountain goats (*Oreamnos americanus*), and bear (*Ursus* spp.) also feed on hypogeous fungi in North America, providing possibilities for longer dispersal distances than by the small rodents. The Australian equivalents are wallabies (*Wallabia* spp), medium-sized kangaroos that can travel substantial distances in the course of a day. Through examinations of fecal material, Claridge et al. (2001) found that wallabies frequently eat truffles. Scats of numerous larger marsupials such as wombats (*Vombatus* spp.) and gray kangaroos (*Macropus fuliginosus*) were also checked, but none have been found to contain truffle spores.

The interactions of trees, fungi, small mammals, and predators in North America are epitomized in the mature conifer forests of the Pacific northwestern United States. There hypogeous fungi depend on their ectomycorrhizal tree hosts for energy, the northern flying squirrel depends on the fruitbodies as a primary food source, and, at the same time, is the primary prey of the threatened northern spotted owl. Analogous interactions seem likely in Australia (Claridge and May 1994). For example, in mature eucalypt forests hypogeous fungi form ectomycorrhizae; the smoky mouse eats their sporocarps, and the threatened sooty owl (*Tyto tenebricosa*) preys on the mice.

These independently evolved parallels between the northern and southern hemispheres may seem remarkable. However, they simply evidence the success of the hypogeous fungal fruiting habit in conjunction with mycophagy for increasing fitness of the plants, fungi, and animals participating in the system.

Nutritional value of hypogeous sporocarps

Sporocarps of ectomycorrhizal fungi generally contain much higher concentrations of minerals than do the leaves and fruits of plants. Phosphorus and zinc, for example, are 20 to 50 times more concentrated in sporocarps than in leaves of plants commonly browsed by animals (Stark 1972). Trace elements, too, can be concentrated at relatively high rates in the fungi; for example, copper and selenium. Large animals such as bear, deer, and wallabies may ingest sporocarps more for the mineral content than for other nutritional needs, using the fungi as a type of salt lick (Fogel and Trappe 1978).

The small mammals that depend strongly on hypogeous sporocarps for nourishment, in contrast, appear to do so to meet most of their nutritional needs, and mycophagist specialists such as western red-backed voles, northern flying squirrels, and long-footed potoroos have little else in their diet. Stomach content and fecal analyses reveal that these specialists almost always feed on diverse species in a given day. This is true even of the very small rodents. The typical volume of a single *Rhizopogon* sporocarp, for example, would exceed the stomach capacity of a western red-backed vole, yet those voles invariably have pieces of at least three and sometimes as many as 12 species in their stomachs at any given time. We can infer, then, that this diversity in their diets reflects a nutritional imperative.

As more studies are undertaken, the nutritional value of truffles to mammals is becoming better known. A large portion of a sporocarp is water, suggesting that quantities must be eaten to gain adequate nutrition (Miller and Halls 1969). However, truffles contain substantially higher amounts of nitrogen, phosphorous, potassium, sodium, iron, and aluminum than some epigeous sporocarps (Fogel 1976, Grönwall and Pehrson 1984). Fungi also contain vitamins (Shemakhanova 1967), nonmetallic and metallic elements (Stark 1972), steroids, triterpenes, amines, indoles, and phenols (Catalfomo and Trappe 1970) that could potentially benefit mycophagous animals.

The digestibilities of dry matter, nitrogen, cell wall constituents, and energy in two species of truffles, *Elaphomyces granulatus* and *Rhizopogon vinicolor*, have been studied in detail. In a feeding trial, Cork and Kenagy

(1989) fed captive golden-mantled ground squirrels (*Spermophilus satura-tus*) the fruit bodies of *Elaphomyces granulatus*, a common truffle. They compared the digestibility of the fungus to the digestibility of the leaves of a variety of plant species eaten naturally by the squirrels, as well as cones, pine nuts, leguminous foliage, and grass. A high-quality food, rodent laboratory chow, was used as a reference diet. Squirrels were offered pre-weighed amounts of the different foods. During the experiment, squirrels maintained or gained body mass on two of the food types, pine nuts and rodent chow. Squirrels consuming a high daily intake of only *Elaphomyces* lost weight. The digestibility of nitrogen and energy from *Elaphomyces* was lower than that recorded for nearly all the other diets. Although chemical analyses revealed that the nitrogen content of fruit bodies was relatively high, 80% of it was bound in totally indigestible spores that the squirrels rarely ate. Of the remaining 20%, only half was present as protein nitrogen. Sources of energy were tied up in complex, relatively indigestible cell-wall tissue.

The digestible energy requirement of the squirrels was also estimated (Cork and Kenagy 1989). The overall digestibility of *E. granulatus* fruit bodies fell just below the critical threshold for the squirrels to maintain energy balance. For these squirrels, with a relatively simple digestive tract, *E. granulatus* was seen as a marginal but important dietary item when no alternative was available. Moreover, the truffles were readily detectable and required minimal processing time prior to consumption, unlike some foods such as seeds from cones. The truffles, therefore, yielded more energy and nutrients in relation to foraging effort. They suggested that if squirrels cannot maintain normal energy balances by eating truffles, then the minor incorporation of less abundant, higher quality foods may be all that is needed to achieve a positive energy balance.

Claridge et al. (1999) conducted feeding experiments in Oregon with captive northern flying squirrels and western red-backed voles. When fed only a single species of truffle (*Rhizopogon vinicolor*) neither of the animals could maintain their weight. The digestibilities of *R. vinicolor* sporocarps were lower than those of other food types eaten by other mammals of similar size. Voles digested the various sporocarp components as well as did the squirrels, even though average vole body mass was six-fold smaller than that of the squirrels. This supports the hypothesis that western red-back voles, like other microtine rodents, have morphological and physiological adaptations of the digestive system to permit greater digestion of fibrous diets than predicted on the basis of body size. Neither of the animals drank water during the experiments. Fresh truffles are >70% water by weight (Claridge et al. 1999) evidently enough to meet the water requirements of the mycophagists. Nonetheless, individual species of hypogeous fungi appear to be of only moderate nutritional value for many small mammals. Again, this may account for the animal's habit of eating relatively small amounts of several different species within a day, a behavior that may compensate for differences in digestibility and nutritional quality among truffle species.

Little is known about the nutritional value of fungi for other groups of mycophagous mammals. This extends to the terrestrial marsupials of Australia. In Western Australia, Kinnear et al. (1979) assessed the chemical components of fruit bodies of *Mesophellia*, a significant food for the brush-tailed bettong (*Bettongia penicillata*) a member of the rat-kangaroo family. Analyses of the inner core of the truffles (the portion largely consumed by bettongs) indicated a rich source of lipids (around 40% by dry weight) and crude protein levels of 8% to 10%. Although protein levels were high, analyses revealed they were deficient in certain essential amino acids such as lycine. Other amino acids, particularly cysteine and methionine, were present in large quantities. It was suggested that the imbalances in amino acids in fruit bodies could be largely corrected during digestion. Bettongs and other rat-kangaroos (except the musky rat-kangaroo, *Hypsiprymnodon moschatus*) have special adaptations to the gut, including a large sacciform forestomach. The hind-gut is reduced to a well-developed, though simple, caecum and proximal colon. The enlarged sacculated foregut is designed to culture anaerobic microbes that ferment food and convert fungal nitrogen to a form more available for the host animal. This process is called pre-gastric fermentation. Hume (1989) suggested that the foregut of rat-kangaroos might serve as a food storage area, an advantage to an animal subject to predation and needing to minimize feeding time.

By monitoring the passage of labeled chemical markers, Frappell and Rose (1986) studied the movement of digesta through the gut of captive long-nosed potoroos (*Potorous tridactylus*). Food particles entered the foregut, remained up to one hour and then passed into the hindgut, where they remained for seven to eight hours before being excreted. Rose and Frappell concluded that fungi entering the foregut were indeed subject to rapid microbial fermentation but that the hindgut was clearly important in the digestion process. Hume et al. (1993) discovered that most material eaten by long-nosed potoroos and rufous bettongs (*Aepyprymnus rufescens*) bypassed the foregut, but that which did not was retained for periods up to four days. Hindgut digestion was also seen by Richardson (1989) in brush-tailed bettongs, although retention time was only a few hours. It seems that the role of the foregut is to help digest fungi while that of the hindgut is to process lower quality food.

Claridge and Cork (1994) provided the first real evidence that fungal fruit bodies were nutritious for rat-kangaroos. In a controlled feeding trial, captive long-nosed potoroos were fed known amounts of fruit-bodies of two species of truffles, *Mesophellia glauca* and *Rhizopogon luteolus*. Chemical analyses revealed that although the nitrogen concentration was high in both fungi, much of it was in non-protein form or associated with cell walls and was thus presumably of low nutritional value or protected from digestive enzymes. The concentration of cell-wall constituents (fiber) was high in both fungi, suggesting low availability of digestible energy. Nonetheless, potoroos lost little weight and digested much of the dry matter, nitrogen, and energy in the pure fungal diets. Consequently, animals maintained positive nitrogen balances and high intakes of digestible and metabolizable energy. Most other mycophagous mammals in Australia lack an enlarged foregut and most food is digested in the hindgut. The lack of this digestive system may help explain why hindgut-fermenters such as rats and bandicoots seldom rely wholly upon fungi but commonly eat other foods such as seeds and invertebrates.

Studying the reproductive energetics of the Tasmanian bettong (*Bettongia gaimardi*) in a eucalypt woodland, Johnson (1994a) found that when production of truffles was highest, the bettongs were almost entirely mycophagous, whereas at times of low fruit body production the bettongs mainly consumed other foods such as leaves and fruits. Body condition of adult bettongs tended to benefit with increasing amounts of fungi in the diet. When production of truffles increased, energy turnover in adult females and growth rates of pouch young increased concomitantly, suggesting that the fungi provided animals with a surplus of energy, perhaps used in lactation. McIlwee and Johnson (1997) used stable isotopes to determine that nearly all nitrogen assimilated into body tissue by northern bettongs (*Bettongia tropica*) was from fungi. In contrast, the sympatric northern brown bandicoot (*Isoodon macrourus*) derived much of its nitrogen from invertebrates and practically none from fungi. This finding was mirrored by patterns in the diet of the same animals.

Current issues

Little information is available on the relationship of EMF sporocarp abundance and species composition to the diets and population abundances of small mammals. Small mammal population densities are highly variable across stands and landscapes (Carey et al. 1992, Rosenberg and Anthony 1992, Witt 1992) and the species composition and abundance of mushrooms and truffles may influence the ability of forests to provide habitat for small mammals (Waters and Zabel 1995).

Effects of disturbance

Studies from the Pacific Northwest indicate that forest management activities can reduce EMF and forest regeneration success (Amaranthus and Perry 1987, 1989, Amaranthus et al. 1990). In these studies, the abundance and rapidity of ectomycorrhiza formation was critical to seedling survival and growth, especially on harsh sites. However, across the Pacific Northwest the degrees of reduction of EMF and impacts on forest regeneration vary widely and depend on many factors. Ectomycorrhizal fungus species vary in their abilities to provide particular benefits to their hosts, and presence and abundance of EMF species change during forest succession (Trappe 1977, Mason et al. 1983, Trappe 1987, Visser 1995). The abundance and composition of truffle production may also change following natural disturbance (Luoma 1988, Luoma et al. 1991, Waters et al. 1997).

Many vegetational and structural changes during succession in Douglas-fir forests are documented (Franklin et al. 1981, Spies et al. 1988). Yet, despite the importance of ectomycorrhizal fungi to ecosystem processes, little is known about their community structure and dynamics in managed stands. Such data are essential to predict impacts of disturbance and management on ecosystem productivity. Integration of vegetation, wildlife, and landscape responses with knowledge of EMF and underground functions is needed to elucidate critical aspects of the ecology of EMF that have strong management implications. Only integrated research can provide the extended perspective needed to produce information on interactions among mycorrhizal fungi, small mammals, and a range of forest management practices.

Young managed stands may have a different composition of truffle species than old-growth or natural mature stands. Among *Tsuga heterophylla* dominated forests of varying ages studied across northwest Washington by North et al. (1997) truffle species richness was highest in the old-growth stands. The standing crop of truffles was much higher in the natural mature and old-growth stands than in the young managed stands. This was largely due to the presence of large clusters of *Elaphomyces granulatus* in the natural stands. A similar trend was found in *Abies amabilis* stands of western Washington. The annual production of truffles was only $1 \mathrm{kg} \cdot \mathrm{ha}^{-1} \cdot \mathrm{year}^{-1}$ in 23-year-old stands, whereas in the 180-year-old stands, production was $380 \mathrm{kg} \cdot \mathrm{ha}^{-1} \cdot \mathrm{year}^{-1}$ (Vogt et al. 1981). Stands that had been altered by management prescriptions such as slash burning and soil scarification showed marked reductions in their ability to provide truffles for small mammals when compared to unmanaged stands (North and Greenburg 1998).

Clear-cutting

Ectomycorrhizal fungi and the production of truffles are closely linked to host trees. When the trees are removed or the composition of a stand changes, the composition, species richness, or abundance of truffles in the stand changes as well (Amaranthus et al. 1994, Clarkson and Mills 1994, North et al. 1997, Colgan et al. 1999). Clear-cutting forests is especially detrimental to EMF diversity and abundance because all potential hosts are removed. The removal of the host tree cuts off the supply of carbon to the fungus and prevents it from producing truffles (Amaranthus et al. 1994). Additionally, soil temperature or moisture changes and soil compaction will heavily impact the production of truffles (Fogel 1976, Waters and Zabel 1995, Cázares et al. 1998).

A positive relationship between truffle production and coarse woody debris (CWD) was found by several researchers (Luoma 1988, Amaranthus et al. 1994, Waters et al. 1997). Many forest management practices impact the amount and decay class of CWD in a stand. This too, may affect truffle production, abundance, or diversity. Older forests tend to have more CWD in the later stages of decomposition than do younger or recently clear-cut stands (Harmon et al. 1986). Late-seral forest remnants in southwestern Oregon had 20–40 times more sporocarps than the surrounding 10- to 27-year-old clear-cuts (Clarkson and Mills 1994). Out of the 80 sample plots placed within clear-cuts, only one truffle was found. Within the late-seral stands, truffles were four times more numerous in plots with CWD than without.

When comparing the numbers of truffles and truffle dry weight between Douglas-fir forest fragments and the clear-cuts surrounding them, Amaranthus et al. (1994) also found an association between stand age, amount of CWD, and truffle production. A greater number, diversity, and total dry weight of truffles were found in the mature stands than in the plantations. Thirteen of the 21 truffle species were found only in the mature stands and eight species were found only under CWD. The effect of CWD upon truffle production was evident only in the mature stands. In the mature forest fragments, there were more truffles and greater truffle biomass in CWD compared to soil (Amaranthus et al. 1994). Since well-decayed CWD retains water, truffle production may be limited to areas in and around well-decayed CWD during times of drought. Retention of mature forest fragments in the managed landscape can help to maintain a diverse food source for small mammals that may not be available in younger stands during critical times (Amaranthus et al. 1994). Though Amaranthus et al. (1994) demonstrated that previously clear-cut young stands produce fewer sporocarps than intact mature forest fragments, only limited information is now becoming available for a range of partial forest harvests and silvicultural systems.

Thinning

Thinning is a common silvicultural practice throughout the world. Unlike clear-cutting, thinning retains residual trees that can act as refuges for EMF. However, thinning still alters the community structure, diversity or composition of EMF in a stand (Waters et al. 1994, Colgan et al. 1999). The effects of variable density thinning (VDT) on truffle production during the first years following thinning was examined by Colgan et al. (1999). Douglas-fir stands were comprised of a mosaic of patches thinned to different densities of standing live trees. The mosaic was divided into two thinning categories, lightly and heavily thinned. Total standing crop truffle biomass was significantly lower in VDT stands compared to control stands. The abundance of *Gautieria* and *Hysterangium* species was lower in thinned stands, while *Melanogaster* species diversity and productivity was highest in VDT stands.

Initial effects of thinning appear to include reduced truffle biomass, reduced frequency of sporocarps, and shifts species dominance (Colgan et al. 1999). Total truffle biomass and frequency of sporocarps may recover 10–17 years after thinning, while shifts in species dominance persist longer (Waters et al. 1994). Potential effects of shifts in truffle species composition include alteration of tree regeneration composition or impacts on mycophagous animals by altering the nutritional balance of their diets.

Green-tree retention

The retention of green trees during commercial timber harvest can moderate the impact of host loss by providing a refuge for EMF diversity. Under the auspices of the Demonstration of Ecosystem Management Options (DEMO) project, Stockdale (2000) examined the initial response of ectomycorrhizae to a 15% basal-area retention treatment. Ectomycorrhizal root tips beneath the crown of retained trees and in open areas away from the retained trees were evaluated. Ectomycorrhizal fungus richness was reduced by as much as 50% in open areas compared to within the dripline of retention trees. Species composition differed between the open areas and within the dripline as well. These results provide evidence that green trees act as refuges for legacy species and are important in maintaining EMF diversity in managed stands. In the longer-term, the DEMO project will be able to examine the role of green-tree retention in the recovery of truffle production after disturbance (Cázares et al. 1999, Luoma and Eberhart 2001).

McIntire (1984) examined the effects of slash burning on mycophagy within a shelterwood-logged coniferous forest in southwest Oregon. The slash treatment was associated with a reduction of spores in fecal samples from Siskiyou chipmunks (*Eutamias siskiyou*) on the site. Waters and Zabel (1995), working in northeastern California, found that heavy logging (shelterwood stands) and intensive site preparation negatively affected flying squirrel populations and truffle frequency.

Disturbance effects on mycophagy and spore dispersal

Many of these studies found a shift in fungal species composition, diversity, and dominance in managed stands. Some also found a difference in biomass and sporocarp frequency between managed and unmanaged stands. Taken as a whole, these studies indicate that forest management practices have a profound effect on EMF communities.

As the diversity, composition, and abundance of truffles in a forest changes, the ability of small mammals to find an adequate amount and diversity of food may be affected. This, in turn, may affect small mammal population numbers or species composition. Pyare and Longland (2001*b*) suggest that different small mammal species may disperse fungal spores in "ecologically nonredundant ways." Thus, a change in small mammal population composition may reduce dispersal or change dispersal patterns for various fungi.

Evidence also suggests that small mammals of different species compete with each other for the truffle food base (Pyare and Longland 2001b). As truffle abundance is reduced or species become less diverse, those animals heavily reliant upon fungi in their diet may have difficulty finding adequate numbers of truffles. If the small mammal community then changes, predators dependent on small mammals as prey may be impacted (Pyare and Longland 2001b). During periods of low truffle production or when other food sources are not available, the impacts of clearcutting and thinning on the food supply of small mammals may be more severe (Tevis 1952). These effects may resonate throughout the tightly knit relationship between trees, truffles, small mammals, and predators.

Disturbance down under

Australia has a long history of fire, both wild and deliberately applied, that is thought to have shaped the nature of much of the vegetation and in turn the animals utilizing it. Several studies have focused on the effects of fire on foraging behavior and food resources of rat-kangaroos. In Western Australia, Christensen and Maisey (1987) noted that foraging by brush-tailed bettongs increased dramatically in recently burnt areas, apparently in response to truffle formation. Similarly, in northern Tasmania, Taylor (1992, 1993) compared the foraging activity of Tasmanian bettongs in plots within unburned and recently burned woodland. The density of forage-diggings of animals on burned plots on lateritic soils was tenfold higher than elsewhere across the study area, and this activity began a few days after the fire event. Most forage-diggings in the burned plots had the remains of truffles, entirely from species within the family Mesophelliaceae. Taylor speculated that because of a gap of several days before forage-diggings were made, stimulation of fruiting had actually occurred.

To explain the responses observed by Christensen and Maisey (1987) and Taylor (1992, 1993), Johnson (1994b, 1995) set up a series of experimental burns in a Tasmanian woodland dominated by *Eucalyptus tenuiramis*, which was the preferred habitat of a local population of Tasmanian bettongs. He established a series of matched sites, some subject to deliberately applied (prescribed) fire, the others remaining as unburned or control sites.

In the first set of small-scale fires, Johnson (1994b) found that the density of forage-diggings of bettongs increased three- and ninefold within one month post-fire, then returned to pre-fire levels within four months and matched forage-digging densities on control sites. A similar result was recorded during the second experiment, with the density of forage-diggings increasing eightfold (Johnson 1995). The relative abundance of fungal fruit bodies changed somewhat, but these changes differed in relation to the intensity of fire. The overall result was that long-term productivity of fungal fruit bodies did not differ significantly across all sites, burned or unburned.

In the second experiment, in which truffles were collected from within exclosures, there were no significant increases in truffles around trees on burned plots (Johnson 1995). Instead, fungal productivity on the burned site remained stable. This contrasted with the unburned control site, where fungal productivity decreased after one month post-fire, but later recovered and surpassed that on the burned site seven to ten months post-fire.

In summary, Johnson (1994b, 1995) concluded that fire had actually triggered the fungi to fruit, but the data indicated that the fungal response was complicated and perhaps unpredictable. Whatever the cause, given that increases in foraging activity occurred mostly within two to three days post-fire, the response was extremely rapid. The ability of rat-kangaroos to rapidly enter burned areas and consume fungi may lead to efficient dispersal of fungal spores that later germinate and form ectomycorrhizal associations on host plant roots. These interactions are critically important to plant recovery post-disturbance in Australia's fire-prone and fire-adapted *Eucalyptus* forests.

More recent experiments by Vernes and Hayden (2001) and Vernes et al. (2001) at Davies Creek in northeastern Queensland largely corroborate the patterns observed by Johnson (1994a, 1994b, 1995) in Tasmania. They found that significantly more animals within the study population chose to forage in burnt habitat than in unburnt habitat. Moreover, they found that foraging success (i.e., probability of recovering a truffle) of animals was higher in burnt than unburnt sites (Vernes and Hayden 2001, Vernes et al. 2001).

The longer-term influences of disturbances on fungal populations are largely undocumented. Johnson (1995) compared diversity and relative abundance of truffles at six sites matched for soil, vegetation, and climate but differing in time since last fire (1, 2, 4, 10, 25, and 50 years). Notably, all fungal taxa present in recently burned sites were also present at sites long unburned. In conclusion, Johnson (1995) suggested that too frequent fire may be detrimental to the fungi and, in turn, animals such as the Tasmanian bettong, and that intervals around ten years between fires might be optimum. Relationships among time-since-disturbance, litter accumulation, and diversity of ectomycorrhizal mushrooms have been found in other studies (Dighton and Mason 1985, Dighton et al. 1986).

Differences in fungal diversity between sites varying in time since disturbance were also noted by Claridge et al. (1993) in southeastern mainland Australia. There, long-nosed potoroos fed on a higher diversity of fungi in a site never logged than in a site with recent intensive logging. Moreover, the relative abundance of different fungi also varied between the two stands, with the spores of some taxa (Mesophelliaceae) more commonly represented in the scats of animals from the recently logged site than the site never logged. Recent studies by Green and Mitchell (1997) of the long-footed potoroo, an obligate fungivore, suggest that animals in sites with minimal recent disturbance may have higher fecundity, and forage for significantly shorter periods, than animals in habitats with evidence of recent disturbance.

Potoroos, in contrast to bettongs, prefer dense understory vegetation. The long-footed potoroo and long-nosed potoroo also apparently benefit from lack of recent disturbance to their habitats. Recent work by Claridge and Barry (2000) in southeastern mainland Australia showed that the probability of occurrence of detecting forage-diggings of these animals was higher in sites burned more than 20 years previously, compared to sites burned 0–10 and 11–20 years previously.

Summary

Truffle fungi are primarily disseminated by small mammals that eat the sporocarps and subsequently disperse spore-packed fecal pellets. The spores can germinate to form new fungal mycelia or fuse with existing fungi, thus colonizing new areas or increasing the genetic diversity of existing fungal populations. Although small mammals consume both mushrooms and truffles, forest-dwelling mycophagists in western North America predominantly utilize a diverse array of truffle genera in their diets.

Animal mycophagy studies are mainly based on stomach content or fecal pellet analyses. These analyses can provide an accurate record of an animal's recent meals. Fecal samples provide a non-lethal method useful for long-term and integrated studies of diet habits. Assessment of the relative frequency of spore types and other dietary items is commonly used to rank the importance of food items. Truffle spores remain viable after passage through the animal's digestive tract. The spores of some truffle species require passage through an animal's digestive tract to improve germination.

The trees, fungi, and mammals of the northern and southern hemisphere have evolved in striking ecological parallelism since their separation by continental drift. Truffles are abundant in Australian *Eucalyptus* forests, as are mycophagous marsupials. The effects of fire on truffle production and mycophagy have been more extensively investigated in Australia than in North America. Consumption of truffles in the family Mesophelliaceae increased after fire and five alternative hypotheses were proposed to account for the phenomenon. First, fire may remove ground cover, thereby improving access to the soil and hence truffle resources. Second, animals may dig more often for fungi in recently burned habitats because other foods become unavailable. Third, fire may actually stimulate the truffles to form. And fourth, heating of existing truffles may change their odor such that they become more detectable.

Knowledge of the nutritional value of truffles to mammals is increasing. A large portion of a sporocarp is water, suggesting that quantities must be eaten to gain adequate nutrition. The digestibilities of dry matter, nitrogen, cell-wall constituents, and energy suggest that, as single species, truffles are of moderate nutritional value. Tests of multiple-species diets have not been published. Small mammals that depend strongly on truffles for nourishment appear to use a diversity of species to meet their nutritional needs. The anatomy of the digestive system may influence the degree to which small mammals can rely on fungi as a food source.

As the diversity, composition, and abundance of truffles in a forest changes, the ability of small mammals to find an adequate amount and diversity of food may be affected. This, in turn, may affect small mammal population numbers or species composition. Alternatives to clear-cutting may mitigate the loss of EMF in managed forest ecosystems. Various squirrels, chipmunks, and voles are ecologically important mycophagists that disperse truffles in forested habitats, contribute to the recovery of truffle species in second-growth stands, and provide part of the prey base for a variety of predators. Increased consideration of these ecological relationships and functions will aid in achieving ecosystem management objectives.

Research needs

Questions relevant to small mammal mycophagy in western coniferous forests abound. The fundamental dilemma posed by Parks (1919) has yet to be fully addressed: to what extent are small mammals dependent on truffles (and hence occur at greater densities in truffle-rich habitats) and to what extent do truffle-producing EMF species depend on small mammals for spore dispersal and dominance in the fungal community (and hence attain high levels of sporocarp production)? The nutritional value of a wide variety of truffle species needs to be determined. Ecosystem responses to disturbance are important concerns. What is the rate of re-establishment of these linked ecosystem components (small mammals and truffles) following disturbance? Studies of the role of fire are particularly needed in western coniferous ecosystems. Experimental research will be necessary to test the applicability of conclusions and hypotheses generated by research in Australian fire-adapted ecosystems. What are the processes that facilitate re-establishment of truffles and small mammals in a disturbed ecosystem? What are the functions of coarse woody debris in these processes? What can managers do to maintain fungal resources in ecosystems and how can recovery from disturbance be facilitated?

Whatever questions researchers choose to pursue, broad, important, commonalities can be fostered. Interdisciplinary experiments with long-term objectives and stable funding sources should be emphasized. Integration of stand-level studies with landscape research is desirable. Iterative rounds of data acquisition and modeling will be necessary to advance ecosystem-science-based forest management.

Literature cited

Alexander, L.F., and B.J. Verts. 1992. *Clethrionomys californicus. Mammalian Species* **406**:1–6.

Amaranthus, M.P., and D.L. Luoma. 1997. Diversity of ectomycorrhizal fungi in forest ecosystems: importance and conservation. Pages 99–105 *in* M.T. Martins, M.I.Z. Sato, J.M. Tiedje, L.C.N. Hagler, J. Döbereiner, and P.S. Sanchez, editors. *Progress in Microbial Ecology*. Proceedings of the Seventh International Symposium on Microbial Ecology. Brazilian Society for Micobiology, Santos, Brazil.

Amaranthus, M.P., and D.A. Perry. 1987. Effect of soil transfer on ectomycorrhiza formation and the survival and growth of conifer seedlings on old, nonreforested clear-cuts. *Canadian Journal of Forest Research* **17**:944–950.

Amaranthus, M.P., and D.A. Perry. 1989. Interaction effects of vegetation type and Pacific madrone soil inocula on survival, growth, and mycorrhiza formation of Douglas-fir. *Canadian Journal of Forest Research* **19**:550–556.

Amaranthus, M.P., J.M. Trappe, and R.J. Molina. 1990. Long-term forest productivity and the living soil. Pages 36–52 *in* D.A. Perry, R. Meurisse, B. Thomas, R. Miller, J. Boyle, J. Means, C.R. Perry, and R.F. Powers, editors. *Maintaining the Long-Term Productivity of Pacific Northwest Forest Ecosystems.* Timber Press, Portland, OR, USA.

Amaranthus M.P., J.M. Trappe, L. Bednar, and D. Arthur. 1994. Hypogeous fungal production in mature Douglas-fir forest fragments and surrounding plantations and its relation to coarse woody debris and animal mycophagy. *Canadian Journal of Forest Research* **24**:2157–2165.

Aubry, K.B., J.P. Hayes, B.L. Biswell, and B.G. Marcot. 2003. The ecological role of tree-dwelling mammals in western coniferous forests. Pages 405–443 *in* C.J. Zabel and R.G. Anthony, editors. *Mammal Community Dynamics. Management and Conservation in the Coniferous Forests of Western North America.* Cambridge University Press, Cambridge, UK.

Buller, A.H.R. 1919. The red squirrel of North America as a mycophagist. *Transactions of the British Mycological Society* **6**:355–362.

Buller, A.H.R. 1922. *Researches on Fungi.* Volume 2. Longmans, Green and Co., New York, NY, USA.

Carey, A.B. 1991. *The Biology of Arboreal Rodents in Douglas-fir Forests.* USDA Forest Service, General Technical Report PNW-276. Portland, OR, USA.

Carey, A.B. 1995. Sciurids in Pacific Northwest managed and old-growth forests. *Ecological Applications* **5**:648–661.

Carey, A.B. 2000. Effects of new forest management strategies on squirrel populations. *Ecological Applications* **10**:248–257.

Carey, A.B. 2001. Experimental manipulation of spatial heterogeneity in Douglas-fir forests: effects on squirrels. *Forest Ecology and Management* **152**:13–30.

Carey, A.B., and C.A. Harrington. 2001. Small mammals in second-growth forests: implications for management for sustainability. *Forest Ecology and Management* **154**:289–309.

Carey, A.B., and S.M. Wilson. 2001. Induced spatial heterogeneity in forest canopies: responses of small mammals. *Journal of Wildlife Management* **65**:1014–1027.

Carey, A.B., S.P. Horton, and B.L. Biswell. 1992. Northern spotted owls: influence of prey base and landscape character. *Ecological Monographs* **62**:223–250.

Carey, A.B., J. Kershner, B. Biswell, and L. Dominguez de Toledo. 1999. Ecological scale and forest development: squirrels, dietary fungi, and vascular plants in managed and unmanaged forests. *Wildlife Monographs* **142**:1–71.

Carey, A.B., W. Colgan III, J.M. Trappe, and R. Molina. 2002. Effects of forest management on truffle abundance and squirrel diets. *Northwest Science* **76**:148–157.

Castellano M.A., J.M. Trappe, Z. Maser, and C. Maser. 1989. *Key to Spores of the Genera of Hypogeous Fungi of North Temperate Forests with Special Reference to Animal Mycophagy.* Mad River Press, Eureka, CA, USA.

Catalfomo, J., and J.M. Trappe. 1970. Ectomycorrhizal fungi: a phytochemical survey. *Northwest Science* **44**:19–24.

Cázares, E., D.L. Luoma, J.L. Eberhart, M.P. Amaranthus, C. Cray, M.P. Dodd, and M. McArthur. 1998. Hypogeous fungal diversity and biomass following salvage logging in Mt. Hood National Forest, Oregon, USA. Abstracts. Second International Conference on Mycorrhizae. Uppsala, Sweden. Pages 39–40, *in* U. Ahonen-Jonnarth, E. Danell, P. Fransson, O. Kårén, B. Lindahl, I. Rangel, and

R. Finlay, editors. Programme and abstracts of the second international conference on mycorrhizae. Swedish University of Agricultural Sciences, Uppsala, Sweden.

Cázares, E., D.L. Luoma, M.P. Amaranthus, C.L. Chambers, and J.F. Lehmkuhl. 1999. Interaction of fungal sporocarp production with small mammal abundance and diet in Douglas-fir stands of the southern Cascade Range. *Northwest Science* **73**:64–76.

Christensen, P., and K. Maisey. 1987. The use of fire as a management tool in fauna conservation reserves. Pages 323–329 *in* D.A. Saunders, G.W. Arnold, A.A. Burgidge, and A.J. Hopkins, editors. *Nature Conservation: The Role of Remnants of Native Vegetation*. Surrey Beatty & Sons, Sydney, Australia.

Claridge, A.W. and S.C. Barry. 2000. Factors influencing the distribution of medium-sized ground-dwelling mammals in south-eastern mainland Australia. *Austral Ecology* **25**:678–688.

Claridge, A.W. and S.J. Cork. 1994. Nutritional value of hypogeal fungal sporocarps for the long-nosed potoroo (*Potorous tridactylus*), a forest-dwelling mycophagous marsupial. *Australian Journal of Zoology* **42**:701–710.

Claridge, A.W., and T.W. May. 1994. Mycophagy among Australian mammals. *Australian Journal of Ecology* **19**:251–275.

Claridge, A.W., M.T. Tanton, J.H. Seebeck, S.J. Cork, and R.B. Cunningham. 1992. Establishment of ectomycorrhizae on the roots of two species of eucalyptus from fungal spores contained in the feces of the long-nosed potoroo (*Potorous tridactylus*). *Australian Journal of Ecology* **17**:207–217.

Claridge, A.W., M.T. Tanton, and R.B. Cunningham. 1993. Hypogeal fungi in the diet of the long-nosed potoroo (*Potorous tridactylus*) in mixed-species and regrowth eucalypt stands in south-eastern Australia. *Wildlife Research* **20**:321–337.

Claridge, A.W., J.M. Trappe, S.J. Cork, and D.L. Claridge. 1999. Mycophagy by small mammals in the coniferous forests of North America: nutritional value of sporocarps of *Rhizopogan vinicolor*, a common hypogeous fungus. *Journal of Comparative Physiology. B: Biochemical, Systemic, and Environmental Physiology* **169**:172–178.

Claridge, A.W., S.C. Barry, S.J. Cork, and J.M. Trappe. 2000. Diversity and habitat relationships of hypogeous fungi. II. Factors influencing the occurrence and number of taxa. *Biodiversity and Conservation* **9**:175–199.

Claridge, A.W., J.M. Trappe, and D.L. Claridge. 2001. Mycophagy by the swamp wallaby (*Wallabia bicolor*). *Wildlife Research* **28**:643–645.

Clarkson, D.A. and L.S. Mills. 1994. Hypogeous sporocarps in forest remnants and clear-cuts in southwest Oregon. *Northwest Science* **68**:259–265.

Colgan, W. III, A.B. Carey, and J.M. Trappe. 1997. A reliable method of analyzing dietaries of mycophagous small mammals. *Northwest Naturalist* **78**:65–69.

Colgan, W. III, A.B. Carey, J.M. Trappe, R. Molina, and D. Thysell. 1999. Diversity and productivity of hypogeous fungal sporocarps in a variably thinned Douglas-fir forest. *Canadian Journal of Forest Research* **29**:1259–1268.

Cooke, M.C. 1890. Animal mycophagists. *Grevillea* **19**:54.

Cork, S.J., and G.J. Kenagy. 1989. Nutritional value of a hypogeous fungus for a forest-dwelling ground squirrel. *Ecology* **70**:577–586.

Currah, R.S., E.A. Smreciu, T. Lehesvirta, M. Niemi, and K.W. Larsen. 2000. Fungi in the winter diets of northern flying squirrels and red squirrels in the boreal mixed wood forest of northeastern Alberta. *Canadian Journal of Botany* **78**:1514–1520.

Dighton, J., and P.A. Mason. 1985. Mycorrhizal dynamics during forest tree development. Pages 117–139 *in* D. Moore, L.A. Casselton, D.A. Wood, and J.C. Frankland, editors. *Developmental Biology of Higher Fungi*. British Mycological Society, Manchester, England.

Dighton, J., J.M. Poskitt, and D.M. Howard. 1986. Changes in the occurrence of Basidiomycete fruitbodies during forest stand development with specific reference to mycorrhizal species. *Transactions of the British Mycological Society* **87**:163–171.

Donaldson, R., and M. Stoddart. 1994. Detection of hypogeous fungi by Tasmanian bettong (*Bettongia gaimardi*: Marsupialia; Macropodoidea). *Journal of Chemical Ecology* **20**:1201–1207.

Fogel, R. 1976. Ecological studies of hypogeous fungi. II. Sporocarp phenology in a western Oregon Douglas-fir stand. *Canadian Journal of Botany* **54**:1152–1162.

Fogel, R. 1981. Quantification of sporocarps produced by hypogeous fungi. Pages 553–568 *in* D.T. Wicklow and G.C. Carroll, editors. *The Fungal Community: Its Organization and Role in the Ecosystem*. Marcel Dekker, New York, NY, USA.

Fogel, R., and G. Hunt. 1979. Fungal and arboreal biomass in a western Oregon Douglas-fir ecosystem: distribution patterns and turnover. *Canadian Journal of Forest Research* **9**:245–256.

Fogel, R., and J.M. Trappe. 1978. Fungus consumption (mycophagy) by small animals. *Northwest Science* **52**:1–31.

Forsman, E.D., E.C. Meslow, and H.M. Wight. 1984. Distribution and biology of the spotted owl in Oregon. *Wildlife Monographs* **87**:1–64.

Franklin, J.F., K. Cromack, Jr., W. Denison, A. McKee, C. Maser, J. Sedell, F. Swanson, and G. Juday. 1981. Ecological characteristics of old-growth Douglas-fir forests. USDA Forest Service, General Technical Report PNW-GTR-118. Portland, OR, USA.

Frappell, P.B. and R.W. Rose. 1986. A radiographic study of the gastrointestinal tract of *Potorous tridactylus*, with a suggestion as to the role of the foregut and hindgut in potoroine marsupials. *Australian Journal of Zoology* **34**:463–471.

Gashwiler, J.S. 1970. Plant and mammal changes on a clear-cut in west central Oregon. *Ecology* **51**:1018–1026.

Getz, L.L. 1968. Influence of water balance and microclimate on the local distribution of the red-backed vole and white-footed mouse. *Ecology* **49**:276–286.

Grayson, D.K., C. Maser, and Z. Maser. 1990. Enamel thickness of rooted and rootless microtine molars. *Canadian Journal of Zoology* **68**:1315–1317.

Green, K. and A.T. Mitchell. 1997. Breeding of the long-footed potoroo, *Potorous longipes* (Marsupialia: Potoroidae), in the wild: behaviour, births and juvenile independence. *Australian Mammalogy* **20**:1–7.

Grenfell, W.E. and M. Fasenfest. 1979. Winter food habits of fishers, *Martes pennanti*, in northwestern California. *California Fish and Game* **65**:186–189.

Grönwall, O., and A. Pehrson. 1984. Nutrient content in fungi as a primary food of the red squirrel *Sciurus vulgaris* L. *Oecologia* **64**:230–231.

Hall, D.S. 1991. Diet of the northern flying squirrel at Sagehen Creek, California. *Journal of Mammalogy* **72**: 615–617.

Hardy, G.A. 1949. Squirrel cache of fungi. *Canadian Field-Naturalist* **63**:86–87.

Harmon, M.E., J.F. Franklin, F.J. Swanson, P. Sollins, S.V. Gregory, J.D. Lattin, N.H. Anderson, S.P. Cline, N.G. Aumen, J.R. Sedell, G.W. Lienkaemper, K. Cromack Jr., and K.W. Cummins. 1986. Ecology of coarse woody debris in temperate ecosystems. *Advances in Ecological Research* **15**:133–302.

Hastings, S., and J.C. Mottram. 1916. Observations upon the edibility of fungi for rodents. *Transactions of the British Mycological Society* **5**:364–378.

Hayes, J.P., S.P. Cross, and P.W. McIntire. 1986. Seasonal variation in mycophagy by the western red-backed vole, *Clethrionomys californicus*, in southwestern Oregon. *Northwest Science* **60**:250–256.

Hume, I.D. 1989. Optimal digestive strategies in mammalian herbivores. *Physiological Zoology* **62**:1145–1163.

Hume, I.D., K.R. Morgan, and G.J. Kenagy. 1993. Digesta retention and digestive performance in sciurid and microtine rodents: effects of hindgut morphology and body size. *Physiological Zoology* **66**:396–411.

Hunt, G.A. and J.M. Trappe. 1987. Seasonal hypogeous sporocarp production in a western Oregon Douglas-fir stand. *Canadian Journal of Botany* **65**:438–445.

Ingold, C.T. 1973. The gift of a truffle. *Bulletin of the British Mycological Society* **7**:32.

Jacobs, K.M. 2002. Response of small mammal mycophagy to varying levels and patterns of green-tree retention in mature forests of Western Oregon and Washington. M.S. Thesis, Oregon State University. Corvallis, Oregon, USA.

Jacobs, K.M., M.A. Castellano, D.L. Luoma, E. Cázares, and J.M. Trappe. 2003. Genera of sequestrate sporocarps of north temperate forests with special reference to animal mycophagy. USDA Forest Service, General Technical Report **PNW-GTR-xxx**. Portland, Oregon, USA.

Johnson, C.N. 1994a. Nutritional ecology of a mycophagous marsupial in relation to production of hypogeous fungi. *Ecology* **75**:2015–2021.

Johnson, C.N. 1994b. Feeding activity of the Tasmanian bettong (*Bettongia gaimardi*) in relation to vegetation patterns. *Wildlife Research* **21**:249–255.

Johnson, C.N. 1995. Interactions between fire, mycophagous mammals, and dispersal of ectomycorrhizal fungi in Eucalyptus forests. *Oecologia* **104**:467–475.

Kinnear, J.E., A. Cockson, P.E.S. Christensen, and A.R. Main. 1979. The nutritional biology of ruminants and ruminant-like mammals – a new approach. *Comparative Biochemistry and Physiology* **64A**:357–365.

Kotter, M.M. and R.C. Farentinos. 1984a. Tassel-eared squirrels as spore dispersal agents of hypogeous mycorrhizal fungi. *Journal of Mammology* **65**:684–687.

Kotter, M.M. and R.C. Farentinos. 1984b. Formation of ponderosa pine ectomycorrhizae after inoculation with feces of tassel-eared squirrels. *Mycologia* **76**:758–760.

Lamont, B.B., C.S. Ralph, and P.E.S. Christensen. 1985. Mycophagous marsupials as dispersal agents for ectomycorrhizal fungi on *Eucalyptus calophylla* and *Gastrolobium bilobum*. *New Phytologist* **101**:651–656.

Laurance, W.F. 1997. A distributional survey and habitat model for the endangered northern bettong (*Bettongia tropica*) in tropical Queensland. *Biological Conservation* **82**:47–60.

Luoma, D.L. 1988. Biomass and community structure of sporocarps formed by hypogeous ectomycorrhizal fungi within selected forest habitats of the H.J. Andrews Experimental Forest, Oregon. Ph.D. Dissertation. Oregon State University, Corvallis, OR, USA.

Luoma, D.L. 1991. Annual changes in seasonal production of hypogeous sporocarps in Oregon Douglas-fir forests. Pages 83–89 *in* L.F. Ruggiero, K.B. Aubry, A.B. Carey, and M.H. Huff, technical coordinators. Wildlife habitat relationships in old-growth Douglas-fir forests. USDA Forest Service, General Technical Report **PNW-GTR-285**. Portland, OR, USA.

Luoma, D.L., and J.L. Eberhart. 2001. Consideration of ectomycorrhizal fungi in sustainable forestry: scaling in ecosystems from root tips to flying squirrels. Abstracts. Third International Conference on Mycorrhizae. Adelaide, Australia.

Luoma, D.L., R.E. Frenkel, and J.M. Trappe. 1991. Fruiting of hypogeous fungi in Oregon Douglas-fir forests: seasonal and habitat variation. *Mycologia* **83**:335–353.

Luoma, D.L., J.L. Eberhart, and M.P. Amaranthus. 1997. Biodiversity of ectomycorrhizal types from Southwest Oregon. Pages 249–253 *in* T.N. Kaye, A. Liston, R.M. Love, D.L. Luoma, R.J. Meinke, and M.V. Wilson, editors. *Conservation and Management of Native Plants and Fungi*. Native Plant Society of Oregon, Corvallis, OR, USA.

Marin, A.B., and M.R. McDaniel. 1987. An examination of hedonic response to *Tuber gibbosum* and three other native Oregon truffles. *Journal of Food Science* **52**:1305–1307.

Martin, K.J. and Anthony, R.G. 1999. Movements of northern flying squirrels in different-aged forest stands of western Oregon. *Journal of Wildlife Management* **63**:291–297.

Maser, C. and Z. Maser. 1988*a*. Interactions among squirrels, mycorrhizal fungi, and coniferous forests in Oregon. *Great Basin Naturalist* **48**:358–369.

Maser, C. and Z. Maser. 1988*b*. Mycophagy of red-gacked voles, *Clethrionomys californicus* and *C. gapperi*. *Great Basin Naturalist* **48**:269–273.

Maser, C., J.M. Trappe, and R.A. Nussbaum. 1978*a*. Fungal-small mammal interrelationships with emphasis on Oregon coniferous forests. *Ecology* **59**:779–809.

Maser, C.M., J.M. Trappe, and D. Ure. 1978b. Implications of small mammal mycophagy to the management of western coniferous forests. *Transactions of the 43rd North American Wildlife and Natural Resources Conference* **43**:78–88.

Maser, Z., C. Maser, and J.M. Trappe. 1985. Food habits of the northern flying squirrel (*Glaucomys sabrinus*) in Oregon. *Canadian Journal of Zoology* **63**:1084–1088.

Maser, C., J.W. Witt, and G. Hunt. 1986. The northern flying squirrel: a mycophagist in southwestern Oregon. *Canadian Journal of Zoology* **64**:2086–2089.

Mason, P.A., J. Wilson, F.T. Last, and C. Walkem. 1983. The concept of succession in relation to the spread of sheathing mycorrhizal fungi on inoculated tree seedlings growing in unsterile soils. *Plant and Soil* **71**: 247–256.

McIlwee, A.P., and C.N. Johnson. 1997. The contribution of fungus to the diets of three mycophagous marsupials in *Eucalyptus* forests, revealed by stable isotope analysis. *Functional Ecology* **12**:223–231.

McIntire, P.W. 1984. Fungus consumption by the Siskiyou chipmunk within a variously treated forest. *Ecology* **65**:137–149.

McIntire, P.W., and A.B. Carey. 1989. A microhistological technique for analysis of food habits of mycophagous rodents. USDA Forest Service, Pacific Northwest Research Station, Research Paper PNW-RP4-404. Portland, Oregon, USA.

McKeever, S. 1960. Food of the northern flying squirrel in northeastern California. *Journal of Mammalogy* **41**:270–271.

Mehus, H. 1986. Fruit body production of macrofungi in some north Norwegian forest types. *Nordic Journal of Botany* **6**:679–701.

Miller, H.A., and L.K. Halls. 1969. Fleshy fungi commonly eaten by southern wildlife. USDA Forest Service, Research Paper SO-49. New Orleans, LO, USA.

Miller, S.L., P. Torres, and T.M. McClean. 1994. Persistence of basidiospores and sclerotia of ectomycorrhizal fungi and *Morchella* in soil. *Mycologia* **86**:89–95.

Mills, L.S. 1995. Edge effects and isolation: red-backed voles on forest remnants. *Conservation Biology* **9**:395–403.

North, M., and J. Greenburg. 1998. Stand conditions associated with truffle abundance in western hemlock/Douglas-fir forests. *Forest Ecology and Management* **112**: 55–66.

North, M., J.M. Trappe, and J. Franklin. 1997. Standing crop and animal consumption of fungal sporocarps in Pacific Northwest forests. *Ecology* **78**: 1543–1554.

Parks, H.E. 1919. Notes on California fungi. *Mycologia* **11**:15–20.

Pyare, S., and W.S. Longland. 2001*a*. Mechanisms of truffle detection by northern flying squirrels. *Canadian Journal of Zoology* **79**:1007–1015.

Pyare, S., and W.S. Longland. 2001*b*. Patterns of ectomycorrhizal-fungi consumption by small mammals in remnant old-growth forests of the Sierra Nevada. *Journal of Mammalogy* **82**:681–689.

Reess, M., and C. Fisch. 1887. Untersuchungen unter bau und lebensgeschichte der hirschtrüffel, *Elaphomyces*. *Bibliotheca Botanica* **7**:1–24.

Richardson, K.C. 1989. Radiographic studies on the form and function of the gastrointestinal tract of the woylie (*Bettongia penicillata*). Pages 205–215 *in* G. Grigg, P. Jarman and I. Hume, editors. *Kangaroos, Wallabies and Rat-kangaroos*. Surrey Beatty & Sons, Sydney, Australia.

Rosenberg, D.K. 1990. Characteristics of northern flying squirrel and Townsend's chipmunk populations in second- and old-growth forests. M.S. Thesis. Oregon State University, Corvallis, OR, USA.

Rosenberg, D.K., and R.G. Anthony. 1992. Characteristics of northern flying squirrel populations in young second- and old-growth forests in western Oregon. *Canadian Journal of Zoology* **70**:161–166.

Rosentreter, R., G.D. Hayward, and M. Wicklow-Howard 1997. Northern flying squirrel seasonal food habits in the interior conifer forests of central Idaho, USA. *Northwest Science* **71**:97–102.

Shemakhanova, N.M. 1967. *Mycotrophy of Woody Plants [Mikotrofiia drevesnykh porod]*. Translated from Russian by S. Nemchonok. Israel Program for Scientific Translations, Jerusalem, Israel.

Smith, S.E., and D.J. Read. 1997. *Mycorrhizal Symbiosis*. Academic Press, London, England.

Smith, W.P., R.G. Anthony, J.R. Waters, N.L. Dodd, and C.J. Zabel. 2003. Ecology and conservation of arboreal rodents of western coniferous forests. Pages 157–206 *in* C.J. Zabel and R.G. Anthony, editors. *Mammal Community Dynamics. Management and Conservation in the Coniferous Forests of Western North America*. Cambridge University Press, Cambridge, UK.

Spies, T., J.F. Franklin, and T. Thomas. 1988. Coarse woody debris in Donglas-fir forests of Western Oregon and Washington. *Ecology* **69**:1689–1702.

Stark, N. 1972. Nutrient cycling pathways and litter fungi. *Bioscience* **22**:355–360.

States, J.S. and W.S. Gaud. 1997. Ecology of hypogeous fungi associated with ponderosa pine. I. Patterns of distribution and sporocarp production in some Arizona forests. *Mycologia* **89**:712–721.

States, J.S. and P.J. Wettstein. 1998. Food habits and evolutionary relationships of the tassel-eared squirrel (*Sciurus aberti*). Pages 185–194 *in* M.A. Steele, J.F. Merritt, and D.A. Zegers, editors. *Ecology and Evolutionary Biology of Tree Squirrels*. Virginia Museum of Natural History, Special Publication Number 6. Martinsville, VA, USA.

Stockdale, C. 2000. Green-tree retention and ectomycorrhiza legacies: the spatial influences of retention trees on mycorrhiza community structure and diversity. M.S. Thesis, Oregon State University, Corvallis, OR, USA.

Taylor, R.J. 1992. Distribution and abundance of fungal sporocarps and diggings of the Tasmanian bettong, *Bettongia gaimardi*. *Australian Journal of Ecology* **17**:155–160.

Taylor, R.J. 1993. Habitat requirements of the Tasmanian bettong (*Bettongia gaimardi*), a mycophagous marsupial. *Wildlife Research* **20**:699–710.

Termorshuizen, A.J. 1991. Succession of mycorrhizal fungi in stands of *Pinus sylvestris* in the Netherlands. *Journal of Vegetation Science* **2**:555–564.

Tevis, L., Jr. 1952. Autumn foods of chipmunks and golden-mantled ground squirrels in the northern Sierra Nevada. *Journal of Mammalogy* **33**:198–205.

Tevis, L., Jr. 1953. Stomach contents of chipmunks and mantled squirrels in northeastern California. *Journal of Mammalogy* **34**:3L6–324.

Tevis, L., Jr. 1956. Responses of small mammal populations to logging of Douglas-fir. *Journal of Mammalogy* **37**:189–196.

Thiers, H.D. 1984. The secotioid syndrome. *Mycologia* **76**:1–8.

Thomas, J.W., E.D. Forsman, J.B. Lint, E.C Meslow, B.R. Noon, and J. Verner. 1990. A conservation strategy for the Northern Spotted Owl: a report of the Interagency Scientific Committee to Address the Conservation of the Northern Spotted Owl. USDA Forest Service and USDI Bureau of Land Management, Fish and Wildlife Service, and National Park Service, Portland, OR, USA.

Thompson, R.L. 1996. Home range and habitat use of western red-backed voles in mature coniferous forests in the Oregon Cascades. M.S. Thesis. Oregon State University, Corvallis, OR, USA.

Trappe, J.M. 1977. Selection of fungi for ectomycorrhizal inoculation in nurseries. *Annual Review of Phytopathology* **15**:203–222.

Trappe, J.M. 1987. Phylogenetic and ecologic aspects of mycotrophy in the angiosperms from an evolutionary standpoint. Pages 2–25 *in* G.R. Safir, editor. *Ecophysiology of VA Mycorrhizal Plants*. CRC Press, Boca Rotan, FL, USA.

Trappe, J.M., and D.L. Luoma. 1992. Chapter 2. The ties that bind: fungi in ecosystems. Pages 17–27 *in* G.C. Carroll, and D.T. Wicklow, editors. *The Fungal Community: Its Organization and Role in the Ecosystem*. Second edition. Marcel Dekker, Inc., New York, NY, USA.

Trappe, J.M., and C. Maser. 1976. Germination of spores of *Glomus macrocarpus* (Endogonaceae) after passage through a rodent digestive tract. *Mycologia* **68**:433–436.

Trappe, J.M., and C. Maser. 1977. Ectomycorrhizal fungi: interactions of mushrooms and truffles with beasts and trees. Pages 165–179 *in* T. Walters, editor. *Mushrooms and Maro, an Interdisciplinary Approach to Mycology*. Linn-Benton Community College, Albany, OR, USA.

Ure, D.C., and C. Maser. 1982. Mycophagy of red-backed voles in Oregon and
    Washington. *Canadian Journal of Zoology* **60**:3307–3315.
Vernes, K., and D.T. Haydon. 2001. Effects of fire on northern bettong (*Bettongia tropica*)
    foraging behaviour. *Austral Ecology* **26**:649–656.
Vernes, K., M. Castellano, and C.N. Johnson. 2001. Effects of season and fire on the
    diversity of hypogeous fungi consumed by a tropical mycophagous marsupial.
    Journal of Animal Ecology **70**:945–954.
Villeneuve, N., F. Le Tacon, and D. Bouchard. 1991. Survival of inoculated *Laccaria bicolor*
    in competition with native ectomycorrhizal fungi and effects on the growth of
    outplanted Douglas-fir seedlings. *Plant and Soil* **135**:95–107.
Visser, S. 1995. Ectomycorrhizal fungal succession in jack pine stands following
    wildfire. *New Phytologist* **129**:389–401.
Vogt, K.A., R.L. Edmonds, and C.C. Grier. 1981. Biomass and nutrient concentrations of
    sporocarps produced by mycorrhizal and decomposer fungi in *Abies amabilis*
    stands. *Oecologia* **50**:170–175.
Waters, J.R., and C.J. Zabel. 1995. Northern flying squirrel densities in fir forests of
    northeastern California. *Journal of Wildlife Management* **59**:858–866.
Waters, J.R., K.S. McKelvey, C.J. Zabel, and W.W. Oliver. 1994. The effects of thinning
    and broadcast burning on sporocarp production of hypogeous fungi. *Canadian
    Journal of Forest Research* **24**:1516–1522.
Waters, J.R., K.S. McKelvey, D.L. Luoma, and C.J. Zabel. 1997. Truffle production in
    old-growth and mature fir stands in northeastern California. *Forest Ecology and
    Management* **96**:155–166.
Wells-Gosling, N. and L.R. Heaney. 1984. *Glaucomys sabrinus*. *Mammalian Species* **229**:1–8.
Wilson, S.M., and A.B. Carey. 2000. Legacy retention versus thinning: influences on
    small mammals. *Northwest Science* **74**:131–145.
Witt, J.W. 1992. Home range and density estimates for the northern flying squirrel,
    Glaucomys sabrinus, in western Oregon. *Journal of Mammalogy* **73**:921–929.
Zabel, C.J. and J.R. Waters. 1997. Food preferences of captive northern flying squirrels
    from the Lassen National Forest in northeastern California. *Northwest Science*
    71:103–107.
Zabel, C.J., K. McKelvey, and J.P. Ward, Jr. 1995. Influence of primary prey on
    home-range size and habitat-use patterns of northern spotted owls (*Strix
    occidentalis caurina*). *Canadian Journal of Zoology* **73**:433–439.
Zielinski, W.J., N.P. Duncan, E.C. Farmer, R.L. Truex, A.P. Clevenger, and R.H. Barrett.
    1999. Diet of fishers (*Martes pennanti*) at the southernmost extent of their range.
    Journal of Mammalogy **80**:961–971.

11
_____

Ecology of coarse woody debris and its role as habitat for mammals

Introduction

A large western redcedar (*Thuja plicata*) may live to be 300 years old, and then may take another 300 years or more to decay (Embry 1963). Throughout its life and after its death, a tree can play a role in contributing to habitat quality for a succession of organisms (Maser et al. 1979) and in contributing to the quality of a site for future tree growth (Harmon et al. 1986). Trees provide cover, foraging sites, and food for many organisms, but the value of a tree to some species of mammals may only begin after the tree has died (Maser and Trappe 1984, Bartels et al. 1985, Hagan and Grove 1999).

Dead wood is a product of disturbances in forested ecosystems as well as a result of stand development (Fig. 11.1). Stands of trees typically develop following a disturbance that provides the opportunity for establishment of often more than 1000 tree seedlings per hectare (Spies and Franklin 1988). As stands develop, inter-tree competition results in mortality among those trees that are intolerant of shade or drought (Drew and Flewelling 1977, Oliver and Larson 1996). It is not uncommon to see over 90% of the trees in a stand die during the first few decades following stand establishment (Oliver and Larson 1996). Inter-tree competition mortality will continue for decades until some exogenous disturbance "thins" the stand providing the remaining trees with sufficient growing space (Agee 1999). At any point in stand development, exogenous disturbances such as fire, wind, ice storms, or insect infestations can lead to mortality that is density independent, creating a pulse of coarse woody debris (CWD) (Fischer and Bradley 1987, Spies et al. 1988, Agee 1999). Consequently, two processes contribute to CWD recruitment in a stand over time:

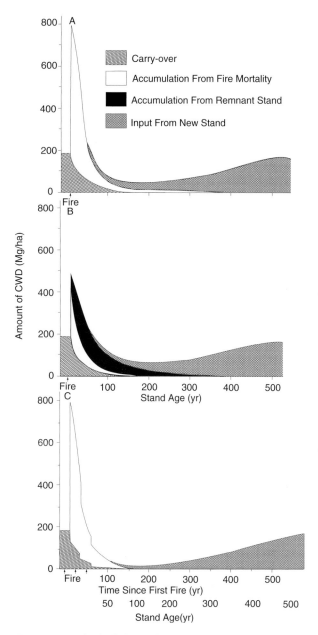

Fig. 11.1. Hypothesized change in coarse woody debris (CWD) biomass per hectare over time since: (A) a catastrophic fire, (B) a partial burn, and (C) subsequent fires following a catastrophic fire (Spies et al. 1988).

(1) the number of trees dying increases rapidly shortly after stand establishment, then declines in a negative exponential manner through the period of "self-thinning" (Drew and Flewelling 1977, Oliver and Larson 1996), and (2) the biomass of CWD increases immediately after an intense disturbance (unless biomass is removed during logging), declines slowly over time, then recovers as large trees die late in stand development (Spies et al. 1988, Fig. 11.1). Woody debris biomass accumulates when inputs of dead wood are greater than decomposition losses. Inputs (suppression mortality or exogenous disturbance) and losses (decomposition or fire) interact to produce the "U"-shaped trend in CWD biomass over time seen in forest types throughout North America (Gore and Patterson 1986, Spies et al. 1988, Van Lear and Waldrop 1994, Spetich et al. 1999).

The accumulation of dead wood and its subsequent decomposition are essential forest ecosystem processes. Dead wood provides a mechanism for energy flow into a detrital-based ecosystem (Harmon et al. 1986). Most energy in a forest is in the form of cellulose that is relatively non-digestible to many animal species. Energy in cellulose is either released to secondary production primarily through decomposition or is lost as heat through fire (Harmon et al. 1986). In addition to facilitating energy flow in forested ecosystems, dead wood along with other features on the forest floor can provide resting, reproduction, and thermal habitat for many mammal species (Maser and Trappe 1984).

Temporal variability in coarse woody debris abundance

The intensity and frequency of disturbances interact to influence the amount and condition of the CWD that is added to a site. Intense disturbances such as stand-replacement fires are usually infrequent and kill most trees in a stand. Low-intensity ground fires are more frequent but may kill only scattered trees (Spies et al. 1988). Intense fires add a pulse of woody debris to a site, while low-intensity ground fires may add little CWD to a site, and instead consume CWD already on the ground, resulting in a net decrease in CWD abundance (Spies et al. 1988). Similarly intense but infrequent wind storms, such as the Columbus Day storm of 1962 in western Oregon and Washington, produce a huge pulse of CWD, while less intense storms lead to windthrow of scattered trees or small clumps. Inter-tree competition represents a very frequent low-intensity disturbance that may produce many dead trees per hectare per year, but changes CWD biomass in a stand only slightly, especially early in stand development (Spies et al. 1988, Oliver and Larson 1996).

Disturbance frequencies and intensities vary among forest types. In low-elevation ponderosa pine (*Pinus ponderosa*) forests fire frequency is relatively high, but intensity is low (Agee 1999), so woody debris inputs are often low and occur every 5–20 years. In the higher-elevation mixed-conifer forests where fire may be less frequent, disturbance intensity is high (Agee 1999), and pulses of woody debris will occur every 50–150 years. The climatic, vegetative and physiographic variables that influence the range of frequencies and intensities of disturbances will dictate the periodicity of pulses of CWD input to a site.

Stand density can also influence CWD input to the site. High-density stands produce many small dead stems early in stand development, but they are of small size (Hester et al. 1989). Competition mortality commences later in stand development in lower density stands, allowing trees to grow rapidly for many years prior to competition (Hester et al. 1989). In these stands, little CWD is added to the site during early stages of stand development, but larger pieces of CWD would be added to the stand later in stand development. Indeed, precommercial thinning to reduce stand density in plantations may benefit the production of CWD of large sizes later in stand development (Hester et al. 1989).

Tree death and decomposition rates interact to dictate CWD biomass fluctuations on a site. Decomposition rates are generally described using decay rate constants (Olson 1963):

$$D_t = D_0 e^{-kt} \text{ where } D = \text{wood density}, t = \text{time (years)},$$
$$k = \text{a decay rate constant.}$$

Decay rate constants for coniferous CWD in the west approximate 0.03 per year (Sollins 1982, Spies et al. 1988). Decay rates of conifers in the western U.S. seem to generally be inversely related to annual precipitation (Harmon et al. 1987) and rates vary among species, with conifers more decay resistant than hardwoods (Harmon and Hua 1991, Table 11.1). As decay proceeds within a bole of wood, the bole becomes subject to fragmentation (Harmon et al. 1986, Tyrell and Crow 1994, Fig. 11.2). As much as 57% of wood loss is the result of fragmentation (Graham 1982). Consequently the CWD biomass on a site at any one time will be dependent on a number of factors. These include site quality and tree species composition, the disturbance regime for the site, and the climatic factors that influence tree growth and decomposition (Muller and Liu 1991). The size and species composition of the live trees influence the potential CWD production on the site. Hardwood forests generally have less CWD than

Table 11.1. *Comparison of decay constants (k) among tree species in various parts of North America (listed from slowest to fastest decay rates)*

Taxon	Location	k (per year)[a]	Citation
Douglas-fir (*Pseudotsuga menziesii*)	Oregon	**0.005–0.10**	Harmon and Hua (1991)
Douglas-fir	Oregon	**0.0063**	Means et al. (1985)
Balsam fir (*Abies balsamea*)	New Hampshire	0.011	Lambert et al. (1980)
Western hemlock (*Tsuga heterophylla*)	Oregon	**0.012**	Greir (1978)
Western hemlock	Oregon	**0.016–0.018**	Harmon and Hua (1991)
Mixed oaks (*Quercus* spp.)	Indiana	0.018	MacMillan (1988)
Western hemlock	Oregon	**0.021**	Graham (1982)
Eastern hemlock (*Tsuga canadensis*)	Wisconsin	0.021	Tyrell and Crow (1994)
Red Spruce (*Picea rubens*)	New Hampshire	0.033	Foster and Lang (1982)
Jack pine (*Pinus banksiana*)	Minnesota	0.042	Alban and Pastor (1993)
Mixed maples Indiana (*Acer* spp.)		0.045	MacMillan (1988)
Red pine (*Pinus resinosa*)	Minnesota	0.055	Alban and Pastor (1993)
White spruce (*Picea glauca*)	Minnesota	0.071	Alban and Pastor (1993)
Trembling aspen (*Populus tremuloides*)	Minnesota	**0.080**	Alban and Pastor (1993)
Mixed hardwoods	New Hampshire	0.096	Arthur et al. (1993)
Mixed hardwoods	Tennessee	0.110	Onega and Eickmeier (1991)

Species occurring in western forests are in bold.

[a] k = a decay rate constant when calculating decay rates as $D_t = D_0 e^{-kt}$ where D = wood density, t = time (years).

conifer forests (Harmon et al. 1986, Harmon and Hua 1991). Eastern hardwood forests may support 11–112 m³/ha of CWD (Tyrell and Crow 1994), but western coniferous forests may have 376–1421 m³/ha of CWD (Huff 1984). Variability in the amounts and distribution of both standing and fallen CWD is considerable (Everett et al. 1999). Indeed, managing CWD to reflect variability among sites over a landscape may be a more meaningful approach than mandating a minimum retention level in managed stands (Everett et al. 1999).

Spatial variability and coarse woody debris

CWD historically was distributed across the landscape as a result of the distribution of patch sizes, conditions, and recovery rates following the

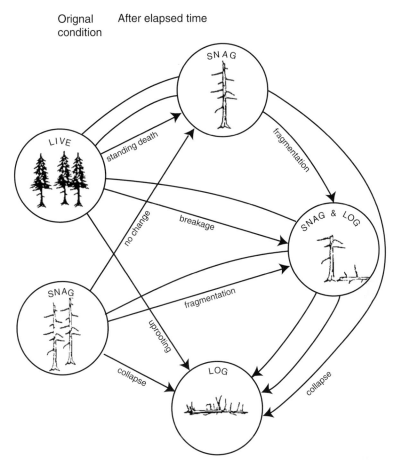

Fig. 11.2. Process of coarse woody debris recruitment and decay (from Tyrell and Crow 1994).

dominant disturbance regimes for the area. A useful frame of reference when considering the spatial and temporal patterns of CWD in western coniferous forests is the historic range of variability (HRV) that occurred under natural disturbance regimes (Landres et al. 1999). By comparing the HRV in CWD to what might be expected in managed stands and landscapes, we can begin to identify where current or future deficiencies may exist.

Over the past 3000 years, old-growth forest probably covered 25% to 75% of Oregon Coast Range at any one time (Wimberly et al. 2000, Fig. 11.3),

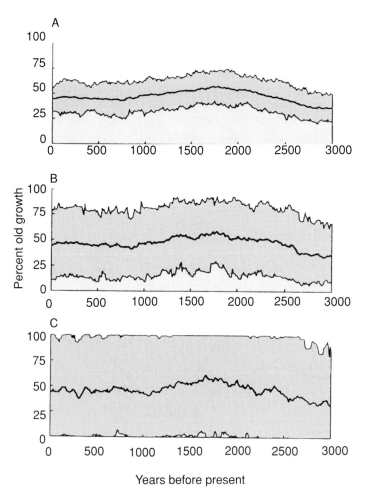

Fig. 11.3. Simulated temporal variability in percent of old-growth forest over the past 3000 years in the Oregon Coast Range at three spatial scales: (A) 2 250 000 ha; (B) 302 500 ha; and (C) 40 000 ha (Wimberly et al. 2000). Shaded area indicates 95% quantiles.

and patch sizes were probably large given the infrequent, high-intensity fires that likely occurred in that environment. Consequently, cohorts of woody debris probably occurred as infrequent pulses into the system over very large areas, followed by decomposition, and finally increases in woody debris recruitment through chronic inputs from individual and group tree death (i.e., from wind, root rot, senescence, Fig. 11.1). For

Table 11.2. *Likely ranges of coarse woody debris (CWD) variability among forest age classes in the Oregon Coast Range under the historic range of variability (HRV), based on Wimberly et al. (2000)*

Age class (years)	Coast Range (%)	Coast Range (ha × 100000)	Number of patches (m³/ha)	CWD range
0–30	4–11	0.9–2.5	1–4	376–1421[a]
31–80	6–19	1.4–4.3	1–6	163–305[b]
80–200	15–45	3.4–10.1	2–14	93–165[b]
>200	25–75	5.6–16.9	4–24	219–324[b]

[a] From Huff (1984).
[b] From Spies and Franklin (1991).

instance, consider the following CWD biomass ranges for the Cascades and Coast Ranges of Washington and Oregon:

- Post-fire or wind disturbance (<30 years) = 376–1421 m³/ha (Huff 1984).
- Young (30–80 years) = 163–305 m³/ha (Spies and Franklin 1991)
- Mature (80–200 years) = 93–165 m³/ha (Spies and Franklin 1991)
- Old-growth (>200 years) = 219–324 m³/ha (Spies and Franklin 1991)

To illustrate the likely distribution of CWD across the Oregon Coast Range, I used Wimberly et al.'s (2000) estimates of historic old-growth area to estimate the likely 95% confidence intervals of CWD biomass among several age classes (Fig. 11.3). If 25% to 75% of Coast Range was old-growth over the past 3000 years, then 25% to 75% of the remaining landscape was in 0- to 200-year-old age classes. Based on probability, approximately 15% of the non-old-growth would be <30 years old (30/200 = 0.15) at any one time, 25% would be 30–80 years old, and 60% would be 80–200 years. Because fires are infrequent but intense, patch sizes typically were large (70600–141200 ha) (Wimberly et al. 2000). The distribution of patches across the Coast Range containing CWD levels under this hypothesized HRV (Table 11.2) raise questions regarding current levels and distribution of CWD across the Coast Range. Recent and young harvested stands now cover over 65% of the Oregon Coast Range (Table 11.3), and some of these stands have CWD levels below that expected based on the HRV (Butts and McComb 2000). During periods of high utilization, CWD retention in plantations is low; when low-quality timber is left on the site, CWD retention in plantations is high (Fig. 11.4). It seems likely that there is the potential for large areas

Table 11.3. *Estimated percent of the Oregon Coast Range in each of four age classes derived from Wimberly et al. (2000; see Table 11.2), and conditions during the period 1984–1996, from Ohmann and Waddell (in press)*

Age class	Estimated historic range of variability (%)	Current unharvested (%)	Current harvested (%)	Current total (%)
Recent (<30 years)	4–11	3.8	35.3	**39.2**
Young (30–80 years)	6–19	10.7	30.1	**40.8**
Mature (80–200 years)	15–45	7.1	4.5	**11.6**
Old-growth (200+ years)	25–75	2.9	1.3	**4.2**
Total[a]		24.6	71.2	95.8

Conditions out of the historic range of variability are in bold.
[a] Total is less than 100% because areas that were salvage logged were excluded from the total.

Fig. 11.4. A 30-year-old Douglas-fir plantation in the Oregon Coast Range established during a period of high wood utilization standards.

west of the Cascades to have CWD levels below that expected within the HRV.

Based on this cursory analysis for this region of the west, there seems to be the potential for adverse effects on species and processes reliant on

CWD in all stand age classes. It is important to keep in mind that these estimates apply to the 2 250 000 ha of the Oregon Coast Range as affected by stand-replacement fire. Estimates of the percent of smaller areas occupied by each age class are much more variable over time (Wimberly et al. 2000; Fig. 11.3), and estimates from other disturbances (wind, disease) would create greater spatial variability. Similar approaches may need to be developed for the Cascades and Interior forests using information in fire regimes as the basis for estimates (Morrison and Swanson 1990, Agee 1999).

Functions of coarse woody debris as habitat for mammals

Habitat is the integration of resources necessary to support a population over space and through time (McComb 2001). Each species has its own habitat requirements, and in forest systems these are often related to vegetative structure and composition, forest floor and below-ground structure and complexity, availability of water, and the spatial arrangement of these attributes across a landscape (McComb 2001). CWD provides habitat for some mammal species. It is clear that many species use CWD as a habitat element within their home range (Table 11.4). There are several key attributes of CWD that influence its value to mammals: piece size and condition (decay stage), biomass or areal cover, and the successional stage in which it occurs.

Piece size can be important to mammals for a number of reasons. Large-diameter logs provide more cover per piece than small diameter logs (Maser et al. 1979, Fig. 11.5). Western red-backed voles (*Clethrionomys californicus*) select large logs as cover (Hayes and Cross 1987), and CWD provides cover and a source of fungi for food for southern red-backed voles (*C. gapperi*) (Buckmaster et al. 1996). A wide variety of other species also reportedly use logs as cover: shrews (*Sorex* spp.), weasels (*Mustela* spp.), mink (*Mustela vison*), and northern river otters (*Lontra canadensis*), among others (Maser et al. 1981, Table 11.4). Further, long logs provide more connectivity across the forest floor than short logs. Connectivity throughout a home range theoretically can influence animal fitness because an individual can remain under cover during movements, thereby reducing the risk of predation while also possibly providing microclimatic advantages to the organism.

The distribution of piece sizes in a forest generally reflects the site quality for tree growth, stage of stand development, and sources of mortality that in the past led to tree death. Trees dying from suppression mortality

Table 11.4. *Species listed by O'Neil et al. (2001) as using coarse woody debris in Oregon and Washington*

Common name	Use [a]
Virginia opossum (*Didelphis virginiana*)	Denning
Masked shrew (*Sorex cinereus*)	Foraging
Montane shrew	Foraging, nesting
Baird's shrew (*S. bairdii*)	Foraging
Fog shrew	Foraging
Pacific shrew	Foraging
Water shrew	Foraging
Marsh shrew	Foraging
Trowbridge's shrew	Foraging, nesting
Shrew-mole	Foraging, nesting
Coast mole (*Scapanus orarius*)	Foraging
Long-eared myotis (*Myotis evotis*)	Roosting
Mountain beaver	Hiding
Least chipmunk (*Tamias minimus*)	Nesting, hiding, cache site
Yellow-pine chipmunk (*T. amoenus*)	Nesting, hiding, cache site
Townsend's chipmunk (*T. townsendii*)	Nesting, hiding, cache site
Siskiyou chipmunk (*T. siskiyou*)	Nesting, hiding, cache site
Red-tailed chipmunk (*T. ruficaudus*)	Nesting, hiding, cache site
Yellow-bellied marmot (*Marmota flaviventris*)	Hiding
Columbian ground squirrel (*Spermophilus columbianus*)	Nesting, hiding
Golden-mantled ground squirrel (*S. lateralis*)	Nesting, hiding, cache site
Red squirrel (*Tamiasciurus hudsonicus*)	Cache site
Douglas squirrel	Foraging, cache site
Northern flying squirrel (*Glaucomys sabrinus*)	Foraging
Northern pocket gopher (*Thomomys talpoides*)	Hiding
Western harvest mouse (*Reithrodontomys megalotis*)	Hiding
Deer mouse	Nesting, hiding
Northwestern deer mouse (*Peromyscus keeni*)	Nesting, hiding
Dusky-footed woodrat (*Neotoma fuscipes*)	Nesting, hiding, cache site
Bushy-tailed woodrat	Nesting, hiding, cache site
Southern red-backed vole	Nesting, hiding, foraging
Western red-backed vole	Nesting, hiding, foraging
Creeping vole	Hiding
Water vole (*Microtus richardsoni*)	Hiding
Western jumping mouse (*Zapus princeps*)	Hiding
Pacific jumping mouse (*Zapus trinotatus*)	Hiding
Porcupine (*Erithizon dorsatum*)	Denning
Gray fox (*Urocyon cineroargenteus*)	Denning, foraging
Black bear	Denning, foraging
Grizzly bear (*Ursus arctos*)	Foraging
Ringtail (*Bassariscus astutus*)	Denning, foraging
Raccoon (*Procyon lotor*)	Denning, foraging
American marten	Denning, foraging
Fisher (*Martes pennanti*)	Denning, foraging
Western spotted skunk	Denning, foraging
Striped skunk (*Mephitis mephitis*)	Denning, foraging
Northern river otter	Denning
Mountain lion (*Felis concolor*)	Foraging
Lynx (*Lynx canadensis*)	Foraging
Elk	Hiding
Black-tailed deer	Hiding

[a] Use based on Thomas (1979), Maser et al. (1981), Brown (1985), and personal observations.

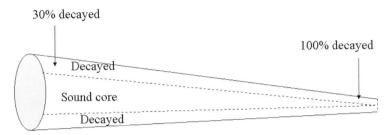

Fig. 11.5. Log condition as a function of piece size. A large log can provide more area of cover for a wider range of animal sizes. Many small logs may not function the same as one large log.

are typically 50% the diameter (but similar in length) of dominant and co-dominant trees in the stand (based on numerous projections using ORGANON, Hester et al. 1989). Because large-diameter pieces take longer to fully decay than small diameter pieces of the same species (a decay constant is applied to a larger mass), large piece sizes may provide an advantage in longevity as functional CWD. The desired size class distribution for the suite of mammal species being managed in the stand or landscape should be determined by the species requiring the largest piece size.

The areal cover or biomass of CWD may influence the function of the wood to mammals. Mean CWD volumes at American marten (*Martes americana*) foraging and resting locations on Vancouver Island, British Columbia were higher for females (638 ± 198.7 m³/ha to 645 ± 93 m³/ha; 9118–9218 ft³/acre) than for males (445 ± 87 m³/ha to 492 ± 48 m³/ha; 6359–7031 ft³/acre) (Baker 1992). Black bears also use stands with high frequencies of CWD for foraging in British Columbia (Davis 1996).

The physical structure of the log is important to some species. Maser et al. (1979) described stages of log decay that are similar to those used to describe snag decay stages (Fig. 11.6). Each stage of decomposition can provide different resources to a suite of organisms (Maser and Trappe 1984). The decomposition process releases energy to other trophic levels from wood through saprophytic fungi and wood-feeding insects (Maser and Trappe 1984). Early in the decay process, sloughing bark and infestation by bark beetles (*Dendroctonus* spp.), carpenter ants (*Camponotus* spp.), and termites provide food and cover resources for small mammals, bears, and woodpeckers (Maser and Trappe 1984, Torgersen and Bull 1995). Once the wood has softened and fragments, vertebrates can begin to excavate

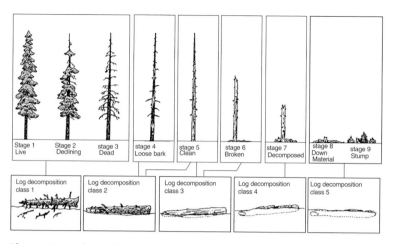

Fig. 11.6. Decay classes of logs and snags (from Maser et al. 1979).

into the wood to extract insects and/or build nests. Western red-backed voles and montane shrews (*S. monitcolus*) use very decayed logs as nest sites (Zeiner et al. 1990, Tallmon and Mills 1994, Thompson 1996) that provide cryptic, dry, and thermally stable environments for their young. Thompson (1996) found that 80% of 746 telemetry locations for 24 western red-backed voles coincided with down logs, indicating the selection of logs as an important habitat element. Raphael and Jones (1997) reported that 32% of American marten breeding and denning sites were in large logs of intermediate decay classes. Eventually the structural integrity of the log is so severely compromised by the fungal infection that the log loses value as a potential nest site or feeding site.

Some mammals such as bushy-tailed woodrats (*Neotoma cinerea*), coyotes (*Canis latrans*), red foxes (*Vulpes vulpes*), black bears, western spotted skunks (*Spilogale gracilis*), and Douglas squirrels (*Tamiasciurus douglasii*) also use hollow logs as dens or food-caching sites (Maser et al. 1981, Zeiner et al. 1990, Timossi et al. 1995, Maser and Gashwiler 1978). Logs do not become hollow after they have fallen, they only occur after a hollow tree falls to the ground. Hollow trees form because trees compartmentalize rot into a column of decay that extends up and down the bole from the wound (Shigo 1984). Recruiting hollow logs into managed stands requires the identification and retention of injured and decaying trees, allowing them to grow to sufficient size or to decay to an acceptable extent, then allowing or promoting their death. Black bears use hollow logs averaging

106 cm (42 inches) in diameter for winter denning in British Columbia (Davis 1996), so recruitment of potential den sites for bears may take centuries. It is apparent that CWD can function as a habitat element for many mammal species in all successional stages of coniferous forests in the western U.S. Based on these functional relationships, we would expect to see numerical relationships between mammals and CWD.

Numerical relationships between CWD and mammals

Numerical relationships between CWD and mammals may emerge at one or more spatial scales meaningful to the species of interest. Sampling of mammals and CWD often occurs at one or more spatial scales to explore associations: (1) selection of habitat elements or patches within a species home range, (2) correlation of mammal abundance with CWD characteristics, or (3) correlation of mammal distributions with CWD across heterogeneous landscapes.

Use of CWD as a habitat element

Runde et al. (1999) and O'Neil et al. (2001) summarized many examples of use of CWD by vertebrates in western coniferous forests. Clearly CWD is used if it is available, and may contribute to the diversity of food and cover resources in an area for a species. The ecological advantage of CWD to these species is less clear. One way to assess if there might be an ecological advantage to a species is to determine if the species selects CWD in higher proportion than its availability. Selection of CWD pieces or patches has been demonstrated for a few species of mammals. For instance, Butts and McComb (2000) found that capture sites of Trowbridge's shrews (*S. trowbridgii*) had a higher percent cover by logs than non-capture sites, but no relationships were detected for fog shrews (*Sorex sonomae*). Montane shrews select areas with concentrations of fallen logs and shrubs during the winter, switching to areas with herbaceous ground cover in spring and summer (Belk et al. 1988). Capture probability was positively related to CWD for marsh shrews (*Sorex bendirii*) but negatively associated with CWD for vagrant shrews (*S. vagrans*) in the Oregon Coast Range (Martin 1998).

CWD is an important element in selection of re-colonization sites by mountain beaver (*Aplodontia rufa*) following clear-cutting (Hacker and Coblentz 1993), and the highest densities of mountain beaver are often associated with CWD in early-seral forests (Carraway and Verts 1993).

Beavers (*Castor canadensis*) select stream reaches with CWD as an anchor for a dam (Leidholt-Bruner 1990), and northern river otters select CWD and beaver lodges as resting sites (Melquist and Hornocker 1983). Capture probability was positively related to CWD for western red-backed voles and negatively associated with CWD for Pacific jumping mice (*Zapus trinotatus*) (Martin 1998). American marten selected logs as resting sites that were larger (mean diameter = 96.7 cm (38 in), mean length = 14.6 m (48 ft)) than the average available (47.2 cm (19 in) diameter, 8.2 m (27 ft) long) in northern California (Martin and Barrett 1991).

Correlations between CWD and mammals
There is evidence, largely through correlational studies, that the abundance or fitness of some species is related to CWD (Table 11.5).

Insectivores
Trowbridge's shrews are one of the most abundant insectivores in western coniferous forests. Their abundance and reproductive rates increase with increasing abundance of CWD (Table 11.5). Positive correlations between CWD and this species were reported with only one exception: a negative association with decay class 1–2 logs (Corn et al. 1988). Montane shrews have higher reproductive rates and abundance in areas with high amounts of CWD (Table 11.5), and their abundance has been correlated with percent cover by stumps in western Washington (West 1991). Capture rates of water shrews (*S. palustris*), marsh shrews (*S. bendirii*), and Townsend's moles (*Scapanus townsendii*) were higher in areas with abundant logs than in areas with few logs (Corn et al. 1988, Gilbert and Allwine 1991). Not all insectivore species are positively associated with CWD, as vagrant shrews were negatively correlated with various aspects of CWD (Table 11.5).

Rodents
Log volume averaged twice as high in the core area of western red-backed vole home ranges than in random locations in the Oregon Cascades (Thompson 1996). Gilbert and Allwine (1991) reported positive associations between cover by logs and captures of western red-backed voles, red tree voles (*Arborimus longicaudus*), and Pacific jumping mice (*Zapus trinotatus*) in the Oregon Cascades. Southern red-backed vole (*Clethrionomys gapperi*) abundance also is related to abundance of CWD (Table 11.5), but West (1991) reported a negative relationship between log cover and southern red-backed voles. He also pointed out that correlations between capture

rates and microhabitat variables were generally weak (West 1991). Not all rodents are associated with CWD. Creeping vole captures in western Washington forests were up to 26 times higher in intensively managed second-growth forests with little CWD than in younger unthinned second-growth with a high amount of CWD but less understory cover (Wilson and Carey 1996). Associations between CWD and deer mouse (*Peromyscus maniculatus*) abundance are inconsistent among three different studies (Table 11.5).

Relationships between mammals and CWD over landscapes

Studies have been conducted at the patch and landscape level to assess the potential for effects of patch isolation and landscape pattern to be associated with genetic heterogeneity and animal abundance, respectively (Mills 1995, Martin 1998). Martin (1998) sampled 30 250- to 300-ha landscapes representing a range of old-forest area and patch configurations. He found that capture rates for six mammal species were associated with area of old forest: Trowbridge's shrew, Pacific shrew (*Sorex pacificus*), fog shrew, shrew mole (*Neurotrichus gibbsii*), western red-backed vole, and red tree vole. Patch configuration was not associated with capture rates for any of these six species, and CWD biomass was not related to capture rates at this scale. Western red-backed voles were negatively associated with edges between young and old conifer forests in Oregon, but CWD biomass did not seem to influence this relationship (Mills 1995). The only significant relationships between capture rates and CWD that Martin (1998) detected were at the trap site scale. Microhabitat relationships may be stronger for these species than patch- or landscape-scale relationships.

CWD obligates and associates

Based on the literature reported in Table 11.5 and previous literature reviews by Runde et al. (1999) and O'Neil et al. (2001), there do not seem to be any mammal species that can be considered CWD obligates, but the following species could be considered CWD associates: Trowbridge's shrew, Pacific shrew, montane shrew, marsh shrew, shrew mole, Townsend's mole, mountain beaver, American beaver, southern red-backed vole, western red-backed vole, American marten, northern river otter, and black bear. There probably are other species associated with CWD, but relationships have not yet been detected. There are a few species that seem to avoid areas with high amounts of CWD including vagrant shrews, Oregon voles

Table 11.5. *Examples of statistical relationships between mammals and coarse woody debris (CWD) in forests of the western U.S.*

Species	Dependent variable[a]	Independent variable[b]	Relationship[c]	Citation
Vagrant shrew	Abundance	Amount	−	Lee (1995)
Vagrant shrew	Abundance	Decay 4,5	−0.55	Corn et al. (1988)
Vagrant shrew	Presence	Amount	−	Martin (1998)
Trowbridge's shrew	Abundance	Amount	+	Lee (1995)
Trowbridge's shrew	Abundance	Decay 1,2	−0.53	Corn et al. (1988)
Trowbridge's shrew	Abundance	Cover (%)	+0.35	Butts and McComb (2000)
Trowbridge's shrew	Abundance	Cover (%)	+	Taylor et al. (1988)
Trowbridge's shrew	Abundance	Cover (%)	+	Carey and Johnson (1995)
Pacific shrew	Repro.	Amount	+	Lee (1995)
Montane shrew	Abundance	Amount	+	Raphael (1988)
Montane shrew	Repro.	Amount	+	Lee (1995)
Montane shrew	Abundance	Cover (%)	+0.30–0.49	Mason (1989)
Montane shrew	Abundance	Decay 1,2	−0.47	Corn et al. (1988)
Marsh shrew	Abundance	Decay 1,2	−0.50	Corn et al. (1988)
Marsh shrew	Abundance	Decay 1–3	+0.26	Gilbert and Allwine (1991)
Marsh shrew	Presence	Amount	+	Martin (1998)
Shrew mole	Abundance	Cover (%)	+0.37	Butts and McComb (2000)
Shrew mole	Abundance	Cover (%)	+0.59	Carey and Johnson (1995)
Townsend's mole	Abundance	Decay 2–5	+0.30	Gilbert and Allwine (1991)
Coast mole	Abundance	Decay 3	−0.43	Corn et al. (1988)
Beaver	Presence	Amount	+	Leidholt-Bruner (1990)
Douglas squirrel	Abundance	Amount	+	Raphael (1988)
Townsend's chipmunk	Presence	Cover (%)	−	Morrison and Anthony (1989)
Deer mouse	Abundance	Amount	o	Hayes and Cross (1987)
Deer mouse	Abundance	Cover (%)	+0.28–0.48	Mason (1989)
Deer mouse	Abundance	Cover (%)	+	Carey and Johnson (1995)

Forest deer mouse	Abundance	Decay 3–5	+0.17	West (1991)
Pacific jumping mouse	Abundance	Decay 3,4	+	Gilbert and Allwine (1991)
Pacific jumping mouse	Presence	Amount	–	Martin (1998)
Red tree vole	Abundance	Decay 2	+0.27	Gilbert and Allwine (1991)
So. red-backed vole	Abundance	Amount	+	Belk et al. (1988)
So. red-backed vole	Abundance	Cover (%)	+0.40–0.49	Mason (1989)
So. red-backed vole	Abundance	Decay 1–2	–0.23	West (1991)
So. red-backed vole	Abundance	Cover (%)	+	Carey and Johnson (1995)
Western red-backed vole	Abundance	Amount	–	Rosenberg et al. (1994)
Western red-backed vole	Abundance	Amount	–0.89	Thompson (1996)
Western red-backed vole	Abundance	Decay 5	+0.31	Gilbert and Allwine (1991)
Western red-backed vole	Abundance	Decay 1,2	–0.53	Corn et al. (1988)
Western red-backed vole	Presence	Log size	+	Hayes and Cross (1987)
Western red-backed vole	Presence	Decay 3,4	+	Tallmon and Mills (1994)
Western red-backed vole	Presence	Amount	+	Martin (1998)
Creeping vole	Abundance	Amount	–	Wilson and Carey (1996)
Creeping vole	Abundance	Decay 4,5	–0.43–0.58	Corn et al. (1988)
Long-tailed vole	Abundance	Cover (%)	+0.28–0.40	Mason (1989)
Ermine	Presence	Amount	+	Wilson and Carey (1996)
Marten	Presence	Size	+	Martin and Barrett (1991)
River otter	Presence	Amount	+	Melquist and Hornocker (1983)
Black bear	Presence	Amount	+	Davis (1996)
Black-tailed deer	Presence	Amount	–	Nyberg (1990)
Elk	Presence	Amount	–	Nyberg (1990)

[a] For most species "Presence" represents more than one capture or observation; "Abundance" represents capture rates in live traps, snap traps or pitfalls; "Repro." represents reproductive rates.

[b] "Amount" may represent volume or number of pieces; "decay" represents the amount within specific decay classes.

[c] Where the study conducted regression or correlation analyses, r-values are presented, or the direction of the relationship is presented (in the case of use/availability studies, logistic regression, or multiple linear regression analyses).

(*Microtus oregoni*), ermine (*Mustela erminea*), mule deer (*Odocoileus hemionus*), and elk (*Cervus elaphus*) (Table 11.5).

The ability of investigators to detect relationships can be affected by a number of factors. Relationships between mammals and CWD are clearly influenced by sampling techniques (Taylor et al. 1988). In addition, populations of many small mammal species fluctuate from year to year. If habitat selection is density dependent (van Horne 1983), then observed relationships during periods of low populations may not be detectable when populations are high because even marginal habitat is being used in systems where CWD is abundant (e.g., mixed-conifer forests of the interior west). Finally, most studies conducted to date have been designed to assess relationships between vegetative patterns and mammal abundance. Few have focused primarily on CWD relationships. Both CWD and mammal populations are patchily distributed and inherently variable. Studies focused on sampling a range of CWD levels over spatial scales encompassing multiple home ranges may be needed to detect associations at the level of precision needed to develop CWD management guidelines for CWD associates.

Functional redundancy in forest systems

Mammalian ecologists have noticed for years that many species use CWD (Table 11.4). Attempts to document quantitative relationships between CWD and animal occurrence, abundance or fitness have produced inconsistent, weak or counterintuitive relationships for some species. An explanation for inconsistency of results is that CWD as a form of cover may play an important role in some systems but not others, because CWD may be one of several structures that can be used by an organism. For many species other forms of cover are available, such as rocks, burrows, dense vegetation, and leaf litter. Indeed, a review of the seven studies reveals strong relationships between mammals and characteristics of the leaf litter, vegetation composition, distance to water, seral stage, or topographic features (Morrison and Anthony 1989, Corn and Bury 1991, Gilbert and Allwine 1991, West 1991, McComb et al. 1993*a*, *b*, Thompson 1996, Gomez and Anthony 1998). Alternative structures often are available to CWD associates. For some species, food may be more abundant where CWD is abundant, but food may not be a limiting factor for some populations. So under what conditions is CWD likely to be critical for mammals? Based on the physiological requirements of forest floor organisms that use

CWD, I would expect relationships to be strongest under the following conditions:

- when sources of cover other than CWD are inadequate, or
- when food is inextricably linked to dead wood.

Even if consistent relationships between CWD and mammals are not found in all studies, CWD could be important to populations under these two conditions. Further, there also are indirect effects of CWD on habitat quality for some mammal populations. Decomposition and mineralization processes that occur in dead wood have the potential to influence the composition and structure of a forest (Harmon et al. 1986), and hence the vegetative conditions important to some mammal species. CWD influences soil moisture and chemistry thereby impacting regeneration and early seedling survival (Sollins et al. 1987). There also is the potential for changes in site productivity (Harmon et al. 1986). For species dependent on aquatic environments, CWD in streams can influence the distribution of fish, stream geomorphology, and production of aquatic invertebrates (Bisson et al. 1987, Maser et al. 1988).

Spatial and temporal scaling of CWD must be considered when interpreting relationships between CWD and mammals. Because coniferous forests are detrital-based systems, a reduction in CWD on a site may lead to reduced energy transfer from dead wood to secondary producers (Harmon et al. 1986). If this phenomenon occurs over a large area, and populations are monitored over multiple generations as wood levels decline, then effects might be observed many years in the future. Unfortunately if effects were detected, it would then take decades or centuries to restore the CWD in the system.

Coarse woody debris management

Despite the lack of consistent quantitative relationships between mammalian abundance and CWD, managers are required to make decisions regarding CWD retention and management. There are two general approaches to CWD management, and they represent two complementary philosophies. First, a manager might ask, "Do the levels of CWD biomass, piece size, and condition over large areas fall within the HRV?" If the answer is "no", questions should be raised about the species and processes that may be adversely affected by this departure from the HRV, and if management actions should be taken to move stands or landscapes back into the HRV. If impacted species and processes are adequately addressed elsewhere in the landscape, then allowing some stands or landscapes to fall outside the HRV may be an acceptable risk. If, however, the addition of another stand or landscape to areas that already fall outside the HRV mean that there is a likelihood of cumulative adverse effects over space and time, then the manager may wish to take actions that contribute to goals related to the HRV (Landres et al. 1999). This approach is based on risk assessment, where the consequences of managing outside the HRV are largely unknown, but, if an error is made, desirable conditions could be costly to restore.

Alternatively, the manager can assess functional relationships between animals and CWD and manage for these conditions as part of a desired future condition (McComb and Lindenmayer 1999). These functional relationships are not clear for most species, but they can be hypothesized and tested in an adaptive management approach. Indeed, some habitat relationship models already include estimates of the abundance of CWD as one component contributing to habitat quality for a species (e.g., Allen 1983). The compilations of these relationships for desired species in future landscapes would dictate the CWD goals. I discourage the use of a guild approach in CWD management because species within a guild often do not respond to management actions in a similar manner across all life functions (Mannan et al. 1984).

Regardless of the CWD management approach chosen, managers should identify high priority sites for CWD management (Bettinger et al. 2001). Intensively managed plantations might fall within this group because they have CWD levels that likely fall outside of the HRV (Butts and McComb 2000). Modest inputs of CWD to these stands may make a greater impact on animal habitat and/or ecological processes than a similar treatment in stands that already contain CWD.

The steps in the management process that I recommend are:

1. Inventory CWD at the desired scale at time 0 (see Harmon and Sexton 1996 for inventory techniques).
2. Compare CWD levels to the HRV estimates for the region and/or compare estimates to habitat goals for species in the desired future condition.
3. Conduct this analysis across the planning area, and prioritize stands for CWD management based on the risk of not meeting future CWD goals.

4. Beginning with the highest priority stands, determine if there are trees of sufficient size that could be felled or killed now to fulfill the CWD goal.

5. If trees in the current stand are not appropriate for meeting CWD goals, then silvicultural actions should be considered to achieve goals. Thinning from below to allow dominant and co-dominant trees to grow more rapidly may be preferable to allowing an overstocked stand to grow slowly and to contribute small amounts of CWD to the stand.

6. Monitor species of highest concern prior to and following active management and assess if populations decline. Given the long-term nature of wood decay and the habitat functions that develop throughout decay processes, monitoring may need to occur periodically for decades.

7. Assess monitoring results and decide if changes should be made to desired future conditions and CWD goals for the area.

Forest Practices Act guidelines (e.g., Oregon Department of Forestry 1996) provide a mechanism for ensuring that minimum retention levels fall within the HRV. Currently in Oregon, the Forest Practices Act requires approximately 1.5 m³/ha of CWD be retained within harvest units (Oregon Department of Forestry 1996). Under the Northwest Forest Plan, approximately 37 m³/ha is required (U.S. Forest Service and U.S. Bureau of Land Management 1994). Both values are below the 95% confidence intervals for what would be expected in unmanaged stands (Spies and Franklin 1991).

Several factors come into play when regulating CWD levels in managed stands. First, the minimum level of the range chosen for regulation is usually the only level that managers will strive to retain in stands. Further, providing one CWD level in all managed stands homogenizes that condition over managed landscapes. It should be clear from the ranges listed, that current practices will not result in the variability in CWD conditions found naturally (Everett et al. 1999). Although current CWD guidelines could be re-written to ensure that CWD levels fall **within** the HRV, it is much more difficult to develop regulations that will lead to CWD levels that represent the HRV for the region. Incentives such as CWD credits provided to landowners by local, state, or federal agencies may allow better representation of the HRV in CWD conditions across landscapes.

Clearly such management actions will require a commitment of time and money to CWD. Costs can be modest if management is for one or a few species, higher if CWD is managed to fall **within** the HRV, and higher still to manage the full range of historic variability. CWD guidelines should be scale-dependent, however. CWD biomass among many stands should collectively contribute to landscape goals. Landscapes should also represent variability in CWD levels, but collectively contribute to regional goals.

Temporal variability is also important. A delay in CWD management in a stand with low levels of CWD now may result in a gap in CWD availability in the future. Certainly a few stands with low CWD levels in an area with otherwise high levels may be relatively unimportant, unless overall CWD levels decline over time and no action is taken now to ensure that advanced decay class (class 5) logs will occur in the stands 50 years from now.

Information and research needs

Understanding the HRV of a variety of aspects of vegetation structure and composition, including CWD, can provide a useful basis for making management decisions. This approach provides managers with a point of reference and allows them to establish CWD goals until more detailed mammal–habitat relationships are quantified. The approach that I have taken in this chapter is elementary and does not include the stochastic disturbance processes that must be adequately represented to fully understand the HRV in a system. Estimation of HRV in CWD and other habitat elements must be spatially explicit and must represent a range of spatial scales relevant to the species and processes of interest (Wimberly et al. 2000). Having estimates of the HRV for CWD among the ecoregions of the west is a high priority research need.

To complement the information derived from estimates of the HRV in CWD across the region, species-specific information on CWD–mammal relationships needs additional attention. Although the following experimental design is idealistic, it can provide the basis for a series of studies that might lend themselves well to meta-analysis. Three characteristics of CWD seem to be associated with mammals: piece size, piece decay class, and abundance (biomass or area). An analysis of variance with three main effects could provide information on the relative importance of each of these characteristics as well as interactions among them. For instance, an idealized design might represent three size classes (<30 cm diameter, 30–50 cm, and >50 cm), five decay classes (Maser et al. 1979), and three levels of abundance (high = >95% confidence interval of HRV, medium = ± one standard deviation of HRV, and low = <95% confidence interval of HRV). With a minimum of three replicates per treatment, 135 sites would be needed. Sampling would have to be conducted over areas representing multiple home ranges for the species of interest. If conducted in plantations with low initial levels of CWD, then initial conditions within each sampling unit may be sufficiently uniform prior to additions of the various levels of CWD. Further, pre- and post-treatment sampling would allow establishment of cause and effect relationships. Monitoring would have to be conducted over multiple generations of the species of concern. Given the spatial and temporal scales associated with such a design, scientists could begin working with managers to conduct an experiment along these lines within an adaptive management framework.

In addition, there is much that we do not know regarding the function of CWD in relation to landscape structure and composition. For instance, I hypothesize that isolated patches of habitat for CWD associates that are surrounded by a relatively impermeable matrix condition could be connected by adding CWD to the matrix. Such an experiment that also included alternative structures might elucidate the relative importance of CWD, shrubs, leaf litter, talus, and other forest-floor features in facilitating movement across landscapes.

Even if scientists and managers can hypothesize the functional relationships between CWD and habitat quality, they must make decisions regarding the amount to create or retain in a managed forest. Completion of a CWD decision-support system such as the DecAID model (Rose et al. 2001) would provide managers with the basis for making management decisions. Monitoring the results of management actions clearly would be required to allow testing of these management hypotheses. In addition, an economic analysis of the marginal gains in CWD function per unit cost of retention may allow managers to make both ecologically and economically informed decisions. Such an approach can also become the basis for optimizing the effectiveness of management actions across managed forests, while ensuring a profit (Bettinger et al. 2001), an outcome that will be important if private forest landowners are to participate in CWD management. Further, such an analysis would provide the basis for identifying the amounts and types of incentives that might encourage private landowner participation in CWD management.

Conclusions

The amounts, sizes, and conditions of CWD observed in western coniferous forests are dependent on the production potential of the forest and the suite of disturbances that lead to dead wood input. Current regulations tend to homogenize CWD amounts over space and time, contrary to the historical spatial and temporal availability of CWD. Wood utilization standards and regulations will dictate the amounts of CWD that will be retained in managed stands in the future, and current regulations will likely place many managed stands outside the HRV. I do not know the level of risk associated with the departure from the HRV.

Associations between CWD amount and mammals vary among species and studies. For many species, CWD serves as an important form of cover or feeding site if other structures are limited or absent. Large-scale, long-term manipulative experiments are needed to test cause and effect relationships between CWD levels and animal fitness.

Management of CWD should consider the range of conditions over space and time in addition to the average conditions within a physiographic province. Some managed stands in western Oregon harvested under high utilization standards and prior to regulations regarding CWD retention fall well below the HRV of CWD in natural stands (Butts and McComb 2000). In contrast, stands in eastern Oregon and Washington that have experienced record insect infestations as a consequence of past and present management actions likely fall above the HRV of CWD in stands under historic fire regimes. The challenge to managers is to develop desired future conditions for stands and landscapes that represent a range of CWD conditions (means and variances) estimated to have occurred under historic disturbance patterns, or at least consider the risks of not doing so.

Literature cited

Agee, J. 1999. Disturbance effects on landscape fragmentation in Interior West Forests. Pages 43–60 *in* J.P. Rochelle, L.A. Lehman, and J. Wisniewski, editors. *Forest Fragmentation: Wildlife and Management Implications*. Brill Press, The Netherlands.

Alban, D.H. and J. Pastor. 1993. Decomposition of aspen, spruce, and pine boles on two sites in Minnesota. *Canadian Journal of Forest Research* **23**:1744–1749.

Allen, A.W. 1983. Habitat suitability index models: southern red-backed vole (Western United States). U.S. Fish and Wildlife Service FWS/OBS-82/10.42. Washington DC, USA.

Arthur, M.A., L.M. Tritton, and T.J. Fahey. 1993. Dead bole mass and nutrients remaining 23 years after clear-felling of a northern hardwood forest. *Canadian Journal of Forest Research* **23**:1298–1305.

Baker, J.M. 1992. Habitat use and spatial organization of pine marten on southern Vancouver Island, British Columbia. Thesis, Simon Fraser University, Burnaby, British Columbia, Canada.

Bartels, R., J.D. Dell, R.L. Knight, and G. Schaefer. 1985. Dead and down woody material. Pages 171–186 *in* E.R. Brown, technical editor. *Management of Wildlife and Fish Habitats in Forests of Western Oregon and Washington*. U.S. Forest Service Publication R6-F&WL-192-1985.

Belk, M.C., H.D. Smith, and J. Lawson. 1988. Use and partitioning of montane habitat by small mammals. *Journal of Mammalogy* **69**:688–695.

Bettinger, P., K. Boston, J. Sessions, and W.C. McComb. 2001. Integrating wildlife species habitat goals and quantitative land management planning processes. Pages 567–579 *in* D. H. Johnson and T. A. O'Neill, managing editors. *Wildlife Habitat Relationships in Oregon and Washington*. Oregon State University Press, Corvallis, Oregon, USA 768pp.

Bisson, P.A., R.E. Bilby, M.D. Bryant, C.A. Dolloff, G.B. Grette, R.A. House, M.L. Murphy, K.V. Koski, and J.R. Sedell. 1987. Large woody debris in forested streams in the Pacific Northwest: past, present, and future. Pages 143–190 *in* E.O. Salo and T.W. Cundy, editors. *Streamside Management: Forestry and Fishery Interactions*. University of Washington, Institute of Forest Resources, Seattle, Washington, USA. Contribution No. 57.

Brown, E.R., technical editor. 1985. *Management of Wildlife and Fish Habitats in Forests of Western Oregon and Washington*. U.S. Forest Service Publication R6-F&WL-192-1985.

Buckmaster, G., W. Bessie, B. Beck, J. Beck, M. Todd, R. Bonar, and R. Quinlan. 1996. Southern red-backed vole (*Clethrionomys gapperi*) year-round habitat. Draft habitat suitability index (HSI) model. Pages 229–234 *in* B. Beck, J. Beck, W. Bessie, R. Bonar and M. Todd, editors. Habitat suitability index models for 35 wildlife species in the Foothills Model Forest: draft report. Canadian Forest Service, Edmonton, Alberta, Canada. (URL: ftp://owlnut.rr.ualberta.ca/pub/barb/hsi.models).

Butts, S.R. and W.C. McComb. 2000. Associations of forest-floor vertebrates with coarse woody debris in managed forests of western Oregon. *Journal of Wildlife Management* **64**:95–104.

Carey, A.B. 1991. Biology of arboreal rodents in Douglas-fir forests. U.S. Forest Service General Technical Report **PNW-276**.

Carey, A.B. and M.L. Johnson. 1995. Small mammals in managed, naturally young and old-growth forests. *Ecological Applications* **5**:336–352.

Carraway, L.N. and B.J. Verts. 1993. Mammalian Species *Aplodontia rufa. American Society of Mammalogists* **431**:1–10.

Corn, P.S. and R.B. Bury. 1991. Small mammal communities in the Oregon Coast Range. Pages 241–253 *in* L.F. Ruggiero, K.B. Aubry, A.B. Carey, and M.H. Huff, technical coordinators. Wildlife and vegetation of unmanaged Douglas-fir forests. U.S. Forest Service General Technical Report **PNW-GTR-285**.

Corn, P.S., R.B. Bury, and T.A Spies. 1988. Douglas-fir forests in the Cascade mountains of Oregon and Washington: is the abundance of small mammals related to stand age and moisture? Pages 340–352 *in* R.C. Szaro, K.E. Severson, and D.R. Patton, technical coordinators. Management of amphibians, reptiles and small mammals in North America. U.S. Forest Service General Technical Report **RM-166**.

Davis, H. 1996. Characteristics and selection of winter dens by black bears in coastal British Columbia. Thesis, Simon Fraser University, Burnaby, British Columbia, Canada.

Drew, T.J. and J.W. Flewelling. 1977. Some Japanese theories of yield-density relationships and their application to Monterey pine plantations. *Forest Science* **23**:517–534.

Embry, R.S. 1963. Estimating how long western hemlock and western redcedar trees have been dead. U.S. Forest Service Research Note NOR-2.

Everett, R., J. Lehmkuhl, R. Schellhaas, P. Ohlson, D. Keenum, H. Reisterer, and D. Spurbeck. 1999. Snag dynamics in a chronosequence of 26 wildfires on the east slope of the Cascade Range in Washington State, USA. *International Journal of Wildfire* **9**:223–234.

Fischer, W.C. and A.F. Bradley. 1987. Fire ecology of western Montana forest habitat types. U.S. Forest Service General Technical Report **INT-223**.

Foster, F.R. and G.E. Lang. 1982. Decomposition of red spruce and balsam fir boles in the White Mountains of New Hampshire. *Canadian Journal of Forest Research* **12**:617–626.

Gilbert, F.F. and R. Allwine. 1991. Small mammal communities in the Oregon Cascade Range. Pages 257–267 *in* L.F. Ruggiero, K.B. Aubry, A.B. Carey, and M.H. Huff, technical coordinators. Wildlife and vegetation of unmanaged Douglas-fir forests. U.S. Forest Service General Technical Report **PNW-GTR-285**.

Gomez, D.M. and R.G. Anthony. 1998. Small mammal abundance in riparian and upland areas of five seral stages in Oregon. *Northwest Science* **72**:293–302.

Gore, J.A. and W.A. Patterson III. 1986. Mass of downed wood in northern hardwood forests in New Hampshire: potential effects of forest management. *Canadian Journal of Forest Research* **16**:335–339.

Graham, R.L. 1982. Biomass dynamics of dead Douglas-fir and western hemlock boles in mid elevation forests of the Cascade Range. PhD Dissertation. Oregon State University, Corvallis, Oregon, USA.

Greir, C.C. 1978. A *Tsuga heterophylla – Picea stichensis* ecosystem of coastal Oregon: decomposition and nutrient balances of fallen logs. *Canadian Journal of Forest Research* **8**:198–206.

Hacker, A.L. and B.E. Coblentz. 1993. Habitat selection by mountain beavers recolonizing Oregon Coast Range clear-cuts. *Journal of Wildlife Management* **57**:847–853.

Hagan, J.M. and S.L. Grove. 1999. Coarse woody debris. *Jouranl of Forestry* **97**(1):6–11.

Harmon, M.E. and C. Hua. 1991. Coarse woody debris dynamics in two old-growth ecosystems. *BioScience* **41**:604–610.

Harmon, M.E. and J. Sexton. 1996. Guidelines for measurements of woody detritus in forest ecosystems. U.S. Long Term Ecological Research Publication No. 20. University of Washington, Seattle, Washington, USA.

Harmon, M.E., J.F. Franklin, F.J. Swanson, P. Sollins, S.V. Gregory, J.D. Lattin, N.H. Anderson, S.P. Cline, N.G. Aumen, J.R. Sedell, G.W. Lienkaemper, K. Cromack, Jr., and K.W. Cummins. 1986. Ecology of coarse woody debris in temperate ecosystems. *Advances in Ecological Research* **15**.

Harmon, M.E., K. Cromack, Jr., and B.G. Smith. 1987. Coarse woody debris in mixed-conifer forests, Sequoia National Park, California. *Canadian Journal of Forest Research* **17**:1265–1272.

Hayes, J.P. and S.P. Cross. 1987. Characteristics of logs used by western red-backed voles, *Clethrionomys californicus*, and deer mice, *Peromyscus maniculatus*. *Canadian Field-Naturalist* **101**:543–546.

Hester, A.S., D.W. Hann, and D.R. Larsen. 1989. *ORGANON: Southwest Oregon Growth and Yield Model User Manual*, ver. 2.0. Oregon State University Forest Research Laboratory, Corvallis, Oregon, USA.

Huff, M.H. 1984. Post-fire succession in the Olympic Mountains, Washington: Forest vegetation, fuels, avifauna. Dissertation, University of Washington, Seattle, Washington, USA.

Lambert, R.C., G.E. Lang, and W.A. Reiners. 1980. Loss of mass and chemical change in decaying boles of a subalpine balsam fir forest. *Ecology* **61**:1460–1473.

Landres, P.B., P. Morgan, and F.L. Swanson. 1999. Overview of the use of natural variability concepts in managing ecological systems. *Ecological Applications* **9**:1179–1188.

Lee, S.D. 1995. Comparison of population characteristics of three species of shrews and shrew-mole in habitats with different amounts of coarse woody debris. *Acta Theriologica* **40**(4):415–424.

Leidholt-Bruner, K. 1990. Effects of beaver on streams, streamside habitat, and coho salmon fry populations in two coastal Oregon streams. Thesis. Oregon State University, Corvallis, Oregon, USA.

MacMillan, P.C. 1988. Decomposition of coarse woody debris in an old-growth Indiana forest. *Canadian Journal of Forest Research* **18**:1353–1362.

Mannan, R.W., M.L. Morrison, and E.C. Meslow. 1984. Comment: the use of guilds in forest bird management. *Wildlife Society Bulletin* **12**:426–430.

Martin, K.J. 1998. Habitat associations of small mammals and amphibians in the central Oregon Coast Range. Dissertation. Oregon State University, Corvallis, Oregon, USA. 88 pp.

Martin, S.K. and R.H. Barrett. 1991. Resting site selection by marten at Sagehen Creek, California. *Northwestern Naturalist* **72**(2):37–42.

Maser, C. and J.S. Gashwiler. 1978. Interrelationships of wildlife and western juniper. Pages 37–82 *in* Proceedings of the western juniper ecology and management workshop. U.S. Forest Service General Technical Report **PNW-74**.

Maser, C. and J.M. Trappe, technical editors. 1984. The seen and unseen world of the fallen tree. U.S. Forest Service General Technical Report **PNW-164**.

Maser, C., R.G. Anderson, and K. Cromack, Jr. 1979. Dead and down woody material. Pages 78–95 *in* J.W. Thomas, technical editor. *Wildlife Habitats in Managed Forests: the Blue Mountains of Oregon and Washington. U.S. Forest Service Agricultural Handbook* No. 553.

Maser, C., B.R. Mate, J.F. Franklin, and C.T. Dyrness. 1981. Natural history of Oregon Coast mammals. U.S. Forest Service General Technical Report **PNW-133**.

Maser, C., R.F. Tarrant, J.M. Trappe, and J.F. Franklin, editors. 1988. From the forest to the sea: a story of fallen trees. U.S. Forest Service General Technical Report **PNW-GTR-229**.

Mason, D.T. 1989. Small mammal microhabitats influenced by riparian woody debris. Pages 697–709 *in* R. R. Sharitz and J.W. Gibbons, editors. Freshwater wetlands and wildlife. USDOE Office of Scientific and Technical Information. *DOE Symposium Series* No. 61. Oak Ridge, Tennessee, USA.

McComb, W.C. 2001. Management of within-stand features in forested habitats. Pages 140–153 *in* D.H. Johnson and T.A. O'Neill, managing directors. *Wildlife Habitat Relationships in Oregon and Washington*. Oregon State University Press, Corvallis, Oregon, USA.

McComb, W.C. and D. Lindenmayer. 1999. Dying, dead, and down trees. Pages 335–372, *in* Hunter, M.L., Jr., *Maintaining Biodiversity in Forest Ecosystems*. Cambridge University Press, Cambridge, UK.

McComb, W.C., C.L. Chambers, and M. Newton. 1993*a*. Small mammal and amphibian communities and habitat associations in red alder stands, central Oregon Coast Range. *Northwest Science* **67**:181–208.

McComb, W.C., K. McGarigal, and R.G. Anthony. 1993*b*. Small mammal and amphibian abundance in streamside and upslope habitats of mature Douglas-fir stands, western Oregon. *Northwest Science* **67**:7–15.

Means, J.E., K. Cromack, and P.C. MacMillan. 1985. Comparison of decomposition models using wood density of Douglas-fir logs. *Canadian Journal of Forest Research* **15**:1092–1098.

Melquist, W.E. and M.G. Hornocker. 1983. Ecology of river otters in west central Idaho. *Wildlife Monograph* **83**.

Mills, L.S. 1995. Edge effects and isolation: red-backed voles on forest remnants. *Conservation Biology* **9**:395–403.

Morrison, M.L. and R.G. Anthony. 1989. Habitat use by small mammals on early-growth clear-cuttings in western Oregon. *Canadian Journal of Zoology* **67**:805–811.

Morrison, P.H. and F.J. Swanson. 1990. Fire history and pattern in a Cascade Range landscape. U.S. Forest Service General Technical Report **PNW-GTR-254**.

Muller, R.N. and Y. Liu. 1991. Coarse woody debris in an old-growth forest on the Cumberland Plateau, southeastern Kentucky. *Canadian Journal of Forest Research* **21**:1567–1572.

Nyberg, J.B. 1990. Interactions of timber management with deer and elk. Pages 99–131 *in* J.B. Nyberg and J.W. Janz, technical editors. *Deer and Elk Habitats in Coastal Forests of Southern British Columbia*. B.C. Ministry of Forests Special Report Series No. 5. Vancouver, BC, Canada.

Ohmann, J.L. and K.L. Waddell. 2003. Regional patterns of dead wood in forested habitats of Oregon and Washington. Pages xxx–xxx *in* Ecology and management of dead wood in western forests. Proceedings of the Conference. U.S. Forest Service General Technical Report **PSW-GTR-xxx**. In press.

Oliver, C.D. and B.C. Larson. 1996. *Forest Stand Dynamics*. Update Edition. John Wiley and Sons, New York, New York, USA.

Olson, J.S. 1963. Energy storage and the balance of producers and decomposers in ecological systems. *Ecology* **44**:322–331.

Onega, T.L. and W.G. Eickmeier. 1991. Woody detritus inputs and decomposition kinetics in a southern temperate deciduous forest. *Bulletin of the Torrey Botanical Club* **118**:52–57.

O'Neil, T.A., D.H. Johnson, C. Barrett, M. Trevithick, K.A. Bettinger, C. Kiilsgaard, M. Vander Heyden, E.L. Greda, D. Stinson, B.G. Marcot, P.J. Doran, S. Tank, and L. Wunder. 2001. Matrixes for wildlife-habitat relationship in Oregon and Washington *in* D.H. Johnson and T.A. O'Neil, managing directors. *Wildlife–Habitat Relationships in Oregon and Washington*. Oregon State University Press, Corvallis, Oregon, USA.

Oregon Department of Forestry. 1996. Oregon Department of Forestry: Oregon Forest Practices Act. Oregon Department of Forestry, Salem, Oregon, USA.

Raphael, M.G. 1988. Long-term trends in abundance of amphibians, reptiles and mammals in Douglas-fir forests of northwestern California. Pages 23–31 *in* R.C. Szaro, K.E. Severson, and D.R. Patton, technical coordinators. Proceedings of the symposium on management of amphibians, reptiles, and small mammals in North America. U.S. Forest Service General Technical Report **RM-166**.

Raphael, M.G. and L.L.C. Jones. 1997. Characteristics of resting and denning sites of American martens in central Oregon and western Washington. Pages 146–165 *in* G. Proulx, H.N. Bryant and P.M. Woodard, editors. Martes: *Taxonomy, Ecology, Techniques and Management*. Provincial Museum of Alberta, Edmonton, Alberta, Canada.

Rose, C.L., B.G. Marcot, T.K. Mellen, J.L. Ohmann, K.L. Waddell, D.L. Lindley, and B. Schrieber. 2001. Decaying wood in Pacific Northwest forests: concepts and tools for habitat management. Pages 580–623 *in* D.H. Johnson and T.A. O'Neil, managing directors. *Wildlife–Habitat Relationships in Oregon and Washington*. Oregon State University Press, Corvallis, Oregon, USA.

Rosenberg, D.K., K.A. Swindle, and R.G. Anthony. 1994. Habitat associations of California red-backed voles in young and old-growth forests of western Oregon. *Northwest Science* **68**:266–272.

Runde, D.E., L.A. Dickson, S.M. Desimone, and J.B. Buchanan. 1999. *Notes on Habitat Relationships for Selected Forest Wildlife Species in Southwestern Washington*. Weyerhaeuser, Tacoma, Washington, USA.

Runkle, J.R. 1990. Eight years change in an old *Tsuga canadensis* woods affected by beech bark disease. *Bulletin of the Torrey Botanical Club* **117**:409–419.

Shigo, A.L. 1984. Compartmentalization: a conceptual framework for understanding how trees defend themselves. *Annual Review of Phytopathology* **22**:189–214.

Sollins, P. 1982. Input and decay of coarse woody debris in coniferous stands in western Oregon and Washington. *Canadian Journal of Forest Research* **12**:18–28.

Sollins, P., S.P. Cline, T. Verhoeven, D. Sachs and G. Spycher. 1987. Patterns of log decay in old-growth Douglas-fir forests. *Canadian Journal of Forest Research* **17**:1585–1595.

Spetich, M.A., S.R. Shifley, and G.R. Parker. 1999. Regional distribution and dynamics of coarse woody debris in midwestern old-growth forests. *Forestry Science* **45**:302–313.

Spies, T.A. and J.F. Franklin. 1988. Old-growth and forest dynamics in the Douglas-fir region of western Oregon and Washington. *Natural Areas Journal* **8**:190–201.

Spies, T.A. and J.F. Franklin 1991. The structure of natural young, mature and old-growth Douglas-fir forests in Oregon and Washington. Pages 91–109 *in* L.F. Ruggiero, K.B. Aubry, A.B. Carey, and M.H. Huff, technical coordinators. Wildlife and vegetation of unmanaged Douglas-fir forests. U.S. Forest Service General Technical Report **PNW-GTR-285**.

Spies, T.A., J.F. Franklin, and T.B. Thomas. 1988. Coarse woody debris in Douglas-fir forests of western Oregon and Washington. *Ecology* **69**:1689–1702.

Tallmon, D.A. and L.S. Mills. 1994. Use of logs within home ranges of California red-backed voles on a remnant of forest. *Journal of Mammalogy* **75**:97–101.

Taylor, C.A., C.J. Ralph, and A.T. Doyle. 1988. Differences in the ability of vegetation models to predict small mammal abundance in different aged Douglas-fir forests. Pages 368–374 *in* R.C. Szaro, K.E. Severson, and D.R. Patton, technical coordinators. Management of amphibians, reptiles and small mammals in North America. U.S. Forest Service General Technical Report **RM-166**.

Thomas, J.W., technical editor. 1997. Wildlife habitats in managed forests: the Blue Mountains of Oregon and Washington. *U.S. Forest Service Agricultural Handbook* No. 553.

Thompson, R.L. 1996. Home range and habitat use of western red-backed voles in mature coniferous forests in the Oregon Cascades. MS Thesis. Oregon State University, Corvallis, Oregon, USA.

Timossi, I.C., E.L. Woodard, and R.H. Barrett. 1995. Habitat suitability models for use with ARC: Douglas' squirrel. California Wildlife Habitat Relationships Program Technical Report 5. California Department of Fish and Game, Sacramento, California, USA.

Torgersen, T.R. and E.L. Bull. 1995. Down logs as habitat for forest-dwelling ants – the primary prey of pileated woodpeckers in northeastern Oregon. *Northwest Science* **69**:294–303.

Tyrell, L.E., and T.R. Crow. 1994. Dynamics of dead wood in old-growth hemlock-hardwood forests of northern Wisconsin and northern Michigan. *Canadian Journal of Forest Research* **24**:1672–1683.

U.S Forest Service and U.S. Bureau of Land Management. 1994. Record of Decision for amendments to Forest Service and Bureau of Land Management planning documents within the range of the northern spotted owl. Standards and guidelines for management of habitat for late-successional and old-growth forest related to species within the range of the northern spotted owl. U.S. Forest Service, U.S. Bureau of Land Management, USA.

van Horne, B. 1983. Density as a misleading indicator of habitat quality. *Journal of Wildlife Management* **47**:893–901.

Van Lear, D.H. and T.A. Waldrop. 1994. Coarse woody debris considerations in southern silviculture. Pages 63–72 *in* Proceedings of the Eighth Biennial Southern Silvicultural Research Conference, Auburn, Alabama, USA.

Van Sickle, J. and S.V. Gregory. 1990. Modeling inputs of large woody debris to streams from falling trees. *Canadian Journal of Forest Research* **20**:1593–1601.

Van Wagner, C.E. 1968. The line intersect method in forest fuel sampling. *Forest Science* **14**:20–26.

West, S.D. 1991. Small mammal communities in the southern Washington Cascade Range. Pages 269–283 *in* L.F. Ruggiero, K.B. Aubry, A.B. Carey, and M.H. Huff, technical coordinators. Wildlife and vegetation of unmanaged Douglas-fir forests. U.S. Forest Service General Technical Report **PNW-GTR-285**.

Wilson, T.M. and A.B. Carey. 1996. Observations of weasels in second-growth Douglas-fir forests in the Puget Trough, Washington. *Northwestern Naturalist* **77**:35–39.

Wimberly, M.C., T.A. Spies, C.J. Long, and C. Whitlock. 2000. Simulating the historical variability in the amount of old forests in the Oregon Coast Range. *Conservation Biology* **14**:167–180.

Zeiner, D.C., W.F. Laudenslayer, Jr., K.E. Mayer, and M. White. 1990. *California's Wildlife.* Volume III. *Mammals. California Statewide Wildlife Habitat Relationships System.* California Department of Fish and Game, Sacramento, California, USA.

KEITH B. AUBRY, JOHN P. HAYES, BRIAN L. BISWELL
AND BRUCE G. MARCOT

12

The ecological role of tree-dwelling mammals in western coniferous forests

Three groups of mammals that occur in coniferous forests of western North America are closely associated with large healthy, decaying, or dead trees: bats, arboreal rodents, and forest carnivores. Detailed descriptions of the ecological relations of these species are presented elsewhere in this book (Buskirk and Zielinski 2003, Hayes 2003, Smith et al. 2003). Although many other kinds of mammals use large vertical forest structures to some degree, these are the species groups that depend on them to meet their life history requirements. Consequently, these are also the mammals that are most likely to suffer population declines in forests where these structures are reduced in abundance. The need to provide for large snags and decadent trees in managed forests to maintain populations of cavity-using birds and mammals has received much attention in the literature (e.g., Balda 1975, Thomas 1979, Hoover and Wills 1984, Brown 1985). However, the perceived consequences of providing inadequate numbers and sizes of these structures in managed forests have generally been limited to the decline or loss of the wildlife species that depend on them. The broader ecological consequences that could also result from the loss or decline of tree-dwelling birds and mammals have received relatively little attention (but see Machmer and Steeger 1995, Aubry and Raley 2002).

Managing forests primarily for timber production often involves not only the removal of a substantial proportion of large, healthy trees from each harvest unit, but also the elimination of large dead or decadent trees. Retaining tall, large-diameter snags or decaying live trees during timber harvest operations involves substantial safety risks (Styskel 1983, Hope and McComb 1994) that often result in the removal of these structures, even when options are available for retaining them in leave patches (Aubry and Raley 2002). Consequently, although forest-management plans may involve green-tree retention harvests and prescriptions for preserving key dead and decadent structures in harvest units, the latter goals may be very difficult to attain. Thus, mammals that are closely associated with large, vertical forest structures may suffer population declines or extirpation in intensively managed forest landscapes.

The recent emergence of ecosystem management as a guiding principle for conservation (e.g., Boyce and Haney 1997) has led some ecologists and resource managers to move away from focusing only on the management of habitat to maintain populations of key wildlife species. Increasingly, wildlife managers are being encouraged to go beyond this narrow focus and consider how ecosystems could be managed to provide for ecological integrity, not simply to provide habitat for one or several sensitive wildlife species (Marcot 1996, Marcot et al. 1998, Marcot and Aubry 2003). For example, a recently published synthesis of wildlife biology, ecology, and resource management in the Pacific Northwest (Johnson and O'Neil 2001) was much more than an update of previous summaries of wildlife–habitat relationships (Thomas 1979, Brown 1985); it also included explicit consideration of the ecological roles that wildlife species may play in the ecosystems they occupy, i.e., their "key ecological functions". However, the matrices of key ecological functions included in this synthesis (O'Neil et al. 2001) were based partly on professional judgments and expert opinion (Marcot and Vander Hayden 2001), and represent only a categorical summarization of known and hypothesized ecological roles that vertebrate and invertebrate organisms may play in northwestern ecosystems. No comprehensive assessment of the contributions that tree-dwelling mammals may make to the composition, structure, and function of western coniferous forest ecosystems has been conducted.

The purpose of this chapter is to present what we believe are the most important ecological roles that each group of tree-dwelling mammals may play in western forests, and review available literature that provides empirical support for each hypothesized role. We consider 14 species of bats, 11 species of arboreal rodents, and six species of carnivores to be associated with large healthy, decaying, or dead trees in coniferous forests of western North America (Table 12.1). The chapter is divided into four major sections: in the first three sections, we discuss the primary ecological

Table 12.1. *Mammals associated with large trees in coniferous forests of western North America*

Bats	Arboreal rodents	Forest carnivores
Silver-haired bat (*Lasionycteris noctivagans*)	Northern flying squirrel (*Glaucomys sabrinus*)	Black bear (*Ursus americanus*)
Hoary bat (*Lasiurus cinereus*)	Douglas squirrel (*Tamiasciurus douglasii*)	Raccoon (*Procyon lotor*)
Western red bat (*L. borealis*)[a]	Red squirrel (*T. hudsonicus*)	Ringtail (*Bassariscus astutus*)
Big brown bat (*Eptesicus fuscus*)	Western gray squirrel (*Sciurus griseus*)	Fisher (*Martes pennanti*)
Pallid bat (*Antrozous pallidus*)	Arizona gray squirrel (*S. arizonensis*)[a]	American marten (*M. americana*)
Townsend's big-eared bat (*Corynorhinus townsendii*)[b]	Abert's squirrel (*S. aberti*)[a]	Western spotted skunk (*Spilogale gracilis*)
Long-eared myotis (*Myotis evotis*)	Dusky-footed woodrat (*Neotoma fuscipes*)	
Keen's myotis (*M. keenii*)	Bushy-tailed woodrat (*N. cinerea*)	
Northern long-eared myotis (*M. septentrionalis*)	Red tree vole (*Arborimus longicaudus*)	
Long-legged myotis (*M. volans*)	Sonoma tree vole (*A. pomo*)[a]	
California myotis (*M. californicus*)	Porcupine (*Erethizon dorsatum*)	
Little brown myotis (*M. lucifugus*)		
Fringed myotis (*M. thysanodes*)		
Yuma myotis (*M. yumanensis*)		

[a] Species does not occur in Washington and Oregon.
[b] Townsend's big-eared bats are listed as associated with large trees in this chapter based on their apparent use of hollow redwoods (Gellman and Zielinski 1996).

roles that each species group may play in western coniferous forests; and in the last, we focus on the subset of tree-dwelling mammals that occur in coniferous forests of Washington and Oregon, and use the key ecological functions provided by O'Neil et al. (2001) to present an example of how information on the ecological contributions of individual species could be used by managers to compare the effects of alternative forest-management strategies on ecosystem function.

Bats

Bats comprise a substantial amount of the mammalian species richness in most western coniferous forests. Although quantitative estimates of their population densities in western forests are not available (see Hayes 2003), bats appear to be locally abundant in many areas. As a consequence of their abundance and specialized ecological niches, it is likely that bats play significant ecological roles in western coniferous forests (Marcot 1996). Despite substantial recent advances in our understanding of the habitat ecology of bats (see Hayes 2003), relatively little work has been conducted that directly examines the functional roles played by bats in forests. Consequently, the magnitude of their contributions to the functioning of western coniferous forest ecosystems remains speculative. We consider three potential ecological roles of bats in western coniferous forests: as predators of insects, as prey of other vertebrates, and as agents of nutrient transport.

Bats as predators of insects

Probably the most significant functional role played by bats in western coniferous forests is as predators of insects. Bats are often purported to play an important role in controlling insect populations (especially in agricultural systems; e.g., Whitaker 1993, 1995, Long 1996, Long et al. 1998). Although it is likely that bats significantly impact insect populations, this conclusion is based largely on expert opinion and logical argument, not scientific research. We are not aware of any studies that have directly examined the influence of bat predation on abundance, population dynamics, or demographics of insects. Assertions that bats play an important ecological role in predation of insects in western coniferous forests are based primarily on information about the types and amounts of prey consumed by bats. All of the bats associated with western coniferous forests (see Hayes 2003) are insectivorous (Black 1974, Whitaker et al. 1977, 1981a,b, Barclay 1985, Rolseth et al. 1994, Wilson and Ruff 1999). Most are aerial insectivores that feed on nocturnal flying insects, although some species also regularly glean insects and other invertebrates from the ground or vegetation.

Obtaining accurate estimates of the amounts of prey consumed by bats is challenging. Amounts and types of prey consumed vary with prey availability (Brigham and Saunders 1990, Hickey and Fenton 1996), time of night (Whitaker et al. 1996, Best et al. 1997), and with the species (Black 1974, Whitaker et al. 1977, 1981a,b), sex, reproductive status (Kunz 1974, Kurta et al. 1989, 1990, Kunz et al. 1995, McLean and Speakman 1999), and age (Kunz 1974, Rolseth et al. 1994, Hamilton and Barclay 1998) of bats. A variety of approaches have been used to estimate the amount of prey consumed, including direct observation (e.g., Hickey and Fenton 1990, 1996), comparison of pre- and post-flight body mass (e.g., Kunz 1974, Anthony and Kunz 1977, Kunz et al. 1995), use of doubly labeled water (e.g., Kurta et al. 1989, 1990), and fecal sample analysis (e.g., Whitaker and Clem 1992, Whitaker 1995). However, each of these approaches has inherent limitations and assumptions that affect estimates. For example, Barclay et al. (1991) hypothesized that overestimation of digestive efficiency in some studies has resulted in substantial underestimation of the biomass of prey consumed by bats.

Early estimates of foraging intensity by bats were based on the mass of stomach contents in bats shot while foraging, body mass of insects that were considered to be primary prey species, and observations and assumptions about the foraging behavior of bats. These studies estimated the number of insects consumed to be as high as 100 to 500 insects per hour (or 1.6–8.3 insects/minute; Gould 1955, 1959). This rate is comparable to feeding rates estimated by others (1.5–9.5 insects/minute, Griffin et al. 1960; 7 insects/minute, Anthony and Kunz 1977). Because the rate of insect consumption varies both temporally and with the type of prey, however, extrapolating these figures to nightly consumption rates should be done with caution. Estimates of total biomass consumed per night is probably a more reliable approach to estimating total consumption (Table 12.2). The cumulative consumption of insects by bats is impressive. Based on fecal sample analyses, Whitaker and Clem (1992) estimated that a colony of 300 evening bats (*Nycticeius humeralis*) in Indiana would consume 6.3 million insects per year. Whitaker (1995) estimated that a colony of 150 big brown bats in the same region would consume roughly 1.3 million insects per year.

To our knowledge, estimates of the amount of insects consumed have not been made for bats in western coniferous forests, but the number of insects consumed is undoubtedly huge. Bats are the primary predators

Table 12.2. *Estimated nightly consumption of prey by bats*

Species	Prey consumed per night		Method of analysis	Reference
	Biomass (g)	% of body mass consumed		
Hoary bat	17.13	57	Direct observation	Hickey and Fenton (1996)
Western red bat	6.3	42	Direct observation	Hickey and Fenton (1990)
Big brown bat	17.2	99	Doubly labeled water	Kurta et al. (1990)
Little brown myotis	5.5–6.7	61–85	Doubly labeled water	Kurta et al. (1989)
Little brown myotis	2.5–3.7	31–49	Pre-post flight body mass comparison	Anthony and Kunz (1977)
Cave myotis[a,b]	2.0–3.4	17–30	Pre-post flight body mass comparison	Kunz (1974)
Mexican free-tailed bat[c]	4.7–8.6	39–73	Pre-post flight body mass comparison	Kunz et al. (1995)

[a] *Myotis velifer.*
[b] Estimated maximum daily consumption for males and females combined.
[c] *Tadarida brasiliensis.*

of many species of nocturnal, flying insects, and some of the insects fed upon by bats play important ecological roles in western coniferous forests. Some, such as the western spruce budworm (*Choristoneura occidentalis*) and the Douglas-fir tussock moth (*Orygia pseudotsugata*), are considered to be forest pests, and can significantly impact wood fiber production. Although there are no estimates of the impacts of bats on these species, bats may play an important role in forest health by minimizing the impacts or outbreaks of forest pests.

Bats as prey of other vertebrates

Bats are preyed upon by a number of vertebrate predators throughout the world (Gillette and Kimbrough 1970, Speakman 1991*b*, Fenton 1995). Despite the importance of predation on bats in some settings, many of the anecdotal observations of predation on bats by fish, amphibians, birds, and mammals (e.g., Stager 1942, Davis 1951, Kinsey 1961, Martin 1961, Wilks and Laughlin 1961, Elwell 1962, Wiseman 1963, Barr and Norton 1965, Cleeves 1969, Lee 1969, Mumford 1969, Thomas 1974, Kirkpatrick 1982, Wroe and Wroe 1982, Yager and Williams 1988) have been published because of their uniqueness or rarity, and these observations often represent sightings of unusual opportunistic foraging events or specialized foraging strategies developed by individual predators.

Although predation on bats sometimes occurs when bats are foraging or commuting, much of this predation occurs when bats are roosting or emerging from roosts. Large concentrations of bats at roost sites (Kunz 1982), coupled with relatively predictable patterns of emergence from roosts (Erkert 1982), provide significant opportunities for predators to prey on bats in some areas (Fenton et al. 1994). In western coniferous forests, bats typically do not congregate in very large colonies and generally exhibit low fidelity to day roosts (see Hayes 2003), reducing the potential functional significance of bats as prey. Low fidelity to roost sites may be a strategy to minimize risk of predation (Lewis 1995). Furthermore, bats may select times (Jones and Rydell 1994, Kunz and Anthony 1996, Rydell et al. 1996, Duvergé et al. 2000) and patterns (Swift 1980, Brigham and Fenton 1986; but see Speakman et al. 1992, Kalcounis and Brigham 1994) of emergence from roosts to minimize predation. Indeed, nocturnality in bats may have evolved to minimize the risk of predation (Speakman 1991a, 1995).

Regular predation on bats in temperate forests appears to be uncommon and generally restricted to a relatively small group of predators. Although diurnal raptors feed on bats during twilight hours in some parts of the world (e.g., Fenton et al. 1994), nocturnal predation by owls is the most significant predation pressure on bats in temperate regions (Speakman 1991b). Bats generally comprise a relatively small proportion of the diet of most predators. Speakman (1991b) estimated that bats represented only 0.003% of the diet of small falcons and hawks and 0.036% of the diet of owls in Great Britain. However, bats can comprise a substantial amount of the prey taken by predators that have learned to specialize on bats. For example, although bats comprise less than 1% of the diet of spotted owls (*Strix occidentalis*) in most areas (Forsman et al. 1984, Smith et al. 1999), they represented 12% of prey taken in southern Arizona (Duncan and Sidner 1990). Overall, it appears that bats probably do not play a major ecological role as prey over large geographic areas in western coniferous forests. However, there has been very little work investigating interactions between bats and predators in this region.

Bats as agents of nutrient transport
Bats may play an important ecological role in western coniferous forests by transporting nutrients, especially from riparian areas to upland areas

(Cross 1988). To our knowledge, no empirical study has estimated the amount of nutrient transfer by bats. Given the relative mobility of bats and the fact that bats often use different habitats for roosting and foraging (see Hayes 2003), it is reasonable to hypothesize that bats may play a significant role in nutrient transfer within ecosystems. Nutrient transfer has been shown to be facilitated by other species of mammals, and this can influence nutrition and nutrient content of plants (e.g., Ben-David et al. 1998). However, we suspect that the importance of nutrient transfer by bats in overall ecosystem function is probably relatively low, but may influence microsite conditions. We hypothesize that the ecological function of nutrient transfer is probably concentrated in relatively small areas, such as the area in, under, or immediately surrounding roosts.

Arboreal rodents

Rodents are an extremely diverse order of mammals that occupy a variety of ecological niches and comprise a substantial amount of the biomass and diversity in western coniferous forests. Some species exploit the forest floor for nesting and foraging, while others spend much of their time in trees; the latter group are often referred to as "arboreal" rodents (Table 12.1, see Smith et al. 2003). The red and Sonoma tree voles are considered to be the most arboreal mammals in western North America (Carey 1991). Both species are closely associated with large trees and conduct almost all of their activities high in the canopy of Douglas-fir (*Pseudotsuga menziesii*) forests (Maser et al. 1981, Gillesberg and Carey 1991, Meiselman and Doyle 1996). Red tree voles are most abundant in late-successional forests (Corn and Bury 1986, Gillesberg and Carey 1991, Meiselman and Doyle 1996) where large tree canopies interweave. At the low end of the scale of arboreal activity is the bushy-tailed woodrat, which uses cavities in standing and fallen trees. It builds nests of sticks and woody debris on branches of trees, in tree hollows, and on the ground in low-elevation transitional and mixed-conifer forests (Carey 1991). Here, we consider four potential ecological roles of arboreal rodents in western coniferous forests: as prey of other vertebrates, as disseminators of fungal spores and mistletoe seeds, as predators and disseminators of conifer seeds, and as modifiers of forest composition and structure.

Arboreal rodents as prey of other vertebrates
Arboreal rodents are preyed upon by a wide array of vertebrate predators throughout the West. Some avian predators, including the federally

listed northern spotted owl (*S. o. caurina*), consume relatively large and energetically profitable prey such as woodrats, northern flying squirrels, and red tree voles (Forsman et al. 1984, Smith et al. 1999). Northern flying squirrels comprised from 25.1% to 57.5% of the biomass consumed by spotted owls in Douglas-fir and western hemlock (*Tsuga heterophylla*) forests in western Oregon, whereas in the dry mixed-conifer and mixed-evergreen forests in the Klamath Mountains of southern Oregon, dusky-footed woodrats represented up to 69.9% of the biomass in spotted owl diets (Forsman et al. 1984). The abundance of these arboreal prey species can influence the reproductive success and densities of predators. Regional differences in densities of the northern spotted owl have been attributed to regional differences in the abundance of northern flying squirrels and dusky-footed and bushy-tailed woodrats (Carey et al. 1992). There is also a significant relationship between reproductive status and the percent biomass of woodrats in California spotted owl (*S. o. occidentalis*) diets from the San Bernardino Mountains of southern California; successful nesters consumed a greater percent biomass of woodrats than non-nesters (Smith et al. 1999).

Both species of tree voles are locally important prey for the northern spotted owl. The percent occurrence of red tree voles in regurgitated owl pellets varied in different portions of the owl's range, from a low of 3.7% of items in the northern Oregon Cascades to a high of 49.1% in the Douglas-fir/coast redwood (*Sequoia sempervirens*) zone along the southern Oregon Coast (Forsman et al. 1984). Averaging across seven study areas, the red tree vole represented 15.1% of all prey items taken by spotted owl pairs. Due to their small size relative to other arboreal rodents, however, it only provided 2% to 19% of total biomass in the owl's diet (Forsman et al. 1984). Although the northern spotted owl is probably the primary predator of red tree voles, the long-eared owl (*Asio otus*; Reynolds 1970), saw-whet owl (*Aegolius acadicus*; Forsman and Maser 1970), raccoon, marten, fisher, various corvids (Maser et al. 1981), and ringtail (Alexander et al. 1994) also prey upon them. The Steller's jay (*Cyanocitta stelleri*) is a common predator that will systematically destroy nests in search of young voles (Howell 1926).

Diurnal squirrels can be important in the diet of predators, especially during winter. The most frequently occurring item in the winter diet of marten in northwestern Montana (Marshall 1946) and north-central Washington (Newby 1951) was the red and Douglas squirrel, respectively. In lodgepole pine (*Pinus contorta*) forests of Wyoming, northern goshawk (*Accipiter gentilis*) diets may contain up to 50% red squirrels (Squires 2000).

Goshawks may also have a stabilizing influence on Abert's squirrel populations in the southwest (Reynolds 1963, Boal and Mannan 1994). Other occasional predators of diurnal tree squirrels in western coniferous forests include the red-tailed hawk (*Buteo jamaicensis*; Luttich et al. 1970), bald eagle (*Haliaeetus leucocephalus*), and lynx (*Lynx canadensis*; Aubry et al. 2000).

Arboreal rodents as disseminators of ectomycorrhizal fungi and dwarf mistletoe

Ectomycorrhizal fungi form symbiotic relationships with the root systems of conifer trees, shrubs, and other vegetation. The tree provides carbon from photosynthesis to the fungi, and the ectomycorrhizal fungi absorb minerals and nutrients from the soil and transfer them to tree roots (Smith and Read 1997). This process helps conifer trees and other woody plants uptake water and nutrients from the soil, and facilitates the transport of carbohydrates from plants into the mycorrhizosphere. Sporocarps of these hypogeous fungi (truffles) are an important food resource for many forest mammals worldwide (Whitaker 1962, Fogel and Trappe 1978, Maser et al. 1978, Viro and Sulkava 1985, Carey et al. 1992, Carey 1995, see Luoma et al. 2003). Most sporocarps consumed by arboreal rodents in western conifer forests are from ectomycorrhizal fungi that form a symbiotic association with feeder roots of trees in the Pinaceae, Fagaceae, Betulaceae, Myrtaceae, and Salicaceae (Molina et al. 1992, North et al. 1997, see Luoma et al. 2003).

Mammals are attracted to hypogeous fungi by aromatic compounds produced by the maturing sporocarps (Fogel and Trappe 1978). Animals unearth the truffles and consume all or part of the sporocarps including the spores (Trappe and Maser 1976, Fogel and Trappe 1978). Fungal spores pass through the digestive tract unharmed and are deposited in feces in the forest soil at new locations (Trappe and Maser 1976). Although the nitrogen concentration of sporocarps is high, much of this nitrogen is in non-protein forms or is associated with cell walls, suggesting that fungal sporocarps may be low in nutritional value or protected from mammalian digestive enzymes (Claridge et al. 1999). The nutritional value of fungi to tree-dwelling mammals is largely unstudied; however, Cork and Kenagy (1989) found that one species of truffle (*Elaphomyces granulatus*) provided few nutritional benefits to the Cascade golden-mantled ground squirrel (*Spermophilus saturatus*) due to low digestibility of the sporocarps. Although they are apparently low in nutritional benefits, truffles may be consumed simply because they are seasonally abundant and highly detectable due to the strong odor they develop when mature. Sporocarps may also be an important source of water in some regions or during certain seasonal periods because of their high moisture content (Fogel and Trappe 1978).

In western coniferous forests, many species of arboreal rodents, especially the northern flying squirrel and Douglas squirrel, contribute to forest ecosystem processes by consuming fungi and dispersing fungal spores (Maser and Maser 1988). Diets of the northern flying squirrel in the Pacific Northwest include primarily fungi and lichens (Maser et al. 1986, McIntire and Carey 1989). Northern flying squirrels consumed 12 taxa of fungi during a single season (McIntire and Carey 1989) and up to 20 taxa annually, including Basidiomycetes, Ascomycetes, and Zygomycetes (Maser et al. 1986). Both squirrel species were found to be more abundant in old forests than in managed forests (Volz 1986, Carey et al. 1992, Witt 1992, Zabel and Waters 1997) where fungal abundance and diversity were higher. Douglas and red squirrels are adept at detecting hypogeous fungi, and Douglas squirrels are reported to eat a wider array of fungi (89 species) than any other mycophagist (Fogel and Trappe 1978).

Dissemination of the spores of hypogeous fungi can only occur through the foraging activities of animals. Thus, despite the conclusion that truffles are only of moderate nutritional value for most small mammals (Claridge et al. 1999), arboreal rodents appear to be important dispersal agents for these fungi. By doing so, arboreal rodents enhance their own food supply and disseminate the ectomycorrhizal fungi that are important for the growth and survival of conifers in western forests.

Arboreal rodents may also aid in the dispersal of other plant propagules. For example, arboreal rodents have been suggested as potential long-distance dispersal agents for dwarf mistletoe (*Arceuthobium* spp.) seeds (Hawksworth et al. 1987, Hawksworth and Wiens 1996). Parasitism of conifer trees by dwarf mistletoes can have a dramatic effect on the structure and function of both individual trees and forest stands, and sets in motion a complex web of interactions with disease organisms, decay fungi, arthropods, birds, and mammals (Hawksworth and Wiens 1996, Mathiasen 1996). Arboreal rodents often use witches' brooms caused by dwarf mistletoes for nests, rest sites, and cover (Lemons 1978, Smith 1982, Tinnin et al. 1982). Dwarf mistletoes disperse seeds by an explosive mechanism that forcibly expels a single, sticky seed from the fruit as far as 16 m away (Hawksworth and Wiens 1996). Although many species of birds and mammals feed on dwarf mistletoes, seeds lose their viability after being ingested (Hudler et al. 1979). Thus, dispersal of dwarf mistletoes by animals, and long-distance dispersal in general, is believed to occur only via external transport by birds and mammals (Hudler et al. 1979, Hawksworth et al. 1987). Few studies have been conducted to assess the ecological importance of mammals as dispersal agents of dwarf mistletoes, but seeds have been found on the fur of the least chipmunk (*Tamias minimus*), golden-mantled ground squirrel (*Spermophilus lateralis*), red squirrel, northern flying squirrel, and marten (Ostry et al. 1983, Hawksworth et al. 1987). In British Columbia, Canada, one in 15 red squirrels examined had dwarf mistletoe seeds in its fur (Hawksworth et al. 1987), suggesting that dispersal of mistletoe seeds by arboreal rodents may occur relatively frequently.

Arboreal rodents as predators and disseminators of conifer seeds

Douglas and red squirrels feed primarily on conifer seeds and are behaviorally and anatomically adapted to exploit them (Smith 1970, 1981). Their food habits are generally similar except that red squirrels are adapted to feed on serotinous cones that remain closed on the tree, whereas Douglas squirrels are not (Smith 1970). Douglas and red squirrels are so dependent on conifer seeds for survival that their population densities fluctuate with conifer cone crops (Smith 1970). These squirrels do not hibernate and rely on cones stored in large caches to survive the harsh winter conditions that occur in many coniferous forest habitats (Lindsay 1986). The caching behavior of Douglas and red squirrels can have a significant effect on the amount and distribution of conifer seeds available for germination. Red squirrels may collect and cache enough cones to support themselves through one or more years of poor cone crops. Red squirrel caches in Alaska contained up to 8500 cones (Smith 1968), caches in the Rocky Mountains were 30–45 cm deep (Finley 1969), and caches may contain up to 24 bushels of cones in the southwest (Patton and Vahle 1986). These large caches are not covered, and seeds may remain viable for many years (Shaw 1936). Douglas squirrels generally use smaller cone caches, that may include as few as 30 cones (Carey 1991). In California, Douglas squirrels cache only enough cones to last through the winter and early spring (Koford 1982). These cone caches or "middens" function to concentrate nutrients, and probably provide food for a wide variety of organisms. Middens are also reported to be particularly important ecological features for the American marten because they facilitate access to the subnivean layer, provide resting sites that may be energetically important in winter, and serve as natal denning sites (Buskirk and Ruggiero 1994). In lodgepole pine/spruce-fir forests of southern Wyoming, red squirrel middens were used by martens for natal den sites more than any other structure (Ruggiero et al. 1998).

Cone caching by Douglas and red squirrels can seriously interfere with natural re-seeding in some areas because the entire cone crop may be collected in poor cone years (Finley 1969). In ponderosa pine (*Pinus ponderosa*) and lodgepole pine forests, red squirrels may harvest up to 67% of mature ponderosa pine cones and remove an additional 14% before they are ripe (Schmidt and Shearer 1971). In ponderosa pine forests of the Southwest, Abert's squirrel populations fluctuated in response to cycles of ponderosa pine cone production (Farentinos 1972). In addition to seeds from mature ovulate cones, Abert's squirrels also consumed significant numbers of terminal buds and fed heavily on the inner bark of pine shoots (Keith 1965). Abert's squirrels can substantially reduce cone production of ponderosa pine trees through intensive foraging on the inner bark of excised pine shoots that contain immature ovulate cones (Allred et al. 1994). The effects of this feeding behavior may reduce the potential cone crops of these forest stands by up to 20%. By shifting their foraging from the inner bark to the seeds of mature cones, Abert's squirrels effectively reduced the total cone crop by 55% (Allred et al. 1994).

Overall, cone production differs among stands, years, seasons, tree species, and individual trees (Eis et al. 1965, Fowells 1965, Smith and Balda 1979). A tree's ability to produce cones increases with age; trees in late-successional stands may produce up to 14 times more seed than trees in young stands (Buchanan et al. 1990, Carey 1991). Large fluctuations in seed production and the synchrony of cone production among western conifers resulted in parallel fluctuations in Douglas squirrel populations (Smith 1970, Buchanan et al. 1990). Years of cone failures between years of heavy seed production led to dramatic fluctuations in populations of all seed predators and allowed a large percentage of the seeds in productive years to go uneaten and be available for germination (Smith and Balda 1979). Cone caching provides a mechanism for seed transport within a stand of trees and may provide for the storage of viable seeds over several years. Thus, available evidence indicates that tree squirrels are likely to be important vectors for the dispersal of conifer seeds in western forests.

Arboreal rodents as modifiers of forest composition and structure

Numerous studies have been conducted on damage to commercial timber plantations from rodents foraging on conifer seeds and cambium, with variable conclusions depending on the region and forest crop. Although these problems often result in efforts to control animal damage, the same behaviors that cause economic losses to commercial forests have potential functional roles in the ecosystem. Red squirrels consume the vascular tissue of young lodgepole pine by peeling bark from branches and the main trunk from May to early July (Sullivan and Sullivan 1982, Kenward 1983, Sullivan and Vyse 1987) when other natural foods are unavailable. The winter diet of porcupines consists almost exclusively of cambium and phloem of saplings and small conifer trees (Sullivan et al. 1986). Foraging activities by arboreal rodents may also provide food for other species, especially during winter; foliage dropped by porcupines when snow covered the ground in Maine was eaten by both deer and rabbits (Curtis and Kozicky 1944).

Arboreal rodents may have their most significant influence on stand structure and dynamics during the sapling and pole stages of forest development (Van Deusen and Meyers 1962), because animal damage is more likely to cause tree mortality or change the growth form of individual trees in young stands. Feeding damage by arboreal rodents is usually greatest in stands with an average tree diameter less than 6.0 cm (Sullivan and Sullivan 1982). Feeding damage to sapwood and cambium can reduce the growth and vigor of a tree, increase its susceptibility to fungal attack, and eventually kill it (Sullivan and Vyse 1987). Girdling of saplings and small trees may cause significant tree mortality in plantations and naturally decrease tree stocking rates. In situations where trees are sparse and insular in distribution, porcupines may be capable of causing local extirpations of some conifer species (Gill and Cordes 1972).

The effects of foraging by arboreal rodents on the growth form of individual trees can be significant. Stripping bark from the leader causes mortality at the tops of small trees in Douglas-fir plantations. This leader mortality can result in tree death, but may have a greater ecological impact by creating defects in living trees. Re-sprouting at the top of a small conifer often results in multiple leaders that reduce the value of crop trees in commercial forests. However, platforms created by multiple tops provide nest sites for many arboreal rodents and passerine birds; red tree voles, flying squirrels, and Douglas squirrels all build stick nests in the protected forked tops of previously damaged trees (Maser 1966, Maser et al. 1981). The increased availability of nest sites for arboreal rodents may significantly increase the diversity and abundance of beneficial ectomycorrhizal fungi in these stands. Thus, damage by arboreal rodents to young conifer stands, while economically detrimental, may introduce canopy structural diversity that sustains species richness.

Forest carnivores

Carnivores may influence the ecosystems they occupy by affecting the behavior and demography of prey and competitors, facilitating the dispersal of seeds, completing or interrupting the life cycles of pathogens and parasites, and cycling nutrients by scavenging carrion (Buskirk 1999). In addition, forest carnivores transport nutrients and contaminants within ecosystems, concentrate them at den and rest sites, and probably aid in the long-distance dispersal of dwarf mistletoes and hypogeous fungi. However, the functional significance of carnivores in ecosystems has received relatively little attention from researchers, and remains largely conjectural (Estes 1996). To our knowledge, no study of tree-dwelling carnivores in western coniferous forests has directly evaluated the effects of their activities on ecosystem function. Consequently, empirical information on the ecological significance of carnivores in forested habitats is sparse and speculative in nature. Here, we consider three potential ecological roles of tree-dwelling carnivores in western coniferous forests: as predators and competitors, as long-distance disseminators of propagules, and as hosts for parasites.

Forest carnivores as predators and competitors

Food webs are key features of all ecosystems, and predator–prey interactions are the fundamental linkages among species in each food web (Estes 1996). Some mammalian carnivores exert such a strong influence on the structure of prey communities that they function as keystone predators; such a role has been described for the sea otter (*Enhydra lutris*; Estes and Palmisano 1974), coyote (*Canis latrans*; Henke and Bryant 1999), and wolf (*C. lupus*; McLaren and Peterson 1994). Although keystone roles for tree-dwelling forest carnivores have not been proposed, this may simply reflect a lack of research designed to elucidate the functional significance of these predators in the ecosystem.

There is little evidence, however, that tree-dwelling mammalian carnivores exert a strong influence on their prey populations. Although inferring the ecological effects of predation from food habits data is problematic, all species considered here (Table 12.1) consume a broad array of animal and plant foods and appear to be largely opportunistic in their choice of prey (see Buskirk and Zielinski 2003). Thus, the effects of predation by forest carnivores on the population dynamics or community structure of prey are probably minimal. However, there is evidence that predation by fishers on porcupines may have substantial ecological effects in forests where these species co-exist. Long-term data on the densities of fisher and porcupine populations in Michigan during the 1970's provided strong evidence that predation by fishers reduced porcupine populations to lower and more stable levels than occurred in the absence of fishers (Powell and Brander 1977, Earle 1978). Anecdotal observations on the effects of fisher predation on porcupine populations have also been reported from Wisconsin, New York, and Maine (Powell 1993).

Reports of heavy damage by porcupines to tree plantations in the Pacific Northwest was attributed to a dramatic increase in porcupine numbers resulting from the overtrapping of fisher populations (Stone 1952). One of the objectives of fisher re-introductions that occurred in Oregon (Kebbe 1961, Aubry et al. 1996), Idaho (Williams 1962), and Montana (Weckworth and Wright 1968) was to reduce damage to forest plantations by controlling porcupine populations. It has even been proposed that, during re-establishment of fishers in areas where porcupines have become unnaturally dense and where they have not been subjected to fisher predation for several decades, fishers may be capable of exterminating porcupines (Powell and Brander 1977). Porcupines can injure or kill large proportions of young conifer stands (Dodge and Borrecco 1992). Wounds resulting from porcupine feeding may expose trees to fungal infection (Sullivan et al. 1986, Eglitis and Hennon 1997), including the heart-rot decay fungi that are essential for creating hollows in trees and for softening heartwood to facilitate cavity excavation by woodpeckers (Conner et al. 1976, Bull et al. 1997). Consequently, porcupines may influence the availability of structures that are used by a wide array of secondary cavity-using birds and mammals. Sublethal porcupine injury may also affect tree vigor and growth and increase susceptibility to attack by insects (Sullivan et al. 1986). Thus, in some situations,

predation by fishers could be a mediating factor on the effects of porcupine foraging on the structure and composition of coniferous forest
ecosystems.

Competitive interactions among tree-dwelling forest carnivores do not
appear to have an important influence on their populations, even between
the congeneric and ecologically similar marten and fisher. However, competition with fishers may have contributed to the decline of the Humboldt
marten (*Martes a. humboldtensis*) in coastal areas of northern California
(Krohn et al. 1997). Fishers are known to kill martens (de Vos 1952, Raine
1981), and it has been suggested that at very high fisher densities, interference competition may prevent martens from maintaining viable populations (Krohn et al. 1995). In the western mountains of North America,
martens generally occur at higher elevations and in areas with deeper
snowpacks than fishers (Powell and Zielinski 1994, Krohn et al. 1997).
Martens have an energetic advantage over fishers when moving through
soft, deep snow and are more efficient predators of winter-active small
mammals in the subnivean layer (Leonard 1980, Raine 1983). Although
their diets overlap extensively (Martin 1994), the smaller size of martens
enables them to better exploit both subnivean and arboreal microhabitats
and to be more efficient predators of microtine rodents and other small
prey. In contrast, fishers are capable of killing porcupines and are better
adapted to prey on snowshoe hares (*Lepus americanus*) and other medium-
sized prey (Raine 1983). Thus, it appears that marten and fisher populations co-exist at the regional scale by partitioning available habitat according to snow conditions, and co-exist occasionally at the local scale by
partitioning food resources.

Forest carnivores as long-distance disseminators of propagules

Forest carnivores are omnivorous, highly mobile, and generally occupy
large home ranges that encompass a variety of habitat conditions. All six
species of tree-dwelling forest carnivores considered here feed extensively
on both animal and plant foods, especially fruits. During summer and
fall, the diet of black bears is dominated by fruits and mast crops (Pelton
1982). Except during the spring, when animal matter predominates in
the diet, raccoons feed more often on fruits and seeds than on other food
items (Kaufmann 1982). Ringtails are less herbivorous than raccoons, but
fruits are also common in ringtail diets (Toweill and Teer 1977, Alexander
et al. 1994), and are important constituents of late-summer and fall diets

for both the marten (Buskirk and Ruggiero 1994) and fisher (Powell and Zielinski 1994).

Several authors have summarized the prevalence of frugivory among North American mammals (Martin et al. 1951, Halls 1977, Willson 1993), but few studies have addressed the ecological or evolutionary importance of carnivores as dispersal agents. For frugivores to have a positive effect on plant fitness, they must also be "legitimate", "efficient", and "effective" dispersers (Bustamente et al. 1992). Legitimate dispersers defecate seeds that are undamaged and capable of germinating, efficient dispersers defecate seeds in sites where they are likely to germinate and survive, and effective dispersers disseminate a large proportion of the seedlings that are recruited into the population.

Field studies have been conducted in Europe and South America on the ecological role of forest carnivores in fruit dispersal (e.g., Herrera 1989, Bustamente et al. 1992, Pigozzi 1992), but similar research has not been conducted in coniferous forests of western North America. However, a few studies have evaluated the legitimacy of forest carnivores as seed-dispersal agents in conifer forests. Black bears were considered to be legitimate fruit-dispersal agents in northeastern Minnesota because fruits were apparently swallowed whole and seeds were defecated intact (Rogers and Applegate 1983). The authors speculated that because many species of *Prunus* and *Pyrus* contain cyanogenetic glycosides, swallowing fruits whole may reduce the incidence of poisoning. Bears may also be relatively efficient dispersal agents, because germination rates of seeds from feces were higher than for seeds of uneaten fruits for all eight species studied, and significantly higher for five species. Thus, chemical or mechanical scarification in the gut of black bears enhanced the germination rate of fruit seeds. Six species of relatively small-seeded fruits from coniferous forests in southeast Alaska were fed to captive black bears, but only salmonberry (*Rubus spectabilis*) and elderberry (*Sambucus racemosa*) had significantly higher germination rates after gut passage than unpassed seeds (Traveset and Willson 1997). Because they are one of the few species capable of ingesting relatively large-seeded fruit and can travel up to 32 km per day (Rogers and Applegate 1983), black bears may be particularly important long-distance dispersal agents for large-seeded fruit.

Similar feeding trials involving salmonberry and two species of huckleberry (*Vaccinium alaskaense* and *V. ovalifolium*) were conducted on captive martens in southeast Alaska (Hickey et al. 1999). Passage through the gut of martens resulted in no difference in the germination rate for

V. ovalifolium compared to seeds taken from inside berries. Although gut passage decreased the germination rate for *V. alaskaense,* 41% were still viable. However, none of the salmonberry seeds fed to martens germinated. Thus, martens appear to be legitimate dispersers of huckleberry, but not salmonberry. In addition, movement models based on radiotelemetry data indicated that in four to five hours of travelling, martens are capable of transporting viable seeds in feces as far as 3.5 km.

Forest carnivores may also facilitate the long-distance dispersal of other propagules, such as mistletoe seeds and fungal spores. Dwarf mistletoe seeds are forcibly ejected from fruits in late summer or early fall (Hawksworth and Wiens 1996) and have been found on the fur of martens (Hawksworth et al. 1987). In the western U.S., martens and fishers often use witches' brooms caused by dwarf mistletoes as rest sites (Buskirk et al. 1987, Spencer 1987, Jones 1991, Seglund 1995, Parks and Bull 1997, Aubry and Raley 2001), and fishers have been documented using mistletoe brooms for maternal dens (Aubry and Raley 2001). Use of broom rest sites by both species occurs most often during the summer, probably because these sites are relatively exposed and provide poor protection from heat loss when ambient temperatures are low (Spencer 1987, Parks and Bull 1997, K. Aubry and C. Raley unpublished data). Because the timing of seed ejection coincides with the period when martens and fishers are most likely to use mistletoe brooms for rest and den sites, seeds may often be ejected onto their fur. Consequently, American martens and fishers may also facilitate the long-distance dispersal of dwarf mistletoes, due to their large spatial requirements and use of a variety of different structures within their home ranges for rest and den sites.

Forest carnivores may also disseminate fungal spores. Analysis of the stomach contents of eight fishers from northwestern California showed that the most important item by volume in the diet was the spores and tissues of the false truffle (*Rhyzopogon* spp.; Grenfell and Fasenfest 1979). In a study conducted in southeastern California, spores from at least six genera of fungi were found to be ubiquitous in samples taken from 24 fisher scats; in 17 of the scats, fungal spores or tissue comprised 5% to 50% of the sample by volume (Zielinski et al. 1999). These findings indicate that fungal spores in fisher gastrointestinal tracts were not obtained solely by predation on northern flying squirrels, western redbacked voles (*Clethrionomys californicus*), or other mycophagous rodents. Fungi have not been reported in diets of the other forest carnivores considered here (Chapman and Feldhamer 1982), but have been reported several times for the eastern spotted skunk (*Spilogale putorius*; Fogel and Trappe 1978).

Sporocarp tissues are fragile and may be unidentifiable in scats. In addition, most investigators do not examine carnivore scats or stomach contents for the presence of fungal spores. Thus, mycophagy by forest carnivores may be more common than is indicated by food habits studies reported in the literature. We do not know if fungal spores present in the gastrointestinal tracts of mycophagous rodents eaten by carnivores can be passed in a viable condition through carnivore digestive tracts. However, because hypogeous fungi have a highly co-evolved relationship with mycophagous small mammals and apparently rely on them for spore dispersal (Maser et al. 1978), it seems likely that spores would be well adapted to survive passage through the simpler gastrointestinal tracts of carnivores. Thus, either via mycophagy or predation on mycophagous small mammals, forest carnivores may also contribute to the long-distance dispersal of hypogeous fungi.

Forest carnivores as hosts for parasites

Mammals serve as hosts for a vast array of microparasites (e.g., bacteria, viruses, protozoans) and both internal and external macroparasites (e.g., helminths, arthropods; Scott 1988, Samuel et al. 2001). Taxonomists have only described a fraction of the species diversity represented by these organisms, however, and have only recently begun to study their influence on biodiversity and other ecosystem attributes (Windsor 1997, Brooks and Hoberg 2000). Infection by parasites can be relatively benign, as is the case for many species of macroparasites that rely on mammals to complete their life cycles, or it can cause diseases that result in morbidity or death. However, even relatively benign parasitic infections can affect host behavior, vigor, and reproductive output, which may in turn alter population demography, genetic diversity, and community interactions (Scott 1988). Because of their potential influence on the structure and functioning of ecosystems, and as a selective force in evolution, several ecologists have argued for an enhanced awareness of the role of parasites in the conservation of ecological diversity (May and Anderson 1983, Rózsa 1992, Windsor 1997, Murray et al. 1999, Brooks and Hoberg 2000). Although all tree-dwelling mammals perform this ecological function to some degree, we focused our discussion of this topic on forest carnivores because their ecological role as vertebrate predators also functions to complete or interrupt the life cycles of a variety of parasites.

The common macroparasites of forest carnivores have been well described (Samuel et al. 2001), but little effort has been made to understand the role of parasites in the ecology of forest carnivores or the functioning of forest ecosystems. In coniferous forests of the Cascade Range in Washington, American martens are commonly infected by the cestode *Taenia martis americana* (Hoberg et al. 1990), that is host-specific and cannot complete its life cycle in the absence of martens. In coastal areas of Washington, marten populations declined dramatically during the twentieth century and appear to have been extirpated in many areas (Marshall 1994, Zielinski et al. 2001). Consequently, the loss of martens from these areas probably also involved the extirpation of one or more of the parasite species they harbored. Martens are also reported to harbor the host-specific flea *Chaetopsylla floridensis* in the central Sierra Nevada in California (Zielinski 1984); thus, a similar scenario of co-extirpation is possible for ectoparasites of marten. Furthermore, because martens are carriers of the plague bacterium and suffer only brief clinical symptoms after infection (Marchette et al. 1962), they may also be involved in the epidemiology of plague. Because martens have much larger home ranges than the rodent prey species that serve as plague reservoirs, they may facilitate the transmission of infected fleas to uninfected rodent populations (Zielinski 1984).

Diseases do not appear to be an important source of mortality in black bears, American martens, or fishers. All are susceptible to trichinosis, however, and the latter two species carry other diseases such as toxoplasmosis, leptospirosis, and Aleutian disease (Rogers and Rogers 1976, Strickland et al. 1982*a,b*). Spotted skunks and raccoons are important vectors for rabies, histoplasmosis, and canine distemper, and raccoons are susceptible to several forms of encephalitis (Howard and Marsh 1982, Kaufmann 1982). However, the interrelationships among carnivores, parasites, and various ecosystem attributes remain largely unknown.

An assessment of key ecological functions of tree-dwelling mammals in the Pacific Northwest

Here, we assess the ecological functional roles played by native mammals (see Marcot and Aubry 2003) that are closely associated with trees in coniferous forests. For this analysis, we used the database of key ecological functions (KEFs) available for Washington and Oregon (O'Neil et al. 2001) to illustrate the array of KEFs provided by tree-dwelling mammals, and to explore how management actions affecting their habitat elements can influence those functions. The term "key ecological function" refers to the major ecological roles played by each species. KEFs are hierarchically categorized according to the trophic and feeding roles of organisms, as well as their roles in nutrient cycling, and various kinds of organismal, disease, soil, wood structure, water, and vegetation relations (Marcot and Vander Heyden 2001). KEFs of species can influence habitat conditions and resources used by other species, thereby influencing the biodiversity and productivity of ecosystems. Thus, managers may wish to understand the effects of habitat management on KEFs. Many of these KEFs are well known, but their rates and specific effects on ecosystem diversity and productivity generally have not been quantified. Managers can view KEFs as testable management hypotheses about how tree-dwelling mammals may influence a particular ecosystem.

KEFs of mammals associated with specific vertical tree structures

Among the nine forested habitat types that occur in Washington and Oregon, 27 mammal species are associated with vertical tree structures (Table 12.1). Most of these species occur in forests west of the crest of the Cascade Range, and fewer occur in western juniper and mountain mahogany woodlands and in lodgepole pine forest and woodlands (Fig. 12.1). Species composition varies among forest types, and even the low-richness types provide unique habitats for some species.

Collectively, the 27 species participate in 60 categories of KEFs; because tallies of KEFs include both categories and subcategories, there is some minor duplication in the counts. These KEFs in turn extend well beyond the immediate vertical tree structures with which the species are associated. As examples: of the 27 species, 11 species of bats and one carnivore may decrease insect populations through insectivory; one squirrel and two carnivores excavate burrows and create runways that can be used by other species; five rodents and five carnivores help disperse seeds, spores, plants, or animals, including dispersal of fungi, lichens, and fruits; six rodents create nesting structures that can be used by other species; two rodents churn soil by digging, potentially improving soil structure and aeration; and other functions.

Nine habitat elements pertaining to vertical tree structures provide part of the habitat requirements for these 27 species, and thus support the array of KEFs performed by those species (Fig. 12.2). The four vertical

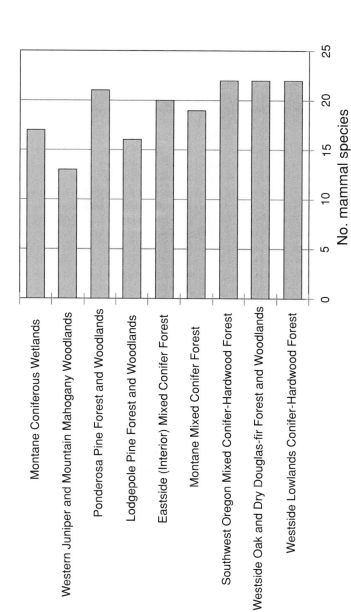

Fig. 12.1. Number of mammal species associated with large tree structures in the nine forested wildlife habitat types occurring in Washington and Oregon (Chappell et al. 2001).

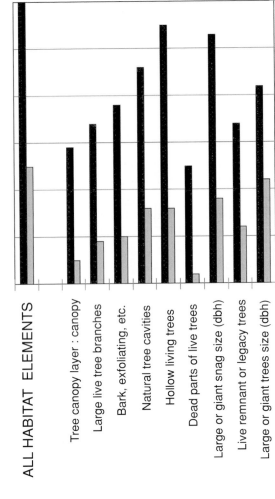

Fig. 12.2. Number of mammal species associated with vertical tree structure habitat elements (vertical axis) and the number of categories of key ecological functions (KEFs) they perform collectively in forests of Washington and Oregon.

tree structures that provide for the greatest number of these 27 species and their associated KEF categories are hollow living trees, large or giant snags, natural tree cavities, and large or giant live trees. However, even those structural elements, such as the dead parts of live trees, that provide for relatively few of these species and KEFs nonetheless make unique contributions to several species, and may play important roles in ecosystem function.

To managers, this means that providing vertical tree structures (as listed in Fig. 12.2) within each of these forest types not only provides some of the habitat needs of these species, but also helps contribute to the "functional web" of their ecological roles within ecosystems. The specific array of KEFs and associated species varies by type of vertical tree structures and forest habitats. Thus, all forest types are necessary to provide the full, collective array of these species and their KEFs. The specific rate of each ecological function, however, and the quantitative influence on ecosystem productivity, are essentially unstudied. Our evaluation provides a qualitative starting point by which to test functional roles as management hypotheses.

Information and research needs

Our review shows there is empirical support for the ecological roles we have hypothesized for tree-dwelling mammals in western coniferous forests. However, research that quantifies how and to what extent this group of mammals influences the composition, structure, or functioning of ecosystems is generally lacking. For example, we know that bats consume large numbers of nocturnal insects, but we do not know if their foraging activities regulate the size of nocturnal insect populations. Similarly, we know that arboreal rodents and forest carnivores transport and probably aid in the long-distance dispersal of fungal spores and dwarf mistletoe seeds, but we do not know whether ectomycorrhizal fungi would be unable to inoculate areas of new regeneration, or if mistletoe infections would be reduced in mature forests, in the absence of tree-dwelling mammals. Field studies designed to answer such questions will be challenging, but are essential for understanding the ecological roles that tree-dwelling mammals play in coniferous forest ecosystems.

Comparing similar ecosystems with and without a given species may be the most useful approach to understanding the ecological importance of that species, even though such studies may suffer from inadequate replication or controls (Estes 1996). Manipulative experiments involving the removal of tree-dwelling mammals from forested ecosystems are neither feasible nor socially acceptable. However, local extirpations and subsequent re-introduction efforts provide unique opportunities to study the ecological importance of tree-dwelling mammals. For example, both martens and fishers were extirpated from large portions of their former range in the Pacific states during the last century (Zielinski et al. 1995, Aubry et al. 1996, Zielinski et al. 2001). In response to these population declines, the Washington Department of Fish and Wildlife has begun to plan the re-introduction of fishers to Washington state, and re-introductions may also occur in portions of Oregon and California where fishers no longer occur. Thus, opportunities currently exist to compare ecosystem processes in western coniferous forests with and without these species, and to conduct such studies before and after re-introductions. Empirical evaluation of the ecological roles we have hypothesized for tree-dwelling mammals would also increase our understanding of the functioning of coniferous forest ecosystems and provide a much stronger scientific basis for evaluating forest management and conservation alternatives.

Summary

Despite the scarcity of empirical evidence, we hypothesize that bats have an important influence on ecosystem function in western coniferous forests because of their abundance and the specialized ecological niches they occupy. Bats serve as prey for aerial predators and as agents of nutrient transport, but we suspect that these functions influence ecosystem processes at relatively small spatial scales, such as at roosting and foraging sites. We predict that the most important ecological role performed by bats is their influence on nocturnal insect populations.

   With the exception of extensive work on the role of arboreal rodents in the dissemination of ectomycorrhizal fungi, few studies have examined the ecological influence of arboreal rodents in western coniferous forests. By dispersing these specialized fungi, arboreal rodents help sustain productivity of both forests and forest commodities in western coniferous forests. In addition, arboreal rodents are important prey for both avian and mammalian predators, and some predator populations may be strongly influenced by the diversity and abundance of arboreal rodents.

   In western coniferous forests, tree-dwelling carnivores may be important agents for long-distance dispersal of propagules, and may contribute to the maintenance of biodiversity by serving as hosts for a variety of parasites. With several exceptions, forest carnivores do not appear to exert strong influences on populations of competitors or prey in western coniferous forests. However, Terborgh (1988) argued that top predators are primarily responsible for the stability and extraordinary diversity of plants and animals in pristine tropical forests. He reasoned that this profound influence on ecosystem structure resulted from the propagation of perturbations through multiple trophic levels in the ecosystem. Thus, effects are felt even in organisms that are far removed, both geographically and taxonomically, from the predator and its prey. Whether forest carnivores exert a similarly strong influence on ecosystem structure in western coniferous forests is unknown.

We have argued that managing forests solely for the persistence of a few key vertebrate species or target vegetative conditions is an overly simplistic approach to forest management, and one that is unlikely to provide for long-term ecosystem sustainability. We believe that resource managers can improve the outcome of management decisions in western coniferous forests by considering the ecological roles that mammals play and evaluating the functional webs to which they contribute. By comparing the array of ecological functions that are predicted to occur under a set of alternative management strategies, managers can select the alternative that will be most likely to provide for long-term ecosystem integrity.

Literature cited

Alexander, L.F., B.J. Verts, and T.P. Farrell. 1994. Diet of ringtails (*Bassariscus astutus*) in Oregon. *Northwestern Naturalist* 75:97–101.

Allred, W.S., W.S. Gaud, and J.S. States. 1994. Effects of herbivory by Abert squirrels (*Sciurus aberti*) on cone crops of ponderosa pine. *Journal of Mammalogy* 75:700–703.

Anthony, E.L.P. and T.H. Kunz. 1977. Feeding strategies of the little brown bat, *Myotis lucifugus*, in southern New Hampshire. *Ecology* 58:775–786.

Aubry, K.B., G.M. Koehler, and J.R. Squires. 2000. Ecology of Canada lynx in southern boreal forests. Pages 373–396 *in* L.F. Ruggiero, K.B. Aubry, S.W. Buskirk, G.M. Koehler, C.J. Krebs, K.S. McKelvey, and J.C. Squires, editors. *Ecology and Conservation of Lynx in the United States.* University Press of Colorado, Boulder, Colorado, USA.

Aubry, K.B., J.C. Lewis, and C.M. Raley. 1996. Reintroduction, current distribution, and ecology of fishers in southwestern Oregon: a progress report. *Martes Working Group Newsletter* 4:8–10.

Aubry, K.B. and C.M. Raley. 2001. *Ecological Characteristics of Fishers in Southwestern Oregon.* Annual report. USDA, Forest Service, Pacific Northwest Research Station, Forestry Sciences Laboratory, Olympia, Washington, USA.

Aubry, K.B. and C.M. Raley. 2002. The pileated woodpecker as a keystone habitat modifier in the Pacific Northwest. Pages 257–274 *in* W.F. Laudenslayer, Jr., P.J. Shea, B.E. Valentine, C.P. Weatherspoon, and T.E. Lisle, technical coordinators. Proceedings of the symposium on the ecology and management of dead wood in western forests; November 2–4, 1999; Reno, Nevada. General Technical Report **PSW-GTR-181**. USDA, Forest Service, Pacific Southwest Research Station, Fresno, California, USA.

Balda, R.P. 1975. The relationship of secondary cavity nesters to snag densities in western coniferous forests. Wildlife Habitat Technical Bulletin **No. 1**. USDA, Forest Service, Southwestern Region, Albuquerque, New Mexico, USA.

Barclay, R.M.R. 1985. Long- versus short-range foraging strategies of hoary (*Lasiurus cinereus*) and silver-haired (*Lasionycteris noctivagans*) bats and the consequences for prey selection. *Canadian Journal of Zoology* **63**:2507–2515.

Barclay, R.M. R., M. Dolan, and A. Dyck. 1991. The digestive efficiency of insectivorous bats. *Canadian Journal of Zoology* **69**:1853–1856.

Barr, T.C., Jr. and R.M. Norton. 1965. Predation on cave bats by the pilot black snake. *Journal of Mammalogy* **46**:672.

Ben-David, M., R.T. Bowyer, L.K. Duffy, D.D. Roby, and D.M. Schell. 1998. Social behavior and ecosystem processes: river otter latrines and nutrient dynamics of terrestrial vegetation. *Ecology* **79**:2567–2571.

Best, T.L., B.A. Milam, T.D. Haas, W.S. Cvilikas, and L.R. Saidak. 1997. Variation in diet of the gray bat (*Myotis grisescens*). *Journal of Mammalogy* **78**:569–583.

Black, H.L. 1974. A north temperate bat community: structure and prey populations. *Journal of Mammalogy* **55**:138–157.

Boal, C.W. and R.W. Mannan. 1994. Northern goshawk diets in ponderosa pine forests on the Kaibab Plateau. *Studies in Avian Biology* **16**:97–102.

Boyce, M.S. and A. Haney, editors. 1997. *Ecosystem Management: Applications for Sustainable Forest and Wildlife Resources*. Yale University Press, New Haven, Connecticut, USA.

Brigham, R.M. and M.B. Fenton. 1986. The influence of roost closure on the roosting and foraging behaviour of *Eptesicus fuscus* (Chiroptera: Vespertilionidae). *Canadian Journal of Zoology* **64**:1128–1133.

Brigham, R.M. and M.B. Saunders. 1990. The diet of big brown bats (*Eptesicus fuscus*) in relation to insect availability in southern Alberta, Canada. *Northwest Science* **64**:7–10.

Brooks, D.R. and E.P. Hoberg. 2000. Triage for the biosphere: the need and rationale for taxonomic inventories and phylogenetic studies of parasites. *Comparative Parasitology* **67**:1–25.

Brown, E.R., technical editor. 1985. Management of wildlife and fish habitats in forests of western Oregon and Washington. Part 1 – Chapter Narratives. **R6-F&WL-192-1985**. USDA, Forest Service, Pacific Northwest Region, Portland, Oregon, USA.

Buchanan, J.B., R.W. Lundquist, and K.B. Aubry. 1990. Winter populations of Douglas' squirrels in different-aged Douglas-fir forests. *Journal of Wildlife Management* **54**:577–581.

Bull, E.L., C.G. Parks, and T.R. Torgersen. 1997. Trees and logs important to wildlife in the Interior Columbia River Basin. General Technical Report **PNW-GTR-391**. USDA, Forest Service, Pacific Northwest Research Station, Portland, Oregon, USA.

Buskirk, S.W. 1999. Mesocarnivores of Yellowstone. Pages 165–187 *in* T.W. Clark, A.P. Curlee, S.C. Minta, and P.M. Kareiva, editors. *Carnivores in Ecosystems: the Yellowstone Experience*. Yale University Press, New Haven, Connecticut, USA.

Buskirk, S.W. and L.F. Ruggiero. 1994. American marten. Pages 7–37 *in* L.F. Ruggiero, K.B. Aubry, S.W. Buskirk, L.J. Lyon, and W.J. Zielinski, technical editors. The scientific basis for conserving forest carnivores: American marten, fisher, lynx, and wolverine in the western United States. General Technical Report **RM-254**. USDA, Forest Service, Rocky Mountain Forest and Range Experiment Station, Fort Collins, Colorado, USA.

Buskirk, S.W. and W.J. Zielinski. 2003. Small and mid-sized carnivores. Pages 207–249 *in* C.J. Zabel and R.G. Anthony, editors. *Mammal Community Dynamics. Management and Conservation in the Coniferous Forests of Western North America*. Cambridge University Press, Cambridge, UK.

Buskirk, S.W., H.J. Harlow, and S.C. Forrest. 1987. Studies on the resting site ecology of marten in the central Rocky Mountains. Pages 150–153 *in* C.A. Troendle, M.R. Kaufmann, R.H. Hamre, and R.P. Winokur, technical coordinators. Management of subalpine forests: building on 50 years of research. General Technical Report **RM-149**. USDA, Forest Service, Rocky Mountain Forest and Range Experiment Station, Fort Collins, Colorado, USA.

Bustamente, R.O., J.A. Simonetti, and J.E. Mella. 1992. Are foxes legitimate and efficient seed dispersers? A field test. *Acta Oecologica* **13**:203–208.

Carey, A.B. 1991. The biology of arboreal rodents in Douglas-fir forests. General Technical Report **PNW-GTR-276**. USDA, Forest Service, Pacific Northwest Research Station, Portland, Oregon, USA.

Carey, A.B. 1995. Sciurids in managed and old growth forest in the Pacific Northwest. *Ecological Applications* **5**:648–661.

Carey, A.B., S.P. Horton, and B.L. Biswell. 1992. Northern spotted owls: influence of prey base and landscape character. *Ecological Monographs* **62**:223–250.

Chapman, J.A. and G.A. Feldhamer, editors. 1982. *Wild Mammals of North America: Biology, Management, and Economics*. The Johns Hopkins University Press, Baltimore, Maryland, USA.

Chappell, C.B., R.C. Crawford, C. Barrett, J. Kagan, D.H. Johnson, M. O'Mealy, G.A. Green, H.L. Ferguson, W.D. Edge, E.L. Greda, and T.A. O'Neil. 2001. Wildlife-habitats: descriptions, status, trends, and system dynamics. Pages 22–114 *in* D.H. Johnson and T.A. O'Neil, managing directors. *Wildlife–Habitat Relationships in Oregon and Washington*. Oregon State University Press, Corvallis, Oregon, USA.

Claridge, A.W., J.M. Trappe, S.J. Cork, and D.L. Claridge. 1999. Mycophagy by small mammals in the coniferous forests of North America: nutritional value of sporocarps of *Rhizopogon vinicolor*, a common hypogeous fungus. *Journal of Comparative Physiology* **169**:172–178.

Cleeves, T.R. 1969. Herring gull catching and eating bat. *British Birds* **62**:333.

Conner, R.N., O.K. Miller, Jr., and C.S. Adkisson. 1976. Woodpecker dependence on trees infected by fungal heart rots. *The Wilson Bulletin* **88**:575–581.

Cork, S.J. and G.J. Kenagy. 1989. Nutritional value of hypogeous fungus for a forest-dwelling ground squirrel. *Ecology* **70**:577–586.

Corn, P.S. and R.B. Bury. 1986. Habitat use and terrestrial activity by red tree voles (*Arborimus longicaudus*) in Oregon. *Journal of Mammalogy* **67**:404–405.

Cross, S.P. 1988. Riparian systems and small mammals and bats. Pages 93–112 *in* K.J. Raedeke, editor. *Streamside Management: Riparian Wildlife and Forestry Interactions.* University of Washington Institute of Forest Resources, Contribution No. 59. Seattle, Washington, USA.

Curtis, J.D. and E.L. Kozicky. 1944. Observations on the eastern porcupine. *Journal of Mammalogy* **25**:137–146.

Davis, W.B. 1951. Bat, *Molossus nigtricans*, eaten by the rat snake, *Elaphe laeta. Journal of Mammalogy* **32**:219.

de Vos, A. 1952. The ecology and management of fisher and marten in Ontario. Technical Bulletin 1. Ontario Dept. of Lands and Forests.

Dodge, W.E. and J.E. Borrecco. 1992. Porcupines. Pages 253–270 *in* H.C. Black, technical editor. Silvicultural approaches to animal damage management in Pacific Northwest forests. General Technical Report **PNW-GTR-287**. USDA, Forest Service, Pacific Northwest Research Station, Portland, Oregon, USA.

Duncan, R.B. and R. Sidner. 1990. Bats in spotted owl pellets in southern Arizona. *Great Basin Naturalist* **50**:197–200.

Duvergé, P.L., G. Jones, J. Rydell, and R.D. Ransome. 2000. Functional significance of emergence timing in bats. *Ecography* **23**:32–40.

Earle, R.D. 1978. The fisher-porcupine relationship in Upper Michigan. Thesis. Michigan Technical University, Houghton, Michigan, USA.

Eglitis, A. and P.E. Hennon. 1997. Porcupine feeding damage in precommercially thinned conifer stands of central southeast Alaska. *Western Journal of Applied Forestry* **12**:115–121.

Eis, S., E.H. Garman, and L.F. Ebell. 1965. Relation between cone production and diameter increment of Douglas-fir (*Pseudotsuga menziesii* [Mirb.] Franco), grand fir (*Abies grandis* [Dougl.] Lindl.) and western white pine (*Pinus monticola* Dougl.). *Canadian Journal of Botany* **43**:1553–1559.

Elwell, A.S. 1962. Blue jay preys on young bats. *Journal of Mammalogy* **43**:434.

Erkert, H.G. 1982. Ecological aspects of bat activity rhythms. Pages 201–242 *in* T.H. Kunz, editor. *Ecology of Bats.* Plenum Press, New York, New York, USA.

Estes, J.A. 1996. Predators and ecosystem management. *Wildlife Society Bulletin* **24**:390–396.

Estes, J.A. and J.F. Palmisano. 1974. Sea otters: their role in structuring nearshore communities. *Science* **185**:1058–1060.

Farentinos, R.C. 1972. Observations on the ecology of the tassel-eared squirrel. *Journal of Wildlife Management* **36**:1234–1239.

Fenton, M.B. 1995. Constraint and flexibility – bats as predators, bats as prey. *Zoological Society of London Symposia* **67**:277–289.

Fenton, M.B., I.L. Rautenbach, S.E. Smith, C.M. Swanepoel, J. Grosell, and J. van Jaarsveld. 1994. Raptors and bats: threats and opportunities. *Animal Behaviour* **48**:9–18.

Finley, R.B., Jr. 1969. Cone caches and middens of *Tamiasciurus* in the Rocky Mountain region. University of Kansas Museum of Natural History. *Miscellaneous Publication* **51**:233–273.

Fogel, R. and J.M. Trappe. 1978. Fungus consumption (mycophagy) by small animals. *Northwest Science* **52**:1–31.

Forsman, E. and C. Maser. 1970. Saw-whet owl preys on tree mice. *Murrelet* **51**:10.

Forsman, E.D., E.C. Meslow, and H.M. Wight. 1984. Distribution and biology of the spotted owl. *Wildlife Monographs* **87**:1–64.

Fowells, H.A. 1965. Silvics of forest trees of the United States. Agricultural Handbook 271. USDA, Forest Service, Washington, D.C., USA.

Gellman, S.T. and W.J. Zielinski. 1996. Use by bats of old-growth redwood hollows on the north coast of California. *Journal of Mammalogy* **77**:255–265.

Gill, D. and L.D. Cordes. 1972. Winter habitat preference of porcupines in the southern Alberta foothills. *Canadian Field-Naturalist* **86**:349–355.

Gillesberg, A.-M. and A.B. Carey. 1991. Arboreal nests of *Phenacomys longicaudus* in Oregon. *Journal of Mammalogy* **72**:784–787.

Gillette, D.D. and J.D. Kimbrough. 1970. Chiropteran mortality. Pages 262–281 *in* B.H. Slaughter and D.W. Walton, editors. *About bats*. Dallas Southern Methodist University Press, Dallas, Texas, USA.

Gould, E. 1955. The feeding efficiency of insectivorous bats. *Journal of Mammalogy* **36**:399–407.

Gould, E. 1959. Further studies on the feeding efficiency of bats. *Journal of Mammalogy* **40**:149–150.

Grenfell, W.E. and M. Fasenfest. 1979. Winter food habits of fishers, *Martes pennanti*, in northwestern California. *California Fish and Game* **65**:186–189.

Griffin, D.R., F.A. Webster, and C.R. Michael. 1960. The echolocation of flying insects by bats. *Animal Behaviour* **8**:141–154.

Halls, L.K., editor. 1977. Southern fruit-producing woody plants used by wildlife. General Technical Report **SO-16**. USDA, Forest Service, Southern Forest Experiment Station, New Orleans, Louisiana, USA.

Hamilton, I.M. and R.M.R. Barclay. 1998. Diets of juvenile, yearling, and adult big brown bats (*Eptesicus fuscus*) in southeastern Alberta. *Journal of Mammalogy* **79**:764–771.

Hawksworth, F.G. and D. Wiens. 1996. Dwarf mistletoes: biology, pathology, and systematics. USDA, Forest Service, Washington DC, USA.

Hawksworth, F.G., T.H. Nicholls, and L.M. Merrill. 1987. Long-distance dispersal of lodgepole pine dwarf mistletoe. Pages 220–226 *in* C.A. Troendle, M.R. Kaufmann, R.H. Hamre, and R.P. Winokur, technical coordinators. Management of subalpine forest: building on 50 years of research. General Technical Report **RM-149**. USDA, Forest Service, Rocky Mountain Forest and Range Experiment Station, Fort Collins, Colorado, USA.

Hayes, J.P. 2003. Habitat ecology and conservation of bats in western coniferous forests. Pages 81–119 *in* C.J. Zabel and R.G. Anthony, editors. *Mammal Community Dynamics. Management and Conservation in the Coniferous Forests of Western North America*. Cambridge University Press, Cabridge, UK.

Henke, S.E. and F.C. Bryant. 1999. Effects of coyote removal on the faunal community in western Texas. *Journal of Wildlife Management* **63**:1066–1081.

Herrera, C.M. 1989. Frugivory and seed dispersal by carnivorous mammals, and associated fruit characteristics, in undisturbed Mediterranean habitats. *Oikos* **55**:250–262.

Hickey, J.M., R.W. Flynn, S.W. Buskirk, K.G. Gerow, and M.F. Willson. 1999. An evaluation of a mammalian predator, *Martes americana*, as a disperser of seeds. *Oikos* **87**:499–508.

Hickey, M.B.C. and M.B. Fenton. 1990. Foraging by red bats (*Lasiurus borealis*): do intraspecific chases mean territoriality? *Canadian Journal of Zoology* **68**:2477–2482.

Hickey, M.B.C. and M.B. Fenton. 1996. Behavioural and thermoregulatory responses of female hoary bats, *Lasiurus cinereus* (Chiroptera: Vespertilionidae), to variations in prey availability. *Ecoscience* **3**:414–422.

Hoberg, E.P., K.B. Aubry, and J.D. Brittell. 1990. Helminth parasitism in martens (*Martes americana*) and ermines (*Mustela erminea*) from Washington, with comments on the distribution of *Trichinella spiralis*. *Journal of Wildlife Diseases* **26**:447–452.

Hoover, R.L. and D.L. Wills, editors. 1984. Managing forested lands for wildlife. Colorado Division of Wildlife in cooperation with USDA, Forest Service, Rocky Mountain Region, Denver, Colorado, USA.

Hope, S. and W.C. McComb. 1994. Perceptions of implementing and monitoring wildlife tree prescriptions on national forests in western Washington and Oregon. *Wildlife Society Bulletin* **22**:383–392.

Howard, W.E. and R.E. Marsh. 1982. Spotted and hog-nosed skunks and allies. Pages 664–673 *in* J.A. Chapman and G.A. Feldhamer, editors. *Wild Mammals of North America: Biology, Management, and Economics*. The Johns Hopkins University Press, Baltimore, Maryland, USA.

Howell, A.B. 1926. Voles of the genus *Phenacomys*, II: life history of the red tree mouse *Phenacomys*. *North American Fauna* **48**:39–66.

Hudler, G., N. Oshima, and F.G. Hawksworth. 1979. Bird dissemination of dwarf mistletoe on ponderosa pine in Colorado. *American Midland Naturalist* **102**:273–280.

Johnson, D.H. and T.A. O'Neil, managing directors. 2001. *Wildlife-Habitat Relationships in Oregon and Washington*. Oregon State University Press, Corvallis Oregon, USA.

Jones, G. and J. Rydell. 1994. Foraging strategy and predation risk as factors influencing emergence time in echolocating bats. *Philosophical Transactions of the Royal Society of London B* **346**:445–455.

Jones, J.L. 1991. Habitat use of fisher in northcentral Idaho. Thesis. University of Idaho, Moscow, Idaho, USA.

Kalcounis, M.C. and R.M. Brigham. 1994. Impact of predation risk on emergence by little brown bats, *Myotis lucifugus* (Chiroptera: Vespertilionidae), from a maternity colony. *Ethology* **98**:201–209.

Kaufmann, J.H. 1982. Raccoon and allies. Pages 567–585 *in* J.A. Chapman and G.A. Feldhamer, editors. *Wild Mammals of North America: Biology, Management, and Economics*. The Johns Hopkins University Press, Baltimore, Maryland, USA.

Kebbe, C.E. 1961. Return of the fisher. *Oregon State Game Commission Bulletin* **16**:3–7.

Keith, J.O. 1965. The Abert squirrel and its dependence on ponderosa pine. *Ecology* **46**:150–163.

Kenward, R.E. 1983. The causes of damage by red squirrels and gray squirrels. *Mammal Review* **13**:159–166.

Kinsey, C. 1961. Leopard frog attacks bat. *Journal of Mammalogy* **42**:408.

Kirkpatrick, R.D. 1982. *Rana catesbeiana* (bullfrog) food. *Herpetological Review* **13**:17.

Koford, R.R. 1982. Mating systems of a territorial tree squirrel (*Tamiasciurus douglasii*) in California. *Journal of Mammalogy* **63**:274–283.

Krohn, W.B., K.D. Elowe, and R.B. Boone. 1995. Relations among fishers, snow, and martens: development and evaluation of two hypotheses. *The Forestry Chronicle* **71**:97–105.

Krohn, W.B., W.J. Zielinski, and R.B. Boone. 1997. Relations among fishers, snow, and martens in California: results from small-scale spatial comparisons. Pages 211–232 *in* G. Proulx, H.N. Bryant, and P.M. Woodward, editors. Martes: *Taxonomy, Ecology, Techniques, and Management*. Provincial Museum of Alberta, Edmonton, Alberta, Canada.

Kunz, T.H. 1974. Feeding ecology of a temperate insectivorous bat (*Myotis velifer*). *Ecology* **55**:693–711.

Kunz, T.H. 1982. Roosting ecology. Pages 1–56 *in* T. H. Kunz, editor. *Ecology of Bats*. Plenum Press, New York, New York, USA.

Kunz, T.H. and E.L.P. Anthony. 1996. Variation in the timing of nightly emergence behavior in the little brown bat, *Myotis lucifugus* (Chiroptera: Vespertilionidae). Pages 225–235 *in* Anonymous, editor. *Contributions in Mammalogy: a Memorial Volume Honoring Dr. J. Knox Jones, Jr.* Museum of Texas Tech University, Lubbock, Texas, USA.

Kunz, T.H., J.O. Whitaker, Jr., and M.D. Wadanoli. 1995. Dietary energetics of the insectivorous Mexican free-tailed bat (*Tadarida brasiliensis*) during pregnancy and lactation. *Oecologia* **101**:407–415.

Kurta, A., G.P. Bell, K.A. Nagy, and T.H. Kunz. 1989. Energetics of pregnancy and lactation in free-ranging little brown bats (*Myotis lucifugus*). *Physiological Zoology* **62**:804–818.

Kurta, A., T.H. Kunz, and K.A. Nagy. 1990. Energetics and water flux of free-ranging big brown bats (*Eptesicus fuscus*) during pregnancy and lactation. *Journal of Mammalogy* **71**:59–65.

Lee, D.S. 1969. Notes on the feeding behavior of cave-dwelling bullfrogs. *Herpetologica* **25**:211–212.

Lemons, D.E. 1978. Small mammal dissemination of dwarf mistletoe seeds. Thesis. Portland State University, Portland, Oregon, USA.

Leonard, R.D. 1980. Winter activity and movements, winter diet and breeding biology of the fisher in southeast Manitoba. Thesis. University of Manitoba, Winnipeg, Manitoba, Canada.

Lewis, S.E. 1995. Roost fidelity of bats: a review. *Journal of Mammalogy* **76**:481–496.

Lindsay, S.L. 1986. Geographic size variation in *Tamiasciurus douglasii*: significance in relation to conifer cone morphology. *Journal of Mammalogy* **67**:317–325

Long, R.F. 1996. Bats for insect biocontrol in agriculture. *The IPM Practitioner* **18**(9):1–6.

Long, R.F., T. Simpson, T.-S. Ding, S. Heydon, and W. Reil. 1998. Bats feed on crop pests in Sacramento Valley. *California Agriculture* **52**:8–10.85744.

Luoma, D.L., J.M. Trappe, A.W. Claridge, K. Jacobs, and E. Cazares. 2003. Relationships among fungi and small mammals in forested ecosystems. Pages 343–373 *in* C.J. Zabel and R.G. Anthony, editors. *Mammal Community Dynamics. Management and Conservation in the Coniferous Forests of Western North America*. Cambridge University Press, Cambridge, UK.

Luttich, S., D.H. Rusch, E.C. Meslow, and L.B. Keith. 1970. Ecology of red-tailed hawk predation in Alberta. *Ecology* **51**:190–203.

Machmer, M.M. and C. Steeger. 1995. The ecological roles of wildlife tree users in forest ecosystems. *Land Management Handbook* **35**. B.C. Ministry of Forests, Victoria, BC, Canada.

Marchette, N.J., D.L. Lungren, D.S. Nicholes, J.B. Bushman, and D. Vest. 1962. Studies on infectious diseases in wild animals in Utah. II. Susceptibility of wild mammals to experimental plague. *Zoonoses Research* **1**:235–250.

Marcot, B.G. 1996. An ecosystem context for bat management: a case study of the interior Columbia River Basin, U.S.A. Pages 19–36 *in* R.M.R. Barclay and R.M. Brigham, editors. Bats and forests symposium; October 19–21, 1995; Victoria, British Columbia. Ministry of Forests, Victoria, BC, Canada.

Marcot, B.G. and K.B. Aubry. 2003. The functional diversity of mammals in coniferous forests of western North America. Pages 631–664 *in* C.J. Zabel and R.G. Anthony, editors. *Mammal Community Dynamics. Management and Conservation in the Coniferous Forests of Western North America.* Cambridge University Press, Cambridge, UK.

Marcot, B.G. and M. Vander Heyden. 2001. Key ecological functions of wildlife species. Pages 168–186 *in* D.H. Johnson and T.A. O'Neil, managing directors. *Wildlife-Habitat Relationships in Oregon and Washington.* Oregon State University Press, Corvallis, Oregon, USA.

Marcot, B.G., L.K. Croft, J.F. Lehmkuhl, R.H. Naney, C.G. Niwa, W.R. Owen, and R.E. Sandquist. 1998. Macroecology, paleoecology, and ecological integrity of terrestrial species and communities of the interior Columbia River Basin and portions of the Klamath and Great Basins. General Technical Report **PNW-GTR-410**. USDA, Forest Service, Pacific Northwest Research Station, Portland, Oregon, USA.

Marshall, D.B. 1994. *Status of the American Marten in Oregon and Washington.* Audubon Society of Portland, Portland, Oregon, USA.

Marshall, W.H. 1946. Winter food habits of the pine marten in Montana. *Journal of Mammalogy* **27**:83–84.

Martin, A.C., H.S. Zim, and A.L. Nelson. 1951. *American Wildlife and Plants: a Guide to Wildlife Food Habits.* Dover Publications, Inc., New York, New York, USA.

Martin, R.L. 1961. Vole predation on bats in an Indiana cave. *Journal of Mammalogy* **42**:540–541.

Martin, S.K. 1994. Feeding ecology of American martens and fishers. Pages 297–315 *in* S.W. Buskirk, A.S. Harestad, M.G. Raphael, and R.A. Powell, editors. *Martens, Sables, and Fishers: Biology and Conservation.* Cornell University Press, Ithaca, New York, USA.

Maser, C. 1966. Life histories and ecology of *Phenacomys albipes, Phenacomys longicaudus,* and *Phenacomys silvicola.* Thesis. Oregon State University, Corvallis, Oregon, USA.

Maser, C. and Z. Maser. 1988. Interactions among squirrels, mycorrhizal fungi, and coniferous forests in Oregon, USA. *Great Basin Naturalist* **48**:358–369.

Maser, C., J.M. Trappe, and R.A. Nussbaum. 1978. Fungal – small mammal interrelationships with emphasis on Oregon coniferous forests. *Ecology* **59**:799–809.

Maser, C., B.R. Mate, J.F. Franklin, and C.T. Dyrness. 1981. Natural history of Oregon Coast mammals. General Technical Report **PNW-133**. USDA, Forest Service, Pacific Northwest Research Station, Portland, Oregon, USA.

Maser, C., Z. Maser, J.W. Witt, and G. Hunt. 1986. The northern flying squirrel: a mycophagist in southwestern Oregon. *Canadian Journal of Zoology* **64**: 2086–2089.

Mathiasen, R.L. 1996. Dwarf mistletoes in forest canopies. *Northwest Science* **70**:61–71.

May, R.M. and R.M. Anderson. 1983. Epidemiology and genetics in the coevolution of parasites and hosts. *Proceedings of the Royal Society of London Series B* **219**:281–313.

McIntire, P.W. and A.B. Carey. 1989. A microhistological technique for analysis of food habits of mycophagous rodents. Research Paper **PNW-RP-404**. USDA, Forest Service, Pacific Northwest Research Station, Portland, Oregon, USA.

McLaren, B.E. and R.O. Peterson. 1994. Wolves, moose, and tree rings on Isle Royale. *Science* **266**:1555–1558.

McLean, J.A. and J.R. Speakman. 1999. Energy budgets of lactating and non-reproductive brown long-eared bats (*Plecotus auritus*) suggest females use compensation in lactation. *Functional Ecology* **13**:360–372.

Meiselman, N. and A.T. Doyle. 1996. Habitat and microhabitat use by the red tree vole, *Phenacomys longicaudus*. *American Midland Naturalist* **135**:33–42.

Molina, R., H.B. Massicotte, and J.M. Trappe. 1992. Specificity phenomena in mycorrhizal symbiosis: community-ecological consequences and practical implications. Pages 357–423 *in* M. Allen, editor. *Mycorrhizal Functioning: an Integrative Plant-Fungal Process*. Chapman & Hall Inc., New York, New York, USA.

Mumford, R.E. 1969. Long-tailed weasel preys on big brown bats. *Journal of Mammalogy* **50**:360.

Murray, D.L., C.A. Kapke, J.F. Evermann, and T.K. Fuller. 1999. Infectious disease and the conservation of free-ranging large carnivores. *Animal Conservation* **2**:241–254.

Newby, F.E. 1951. Ecology of the marten in the Twin Lakes area, Chelan County, Washington. Thesis. Washington State University, Pullman, Washington, USA.

North, M.J., J.M. Trappe, and J. Franklin. 1997. Standing crop and animal consumption of fungal sporocarps in Pacific Northwest forests. *Ecology* **78**:1543–1554.

O'Neil, T.A., D.H. Johnson, C. Barrett, M. Trevithick, K.A. Bettinger, C. Kiilsgaard, M. Vander Heyden, E.L. Greda, D. Stinson, B.G. Marcot, P.J. Doran, S. Tank, and L. Wunder. 2001. Matrixes for wildlife-habitat relationships in Oregon and Washington. CD-ROM *in* D.H. Johnson and T.A. O'Neil, managing directors. *Wildlife-Habitat Relationships in Oregon and Washington*. Oregon State University Press, Corvallis, Oregon, USA.

Ostry, M.E., T.H. Nicholls, and D.W. French. 1983. Animal vectors of eastern dwarf mistletoe on black spruce. Research Paper **NC-232**. USDA, Forest Service, North Central Forest Experiment Station, St. Paul, Minnesota, USA.

Parks, C.G. and E.L. Bull. 1997. American marten use of rust and dwarf mistletoe brooms in northeastern Oregon. *Western Journal of Applied Forestry* **12**:131–133.

Patton, D.R. and J.R. Vahle. 1986. Cache and nest characteristics of the red squirrel in Arizona mixed-conifer forest. *Western Journal of Applied Forestry* **1**:48–51.

Pelton, M.R. 1982. Black bear. Pages 504–514 *in* J.A. Chapman and G.A. Feldhamer, editors. *Wild Mammals of North America: Biology, Management, and Economics*. The Johns Hopkins University Press, Baltimore, Maryland, USA.

Pigozzi, G. 1992. Frugivory and seed dispersal by the European badger in a Mediterranean habitat. *Journal of Mammalogy* **73**:630–639.

Powell, R.A. 1993. *The Fisher: Life History, Ecology, and Behavior*. Second Edition. University of Minnesota Press, Minneapolis, Minnesota, USA.

Powell, R.A. and R.B. Brander. 1977. Adaptations of fishers and porcupines to their predator prey system. Pages 45–53 *in* R.L. Phillips and C. Jonkel, editors. Proceedings of the 1975 Predator Symposium, University of Montana, Missoula, Montana, USA.

Powell, R.A. and W.J. Zielinski. 1994. Fisher. Pages 38–73 *in* L.F. Ruggiero, K.B. Aubry, S.W. Buskirk, L.J. Lyon, and W.J. Zielinski, technical editors. The scientific basis for conserving forest carnivores: American marten, fisher, lynx, and wolverine in the western United States. General Technical Report **RM-254**. USDA, Forest Service, Rocky Mountain Forest and Range Experiment Station, Fort Collins, Colorado, USA.

Raine, R.M. 1981. Winter food habits, responses to snow cover and movements of fisher (*Martes pennanti*) and marten (*Martes americana*) in southeastern Manitoba. Thesis. University of Manitoba, Winnipeg, Manitoba, Canada.

Raine, R.M. 1983. Winter habitat use and responses to snow cover of fisher (*Martes pennanti*) and marten (*Martes americana*) in southeastern Manitoba. *Canadian Journal of Zoology* **61**:25–34.

Reynolds, H.G. 1963. Western goshawk takes Abert squirrel in Arizona. *Journal of Forestry* **61**:839.

Reynolds, R.T. 1970. Nest observations of the long-eared owl (*Asio otus*) in Benton County, Oregon, with notes on their food habits. *Murrelet* **51**:8–9.

Rogers, L.L. and R.D. Applegate. 1983. Dispersal of fruit seeds by black bears. *Journal of Mammalogy* **64**:310–311.

Rogers, L.L. and S.M. Rogers. 1976. Parasites of bears: a review. Pages 411–430 *in* M.R. Pelton, J.W. Lentfer, and G.E. Folk, editors. *Bears – their Biology and Management*. IUCN Publications new series **40**. International Union for Conservation of Nature and Natural Resources, Morges, Switzerland.

Rolseth, S.L., C.E. Koehler, and R.M.R. Barclay. 1994. Differences in the diets of juvenile and adult hoary bats, *Lasiurus cinereus*. *Journal of Mammalogy* **75**: 394–398.

Rózsa, L. 1992. Endangered parasite species. *International Journal for Parasitology* **22**:265–266.

Ruggiero, L.F., D.E. Pearson, and S.E. Henry. 1998. Characteristics of American marten den sites in Wyoming. *Journal of Wildlife Management* **62**:663–673.

Rydell, J., A. Entwistle, and P.A. Racey. 1996. Timing of foraging flights of three species of bats in relation to insect activity and predation risk. *Oikos* **76**: 243–252.

Samuel, W.M., M.J. Pybus, and A.A. Kocan, editors. 2001. *Parasitic Diseases of Wild Mammals*. Second Edition. Iowa State University Press, Ames, Iowa, USA.

Schmidt, W.C. and R.C. Shearer. 1971. Ponderosa pine seed, for animals or trees? Research Paper **INT-112**. USDA, Forest Service, Intermountain Forest and Range Experiment Station, Ogden, Utah, USA.

Scott, M.E. 1988. The impact of infection and disease on animal populations: implications for conservation biology. *Conservation Biology* **2**:40–56.

Seglund, A.E. 1995. The use of resting sites by the Pacific fisher. Thesis. Humboldt State University, Arcata, California, USA.

Shaw, W.T. 1936. Moisture and its relation to the cone-storing habit of the western pine squirrel. *Journal of Mammalogy* **17**:337–349.

Smith, C.C. 1970. The coevolution of pine squirrels (*Tamiasciurus*) and conifers. *Ecological Monographs* **40**:349–371.

Smith, C.C. 1981. The indivisible niche of *Tamiasciurus*: an example of nonpartitioning of resources. *Ecological Monographs* **51**:343–363.

Smith, C.C. and R.P. Balda. 1979. Competition among insects, birds, and mammals for conifer seeds. *American Zoologist* **19**:1065–1083.

Smith, G.W. 1982. Habitat use by porcupines in a ponderosa pine/Douglas-fir forest in northeastern Oregon. *Northwest Science* **56**:236–240.

Smith, M.C. 1968. Red squirrel responses to spruce cone failure in interior Alaska. *Journal of Wildlife Management* **32**:306–316.

Smith, R.B., M.Z. Peery, R.J. Gutiérrez, and W.S. Lahaye. 1999. The relationship between spotted owl diet and reproductive success in the San Bernardino Mountains, California. *Wilson Bulletin* **111**:22–29.

Smith, S.E. and D.J. Read. 1997. *Mycorrhizal Symbiosis*. Academic Press, London, UK.

Smith, W.P., R.G. Anthony, J.R. Waters, N.L. Dodd, and C.J. Zabel. 2003. Ecology and conservation of arboreal rodents in western coniferous forests. Pages 157–206 *in* C.J. Zabel and R.G. Anthony, editors. *Mammal Community Dynamics. Management and Conservation in the Coniferous Forests of Western North America.* Cambridge University Press, Cambridge, UK.

Speakman, J.R. 1991*a*. Why do insectivorous bats in Britain not fly in daylight more frequently? *Functional Ecology* **5**:518–524.

Speakman, J.R. 1991*b*. The impact of predation by birds on bat populations in the British Isles. *Mammal Review* **21**:123–142.

Speakman, J.R. 1995. Chiropteran nocturnality. *Symposium of the Zoological Society of London* **67**:187–201.

Speakman, J.R., D.J. Bullock, L.A. Eales, and P.A. Racey. 1992. A problem defining temporal pattern in animal behaviour: clustering in the emergence behaviour of bats from maternity roosts. *Animal Behaviour* **43**:491–500.

Spencer, W.D. 1987. Seasonal rest-site preferences of pine martens in the northern Sierra Nevada. *Journal of Wildlife Management* **51**:616–621.

Squires, J.R. 2000. Food habits of northern goshawks nesting in south central Wyoming. *Wilson Bulletin* **112**:536–539.

Stager, K.E. 1942. The cave bat as the food of the California lyre snake. *Journal of Mammalogy* **23**:92.

Stone, J.H. 1952. Porcupine damage to trees serious in Northwest. *Journal of Forestry* **50**:891.

Strickland, M.A., C.W. Douglas, M. Novak, and N.P. Hunziger. 1982*a*. Fisher. Pages 586–598 *in* J.A. Chapman and G.A. Feldhamer, editors. *Wild Mammals of North America: Biology, Management, and Economics*. The Johns Hopkins University Press, Baltimore, Maryland, USA.

Strickland, M.A., C.W. Douglas, M. Novak, and N.P. Hunziger. 1982*b*. Marten. Pages 599–612 *in* J.A. Chapman and G.A. Feldhamer, editors. *Wild Mammals of North America: Biology, Management, and Economics*. The John Hopkins University Press, Baltimore, Maryland, USA.

Styskel, E.W. 1983. Problems in snag management implementation – a case study. Pages 24–27 *in* Snag habitat management: proceedings of the symposium; June 7–9, 1983; Flagstaff, Arizona. General Technical Report **RM-99**. USDA, Forest Service, Rocky Mountain Forest and Range Experiment Station, Albuquerque, New Mexico, USA.

Sullivan, T.P. and D.S. Sullivan. 1982. Barking damage by snowshoe hares and red squirrels in lodgepole pine stands in central British Columbia. *Canadian Journal of Forest Research* **12**:443–448.

Sullivan, T.P. and A. Vyse. 1987. Impact of red squirrel feeding damage on spaced stands of lodgepole pine in the Cariboo Region of British Columbia. *Canadian Journal of Forest Research* **16**:1145–1149.

Sullivan, T.P., W.T. Jackson, J. Pojar, and A. Banner. 1986. Impact of feeding damage by the porcupine on western hemlock – Sitka spruce forests of north-coastal British Columbia. *Canadian Journal of Forest Research* **16**:642–647.

Swift, S.M. 1980. Activity patterns of pipistrelle bats (*Pipistrellus pipistrellus*) in north-east Scotland. *Journal of Zoology (London)* **190**:285–295.

Terborgh, J. 1988. The big things that run the world – a sequel to E.O. Wilson. *Conservation Biology* **2**:402–403.

Thomas, J.W., technical editor. 1979. Wildlife habitats in managed forests: the Blue Mountains of Oregon and Washington. Agriculture Handbook **No. 553**. USDA, Forest Service, Washington DC, USA.

Thomas, M.E. 1974. Bats as a food source for *Boa constrictor. Journal of Herpetology* **8**:188.

Tinnin, R.O., F.G. Hawksworth, and D.M. Knutson. 1982. Witches' broom formation in conifers infected by *Arceuthobium* spp.: an example of parasitic impact upon community dynamics. *American Midland Naturalist* **107**:351–359.

Toweill, D.E. and J.G. Teer. 1977. Food habits of ringtails in the Edwards Plateau region of Texas. *Journal of Mammalogy* **58**:661–663.

Trappe, J.M. and C. Maser. 1976. Germination of spores of *Glomus macrocarpus* (Endogonaceae) after passage through a rodent digestive tract. *Mycologia* **67**:433–436.

Traveset, A. and M.F. Willson. 1997. Effect of birds and bears on seed germination of fleshy-fruited plants in temperate rainforests of southeast Alaska. *Oikos* **80**:89–95.

Van Deusen, J.L. and C.A. Meyers. 1962. Porcupine damage in immature stands of ponderosa pine in the Black Hills. *Journal of Forestry* **6**:811–813.

Viro, P. and S. Sulkava. 1985. Food of the bank vole in northern Finnish spruce forests. *Acta Theriologica* **30**: 259–266.

Volz, K. 1986. Habitat requirements of northern flying squirrels in west-central Oregon. Thesis. Washington State University, Pullman, Washington, USA.

Weckworth, R.P. and P.L. Wright. 1968. Results of transplanting fisher in Montana. *Journal of Wildlife Management* **32**:977–979.

Whitaker, J.O., Jr. 1962. Endogone, hymenogaster, and melanogaster as small mammal foods. *American Midland Naturalist* **62**:152–156.

Whitaker, J.O., Jr. 1993. Bats, beetles and bugs. More big brown bats mean less agricultural pests. *Bats* **11**:23.

Whitaker, J.O., Jr. 1995. Food of the big brown bat *Eptesicus fuscus* from maternity colonies in Indiana and Illinois. *American Midland Naturalist* **134**:346–360.

Whitaker, J.O., Jr. and P. Clem. 1992. Food of the evening bat *Nycticeius humeralis* from Indiana. *American Midland Naturalist* **127**:211–217.

Whitaker, J.O., Jr., C. Maser, and L.E. Keller. 1977. Food habits of bats of western Oregon. *Northwest Science* **51**:46–55.

Whitaker, J.O., Jr., C. Maser, and S.P. Cross. 1981*a*. Foods of Oregon silver-haired bats, *Lasionycteris noctivagans. Northwest Science* **55**:75–77.

Whitaker, J.O., Jr., C. Maser, and S.P. Cross. 1981*b*. Food habits of eastern Oregon bats, based on stomach and scat analyses. *Northwest Science* **55**:281–292.

Whitaker, J.O., Jr., C. Neefus, and T.H. Kunz. 1996. Dietary variation in the Mexican free-tailed bat (*Tadarida brasiliensis mexicana*). *Journal of Mammalogy* **77**:716–724.

Wilks, B.J. and H.E. Laughlin. 1961. Roadrunner preys on a bat. *Journal of Mammalogy* **42**:98.

Williams, O. 1962. A technique for studying microtine food habits. *Journal of Mammalogy* **43**:365–368.

Willson, M.F. 1993. Mammals as seed-dispersal mutualists in North America. *Oikos* **67**:159–176.

Wilson, D.E. and S. Ruff. 1999. *The Smithsonian Book of North American Mammals*. Smithsonian Institution Press, Washington DC, USA.

Windsor, D.A. 1997. Stand up for parasites. *Trends in Ecology and Evolution* **12**:32.

Wiseman, J.S. 1963. Predation by the rat snake on the hoary bat. *Journal of Mammalogy* **44**:581.

Witt, J.W. 1992. Home range and density estimates for the northern flying squirrel, *Glaucomys sabrinus*, in western Oregon. *Journal of Mammalogy* **73**:921–929.

Wroe, D.M. and S. Wroe. 1982. Observation of bobcat predation on bats. *Journal of Mammalogy* **63**:682–683.

Yager, J. and D. Williams. 1988. Predation by gray snapper on cave bats in the Bahamas. *Bulletin of Marine Science* **43**:102–103.

Zabel C. and J.R. Waters. 1997. Food preferences of captive northern flying squirrels from the Lassen National Forest in northern California. *Northwest Science* **71**:103–107.

Zielinski, W.J. 1984. Plague in pine martens and the fleas associated with its occurrence. *Great Basin Naturalist* **44**:170–175.

Zielinski, W.J., T.E. Kucera, and R.H. Barrett. 1995. Current distribution of fishers, *Martes pennanti*, in California. *California Fish and Game* **81**:104–112.

Zielinski, W.J., N.P. Duncan, F. Emma, C.R.L. Truex, et al. 1999. Diet of fishers (*Martes pennanti*) at the southernmost extent of their range. *Journal of Mammalogy* **80**:961–971.

Zielinski, W.J., K.M. Slauson, C.R. Carroll, C.J. Kent, and D.G. Kudrna. 2001. Status of American martens in coastal forests of the Pacific states. *Journal of Mammalogy* **82**:478–490.

FRANCIS J. SINGER, GUIMING WANG AND
N. THOMPSON HOBBS

13
_____

The role of ungulates and large predators on plant communities and ecosystem processes in western national parks

Introduction

Human activities have caused fundamental changes in relationships between predators and large herbivores in ecosystems of western North America. In particular, large predators, especially wolves (*Canis lupus*) and grizzly bears (*Ursus arctos*), have been eliminated from most of the region. Human developments, as well as agricultural and livestock grazing activities have altered or eliminated many migration routes and habitats of ungulates. Developments such as towns and paved areas have eliminated habitat. Alternatively, human modifications to the landscape, in some cases, have created new and more fertile habitats for ungulates. Habituated ungulates use lawns, golf courses, and agriculture fields that are rich in nutrients in many of the Rocky Mountain states and provinces (Thompson and Henderson 1998).

No magic formula exists for setting goals for the appropriate number of ungulates in a national park or in a more managed ecosystem. National Park Service policy calls for managing for natural processes within the parks, where high ungulate populations may be managed if the concentrations are due to human effects (National Park Service 2001; NPS-77). Unfortunately, most current-day park ecosystems are extensively altered by many human factors, such as the extirpation of large predators, abbreviated or lost migrations of ungulates, and concentrations of ungulates on artificially rich human habitats, such as restored agriculture sites (in many eastern and mid-western parks) or lawns and golf courses within or adjacent to the parks. Determining what constituted natural conditions and natural processes under these altered states can be difficult.

[444]

We focus our analyses on the effects of the larger grazing ungulates, particularly elk (*Cervus elaphas*) and American bison (*Bison bison*). First, we review the effects of ungulates on plant size, morphology, density, biomass production, seed production, and recruitment rates in the parks. Second, we describe the potential for limitation of ungulates by multiple predators by modeling the predicted northern Yellowstone elk (*Cervus elaphus*) population without wolves through 1999, and we compare this model to the observed elk population size and recruitment rate, 1995–2000, when wolves were present. Last, we project the effects on plants and processes that ungulates might have if wolves were recovered into the ecosystems.

For our case studies, we selected two ungulate-plant grazing systems located in two national parks (Yellowstone and Rocky Mountain National Parks, YNP and RMNP) where high ungulate densities were at or near food-limited ecological carrying capacity (hereafter K) at the time of our studies. Human alterations of the landscape have been held to a minimum in the parks, thus providing baseline candidate natural systems where the effects of ungulates on plants and ecosystem processes could be compared to those in manipulated systems and harvested populations located outside of the parks. Our goal was to present the reader with the types of effects that might occur on plant morphology, community structure, and ecosystem processes by high densities of ungulates at or near food-limited K, versus effects of lower densities of ungulates that are limited by predators and/or sport hunting.

Grasslands are abundant in both of the two national parks (55% of the ungulate winter range in YNP; 26% in RMNP, Marr 1961, Peet 1991), and graminoids (86% in YNP, 72% in RMNP) predominate in the winter diet of elk in both parks (Houston 1982, Singer and Norland 1994). Bison winter diets in YNP are even higher in graminoids (97%, Singer and Norland 1994) compared to elk. Thus, our attention in this chapter logically focuses on ungulate influences in herbaceous communities, but includes information on the effects on sensitive riparian trees and shrubs due to concerns over their declines in the parks (Kay and Wagner 1994, Wagner et al. 1995, Baker et al. 1997).

Views of ungulate abundance

A number of ecological approaches exist to assist managers in evaluating the most appropriate numbers of ungulates either in national parks or in managed ecosystems. We describe four of the most common views and their associated key measures that will be useful in our review of the abundance of ungulates and their ecosystem effects in national parks.

Predator-limitation of ungulates

Analyses of ungulate population dynamics and growth might be central to a manager's assessment of ungulate abundance. In the absence of predators, negative feedback from effects of grazing on plants is presumed to cause ungulate populations to reach quasi-steady states. Ungulate abundance, in the absence of any predator limitation, is thought to be limited by density-dependent population processes mediated through a lower per capita food consumption and lowered body condition of individuals (Caughley 1976, Fryxell 1987, Dublin et al. 1990, Coughenour and Singer 1995). Annual variation in the strength of density dependence and hence in animal population size may also result from variation in weather. Ecological carrying capacity (hereafter food-limited K) can be determined from: (1) linear and non-linear density feedback relationships (Caughley 1976, Houston 1982, Boyce 1989); or (2) nutritional, landscape-level forage-based methods (Hobbs et al. 1982, Coughenour and Singer 1996). Populations at food-limited K are assumed to be regulated within some range by density-dependent processes caused by per capita restrictions in food availability (Caughley 1976, Dublin et al. 1990). The process of forage restriction on food-limited K may result in reductions in plant cover and production, plant species alterations, reduced body condition and survival rate of ungulates, and a new equilibrium between ungulates and forage conditions (Caughley 1976, Sinclair et al. 1985).

Alternatively, predators may limit ungulates below food-limited K under natural conditions. Predators may either limit ungulates slightly below (7% to 30%) food-limited K (Boutin 1992, Boyce 1993, Mack and Singer 1993) or well below (40% to 60%) food-limited K (Gasaway et al. 1992, Lime et al. 1993, Messier 1994), or the limitation may be highly variable.

The biodiversity-based approach

Ungulate abundance and grazing effects may also be assessed using plant diversity measures. Management of much of the world's grazing land is based primarily on changes in plant-species composition (Milchunas and Lauenroth 1992). In particular, dominant individual species have been used as indicators of range condition under the increaser-decreaser-invader concept. Some hold the view that ungulates should have no effect on plant species composition (Kay and Wagner 1994, Wagner et al. 1995:49, 52–53, Berry et al. 1997). For example, totally ungrazed sites inside exclosures were used as the primary benchmark for the multiple comparison examples presented by Wagner et al. (1995:52–59) as evidence of too much ungulate herbivory. National Park Service (2001) policy states, however, that the natural abundance and diversity of plants and animals should be maintained within the bounds of natural processes, not that there should be absolutely no influences on plants due to ungulates. But what effect should be acceptable on plant diversity under natural conditions in a particular park ecosystem by ungulates? This answer may be challenging, since ungulate herbivory has alternatively been shown to decrease (Rummel 1951, Chew 1982), result in no change (Gough and Grace 1998, Stohlgren et al. 1999), or to increase plant species diversity (Grime 1973, Mueggler 1984, Chadde and Kay 1988).

Grazing optimization and the sustainability views

The sustainability of the ecosystem to the ungulate grazing may be the central focus of the assessment of ungulate effects. The central goal of the sustainability view is to maintain plant production and soil fertility in grazing systems (McNaughton 1979, 1993, Frank and Groffman 1998), while shifts in the abundance of individual plant species may be less of a concern (McNaughton 1979, Milchunas and Lauenroth 1993).

Compensatory responses in grasses may enable moderate levels of herbivory to be sustained (Biondini et al. 1998, Mazancourt et al. 1998). These compensatory processes, at moderate levels of herbivory, may result in production on grazed sites (net primary productivity or NPP) to be maximized over ungrazed controls (Fig. 13.1). This is referred to as

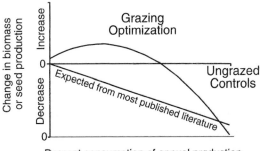

Fig. 13.1. The grazing optimization model predicts a peak in net primary or seed production at some moderate level of herbivory above the level for ungrazed controls (McNaughton 1979, Hilbert et al. 1981, Hamilton et al. 1982, McNaughton 1986, Paige and Whitham 1987, Dyer et al. 1993), but alternatively no optimization and linear decreases are predicted by other authors (Owen and Wiegert 1981, Painter and Belsky 1993, Belsky 1986, Verkaar 1986).

grazing optimization (McNaughton 1979, 1983, McNaughton et al. 1983; Frank and McNaughton 1993, McNaughton 1993, Mazancourt et al. 1998). In contrast, several authors have found no evidence for grazing optimization in grasslands across a wide range of herbivory rates by domestic and wild ungulates (Lacey and Van Poolen 1981, Belsky 1986, Verkaar 1992, Painter and Belsky 1993). In a review of 236 grazing studies worldwide, Milchunas and Lauenroth (1993) reported grazing-related increases in net aboveground plant production (i.e., grazing optimization) in only 17% of the cases. Detling (1988) concluded that most well-documented cases of grazing optimization came from tropical African grasslands (McNaughton 1983, Coughenour 1985, McNaughton 1993). He also found considerably less evidence for grazing optimization from North American grasslands (Detling et al. 1979, Detling and Painter 1983), although there is now more recent evidence of grazing optimization for ungulates from North America by ungulates (Hik and Jefferies 1990, Turner et al. 1993, Green and Detling 2000), especially in areas of high nutrient availability (Mazancourt et al. 1998).

The allowable-use approach

Perhaps the most commonly used criteria in the U.S. west for evaluating the appropriate number of ungulates are allowable-use measurements. First, aboveground plant biomass and the percentage of that biomass consumed by ungulates are measured. A judgment is then made as to what level of percentage use of annual biomass production, or annual consumption, is allowable in that system.

Effects of ungulates on plant communities and ecosystem processes

Effects on plant sizes and morphology

Herbivory by native ungulates typically resulted in shorter, more prostrate plants with smaller canopy sizes (Detling and Painter 1983, McNaughton 1984, Jaramillo and Detling 1988). Declines in root biomass in grassland ecosystems often occurred due to grazing (Pearson 1965, Belsky 1986, Detling 1988). But total herbaceous root biomass was consistently unaffected by ungulate herbivory in RMNP and YNP (Coughenour 1991; Zeigenfuss et al. unpubl. data) and in African parks (McNaughton et al. 1998). Because total root biomass is unchanged in the parks, and aboveground total mass (current and all accumulated prior years) is reduced, root:shoot ratios of individual plants are typically higher in most grazed areas (Detling et al. 1979, Coughenour 1985, Detling 1988). Although entire root systems of willows were not excavated and weighed in the parks, improved rate of water and nitrogen (N) uptake of grazed plants also suggested that root:shoot ratios were higher for grazed compared to ungrazed willow plants (Ruess 1984, Welker and Menke 1990, Alstad et al. 1999).

We observed grazed/browsed plants to be shorter than those plants that were not fed upon. Heights of grazed plants were less in 15 forage species in both parks (Singer et al. 1994, Singer and Renkin 1995). Crown diameters were also consistently smaller, with the single exception of Junegrass (*Koeleria macrantha*) in YNP, whose diameters were 11% larger on grazed sites (Fig. 13.2). Densities (plants per unit area) of smaller grazed plants were higher on grazed than ungrazed sites for seven of ten plant species in YNP. For example, on average there were 26% more grass clumps per plot (bluebunch wheatgrass, Junegrass, Idaho fescue), 21% more big sagebrush individuals per plot, and 100% or more willow clumps per plot on grazed sites in YNP. However, most heavily browsed willow clumps in YNP were apparently root suckers of the same individual plants and not different individuals. Disparity in heights of grazed versus ungrazed herbaceous plants was even greater in the Wind Cave NP where grazed individuals averaged half the height of controls (Detling 1988). Browsed willow plants were dramatically shorter (20% to 50% as tall), dramatically smaller in diameter, and willow aerial cover was reduced dramatically 50% to 80% in browsed patches in the two parks (Singer et al. 1994, 1998*b*, Zeigenfuss et al. 2002).

The presence of shorter and smaller plants on grazed sites can alter several aspects of plant community dynamics including: (1) less competition for space, light and other resources (Vesey-Fitzgerald 1973, McNaughton 1984, Detling 1988); (2) greater solar interception to the lower portions of plants; and (3) more bare ground between the smaller plants (in conjunction with reduced litter on most grazed sites) that creates open space for new seedling establishment. In the tropical grasslands of Africa, ungulate herbivory results in a dense, bushy geometry of shrubs and dense mats of grasses (the latter are called grazing lawns)(Vesey-Fitzgerald 1973, McNaughton 1983, 1993). Increases in bare ground are associated with the reduction of standing and accumulated litter that result from the grazing process, and a reduction in the average size of most grazed plants due to herbivory. Greater soil heating, and thus less soil moisture, are

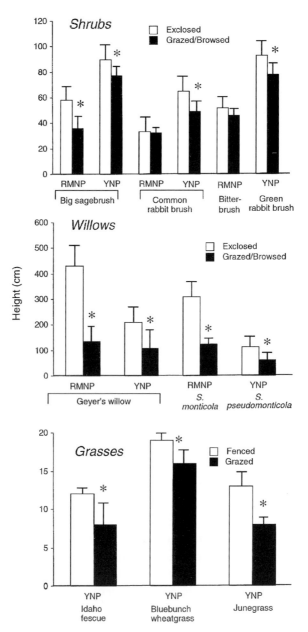

Fig. 13.2. Heights of riparian shrubs, upland shrubs, and grasses were substantially less on grazed compared to ungrazed sites in Yellowstone National Park (YNP), Wyoming, and Rocky Mountain National Park (RMNP), Colorado. * Denotes statistical significance, $p < 0.05$, in all figures.

also predicted on grazed sites due to greater solar interception and more open ground (Belsky 1986, McNaughton 1993), but the empirical findings on this topic are equivocal. Higher soil temperatures and lower soil moisture have been reported due to ungulate herbivory in some areas (Archer and Detling 1984). In YNP, no effects of native ungulates on either soil temperatures or moisture were reported in four studies (Frank 1990, Coughenour 1991, Singer and Harter 1996), but in other studies lower soil moisture on grazed sites was reported (Merrill et al. 1994, Frank and Groffman 1998). Ungulates are heavy animals (elk – 240 kg; bison – 450 kg; mule deer – 60 kg) and their hoof action commonly increased soil surface compaction and soil bulk density. Consequently, soil infiltration rates may be reduced on grazed sites (Fig. 13.3).

Bare ground was increased by wild ungulate grazing, but the extent of the increase varied. The increase may be rather minor, as is apparently the case in RMNP, where there was an increase of only 4.6% more total area of bare ground due to herbivory (a 148% increase from 3.1% bare ground

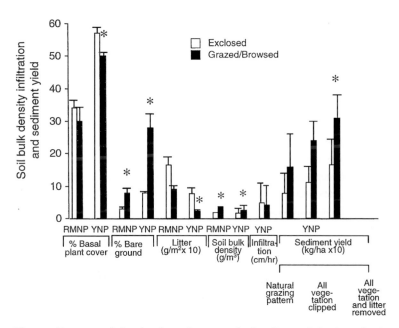

Fig. 13.3. Responses of plant basal area, bare ground, plant litter, soil density and sediment yield to ungulate herbivory, in RMNP Colorado, and YNP Wyoming. Data from Singer et al. (1994), Singer (1995), Singer and Renkin (1995), Lane and Montagne (1996), and Binkley et al. (2001). Data were not available for infiltration rates or sediment yield for RMNP. * Denotes statistical significance, $p < 0.05$.

on grazed sites to 7.7% bare ground on exclosed sites). However, larger increases in total area of bare ground due to ungulate grazing were observed in YNP (a 31% increase in bare ground: 4% bare ground on exclosed sites versus 35% on ungrazed sites) (Fig. 13.3) (Singer et al. 1998a). Soil surface alterations due to ungulate activity, if of sufficient magnitude, may result in accelerated rates of sediment yield from grazed slopes. In one study, there was preliminary evidence for increased sediments in YNP (average sediment yield across six sites was 56% greater on grazed sites) but, due to high variances, the differences were not statistically significant (Lane and Montagne 1996). Increased sediment yields were observed, but only for the treatments where 100% of all grasses and other herbaceous materials were artificially clipped to the ground surface (Lane and Montagne 1996). However, this was a highly artificial treatment, five times more severe than the average native ungulate herbivory removal of 28% for the area (Frank 1990).

Willow and mountain big sagebrush current annual growth was reduced 74% and 77% in core portions (the core 8%) of elk winter range in RMNP where average elk densities were 30–110 elk/km^2 over the six-month winter period (Lubow et al. 2002). Winter elk densities averaged less ($<$10/km^2) on the remaining 82% of the winter range, and shrub production was generally not reduced in those areas. Mountain big sagebrush is an uncommon shrub in RMNP. Three other more common upland shrubs (bitterbrush, rabbitbrush, *Ribes* spp.) were not negatively influenced by ungulates. Willows are more common in RMNP than in YNP, they dominate only 4% of the landscape of the entire winter range in RMNP and 30% of the core elk winter range (Singer et al. 1998b), and willows were negatively influenced by ungulates in both parks (Fig. 13.4).

Ungulates had negative effects on plant production of only a few plant species, in general (Fig. 13.4). In RMNP, ungulate herbivory resulted in no decline in vegetative production in 84% of the plants sampled (36 of 43). However, there was a decline in aboveground production of the remaining 16% of the species (7 of 43), including declines in production in one upland shrub (*Artemisia tridentata tridentata*), in all three willows sampled (*Salix monticola, S. geyeriana, S. planifolia*) and in three forbs (*Mertensia ciliata, Artemisia ludoviciana, Eriogonum umbellatum*). One other forb (*Solidago* spp.) increased on grazed sites. In YNP, vegetative production of 12% of the herbaceous species (16 of 128 plant species) was influenced by ungulates, but the effects included both increases and declines. Aboveground annual vegetative production of ten species was higher on grazed sites, production for six other species was less on grazed sites, while there was

Aboveground Production

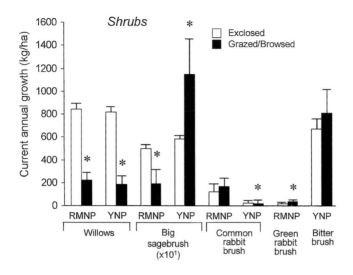

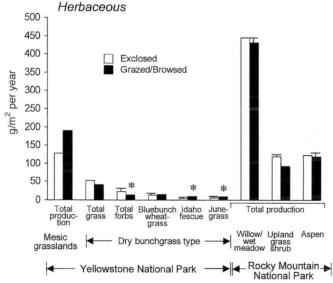

Fig. 13.4. Responses of aboveground production in plants to ungulate herbivory in RMNP and YNP. * Denotes statistical difference.

no difference for the remaining 112 (84%) plant species (Singer et al. 1998*a*). Vegetative production of one grass, (Sandberg's bluegrass (*Poa sandbergii*)), and three forbs (goatsbeard (*Tragopogon dubius*), Holboell's rockcress (*Arabis holboelii*), and wild onion (*Allium cernum*)) was consistently less abundant on grazed sites. However, three grasses (Idaho fescue, bluebunch wheatgrass, thick-spiked wheatgrass (*Agropyron dasystachum*)) were consistently more abundant on grazed sites. Wyoming big sagebrush (*Artemisia tridentata wyomingensis*) was also negatively affected by ungulates. The subspecies of big sagebrush was consumed at a high rate by several ungulates on the lower 11% of the winter range (68% of current annual growth was removed), seed production per plant was reduced 80%, current annual growth was reduced 63%, and density of plants was reduced 63% due to ungulate browsing (Singer and Renkin 1995), although the mountain big sagebrush (*A. t. vaseyana*) found on elk-only winter range (89% of the winter range) was consumed only at the 8% level and was healthy, vigorous, and even increasing.

Compensatory responses of plants to ungulate herbivory

Six of ten plant species in YNP strongly compensated for intermediate levels of ungulate herbivory by greater aboveground biomass production, higher seed production, and, ultimately, greater recruitment. These variables followed the classic curvelinear function (Fig. 13.5) predicted by the grazing optimization hypothesis (Fig. 13.1). Production levels fell below values for ungrazed plants at higher levels of plant consumption by ungulates in four plant species (Figs. 13.5, 13.6). Willows in RMNP also followed the grazing optimization pattern (Fig. 13.7), but we observed no compensatory responses by willows in YNP (Singer et al. 1994, 1997, 1998*b*). Current annual growth (CAG) of willows in RMNP peaked at intermediate browse levels at about 21% consumption of CAG, but CAG of willows declined for those patches of willows that were consumed at 37% or greater (Fig. 13.7). In our YNP study areas, intermediate herbivory levels, where optimization occurred, were about 20% for shrubs and 45% for the graminoids. Excessive consumption levels for shrubs ranged from 32% to 75%.

The literature suggests that willows, a seral shrub species, would be able to sustain higher consumption rates by ungulates than we observed (see Krefting et al. 1966, Willard and Pickel 1978, Wolfe et al. 1983, Bergström and Danell 1987). However, these studies document effects due to artificial clipping. We concluded that ungulate browsing was more

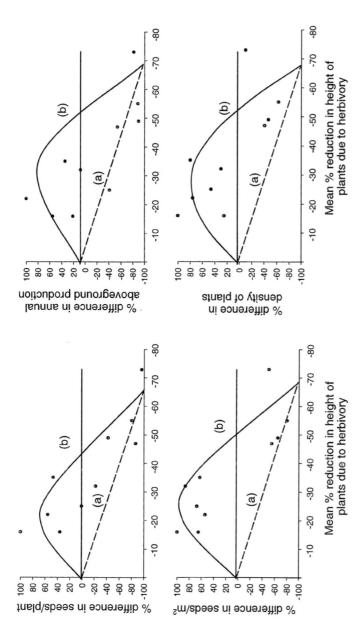

Fig. 13-5. Plant compensatory responses in annual biomass production, seed production, and recruitment in Yellowstone National Park, Wyoming, at moderate levels of herbivory and height reductions (i.e., grazing optimization was observed). At higher levels of herbivory, seed, biomass production and densities of plants declined, over ungrazed controls (b). The flat horizontal line (b) reflects values for ungrazed controls, the humped curvilinear line reflects grazing optimization for YNP grazed plants, while a linear decline in plant production was predicted by most published literature (Belsky 1986; Belsky et al. 1993).

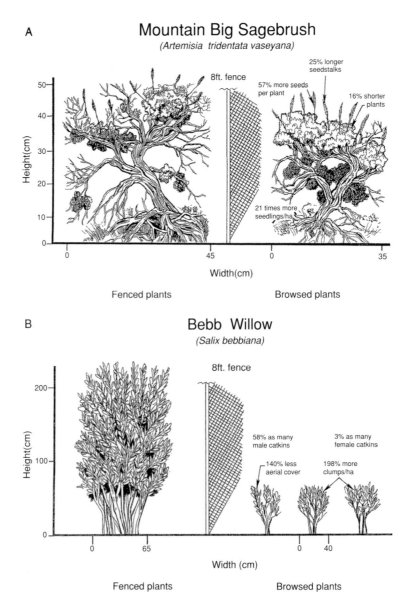

Fig. 13.6. Comparison of annual aboveground biomass, seed production, and recruitment rates in grazed and ungrazed plants in YNP, Wyoming. These drawings reflect average values for typical plants located inside and outside of grazing exclosures, where plants have been fenced for 41 or 45 years.

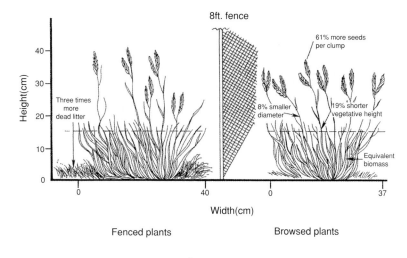

C

Bluebunch Wheatgrass
(Pseudorognaria spicata)

8ft. fence

61% more seeds per clump

Height(cm)

40

30

20

10

0

Three times more dead litter

8% smaller diameter

19% shorter vegetative height

Equivalent biomass

0   40   0   37

Width(cm)

Fenced plants   Browsed plants

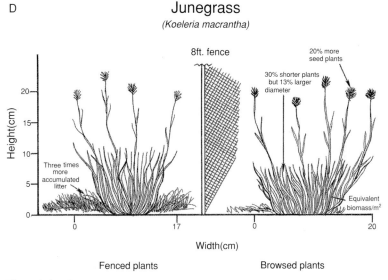

D

Junegrass
(Koeleria macrantha)

8ft. fence

20% more seed plants

30% shorter plants but 13% larger diameter

Height(cm)

20

15

10

5

0

Three times more accumulated litter

Equivalent biomass/m²

0   17   0   20

Width(cm)

Fenced plants   Browsed plants

Fig. 13.6. (*cont.*)

detrimental to plants than was clipping due to the rough breakage and shattering of browsed stems by ungulates.

Evidence for compensatory growth and relatively unique verification of grazing optimization that we documented may be a direct consequence

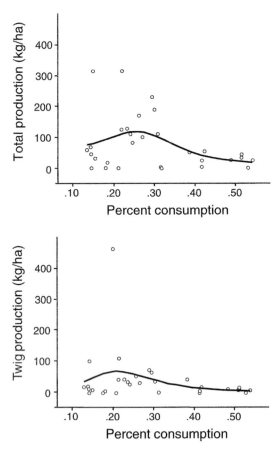

Fig. 13.7. A hump-shaped curve and peak values of production of willows above levels for ungrazed willows at moderate levels of herbivory strongly suggested grazing optimization was occurring in the willows of RMNP, Colorado.

of the smaller plant sizes, reduced competition, and greater solar interception by the remaining live plant tissues on grazed sites. Compensation in plant growth of grazed plants may also be due to increased rates of N uptake by roots, greater allocation of N to shoots versus roots, and increased rates of photosynthesis and tillering (Caldwell et al. 1981, Ruess et al. 1983, Jaramillo and Detling 1988).

Compensatory production responses by grazed plants also included consistently heavier and longer shoots in seven of eight shrubs and in two of three grasses in the two parks (Fig. 13.8). More willow shoots and leaves were produced per unit of total plant biomass and more current year's biomass was produced per unit of the previous year's biomass in RMNP

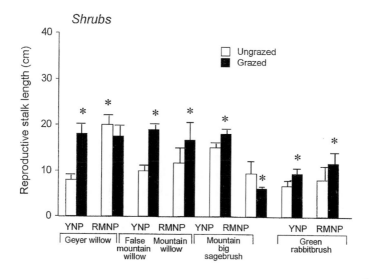

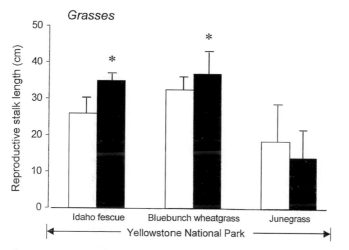

Fig. 13.8. Compensatory responses of shrubs and grasses included longer shoots on grazed shrubs and longer reproductive stalks on grazed grasses.

(Peinetti et al. 2001). A higher proportion of total willow plant N was allocated to the new leaves and shoots in grazed willow plants (Peinetti et al. 2001).

Compensatory responses were sufficient to maintain annual above-ground production of plants in nearly all grazed sites at levels comparable to ungrazed sites in YNP, with the exception of willows (Coughenour 1991, Frank and McNaughton 1992, Singer et al. 1994, Singer 1995, Singer et al. 1998*b*). Grazing even stimulated aboveground production of some grasses in YNP compared to ungrazed controls (Frank and McNaughton 1993). Grazed plants with higher N concentrations grew longer into the growing season in both parks, enabling grazed grasses in YNP and some willows in RMNP to "catch up" in aboveground production with their ungrazed counterparts by about July of each growing season (Merrill et al. 1994, Peinetti et al. 2001).

A number of compensatory plant responses were directly attributed to the elevated N concentrations found in grazed plants in YNP. Maschinski and Whitham (1989) also concluded that compensation to herbivory by plants was more likely to occur if nutrient availability was high. Six grazed plants in YNP that possessed higher concentrations of N also produced longer reproductive stalks (Fig. 13.8), heavier individual seeds more seed per stalk, more seeds per plant and more seeds per unit area of plant stalks (Figs. 13.5, 13.6). These results seem paradoxic because it is widely held that any level of herbivory reduces, rather than increases, seed production in plants (Jameson 1963, Belksy 1986). Upon closer inspection, however, these studies were either: (1) controlled clipping experiments conducted out of the context of the potentially compensatory influences of herbivore actions on the ecosystem, such as ungulate deposition of feces and urine and reduced competition due to ungulate effects on plant morphology and structure, or (2) conducted during the growing season when plants are more sensitive to herbivory. Most of the herbivory in YNP and RMNP occurred during the winter period of plant dormancy.

Increases in seed production following herbivory have been rarely reported (Hamilton et al. 1982, McNaughton 1986, Elmquist et al. 1987, Paige and Whitham 1987) (but see Paige 1999 for an exception), and reports of elevated seed production often include only plants growing under the best of water and growth conditions (see reviews by Belsky 1986, Belsky et al. 1993). The unusual compensatory response in seed production in plants observed in YNP followed the classic grazing optimization hypothesis (i.e., seed production was optimized for moderately, but not for intensely, grazed plants) (Figs. 13.5, 13.6). Individual seeds produced by grazed compared to ungrazed grasses were both heavier and more viable.

All of the plants with higher seed production on grazed sites were common species found in the core of their local ranges and thus presumably were growing on favorable growing sites. We observed less seed production in four shrub species that were grazed at apparently excessively high

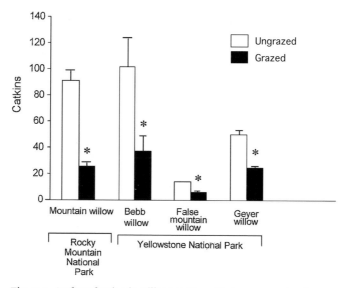

Fig. 13.9. Seed production in willows compared between grazed and ungrazed sites in both Rocky Mountain National Park, Colorado, and Yellowstone National Park, Wyoming. * Denotes statistical differences, $p < 0.05$.

levels and, in the case of the willows (Fig. 13.9), were apparently growing on less than optimal sites in YNP (Singer et al. 1994). Also, seed production was dramatically reduced for grazed Wyoming big sagebrush growing on the lowest elevation boundary line area of YNP (i.e., on ~9% of the YNP winter range) where that species also grows on the periphery of its local distribution. Maschinski and Whitham (1989) found that the degree to which a plant species compensates to herbivory will vary according to local conditions.

Effects of ungulates on ecosystem processes

Ungulate grazers were major processors and active regulating agents of N in the national park study areas (McNaughton 1979, Detling 1988, Day and Detling 1990, Holland and Detling 1990, Frank and McNaughton 1992). N is an essential nutrient that may limit ecosystem productivity in many areas. N availability is closely tied to the amount of organic matter, and soil organic matter may determine the ability of the soil to hold moisture. Any depletion of N, and also depletion of organic matter, can reduce the long-term productivity of the system. Depletion of N can also alter plant diversity, because many species have specific N needs, which, if not met, leads to impaired competitive abilities (Ritchie et al. 1998). Higher N

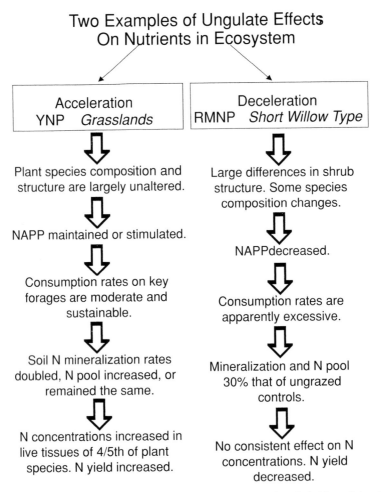

Fig. 13.10. The nutrient accelerating and decelerating scenarios of Ritchie et al. (1998) for herbivores in the Rocky Mountain National Park and Yellowstone National Parks' ecosystems.

concentrations in grazed plants can also contribute to higher decomposition rates and higher N mineralization rates (Irons et al. 1991, McInnes et al. 1992, Pastor et al. 1993, Ritchie et al. 1998). Higher N mineralization rates due to ungulate herbivory produces N in a more useable or labile form for more rapid uptake by plants (Ruess et al. 1983).

Ritchie et al. (1998) proposed that ungulates may either: (1) accelerate or (2) decelerate nutrient cycling (Fig. 13.10). Nutrient cycling may be accelerated if ungulates prefer to forage on groups of plants that have high tissue nutrient concentrations and, most importantly, tolerate the herbivory. The grazed plants may compensate for herbivory with faster nutrient uptake, faster relative growth rates, and higher tissue concentrations of N. Higher litter quality and feces/urine deposition by ungulates on grazed sites then result in higher decomposition rates and subsequently in higher mineralization rates that make N more available to plants. In turn, plants develop higher tissue concentrations of N.

Examples of ecosystems where acceleration of nutrient cycling following herbivore grazing (accelerating scenario) has been documented include the African Serengeti tropical grasslands (McNaughton 1976, 1979, Ruess et al. 1983), grasses growing in coastal salt marshes (Hik and Jefferies 1990), the mixed-grass prairie of Wind Cave NP, South Dakota (Holland et al. 1992, Green and Detling 2000), the bunchgrass communities and wet meadow communities of YNP, Wyoming (Frank et al. 1994, Frank and Groffman 1998) and the tall grass prairie of the U.S. (Turner et al. 1993). All of these examples include predominantly grassland communities.

But in the decelerating nutrient cycling scenario of Ritchie et al. (1998), ungulate herbivores selectively feed on more grazing-sensitive plants. As a result, herbivory increases the less palatable plants whose tissues are typically nutrient-poor, are more defended by secondary compounds, and are less palatable to the ungulates. This dominance shift to less palatable plants produces a lower quality and less decomposable litter, which in turn results in lower decomposition rates, lower soil mineralization rates, and a decline in total soil N pools (Fig. 13.10).

Examples of ecosystems where nutrient deceleration due to herbivores has been documented include the birch-snowshoe hare browsing system in Alaska (Bryant 1981), Great Lakes forest shrub and tree sapling understories that are browsed by moose (Pastor et al. 1993), and Minnesota oak savanna that is browsed by white-tailed deer (Ritchie et al. 1998). All reported examples of the declining nutrients include shrubs, tree saplings, or forbs growing in forest understories, seral forests, or in savannas. None of the declining nutrient examples included either grassland ecosystems or grasses growing in other systems, except that Biodini et al. (1998) reported declining nutrients at very high offtake (90% consumption of annual biomass) levels in the mixed-grass prairie of South Dakota.

We concluded that the grasslands of YNP fit the accelerating nutrient cycling scenario for ungulates. Soil nitrogen mineralization rates were about twice as high on grazed versus ungrazed sites in YNP (Frank and Groffman 1998) and soil organic matter was 37% higher on grazed sites (Lane and Montagne 1996). As an apparent consequence, about 80% of the grazed plants in YNP possessed an average of about 26% higher N concentrations in the current year's plant tissues (Frank 1990, Coughenour 1991, Singer 1995, Singer and Harter 1996), and aboveground N yields were higher except in drought years (Singer et al. 1998a). Ungulates played a large role in the annual grassland nutrient cycle, and ungulate urine and feces constituted one-half the total N mineralized each year (Frank et al. 1994).

We are uncertain as to why grazing by native ungulates results in a large increase in soil mineralization rates in YNP but not in RMNP. The increase in readily available N resulted in the substantial ability of plant species in YNP to compensate for the tissue losses resulting from herbivory. Possibly the ungulates lose more weight over the course of the winter in YNP versus RMNP since the winters are more severe (colder temperatures, deeper snow) than in RMNP, and, thus, ungulates would transpose more net N to Yellowstone's winter range. We concluded that net movement of N from the summer range in RMNP to the winter range resulted in the input of 0.16 gN/m^2 per year, based on body mass gains while elk were on summer range and mass losses while elk were on winter range. Frank et al. (1994) also concluded that there was a net movement of N from summer ranges (where ungulates gained weight) to winter ranges (where the ungulates lost weight). But to test this hypothesis, weight losses for elk and the same calculations would have to be made for YNP.

Additionally, elk densities were higher and elk herbivory was more substantial in the RMNP system. Elk densities were three times higher in RMNP than in YNP (30–110 elk/km^2 concentrated on 7% of the winter range in RMNP versus a more uniform and lower elk density of 16–21 elk/km^2 on Yellowstone's northern range). Elk grazed the winter range longer in RMNP than YNP (210 versus 180 days) and consumed more of the available annual vegetation biomass. Average annual consumption of herbaceous vegetation was double in RMNP: 56% in RMNP versus 28% in YNP. Elk consumed nearly double the N yield per unit area (2.4 versus 1.10 gN/m^2 per year in RMNP compared to YNP) and this may explain the negative influence on N mineralization in RMNP. This speculation is supported by the findings of Biondini et al. (1998), who also reported that mineralization rates declined as grazing intensity increased.

Effects of ungulates on plant community abundance and composition

The intermediate disturbance hypothesis predicts that maximum species diversity will be observed at some intermediate levels of disturbance (Grime 1973, Horn 1975). Grime (1973) obtained empirical evidence for the relationship for plant disturbances in British pastures. Horn (1975) then derived the relationship for Markov-chain models of forest succession, but the hypothesis has rarely been tested for herbivore disturbances. Milchunas et al. (1988) predicted a bell-shaped curve for the relationship between plant species diversity and grazing intensity for ecosystems with a long history of large ungulate herbivory such as North American grasslands. They predicted a left humped (right skewed) curve for grasslands with a brief history of grazing.

The intermediate disturbance hypothesis is based on the premise that changes in plant community structure encourage compensatory responses in the plants and the system. Competition severely limits plant species diversity in undisturbed communities (Huston 1979), and large ungulate herbivory both reduces plant competition and increases the amount of light reaching the soil surface. Additionally, herbivory can enhance plant species diversity by differentially altering growth rates of plant species (McNaughton 1979, Belsky 1986), differentially altering seed production and the reproductive fitness of plants (Jameson 1963, Dyer 1975), and by reducing leaf areas, opening up the canopy, creating small disturbances and, thus, further altering relative species abundances (Edroma 1981).

Ungulates often alter plant species diversity even in pristine, naturally functioning ecosystems (McNaughton 1979, Pastor et al. 1993, Stohlgren et al. 1999), but there are no universal rules to predict the direction of the differences due to ungulates. The effect of herbivores on plant diversity apparently is related to impacts on the dominant plant species, or plant regeneration opportunities and propagule transport (Ritchie and Olff 1999). In more productive grasslands, herbivores are more likely to increase plant diversity (Belsky 1992, Fahnestock and Knapp 1994, McNaughton 1994), but in areas with poor or saline soils herbivores are likely to decrease plant diversity (Milchunas et al. 1988, Hobbs and Huenneke 1992, Ritchie and Olff 1999).

Ungulate herbivory caused a decline in community-level diversity in the two parks by contributing to a decline in both the abundance and the size and structure of aspen and willow patches. Willow and aspen patches have declined >50% since the 1930s in YNP, and willows have declined 19% to 22% since 1946 in RMNP (Houston 1982, Singer et al. 1998*b*; R. Peinetti et al. unpubl. data, Natural Resource Ecology Laboratory, Fort Collins, Colorado, USA), although the declines are likely explained by a multiple-factor response to succession, fire suppression, and climate change (Romme et al. 1995, Singer et al. 1998*b*). Tall willow patches are being converted to short, hedged willow patches in both parks (Houston 1982; Singer et al. 1994). This loss of tall willow patches may eliminate necessary habitats for songbirds, some insects, and small mammals. For example, higher songbird diversity was found in tall willow patches where the number of browsing moose (*Alces alces*) and browsing levels were reduced by sport hunting in the Jackson Valley than in willow patches in the adjacent Grand Teton NP where moose hunting was not allowed and the large predator fauna was depleted (Berger et al. 2001).

Plant species diversity (evenness or richness) was not consistently influenced by ungulate herbivory in RMNP (Singer et al. 2001). In YNP, ungulates resulted in 16% more total herbaceous species on grazed (80 species) versus ungrazed (69 species) plots. Observations of higher herbaceous plant diversity on moderately grazed sites are consistent with the intermediate disturbance hypothesis (Grime 1973). However, there were no significant effects on shrub diversity.

Decline in tree and shrub communities
Willow shrub sizes were reduced by ungulate herbivory by roughly one-half in RMNP and roughly four to five times in YNP. Such large declines in willow heights and crown sizes can dramatically alter the amount of solar interception, temperatures, nutrients, and microclimates of the willow communities. Essentially, many of the willow patches in YNP and some patches in RMNP were being converted to wet meadows. Herbaceous plant diversity might increase (Chadde and Kay 1988) during this process, but avian and other faunal diversity declined (Berger et al. 2001).

Ungulates may be contributing to the decline in community dominance of some woody species such as willows, aspen, and perhaps cottonwood in the two parks (Olmstead 1987, Singer et al. 1994, Wagner et al. 1995, Baker et al. 1997, Meagher and Houston 1998, Singer et al. 1998*b*), although willows and aspen did not decline on all areas of the winter ranges (Singer et al. 1994, Suzuki et al. 1999). Rather than a one-factor explanation, the woody declines are more likely a multifactor response, due not only to the elk increases but also to warmer, drier climates, fewer fires due to suppression of firestarts, and lower local water tables, and reduced recruitment sites due to large-scale declines in beaver numbers (Romme et al. 1995, Singer et al. 1998b). For example, declines in active beaver colonies have been in excess of 90% in both parks over their peak observed densities earlier last century (Jonas 1955, Consolo-Murphy and Hanson 1993, Zeigenfuss et al. 2002). As a direct consequence of fewer beaver dams, stream channels are straighter, less braided and there is less stream length (56% to 69% less stream length) in 1996 compared to 1946 in RMNP (Peinetti et al. unpubl. data). As a result of greater energy, some streams have visibly downcut 1–1.5 m below the levels that had been maintained by the now abandoned beaver dams in YNP (Singer et al. 1994, Singer and Cates 1995), although no such downcutting was observed in RMNP. Willows are now co-dominant with such dryland shrubs as big sagebrush on downcut stream sites in YNP (Singer et al. 1994, Singer and Cates 1995). Since these multiple factors co-varied in both parks, it is difficult to determine which factor contributed more to the willow declines (Romme et al. 1995, Singer and Cates 1995, Singer et al. 1998b). Experiments are needed to identify the most significant factors.

Ungulate herbivory is affecting some woody plant communities with shorter shrubs and a lower density of shrubs. Willows and aspen are substantially shorter, and in some cases reduced to the heights of the herbaceous plants, where the shrubs may persist for decades as annual suckers from roots (Renkin and Despain 1994, Singer et al. 1994). Browse-suppressed stands support reduced avifauna and other altered biota (Berger et al. 2001), although herbaceous species richness may be increased due to greater light interception and less competition from willows (Chadde and Kay 1988). N processes, N pools, and depth of the rooting zone may be reduced in short willow stands, and N inputs may be reduced in non-recruiting aspen stands (Alstad et al. 1999), thus further increasing the stands' vulnerability to additional ungulate herbivory. This downward spiraling decline of reduced woody riparian shrubs → fewer beavers, → lowered water tables → more arid climate all contribute to further woody browse declines. Elk declines alone may not be sufficient for woody riparian ecosystem recovery, but a combination of fewer elk, cooler, wetter conditions, and the return of beavers may be necessary (Romme et al. 1995, Singer et al. 1998b).

Top-down control of ungulate populations: the effect of wolf restoration

Ungulates are presumed to reach an upper threshold in numbers determined by food-limitation and density-dependent population processes brought on by a lower per capita consumption of forages (Caughley 1976, Fryxell 1987, Dublin et al. 1990, Coughenour and Singer 1995). Food-limited ungulate populations usually have lower body condition of individuals, lower pregnancy rates, lower recruitment, and, sometimes, negative effects on their own food supplies (Caughley 1976, Sinclair 1977, Dublin et al. 1990, Coughenour and Singer 1996). However, evidence has accumulated that large predators may limit ungulates below food-limited K (Messier 1991, 1994, Van Ballenberghe and Ballard 1994, Orians et al. 1997). Limitation is the process that lowers an equilibrium point of a population through a change in population production or a loss of individuals (Sinclair 1989, Ballard 1992, Boutin 1992). Limiting factors are defined as any density-dependent or density-independent factor that causes a lowering in reproduction or survival (Boutin 1992, Sinclair and Pech 1996). By contrast, regulation is the process by which a population returns to an equilibrium density, after perturbation, and thus regulating factors are always density-dependent (Sinclair 1989, Messier 1991). Predation may also be regulatory in some cases (Van Ballenberghe and Ballard 1994).

Gray wolves have dispersed to, or have been re-introduced into, several sites in the western U.S. in the past 5–15 years (Bangs and Fritts 1996). Wolves dispersed to northwestern Montana in the early 1980's and into north-central Washington in the 1990's from Canada. Wolves were re-introduced into YNP in 1995. The YNP wolves currently number in excess of 150 animals, and about 60 wolves kill an estimated 1000+ elk annually from the northern Yellowstone elk population (Douglas Smith et al. Unpublished annual reports, 1997 through 2000, Yellowstone wolf project, Yellowstone Center for Resources, Yellowstone National Park, Wyoming). Wolves were re-introduced into Idaho in 1995 and now number more than 100 animals, and into Arizona in 1998. At the same time, black bears and coyotes (*Canis latrans*) were common in many of these areas, while grizzly bears have recovered in parts of Montana and in northwestern Wyoming. A number of western ecosystems now support populations of these multiple large predators. Expansion of wolves and continued human-assisted restoration into other large western U.S. national parks and additional western U.S. wilderness areas are highly likely, although the ultimate extent of the distribution of gray wolves in the U.S. west is conjectural.

Limitation of ungulates is more likely when there are multiple predators. Wolves in concert with a bear species limited moose populations to approximately one-fifth of food-limited K (to 0.4 moose/km²) across large areas of the arctic and boreal forest zones (Gasaway et al. 1992, Messier and Crête 1985, Messier 1994). Small woodland caribou populations were also apparently sharply limited in several instances where there were multiple predators and also where there was one or more abundant additional species of ungulate(s) (e.g., moose or deer) present to support a high predator density (Messier 1991, Gasaway et al. 1992, Messier 1994, Orians et al. 1997). Messier (1994) and McLaren and Peterson (1994) concluded that predator limitation of ungulates may have top-down effects on lower trophic levels, including effects on plant communities and other vertebrates utilizing the vegetation.

However, findings of limitation are mostly derived from moose and caribou systems in deep-snow environments located in northern Canada and Alaska. It remains to be seen if they can be reliably applied to milder, more productive, and diverse mammal communities of the Rocky Mountains. For example, many U.S. western grassland-conifer mosaic communities support ungulate guilds consisting of five to six species of ungulates (elk, bison, mule deer, pronghorns, bighorns, mountain goats) and many of the Rocky Mountain winter ranges are nearly snow-free, especially on wind-swept hills.

Top-down controls of moose, for example, may be less likely in more productive environments (Crête and Manseau 1996) possibly due to greater interference competition among a more diverse fauna of predators (Abrams 1994). Limitation was more severe (i.e., 0.1 moose/km²) in boreal ecosystems located further to north than in more productive mixed-deciduous forests located to the south (Crête and Courtois 1997). Forage production may also have been a factor in the lesser limitation. Forage production/ha was 8–14 times higher in the more southern mixed forests versus the more northern boreal moose ranges (Crête and Courtois 1997).

Limitation of the northern Yellowstone elk following wolf restoration

We modeled elk population dynamics using models developed for the northern Yellowstone elk population (Mack and Singer 1993, Coughenour and Singer 1995, Singer et al. 1997, Singer and Mack 1999). Prior elk population model efforts demonstrate the effects of density dependence (Taper and Gogan 2002), harvests, and weather effects on the elk population that were in close agreement with observed elk population dynamics (Singer and Mack 1999).

We modeled elk population sizes with a deterministic logistic equation:

$$N_{[t+1]} = N_t + N_t \cdot R_t \cdot (1 - N_t/K), \tag{1}$$

Where $N_{[t+1]}$ is the estimate of elk population size of year $t + 1$, N_t the estimate of population size of year t, K the equilibrium population size or carrying capacity, and R_t annual population growth rate. R_t was modeled by equation:

$$R_t = a_0 + a_1 \cdot WSI_t, \tag{2}$$

where a_0 is the intercept, WSI_t the winter severity index of year t (Farnes 1996), and a_1 the coefficient for WSI_t. Calf:cow ratios were modeled by a simplified equation:

$$Y_{[t+1]} = b_0 + [b_1 \cdot N_t/(1 + b_1 \cdot N_t)] + [b_2 \cdot WSI_t/(1 + b_2 \cdot WSI_t)] \tag{3}$$

where $Y_{[t+1]}$ is the calf:cow ratio of year $t + 1$, N_t the estimate of population size of year t, WSI_t the winter severity index of year t, b_0 the intercept, b_1 the coefficient of N_t, and b_2 the coefficient for WSI_t. We corrected the calf ratio for yearling cows (assumed to be 0, Houston 1982) by assuming that the ratio of yearling:adult cows was similar to classifications of yearling males (observed spike bulls:adult cows).

We fit models $(1) + (2)$ to the observed data on the elk population sizes from 1969 to 1995 to estimate a_0, a_1, and K with the non-linear least-squares method. Coefficients b_0, b_1, and b_2 were estimated by fitting the model (3) to data of 1986–1992. We predicted population sizes for 1996–1999, and calf:cow ratios for 1969–1981 and 1996–1999. In a t-test, we did not find a significant difference in the calf:cow ratios from 1969 to 1981 between the observed and predicted values ($p > 0.05$). However, the observed calf:cow ratios from 1996 to 1999 were significantly lower ($p < 0.05$) than the values predicted by model (3). We interpreted the differences in population size and calf:cow ratio between the observed and predicted as the effect of wolf predation (Fig. 13.11).

In order to further assess the potential limitation of the northern Yellowstone elk population by wolves, we developed a stochastic model of the elk population:

$$N_{t+1} = N_t \cdot \exp(r - b \cdot N_t + V_e \cdot s), \tag{4}$$

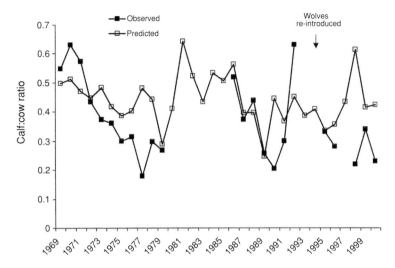

A Non-linear regression of the calf:cow ratio on winter severity index and population size

B Non-linear regression of elk population size on winter severity index at YNP using discrete logistic model

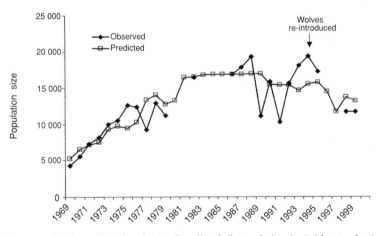

Fig. 13.11. (A) Observed (with wolves) and predicted elk population size (without wolves), (B) and elk recruitment rates for the Northern Yellowstone elk population. The predicted values (without wolves) were based upon relations to the observed weather (especially the winter severity index) patterns, and well-established density dependence relations for the elk population minus known hunter removals were then compared to the observed ungulate patterns that occurred with wolves present. The effects of wolf re-introduction to the ecosystem in 1995 and the potential limitation of elk by the multiple predators present in the system (wolves, bears, coyotes, humans) at both low and high wolf predation rates are demonstrated in (C) ($p < 0.05$).

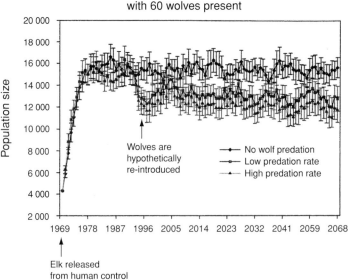

Fig. 13.11. (*cont.*)

where N_{t+1} and N_t are the population sizes at the time of $t + 1$ and t, respectively; r is the maximum population growth rate, b is the density regulation coefficient, V_e is the environmental stochasticity, and s is the random variable of the standard normal distribution. Parameters r, b, and V_e were estimated. We then simulated the predation of wolves on elk in YNP by adding a predation term into the model (4):

$$N_{t+1} = N_t \cdot \exp(r - b \cdot N_t + V_e \cdot s) - g \cdot N_w, \qquad (5)$$

where g is the annual predation rate of wolves on elk (the number of elk eaten by a wolf per year) and N_w the population size of wolves. Low (15.98 elk/wolf per year) and high (19.97 elk/wolf per year) predation rates were used in the simulations, respectively. We simulated the elk population dynamics for 100 years with the initial population size of 4309 that was the estimate of the elk population size in 1969. Wolves were introduced into the system at year 26 to simulate the wolf predation from 1995. We assumed that population size and predation of wolves rate were constant. We ran the simulation 50 times for each predation scenario, and computed the mean and 95% confidence intervals (vertical lines in Fig. (13.11C) for each year.

Our model effort suggests potential limitation of the northern Yellowstone elk population since the restoration of wolves in 1995 (Fig. 13.11A,B). Our simulations projected that the predation of 60 wolves on the northern range at the low and high predation rates depressed the population sizes of northern range elk ($p < 0.05$, Fig. 13.11). Both observed elk population growth trajectory and calf recruitment since the restoration of wolves in 1995 were significantly lower than the "expected" elk population size and recruitment values that we predicted from elk, weather, and density-dependent relationships for the same period. The dynamics of the northern Yellowstone elk population have been extensively studied, so we are confident in these predicted or "expected" relationships. The density-dependent responses of the Yellowstone elk population and weather relations we modeled were well established from prior model efforts for the population (Houston 1982, Boyce 1993, Coughenour and Singer 1995, Singer and Mack 1999, Taper and Gogan 2002).

We do not attribute this apparent limitation of the northern Yellowstone elk entirely to wolves. The losses of elk to other sources of mortality are likely additive versus any potential compensatory responses to the new source of mortality due to wolves. Prior to wolf recovery, survival of elk calves to their first year was only 0.58. Nearly half of first-year mortality was due to predation mostly by bears and coyotes during the first six to eight weeks of life of the calves and the other half was due to winter malnutrition losses (Singer et al. 1997). While some of the winter malnutrition losses might be compensatory with wolf predation (i.e., wolves might kill weak calves that would have otherwise died from winter malnutrition) (Singer and Mack 1999), predator losses to grizzly bears and coyotes during the early summer are unlikely to be compensatory with wolf predation. Prior computer models suggested that if the projected wolf predation was largely additive, and if hunter harvests of female elk were not substantially reduced, then the elk population would decline at a rapid rate (Singer and Mack 1999). The model exercise suggested if 78 wolves occupied the area and the kill rate was 12 elk/wolf per year, then the antlerless elk harvest would have to be reduced 18% if there was some compensation in the new mortality due to wolves, or 27% if the new mortality was entirely additive (Singer and Mack 1999).

These earlier model projections have apparently come true. Sport hunting harvests of the northern Yellowstone elk have increased about 64% to about 1459 ± 815 elk or about 8% of the population per year since 1988, from an earlier average of about 892 ± 506 elk (5% of the population per year) (Coughenour and Singer 1996, Yellowstone NP 1997, Lemke et al. 1998). Increased harvest also contributes to the multiple predator limitation of the elk population. Reduced calf recruitment rates since 1997 also suggest that wolves are strongly focusing predation on the calf component of the population.

We predict that limitation of the Yellowstone elk will be far less than the more severe limitation to one-fifth of food-limited K that is reported for moose across large areas of northern, less productive boreal forest (Messier and Crête 1985, Gasaway et al. 1992, Crête and Manseau 1996) because: (1) the multiple predators (wolves, grizzly bears, black bear, mountain lion, coyotes) may experience interference competition among the diverse predator fauna (Abrams 1994); (2) plant production in the ecosystem is higher and the ungulate fauna is more diverse (six common species of ungulates occur in many areas) (Crête and Courtois 1997, Crête 1999); and (3) the elk population is seasonally migratory (20 km separates most winter and summer ranges, range = 10–55 km), and predator limitation of migratory ungulates is generalized to be less extreme than for non-migratory ungulates (Fryxell et al. 1988). These authors concluded that 1000 predators could limit a non-migratory population of wildebeest in Serengeti NP, but not if the wildebeest were migratory (Fryxell et al. 1988).

Information needs
Our model projections of wolf/other predator limitation of ungulate numbers are based upon current information on kill rates and current wolf densities in YNP. But these values may change. These parameters need to be closely monitored, as well as any behaviorally mediated effects of wolves on elk, or other ungulates. Ultimately, the effects of wolves on elk, or other ungulates, might exceed our projections if these parameters prove to be substantially different, or if wolves further increase.

Our studies of the effects of ungulate herbivory on plants tended to be mostly oriented to plant-species-specific responses, although we also extended our work to influences on ecosystem process, soil fertility, and plant species diversity (mostly richness). The structure of a few shrub communities (willow, aspen, some upland shrub patches) is being greatly altered by current rates of ungulate herbivory in the two parks. More research needs to be focused at structural and community levels and other shrub-community-dependent trophic-level responses to ungulate herbivory (e.g., Berger et al. 2001). At the same time, recent height releases of some willow and aspen patches on the northern winter range in YNP, apparently related to the re-introduction of wolves, need to also be incorporated into community-related research.

Summary and concluding remarks

Our studies concluded, in agreement with much other published work, that there were several consistent effects of ungulates on plant structure and ecosystem. These effects included: (1) height and diameter reductions in the grazed plants; (2) increases in bare ground that varied from very minor (3% total increase) in the most productive grasslands to substantial (28% total increase) in less productive bunchgrass communities; (3) increases in compaction of the soil surface and a resultant potential tendency to higher surface bulk density and higher soil compaction; and (4) a substantial reduction in the sizes and community structure of willow communities and some, but not all, upland shrub patches.

Soil density changes due to ungulates resulted in the potential for increased sediment yield, but apparently only if the increases in bare ground, soil bulk density, and surface compaction increased beyond certain threshold values (see review by Packer 1963). Increased sediment yield was not statistically documented in any of these parks, although the potential for increased sediment losses clearly existed in YNP.

Grazing effects on herbaceous root biomass, however, were variable. Lower root:shoot ratios were documented or suspected for shrubs in YNP and RMNP, no evidence was found for declines in herbaceous root biomass, while root biomass of some grazed plants "caught up" with ungrazed counterparts later during the growing season.

We concluded that there was substantial deposition of urine and feces by ungulates in grasslands in each of the parks, and plants responded strongly to these nutrient additions. Higher N concentrations in plants were typically associated with higher plant digestibility and lower fiber and lignin and, thus, more concentrated biomass, high plant nutrient concentrations, and higher N yield on previously grazed patches. Other responses included higher rates of N uptake by the grazed plants and higher rates of photosynthesis in new plant tissues, that likely occurred in concert with decreased N uptakes and decreased growth rates by the plant roots (Ruess et al. 1983, Detling 1988, Jaramillo and Detling 1988).

These conclusions apply mostly to grasslands. However, ungulate effects on woody riparian shrubs and trees were more substantial. Aspen and willow have declined as a community type on the core winter range of both parks, and there has also been a shift from tall to short aspen and willow communities in both parks. For willow and aspen patches on the core elk winter range, N inputs, N mineralization rates, N pools, and thus forage biomass available to elk were declining. The reduced canopy area of heavily browsed willows resulted in increased herbaceous richness (Chadde and Kay 1988), but vertebrate fauna richness has been reduced (Berger et al. 2001).

Wolf restoration is occurring in many areas of the U.S. west (Bangs and Fritts 1996). We do not predict a stable, uniform predator limitation following wolf restoration. Perhaps an average limitation may hold for a number of years, but periodically ungulates may be held to even lower levels in conjunction with unfavorable stochastic events, such as epizootics or severe winters (Van Ballenberghe 1985, Peterson et al. 1998). Alternatively, ungulates may also periodically escape the effects of the predators and reach food-limited K without any predator limitation for a number of years (Singer 1987, McLaren and Peterson 1994).

Preservation or restoration of natural processes is a primary goal in national parks. Two of the most significant missing natural processes that influenced ungulate dynamics in the U.S. parks are elimination of the large predators and elimination or abbreviation of ungulate migrations. These processes should be restored to the largest extent feasible or their effects should be duplicated by management.

Evidence is rapidly accumulating in YNP that there will be significant, top-down controls and cascading effects of the wolves on ungulates, their forage base, and other faunal components. We predict that limitation will ultimately be at least 20% to 30% below food-limited K in the case of elk in YNP, since wolves have continued to expand, and the eventual degree of limitation might be even more substantial. Ungulate declines will result in overall fewer negative effects by ungulates on plant growth, recruitment and on ecosystem processes. Projected declines in elk numbers attributable to predator limitation alone, however, are unlikely to be large enough to result in positive responses in willows in YNP or in aspen stands in many areas. However, recent information suggests that a number of aspen and willow stands have less browse pressure, are taller, and are being released from browsing suppression since wolf restoration in 1995 in YNP (Ripple and Larsen 2000; F. Singer and N.T. Hobbs, unpubl. data). The current limitation of elk numbers (~15% fewer elk) by the predator guild in YNP is likely presently too small to explain the height releases of woody browse. Substantial alterations of habitat selection and distributions of ungulates away from the released patches must also be occurring. Alteration of distributions and spatial-use patterns of elk by wolves has been reported that may release browse patches (Dekker et al. 1995, Ripple and Larsen 2000, Ripple et al. 2001, White and Feller 2001). Mean aspen sucker heights were higher in the high wolf use areas of northern YNP than in low wolf use areas (Ripple et al. 2001). Lower densities of elk pellet groups in the high wolf use areas suggested that wolves may have an indirect benefit to aspen regeneration by altering elk movements, browsing patterns, and foraging behavior due to predation risk effects (Ripple et al. 2001). Schmitz et al. (1997) also pointed out that predation risk to herbivores may result in behaviorally mediated effects of the herbivores on trophic cascades. Ultimately, behaviorally mediated ecosystem effects of wolves on ungulates may equal, or even exceed, effects from predator limitation of elk numbers.

Acknowledgments

We thank the staff of YNP, especially John Mack and Glenn Plumb, for access to unpublished elk counts and elk demographic information, 1998–2000. John Varley, Director of the Yellowstone Center for Resources, provided the opportunity and the ideal research environment for the senior author, who studied ungulate-vegetation relations in Yellowstone NP, 1986–1993. We also thank Craig Axtell, Therese Johnson, James Thompson, and Homer Rouse for extensive support, including constructing exclosures and radiocollaring elk in RMNP during the 1994–1999 studies. The research was funded by the U.S. Geological Survey, Midcontinent Ecological Science Center, Fort Collins, and the Washington Office of the U.S. Geological Survey, Biological Resources Division.

Literature cited

Abrams, P.A. 1994. The fallacies of "ratio-dependent" predation. *Ecology* **75**:1842–1850.

Alstad, K.P., J.M. Welker, S.A. Williams, and M.J. Trlica. 1999. Carbon and water relations of Salix monticola in response to winter browsing and changes in surface water hydrology: an isotopic study using $\delta^{13}C$ and $\delta^{18}O$. *Oecologia* **120**:375–385.

Archer, S. and J.K. Detling. 1984. The effects of defoliation and competition on regrowth of tillers of two North American mixed grass prairie graminoids. *Oikos* **43**:351–357.

Baker, W.L., J.A. Munroe, and A.E. Hessl. 1997. The effects of elk on aspen in the winter range in Rocky Mountain National Park. *Ecography* **20**:155–165.

Ballard, W.B. 1992. Bear predation on moose: a review. *Alces Supplement* **1**:162–176.

Bangs, E.E. and S.E. Fritts. 1996. Reintroducing the gray wolf to central Idaho and Yellowstone National Park. *Wildlife Society Bulletin* **24**:402–413.

Belsky, A.J. 1986. Does herbivory benefit plants? A review of the evidence. *American Naturalist* **127**:870–892.

Belsky, A.J. 1992. Effects of grazing, competition, disturbance and fire on species composition and diversity in grassland communities. *Journal of Vegetation Science* **3**:187–200.

Belsky, A.J., W.P. Carson, and C.L. Jensen. 1993. Overcompensation by plants: herbivore optimization or red herring? *Evolutionary Ecology* **7**:109–121.

Berger, J., P.B. Stacey, L. Bellis, and M.P. Johnson. 2001. A mammalian predator–prey imbalance: grizzly bear and wolf extinction affect avian neotropical migrants. *Ecological Applications* **11**:947–960.

Bergström, R. and K. Danell. 1987. Effects of simulated winter browsing by moose on morphology and biomass of two birch species. *Journal of Ecology* **75**: 533–544.

Berry, J., D. Decker, J. Gordon, R. Heitschmidt, D. Huff, D. Knight, W. Romme, and D. Swift. 1997. Science-Based Assessment of Vegetation Management Goals of Elk Winter Range – *Rocky Mountain National Park*. Environment and Natural Resources Policy Institute, Colorado State University, Fort Collins. 16 pp.

Binkley, D., F.J. Singer, M. Kaye, and R. Rochelle. 2001. Influence of elk grazing on soil and nutrients in Rocky Mountain National Park. *In* F.J. Singer, editor. *Ecological Evaluation of the Abundance and Effects of Elk in Rocky Mountain National Park, Colorado*. Final report to National Park Service, U.S. Geological Survey, Fort Collins, Colorado, USA.

Biondini, M., B.D. Patton, and P.E. Nyren. 1998. Grazing intensity and ecosystem processes in a northern mixed-grass prairie, USA. *Ecological Applications* **8**:469–479.

Boutin, S. 1992. Predation and moose population dynamics: a critique. *Journal of Wildlife Management* **56**:116–127.

Boyce, M.S. 1989. *The Jackson Elk Herd: Intensive Wildlife Management in America*. Cambridge University Press, United Kingdom.

Boyce, M.S. 1993. Predicting the consequences of wolf recovery in Yellowstone National Park. Pages 234–269 *in* R.S. Cook, editor. *National Park Service Monograph* **No. 22**. USDI, National Park Service, Denver, Colorado, USA.

Bryant, J.P. 1981. Phytochemical deterrence of snowshoe hare browsing by adventitious shoots of four Alaskan trees. *Science* **231**:889–890.

Caldwell, M.M., J.H. Richards, and D.A. Johnson. 1981. Coping with herbivory: photosynthetic capacity and resource allocation in two semiarid Agropyron grasses. *Oecologia* **50**:14–24.

Caughley, G. 1976. Wildlife management and the dynamics of ungulate populations. Pages 183–246 *in* T.H. Coaker, editor. *Applied Biology*. Volume I. Academic Press, New York, New York, USA.

Chadde, S. and C. Kay. 1988. Willows and moose: a study of grazing pressure. Montana Forest and Conservation Experiment Station, Research Note 24.

Chew, R.M. 1982. Changes in herbaceous and suffrutescent perennials in grazed and ungrazed desertified grassland in southeastern Arizona. 1958–78. *American Midland Naturalist* **108**:159–169.

Consolo-Murphy, S. and D.D. Hanson. 1993. Distribution of beaver in Yellowstone National Park, 1988–89. Pages 38–48 *in* R.S. Cook, editor. Ecological issues on reintroducing wolves into Yellowstone National Park. National Park Service, Science Monograph No. 22. National Park Service, Denver, Colorado, USA.

Coughenour, M.B. 1985. Graminoid response to grazing by range herbivores: adaptations, exaptations and interacting processes. *Annuals of Missouri Botanical Gardens* **72**:852–863.

Coughenour, M.B. 1991. Biomass and nitrogen responses to grazing of upland step on Yellowstone's northern winter range. *Journal of Applied Ecology* **28**:71–82.

Coughenour, M.B. and F.J. Singer. 1995. Elk population processes in Yellowstone National Park under the policy of natural regulation. *Ecological Applications* **6**:573–593.

Crête, M. 1999. The distribution of deer biomass in North America supports the hypothesis of exploitation ecosystem. *Ecology Letters* **2**:223–227.

Crête, M. and R. Courtois. 1997. Limiting factors might obscure population regulation of moose (*Cervidae: Alces alces*) in unproductive boreal forests. *Journal of Zoology (London)* **242**:765–781.

Crête, M. and M. Manseau. 1996. National regulation of cervidae along a 1000km latitudinal gradient: changes in trophic dominance. *Evolutionary Ecology* **10**:51–62.

Day, T.A. and J.K. Detling. 1990. Grassland patch dynamics and herbivore grazing preference following urine deposition. *Ecology* **71**:180–188.

de, Mazancourt, G., M. Loreau, and L. Abbadie. 1998. Grazing optimization and nutrient cycling: when do herbivores enhance plant production? *Ecology* **79**:2242–2252.

Dekker, D., W. Bradford, and J.R. Gunson. 1995. Elk and wolves in Jasper National Park, Alberta, from historical to 1992. Pages 85–94 *in Ecology and Conservation of Wolves in a Changing World*. Canadian Circumpolar Institute, University of Alberta, Canada.

Detling, J.K. 1988. Grassland savannas: regulation of energy flora and nutrient cycling by herbivores. Pages 131–148 *in* L.R. Pomeroy and J.J. Albert, editors. *Concepts of Ecosystem Ecology*. Springer-Verlag, Berlin.

Detling, J.K. and E.L. Painter. 1983. Defoliation responses of western wheat grass populations with disease histories of prairie dog grazing. *Oecologia* **57**: 65–71.

Detling, J.K., M.I. Dyer, and D.T. Winn. 1979. Net photosynthesis, cost respiration, and regrowth of Bouteloua gracilis following simulated grazing. *Oecologia* **41**:172–134.

Dublin, H.T., A.R.E. Sinclair, S. Boutin, E. Anderson, M. Jago, and P. Arcese. 1990. Does competition regulate ungulate populations? *Oecologia* **82**:283–288.

Dyer, M.I. 1975. The effects of red-winged blackbirds on biomass of corn grains. *Journal of Applied Ecology* **12**:719–726.

Dyer, M.I., C.L. Turner, and T.R. Seastedt. 1993. Herbivory and its consequences. *Ecological Applications* **3**:10–16.

Edroma, E.L. 1981. The role of grazing in maintaining high species-composition on Imperata grassland in Rwenzori National Park, Uganda. *African Journal of Ecology* **19**:215–233.

Elmquist, T., L. Ericson, K. Danell, and A. Salomonson. 1987. Flowering, shoot production, and role herbivory in a boreal willow. *Ecology* **68**:1623–1629.

Fahnestock, J.T. and A.K. Knapp. 1994. Responses of grasses to selective herbivory by bison: interactions between herbivory and water stress. *Vegetation* **115**: 123–131.

Farnes, P. 1996. A winter severity index for ungulates of Yellowstone National Park. *in* F.J. Singer, editor. Effects of grazing by wild ungulates in Yellowstone National Park. Technical Report **NPS 96-01**. Natural Resource Information.

Frank, D.A. 1990. Interactive ecology of plants, large mammalian herbivores, and drought in Yellowstone National Park. PhD Dissertation. Syracuse University, Syracuse, New York, USA.

Frank, D.A. and P.M. Groffman. 1998. Ungulate vs. landscape control of soil C and N processes in grasslands of Yellowstone National Park. *Ecology* **79**:2229–2241.

Frank, D.A. and S.J. McNaughton. 1992. The ecology of plants, large mammalian herbivores, and drought in Yellowstone National Park. *Ecology* **73**:2034–2058.

Frank, D.A. and S.J. McNaughton. 1993. Evidence for promotion of aboveground grassland production by native large herbivores in Yellowstone National Park. *Oecologia* **96**:157–161.

Frank, D.A., R.S. Inouye, N. Huntley, G.W. Minshall, and J.E. Anderson. 1994. Biogeochemistry of a north-temperate grassland with native ungulates; nitrogen dynamics in Yellowstone National Park. *Biogeochemistry* **26**:163–188.

Fryxell, J.M. 1987. Food limitation and demography of a migratory antelope, the white-eared Kob. *Oecologia* **72**:83–91.

Fryxell, J.M., J. Greever, and A.R.E. Sinclair. 1988. Why are migratory ungulates so abundant? *American Naturalist* **131**:781–798.

Gasaway, W.C., R.D. Boertje, D.V. Grangaard, D.G. Kelleyhouse, R.O. Stephenson, and D.G. Larsen. 1992. The role of predation in limiting moose at low densities in Alaska and Yukon and implications for conservation. *Wildlife Monographs* **120**:1–59.

Gough, L. and J.B. Grace. 1998. Herbivore effects on plant species diversity at varying productivity levels. *Ecology* **79**:1586–1594.

Green, R.A. and J.K. Detling. 2000. Defoliation-induced enhancement of total aboveground yield of grasses. *Oikos* **91**:280–284.

Grime, J.P. 1973. Control of species density in herbaceous vegetation. *Journal of Environmental Management* **1**:151–167.

Hamilton, R.I., B. Subramanian, M.N. Reddy, and C.H. Rao. 1982. Comparison in grain yield components in a panicle of rain-fed sorghum. *Annals of Applied Biology* **101**:119–125.

Hik, D.S. and R.L. Jefferies. 1990. Increases in net above-ground primary production of a salt-marsh forage grass: a test of the predictions of herbivore-optimization model. *Journal of Ecology* **78**:180–195.

Hilbert, D.W., D.M. Swift, J.K. Detling, and M.I. Dyer. 1981. Relative growth rates and the grazing optimization hypothesis. *Oecologia* **51**:14–18.

Hobbs, N.T. and L.F. Huenneke. 1992. Disturbance, diversity and invasions: implications for conservation. *Conservation Biology* **6**:324–337.

Hobbs, N.T., D.L. Baker, J.E. Ellis, D.M. Swift, and R.A. Green. 1982. Energy and nitrogen based estimates of elk winter range carrying capacity. *Journal of Wildlife Management* **46**:12–21.

Holland E.A. and J.K. Detling. 1990. Plant response to herbivory and belowground nitrogen cycling. *Ecology* **71**:1040–1049.

Holland, E.A., W.J. Parton, J.K. Detling, and D.L. Coppock. 1992. Physiological responses of plant populations to herbivory and their consequences for ecosystem nutrient flow. *American Naturalist* **140**:685–706.

Horn, H.S. 1975. Markovian properties of forest succession. Pages 196–211 *in* M.L. Cody and J. M. Diamond, editors. *Ecology and Evolution of Communities*. Belknap, New York, New York, USA.

Houston, D.B. 1982. *The Northern Yellowstone elk: Ecology and Management*. MacMillan Publishing, New York, New York, USA.

Huston, M.A. 1979. A general hypothesis of species diversity. *American Naturalist* **113**:81–101.

Irons, J.G. III., J.P. Bryant, and M.W. Oswood. 1991. Effects of moose browsing on decomposition rates of birch leaf litter in a subarctic stream. *Canadian Journal of Fisheries and Aquatic Sciences* **48**:442–444.

Jameson, D.A. 1963. Responses of individual plants to harvesting. *Botanical Review* **29**:532–594.

Jaramillo, V.J. and J.K. Detling. 1988. Grazing history, defoliation, and competition: effects on shortgrass production and nitrogen accumulation. *Ecology* **69**:1599–1608.

Jonas, R.J. 1955. A population and ecological study of the beaver (*Castor canadensis*) of Yellowstone National Park. Thesis, University of Idaho, Moscow, Idaho, USA.

Kay, C.E. and F.H. Wagner. 1994. Historical condition of woody vegetation on Yellowstone's northern range. Pages 151–169 *in* D.G. Despain, editor. Plants and their environments. National Park Service, Technical Report 93. Denver, Colorado, USA.

Krefting, L.W., M.H. Stenlund, and R.K. Seemel. 1996. Effects of simulated and natural browsing on mountain maple. *Journal of Wildlife Management* **42**:514–519.

Lacey, J.R. and H.W. Van Poolen. 1981. Comparison of herbage production on moderately grazed and ungrazed western rangelands. *Journal of Range Management* **34**:210–292.

Lane, J.R. and C. Montagne. 1996. Comparison of soils inside and outside of grazing exclosures in Yellowstone's National Park northern winter range. Pages 63–70 *in* F.J. Singer, editor. Effects of grazing by wild ungulates in Yellowstone National Park. U.S. Department of the Interior, National Park Service. Technical Report NPS/96–01.

Lemke, T.O., J.A. Mack, and D.B. Houston. 1998. Winter range expansion by the northern Yellowstone elk herd. *Intermountain Journal of Sciences* **4**:1–9.

Lime, D.W., B.A.D.W.B.A. Koth, and J.C. Vlaming. 1993. Effects of restoring wolves on Yellowstone area big game and grizzly bears: opinions of scientists. Pages 306–326 *in* R.S. Cook, editor. Ecological issues on reintroducing wolves into Yellowstone National Park. *National Park Service Science Monographs* **92/22**.

Lubow, B.C., F.J. Singer, T.L. Johnson, and D.C. Bowden. 2002. Dynamics of interacting elk populations within and adjacent to Rocky Mountain National Park. *Journal of Wildlife Management*, in press.

Mack, J.A. and F.J. Singer. 1993. Predicted effects of wolf predation on northern range elk using Pop II models. Pages 49–74 *in* R. Cook, editor. *National Park Service Science Monograph* **No. 22**. National Park Service, Denver, Colorado, USA.

Marr, J.W. 1961. Ecosystem of the east slope of the Front Range of Colorado. *University of Colorado Studies, Series in Biology* **No. 8**.

Maschinski, J. and T.G. Whitham. 1989. The continuum of plant responses to herbivory: the influence of plant association, nutrient availability, and timing. *American Naturalist* **134**:1–19.

McInnes, P.F., R.J. Naiman, J. Pastor, and Y. Cohen. 1992. Effects of more browsing on vegetation and litter of the boreal forest, Isle Royale, Michigan, USA. *Ecology* **73**:2059–2075.

McLaren, B.E. and R.O. Peterson. 1994. Wolves, moose, and tree rings on Isle Royale. *Science* **266**:1555–1558.

McNaughton, S.J. 1976. Serengeti migratory wildebeest: facilitation of energy flow by grazing. *Science* **191**:92–94.

McNaughton, S.J. 1979. Grazing as an optimization process: grass–ungulate relationships in the Serengeti. *American Naturalist* **11**:691–703.

McNaughton, S.J. 1983. Compensatory plant growth as a response to herbivory. *Oikos* **40**:329–336.

McNaughton, S.J. 1984. Grazing lawns: animals in herds, plant form and co-evolution. *American Naturalist* **124**:863–886.

McNaughton, S.J. 1986. Grazing lawns: on domesticated and wild grazers. *American Naturalist* **128**:937–939.

McNaughton, S.J. 1993. Grasses and grazers, science and management. *Ecological Applications* **3**:17–20.

McNaughton, S.J. 1994. Biodiversity and function of grazing ecosystems. Pages 361–383 *in* E.-D. Schulze and H.M. Moony, editors. *Biodiversity and Ecosystem Function*. Springer-Verlag, New York, New York, USA.

McNaughton, S.J., L.L. Wallace, and M.B. Coughenour. 1983. Plant adaptation in an ecosystem context: effects of defoliation, nitrogen, and water on growth of an African C$_4$ sedge. *Ecology* **64**:307–318.

McNaughton, S.J., F.F. Banyikwa, and M.M. McNaughton. 1998. Root biomass and productivity in a grazing ecosystem: the Serengeti. *Ecology* **79**: 587–592.

Meagher, M.M. and D.B. Houston. 1998. *Yellowstone and the Biology of Time*. Oklahoma State University Press, Norman, Oklahoma, USA. 287 pp.

Merrill, E.H., N.L. Stanton, and J.C. Hak. 1994. Response of bluebunch wheatgrass, Idaho fescue and nematodes to ungulate grazing in Yellowstone National Park. *Oikos* **69**:231–240.

Messier, F. 1991. The significance of limiting and regulating factors on the demography of moose and white-tailed deer. *Journal of Animal Ecology* **60**:377–393.

Messier, F. 1994. Ungulate population models with predation: a case study with the North American moose. *Ecology* **75**:478–488.

Messier, F. and M. Crête. 1985. Moose-wolf dynamics and the natural regulation of moose populations. *Oecologia* **65**:503–512.

Milchunas, D.G. and W.K. Lauenroth. 1992. Quantitative effects of grazing on vegetation and soils over a global range of environments. *Ecological Monographs* **63**:327–366.

Milchunas, D.G., O.E. Sala, and W.K. Lauenroth. 1988. A generalized model of the effects of grazing by large herbivores on grassland community structure. *American Naturalist* **132**:87–106.

Mueggler, W.F. 1984. Diversity of western rangelands. Pages 211–217 *in* J.L. Cooley and J.H. Cooley, editors. *Natural Diversity in Forest Ecosystems Proceedings Workshop.* University of Georgia, Athens.

National Park Service. 2001. *Management Policies.* U.S. Department of the Interior, Washington DC, USA. 137 pp.

Olmstead, C.E. 1987. The ecology of aspen with reference to utilization by large herbivores in Rocky Mountain National Park. Pages 89–97 *in* M.S. Boyce and L.D. Hayden-Winger, editors. *North American Elk: Ecology, Behavior, and Management.* Laramie, University of Wyoming, Wyoming, USA.

Orians, G., D.A. Chochran, J.W. Duffield, T.K. Fulla, R.J. Gutierrez, W.M. Hanemann, F.C. James, P. Karieva, S.R. Kellert, D. Klein, B.N. McLellan, P.D. Olson, and G. Yaska. 1997. *Wolves, Bears and their Prey in Alaska.* National Academy of Sciences Press, Washington DC, USA.

Owen, D.F. and R.G. Wiegert. 1981. Mutualism between grasses and grazers: an evolutionary hypothesis. *Oikos* **36**:376–378.

Packer, P.E. 1963. Soil, stability requirements for the Gallatin elk winter range. *Journal of Wildlife Management* **27**:401–410.

Paige, K.N. 1999. Regrowth following ungulate herbivory in *Impomopsis aggregotis*: geographic evidence for overcompensation. *Oecologia* **118**:316–323.

Paige, K.N. and T.G. Whitham. 1987. Overcompensation in response to mammalian herbivory: the advantage of being eaten. *American Naturalist* **129**:407–416.

Painter, E.L. and A.J. Belsky. 1993. Application of herbivore optimization theory to rangelands of the western United States. *Ecological Applications* **3**:2–9.

Pastor, J., B. Dewey, R.J. Naiman, P.F. McInnes, and Y. Cohen. 1993. Moose browsing and fertility in the boreal forests of Isle Royale National Park. *Ecology* **74**: 467–480.

Pearson, L.C. 1965. Primary production in grazed and ungrazed desert communities of eastern Idaho. *Ecology* **46**:278–285.

Peet, R.K. 1991. Forest vegetation of the Colorado Front Range. *Vegetatio* **45**:3–71.

Peinetti, H.R., R.S.C. Menezes, and M.B. Coughenour. 2001. Changes induced by elk browsing in the aboveground biomass production and distribution of willow (*Salix monticola*): their relationships with plant water, carbon, and nitrogen dynamics. *Oecologia* **127**:334–342.

Peterson, R.O., N.J. Thomas, J.M. Thurber, J.A. Vucetich, and T.A. Waite. 1998. Population limitation and the wolves of Isle Royale. *Journal of Mammalogy* **79**:828–841.

Renkin, R. and D. Despain. 1994. Preburn root biomass/basal area influences on the response of aspen to fire and herbivory. Pages 95–104 *in* J. Greenlee, editor. *The Ecological Implications of Fire in the Greater Yellowstone Ecosystem.* International Association for Wildland Fire, Fairfield, Washington, USA.

Ripple, W.J. and E.J. Larsen. 2000. Historic aspen recruitment, elk, and wolves in northern Yellowstone National Park, USA. *Biological Conservation* **95**:361–370.

Ripple, W.J., E.J. Larsen, R.A. Renkin, and D.W. Smith. 2001. Trophic cascades among wolves, elk and aspen on Yellowstone National Park's northern range. *Biological Conservation* in press.

Ritchie, M.E. and H. Olff. 1999. Herbivore diversity and plant dynamics: compensatory and additive effects. *In* H. Olff, V.K. Brown, and R.H. Drent, editors. *Herbivores: Between Plants and Predator.* Blackwell Science Press, Oxford, England.

Ritchie, M.E., D. Tilman, and J.M.H. Knops. 1998. Herbivore effects on plant and nitrogen dynamics on oak savanna. *Ecology* **79**:165–177.

Romme, W.H., M.G. Turner, L.L. Wallace, and J.S. Walker. 1995. Aspen, fire and elk in northern Yellowstone National Park. *Ecology* **76**:2097–2104.

Ruess, R.W. 1984. Nutrient movement and grazing: experimental effects of clipping and nitrogen source on nutrient uptake in *Kyllinga nervosa*. *Oikos* **43**:183–188.

Ruess, R.W., S.J. McNaughton, and M.B. Coughenour. 1983. The effects of clipping, nitrogen source, and nitrogen concentration on growth response and nitrogen uptakes of an East African sedge. *Oecologia* **59**:253–261.

Rummel, R.S. 1951. Some effects of grazing on ponderosa pine forest and range in central Washington. *Ecology* **32**:594–607.

Schmitz, O.J., A.P. Beckerman, and K.M. O'Brien. 1997. Behaviorally mediated trophic cascades: effects of predation risk on food web interactions. *Ecology* **78**:1388–1399.

Sinclair, A.R.E. 1977. *The African Buffalo, A Study of Resource Limitation of Populations.* University of Chicago Press, Chicago, USA. 355 pp.

Sinclair, A.R.E. 1989. Population regulation in arrivals. Pages 197–241 *in* J.M. Cherret, editor. *Ecological Concepts.* Blackwell Scientific Publications, Oxford, England.

Sinclair, A.R.E. and R.P. Pech. 1996. Density dependence, stochasticity, compensation and predator regulation. *Oikos* **75**:164–173.

Sinclair, A.R.E., H. Dublin, and M. Borner. 1985. Population regulation of Serengeti wildebeest: a test of the food hypothesis. *Oecologia* **65**:266–268.

Singer, F.J. 1987. Dynamics of caribou and wolves in Denali National Park. Pages 117–157 *in Fourth Triennial Conference on Science in the National Parks.* Volume 2. George Wright Society. Houghton, Michigan, USA.

Singer, F.J. 1995. Effects of grazing by ungulates on upland bunchgrass communities of the northern range of Yellowstone National Park. *Northwest Science* **69**:191–203.

Singer, F.J. and R.G. Cates. 1995. Response to comment: ungulate herbivory on willows on Yellowstone's northern winter range. *Journal of Range Management* **48**: 563–565.

Singer, F.J. and M.K. Harter. 1996. Comparative effects of elk herbivory and the 1988 fires on northern Yellowstone National Park grasslands. *Ecological Applications* **6**:185–199.

Singer, F.J. and J.A. Mack. 1999. Predicting the effects of wildfire and carnivore predation on ungulates. Pages 189–237 *in* T.W. Clark, A.P. Curlee, S.C. Minta, and P.M. Kareiva, editors. *Carnivores in Ecosystems: The Yellowstone Experience.* Yale University Press, New Haven, Connecticut, USA.

Singer, F.J. and J.E. Norland. 1994. Niche relationships within a guild of ungulates following release from artificial controls. *Canadian Journal of Zoology* **72**:1383–1394.

Singer, F.J. and R.A. Renkin. 1995. Effects of browsing by native ungulates on the shrubs in big sagebrush communities in Yellowstone National Park. *Great Basin Naturalist* **55**:201–212.

Singer, F.J., L. Mack, and R.G. Cates. 1994. Ungulate herbivory of willows on Yellowstone's northern winter range. *Journal of Range Management* **47**:435–443.

Singer, F.J., A. Harting, K.K. Symonds, and M.B. Coughenour. 1997. Density
dependence, compensation, and environmental effects on calf mortality. *Journal
of Wildlife Management* **61**:12–25.

Singer, F.J., D.M. Swift, M.B. Coughenour, and J.D. Varley. 1998*a*. Thunder on the
Yellowstone revisited: an assessment of management of native ungulates by
natural regulation, 1968–93. *Wildlife Society Bulletin* **26**:375–390.

Singer, F.J., L.C. Zeigenfuss, R.G. Cates, and D.T. Barnett. 1998b. Elk, multiple factors,
and persistence of willows in national parks. *Wildlife Society Bulletin* **26**:419–428.

Singer, F.J., L.C. Zeigenfuss, B. Lubow, and M.J. Rock. 2002. Ecological evaluation of
potential overabundance of ungulates in U.S. National Parks: a case study. Pages
205–248 *in* U.S. Geological Survey, Open File Report 01-208, U.S. Geological
Survey, Fort Collins, Colorado. 268 pp.

Stohlgren, T.J., L.D. Schell, and B.V. Heuvel. 1999. How grazing and soil quality affect
native and exotic plant diversity in Rocky Mountain Grasslands. *Ecological
Applications* **9**:45–64.

Suzuki, K., and H. Suzuki, D. Binkley, and T.J. Stohlgren. 1999. Aspen regeneration in
the Colorado front range: differences at local and landscape scales. *Landscape
Ecology* **14**:321–237.

Taper, M.L. and P.J.P. Gogan. 2002. The northern Yellowstone elk: density dependence
and climatic conditions. *Journal of Wildlife Management* **66**:106–122.

Thompson, M.J. and R.E. Henderson. 1998. Elk habituation as a credibility challenge
for wildlife professionals. *Wildlife Society Bulletin* **26**:477–483.

Turner, C.L., T.R. Seastedt, and M.I. Dyer. 1993. Maximization of above ground
grassland production: the role of defoliation frequency, intensity, and history.
Ecological Applications **3**:175–186.

Van Ballenberghe, V. 1985. Wolf predation on caribou: the Nelchina herd case history.
Journal of Wildlife Management **49**:711–720.

Van Ballenberghe, V. and W. Ballard. 1994. Limitation and regulation of moose
populations: the role of predation. *Canadian Journal of Zoology* **72**:2071–2077.

Verkaar, H.J. 1986. When does grazing benefit plants? *Trends in Ecology and Evolution*
1:168–169.

Verkaar, H.J. 1992. When does grazing benefit plants? *Trends in Ecology and Evolution*
1:168–169.

Vesey-Fitzgerald, V. 1973. *East African Grasslands*. East African Publishing House,
Nairobi, Kenya. 265 pp.

Wagner, F.H., R. Foresta, R.B. Gill, D.R. McCullough, M.R. Pelton, W.F. Porter, and
H. Salwasser. 1995. *Wildlife policies in the U.S. National Parks*. Island Press,
Washington DC, USA.

Welker, J.M. and J.W. Menke. 1990. The influence of simulated browsing on tissue water
relations, growth and survival of *Quercus douglasii* seedlings under slow and rapid
rates of soil drought. *Functional Ecology* **4**:807–817.

White, C.A. and M.C. Feller. 2001. Predation risk and elk-aspen foraging patterns.
Sustaining aspen in the western landscape. USDA Forest Service Proceedings
RMRS-P-18, Fort Collins, Colorado, USA.

Willard, E.E. and C.M. McKello. 1978. Responses of shrubs to simulated browsing on
Drummond's willow. *Journal of Wildlife Management* **42**:514–519.

Wolfe, M.L., W.H. Babcock, and R.M. Welch. 1983. Effects of simulated browsing on
Drummond's willow. *Alces* **19**:14–35.

Yellowstone National Park. 1997. Yellowstone's northern range: complexity and change in a wildland ecosystem. National Park Service, Mammoth Hot Springs, Wyoming. 148 pages.

Zeigenfuss, L.C., F.J. Singer, S. Williams and T.L. Johnson. 2002. Influence of herbivory and water additions to willow communities on elk winter range. *Journal of Wildlife Management*, in press.

STAN BOUTIN, CHARLES J. KREBS, RUDY BOONSTRA AND
ANTHONY R.E. SINCLAIR

14

The role of the lynx–hare cycle in boreal forest community dynamics

Snowshoe hare (*Lepus americanus*) and lynx (*Lynx canadensis*) populations show 9- to 11-year cycles throughout the boreal coniferous forests of Canada and Alaska. Lynx cycles, recorded in Hudson Bay trapping records covering many decades, have been analyzed by population ecologists interested in understanding the mechanisms underlying the cycle (Royama 1992, Ranta et al. 1997). Long-term records of hare cycles are less common but there have been two major field programs that have measured demographic changes over at least one full cycle. This includes the work based at Rochester, Alberta by Lloyd Keith and co-workers (Keith et al. 1984, Keith 1990) and our work based at Kluane, Yukon (Krebs et al. 1986, Sinclair et al. 1988, Krebs et al. 1995). These field studies have established the demographic mechanisms that drive changes in hare numbers and have tested mechanistic hypotheses about factors that stop population increase and cause the hare population decline (Keith 1990, Krebs et al. 2001*b*).

It is clear that the lynx and hare are tightly linked (Royama 1992, Stenseth et al. 1998). Lynx show strong functional and numerical responses to changes in hare density (Keith et al. 1977, O'Donoghue et al. 1998). In turn, the dominant proximate cause of hare mortality is predation (Keith et al. 1984, Boutin et al. 1986) with lynx being one of a suite of hare predators. Statistical analyses of time series data suggest that lynx cycles can be explained by a two trophic level interaction whereas hare cycles require a three trophic level interaction (Stenseth et al. 1997). This simple representation masks a rich set of hypotheses concerning the details of trophic interactions and the role that this predator–prey cycle plays in the broader forest community. Over the past 25 years our research team at Kluane has attempted to test hypotheses explaining hare

[487]

cycles and to place the hare cycle in the broader boreal forest community. The work is unique in that it has combined the case study approach similar to the studies of Serengeti (Sinclair and Arcese 1995), Yellowstone (Houston 1982) and Bialowieza (Jedrzejewska and Jedrzejewski 1998) with large-scale experiments designed to test hypotheses about community organization.

In this chapter we focus on how the hare cycle affects the forest community including its vegetation, vertebrate herbivores, and predators as revealed by the Kluane research. We will address the following questions:

1.   How many species and what processes are affected by the lynx–hare cycle?
2.   Are trophic levels limited by bottom-up or top-down effects?
3.   Is the lynx or hare a keystone species?

We take advantage of the major natural perturbation caused by the cycle to study the response of the community. In addition, we performed large-scale experiments designed to alter each trophic level and monitored how these alterations affected other trophic levels. The boreal coniferous forests of Canada and Alaska represent some of the largest tracts of natural forest left in the world but this is changing rapidly as commercial forestry and oil and gas development press northward. We will end by discussing how these human-caused changes might affect the lynx, hare, and their associated community.

Study area and approach

Details for the Kluane study area and the methods used can be found in Krebs et al. (2001b). The 350 km² study area was located in the southwestern Yukon at the base of a broad glacial valley. The forest was roughly an equal mix of willow (*Salix glauca*) shrub communities and mature open and closed (>50% canopy cover) white spruce (*Picea glauca*). The dominant shrubs were willow (*Salix glauca*) and dwarf birch (*Betula glandulosa*). All of the communities were of fire origin but there have been no fires in the past 50 years and over 90% of the forest is more than 100 years old. There has been no commercial forestry in the study site, which is bisected by a single major road, the Alaska Highway.

The northern location and relatively high elevation of the Kluane area creates a highly seasonal environment with below average productivity. The growing season does not begin before the third week of May and killing frosts occur as early as mid August. The timing of arrival of snow is highly variable and accumulation is normally around 1 m. Snowmelt is usually not complete before mid May.

We have been studying snowshoe hares at Kluane since 1977 and the work we describe here occurred from 1986 to 1996. There were two basic components: description of the vertebrate food web, including standing biomass and energy flow, and experimental manipulation of trophic levels. For the first component we used live-trapping and mark-recapture to estimate the abundance of snowshoe hares, deer mice (*Peromyscus maniculatus*) and voles (*Clethrionomys rutilus, Microtus* spp.), red squirrels (*Tamiasciurus hudsonicus*) and Arctic ground squirrels (*Spermophilus parryii.*) We did not estimate moose (*Alces alces*) density, but it was low (0.12/km^2, Gasaway et al. 1992). Lynx and coyote (*Canis latrans*) densities were estimated by a combination of snow track transects, intensive snow tracking, and radiotelemetry monitoring of individuals. We followed changes in relative abundance of other predators including red fox (*Vulpes vulpes*), wolf (*Canis lupus*), wolverine (*Gulo gulo*), and weasel (*Mustela* spp.) by snow track transects but we did not estimate their abundance. We used a combination of telemetry monitoring of prey and snow tracking of predators to estimate kill rates by predators. Marten (*Martes americana*) and fisher (*Martes pennanti*) were absent from our study area.

We performed a series of large-scale experiments designed to "kick" each trophic level (see Krebs et al. 2001b for details). We fertilized the forest each spring with nitrogen, phosphorus, and potassium to pulse the plant trophic level. Addition of rabbit chow ad libitum was used to increase the herbivore trophic level (hares, ground squirrels, mice, and voles). We erected a large (1 km^2) predator exclosure by using chickenwire combined with electric wires to stop predators from digging under or climbing over the fence. All small herbivores, including snowshoe hares, could move freely through the fence. We also erected a small hare exclosure. Finally treatments were combined to produce a food addition plus predator exclusion treatment and a hare exclusion plus fertilization treatment. These treatments are summarized in Table 14.1.

Changes in the vertebrate food web over the cycle

During our study snowshoe hares underwent a typical cycle whereby they increased to peak autumn densities of 2–3 hares/ha in 1989 and 1990 and dropped to 0.01–0.1 hares/ha in 1993 (Boutin et al. 1995, Krebs et al.

Table 14.1. *Experimental treatments used in the Kluane Boreal Forest Ecosystem Project, 1986–1996*

Treatment	Replicates	Details
Food addition	2	Commercial rabbit chow supplied year round ad libitum, spread over 35 ha
Mammalian predator exclosure	1	Electric fence and chickenwire around 1 km^2
Predator exclosure + food addition	1	Electric fence around 1 km^2 plus rabbit chow spread on central 35 ha
Fertilization	2	Nitrogen, phosphorus, potassium added each spring on 1 km^2
Hare exclosure	1	4 ha
Hare exclosure + fertilizer	1	4 ha inside fertilizer replicate

1995, Hodges et al. 2001). The amplitude (minimum spring to maximum autumn densities) was roughly 20- to 40-fold. The peak densities of hares observed at Kluane were relatively low compared to other studies (Hodges et al. 2001). Figure 14.1A shows the standing biomass of the vertebrate herbivore community over the cycle. There are three important things to note. First, hares dominated the vertebrate herbivore biomass at the cyclic peak, comprising 65% of the total biomass. This dropped to about 10% during the low. Over the entire cycle, hares represented roughly 50% of the standing herbivore biomass. Second, total herbivore biomass varied roughly three fold over the cycle. Finally, apart from hares, red squirrels and ground squirrels were the only substantial alternative herbivore biomass available to predators. Because ground squirrels hibernate for eight months of the year, red squirrels are the only alternative prey from September to April. Mice and voles were never >5% of the vertebrate herbivore community (Boonstra et al. 2001*b*).

Predator biomass roughly doubled over the cycle with lynx and coyote being the predominant species (Fig. 14.1B). Fox, wolf, and wolverine were occasional visitors but we think that fox occupied the alpine areas and avoided the valley because of coyotes (O'Donoghue et al. 2001). One pack of wolves passed through our study area occasionally but their numbers were low. During the low phase, great horned owls (*Bubo virginianus*) and red-tailed hawks (*Buteo j. harlani*) equaled the biomass of lynx and coyotes.

Complete food web linkages are outlined in Krebs et al. (2001*a*). Hares were the dominant component of the diet of lynx, coyotes, and owls even

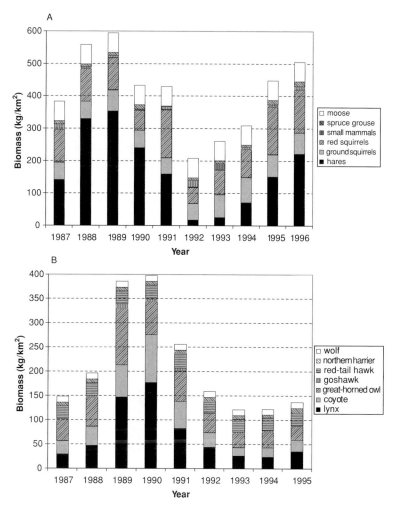

Fig. 14.1. Autumn biomass of vertebrate herbivores (A) and predators (B) at Kluane Lake in 1987–1996.

during the low phase of the cycle (O'Donoghue et al. 1998, Rohner et al. 2001). All of the predators consumed grouse, voles, ground squirrels, and red squirrels. Hares browsed all of the shrubs and trees in the study area but birch was the preferred species (Smith et al. 1988). Birch comprised roughly 10% of the shrub biomass and it was not present on all of our study sites (Krebs et al. 2001c). Hares tended to feed on twigs <5 mm in diameter (Pease et al. 1979, Smith et al. 1988) and these made up roughly 10% of the standing shrub biomass (Krebs et al. 2001c).

The plant trophic level

Effects of nutrients

We expected the plant trophic level to be strongly limited by nutrient availability as control sites had extremely low nitrogen values (0.97 ppm, Turkington et al. 1998). Fertilization had strong positive effects on grass and herb growth and biomass (Turkington et al. 1998). Fertilization increased the growth of small willow twigs (<5 mm) by 30% overall (Fig. 14.2). Birch showed similar increases except for the food addition treatment, where they were lower. There were similar increases on the hare exclosure plus fertilization plot, and even higher growth rates on the food addition plot and the predator exclosure plus food plot (Krebs et al. 2001c). Growth rates of birch peaked in 1992 on control sites whereas willow tended to decline to the low point of the hare cycle (Krebs et al. 2001c). Standing biomass of small twigs and shrubs reached their highest levels in 1992 and there was no difference in relative standing crop on fertilizer versus control grids.

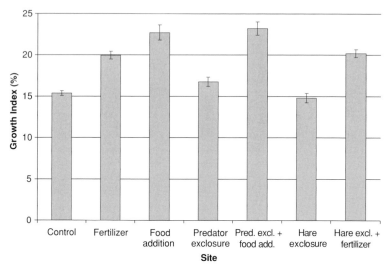

Fig. 14.2. Average growth index (%) of small twigs (<5 mm) of gray willow (*Salix glauca*) on each of the study treatments, with 95% confidence limits. The average was calculated over 1988–1995. The growth index was calculated as:

$$\frac{\text{Dry weight of current annual growth on 5-mm twig}}{\text{Dry weight of complete 5-mm twig}}$$

Effects of herbivores

There was little evidence that mammalian herbivores in the Kluane system affected the composition or abundance of herbs and grasses present (John and Turkington 1995, Turkington et al. 2001). In the case of shrubs, the impact was somewhat different. Browse rates of hares were monitored by following the fates of a tagged sample of twigs. Browsing of birch twigs rapidly increased to >50% in the late increase phase of the cycles and was as high as 80% at the peak and early decline (Fig. 14.3A). Once the hare numbers had crashed, virtually none of the twigs were clipped (Krebs et al. 2001c). The pattern of browsing on willow was similar but the percentage clipped on control areas was never >20% (Fig. 14.3B). During high hare densities, birch could not compensate for the loss of twigs due to hare browsing and twig biomass was less than one-fifth that of the previous year (Fig. 14.3A). In contrast, willow was able to fully compensate for the increased clipping by hares.

The density of hares was increased by 5- to 11-fold on the food addition and food addition plus predator exclusion grids (Krebs et al. 1995), and this led to a substantial increase in browse rates of willow. These levels were such that willow was not able to compensate and biomass was reduced. Averaged over the entire cycle, biomass of twigs on the predator exclosure plus food addition treatment was one-third that of controls (Sinclair et al. 2001). Hares had a significant effect on birch during the peak but browsing decreased rapidly thereafter such that total biomass actually increased to a peak in 1992. Browsing by hares on willow had little effect on standing crop. Overall, biomass of shrubs showed a clear cycle but, contrary to expectation, biomass increased from the start of the study to a peak in 1991 and 1992 before declining again (Krebs et al. 2001c).

The herbivore trophic level

Effects of plants

Herbivores had relatively little effect on the primary producer trophic level but can the same be said for the effect of plants on herbivores? Growth of grasses, herbs, and shrubs was increased roughly 30% by fertilizing (Krebs et al. 2001c, Turkington et al. 2001) but this did not result in higher herbivore biomass (Fig. 14.4). Ground squirrels and red squirrels actually declined on the fertilizer plots whereas meadow voles increased by 40% (Boonstra et al. 2001b). Food supplementation produced much stronger

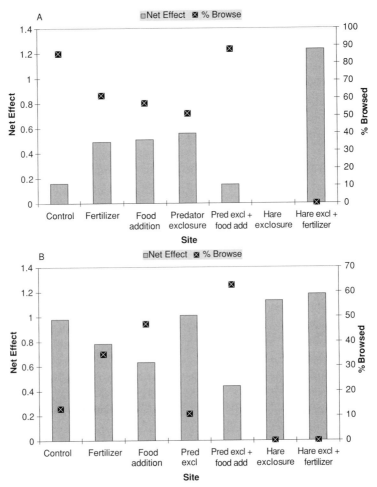

Fig. 14.3. Percentage of bog birch (*Betula glandulosa*) (A) and gray willow (*Salix glauca*) (B) browsed by hares in 1990–1991. The net effect of winter browsing is also shown and was calculated as $(1 - h)(1 + g)$ where h = biomass loss to herbivory and g = growth rate (see Sinclair et al. 2001 for details). Values <1 represent when browsing exceeded growth.

effects than did fertilization. Hare and ground squirrel densities increased 2- to 5-fold on average and this produced an average biomass increase over the cycle of 2.3-fold (Fig. 14.4). In the case of hares, much of the population increase was due to immigration (Krebs et al. 2001*a*). Red squirrels,

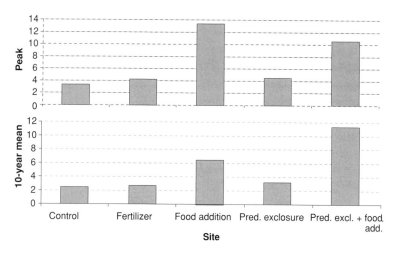

Fig. 14.4. Total herbivore biomass (kg/ha) on experimental treatments in the year of peak biomass and as a mean from 1987 to 1996.

mice, and voles showed no response to the food addition (Boonstra et al. 2001*a*,*b*).

Effects of predators

Predation was a dominant limiting factor for the herbivore trophic level. We recorded 866 mortalities of radiocollared hares on grids excluding the fenced areas (Hodges et al. 2001). Cause of death could be identified in roughly 85% of these cases and >90% of these were attributed to predation. Lynx and coyotes were responsible for 65% to 75% of these mortalities. Figure 14.5 shows the percentage of the autumn hare population removed by lynx and coyotes over winter as determined by radiotelemetry monitoring of hares and calculation of the functional and numerical response of lynx and coyotes from snowtracking. Losses as assessed by telemetry were consistently higher than those determined by snowtracking and both techniques have potential biases (see O'Donoghue et al. 2001). If true losses fell between the two estimates it is clear that lynx and coyote removed a substantial portion of the hare population during the peak and early decline. During the hare low, predators removed 10% to 40%.

We also followed the fates of juvenile hares during their first month of life by radiotelemetry. Mortality rates were high during this period but the striking finding was that lynx and coyotes took very few of these

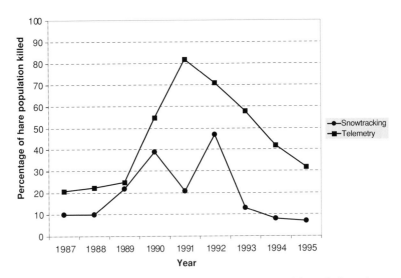

Fig. 14.5. Estimated total impact of predation by coyotes and lynx during winter (October through April) on snowshoe hares from 1987–1988 through 1995–1996 at Kluane. Estimates from snowtracking combined functional and numerical responses of the predators while estimates from telemetry used mortality rates of radiocollared hares.

small animals. Instead red squirrels and ground squirrels were the primary predators. When hares were abundant, up to 40% of the young leverets were killed by squirrels (O'Donoghue 1994, Stefan 1998, Hodges et al. 2001).

We also monitored the fates of ground squirrels and red squirrels via radiotelemetry. Ground squirrels were monitored from 1992 to 1995 and of 130 mortalities, 96% were due to predation (Hubbs and Boonstra 1997, Boonstra et al. 2001a). Lynx and coyote were responsible for only 20% to 30% of these losses with the rest caused by raptors. Red squirrels were also taken by coyotes and lynx but losses had little effect on red squirrel densities (Stuart-Smith and Boutin 1995). Winter losses were never >5% (O'Donoghue et al. 2001).

Exclusion of lynx and coyotes improved hare survival significantly during the population peak and early decline but this resulted in a relatively small increase in hare density (Hodges et al. 2001). This was partially due to the fact that hares could move freely through the fence. Once they left the enclosure, they suffered losses equal to hares living outside of the fence.

The combined treatment of food addition plus predator exclusion produced the largest increase in herbivore biomass. The increase over the controls was 6-fold over the entire cycle (Fig. 14.4) and levels reached as high as 18 kg/ha in the spring of the last year of the study. This compared to an average of 2.48 kg/ha for controls.

Linkages between herbivores

We were interested in whether other herbivores were influenced by the hare cycle either indirectly through predator switching or possibly through direct competition. Changes in ground squirrels, forest grouse (spruce grouse: *Falcipennis canadensis*, ruffed grouse: *Bonasa umbellus*), willow ptarmigan (*Lagopus lagopus*), and muskrats (*Ondatra zibethicus*) were directly correlated with changes in hare numbers and these species declined in synchrony with hares (Boutin et al. 1995, see also appropriate chapters in Krebs et al. 2001*b*). Redbacked and meadow vole numbers were inversely correlated with hare numbers (Boonstra et al. 2001*b*). Red squirrel numbers were largely unaffected by changes in hare numbers (Stuart-Smith and Boutin 1995, Boonstra et al. 2001*a*).

The predator trophic level

Effects of herbivores

It is clear that predators had a strong influence on some but not all of the herbivores in the system. The large changes in hare densities over the cycle had a strong influence on both lynx and coyotes. With the exception of the extreme cyclic low, hares represented virtually 100% of the winter diet of both predators (O'Donoghue et al. 1998). Even when hares were very scarce, they still comprised >50% of the biomass of kills. Red squirrels made up as much as 40% of the lynx diet in 1 year of the low but they were never >13% of the coyote's diet (O'Donoghue et al. 1998). Coyotes and lynx made many kills of small mammals in 1 year but mice and voles never formed a large component of the biomass consumed.

The number of lynx and coyotes tracked hare densities with a 1-year lag (Fig. 14.6). Lynx peaked at 17/100 km² and coyotes at 9/100 km². The amplitude of change was 7.5-fold for lynx and 6-fold for coyotes. Predators declined rapidly in the second winter of the hare crash (1991–1992). This was due to a combination of complete recruitment failure (no tracks of family groups) and relatively high emigration. In 1992–1993 winter survival of lynx plummeted to less than 10% (O'Donoghue et al. 1997, 2001). During this time, a number of lynx were killed by other predators including wolves, wolverine, and another lynx (O'Donoghue et al. 2001). In

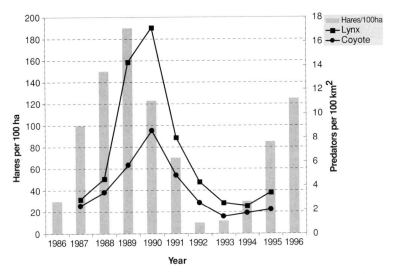

Fig. 14.6. Estimates of densities of snowshoe hares (mean of autumn and late winter estimates) and coyotes and lynx (early winter) for 1986–1997.

contrast, few coyotes died in our study area but 13 of 21 collared animals dispersed. Lynx and coyote populations did not begin to increase until hare numbers began to increase. Family groups did not reappear until 1995–1996.

The results at Kluane are consistent with findings from other northern studies over roughly the same time period. In south central Yukon, Slough and Mowat (1996) found a 10- to 17-fold change in lynx numbers with peak densities of 50/100 km². In the Northwest Territories, Poole (1994) observed a 10-fold change and peak densities of 30/100 km². In both of these areas coyotes were largely absent. These studies and observations of the previous population cycle at Kluane (Ward and Krebs 1985) also reported high emigration rates of lynx during the hare crash. In north central Alberta, lynx and coyote densities varied 3- to 4-fold, with lynx numbers peaking at 10/km² and coyotes at 44/100 km² (Keith et al. 1977, Todd et al. 1981).

Competitive interactions

Both coyotes and lynx relied heavily on snowshoe hares during our study. The question is how do these two similar-sized carnivores co-exist on a relatively limited resource base? Lynx have long been viewed as a snowshoe hare specialist. Coyotes are relative newcomers to the boreal forest and they are considered to be generalists (O'Donoghue et al. 2001). This was not the case at Kluane. Overlap in diet and habitat use remained extremely high between lynx and coyotes even during the hare low. It seemed that the low diversity of alternative prey at Kluane led to coyotes having no alternative but to specialize on hares (O'Donoghue et al. 2001).

Community organization

In general, the Kluane study provides the following insights into the major processes that operate on the vertebrate community. The snow-shoe hare dominates the vertebrate herbivore biomass and a broad suite of mammal and raptor species feed primarily on hares. Hares, in turn, inhibit the growth of their main winter food plants, willow and birch, during the hare peak. This effect is short-lived and there appears to be some stimulation of shrub growth one to two years after the hare peak. It appears that the hare is regulated from both above and below with the predator–hare interactions being stronger than the vegetation–hare interactions. These findings are in accordance with statistical time-series analyses (Royama 1992, Stenseth et al. 1997). Predators, on the other hand, are regulated from below (Stenseth et al. 1997). The hare cycle appears to create cycles in most of the small herbivores present either through predators switching to alternative prey as hares decline or through stimulation of berry production through increased nutrient cycling (Boutin et al. 1995, Boonstra et al. 2001b). The notable exception to this is the red squirrel whose populations do not appear to be coupled to hares. Red squirrels represent the only alternative prey available to predators in winter.

The Kluane study was designed to test a number of hypotheses related to the regulation of each trophic level (Sinclair et al. 2000, 2001). The focus was on whether the size (biomass) of each level was controlled by bottom-up (White 1978, 1984, Polis and Strong 1997), top-down (Hairston et al. 1960, Menge and Sutherland 1976), or reciprocal effects (Power 1984, Benndorf and Horn 1985). Our results suggest that there is not one pure model that applies to all trophic levels. Instead, we found strong reciprocal effects at the upper trophic levels. Hares and the main predators in the system (lynx, coyote, great horned owl) were tightly linked with predators limiting hare numbers and vice versa. The reciprocal interaction was weaker between herbivores and vegetation. Although hares had some effect on the small twig biomass of the most preferred shrub species, they had little overall effect on the total biomass or species composition of any of the major vegetation compartments. Vegetation was limited by bottom-up nutrient availability.

Despite the fact that we increased vegetation production by 30% through fertilization, we saw very little increase in herbivore biomass. In contrast, food addition to herbivores led to a 3-fold increase overall. We attribute these results to the fact that laboratory rabbit chow represented a major increase in quality as well as quantity of food for hares and ground squirrels. Fertilization served to increase the quantity and quality (Sinclair et al. 2001) of natural vegetation only slightly and this was not enough to overcome other limitations on the hare population. Hares are capable of reaching much higher densities in other habitat types. For example, Slough and Mowat (1996) recorded hare densities twice as high as ours in regenerating pine (*Pinus contorta*) stands some 300 km from the Kluane study site. It would take many years of fertilization at the Kluane site to produce these sorts of hare densities because of the slow response of the plants to increased nutrients.

A 3-fold increase in herbivore densities with food addition did not lead to major changes in shrub vegetation. Although birch could not compensate for the higher browsing during the peak and early decline, it quickly recovered thereafter. It was only when hare densities were pushed to even higher levels and were maintained at or above peak control densities over an entire cycle (food addition plus predator exclusion, Fig. 14.4) that shrub biomass showed significant declines. It appears that chronic high hare densities would be required to change shrub cover in these hare-dominated systems.

Despite overwhelming evidence that predation is the proximate cause of mortality of most herbivores in the Kluane system, and population declines in part are due to lower survival, we did not observe a large increase in herbivore density on the mammalian predator exclusion grid. This was partially due to the fact that avian predation compensated somewhat for the reduced mammalian predation (Hodges et al. 2001). More importantly though, hares moved out of the predator exclosure when their numbers reached high densities (Hodges et al. 2001). We do not know why this occurred, but it may have been that hares were driven to forage more widely to find patches of unbrowsed birch or willow (Hik 1995).

Lynx and coyotes relied heavily on snowshoe hares even during the population crash. In fact, they had few options, as red squirrels represented the only alternative prey during winter. This situation appeared to trigger long-range emigration by lynx and possibly coyotes after the hare crash. This phenomenon has been observed a number of times (Ward and Krebs 1985, Poole 1994, Slough and Mowat 1996, O'Donoghue et al. 2001) and is a unique feature of the boreal forest created by the hare cycles.

Finally, we could not disrupt the hare cycle completely. Single-factor manipulations altered the amplitude of the cycle but only the combination of food addition plus predator exclusion affected both the amplitude and timing of the hare decline. We were able to maintain hare densities at or above peak control levels from 1989 to 1995 on the food addition plus predator exclusion treatment. However, the cycle was still apparent although the amplitude was reduced to a 5-fold change as opposed to 20- to 40-fold changes on controls. It appears that feedbacks between the vegetation–herbivore–predator trophic levels make it difficult to disrupt the cycle completely (see also Royama 1992). The Kluane work did not address the reasons for the broad scale synchrony observed in the long-term Hudson Bay lynx trapping records and snowshoe hare questionnaires (see Ranta et al. 1997).

Is the hare or lynx a keystone species?

Although the concept of keystone species has become overused (Mills et al. 1993), we think that it is applicable in the case of the snowshoe hare for the following reasons. Hares represent the main food for many of the predators in the boreal forest and the elimination of hares would lead to extinction of lynx and coyotes and great horned owls. None of the resident predators were capable of successful recruitment during the hare crash despite expanding their diet to consume alternative prey. A more appropriate test of the keystone species idea would be to replace hare biomass with an equivalent amount of alternative herbivores. Although this is logistically impossible, we can do the thought experiment. Replacing hares with equivalent amounts of moose or ground squirrels would still lead to the elimination of lynx, coyotes and owls because moose are too big to be killed and ground squirrels are unavailable during winter. The situation would be unlikely to change as well if hare biomass were to be converted into red squirrels. Although both lynx and coyotes hunted red squirrels more often during the hare low, they took only a small percentage of the available population. Red squirrels are active for only a small portion of the day during winter, particularly when temperatures are very low. For the majority of time, they are safely hidden in nests that are inaccessible to the main predators. Voles are likely too small for them to be efficient prey items for lynx and coyotes. Capture rates would have to be two orders of magnitude higher to produce the energetic returns obtained from hunting hares. Grouse seem like the only alternative prey that might function as a substitute for hares. Spruce grouse densities in our study were intermediate ($50/km^2$) relative to published records (Boag and Schroeder 1987, 1992, Martin et al. 2001). Grouse made up a small portion of the diets of lynx and coyotes at Kluane (O'Donoghue et al. 2001) but there is no information available in the literature to indicate whether these predators could specialize on grouse if grouse numbers were increased substantially.

Besides having a major influence on the persistence of predator populations, the hare cycle also acts to entrain cycles in five of the seven other herbivores in the Kluane system (Boutin et al. 1995). In the case of ground squirrels, grouse, ptarmigan, and probably muskrats, this appears to result from predator switching as hares, the primary prey, decline in numbers. Vole numbers appear to cycle in an inverse fashion relative to hares and one hypothesis is that this may be due to a time-lag between increased nutrient cycling from high hare densities and berry production in dwarf shrubs (Boonstra et al. 2001b).

It seems that hares possess some unique features that are not easily substituted by other herbivores in the system. Can the same be said for lynx and other predators? This does not appear to be the case. The predator–hare interaction is an important driver of the hare cycle (Keith et al. 1984, Krebs et al. 1995) but it seems that the key predators, lynx, coyote, and great horned owl, are highly interchangeable (Stenseth et al. 1997). The long-term hare studies by Keith and co-workers in central Alberta occurred where lynx were relatively scarce and coyotes were the dominant predator (Keith et al. 1977, Todd and Keith 1983). Great horned owls were partially able to compensate for the exclusion of lynx and coyotes at our Kluane site (Hodges et al. 2001). It seems that removal of any one of these species of predator would not lead to a change in the hare cycle and associated parts of the community.

Is the snowshoe hare a keystone throughout its range? Unfortunately, information on the role that hares play in community dynamics of the montane western coniferous forests of British Columbia and the northwestern U.S. is limited (see Hodges 2000 for a review). However, information on population densities suggests that hare numbers may continue to cycle in the southern part of their range but at less than half of the densities observed in their northern range (Hodges 2000). Because of this, it is likely that the hare's influence on the community is diminished. In fact, it has been suggested that the presence of lynx in the southern montane portion of the hare's distribution is primarily due to periodic immigration into these areas from more northerly regions following hare crashes (Thiel 1987, Mowat et al. 2000, see also Schwartz et al. 2002).

It has been suggested that the hare cycle is diminished in the southern part of its range because more and more of the available habitat is of lower quality and a broad suite of generalist predators do not allow hare densities to build to levels observed during peaks in the north (Keith et al. 1993). Experiments in western montane forests are required to test this hypothesis but it is likely that lynx and snowshoe hare are secondary components of a more diverse vertebrate community in this region.

Potential effects of human activities

The large tracts of northern coniferous forests in Canada and Alaska are undergoing changes through a variety of human activities: large-scale commercial forestry, widespread oil and gas exploration, and encroachment of agriculture from the south. How resilient are lynx–hare cycles to these activities? We could not completely stop the hare cycle with our relatively small-scale experiments and it is clear that the boreal forest and hare cycles have experienced large-scale natural disturbances such as fire many times in the past. Hare densities are likely higher in young forest and in areas with high shrub cover. Intensive forest management designed to reduce shrub density to favor rapid regeneration of conifer will likely be detrimental to hare and lynx populations. New forest-management approaches designed to approximate patterns of regeneration following natural disturbance such as fire are much less likely to have these negative effects.

Broad changes in the average age of forest cover could possibly affect the amplitude of the cycles but it is unlikely to alter the cycles themselves. The Kluane study suggests that major changes in the hare–predator interaction would be required for this to occur. Conventional forestry leaves a pattern of alternating cut and uncut forest that creates a relatively high amount of edge habitat. It is possible that this could affect the hunting success of predators but not to the degree that the cycle would be disrupted. Hare cycles do disappear near the southern boundary of the species' range and studies by Keith and Bloomer (1993) suggest that this may be due to changes in the proportion of hare habitat versus non-hare habitat. As the amount of useable habitat declines, the probability of hares being killed by a suite of generalist predators in the unfavorable habitat increases. When these rates become high enough the cycle is actually stopped and hares never reach peak densities. It is possible that this could occur in other parts of the hare's range if large tracts of forest were to be converted to unforested areas through agriculture or oil sands mining. However, if the predominant vegetation cover in a region continues to be forest this seems unlikely.

As industrial activities increase in the north, so does the amount of linear corridor development. This in turn, increases public access. These sorts of changes could affect lynx by creating trapper access to refuge areas. Lynx are easily trapped and the numbers of lynx have been greatly reduced in the southern part of their range through habitat loss and heavy trapping pressure. Slough and Mowat (1996) suggested that high numbers of lynx remain in the Yukon despite extensive trapping because large refugia exist. This scenario can change rapidly as oil and gas exploration creates many kilometers of seismic lines that become access routes for trappers. Although these changes may affect lynx negatively, it is likely that they will favor coyotes. Coyotes have expanded their distribution into the boreal forest and it is likely that this has been aided, if not precipitated, by industrial expansion. In the Yukon, coyotes are not widely distributed and they tend to be concentrated in areas where there are major roads. It appears that the hare cycle is robust to relative changes in predator species composition. Whether it would persist if lynx were to disappear completely is another question.

Summary

As mammalian ecologists, it was not hard to be occupied with the lynx–hare cycle when we began our study of the Kluane system. Virtually all of the vertebrate community is linked to the hare in a direct or indirect fashion. The Kluane study has helped to establish the hare–lynx cycle in the broader context of the boreal forest community. The snowshoe hare is the dominant vertebrate herbivore in this system and changes in its numbers have strong effects up the trophic web with lesser effects down the web. Hare cycles directly determine cycles in most vertebrate predators, which indirectly creates cycles in other herbivores through shared predators. Hares do not have major effects on vegetation except on selected species during peak densities. Hares function as a keystone species, and lynx, coyotes, and great horned owls would disappear in the absence of hares. In contrast, it appears that members of the predator guild are highly interchangeable. The Kluane work has provided insights into the strength of the various trophic relationships and their robustness to experimental perturbation. The experiments we conducted were large scale as far as experiments go, but they were very small scale relative to the changes that are about to unfold in the world's northern forests. The lynx–hare cycle has proven to be remarkably robust but it is crucial that ecologists continue to monitor this vertebrate complex as the large experiment of industrial development unfolds on the landscape. There is still much to be learned about this fascinating system.

Acknowledgments

We wish to thank all of the workers and colleagues who participated in the Kluane Project. We thank Dennis Murray and J. David Brittell for helpful comments. Finally we thank Cathy Shier and Ainsley Sykes for support in preparation of the manuscript.

Literature cited

Benndorf, J. and W. Horn. 1985. Theoretical considerations on the relative importance of food limitation and predation in structuring zooplankton communities. Pages 383–396 *in* W. Lampert, editor. *Food Limitation and the Structure of Zooplankton Communities*. E. Schweizerbart'sche Verlagsbuchhandlung, Stuttgart, Germany.

Boag, D.A. and M.A. Schroeder. 1987. Population fluctuations in spruce grouse: what determines their number in spring? *Canadian Journal of Zoology* 65:2430–2435.

Boag, D.A. and M.A. Schroeder. 1992. Spruce grouse (*Dendragapus canadensis*). *In* A. Poole, P. Stettenheim and F. Gill, editors. *The Birds of North America*. The Academy of Natural Sciences, Philadelphia, The American Ornithologists' Union, **No. 5**. Washington DC, USA.

Boonstra, R., S. Boutin, A. Byrom, T. Karels, A. Hubbs, K. Stuart-Smith, M. Blower, and S. Antpoehler. 2001a. The role of red squirrels and Arctic ground squirrels. Pages 179–214 *in* C.J. Krebs, S. Boutin, and R. Boonstra, editors. *Vertebrate Community Dynamics in the Kluane Boreal Forest*. Oxford University Press, New York, New York, USA.

Boonstra, R., C.J. Krebs, S. Gilbert, and S. Schweiger. 2001b. Voles and mice. Pages 215–239 *in* C.J. Krebs, S. Boutin, and R. Boonstra, editors. *Vertebrate Community Dynamics in the Kluane Boreal Forest*. Oxford University Press, New York, New York, USA.

Boutin, S., C.J. Krebs, A.R.E. Sinclair, and J.N.M. Smith. 1986. Proximate causes of losses in a snowshoe hare population. *Canadian Journal of Zoology* 64:606–610.

Boutin, S., C.J. Krebs, R. Boonstra, M.R.T. Dale, J. Hannon, K. Martin, A.R.E. Sinclair, J.N.M. Smith, R. Turkington, M. Blower, A. Byrom, F.I. Doyle, C. Doyle, D. Hik, L. Hofer, A. Hubbs, T. Karels, D.L. Murray, V.O. Nams, M. O'Donoghue,

C. Rohner, and S. Schweiger. 1995. Population changes of the vertebrate community during a snowshoe hare cycle in Canada's boreal forest. *Oikos* **74**:69–80.

Gasaway, W.C., R.D. Boertje, D.V. Grangaard, D.G. Kelleyhouse, R.O. Stephenson, and D.G. Larsen. 1992. The role of predation in limiting moose at low densities in Alaska and Yukon and implications for conservation. *Wildlife Monographs* **120**:1–59.

Hairston, N.G., F.E. Smith, and L.B. Slobodkin. 1960. Community structure, population control and competition. *American Naturalist* **94**:421–425.

Hik, D.S. 1995. Does risk of predation influence population dynamics? Evidence from the cyclic decline of snowshoe hares. *Wildlife Research* **22**:115–129.

Hodges, K.E. 2000. Ecology of snowshoe hares in southern boreal and montane forests. Pages 163–206 *in* L.F. Ruggiero, K.B. Aubry, S.W. Buskirk, G.M. Koehler, C.J. Krebs, K.S. McKelvey, and J.R. Squires, editors. *Ecology and Conservation of Lynx in the United States.* University Press of Colorado, Boulder, Colorado, USA.

Hodges, K.E., C.J. Krebs, D.S. Hik, C.I. Stefan, E.A. Gillis, and C.E. Doyle. 2001. Snowshoe hare demography. Pages 141–178 *in* C.J. Krebs, S. Boutin, and R. Boonstra, editors. *Vertebrate Community Dynamics in the Kluane Boreal Forest.* Oxford University Press, New York, New York, USA.

Houston, D.B. 1982. *The Northern Yellowstone Elk: Ecology and Management.* MacMillan, London, United Kingdom.

Hubbs, A.H. and R. Boonstra. 1997. Population limitation in Arctic ground squirrels: effects of food and predation. *Journal of Animal Ecology* **66**:527–541.

Jedrzejewska, B. and W. Jedrzejewski. 1998. Predation in vertebrate communities: the Białowieża primeval forest as a case study. *Ecological Studies.* Volume 135. Springer Verlag, Berlin, Germany.

John, E. and R. Turkington. 1995. Herbaceous vegetation in the understory of the boreal forest: does nutrient supply or snowshoe hare herbivory regulate species composition and abundance? *Journal of Ecology* **83**:581–590.

Keith, L.B. 1990. Dynamics of snowshoe hare populations. Pages 119–195 *in* H.H. Genoways, editor. *Current Mammalogy.* Plenum Press, New York, New York, USA.

Keith, L.B. and S.E.M. Bloomer. 1993. Differential mortality of sympatric snowshoe hares and cottontail rabbits in central Wisconsin. *Canadian Journal of Zoology* **71**:1694–1697.

Keith, L.B., A.W. Todd, C.J. Brand, R.S. Adamcik, and D.H. Rusch. 1977. An analysis of predation during a cyclic fluctuation of snowshoe hares. *Proceedings of the International Congress of Game Biologists* **13**:151–175.

Keith, L.B., J.R. Cary, O.J. Rongstad, and M.C. Brittingham. 1984. Demography and ecology of a declining snowshoe hare population. *Wildlife Monographs* **90**:1–43.

Keith, L.B., S.E.M. Bloomer, and T. Willebrand. 1993. Dynamics of a snowshoe hare population in fragmented habitat. *Canadian Journal of Zoology* **71**:1385–1392.

Krebs, C.J., B.S. Gilbert, S. Boutin, A.R.E. Sinclair, and J.N.M. Smith. 1986. Population biology of snowshoe hares. I. Demography of food-supplemented populations in the southern Yukon, 1976–84. *Journal of Animal Ecology* **55**:963–982.

Krebs, C.J., S. Boutin, R. Boonstra, A.R.E. Sinclair, J.N.M. Smith, and M.R.T. Dale, K. Martin, and R. Turkington. 1995. Impact of food and predation on the snowshoe hare cycle. *Science* **269**:1112–1115.

Krebs, C.J., R. Boonstra, S. Boutin, and A.R.E. Sinclair. 2001*a*. What drives the 10-year cycle of snowshoe hares? *BioScience* **51**(1):25–35.

Krebs, C.J., S. Boutin, and R. Boonstra, editors. 2001*b*. *Vertebrate Community Dynamics in the Kluane Boreal Forest*. Oxford University Press, New York, New York, USA.

Krebs, C.J., M.R.T. Dale, V.O. Nams, A.R.E. Sinclair, and M. O'Donoghue. 2001*c*. Shrubs. Pages 92–115 *in* C.J. Krebs, S. Boutin, and R. Boonstra, editors. *Vertebrate Community Dynamics in the Kluane Boreal Forest*. Oxford University Press, New York, New York, USA.

Martin, K., C. Doyle, S. Hannon, and F. Mueller. 2001. Forest grouse and ptarmigan. Pages 240–260 *in* C.J. Krebs, S. Boutin, and R. Boonstra, editors. *Vertebrate Community Dynamics in the Kluane Boreal Forest*. Oxford University Press, New York, New York, USA.

Menge, B.A. and J.P. Sutherland. 1976. Species diversity gradients: synthesis of the roles of predation, competition and temporal heterogeneity. *American Naturalist* **110**:351–369.

Mills, S.L., M.E. Soulé, and D.F. Doak. 1993. The keystone-species concept in ecology and conservation. *BioScience* **43**(4):219–224.

Mowat, G., K.G. Poole, and M. O'Donoghue. 2000. Ecology of lynx in northern Canada and Alaska. Pages 265–306 *in* L.F. Ruggiero, K.B. Aubry, S.W. Buskirk, G.M. Koehler, C.J. Krebs, K.S. McKelvey, and J.R. Squires, editors. *Ecology and Conservation of Lynx in the United States*. University Press of Colorado, Boulder, Colorado, USA.

O'Donoghue, M. 1994. Early survival of juvenile snowshoe hares. *Ecology* **75**:1582–1592.

O'Donoghue, M., S. Boutin, C.J. Krebs, and E.J. Hofer. 1997. Numerical responses of coyotes and lynx to the snowshoe hare cycle. *Oikos* **80**:150–162.

O'Donoghue, M., S. Boutin, C.J. Krebs, G. Zuleta, D.L. Murray, and E.J. Hofer. 1998. Functional responses of coyotes and lynx to the snowshoe hare cycle. *Ecology* **79**:1193–1208.

O'Donoghue, M., S. Boutin, D.L. Murray, C.J. Krebs, E.J. Hofer, U. Breitenmoser, C. Breitenmoser-Wuersten, G. Zuleta, C. Doyle, and V.O. Nams. 2001. Coyotes and lynx. Pages 275–323 *in* C.J. Krebs, S. Boutin, and R. Boonstra, editors. *Vertebrate Community Dynamics in the Kluane Boreal Forest*. Oxford University Press, New York, New York, USA.

Pease, J.L., R.H. Vowles, and L.B. Keith. 1979. Interaction of snowshoe hares and woody vegetation. *Journal of Wildlife Management* **43**:43–60.

Polis, G.A. and D.R. Strong. 1997. Food web complexity and community dynamics. *American Naturalist* **147**:813–846.

Poole, K.G. 1994. Characteristics of an unharvested lynx population during a snowshoe hare decline. *Journal of Wildlife Management* **58**:608–618.

Power, M.E. 1984. Depth distributions of armoured catfish: predator-induced resource avoidance? *Ecology* **65**:523–528.

Ranta, E., V. Kaitala, and J. Lindstrom. 1997. Dynamics of Canadian lynx populations in space and time. *Ecography* **20**:454–460.

Rohner, C., F.I. Doyle, and J.N.M. Smith. 2001. Great horned owls. Pages 339–376 *in* C.J. Krebs, S. Boutin, and R. Boonstra, editors. *Vertebrate Community Dynamics in the Kluane Boreal Forest*. Oxford University Press, New York, New York, USA.

Royama, T. 1992. *Analytical Population Dynamics*. Chapman & Hall, London.

Schwartz, M.K., L.S. Mills, K.S. McKelvey, L.F. Ruggiero, and F.W. Allendorf. 2002. DNA reveals high dispersal synchronizing the population dynamics of Canada lynx. *Nature* **415**:520–522.

Sinclair, A.R.E. and P. Arcese, editors. 1995. *Serengeti II: Dynamics, Management and Conservation of an Ecosystem*. University of Chicago Press, Chicago, Illinois, USA.

Sinclair, A.R.E., C.J. Krebs, J.N.M. Smith, and S. Boutin. 1988. Population biology of snowshoe hares. III. Nutrition, plant secondary compounds and food limitation. *Journal of Animal Ecology* **57**:787–806.

Sinclair, A.R.E., C.J. Krebs, J.M. Fryxel, R. Turkington, S. Boutin, R. Boonstra, P. Seccombe-Hett, P. Lundberg, and L. Oksanen. 2000. Testing hypotheses of trophic level interactions: a boreal forest ecosystem. *Oikos* **89**(2):313–328.

Sinclair, A.R.E., C.J. Krebs, R. Boonstra, S. Boutin, and R. Turkington. 2001. Testing hypotheses of community organization for the Kluane ecosystem. Pages 407–436 *in* C.J. Krebs, S. Boutin, and R. Boonstra, editors. *Vertebrate Community Dynamics in the Kluane Boreal Forest*. Oxford University Press, New York, New York, USA.

Slough, B.G. and G. Mowat. 1996. Lynx population dynamics in an untrapped refugium. *Journal of Wildlife Management* **60**:946–961.

Smith, J.N.M., C.J. Krebs, A.R.E. Sinclair, and R. Boonstra. 1988. Population biology of snowshoe hares. II. Interactions with winter food plants. *Journal of Animal Ecology* **57**:269–286.

Stefan, C.I. 1998. Reproduction and pre-weaning juvenile survival in a cyclic population of snowshoe hares. MSc Thesis. University of British Columbia, Vancouver, British Columbia, Canada.

Stenseth, N.C., W. Falck, O.N. Bjørnstad, and C.J. Krebs. 1997. Population regulation in snowshoe hare and Canadian lynx: asymmetric food web configurations between hare and lynx. *Proceedings of the National Academy of Sciences of the United States of America* **94**:5147–5152.

Stenseth, N.C., W. Falck, K.S. Chan, O.N. Bjørnstad, M. O'Donoghue, H. Tong, R. Boonstra, S. Boutin, C.J. Krebs, and N.G. Yoccoz. 1998. From patterns to processes: phase and density dependencies in the Canadian lynx cycle. *Proceedings of the National Academy of Sciences of the United States of America* **95**:15430–15435.

Stuart-Smith, A.K. and S. Boutin. 1995. Predation on red squirrels during a snowshoe hare decline. *Canadian Journal of Zoology* **73**:713–722.

Thiel, R.P. 1987. The status of Canada lynx in Wisconsin, 1865–1980. *Wisconsin Academy of Sciences, Arts and Letters* **75**:90–96.

Todd, A.W. and L.B. Keith. 1983. Coyote demography during a snowshoe hare decline in Alberta. *Journal of Wildlife Management* **47**:394–404.

Todd, A.W., L.B. Keith, and C.A. Fischer. 1981. Population ecology of coyotes during a fluctuation of snowshoe hares. *Journal of Wildlife Management* **45**:629–640.

Turkington, R., E. John, C.J. Krebs, M.R.T. Dale, V.O. Nams, R. Boonstra, S. Boutin, K. Martin, A.R.E. Sinclair, and J.N.M. Smith. 1998. The effects of NPK fertilization for nine years on boreal forest vegetation in northwest Canada. *Journal of Vegetation Science* **9**:333–346.

Turkington, R., E. John, and M.R. Dale. 2001. Herbs and grasses. Pages 69–91 *in* C.J. Krebs, S. Boutin, and R. Boonstra, editors. *Vertebrate Community Dynamics in the Kluane Boreal Forest*. Oxford University Press, New York, New York, USA.

Ward, R.M.P. and C.J. Krebs. 1985. Behavioural responses of lynx to declining snowshoe hare abundance. *Canadian Journal of Zoology* **63**:2817–2824.

White, T.C.R. 1978. The importance of a relative shortage of food in animal ecology. *Oecologia* **3**:71–86.

White, T.C.R. 1984. The abundance of invertebrate herbivores in relation to the availability of nitrogen in stressed food plants. *Oecologia* **63**:90–105.

ROBERT G. ANTHONY, MARGARET A. O'CONNELL,
MICHAEL M. POLLOCK AND JAMES G. HALLETT

15
_____

Associations of mammals with riparian ecosystems in Pacific Northwest forests

Introduction

The aquatic and terrestrial components of riparian systems provide eco-logical opportunities for many species of mammals. The importance of riparian habitat to wildlife populations has been documented in a wide range of habitats in North America: the midwestern United States (Stauffer and Best 1980), desert southwest (England et al. 1984), Rocky Mountains (Knopf 1985), Oregon (Anthony et al. 1987, Doyle 1990, McComb et al. 1993, Gomez and Anthony 1998, Kauffman et al. 2001), Washington (O'Connell et al. 1993, Kelsey and West 1998, Kauffman et al. 2001), and the Okanogan Highlands of British Columbia (Gyug 2000). These studies indicate that wildlife species richness is high in these ecosys-tems, and use of riparian zones by some species is disproportionately higher than in other areas. Although this is especially true in the more arid regions of North America (Johnson and Jones 1977, Brinson et al. 1981), this pattern can also be found in mesic forests of the Pacific Northwest. For example, Thomas et al. (1979) report that 285 of the 378 terrestrial wildlife species in the Blue Mountains of Oregon and Washington are found exclusively or more commonly in riparian areas, and Oakley et al. (1985) report similar patterns of 359 of the 414 wildlife species using riparian zones of western Washington and Oregon forests. Kauffman et al. (2001) estimate that 53% of the 593 wildlife species that occur in Washington and Oregon use riparian zones, whereas riparian zones and wetlands constitute only 1% to 2% of the landscape. This dispropor-tionate use of riparian zones by wildlife species, including mammals, reflects their response to the diversity of biological and physical features found there.

[510]

Riparian zone management has come under increasing scrutiny over the last two decades (e.g., Young 2000), and wildlife use of riparian areas is one of the key issues influencing management considerations. Management of these areas is complicated because of the variety of human impacts on these systems including forest harvest, livestock grazing, road building, and the range of uses by wildlife species. For mammals, the use of riparian areas varies widely from obligate (e.g., semi-aquatic species) to occasional, and reflects the degree to which mammals are specialized for these areas and the opportunities these areas afford for foraging, cover, or roosting.

In this chapter, we first describe the impacts of physical features, vegetation, and disturbance on the structure and function of riparian systems. We next review patterns of mammalian use of riparian habitats with emphasis on Pacific Northwest forests, and how a keystone species, the beaver, can influence the riparian zone. We then focus on management problems, particularly those related to the use of riparian buffer strips associated with timber harvest. Finally, we consider future directions for research.

Structure and function of riparian zones

Riparian zones are found adjacent to streams, rivers, springs, ponds, lakes, or tidal estuaries and represent areas of great ecological richness and diversity. They can be variously defined in terms of vegetation, topography, hydrology, or ecosystem function as the zone of interaction between the aquatic and terrestrial environments (Swanson et al. 1982, Bilby 1988, Naiman et al. 2000, Kauffman et al. 2001). This definition encompasses the concept that the terrestrial and aquatic systems influence each other. The zone of interaction can be identified as the water's edge or, on a broader scale, as a zone extending from the water through the canopy of vegetation associated with the zone (Swanson et al. 1982). On the latter scale, riparian zones include the relatively mesic vegetation and associated faunas occurring between aquatic and more xeric upland sites (Knopf et al. 1988). The structure of the riparian zone is closely related to the size of the body of water or watercourse. In the Pacific Northwest, most riparian zones are found adjacent to streams and rivers (Oakley et al. 1985), and this is especially true for the forested lands of the region (Swanson et al. 1982).

The interaction between the terrestrial and aquatic environment that occurs in the riparian zone changes with stream size. Small streams produce smaller riparian zones than larger streams, and the effect of the terrestrial system on the aquatic system is inversely related to stream size. Understanding the relation between stream size and the interaction between aquatic and terrestrial systems is important in understanding the structure and function of riparian zones. Consequently, a classification system for the size of streams is necessary. A widely adopted system to describe drainages classifies small, headwater channels with perennial surface waters as first-order streams. Each union of first-order streams forms larger second-order streams, and each union of second-order streams forms still larger third-order streams and so forth (Strahler 1957, Everest et al. 1985).

The structure and function of riparian zones are determined by topography, surface water, soils, microclimate, vegetation, and disturbance patterns (Cummins 1980, Swanson et al. 1982, Brosofske et al. 1997, Kauffman et al. 2001). These elements combine to create features that distinguish riparian zones from upland areas. For example, riparian zones are characterized by increased primary productivity, higher levels of energy transport, and often, more frequent natural disturbance than upland areas. Variation in these key elements results in differences within and among riparian systems and their use by mammals.

Topography
Topography within and adjacent to riparian zones in the Pacific Northwest varies from narrow, entrenched channels that are typically associated with lower order streams to broad floodplains associated with higher order rivers. Riparian zones surrounded by steep upland slopes have soils that are typically shallow and coarse textured, are not exposed to direct sunlight for long periods, have erosion and active transport of material as dominant processes, and often have associated plant and animal communities that are relatively limited. In contrast, riparian zones associated with broad floodplains typically have deep and fine-textured soils, are exposed to sunlight, have deposition of materials as the dominant process, and have associated plant and animal communities that are diverse. The former are confined between rock outcrops and have little, if any, developed floodplain (Brinson et al. 1981). The latter have well-developed floodplains and can change dimensions, shape, and gradient in response to changing water flows. Individual streams often have alternating sections of both topographies along their entire length, which greatly increases the diversity of associated plant and mammal communities.

Hydrology

A common element in all riparian systems that sets them apart from upland areas is the presence of surface water that varies from standing to running and perennial to intermittent. In the Pacific Northwest, perennial streams and rivers exhibit pronounced annual variation in flow levels (Hall 1988). In addition, many riparian zones experience periodic catastrophic floods that are often accompanied by ice flows or debris torrents (Cummins 1980, Hall 1988). The dynamic nature of the water flow shapes the structure of the riparian zone through erosive down-cutting and deposition and is responsible for the high levels of nutrient cycling characteristic of riparian zones. Seasonal variation in water level and flow are important for nutrient recycling in riparian zones, and the expansion and contraction of stream channels with changing flow levels influence the structure and composition of plant communities (Brinson et al. 1981).

Soils

Both the surface water character and topography of the riparian zones have a direct influence on the types of soils found in riparian zones. In general, riparian soils differ from upland soils in the origin of soil mineral content, organic content of soils, and amount of soil litter (Bilby 1988). The increased moisture content of riparian soils relative to upland soils generally results in higher decomposition rates in some systems and therefore increased organic content. However, if riparian soils become saturated with stagnant water, decomposition rates will decrease due to lack of oxygen. The organic content of riparian soils is also determined, to some extent, by redistribution during periodic flooding. Large amounts of organic matter will be flushed from areas with high-energy flows and deposited in other areas of low water flow (Bilby 1988). The organic content of riparian soils can be greater than that of upland soils in part because many riparian plants decompose more easily than upland plants (Edmonds 1980, Bilby 1988). The nitrogen (N) content of the litter can also affect decomposition rates. Elevated N content results in more rapid decomposition and, consequently, increased organic turnover (Swanson et al. 1982). Riparian zones often have exposed soil surfaces, whereas upland areas have greater amounts of terrestrial litter. This is due to the combined effects of deposition and flooding in riparian zones (Bell and Sipp 1975).

Microclimate

Topographic features and presence of surface water often result in microclimatic differences between riparian zones and upland areas that may extend up to 60 m from the stream (Brosofske et al. 1997). Riparian habitats, for example, usually have higher humidity, rates of transpiration, and air movement than upland areas (Thomas et al. 1979). In the Pacific Northwest, these conditions allow the extension of maritime-associated plant species further inland in riparian areas than on hillsides (Wimberly and Spies 2001).

Vegetation

The hydrologic, topographic, substrate, and microclimatic features of riparian zones result in distinct physiological, compositional, and structural features of riparian vegetation (Campbell and Franklin 1979, Franklin et al. 1981, Swanson et al. 1982, Oakley et al. 1985). Studies of the structure and composition of riparian forests describe the mosaic aspect of riparian vegetation and the need to assess both coarse- and fine-scale factors to characterize riparian vegetation (Jonsson 1997, Pabst and Spies 1999, Nierenberg and Hibbs 2000, Russell and McBride 2001).

The hydrology of riparian zones affects the metabolism and growth of vegetation in three primary ways. First, increased soil moisture is important in maintaining riparian forest vegetation, especially in more xeric forests east of the Cascade Mountains. Second, the nutrient supply for riparian vegetation depends, in part, on the transport action of streams. Third, flowing water aerates the soils and roots of riparian plants resulting in more rapid gas exchange. These three factors contribute to faster growth rates and increased primary productivity of riparian plant communities relative to upland communities.

Riparian zones typically have higher plant species diversity than upland areas (Naiman et al. 1993). Variation in the diversity of vegetation among riparian areas is related to a site's size, aspect, soil moisture, amount of coarse woody debris (CWD), and time since disturbance (Gawler 1988, Malanson and Butler 1990). Riparian vegetation is composed of generalized species that inhabit both riparian and upslope sites, but are often more abundant in riparian areas because of favorable conditions. Specialized species are found only in moist riparian areas and are adapted to conditions created by patterns of natural disturbance characteristic of riparian areas (Gawler 1988). A common feature of many Pacific Northwest forests is a shift from the dominance of hardwoods to conifers with increasing distance from streams. Although disturbance patterns (e.g., fire, flooding) contribute to the dominance of hardwoods in riparian forests, the proximity to water (Russell and McBride 2001) and the competitive advantage of hardwoods over conifers in these habitats (Nierenberg and Hibbs 2000) are thought to be the principal determinants.

Riparian vegetation, in turn, has a direct influence on stream structure and function. First, roots of riparian vegetation stabilize stream banks that help define stream morphology and reduce sedimentation (Swanson et al. 1982). Second, riparian vegetation is an important source of CWD in Pacific Northwest streams. CWD is recognized as an integral link between terrestrial and aquatic components of forest ecosystems and might be the primary influence on lower order streams in this area (Swanson et al. 1982, Hyatt and Naiman 2001). It can help define stream structure by retaining gravel and sediment, forming pools, and creating waterfalls. CWD facilitates deposition of sediments in the stream and consequently affects the morphology and energy transport in lower order streams (Keller and Swanson 1979, Swanson et al. 1982, Bilby 1988). Third, standing riparian vegetation has an important effect on stream function. Riparian vegetation influences the chemistry of the stream through nutrient assimilation and transformation. The absence of vegetation in the riparian zone can result in greater export of dissolved materials (Brinson et al. 1981, Bilby 1988). Fourth, shading of streams by riparian vegetation affects water temperatures, and the magnitude of the effect is inversely related to stream size. In smaller streams, riparian vegetation can completely shade the water from sunlight, and these streams typically exhibit stable, cool temperatures year-round. Lastly, riparian vegetation is also important for delivery of leaves, litter, and logs that contribute to the aquatic food chain.

Disturbance

Riparian zones are a product of disturbance (Agee 1988) and an understanding of how natural disturbance affects riparian zone structure and function provides insight into mammalian associations. In Pacific Northwest forests, natural disturbances such as flooding, fire, and wind vary in frequency, magnitude, and relative importance in upland versus riparian areas. Given the topographic constraints inherent to many riparian areas, Wimberly and Spies (2001) suggest that the relationships between disturbance regimes, environmental gradients, and vegetation patterns are more tightly linked in riparian than upslope habitats.

Fluvial disturbance is a major factor shaping the structure and composition of riparian habitats in the forests of the Pacific Northwest (e.g., Naiman et al. 2000), and riparian networks in the region have been classified according to the predominant type of fluvial disturbance

as: (1) debris-flow avalanche channels, (2) fluvial and debris-flow channels, or (3) fluvial channels (e.g., Fetherston et al. 1995). In general, the
impact of fluvial disturbance is greater in riparian than in upslope habitats and the likelihood of debris-flow avalanche disturbances is greater in
lower order streams (Wimberly and Spies 2001). Fluvial disturbances in
forests of the Pacific Northwest can occur as seasonal small-scale events
or episodic large-scale floods. Annual variations in flow make portions
of riparian zones available for plants each dry season as channel width
decreases 16% to 60% (Swanson et al. 1982). Large-scale flooding has a
much greater impact on riparian vegetation, especially in small streams
and if it includes the battering action of debris and/or ice. Deciduous
trees and shrubs (e.g., willows (*Salix* spp.), red alder (*Alnus rubra*), aspens
(*Populus* spp.)) usually dominate post-disturbance riparian zones within
5–10 years. In larger streams, fluvial disturbance might result in a stepped
progression of successional stages from the channel to the upslope forest (Agee 1988). Deciduous trees colonize recent gravel bars and dominate
lower, younger terraces. Older, higher terraces support conifer stands.

Fire is an important determinant of the composition and structure of
western forests (e.g., Agee 1993, Arno 2000). The features that characterize riparian areas have been thought to create conditions that reduce fire
frequency, but increase fire severity (Agee 1994). In general, the higher
humidity and fuel moisture of riparian forests make them less susceptible
to the more frequent, low-intensity fire regimes. These wetter forests,
however, typically support tree species that are fire sensitive. Consequently, when environmental conditions lead to fires (i.e., drought years,
heavy fuel loads), they are often stand-replacement fires (Arno 2000).
Therefore, although fires are not frequent in riparian forests, they are
typically large and severe (Agee 1993). Evidence in support of this is seen
in riparian areas of Oregon's Coastal Range forests (Wimberly and Spies
2001). The number of large remnant old-growth trees that survived a nineteenth century fire was similar (albeit low) between riparian and upslope habitats, suggesting a comparable impact of this stand-replacement
fire. In contrast, the mean age of the riparian forest was more than that
of the adjacent upslope forests, indicating that the lower intensity fires
that had occurred since the stand-replacement fire had not impacted the
riparian as much as the upslope habitats (Wimberly and Spies 2001). In
mixed-coniferous forests of northern California, Russell and McBride
(2001) found that the frequency of fire scars in riparian areas increases
linearly with increasing distance from streams. Olson (2000) compared

fire frequencies between riparian and upland habitats of Oregon's Blue Mountains and southern Cascades. She found that fire frequencies were comparable when forest composition was similar between the riparian and upland habitats, but fire-return intervals were longer in riparian habitats that had more mesic forests than the surrounding uplands.

Disturbance from windthrow can be a major factor contributing to the falling of trees in Pacific Northwest forests (Hairston-Strang and Adams 1998). Although windthrow is often more common in upslope as compared to riparian habitats (e.g., Wimberly and Spies 2001), it can be pronounced in streamside buffers that are left after logging (e.g., Grizzel and Wolff 1998). The susceptibility of a riparian zone to wind disturbance is specific and dependent upon local topography, stream size, soil conditions, and forest structure and composition (Agee 1988). Conditions that increase the likelihood of blowdown in a riparian area include: (1) little topographic depression of the riparian area, (2) poorly drained soils, (3) orientation of the riparian zone across the direction of prevailing winds, and (4) presence of species prone to windthrow (i.e., western hemlock). Consequently, patterns of wind speed in riparian habitats appear to be highly variable and very site-specific (e.g., Brosofske et al. 1997, Ruel 2000). For example, in Douglas-fir and western hemlock forests of western Washington, Brosofske et al. (1997) observed that, before timber harvest, wind speed patterns varied greatly between riparian sites, reflecting topographic and vegetative differences between sites. Although wind speed generally increased after timber harvest, the high levels of intersite variation remained and were not directly related to buffer width.

Mammalian species richness, diversity, and evenness

When examined on a regional scale, overall mammalian species richness is comparatively high in riparian zones of the Pacific Northwest. Kauffman et al. (2001) estimate that 95 of the 147 mammal species of Oregon and Washington use riparian zones. On the local scale, evaluation of richness, diversity, and evenness of mammals in riparian zones has typically involved comparisons with adjacent upland areas. Such comparisons have largely been restricted to small mammals because of the difficulties in effectively sampling across all classes of mammals. In the Pacific Northwest, most species exhibit higher relative abundances in one habitat type or the other (Cross 1985, Anthony et al. 1987, Doyle 1990, McComb et al. 1993, Hallett and O'Connell 2000, West 2000a), but most species can occur in both habitats. Because a few species are restricted to riparian zones, species richness may be greater in these areas (Doyle 1990, West 2000a), although this is not always the case (McComb et al. 1993, Gomez and Anthony 1998, Hallett and O'Connell 2000). Within a region, species richness may vary spatially with local conditions and temporally with peaks in abundance. For example, in northeastern Washington with a potential pool of 15 species, mean species richness along third-order streams varied from 4.1 to 6.2 species at low abundance and from 7 to 9.5 species at high abundance (Hallett and O'Connell 2000).

Species diversity measures also have not yielded consistent differences in mammalian communities between riparian and upland areas. McComb et al. (1993) reported higher species diversity in riparian than upland habitats for mature Douglas-fir forests in western Oregon, whereas Hallett and O'Connell (2000) found no differences in mixed-coniferous, second-growth forests in northeastern Washington. Evenness or equitability may be low in upland areas that are dominated by a few abundant species, such as Trowbridge's shrew (*Sorex trowbridgii*), deer mouse (*Peromyscus maniculatus*), or southern red-backed vole (*Cleithrionomys gapperi*) (McComb et al. 1993), but may also be low in riparian areas because of the presence of a few rare species (Doyle 1990). Hallett and O'Connell (2000) observed a significant reduction in evenness in riparian zones when mammalian richness peaked with overall abundance.

Mammalian use of riparian zones

Naturalists have long recognized the high value of riparian habitats to some mammalian species. During the past two decades, quantitative studies have supported these observations and have identified biological and physical attributes of riparian habitats that enhance their value to these species (Brinson et al. 1981, Oakley et al. 1985, Kauffman et al. 2001). Wildlife, in turn, can impact the structure and function of riparian systems. Activities of some species, such as the beaver (*Castor canadensis*), can alter stream flow; selective foraging by beavers and ungulates on plant species can impact the composition of vegetative communities (Schreiner et al. 1996). Foraging by mammals on spawning salmon (*Oncorhynchus* spp.) can enhance nutrient transfer from aquatic to terrestrial systems (Bilby et al. 1996, Ben-David et al. 1998) and plant growth relative to upland areas (Naiman et al. 2000).

For the purposes of this chapter, it is helpful to recognize four types of mammalian species that inhabit riparian zones. First, *riparian obligate* species are those that require free water or riparian vegetation for some aspect of their natural history and must inhabit the riparian zone. They will be found exclusively in the riparian zone or decline dramatically in abundance with distance from it. Second, *riparian-associated* species are those that are statistically more abundant in riparian zones than upland areas or show positive correlations to riparian areas in their movements or location of home range or activity centers. These species may have higher physical condition or reproductive fitness in riparian areas. The third group of species consists of those associated with *early successional* stages. They have an interesting relation to riparian zones in that these areas almost always provide habitats to support them as the result of periodic disturbances characteristic of these zones. They may inhabit riparian zones embedded within old forest in small but persistent numbers. Should the forest be harvested, burned, or an episodic flood occur, the forest successional sequence will be initiated, and these species will colonize the area. Lastly, numerous *generalist* species may be found in riparian zones, but are not statistically more abundant there than in upland areas, nor are their movements or home ranges positively correlated with riparian areas. Sometimes *generalist* species are more abundant in upland areas than riparian areas. In addition, upland species are found less frequently than expected in riparian areas and appear to avoid these areas (McComb et al. 1993, Gomez and Anthony 1998). A species' association with riparian and upland habitats can vary regionally.

Riparian obligates

In the Pacific Northwest, several species of mammals from small insectivores to large ungulates are obligate inhabitants of riparian areas, and some species are adapted for locomotion in water. Their strong association with riparian areas and open water is consistent between west and east sides of the Cascade Mountains (Table 15.1). Among small mammals, water shrews (*Sorex palustris*) are semi-aquatic and are captured exclusively next to water (Bailey 1936, Conaway 1952, Verts and Carraway 1998) where they can be found in association with debris piles (Steel et al. 1999) and are known to feed upon salmon carcasses (Cederholm et al. 1989). Marsh shrews (*Sorex benderii*) are generally captured adjacent to running water with captures declining sharply beyond 25 m from streams in the Oregon Coast Range (McComb et al. 1993, Gomez and Anthony 1998). Beavers (Hill 1982) are a keystone species that have profound influence on the hydrology, vegetation, and nutrient cycling in areas they inhabit (see below).

Muskrats (*Ondatra zibethicus*) (Willner et al. 1980), water voles (*Microtus richardsoni*) (Bailey 1936, Anthony et al. 1987, Doyle 1990), and the exotic nutria (*Myocastor coypus*) (Verts and Carraway 1998) are also restricted to sites near water and usually are not found more than a few meters from permanent water. Water shrews, beaver, muskrat, and nutria are semi-aquatic and all are excellent swimmers and divers. Water shrews and water voles require moving water (Beneski and Stinson 1987), whereas muskrats use water that is lentic or slightly lotic (Perry 1982).

Many species of carnivores are associated with riparian habitats in the Pacific Northwest, but river otters (*Lontra canadensis*) and mink (*Mustela vison*) are most closely associated with open water in riparian areas. Riparian areas are very important to foraging otters because aquatic animals are their favored prey, and they hunt near undercut banks and logs or other debris in small streams and among log jams in deep, slow-moving pools (Melquist and Hornocker 1983). River otters are known to select stream-associated habitats but will use lakes, reservoirs, and ponds in winter and mud flats and associated open marshes, swamps, and backwater sloughs in summer months (Melquist and Hornocker 1983). Mink inhabit all types of wetlands such as riverbanks, streams, lakes, ditches, swamps, marshes, and backwater areas (Chapman and Feldhamer 1982). In the Yukon, the highest density of mink occurred in swampy habitats surrounding large bodies of water that supported large numbers of fish (Burns 1964). Mink are more generalist predators than otters and generally feed on aquatic animals such as muskrats, frogs, ducks, other birds, mice, insects, and fish (Errington 1943, Sealander 1943, Errington 1954, Wilson 1954, Korschgen 1958, Waller 1962, Erlinge 1969, Gerell 1970, Eberhardt 1973, Eagle and Whitman 1987). Both river otters and mink feed upon salmon carcasses (Cederholm et al. 1989) and are important links in the transfer of nutrients from aquatic to terrestrial systems.

Columbian white-tailed deer (*Odocoileus virginianus leucurus*) are found in western Washington and Oregon along the lower Columbia River and in southwestern Oregon along the Umpqua River where they are restricted to bottomland riparian areas. Gavin (1984) describes sightings until the 1940's of the species, and they all appeared to be within riparian zones of the Columbia River systems. Both historical information and recent research (Gavin et al. 1984, Smith 1985, Ricca 1999) indicate that the species is highly associated with riparian areas and may be restricted to these systems. Because its present range is limited to habitat along large rivers, its conservation might be more affected by agricultural practices than by forestry.

Table 15.1. Associations of mammals with riparian zones in coniferous forests of the Pacific Northwest

Species	Riparian obligate	Riparian associate	Early seral	Non-riparian associate	Reference
Didelphimorphia: Opossums					
Didelphidae: Opossums					
Virginia opossum (*Didelphis virginiana*)				■	Verts and Carraway (1998)
Insectivora: Insectivores					
Soricidae: Shrews					
Baird's shrew (*Sorex bairdi*)	W			W	Verts and Carraway (1998)
Marsh shrew (*Sorex bendirii*)			W		Gomez and Anthony (1998), West (2000a)
Masked shrew (*Sorex cinereus*)				E	West (2000a), Hallett and O'Connell (2000)
Pygmy shrew (*Sorex hoyi*)				E	Hallett and O'Connell (2000)
Montane shrew (*Sorex monticolus*)		W		E	West (2000a), Hallett and O'Connell (2000)
Pacific shrew (*Sorex pacificus*)		W			Verts and Carraway (1998)
Water shrew (*Sorex palustris*)	■				Anthony et al. (1987), Hallett and O'Connell (2000), Peffer (2001)
Fog shrew (*Sorex sonomae*)		W			Verts and Carraway (1998)
Trowbridge's shrew (*Sorex trowbridgii*)				■	West (2000a), O'Neil et al. (2001), Peffer (2001)
Vagrant shrew (*Sorex vagrans*)			E	W	Hallett and O'Connell (2000), West (2000b)
Talpidae: Moles					
Shrew mole (*Neurotrichus gibbsii*)		■			Gomez and Anthony (1998), Verts and Carraway (1998), Peffer (2001)
Coast mole (*Scapanus orarius*)				■	West (2000a)

(cont.)

Table 15.1. (cont.)

Species	Riparian obligate	Riparian associate	Early seral	Non-riparian associate	Reference
Chiroptera: Bats					
Vespertilionidae: Plainnose bats					
California myotis (Myotis californicus)		■			Hayes (2003)
Western small-footed myotis (Myotis ciliolabrum)		E			Hayes (2003)
Long-eared myotis (Myotis evotis)		■			Hayes (2003)
Little brown bat (Myotis lucifugus)		■			Hayes (2003)
Fringed myotis (Myotis thysanodes)		■			O'Farrell and Studier (1980)
Long-legged myotis (Myotis volans)		■			Hayes (2003)
Yuma myotis (Myotis yumanensis)		■			Hayes (2003)
Hoary bat (Lasiurus cinereus)		■			Hayes (2003)
Silver-haired bat (Lasionycteris noctivagans)		■			Hayes (2003)
Big brown bat (Eptesicus fuscus)		■			Hayes (2003)
Townsend's big-eared bat (Corynorhinus townsendii)		■			Hayes (2003)
Lagomorpha: Hares and rabbits					
Leporidae: Hares and rabbits					
Snowshoe hare (Lepus americanus)				■	Verts and Carraway (1998)
Mountain cottontail (Sylvilagus nuttallii)				■	Verts and Carraway (1998)
Rodentia: Rodents					
Aplodontidae: Mountain beaver					
Mountain beaver (Aplodontia rufa)				W	Verts and Carraway (1998)
Sciuridae: Squirrels and chipmunks					
Yellow-pine chipmunk (Tamias amoenus)				■	Verts and Carraway (1998), Hallett and O'Connell (2000)

(cont.)

Species			References
Red-tailed chipmunk (*Tamias ruficaudus*)		E	Hallett and O'Connell (2000)
Allen's chipmunk (*Tamias senex*)		E	Waters et al. (2001)
Siskiyou chipmunk (*Tamias siskiyou*)		W	Verts and Carraway (1998)
Townsend's chipmunk (*Tamias townsendii*)		■	Verts and Carraway (1998)
Columbian ground squirrel (*Spermophilus columbianus*)	W	■	O'Neil et al. (2001)
Golden-mantled ground squirrel (*Spermophilus lateralis*)		■	Verts and Carraway (1998)
Cascade golden-mantled ground squirrel (*Spermophilus saturatus*)		■	Verts and Carraway (1998)
Western gray squirrel (*Scurus griseus*)		■	Verts and Carraway (1998)
Douglas squirrel (*Tamiasciurus douglasii*)		■	Steele (1999), Verts and Carraway (1998)
Red squirrel (*Tamiasciurus hudsonicus*)		■	Verts and Carraway (1998)
Northern flying squirrel (*Glaucomys sabrinus*)			Waters and Zabel (1995), Rosenberg and Anthony (1992), Verts and Carraway (1998)
Geomyidae: Pocket gophers			
Western pocket gopher (*Thomomys mazama*)		■	Verts and Carraway (1998)
Northern pocket gopher (*Thomomys talpoides*)		■	Verts and Carraway (1998)
Castoridae: Beaver			
Beaver (*Castor canadensis*)	■		Jenkins (1979), Naiman et al. (1988)
Muridae: Mice and rats			
Deer mouse (*Peromyscus maniculatus*)		■	Anthony et al. (1987), West (2000a), Hallett and O'Connell (2000), Peffer (2001)
Forest mouse (*Peromyscus oreas*)	E	W	West (2000a), Peffer (2001)
Pinyon mouse (*Peromyscus truei*)	W	W	Verts and Carraway (1998)
Bushy-tailed woodrat (*Neotoma cinerea*)	W	E	O'Neil et al. (2001)
Dusky-footed woodrat (*Neotoma fuscipes*)			O'Neil et al. (2001)

Table 15.1. (*cont.*)

Species	Riparian obligate	Riparian associate	Early seral	Non-riparian associate	Reference
Western red-backed vole (*Clethrionomys californicus*)				W	Gomez and Anthony (1998), Anthony et al. (1987)
Southern red-backed vole (*Clethrionomys gapperi*)		E		W	West (2000a), Hallett and O'Connell (2000)
Western heather vole (*Phenacomys intermedius*)		W			Hallett and O'Connell (2000)
White-footed vole (*Arborimus albipes*)				■	Gomez and Anthony (1998)
Red tree vole (*Arborimus longicaudus*)				W	Verts and Carraway (1998), Meiselman and Doyle (1996)
Sonoma tree vole (*Arborimus pomo*)				W	Meiselman and Doyle (1996)
Long-tailed vole (*Microtus longicaudus*)		W		E	Gomez and Anthony (1998), West (2000a), Hallett and O'Connell (2000)
Montane vole (*Microtus montanus*)				E	Hallett and O'Connell (2000)
Meadow vole (*Microtus pennsylvanicus*)				E	Hallett and O'Connell (2000)
Oregon vole (*Microtus oregoni*)			■		McComb (1993), Gomez and Anthony (1998), Anthony et al. (1987)
Water vole (*Microtus richardsoni*)	■				Anthony et al. (1987), Hallett and O'Connell (2000)
Northern bog lemming (*Synaptomys borealis*)		■			Wilson et al. (1980), Hallett and O'Connell (2000)
Muskrat (*Ondatra zibethicus*)	■				Verts and Carraway (1998)
Dipodidae: Jumping mice					
Western jumping mouse (*Zapus princeps*)			■		Verts and Carraway (1998), Hallett and O'Connell (2000)
Pacific jumping mouse (*Zapus trinotatus*)			■		Gomez and Anthony (1998), West (2000a)

	C1	C2	C3	C4	Reference
Erethizontidae: Porcupines					
Porcupine (*Erethizon dorsatum*)				■	Verts and Carraway (1998)
Myocastoridae: Nutria					
Nutria (*Myocastor coypus*)	W				Verts and Carraway (1998)
Carnivora: Carnivores					
Ursidae: Bears					
Black bear (*Ursus americanus*)		■	■		Unsworth et al. (1989)
Grizzly bear (*Ursus arctos*)					LeFranc et al. (1987)
Procyonidae: Raccoons					
Raccoon (*Procyon lotor*)		■			Verts and Carraway (1998)
Ringtail (*Bassariscus astutus*)			W		Verts and Carraway (1998)
Mustelidae: Weasels					
River otter (*Lontra canadensis*)		■			Lavriviere and Walton (1999)
Marten (*Martes americana*)		■			Spencer et al. (1983), Buskirk and Powell (1994)
Fisher (*Martes pennanti*)		■	■		Jones and Garton (1994), Buck et al. (1994)
Mink (*Mustela vison*)			W	E	Verts and Carraway (1998)
Ermine (*Mustela erminea*)			W	■	Doyle (1990), Kelsey and West (1998)
Long-tailed weasel (*Mustela frenata*)				E	O'Neil et al. (2001)
Wolverine (*Gulo gulo*)				E	Verts and Carraway (1998)
Mephitidae: skunks					
Western spotted skunk (*Spilogale gracilis*)				■	Verts and Carraway (1998)
Canidae: Dog-like carnivores					
Red fox (*Vulpes vulpes*)				■	Verts and Carraway (1998)
Gray fox (*Urocyon cinereoargenteus*)			W		Fuller (1978)
Felidae: Cats					
Bobcat (*Lynx rufus*)				■	Verts and Carraway (1998)
Lynx (*Lynx canadensis*)				■	Verts and Carraway (1998)
Mountain lion (*Felis concolor*)				■	Verts and Carraway (1998)

(cont.)

Table 15.2. (cont.)

Species	Riparian obligate	Riparian associate	Early seral	Non-riparian associate	Reference
Artiodactyla: Even-toed ungulates					
Cervidae: Deer					
Rocky Mt. elk (*Cervus elaphus nelsoni*)				■	Verts and Carraway (1998)
Roosevelt elk (*Cervus elaphus roosevelti*)		W			Cole et al. (1997), Jenkins and Starkey (1984)
Black-tailed deer (*Odocoileus hemionus columbianus*)				■	Verts and Carraway (1998)
Mule deer (*Odocoileus hemionus hemionus*)				E	Verts and Carraway (1998)
Columbian white-tailed deer (*Odocoileus virginianus leucurus*)	W				Gavin (1984), Smith (1985), Ricca (1999)
White-tailed deer (*Odocoileus virginianus ochroura*)				E	Verts and Carraway (1998), Peek (1984), Dusek et al. (1989)
Moose (*Alces alces*)		■			O'Neil et al. (2001)
Woodland caribou (*Rangifer tarandus*)				E	Verts and Carraway (1998)

Habitat associations are designated by geographic location east (E) or west (W) of the Cascade Crest or both (■) for species that are not broadly distributed, and for species that have habitat associations that change geographically. Riparian obligate = species that require free water for some aspect of their natural history and must inhabit riparian zones. Riparian associate = species that are statistically more abundant in riparian zones, or show positive correlations to riparian habitats in their movements or location of home ranges or activity centers, but can also be found in upland habitats. Early seral = species that are usually more abundant in riparian zones because of frequent disturbance there and their association with early seral vegetative communities, especially herbaceous vegetation. Non-riparian associate = species that occur in riparian zones but are not statistically more abundant there than in upland habitats, or species that are more abundant in upland habitats.

Riparian associates

Many species of mammals occur in both upland and riparian habitats in northwestern forests, but some are significantly more abundant in riparian areas, their movements are highly associated with riparian areas, or they have higher physical condition or reproductive fitness in these areas. The degree of association for some species varies spatially between the east and west slopes of the Cascade Mountains (Table 15.1). This may be due to the more xeric conditions east of the Cascade Crest that create a more defined riparian zone. The degree of association also appears to vary among different forest types from early- to late-seral stages (McComb et al. 1993, Gomez and Anthony 1998). Consequently, it is challenging to designate some species as being associated with riparian zones or not.

Of the insectivores, Pacific shrews (*Sorex pacificus*) (Anthony et al. 1987), montane shrews (*Sorex monticolus*) (Doyle 1990), and shrew moles (*Neurotrichus gibbsii*) (Terry 1981, Cross 1985, Gomez and Anthony 1998) appear to be strongly associated with riparian zones in Oregon and Washington (Table 15.1) because their relative abundance is greater in riparian areas or the number of captures declined significantly along transriparian transects. Pygmy shrews (*Sorex hoyi*) are considered riparian species by some authors (Long 1974, Stinson and Gilbert 1985), but additional work has not supported this (Hallett and O'Connell 2000). Coast moles (*Scapanus orarius*) have been found to be more abundant in riparian than upland habitats in Washington (Hartman and Yates 1985). However, McComb et al. (1993) and Gomez and Anthony (1998) found coast moles to be equally abundant in riparian and upland habitats in five forest types in the Coast Range of western Oregon. Similarly, Doyle (1990) reported a higher abundance of Trowbridge's shrews (*Sorex trowbridgii*) in riparian zones in the Oregon Cascade Mountains, but this species appears to be equally abundant in riparian and upland habitats in the Oregon Coast Range (McComb et al. 1993, Gomez and Anthony 1998) and west of the Cascade Crest in Washington (West 2000a). Insectivorous shrews and moles that have general habitat associations probably benefit from the abundance of streamside insects and other invertebrates in the moist soils of riparian zones.

The home ranges of bats encompass day and night roosts, foraging areas, and water sources. Riparian areas within forests provide important foraging, drinking, and roosting sites for many species (Cross 1988, Hayes 2003), and all species of bats in the Pacific Northwest have been caught or observed while drinking or foraging over water (Table 15.1). Marcot (1996) considered riparian habitat a key environmental correlate for ten bat species in the Interior Columbia Basin. Most bats feed primarily on flying insects, and aquatic insects are frequently a major component of the diet (Whitaker 1972, Belwood and Fenton 1976, Whitaker et al. 1977, Fenton and Bell 1979, Herd and Fenton 1983). The importance of riparian habitats as foraging areas and sources of free water has been documented in numerous studies (e.g., Lunde and Harestad 1986, Thomas and West 1991, Brigham et al. 1992, Hayes and Adam 1996, Parker et al. 1996, O'Connell and Hallett 2000, Seidman and Zabel 2001). Feeding rates of eight *Myotis* spp. in the Washington Cascade and Oregon Coast Ranges were ten times higher over water than in forested stands (Thomas and West 1991). A comparison of bat activity in riparian, old-growth, clearcuts, and second growth forests of southeast Alaska revealed that riparian habitat had the most bat activity (Parker et al. 1996). The association of roost sites with riparian areas was more variable. At least two species of *Myotis* bats are known to roost near water when suitable roosts are present (Barbour and Davis 1969). Although day roosts of fringed myotis (*Myotis thysanodes*) are not associated with distance to perennial water, roost location is positively associated with distance to the nearest stream channel (Weller and Zabel 2001). In contrast, roost sites of at least some species of bats (e.g., silver-haired bats (*Lasionycterus noctivagans*)) in the Pacific Northwest tend to be removed from riparian habitats, perhaps due to microclimatic characteristics (Betts 1996, Campbell et al. 1996, Frazier 1997, Ormsbee and McComb 1998).

Several species of rodents are associated with riparian zones, and many of these species are locally or broadly endemic to the Pacific Northwest. Bushy-tailed (*Neotoma cinerea*) and dusky-footed woodrats (*N. fuscipes*) are associated with riparian habitats in northern California and parts of southern Oregon (Brown 1985); however, this association is less clear in the northern portion of their geographic ranges where conditions are moister and cooler. Therefore, their association with riparian areas appears to vary across their geographic ranges. This is also the case with species in the genus *Clethrionomys*. Western red-backed voles (*Clethrionomys californicus*) are endemic to western Oregon and extreme northwestern California and are significantly more abundant in upland habitats than riparian areas in the Oregon Coast Range (McComb et al. 1993, Gomez and Anthony 1998) and Cascade Mountains of Oregon (Doyle 1990). Southern red-backed voles (*C. gapperi*) are broadly distributed across northern North America and are not associated with riparian areas west of the crest of the Cascade Mountains in Washington (West 2000*a*). However, they are associated with riparian areas east of the crest of the Cascades where conditions are hotter and drier (Hallett and O'Connell 2000). Long-tailed voles (*Microtus longicaudus*) occupy a variety of moist habitats in the Pacific Northwest, including forests, shrubs, and marshes (Smollen and Keller 1987, Verts and Carraway 1998). Throughout most of their geographic range they are not particularly associated with riparian areas; however, in the Oregon Coast Range they were found to be significantly more abundant in riparian areas than upland habitats (McComb et al. 1993, Gomez and Anthony 1998) and are considered riparian obligates in this region (Kelsey and West 1998). Similarly, Townsend's voles (*Microtus townsendii*) (McComb et al. 1993, Gomez and Anthony 1998) and white-footed voles (*Arborimus albipes*) (Gomez and Anthony 1998) are associated with riparian zones in the Oregon Coast Range. Both species are locally endemic to the Pacific Northwest. Goertz (1964), Maser et al. (1981), and Suzuki (1992) also found Townsend's voles to occupy moist open areas adjacent to streams or springs within forests in western Oregon. Northern bog lemmings (*Synaptomys borealis*) also are associated with riparian areas, particularly marshes and bogs in northern latitudes. Few species of sciurid rodents are associated with riparian areas, which is likely due to their heavy reliance on nuts and seeds for food. An exception to this pattern are Allen's chipmunks (*Tamias senex*), which are more abundant in riparian than upslope areas in northwestern California (Waters et al. 2001).

The association of generalist carnivores such as bears (*Ursus* spp.), raccoons (*Procyon lotor*), ringtails (*Bassariscus astutus*), ermine (*Mustela erminea*), and marten (*Martes americana*) with riparian areas often varies seasonally and in response to changing prey availability. For example, many of these species feed upon salmon carcasses (Cederholm et al. 1989) and are attracted to riparian areas when salmon are available. Black bears (*U. americanus*) generally remain in close proximity to water, feeding and resting in areas <100 m from water during spring, summer, and fall (Unsworth et al. 1989). Riparian areas are used by black bears for foraging in both California and Idaho (Kellyhouse 1980, Young and Beecham 1986). In Idaho, the primary sources of food for black bears in the summer and fall are huckleberries (*Vaccinium* spp.), bitter cherry (*Prunus emarginatus*), and chokecherry (*Prunus virginianus*), which are abundant in riparian zones and mesic aspen stands during these seasons (Unsworth et al. 1989). Grizzly bears (*U. arctos*) will use riparian areas for foraging when runs of salmon are active in the summer and fall or when plants are fruiting, but they are not dependent on these zones for feeding in all seasons (Craighead et al. 1982, LeFranc et al. 1987). Raccoons are strongly aquatic, although less so than river otters and mink. They are strong swimmers and spend most of their life near streams, lakes, or marshes, but are scarce in dry, upland areas (Kaufmann 1982). Grinnell et al. (1937) indicated that ringtails never occurred >0.8 km from water, but more detailed studies of this species are needed to further assess its association with riparian areas. Ermine are more common in riparian than upslope habitats especially during drier months when prey densities might be higher in riparian areas (Doyle 1990, Kelsey and West 1998). Throughout the range of the species, marten are associated with riparian areas. In the northern Sierra Nevada, marten selected riparian areas in lodgepole pine (*Pinus contorta*) forests over upland forest for feeding (Spencer et al. 1983). Marten also are attracted to riparian areas in the Tahoe area of California (Simon 1980, Zielinski 1981) and in the Cascade Mountains of western Washington (Raphael and Jones 1997). Prey are more abundant in riparian zones, and the abundance of CWD in riparian habitats makes prey more available to marten, especially in winter when stumps and large logs provide access to prey living under the snow (Buskirk et al. 1989). In California, gray fox (*Urocyon cinereoargenteus*) select riparian areas when available (Fuller 1978).

Four species of ungulates in the Pacific Northwest are associated with riparian areas because of the free water and forage these habitats provide. However, their dependence on these habitats varies seasonally and with availability of other habitats on the landscape. Selection of riparian habitats has been demonstrated for white-tailed deer (Compton et al. 1988, Dusek et al. 1989) and Rocky Mountain elk (*Cervus elaphus nelsoni*) (Marcum 1976, McCorquodale et al. 1986) east of the Cascade Mountains, Roosevelt elk (*Cervus elaphus roosevelti*) west of the Cascades (Jenkins and Starkey 1984, Cole et al. 1997), and moose (*Alces alces*) in Alaska (LeResche et al. 1974, Coady 1982). The dependence of these ungulates on riparian habitat can vary seasonally. For example, Marcum (1976) found that Rocky Mountain elk on their summer range most frequently selected areas within 320 m of water in dry forests of western Montana, whereas areas >320 m from water were not selected. In the arid shrub-steppe of eastern Washington, natural springs were especially important to lactating female Rocky Mountain elk (McCorquodale et al. 1986). West of the

Cascades, Roosevelt elk cows selected areas less than 200 m from riparian areas especially during the hot dry summer months (Cole et al. 1997).

Early seral species

Riparian zones typically have vegetation that is adapted to high disturbance produced by frequent episodes of flooding, scouring, and deposition of sediment. Such vegetation is usually comprised of grasses, forbs, and deciduous shrubs and trees. In many cases, insectivores and rodents are closely associated with these patches of early successional vegetation in riparian zones. Grassy areas are especially important for vagrant shrews (*Sorex vagrans*), masked shrews, Pacific jumping mice (*Zapus trinotatus*), Oregon voles (*Microtus oregoni*), and meadow voles (*Microtus pennsylvanicus*). Therefore, these species may be found in either riparian areas or upland areas if the conditions are optimal. For example, in western Oregon, Pacific jumping mice are most frequently trapped in riparian zones (McComb et al. 1993, Gomez and Anthony 1998). This species increases in abundance in upland areas following timber harvest and establishment of grasses, forbs, and shrubs (Morrison and Anthony 1989, Gomez and Anthony 1998). Therefore, this species is not dependent on riparian areas *per se*, but is associated with herbaceous vegetation favored in these areas as a result of the high disturbance regime. We believe that a similar relation exists with other rodents and insectivores, including Oregon voles (Carraway and Verts 1985, Cross 1985, Gomez and Anthony 1998), vagrant shrews (Morrison and Anthony 1989, Gomez and Anthony 1998), and masked shrews (Larrison 1976) in western Oregon and Washington, and meadow voles (Getz 1970, Snyder and Best 1988), western jumping mouse (*Zapus princeps*) (Stinson and Gilbert 1985), and western harvest mice (*Reithrodontomys megalotis*) (Stinson and Gilbert 1985) east of the Cascade Mountains in Washington. Although these species are not dependent on riparian zones, they are highly associated with early seral vegetation in riparian areas.

Habitat generalists

Several species of small mammals inhabit riparian areas and are major components of riparian mammal communities, yet they are not dependent on or necessarily more abundant in these areas. For example, deer mice (*Peromyscus maniculatus*) use forested riparian areas, but no strong association has been demonstrated for these areas except in the Cascades

Mountains of Oregon (Doyle 1990) and northwestern California (Waters et al. 2001). Deer mice occupy a wide variety of habitats, including forested uplands, shrub-steppe habitats, and grasslands. Although usually not more abundant in riparian than in upland habitats, deer mice are often a numerically dominant rodent on riparian sites (Anthony et al. 1987, Cross 1988, Doyle 1990). Similarly, Trowbridge's shrews were found to be the most abundant small mammal in riparian areas of the Oregon Coast Range (McComb et al. 1993, Gomez and Anthony 1998), yet were equally abundant in upland and riparian habitats. Other examples of such species can be found throughout the Pacific Northwest. Given their abundance, these species are important components of, and comprise significant biomass in, riparian systems.

Mammalian impacts on riparian system structure and function

Wildlife, especially mammals (Kauffman et al. 2001), can impact the structure and function of riparian systems. Through activities such as predation, herbivory, burrowing, trampling, and stream channel modification, mammals can enhance nutrient transfer and either increase or degrade habitat complexity (e.g., Kauffman et al. 2001).

Many mammalian species will feed on salmon carcasses (*Oncorhynchus* spp.). Cederholm et al. (1989) examined use of 945 coho salmon carcasses in western Washington streams and observed that 15 species of insectivores, rodents, and carnivores consumed the salmon. Willson and Halupka (1995) further suggest that deer also feed upon salmon carcasses and point out that mammalian predation on salmon includes consumption of eggs and juvenile salmon. Foraging by mammals on spawning salmon can enhance nutrient transfer from aquatic to terrestrial systems (e.g., Bilby et al. 1996, Ben-David et al. 1998). For example, Ben-David et al. (1998) found that when predators feed upon salmon at sites removed from the streams, upland plants near these sites have nitrogen isotope levels similar to those in plants found at the stream edge. The increased nitrogen availability from salmon carcasses enhances plant growth (Naiman et al. 2000).

The browsing activities of ungulates can impact the composition of riparian vegetative communities. For example, Schreiner et al. (1996) demonstrated that the selective foraging of deer and elk on hardwood species and certain conifers resulted in the dominance of other conifers.

The subsequent increase in conifer litter corresponded to a decrease in the availability of soil nitrogen (Schreiner et al. 1996). Riparian plants have, in turn, evolved complex interactions to reduce herbivory. In northern California, orchids (*Epipactis gigantea*) are protected from deer browsing by growing in association with sedges (*Carex nudata*) (Levine 2000*a*). Land use practices and population management can contribute to relatively high densities of native ungulates. When populations increase, ungulate browsing can alter the abundance and population structure of woody and herbaceous plants, eliminate especially palatable plants, and reduce rates of recovery in riparian restoration (e.g., Hanley and Taber 1980, Alverson et al. 1988, Case and Kauffman 1997, Opperman and Merenlender 2000, Liang and Seagle 2002). Additionally, heavy use of riparian areas by native ungulates can impact soil and stream channel structure (Kauffman et al. 2001).

Beaver, a keystone species

Although many mammal species impact riparian systems, the effects of beaver can be so fundamental to riparian system structure and function, that this species is considered a keystone species in riparian habitats. Beavers can have an enormous impact on stream and riparian ecosystems by cutting down trees and building dams. Such activities change the physical characteristics of the stream environment thereby impacting the successional dynamics and species composition of riparian flora (Naiman et al. 1988, Johnston and Naiman 1990). The impoundment of water by dams increases sediment loads and retention of organic material within the channel and alters biogeochemical cycles (Naiman et al. 1988, Pollock et al. 1994). Beaver dams create and maintain floristic diversity that enhance habitat for species more typical of wetland environments and modify the composition and productivity of faunal assemblages (Hodkinson 1975, Pollock et al. 1998). The overall result of beaver activities is to increase the spatial and temporal variation in riparian communities (Suzuki 1992, Pollock et al. 1998).

Because beaver populations were quite low for most of the twentieth century, most scientific studies of streams have been done in the absence of the species and have created an image of natural stream systems that is quite different from what existed historically. Historical records refer to many small, low gradient streams and rivers as a series of beaver impoundments characterized by slow water and extensive riparian vegetation (Rudemann and Schoonmaker 1938, Rea 1983, Morgan 1986).

Currently, such streams are rare except in places where beaver are protected. This is the exception, however, and along many streams trapping of beaver continues into the present. Consequently our understanding of how natural low gradient streams function is based on streams that have been fundamentally altered by the removal of beaver and their dams.

Effects on plant communities

Beaver alter riparian plant community composition primarily through two mechanisms: the direct removal of trees and shrubs and indirectly by flooding riparian areas, thereby shifting these communities to more flood-tolerant species. Beaver are known to eat a wide variety of plant species, including species in the genera *Alnus, Nuphar, Betula, Acer, Cornus, Corylus, Sorbus, Thuja,* and *Pinus,* but generally prefer species of *Populus* and *Salix* if they are available (Hall 1960, Nixon and Ely 1968, Aleksiuk 1970, Jenkins 1975, Belovsky 1984, Roberts and Arner 1984). Beaver generally prefer deciduous species to coniferous species and this can result in an acceleration of forest succession in places such as the boreal forest and the western coniferous forests where deciduous trees are generally an early-seral species (Johnston and Naiman 1990, Johnston et al. 1993). Vegetation of beaver-influenced riparian areas can be much more diverse than that of unaffected riparian areas. This is due to the highly variable physical environment over small spatial scales that creates a number of microhabitats to which different plant species are best adapted (Leidholt-Bruner 1990). Such variation includes variation in flood frequency and duration, available light, overhead canopy, and microtopography (Pollock 1995).

Effects on birds

Beaver activity creates highly diverse avian communities and increases the productivity of certain species (Reese and Hair 1976, Dieffenbach and Owen 1989, Grover and Baldassarre 1995, Brown et al. 1996, McCall et al. 1996). Avian communities around beaver ponds can be unique, because the habitat characteristics of these areas are quite different from those of other wetlands due to high numbers of snags, a complex shoreline, and a very complex physical environment. Birds known to benefit from beaver ponds include numerous species of waterfowl such as green-winged teals (*Anas crecca*), mallards (*Anas platyrhynchos*), ring-necked ducks (*Aythya collaris*), Canada geese (*Branta canadensis*), hooded mergansers (*Lophodytes cucullatus*), and American black ducks (*Anas rubripes*), as well as other species such as red-winged blackbirds (*Agelaius phoeniceus*) and Brewer's blackbirds (*Euphagus cyanocephalus*) (Dieffenbach and Owen 1989,

Merendino et al. 1995, Brown et al. 1996, McCall et al. 1996, Edwards and Otis 1999).

Effects on mammals

Studies on the effects of beaver-created habitat on other mammals suggest that, relative to other riparian areas, the diversity of small-mammal species does not increase, and that, for some species, beaver habitat can be very productive. For example, Medin and Clary (1989) found that the small-mammal biomass of beaver ponds was 819 g ha^{-1}, whereas riparian areas unaffected by beaver had an average biomass of just 304 g ha^{-1}. They also found that riparian areas influenced by beaver activity were dominated by herbivores (*Microtus* spp.) and insectivores (*Sorex* spp.), whereas the control riparian areas were dominated by omnivores (*Peromyscus maniculatus*) and herbivores (*Microtus montanus*). Although small-mammal diversity was not increased by beaver activity, abundance of *Microtus* species was higher in beaver-influenced areas of the Oregon Coast Range (Suzuki 1992).

Other mammal species utilize the lodges of beaver. Muskrats may use both active and abandoned lodges, and bobcats (*Lynx rufus*) have been observed using lodges as dens (Lovallo et al. 1993, McKinstry et al. 1997). Additionally, mink and river otter frequently utilize beaver ponds (Newman and Griffin 1994). Finally, beaver provide an important food source for several predators, most notably the gray wolf, but other species such as the black and grizzly bears have been known to eat beaver and destroy their lodges (Shelton and Peterson 1983, Potvin et al. 1992, Smith et al. 1994). The impact of wolf predation on beaver populations can be significant. In one area, beaver populations decreased by 30% following cessation of wolf control measures (Potvin et al. 1992). A study of two Lake Superior islands, one with black bear and one without, suggested that black bear predation was significant enough to cause a decline in the beaver population (Smith et al. 1994). The presence of bear also altered beaver tree selection. Where bear were present, beaver cut few trees >30 m from the water, whereas on the island where bear were absent beaver traveled as far as 200 m from water to forage.

Effects on amphibians

Available research suggests that beavers have a species-specific effect on the distribution and abundance of amphibians. The abundance and diversity of nine amphibian species were not different between stream reaches occupied by beaver and unoccupied reaches in the Oregon Coast Range

(Suzuki 1992). However, rough-skinned newts (*Taricha granulosa*), a pond-breeding amphibian, were more abundant at beaver ponds, whereas tailed frogs (*Ascaphus truei*), a species associated with cold, fast-flowing streams, were less abundant at beaver ponds (Suzuki 1992). In the Piedmont of South Carolina, amphibian richness, abundance, and diversity did not differ among new (<5 years) beaver ponds, old (>5 years) beaver ponds, and unimpounded streams, but several species of anurans were captured predominantly or exclusively at beaver ponds (Russell et al. 1999).

Influence on species composition and abundance of fish
In the Pacific Northwest, the importance of beaver ponds to certain fish species is well established (Bustard and Narver 1975, Peterson 1982*a*, Murphy et al. 1989, Leidholt-Bruner et al. 1992). Fish species that use beaver ponds include coho salmon (*Oncorhynchus kisutch*), chinook salmon (*O. tshawytscha*), sockeye salmon (*O. nerka*), cutthroat trout (*O. clarki*), Dolly Varden char (*Salvelinus malma*), three-spine stickleback (*Gasterosteus aculeatus*), prickly sculpin (*Cottus asper*), western speckled dace (*Rhinichthys osculus*), and the introduced eastern brook trout (*Salvelinus fontinalis*). Less frequently observed species include steelhead trout (*O. mykiss*) and torrent sculpin (*C. rhotheus*) (Peterson 1982*b*).

The importance of beaver ponds to the survival of coho salmon has been well established. Juvenile coho utilize beaver ponds both for summer rearing and overwintering habitat (Bustard and Narver 1975, Peterson 1982*a*, Murphy et al. 1989, Leidholt-Bruner et al. 1992). Beaver ponds not only contain high densities of coho, but coho in beaver ponds also have overwintering survival rates as much as twice that of their instream counterparts (Bustard and Narver 1975, Murphy et al. 1989). Additionally, coho emigrating from beaver ponds are on average larger and have higher growth rates than those emigrating from streams (Swales and Levings 1989). Because size is an important determinant of survival to maturity (Bilton et al. 1982), these studies suggest that coho raised in beaver ponds may constitute a disproportionate percentage of returning adults.

Juvenile chinook salmon also utilize beaver ponds, although not nearly to the same extent as juvenile coho. Swales and Levings (1989) found that overwintering densities of juvenile chinook of the Coldwater River in British Columbia, Canada were highly variable in beaver ponds, ranging from 0.007–0.13/m². In comparison coho ranged from 0.008–0.368/m². Additionally, in the Snake River basin of Idaho at least some juvenile chinook utilized beaver ponds as summer-rearing habitat, and the complex cover created by the dam itself contained high densities of juvenile chinook on the downstream side (M. Pollock, pers. obs.). However data from the Taku River, Alaska indicated that beaver ponds were the habitat type least used by juvenile chinook salmon in summer. These data suggest that juvenile chinook probably used beaver ponds as overwintering habitat to a limited extent, but they were not selected habitats.

Effects of human disturbance in riparian zones

Riparian systems are products of natural disturbance, but they also can be especially susceptible to human disturbance because: (1) humans are attracted to and concentrate their activities in riparian areas, (2) riparian areas generally constitute a relatively smaller amount of area than upland areas, (3) disturbances in upland areas or in headwater streams may be transferred to riparian systems downstream, and (4) riparian habitats support flora that are often susceptible to disturbance (Oakley et al. 1985). Human impacts on riparian habitats vary in space and time and include timber harvesting, livestock grazing, road building, impoundments, channelization, introduction of toxic compounds or sewage, and recreational activities (Brinson et al. 1981, Hall 1988). Such human-induced disturbances differ from natural disturbances in the distribution and abundance of biological legacies that remain after the disturbance (Franklin et al. 2000). Additionally, these activities not only alter current habitat conditions, but also impact natural disturbance regime trajectories (Hessburg et al. 1999). Because of the scope of this book, we will focus primarily on the effects of timber harvest and the role of riparian buffers in maintaining mammalian diversity, but we also discuss road construction and livestock grazing. Many of the effects of timber harvest on specific taxa of mammals are addressed in other chapters (see chapters by Buskirk and Zielinski 2003, Hallett et al. 2003, Hayes 2003, Kie et al. 2003, Smith et al. 2003), so we will not discuss these effects here. However, we focus on the effects of establishing riparian buffer strips.

Use of riparian buffer strips following timber harvest

The impact of timber harvest in riparian and adjacent upland habitats varies from little to substantial with the type of harvest and characteristics of the watershed. Various studies have identified several major stream or habitat changes associated with timber harvest (e.g., Harr et al. 1979, Swanson 1980). Water temperatures increase after tree harvesting due to the reduction of shading. Increased sedimentation often results from logging because logging activities increase input of soil and detritus into streams. Stream flow, especially in smaller streams, can increase significantly after timber harvests. Most important to mammalian species, timber harvest in riparian areas alters the composition and structure of both the overstory and understory plant communities, which alters the dynamics of the food chain and the primary source of CWD in streams. Maintenance of vegetative buffer zones on both sides of streams and retention of CWD in streams can decrease many of these negative impacts (Franklin et al. 1981). The effects of timber harvest on mammal communities depend on a variety of factors including the original plant community; the type, size, and timing of harvest; and onsite treatment of slash and snags. Responses to timber harvest are species specific; populations of some species will increase following clear-cutting or forest-management practices that set back forest succession, whereas populations of others will decrease.

In forest ecosystems, riparian buffer strips or riparian management zones are areas adjacent to streams where trees are retained following timber harvest (Fig. 15.1). Riparian buffers may be no entry areas or selective logging may be permitted depending on the jurisdiction. Although much of the initial impetus for prescribing buffer strips arose from concerns over water quality and fish stocks, recent interest in the benefits that riparian buffers provide to wildlife has led to consideration of their effectiveness in providing sustainable habitat or as possible travel corridors.

In managed forests, riparian buffers can serve two distinct roles as habitat (O'Connell et al. 2000). Historically, when the prevailing successional stage was older, closed-canopy forest, riparian habitat provided refugia for species characteristic of early-successional stages. Aside from the presence of water, the unique features of riparian habitat center on the mixing of early successional characteristics within old forests. The presence of such areas was especially important for the continued existence of species with limited mobility. For example, herbivorous small mammals that survived at low population densities in such areas and could rapidly colonize large areas after forest disturbance needed the small openings supporting grasses and herbs. With the creation of riparian buffers in managed forests, a second function envisioned for riparian buffers was to maintain elements of mature forest in a predominantly young forest landscape.

The benefit provided to wildlife, particularly mammals, depends on the width of the buffer, the amount of tree canopy left, and the specific habitat requirements of the species. To be useful as refugia for mammals,

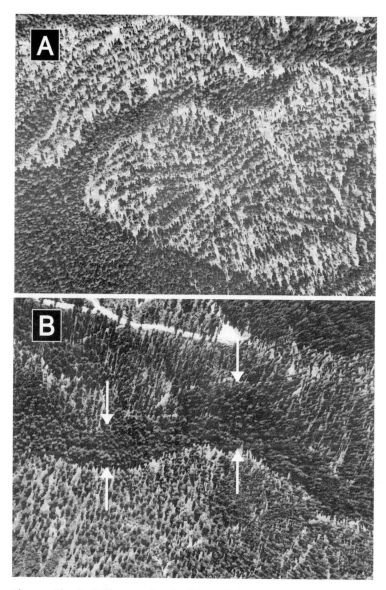

Fig. 15.1. Riparian buffers created under different harvest prescriptions in northeastern Washington. Upland harvest was a selective cut with 6–12 m spacing. (A) Buffer created under state (1991) guidelines for riparian management zones. (B) A modified prescription that included buffers of snags and seeps. The modified prescription resulted in a buffer that is wider on average, but more variable (indicated by arrows).

riparian buffers need to be large enough and to retain sufficient trees and shrubs to allow riparian-associated species to persist until the tree canopy is re-established on adjacent uplands. Riparian buffers associated with stream systems are also habitat patches that form part of a broader network across the landscape. Although the vegetation and environmental characteristics of riparian buffers vary spatially (e.g., with stream size and elevation), these differences may be negligible compared to the contrast between the riparian and adjacent harvested upland habitats (Fig. 15.1). Harris (1984) has argued that riparian buffer strips should provide linkages between fragmented forest stands because of their biotic diversity. Of primary concern is whether mammals use riparian buffers and, if so, what factors affect their use. Research conducted in the forests of the Pacific Northwest (Cross 1985, O'Connell et al. 2000) and other regions (Dickson and Williamson 1988, Darveau et al. 2001, Forsey and Baggs 2001) provide evidence that a variety of terrestrial small mammals, bats, carnivores, and ungulates will utilize riparian buffers.

Given the relatively low mobility of terrestrial small mammals, it is not surprising that buffers can provide habitat for some of these species. Cross (1985) trapped Trowbridge's shrews, Pacific shrews, vagrant shrews, deer mice, bushy-tailed woodrats, Oregon voles, and jumping mice in riparian leave strips that ranged 9–67 m wide in southern Oregon. In an experimental study conducted in western and eastern Washington State, O'Connell et al. (2000) compared the abundance of terrestrial wildlife between intact forest (i.e., controls) and areas logged by two different riparian buffer prescriptions (Fig. 15.1A,B). In both western Washington (West 2000a) and eastern Washington (Hallett and O'Connell 2000), small-mammal species associated with riparian habitats on control sites and prior to logging (Table 15.1) were found in the buffer strips created after logging. Creation of forested buffer strips in a logged landscape can, however, result in the incursion or increase of non-riparian colonizers or generalist species into riparian habitats. For example, Townsend's voles were found in riparian habitats only after logging on western Washington sites (West 2000a). Similar increases of open-habitat species in riparian buffers have been observed in recently logged forests of other regions (Dickson and Williamson 1988, Darveau et al. 2001). In other cases, species such as red squirrels (*Tamiasciurus hudsonicus*) that are associated with the forest interior and not riparian habitats might shift activity to forested riparian buffers following logging of the upland interior forest (e.g., Forsey and Baggs 2001). Studies designed to specifically examine the effects of riparian buffer width on small mammals have concluded that wider buffers support higher overall abundance of small mammals, but responses are species specific and reflect complex interactions between species (e.g., Dickson and Williamson 1988, Darveau et al. 2001). For example, in boreal forests of Quebec, meadow voles increased in riparian buffers after clear-cut harvest adjacent to these areas. This in turn resulted in an increase of southern red-backed voles in 20-m riparian buffers, although this species is not associated with streams in intact forests (Darveau et al. 2001). O'Connell et al. (2000) compared buffers prescribed by Washington State guidelines with site-specific buffers designed to incorporate important habitat features (e.g., snags and seeps) into the buffers. State-prescribed buffers were uniform in width (Fig. 15.1A), whereas site-specific buffers tended to be irregular in shape and established greater linkages with upland habitats (Fig. 15.1B). Upland areas of the western Washington sites were clear-cut, whereas those of the eastern Washington sites were selectively logged. In western Washington, overall small-mammal richness and abundance did not differ statistically between buffer treatments, but post-logging declines of individual species were less pronounced on the site-specific buffers (West 2000a). Small-mammal abundance was higher on sites logged with site-specific buffers than with state-prescribed buffers in eastern Washington (Hallett and O'Connell 2000). The ability of riparian buffers to provide linkages will be of potential importance for the dispersal of small mammals across the landscape.

Genetic approaches may provide some insight into movement along riparian buffers for small mammals. For example, Mech and Hallett (2001) used microsatellite markers to show that forest strips between two closed-canopy patches provided a route of gene flow for southern red-backed voles (closed-canopy specialists), whereas deer mice (habitat generalists) showed no genetic differentiation. This work suggests that the effectiveness of landscape linkages depends on the degree of specialization for that habitat type. Aars et al. (1998), however, reported reduced short-term gene flow (mitochondrial DNA) for female bank voles (*Clethrionomys glareolus*) along river banks. They suggested that territorial behavior of resident females might reduce female dispersal in these linear habitats. The ability of females to establish territories within the riparian zone will be limited by the amount of habitat available.

Bat activity, especially of the larger-bodied silver-haired bats and hoary bats (*Lasiurus cinereus*) increased following clear-cut logging in coastal Pacific Northwest forests (e.g., Erickson and West 1996, Hayes and Adam

1996), and this has been attributed to increased maneuverability in open landscapes. Few studies have examined bat activity in riparian buffers. In an experimental study, activity of *Myotis* spp. increased in riparian buffers following logging of the uplands (O'Connell et al. 2000). West (2000*b*) suggested that the increased activity of bats observed on western Washington sites was associated with the use of edges for foraging by *Myotis* spp. In eastern Washington, O'Connell and Hallett (2000) observed a significant increase in the activity of *Myotis* spp. on the sites harvested by the site-specific riparian buffer prescription as compared to either control sites or state-prescribed buffers. The combination of increased edges and protection of snags on the site-specific buffers might explain this increase.

For more mobile carnivores and ungulates that use riparian habitats as travel corridors, an important role of forested buffers might be to maintain those travel routes. Raedeke et al. (1988) suggested that large mammals use riparian corridors when they provide the only available cover or the easiest route of travel in a watershed. Riparian areas serve as natural travel corridors for ungulates because of their shape, extension from high to low elevations, and their habitat characteristics. Use of riparian habitats as travel routes has been observed for river otters (Melquist and Hornocker 1983), raccoons (Sherfy and Chapman 1980, Riley 1989) black bears in California (Kellyhouse 1980), bobcats (Young 1958), elk in Washington (Taber 1976), and moose (Mastenbrook and Cummings 1989). However, response of carnivores and ungulates to the establishment of riparian buffers has not been widely studied in the Pacific Northwest. Columbian black-tailed deer (*Odocoileus hemionus columbianus*) in northern California utilized timbered strips within clear-cuts as escape routes to more dense cover (Loft et al. 1984). Winter tracking studies of marten and red fox (*Vulpes vulpes*) in Newfoundland revealed greater use of forested buffers compared to recently clear-cut forest (Forsey and Baggs 2001). Riparian buffers need not be large to be used by marten for foraging and travel corridors. Small, scattered old-growth stands may be sufficient for marten if located adjacent to riparian areas (Spencer 1981).

In summary, research indicates that many mammals will use riparian buffers, but responses are variable and species specific. Brinson and VerHoeven (1999) suggest four general principles about the effectiveness of forested riparian buffers: (1) wider is better in all cases, (2) the more intensive the upland activity, the wider the buffer should be, (3) the importance of buffers around open water (i.e., streams, ponds) is irrespective of geomorphology of surrounding area, and (4) buffers are critical for protection of headwater streams. O'Connell et al. (2000) further suggest that the effectiveness of forested riparian buffers is enhanced if buffer design is site specific and linkages between riparian and upland habitats are established or maintained.

Livestock grazing

There has been an almost continuous increase in the prevalence of livestock grazing in the Pacific Northwest since cattle were first brought to the region in 1789 (McIntosh et al. 2000). By the early 1900's, livestock grazing was common throughout the region, and current estimates of grazing stand at 4.2×10^6 Animal Unit Months (AUMs) with cattle representing 96% of the AUMs (McIntosh et al. 2000). Although more common in non-forested floodplains and rangelands, livestock grazing occurs across a range of forest types (Hemstrom et al. 2001) and, in managed forests, is often incorporated into the forest rotation schedule (e.g., Krzic et al. 2001). Livestock grazing has been identified as the most extensive land use in the Pacific Northwest and has had a major impact on all riparian zones throughout the region (Kauffman et al. 2001).

Most research documenting livestock effects on habitat in general, and riparian habitat in particular, has focused on the more arid regions of the west (e.g., Fleischner 1994, Belsky et al. 1999). However, Belsky et al. (1999) cite evidence that suggests the environmental impacts of livestock grazing in more mesic regions, such as the forests of the Pacific Northwest, are similar to those that have been documented in arid regions. Fleischner (1994) identifies three broad ecological impacts of livestock grazing: (1) alteration of species composition of communities, (2) disruption of ecosystem functioning, and (3) alteration of ecosystem structure. Although these impacts are not limited to riparian zones, livestock grazing can have a disproportionately heavy impact on these zones because they represent such a small proportion of the total landscape and livestock often concentrate along streams (e.g., Roath and Krueger 1982, Wales 2001).

Through selection both for and against different plant species, livestock have obvious effects on plant communities (e.g., Fleischner 1994). Grazing removes plant biomass, alters the structure of plant populations, reduces tree and shrub reproduction by seed and seedling browsing, and changes the species composition of plant communities (e.g., Brinson et al. 1981, Case and Kauffman 1997). A critical change in plant community composition has been the introduction of exotic plant species along grazed streams (e.g., Wales 2001). Interestingly, experimental studies of ungrazed riparian plant communities in northern California reveal that the most diverse communities are the most susceptible to invasion by exotic plants

(Levine 2000*b*). Heavy livestock grazing in riparian zones has additional negative impacts including soil compaction, breakdown of stream banks, alterations of channel morphology, increased erosion, lowered water tables, and deterioration of water quality (Thomas et al. 1979, Brinson et al. 1981, Oakley et al. 1985, Hall 1988). For example, McIntosh et al. (2000) compared the frequency of large, deep pools between natural and "commodity" (i.e., grazed, logged) streams of the Columbia River Basin from surveys conducted in 1935–1945 and in 1987–1997. Whereas almost all (96%) of the natural streams exhibited either no change or an increase in the number of pools, the number of pools decreased in the majority (52% to 54%) of the commodity streams (McIntosh et al. 2000). These changes to the vegetation and structure of riparian habitats have consequences on the distribution and abundance of mammals in riparian habitats.

The potential impacts of grazing on native mammals stem from the loss of herbaceous vegetation that serves as both food and cover, increased soil compaction, changes in stream morphology and water quality, and consequent reduction in the prey base (e.g., Fleischner 1994, Kauffman et al. 2001, Wales 2001). Much of the empirical evidence documenting the effects of grazing on mammals is from comparisons of small-mammal populations in grazed versus ungrazed riparian habitats (e.g., Medin and Clary 1989, Leege et al. 1981, Schultz and Leininger 1991, Klaus et al. 1999). For example, capture success, female body mass, and recruitment of young water voles were greater along ungrazed than grazed headwater streams of the Clark's Fork of the Yellowstone River (Klaus et al. 1999). In Idaho, livestock exclosure experiments resulted in increased litter depths and a corresponding increase in small-mammal abundance within the exclosures (Leege et al. 1981).

The impacts of livestock grazing on riparian ecosystems vary depending on the intensity of grazing, season of grazing, and frequency of grazing, as well as on site-specific factors such as species composition, soil characteristics, and slope (Kauffman et al. 2001). Belsky et al. (1999) suggest damages from heavy livestock grazing can be reduced with modified management practices such as herding or fencing cattle from streams, reducing livestock numbers, or increasing periods of rest.

Road construction
Road construction is often associated with logging activities and can have a lasting impact on riparian habitats (Thomas et al. 1979). The construction of roads in riparian habitats changes vegetative composition and

structure, alters microclimatic conditions, can result in debris torrents due to increased erosion, and reduces the size of the riparian zone (Oakley et al. 1985). For example, Jones et al. (2000) showed that road networks in Pacific Northwest forests, especially those below mid slope, have increased the frequency and magnitude of peak flows and debris slides compared to either intact forests or fire-disturbed forests. Jones et al. (2000) also suggest that roads on valley floors will limit access to streams by organisms, thereby limiting the dispersal of colonizers into flood-disturbed areas. In contrast, roads in riparian areas have been linked to the invasion of exotic plants into Pacific Northwest forests (Parendes and Jones 2000). In efforts to mitigate the negative impacts of roads on wildlife, wildlife underpasses have been constructed. However, wildlife response to these structures can be varied. For example, analysis of wildlife use of underpasses in Banff National Park showed that carnivores utilized underpasses near stream drainages, but ungulates avoided these underpasses, presumably because of increased predation risk (Clevenger and Waltho 2000).

Research needs and designs

Most studies on mammals in riparian zones have reported abundance of small mammals from trapping or movements and habitat associations of larger carnivores or ungulates from radiotelemetry. In contrast, there is very little information on reproductive fitness (however see Doyle 1990) or population viability. Consequently, future studies should focus on comparisons of physical condition, reproductive rates, and survival of mammals in riparian zones compared to upland areas. In addition, there have been more studies west than east of the crest of the Cascade Mountains; therefore, there is a need to describe the associations of mammals with riparian habitats east of the Cascade Mountains in the intermountain regions. Some literature suggests that riparian areas are used as travel corridors and, by allowing connectivity between suitable habitat fragments across the landscape, riparian areas might also provide a means for gene flow among populations. Further examination of the extent to which riparian areas are used as travel corridors and their potential to enhance gene flow among populations would provide insight into mammalian use of riparian areas in the larger landscape context.

New approaches to mammalian surveys and to analysis of habitat associations in riparian areas are needed, and designed experiments in conjunction with habitat alterations should be pursued. In view of the difficulties in designing replicated experiments, sampling mammalian habitat relations in narrow riparian zones and the many factors influencing these relations, we suggest two alternate approaches. The first approach attempts to sample the total variability of plant and animal communities in riparian zones along a continuum of many different habitats. Mammalian communities could then be ranked from least to most diverse with multivariate statistical analyses such as principal component analysis or ordination procedures. Then, management schemes or impacts associated with less rich and/or abundant mammal populations could be identified and modified. Such an approach will require an emphasis on point sampling (trap or observation stations) rather than intensive sampling on large quadrants and will allow the sampling of many more points over many different habitat conditions. This approach may allow research to answer more questions and formulate hypotheses with fewer confounding effects and limitations in interpreting results. Second, manipulative experiments are needed for studying habitat associations of mammals or assessing the effects of habitat alterations. Manipulative experiments that create homogeneous vegetation or increase heterogeneity will be more likely to demonstrate effects of habitat alterations than other types of studies.

Small mammals are difficult to sample because of their size and general lack of sign to indicate their presence. As a result, a number of different designs and traps have been used and the selection of a method depends on the objectives of the study. There have been at least three sampling designs for trapping small mammals in riparian zones: grid trapping (Doyle 1990), transect sampling (McComb et al. 1993, Gomez and Anthony 1998, Waters et al. 2001), and paired parallel (Calhoun) trapping lines (Anthony et al. 1987). Grid trapping with capture-recapture methods is common and well described in the literature. Anthony et al. (1987) used parallel trap lines along riparian zones to compare the abundance of species adjacent to the stream versus the riparian fringe and to describe species' use of the ecotone between riparian and upland habitats. McComb et al. (1993) and Gomez and Anthony (1998) used transects arranged in a rectangular design (Fig. 15.2A) to compare relative abundance of small mammals between riparian and upland habitats and to describe the degree of association of species to riparian zones along trans-riparian transects. Waters et al. (2001) used three parallel lines to describe differences in small-mammal captures among streamside, riparian fringe, and upslope transects and to discuss species' associations with riparian areas (Fig. 15.2B).

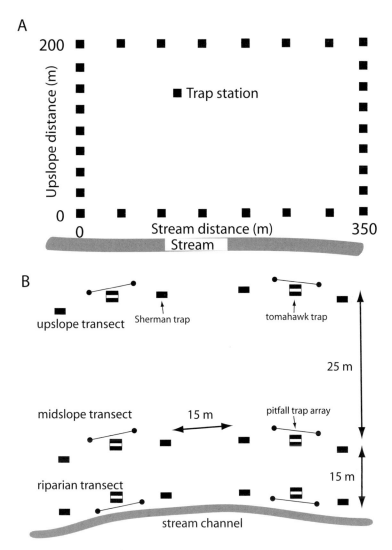

Fig. 15.2. Two designs for investigating the association of small mammals to riparian zones. (A) Sampling scheme for examining changes in small-mammal distribution on gradients extending from riparian into upland habitat as modified from Gomez and Anthony (1996). Two pitfall traps were placed at each trap station. (B) An alternative approach modified from Waters et al. (2001) that employs different trap types on three transects parallel to the stream in riparian, midslope, and upslope areas.

All of these designs are amenable to either live trapping or kill trapping (pitfalls or snap-traps) to describe abundance of small mammals and have potential for use in the future.

More studies on movements and habitat associations of bats, carnivores, and ungulates in relation to riparian areas using radiotelemetry, photographic "traplines", and hair snares are needed. In addition, there is a need to describe the functional benefits of the use of riparian zones with respect to foraging efficiency, energetic requirements, and reproductive fitness. These are not easily researched topics but attempts must be made to address them.

Summary

Riparian zones are found adjacent to streams, rivers, springs, ponds, lakes, or tidal estuaries and represent areas of great ecological richness and diversity. The structure and function of riparian zones are determined by topography, surface water, soils, microclimate, and vegetation, and the vegetation, in turn, has a direct effect on stream structure and function. Riparian zones also are a product of disturbances that affect structure and function of these areas and the associated mammal fauna. As a result of the diversity of biological and physical features found in riparian zones, mammalian species richness is high in these ecosystems and use of riparian zones by some species is disproportionately higher than in other areas. Several species of mammals in the Pacific Northwest are considered obligate inhabitants of riparian areas, and they represent a variety of taxonomic groups from small insectivores to large carnivores. Prominent among these species is the beaver which acts as a keystone species by its ability to alter riparian vegetation and water flows. These activities influence plant and associated mammal communities. Many species of mammals, including shrews, moles, bats, rodents, carnivores, and ungulates, are significantly more abundant in riparian areas, their movements are highly associated with these areas, or they have higher reproductive fitness in these areas. However, the degree of association for some species varies spatially among different forest types and between the east and west slopes of the Cascade Mountains. Riparian zones typically have vegetation that is adapted to high disturbance produced by frequent episodes of flooding, scouring, and deposition of sediment that promotes early-successional vegetation and mammals that are associated with these conditions. Many species of mammals are not dependent on riparian areas nor are they more abundant there; however, they may be major components of riparian mammal communities (i.e., generalist species such as deer mice).

Human disturbances, particularly timber harvest, cattle grazing, and road construction, can have major influences on riparian systems and associated mammal communities. One method of managing for mammals and other wildlife in riparian zones is the retention of buffer strips where some but not all of the trees are harvested within various distances from the stream border. The amount of benefit provided to mammals depends on the width of the buffer, amount of tree canopy left, and the specific habitat requirements of the species. More information is needed on the potential benefit of these buffer strips, because we know very little about the use of these areas as refugia, travel corridors, or their role in gene flow among subpopulations.

Research on mammals, particularly the larger mobile species, in riparian areas is challenging because of the narrow configuration of these systems and the uncertainty in ascertaining habitat selection patterns of these species. More information is needed on physical condition, reproductive rates, and survival of mammals in riparian zones compared to upland areas, as most of the past research has focused on estimating relative abundance and habitat associations. We recommend the use of point or transect sampling, instead of quadrat sampling, and multivariate statistical analyses to reveal more definitive habitat associations and develop testable hypotheses. Lastly, manipulative experiments will be important in future research to study habitat requirements and demonstrate the effects of habitat alterations on mammals.

Acknowledgments

M.A.O'C and J.G.H. acknowledge the Washington Timber, Fish, and Wildlife Program and the Washington Department of Natural Resources, which supported their research on riparian zones. R.G.A. acknowledges the support of the U.S. Geological Survey while writing this chapter. Parts of this paper were modified from O'Connell et al. (1993). We thank W. McComb, M. Raphael, and C. Zabel for valuable comments on earlier drafts of the chapter.

Literature cited

Aars, J., R.A. Ims, H.P. Liu, M. Mulvey, and M.H. Smith. 1998. Bank voles in linear habitats show restricted gene flow as revealed by mitochondrial DNA (mtDNA). *Molecular Ecology* 7:1383–1389.

Agee, J.K. 1988. Successional dynamics in forest riparian zones. Pages 31–43 *in* K.J. Raedeke, editor. *Streamside Management: Riparian Wildlife and Forestry Interactions.* University of Washington Press, Seattle, Washington, USA.

Agee, J.K. 1993. *Fire Ecology of Pacific Northwest Forests.* Island Press, Washington, DC, USA.

Agee, J.K. 1994. Eastside forest ecosystem health assessment. Fire and weather disturbances in terrestrial ecosystems of the eastern Cascades. US Forest Service, General Technical Report **PNW-GTR-320**, Portland, Oregon, USA.

Aleksiuk, M. 1970. The seasonal food regime of arctic beavers. *Ecology* **51**:264–270.

Alverson, W.S., D.M. Waller, and S.L. Solheim. 1988. Forests too deer: edge effects in northern Wisconsin. *Conservation Biology* **2**:348–358.

Anthony, R.G., E.D. Forsman, G.A. Green, G. Witmer, and S.K. Nelson. 1987. Small mammal populations in riparian zones of different-aged coniferous forests. *Murrelet* **68**:94–102.

Arno, S.F. 2000. Fire in western forest ecosystems. Pages 97–120 *in* J.K. Brown and J.K. Smith, editors. Wildland fire in ecosystems: effects of fire on flora. US Forest Service General Technical Report **RMRS-GTR-42**, Volume 2, Ogden, Utah, USA.

Bailey, V. 1936. The mammals and life zones of Oregon. *North American Fauna* **55**:1–416.

Barbour, R.W. and W.H. Davis. 1969. *Bats of America.* University of Kentucky Press, Lexington, Kentucky, USA.

Bell, D.T. and S.K. Sipp. 1975. The litter stratum in the streamside forest ecosystem. *Oikos* **26**:391–397.

Belovsky, G.E. 1984. Summer diet optimization by beaver. *American Midland Naturalist* **111**:209–221.

Belsky, A.J., A. Matzke, and S. Uselman. 1999. Survey of livestock influences on stream and riparian ecosystems in the western United States. *Journal of Soil and Water Conservation* **54**:419–431.

Belwood, J.J. and M.B. Fenton. 1976. Variation in the diet of *Myotis lucifugus* (Chiroptera:Vespertilionidae). *Canadian Journal of Zoology* **54**:1674–1678.

Ben-David, M., T.A. Hanley, and D.M. Schell. 1998. Fertilization of terrestrial vegetation by spawning Pacific salmon: the role of flooding, and predator activity. *Oikos* **83**:47–55.

Beneski, J.T. and D.W. Stinson. 1987. *Sorex palustris. Mammalian Species* **296**:1–6.

Betts, B.J. 1996. Roosting behavior of silver-haired bats (*Lasionycteris noctivagans*) and big brown bats (*Eptesicus fuscus*) in northeast Oregon. Pages 55–61 *in* R.M.R. Barclay and R.M. Brigham, editors. Bats and forests symposium, October 19–21, 1995. Working Paper **23/1996**. British Columbia Research Branch, British Columbia Ministry of Forests, Victoria, British Columbia, Canada.

Bilby, R.E. 1988. Interactions between aquatic and terrestrial systems. Pages 13–29 *in* K.J. Raedeke, editor. *Streamside Management: Riparian Wildlife and Forestry Interactions.* University of Washington Press, Seattle, Washington, USA.

Bilby, R.E., B.R. Fransen, and P.A. Bisson. 1996. Incorporation of nitrogen and carbon from spawning coho salmon into the trophic system of small streams: evidence from stable isotopes. *Canadian Journal of Fisheries and Aquatic Sciences* **53**:164–173.

Bilton, H.T., D.F. Alderdice, and J.T. Schnute. 1982. Influence of time and size of release of juvenile coho salmon (*Oncorhynchus kisutch*) on returns at maturity. *Canadian Journal of Fisheries and Aquatic Science* **39**:426–447.

Brigham, R.M., H.D.J.N. Aldridge, and R.L. Mackey. 1992. Variation in habitat use and prey selection by Yuma bats, *Myotis yumanensis*. *Journal of Mammalogy* **73**:640–645.

Brinson, M.M. and J. VerHoeven. 1999. Riparian forests. Pages 265–269 *in* M.L. Hunter, Jr., editor. *Maintaining Biodiversity in Forest Ecosystems*. Cambridge University Press, New York, New York, USA.

Brinson, M.M., B.L. Swift, R.C. Plantico, and J.S. Barclay. 1981. Riparian ecosystems: their ecology and status. USDI Fish, and Wildlife Service, Biological Services Program, Kearneysville, West Virginia, USA.

Brosofske, K.D., J. Chen, R.J. Naiman, and J.F. Franklin. 1997. Harvesting effects on microclimatic gradients from small streams to uplands in western Washington. *Ecological Applications* **7**:1188–1200.

Brown, D.J., W.A. Hubert, and S.H. Anderson. 1996. Beaver ponds create wetland habitat for birds in mountains of southeastern Wyoming. *Wetlands* **16**:127–133.

Brown, E.R., editor. 1985. Management of wildlife and fish habitats in forests of western Oregon and Washington. US Forest Service, **R-F&WL-192–1985**, Portland, Oregon, USA.

Buck, S.G., C. Mullis, A.S. Mossman, I. Show, and C. Coolahon. 1994. Habitat use by fishers in adjoining heavily and lightly harvested forest. Pages 368–372 *in* S.W. Buskirk, A.S. Harestad, M.G. Raphael, and R.A. Powell, editors. *Martens, Sables, and Fishers: Biology and Conservation*. Cornell University Press, Ithaca, New York, USA.

Burns, J.J. 1964. The ecology, economics, and management of mink in the Yukon-Koskokwim Delta, Alaska. Thesis. University of Alaska, Fairbanks, Alaska, USA.

Buskirk, S.W. and R.A. Powell. 1994. Habitat ecology of fishers and American martens. Pages 283–296 *in* S.W. Buskirk, A.S. Harestad, M.G. Raphael, and R.A. Powell, editors. *Martens, Sables, and Fishers: Biology and Conservation*. Cornell University Press, Ithaca, New York, USA.

Buskirk, S.W. and W.J. Zielinski. 2003. Small and mid-sized carnivores. Pages 207–249 *in* C.J. Zabel and R.G. Anthony, editors. *Mammal Community Dynamics. Management and Conservation in the Coniferous Forests of Western North America*. Cambridge University Press, Cambridge, UK.

Buskirk, S.W., S.C. Forrest, M.G. Raphael, and H.J. Harlow. 1989. Winter resting site ecology of marten in the central Rocky Mountains. *Journal of Wildlife Management* **53**:191–196.

Bustard, D.R. and D.W. Narver. 1975. Aspects of the winter ecology of juvenile coho salmon (*Oncorhynchus kisutch*) and steelhead trout (*Salmo gairdneri*). *Journal of Fisheries Research Board Canada* **32**:667–680.

Campbell, A.G. and J.F. Franklin. 1979. Riparian vegetation in Oregon's western Cascade Mountains: composition, biomass, and autumn phenology. *Coniferous Forest Biome Ecosystem Analysis Studies Bulletin* **14**:1–90.

Campbell, L.A., J.G. Hallett, and M.A. O'Connell. 1996. Conservation of bats in managed forests: roost use by *Lasionycteris noctivagans*. *Journal of Mammalogy* **77**:976–984.

Carraway, L.N. and B.J. Verts. 1985. *Microtus oregoni. Mammalian Species* **233**:1–6.

Case, R.L. and J.B. Kauffman. 1997. Wild ungulate influences on the recovery of willows, black cottonweed, and thin-leaf alder following cessation of cattle grazing in northeastern Oregon. *Northwest Science* **71**:115–126.

Cederholm, C.J., D.B. Houston, D.L. Cole, and W.J. Scarlett. 1989. Fate of coho salmon (*Oncorhynchus kisutch*) carcasses in spawning streams. *Canadian Journal of Fisheries and Aquatic Sciences* **46**:1347–1355.

Chapman, J.A. and G.A. Feldhamer, editors. 1982. *Wild Mammals of North America*. Johns Hopkins University Press, Baltimore, Maryland, USA.

Clevenger, A.P. and N. Waltho. 2000. Factors influencing the effectiveness of wildlife underpasses in Banff National Park, Alberta, Canada. *Conservation Biology* **141**:47–56.

Coady, J.W. 1982. Moose. Pages 902–922 *in* J. Chapman and G.A. Feldhamer, editors. *Wild Mammals of North America*. Johns Hopkins University Press, Baltimore, Maryland, USA.

Cole, E.K., M.D. Pope, and R.G. Anthony. 1997. Effects of road management on movement and survival of Roosevelt elk. *Journal of Wildlife Management* **61**:1115–1126.

Compton, B.B., R.J. Mackie, and G.L. Dusek. 1988. Factors influencing distribution of white-tailed deer in riparian habitats. *Journal of Wildlife Management* **52**:544–548.

Conaway, C.H. 1952. Life history of the water shrew (*Sorex palustris navigator*). *American Midland Naturalist* **48**:219–248.

Craighead, J.J., J.S. Sumner, and G.B. Scaggs. 1982. A definitive system for analysis of grizzly bear habitat and other wilderness resources using Landsat multispectral imaging and computer technology. *University of Montana Monographs* **1**:1–279.

Cross, S.P. 1985. Responses of small mammals to forest riparian perturbations. Pages 269–275 *in* R.R. Johnson, C.D. Ziebell, D.R. Patton, P.F. Ffolliot, and R.H. Hamre, editors. Riparian ecosystems and their management: reconciling conflicting uses. Proceedings of the First North American Riparian Conference. US Forest Service General Technical Report **RM-GTR-120**, Fort Collins, Colorado, USA.

Cross, S.P. 1988. Riparian systems and small mammals and bats. Pages 93–112 *in* K.J. Raedeke, editor. *Streamside Management: Riparian Wildlife and Forestry Interactions*. University of Washington Press, Seattle, Washington, USA.

Cummins, K.W. 1980. The multiple linkages of forest to streams. Pages 191–198 *in* R.H. Waring, editor. Forests: fresh perspectives from ecosystem analysis. Proceedings of the 40th Annual Biology Colloquium. Oregon State University Press, Corvallis, Oregon, USA.

Darveau, M., P. Labbe, P. Beauchesne, L. Belanger, and J. Huot. 2001. The use of riparian forest strips by small mammals in a boreal balsam fir forest. *Forest Ecology and Management* **143**:95–104.

Dickson, J.G. and J.H. Williamson. 1988. Small mammals in streamside management zones in pine plantations. Pages 375–378 *in* R.C. Szaro, K.E. Severson, and D.R. Patton, editors. Management of amphibians, reptiles, and small mammals in North America. US Forest Service General Technical Report **RM-GTR-166**, Fort Collins, Colorado, USA.

Dieffenbach, D.R. and R.B. Owen, Jr. 1989. A model of habitat use by breeding American black ducks. *Journal of Wildlife Management* **53**:383–389.

Doyle, A.T. 1990. Use of riparian and upland habitats by small mammals. *Journal of Mammalogy* **71**:14–23.

Dusek, G.L., R.J. Mackie, J.D. Herriges, and B.B. Compton. 1989. Population ecology of white-tailed deer along the lower Yellowstone River. *Wildlife Monographs* **104**:1–68.

Eagle, T.C. and J.S. Whitman. 1987. Mink. Pages 615–642 *in* M. Novak, J.A. Baker, M.E. Obbard, and B. Malloch, editors. *Wild Furbearer Management and Conservation in North America*. Ontario Ministry of Natural Resources, Toronto, Ontario, Canada.

Eberhardt, R.T. 1973. Some aspects of mink-waterfowl relationships on prairie wetlands. *Prairie Naturalist* **5**:17–19.

Edmonds, R.L. 1980. Litter decomposition and nutrient release in Douglas fir, red alder, western hemlock, and Pacific silver fir ecosystems in western Washington. *Canadian Journal of Forest Resources* **10**:327–337.

Edwards, N.T. and D.L. Otis. 1999. Avian communities and habitat relationships in South Carolina piedmont beaver ponds. *American Midland Naturalist* **141**:158–171.

England, A.S., L.D. Foreman, and W.F. Laudenslayer, Jr. 1984. Composition and abundance of bird populations in riparian systems of the California deserts. Pages 694–705 *in* R.E. Warner and K.M. Hendrix, editors. *California Riparian Systems: Ecology, Conservation, and Productive Management*. University of California Press, Berkeley, California, USA.

Erickson, J.L. and S.D. West. 1996. Managed forests in the western Cascades: the effects of seral stage on bat use patterns. Pages 215–227 *in* R.M.R. Barclay and R.M. Brigham, editors. Bats and forests symposium, October 19–21, 1995. Working Paper **23/1996**, British Columbia Research Branch, British Columbia Ministry of Forests. Victoria, British Columbia, Canada

Erlinge, S. 1969. Food habits of the otter *Lutra lutra* L. and mink *Mustela vison* Schreber in a trout water in southern Sweden. *Oikos* **20**:1–7.

Errington, P.L. 1943. An analysis of mink predation upon muskrats in north-central United States. *Research Bulletin Iowa Agricultural Experimental Station* **320**:797–924.

Errington, P.L. 1954. The special responsiveness of minks to epizootics in muskrat populations. *Ecological Monographs* **24**:377–393.

Everest, F.H., N.B. Armantrout, S.M. Keller, W.D. Parante, J.R. Sedell, T.E. Nickelson, J.M. Johnston, and G.N. Haugen. 1985. Salmonids. Pages 199–230 *in* E.R. Brown, editor. Management of wildlife and fish habitats in forests of western Oregon and Washington Part I: Chapter narratives. US Forest Service **R6-F&WL-192-1985**, Portland, Oregon, USA.

Fenton, M.B. and G.P. Bell. 1979. Echolocation and feeding behavior in four species of *Myotis* (Chiroptera). *Canadian Journal of Zoology* **57**:1271–1277.

Fetherston, K.L., R.J. Naiman, and R.E. Bilby. 1995. Large woody debris, physical process, and riparian forest development in montane river networks of the Pacific Northwest. *Geomorphology* **13**:133–144.

Fleischner, T.L. 1994. Ecological costs of livestock grazing in western North America. *Conservation Biology* **8**:629–644.

Forsey, E.S. and E.M. Baggs. 2001. Winter activity of mammals in riparian zones and adjacent forests prior to and following clear-cutting at Copper Lake, Newfoundland, Canada. *Forest Ecology and Management* **145**:163–171.

Franklin, J.F., K. Cromack, Jr., W. Denison, A. McKee, C. Maser, J. Sedell, F. Swanson, and G. Juday. 1981. Ecological characteristics of old-growth Douglas-fir forests. US Forest Service General Technical Report **PNW-GTR-118**, Portland, Oregon, USA.

Franklin, J.F., D. Lindenmayer, J.A. MacMahon, A. McKee, J. Magnusson, D.A. Perry, R. Waide, and D. Foster. 2000. Threads of continuity. *Conservation Biology in Practice* **1**:8–16.

Frazier, M.W. 1997. Roost site characteristics of the long-legged myotis (*Myotis volans*) in the Teanaway River Valley of Washington. Pages 5.1–5.21 *in* K.B. Aubry, S.D. West, D.A. Manuwal, A.B. Stringer, J. Erickson, and S. Pearson, editors. West-side studies: research results. Volume 2 of Wildlife use of managed forests: a landscape perspective. Final report to the Timber, Fish, and Wildlife Cooperative Monitoring, Evaluation, and Research Committee, **TFW-WL4-98-0023**, Washington Department of Natural Resources, Olympia, Washington, USA.

Fuller, T.K. 1978. Variable home-range sizes of female gray foxes. *Journal of Mammalogy* **59**:446–449.

Gavin, T.A. 1984. Pacific Northwest. Pages 487–496 *in* L. K. Halls, editor. *White-tailed Deer: Ecology and Management*. Stackpole Books, Harrisburg, Pennsylvania, USA.

Gavin, T.A., L.H. Suring, P.A. Vohs, Jr., and E.C. Meslow. 1984. Population characteristics, spatial organization, and natural mortality in Columbian white-tailed deer. *Wildlife Monographs* **91**:1–45.

Gawler, S.C. 1988. Disturbance-mediated population dynamics of *Pedicularis furbishae* S. Wats, a rare riparian endemic. Dissertation. University of Wisconsin, Madison, Wisconsin, USA.

Gerell, R. 1970. Home ranges and movements of the mink *Mustela vison* Schreber in southern Sweden. *Oikos* **21**:160–173.

Getz, L.L. 1970. Influence of vegetation on the local distribution of the meadow vole in Wisconsin. *University of Connecticut Occasional Papers (Biological Science Series)* **1**:213–241.

Goertz, J.W. 1964. Habitats of three Oregon voles. *Ecology* **45**:846–848.

Gomez, D.M. and R.G. Anthony. 1996. Amphibian and reptile abundance in riparian and upslope areas of five forest types in western Oregon. *Northwest Science* **70**:109–111.

Gomez, D.M. and R.G. Anthony. 1998. Small mammal abundance in riparian and upland areas of five seral stages in western Oregon. *Northwest Science* **72**:293–302.

Grinnell, J., J.S. Dixon, and J.M. Linsdale. 1937. *Furbearing Mammals of California*. University of California Press, Berkeley, California, USA.

Grizzel, J.D. and N. Wolff. 1998. Occurrence of windthrow in forest buffer strips and its effect on small streams in northwest Washington. *Northwest Science* **72**:214–223.

Grover, A.M. and G.A. Baldassarre. 1995. Bird species richness within beaver ponds in south-central New York. *Wetlands* **15**:108–118.

Gyug, L.W. 2000. Timber-harvesting effects on riparian wildlife and vegetation in the Okanogan Highlands of British Columbia. Ministry of Environment, Lands and Parks, Victoria, British Columbia, Canada.

Hairston-Strang, A.B. and P.W. Adams. 1998. Potential large woody debris sources in riparian buffers after harvesting in Oregon, USA. *Forest Ecology and Management* **112**:67–77.

Hall, F.C. 1988. Characterization of riparian systems. Pages 7–12 *in* K.J. Raedeke, editor. *Streamside Management: Riparian Wildlife and Forestry Interactions*. University of Washington Press, Seattle, Washington, USA.

Hall, J.G. 1960. Willow and aspen in the ecology of beaver on Sagehen Creek, California. *Ecology* **41**:484–494.

Hallett, J.G. and M.A. O'Connell. 2000. East-side small mammal surveys. Pages 11.1–11.23 *in* M.A. O'Connell, J.G. Hallett, S.D. West, K.A. Kelsey, D.A. Manuwal, and S.F. Pearson, editors. Effectiveness of riparian management zones in providing habitat for wildlife. Final report to Timber, Fish, and Wildlife Program, **TFW-LWAG1-00-001**, Washington Department of Natural Resources, Olympia, Washington, USA.

Hallett, J.G., M.A. O'Connell, and C.C. Maguire. 2003. Ecological relationships of terrestrial small mammals in western coniferous forests. Pages 120–156 *in* C.J. Zabel and R.G. Anthony, editors. *Mammal Community Dynamics. Management and Conservation in the Coniferous Forests of Western North America.* Cambridge University Press, Cambridge, UK.

Hanley, T.A. and R.D. Taber. 1980. Selective plant species inhibition by elk and deer in three conifer communities in western Washington. *Forest Science* **26**:97–107.

Harr, R.D., R.L. Frederickson, and J. Rothacher. 1979. Changes in streamflow following timber harvest in southwest Oregon. US Forest Service Research Paper **PNW-249**, Portland, Oregon, USA.

Harris, L.D. 1984. *The Fragmented Forest: Island Biogeography Theory and the Preservation of Biotic Diversity.* University of Chicago Press, Chicago, Illinois, USA.

Hartman, G.D. and T.L. Yates. 1985. *Scapanus orarius. Mammalian Species* **253**:1–5.

Hayes, J.P. 2003. Habitat ecology and conservation of bats in western coniferous forests. Pages 81–119 *in* C.J. Zabel and R.G. Anthony, editors. *Mammal Community Dynamics. Management and Conservation in the Coniferous Forests of Western North America.* Cambridge University Press, Cambridge, UK.

Hayes, J.P. and M.D. Adam. 1996. The influence of logging riparian areas on habitat utilization by bats in western Oregon. Pages 228–237 *in* R.M.R. Barclay and R.M. Brigham, editors. Bats and forests symposium, October 19–21, 1995. Working Paper **23/1996**, British Columbia Research Branch, British Columbia Ministry of Forests. Victoria, British Columbia, Canada.

Hemstrom, M.A., J.J. Korol, and W.J. Hann. 2001. Trends in terrestrial plant communities and landscape health indicate the effects of alternative management strategies in the interior Columbia River Basin. *Forest Ecology and Management* **153**:105–126.

Herd, R.M. and M.B. Fenton. 1983. An electrophoretic, morphological, and ecological investigation of a putative hybrid zone between *Myotis lucifugus* and *Myotis yumanensis* (Chiroptera:Vespertilionidae). *Canadian Journal of Zoology* **61**:2029–2050.

Hessburg, P.F., B.G. Smith, and R.B. Salter. 1999. Detecting change in forest spatial patterns from reference conditions. *Ecological Applications* **9**:1232–1252.

Hill, E.H. 1982. Beaver (*Castor canadensis*). Pages 256–281 *in* J.A. Chapman and G.A. Feldhamer, editors. *Wild Mammals of North America.* Johns Hopkins University Press, Baltimore, Maryland, USA.

Hodkinson, I.D. 1975. Energy flow and organic matter decomposition in an abandoned beaver pond ecosystem. *Oecologia* **21**:131–139.

Hyatt, T.L. and R.J. Naiman. 2001. The residence time of large woody debris in the Queets River, Washington, USA. *Ecological Applications* **11**:191–202.

Jenkins, K.J. and E.E. Starkey. 1984. Habitat use by Roosevelt elk in unmanaged forests of the Hoh Valley, Washington. *Journal of Wildlife Management* **48**:642–646.

Jenkins, S.H. 1975. Food selection by beavers. *Oecologia* **21**:157–173.

Jenkins, S.H. 1979. *Castor canadensis. Mammalian Species* **120**:1–8.

Johnson, D.H. and T.A. O'Neil, managing directors. 2001. *Wildlife-habitat Relationships in Oregon and Washington*. Oregon State University Press, Corvallis, Oregon, USA.

Johnson, R.R. and D.A. Jones, editors. 1977. Importance, preservation, and management of riparian habitat: a symposium. US Forest Service General Technical Report **RM-GTR-43**, Tempe, Arizona, USA.

Johnston, C.A. and R.J. Naiman. 1990. Browse selection by beaver: effects on riparian forest composition. *Canadian Journal of Forest Research* **20**:1036–1043.

Johnston, C.A., J. Pastor, and R.J. Naiman. 1993. Effects of beaver and moose on boreal forest landscapes. Pages 237–254 *in* R. Haines-Young, D.R. Green, and S.H. Cousins, editors. *Landscape Ecology and Geographic Information Systems*. Taylor and Francis, London, UK.

Jones, J.A., F.J. Swanson, B.C. Wemple, and K.U. Snyder. 2000. Effects of roads on hydrology, geomorphology, and disturbance patches in stream networks. *Conservation Biology* **14**:76–85.

Jones, J.L. and E.O. Garton. 1994. Selection of successional stages by fishers in north-central Idaho. Pages 377–382 *in* S.W. Buskirk, A.S. Harestad, M.G. Raphael, and R.A. Powell, editors. *Martens, Sables, and Fishers: Biology and Conservation*. Cornell University Press, Ithaca, New York, USA.

Jonsson, B.G. 1997. Riparian bryophyte vegetation in the Cascade mountain range, northwest USA: patterns at different spatial scales. *Canadian Journal of Botany* **75**:744–761.

Kauffman, J.B., M. Mahrt, L.A. Mahrt, and W.D. Edge. 2001. Wildlife of riparian habitats. Pages 361–388 *in* D.H. Johnson and T.A. O'Neil, managing directors. 2001. *Wildlife-habitat Relationships in Oregon and Washington*. Oregon State University Press. Corvallis, Oregon, USA.

Kaufmann, J.H. 1982. Raccoon and allies. Pages 567–585 *in* J.A. Chapman and G. Feldhamer, editors. *Wild Mammals of North America*. Johns Hopkins University Press, Baltimore, Maryland, USA.

Keller, E.M. and F.J. Swanson. 1979. Effects of large organic material on channel form and fluvial processes. *Earth Surface Processes* **4**:361–380.

Kellyhouse, D.G. 1980. Habitat utilization by black bears in northern California. *International Conference of Bear Research and Management* **4**:221–227.

Kelsey, K.A. and S.D. West. 1998. Riparian wildlife. Pages 235–258 *in* R.J. Bilby and R.E. Naiman, editors. *River Ecology and Management: Lessons from the Pacific Coastal Ecoregion*. Springer-Verlag, New York, New York, USA.

Kie, J.G., R.T. Bowyer, and K.M. Stewart. 2003. Ungulates in western coniferous forests: habitat relationships, population dynamics, and ecosystem processes. Pages 296–340 *in* C.J. Zabel and R.G. Anthony, editors. *Mammal Community Dynamics. Management and Conservation in the Coniferous Forests of Western North America*. Cambridge University Press, UK.

Klaus, M., R.E. Moore, and E. Vyse. 1999. Impact of precipitation and grazing on the water vole in the Beartooth Mountains of Montana and Wyoming, U.S.A. *Arctic, Antarctic and Alpine Research* **31**:278–282.

Knopf, F.L. 1985. Significant riparian vegetation to breeding birds across an altitudinal cline. Pages 105–111 *in* R.R. Johnson, C.D. Ziebell, D.R. Patton, P.F. Ffolliot, and R.H. Hamre, editors. Riparian ecosystems and their management: reconciling conflicting uses. First North American Riparian Conference. US Forest Service General Technical Report **RM-GTR-120**, Fort Collins, Colorado, USA.

Knopf, F.L., R.R. Johnson, T. Rich, F.B. Samson, and R.C. Szaro. 1988. Conservation of riparian ecosystems in the United States. *Wilson Bulletin* **100**:272–284.

Korschgen, L.T. 1958. December food habits of mink in Missouri. *Journal of Mammalogy* **39**:521–527.

Krzic, M., K. Broersma, R.F. Newman, T.M. Ballard, and A.A. Bomke. 2001. Soil quality on harvested and grazed forest cutblocks in southern British Columbia. *Journal of Soil and Water Conservation* **56**:192–197.

Larrison, E.J. 1976. *Mammals of the Northwest*. Seattle Audubon Society, Seattle, Washington, USA.

Lavriviere, S. and L.R. Walton. 1999. *Lontra canadensis*. *Mammalian Species* **587**:1–8.

Leege, T.A., D.J. Herman, and B. Zamora. 1981. Effects of cattle grazing on mountain meadows in Idaho. *Journal of Range Management* **34**:324–328.

LeFranc, M.N., Jr., M.B. Moss, K.A. Patnode, and W.C. Sugg, III. 1987. Grizzly bear compendium. Interagency Grizzly Bear Committee, National Wildlife Federation, Washington, DC, USA.

Leidholt-Bruner, K. 1990. Effects of beaver on streams, streamside habitat, and coho salmon fingerling populations in two coastal Oregon streams. Thesis. Oregon State University, Corvallis, Oregon, USA.

Leidholt-Bruner, K., D.E. Hibbs, and W.C. McComb. 1992. Beaver dam locations and their effects on distribution and abundance of coho salmon fry in two coastal Oregon streams. *Northwest Science* **66**:218–223.

LeResche, R.E., R.H. Bishop, and J.W. Coady. 1974. Distribution and habitats of moose in Alaska. *Le Naturaliste Canadien* **101**:143–178.

Levine, J.M. 2000*a*. Complex interactions in a streamside plant community. *Ecology* **81**:3431–3444.

Levine, J.M. 2000*b*. Species diversity and biological invasions: relating local process to community pattern. *Science* **288**:852–854.

Liang, S.Y. and S.W. Seagle. 2002. Browsing and microhabitat effects on riparian forest woody seedling demography. *Ecology* **83**:212–227.

Loft, E.R., J.W. Menke, and T.S. Burton. 1984. Seasonal movements and summer habitats of female black-tailed deer. *Journal of Wildlife Management* **48**:1317–1325.

Long, C.A. 1974. *Microsorex hoyi* and *Microsorex thompsoni*. *Mammalian Species* **33**:1–4.

Lovallo, M.J., J.H. Gilbert, and T.M. Gehring. 1993. Bobcat, *Felis rufus*, dens in an abandoned beaver, *Castor canadensis*, lodge. *Canadian Field-Naturalist* **107**:108–109.

Lunde, R.E. and A.S. Harestad. 1986. Activity of little brown bats in coastal forests. *Northwest Science* **60**:206–209.

Malanson, G.P. and D.R. Butler. 1990. Woody debris, sediment, and riparian vegetation of a subalpine river, Montana, USA. *Arctic and Alpine Research* **22**:183–194.

Marcot, B.G. 1996. An ecosystem context for bat management: a case study of the Interior Columbia River Basin, USA. Pages 19–36 *in* R.M.R. Barclay and R.M. Brigham, editors. Bats and forests symposium, October 19–21, 1995. Working Paper **23/1996**. British Columbia Research Branch, British Columbia Ministry of Forests, Victoria, British Columbia, Canada.

Marcum, C.L. 1976. Habitat selection and use during summer and fall months by a western Montana elk herd. Pages 91–96 *in* S.R. Heib, editor. *Elk-logging Symposium*. University of Idaho, Moscow, Idaho, USA.

Maser, C., B.R. Mate, J.F. Franklin, and C.T. Dyrness. 1981. Natural history of some Oregon coast mammals. US Forest Service General Technical Report **PNW-GTR-133**, Portland, Oregon, USA.

Mastenbrook, B. and H. Cummings. 1989. Use of residual strips of timber by moose within cutovers in northwestern Ontario. *Alces* **25**:146–155.

McCall, T.C., T.P. Hodgman, D.R. Diefenbach, and R.B. Owen, Jr. 1996. Beaver populations and their relation to wetland habitat and breeding waterfowl in Maine. *Wetlands* **16**:163–172.

McComb, W.C., K. McGarigal, and R.G. Anthony. 1993. Small mammal and amphibian abundance in streamside and upslope habitats of mature Douglas-fir stands, western Oregon. *Northwest Science* **67**:7–15.

McCorquodale, S.M., K.J. Raedeke, and R.D. Taber. 1986. Elk habitat use patterns in the shrub-steppe of Washington. *Journal of Wildlife Management* **50**:664–669.

McIntosh, B.A., J.R. Sedell, R.F. Thurow, S.E. Clarke, and G.L. Chandler. 2000. Historical changes in pool habitats in the Columbia River Basin. *Ecological Applications* **10**:1478–1496.

McKinstry, M.C., R.R. Karhu, and S.H. Anderson. 1997. Use of active beaver, *Castor canadensis*, lodges by muskrats, *Ondatra zibethicus*, in Wyoming. *Canadian Field-Naturalist* **111**:310–311.

Mech, S.G. and J.G. Hallett. 2001. Evaluating the effectiveness of corridors: a genetic approach. *Conservation Biology* **15**:467–474.

Medin, D.E. and W.P. Clary. 1989. Small mammal populations in a grazed and ungrazed riparian habitat in Nevada. US Forest Service Intermountain Research Station Research Paper **6**, Salt Lake City, Utah, USA.

Meiselman, N. and A.T. Doyle. 1996. Habitat and microhabitat use by the red tree vole (*Phenacomys longicaudus*). *American Midland Naturalist* **135**:33–42.

Melquist, W.E. and M.G. Hornocker. 1983. Ecology of river otters in west central Idaho. *Wildlife Monographs* **83**:1–60.

Merendino, M.T., G.B. McCullough, and N.R. North. 1995. Wetland availability and use by breeding waterfowl in southern Ontario. *Journal of Wildlife Management* **59**:527–532.

Morgan, L.H. 1986. *The American Beaver – A Classic of Natural History and Ecology*. Dover Publications, Toronto, Ontario, Canada.

Morrison, M.L. and R.G. Anthony. 1989. Habitat use by small mammals on early-growth clear-cuttings in western Oregon. *Canadian Journal of Zoology* **67**:805–811.

Murphy, M.L., J. Heifetz, J.F. Thedinga, S.W. Johnson, and K.V. Koski. 1989. Habitat utilization by juvenile Pacific salmon in the glacial Taku River, southeast Alaska. *Canadian Journal of Fisheries and Aquatic Sciences* **46**:1677–1685.

Naiman, R.J., C.A. Johnston, and J.C. Kelley. 1988. Alteration of North American streams by beaver. *BioScience* **38**:753–761.

Naiman, R.J., H. Decamps, and M. Pollock. 1993. The role of riparian corridors in maintaining regional biodiversity. *Ecological Applications* **3**:209–212.

Naiman, R.J., R.E. Bilby, and P.A. Bisson. 2000. Riparian ecology and management in the Pacific Coastal Rain Forest. *BioScience* **50**:996–1011.

Newman, D.G. and C.R. Griffin. 1994. Wetland use by river otters in Massachusetts. *Journal of Wildlife Management* **58**:18–23.

Nierenberg, T.R. and D.E. Hibbs. 2000. A characterization of unmanaged riparian areas in the central Coast Range of western Oregon. *Forest Ecology and Management* **129**:195–206.

Nixon, C. and J. Ely. 1968. Foods eaten by a beaver colony in southeast Ohio. *Ohio Journal of Science* **69**:313–319.

Oakley, A.L., J.A. Collins, L.B. Everson, D.A. Heller, J.C. Howerton, and R.E. Vincent. 1985. Riparian zones and freshwater wetlands. Pages 57–80 *in* E.R. Brown, editor. Management of wildlife and fish habitats in forests of western Oregon and Washington. US Forest Service **R6-F&WL-192-1985**, Portland, Oregon, USA.

O'Connell, M.A. and J.G. Hallett. 2000. East-side bat surveys *in* M.A. O'Connell, J.G. Hallett, S.D. West, K.A. Kelsey, D.A. Manuwal, and S.F. Pearson, editors. Effectiveness of riparian management zones in providing habitat for wildlife. Final report to Timber, Fish, and Wildlife Program. **TFW-LWAG1-00-001**, Washington Department of Natural Resources, Olympia, Washington, USA.

O'Connell, M.A., J.G. Hallett, and S.D. West. 1993. Wildlife use of riparian habitats: a literature review. **TFW-WL1-93-001**, Washington Department of Natural Resources, Olympia, Washington, USA.

O'Connell, M.A., J.G. Hallett, S.D. West, K.A. Kelsey, D.A. Manuwal, and S.F. Pearson, editors. 2000. Effectiveness of riparian management zones in providing habitat for wildlife. Final report to Timber, Fish, and Wildlife Program. **TFW-LWAG1-00-001**, Washington Department of Natural Resources, Olympia, Washington, USA.

O'Farrell, M.J. and E.H. Studier. 1980. *Myotis thysanodes. Mammalian Species* **137**:1–5.

Olson, D.L. 2000. Comparison of fire frequency between riparian and upland stands in the Blue Mountains, Oregon. Thesis. University of Washington, Seattle, Washington, USA.

O'Neil, T.A., D.H. Johnson, C. Barrett, M. Trevithick, K.A. Bettinger, C. Kiilsgarrd, M. Vander Heyden, E.L. Greda, D. Stinson, B.G. Marcot, P.J. Doran, S. Tank, and L. Wunder. 2001. Matrixes for wildlife-habitat relationships in Oregon and Washington. A Companion CD-Rom to D.H. Johnson, and T.A. O'Neil, managing directors. 2001. *Wildlife-habitat Relationships in Oregon and Washington.* Oregon State University Press, Corvallis, Oregon, USA.

Opperman, J.J. and A.M. Merenlender. 2000. Deer herbivory as an ecological constraint to restoration of degraded riparian corridors. *Restoration Ecology* **8**:41–47.

Ormsbee, P.C. and W.C. McComb. 1998. Selection of day roosts by female long-legged myotis. *Journal of Wildlife Management* **62**:596–602.

Pabst, R.J. and T.A. Spies. 1999. Structure and composition of unmanaged riparian forests in the coastal mountains of Oregon, USA. *Canadian Journal of Forest Research* **29**:1557–1573.

Parendes, L.A. and J.A. Jones. 2000. Role of light availability and dispersal in exotic plant invasion along roads and streams in the H.J. Andrews Experimental Forest, Oregon. *Conservation Biology* **14**:64–75.

Parker, D.I., J.A. Cook, and S.W. Lewis. 1996. Effects of timber harvest on bat activity in southeastern Alaska's temperate rainforests. Pages 277–292 *in* R.M.R. Barclay and R.M. Brigham, editors. Bats and forests symposium, October 19–21, 1995. Working Paper **23/1996**. British Columbia Research Branch, British Columbia Ministry of Forests. Victoria, British Columbia, Canada.

Peek, J.M. 1984. Northern Rocky Mountains. Pages 497–504 *in* L.K. Halls, editor. *White-tailed Deer: Ecology and Management*. Stackpole Books, Harrisburg, Pennsylvania, USA.

Peffer, R.D. 2001. Small mammal habitat selection in east slope Cascade Mountain riparian and upland habitats. Thesis. Eastern Washington University, Cheney, Washington, USA.

Perry, H.R., Jr. 1982. Muskrats (*Ondatra zibethicus* and *Neofiber alleni*). Pages 282–325 *in* J.A. Chapman and G.A. Feldhamer, editors. *Wild Mammals of North America*. Johns Hopkins University Press, Baltimore, Maryland, USA.

Peterson, N.P. 1982*a*. Immigration of juvenile coho salmon (*Oncorhynchus kisutch*) into riverine ponds. *Canadian Journal of Fisheries and Aquatic Science* **39**:1308–1310.

Peterson, N.P. 1982*b*. Population characteristics of juvenile coho salmon (*Oncorhynchus kisutch*) overwintering in riverine ponds. *Canadian Journal of Fisheries and Aquatic Science* **39**:1303–1307.

Pollock, M.M. 1995. Patterns of plant species richness in emergent and forested wetlands of southeast Alaska. Dissertation. University of Washington, Seattle, Washington, USA.

Pollock, M.M., R.J. Naiman, H.E. Erickson, C.A. Johnston, J. Pastor, and G. Pinay. 1994. Beaver as engineers: influences on biotic and abiotic characteristics of drainage basins. Pages 117–126 *in* C.G. Jones and J.H. Lawton, editors. *Linking Species to Ecosystems*. Chapman & Hall, New York, New York, USA.

Pollock, M.M., R.J. Naiman, and T.A. Hanley. 1998. Predicting plant species richness in forested and emergent wetlands – a test of biodiversity theory. *Ecology* **79**:94–105.

Potvin, F., L. Breton, C. Pilon, and M. Macquart. 1992. Impact of an experimental wolf reduction on beaver in Papineau-Labelle Reserve, Quebec, Canada. *Journal of Canadian Zoology* **70**:180–183.

Raedeke, K.J., R.D. Taber, and D.K. Paige. 1988. Ecology of large mammals in riparian systems of Pacific Northwest forests. Pages 113–132 *in* K.J. Raedeke, editor. *Streamside Management: Riparian Wildlife and Forestry Interactions*. University of Washington Press, Seattle, Washington, USA.

Raphael, M.G. and L.L.C. Jones. 1997. Characteristics of resting and denning sites of American martens in central Oregon and western Washington. Pages 146–165 *in* G. Proulx, H.N. Bryant, and P.M. Woodward, editors. *Martes: Taxonomy, Ecology, Techniques, and Management*. Provincial Museum of Alberta, Edmonton, Alberta, Canada.

Rea, A.M. 1983. *Once a River: Bird Life and Habitat Change on the Middle Gila*. University of Arizona Press, Tucson, Arizona, USA.

Reese, K.P. and J.D. Hair. 1976. Avian species diversity in relation to beaver pond habitats in the Piedmont region of South Carolina. *Proceedings of the Annual Conference, Southeastern Association of Game and Fish Commissioners* **30**:437–447.

Ricca, M.A. 1999. Movements, habitat associations, and survival of Columbian white-tailed deer in western Oregon. Thesis. Oregon State University, Corvallis, Oregon, USA.

Riley, D.G. 1989. Controlling raccoon damage in urban areas. Pages 85–86 *in* A.J. Bjugstad, D.W. Uresk, and R.R. Hamre, editors. Ninth Great Plains Wildlife Damage Control Workshop Proceedings, US Forest Service General Technical Report **RM-GTR-171**, Fort Collins, Colorado, USA.

Roath, L.R. and W.C. Krueger. 1982. Cattle grazing and behavior on a forested range. *Journal of Range Management* **35**:100–103.

Roberts, T.H. and D.H. Arner. 1984. Food habits of beaver in east-central Mississippi. *Journal of Wildlife Management* **48**:1414–1419.

Rosenberg, D.K. and R.G. Anthony. 1992. Characteristics of northern flying squirrel populations in young second- and old-growth forests. *Canadian Journal of Zoology* **70**:161–166.

Rudemann, R. and W.J. Schoonmaker. 1938. Beaver dams as geological agents. *Science* **88**:523–524.

Ruel, J.C. 2000. Factors influencing windthrow in balsam fir forests: from landscape to individual tree studies. *Forest Ecology and Management* **135**:169–178.

Russell, K.R., C.E. Moorman, J.K. Edwards, B.S. Metts, and D.C. Guynn, Jr. 1999. Amphibian and reptile communities associated with beaver (*Castor canadensis*) ponds and unimpounded streams in the Piedmont of South Carolina. *Journal of Freshwater Ecology* **14**:149–158.

Russell, W.H. and J.R. McBride. 2001. The relative importance of fire and watercourse proximity in determining stand composition in mixed conifer riparian forests. *Forest Ecology and Management* **150**:259–265.

Schreiner, E.G., K.A. Krueger, P.J. Happe, and D.B. Houston. 1996. Understory patch dynamics and ungulate herbivory in old-growth forests of Olympic National Park, Washington. *Canadian Journal of Forest Research* **26**:255–265.

Schultz, T.T. and W.C. Leininger. 1991. Nongame wildlife communities in grazed and ungrazed montane riparian sites. *Great Basin Naturalist* **51**:286–292.

Sealander, J.A. 1943. Winter food habits of mink in southern Michigan. *Journal of Wildlife Management* **7**:411–417.

Seidman, V.M. and C.J. Zabel. 2001. Bat activity along intermittent streams in northwestern California. *Journal of Mammalogy* **82**:738–747.

Shelton, P.C. and R.O. Peterson. 1983. Beaver, wolf, and moose interactions in Isle Royale National Park, USA. *Acta Zoologica Fennica* **174**:265–266.

Sherfy, F.C. and J.A. Chapman. 1980. Seasonal home range and habitat utilization of raccoons in Maryland. *Carnivore* **3**:8–18.

Simon, T.L. 1980. An ecological study of the marten in the Tahoe National Forest, California. Thesis. California State University, Sacramento California, USA.

Smith, D.W., D.R. Trauba, R.K. Anderson, and R.O. Peterson. 1994. Black bear predation on beavers on an island in Lake Superior. *American Midland Naturalist* **132**:248–255.

Smith, W.P. 1985. Current geographic distribution and abundance of Columbian white-tailed deer, *Odocoileus virginianus leucurus* (Douglas). *Northwest Science* **59**:243–251.

Smith, W.P., R.G. Anthony, J.R. Waters, N.L. Dodd, and C.J. Zabel. 2003. Ecology and conservation of arboreal rodents of western coniferous forests. Pages 157–206 *in* C.J. Zabel and R.G. Anthony, editors. *Mammal Community Dynamics. Management and Conservation in the Coniferous Forests of Western North America.* Cambridge University Press, Cambridge, UK.

Smollen, M.J. and B.L. Keller. 1987. *Microtus longicaudus. Mammalian Species* **271**:1–7.

Snyder, E.J. and L.B. Best. 1988. Dynamics of habitat use by small mammals in prairie communities. *American Midland Naturalist* **119**:128–136.

Spencer, W.D. 1981. Pine marten habitat preferences at Sagehen Creek, California. Thesis. University of California, Berkeley, California, USA.

Spencer, W.D., R.H. Barrett, and W.J. Zielinski. 1983. Marten habitat preferences in the northern Sierra Nevada. *Journal of Wildlife Management* **47**:1181–1187.

Stauffer, D.F. and L.B. Best. 1980. Habitat selection by birds of riparian communities: evaluating effects of habitat alterations. *Journal of Wildlife Management* **44**:1–15.

Steel, E.A., R.J. Naiman, and S.D. West. 1999. Use of woody debris piles by birds and small mammals in a riparian corridor. *Northwest Science* **73**:19–26.

Steele, M.A. 1999. *Tamiasciurus douglasii*. *Mammalian Species* **630**:1–8.

Stinson, D.W. and F.F. Gilbert. 1985. *Wildlife of the Spokane Indian Reservation: a Predictive Model*. USDI Bureau of Indian Affairs, Spokane Agency, Wellpinit, Washington, USA.

Strahler, A.N. 1957. Quantitative analysis of watershed geomorphology. *American Geophysics Union Transactions* **83**:913–920.

Suzuki, N. 1992. Habitat classification and characteristics of small mammal and amphibian communities in beaver-pond habitats of the Oregon Coast Range. Thesis. Oregon State University, Corvallis, Oregon, USA.

Swales, S. and C.D. Levings. 1989. Role of off-channel ponds in the life cycle of coho salmon (*Oncorhynchus kisutch*) and other juvenile salmonids in the Coldwater River, British Columbia, Canada. *Canadian Journal of Fisheries and Aquatic Science* **46**:232–242.

Swanson, F.J. 1980. Geomorphology and ecosystems. Pages 159–170 *in* R.H. Warning, editor. *Forests: Fresh Perspectives from Ecosystem Analysis. Proceedings of the 40th Annual Biology Colloquium*. Oregon State University Press, Corvallis, Oregon, USA.

Swanson, F.J., S.V. Gregory, J.R. Sedell, and A.G. Campbell. 1982. Land-water interactions: the riparian zone. Pages 267–291 *in* R.L. Edmonds, editor. *Analysis of Coniferous Ecosystems in the Western United States*. Hutchinson Ross Publishing Company, Stroudsburg, Pennsylvania, USA.

Taber, R.D. 1976. Seasonal landscape use by elk in the managed forests of the Cedar River drainage, western Washington. US Forest Service Report **FS-PNW-14**, College of Forest Resources, University of Washington, Seattle, Washington, USA.

Terry, C.J. 1981. Habitat differentiation among three species of *Sorex* and *Neurotrichus gibbsii* in Washington. *American Midland Naturalist* **106**:119–125.

Thomas, D.W. and S.D. West. 1991. Forest age associations of bats in the Washington Cascades and Oregon Coast Ranges. Pages 295–303 *in* L.F. Ruggiero, K.B. Aubry, A.B. Carey, and M.H. Huff, editors. Wildlife and vegetation in unmanaged Douglas-fir forests. US Forest Service General Technical Report **PNW-GTR-285**, Portland, Oregon, USA.

Thomas, J.W., C. Maser, and J.E. Rodieck. 1979. Riparian zones. Pages 40–47 *in* J.W. Thomas, editor. Wildlife habitats in managed forests: the Blue Mountains of Oregon and Washington. US Forest Service Agricultural Handbook No **553**, Portland, Oregon, USA.

Unsworth, J.W., J.J. Beecham, and L.R. Irby. 1989. Female black bear habitat use in west-central Idaho. *Journal of Wildlife Management* **53**:668–673.

Verts, B.J. and L.N. Carraway. 1998. *Land Mammals of Oregon*. University of California Press, Berkeley, California, USA.

Wales, B.C. 2001. The management of insects, diseases, fire, and grazing and implications for terrestrial vertebrates using riparian habitats in eastern Oregon and Washington. *Northwest Science* **75**:119–127.

Waller, D.W. 1962. Feeding behavior of minks at some Iowa marshes. Thesis. Iowa State University, Ames, Iowa, USA.

Waters, J.R. and C.J. Zabel. 1995. Northern flying squirrel densities in fir forests of northeastern California. *Journal of Wildlife Management* **59**:848–866.

Waters, J.R., C.J. Zabel, K.S. McKelvey, and H.H. Welsh, Jr. 2001. Vegetation patterns and abundances of amphibians and small mammals along small streams in a northwestern California watershed. *Northwest Science* **75**:37–52.

Weller, T.J. and C.J. Zabel. 2001. Characteristics of fringed myotis day roosts in northern California. *Journal of Wildlife Management* **65**:489–497.

West, S.D. 2000*a*. West-side small mammal surveys. Pages 10.1–10.45 *in* M.A. O'Connell, J.G. Hallett, S.D. West, K.A. Kelsey, D.A. Manuwal, and S.F. Pearson, editors. Effectiveness of riparian management zones in providing habitat for wildlife. Final report to Timber, Fish, and Wildlife Program **TFW-LWAG1-00-001**, Washington Department of Natural Resources, Olympia, Washington, USA.

West, S.D. 2000*b*. West-side bat surveys. Pages 12.1–12.21 *in* M.A. O'Connell, J.G. Hallett, S.D. West, K.A. Kelsey, D.A. Manuwal, and S.F. Pearson, editors. Effectiveness of riparian management zones in providing habitat for wildlife. Final report to Timber, Fish, and Wildlife Program **TFW-LWAG1-00-001**, Washington Department of Natural Resources, Olympia, Washington, USA.

Whitaker, J.O., Jr. 1972. Food habits of bats from Indiana. *Canadian Journal of Zoology* **50**:877–883.

Whitaker, J.O., Jr., C. Maser, and L.E. Keller. 1977. Food habits of bats of western Oregon. *Northwest Science* **51**:46–55.

Willner, G.R., G.A. Feldhamer, E.E. Zucker, and J.A. Chapman. 1980. *Ondatra zibethicus*. *Mammalian Species* **141**:1–8.

Willson, M.F. and K.C. Halupka. 1995. Anadromous fish as keystone species in vertebrate communities. *Conservation Biology* **9**:489–497.

Wilson, C., R.E. Johnson, and J.D. Reichel. 1980. New records for the northern bog lemming in Washington. *Murrelet* **61**:104–106.

Wilson, K.A. 1954. Mink and otter as muskrat predators in northeastern North Carolina. *Journal of Wildlife Management* **18**:199–207.

Wimberly, M.C. and T.A. Spies. 2001. Influences of environment and disturbance on forest patterns in coastal Oregon watersheds. *Ecology* **82**:1443–1459.

Young, D.D. and J.J. Beecham. 1986. Black bear habitat use at Priest Lake, Idaho. *International Conference on Bear Research and Management* **6**:73–80.

Young, K.A. 2000. Riparian zone management in the Pacific Northwest: who's cutting what? *Environmental Management* **26**:131–144.

Young, S.P. 1958. *The Bobcat in North America*. Wildlife Management Institute, Washington, DC, USA.

Zielinski, W.J. 1981. Food habits, activity patterns, and ectoparasites of the pine marten at Sagehen Creek, California. Thesis. University of California, Berkeley, California, USA.

Part III

Conservation issues and strategies

16
_____

Small mammals in a landscape mosaic: implications for conservation

Introduction

Johnson (1980) provided a conceptual framework to describe habitat selection that identified four scales of selection: geographic range, home range, resource patches within home ranges, and specific resources necessary for survival (Fig. 16.1). The spatial scales over which these patterns of selection occur vary widely among species of mammals found in northwest coniferous forests. Harris (1984) estimated home range sizes for mammals in the Pacific Northwest and indicated they have a home range size distribution described by a negative exponential function (Fig. 16.2). Natural disturbance regimes common in Pacific Northwest coniferous forests (Spies and Turner 1999; Fig. 16.3) result in a distribution of patch sizes remarkably similar to this home range size distribution with frequent disturbances (sun flecks, small gaps, herbivores, and pathogens) creating small patches, and infrequent disturbances (fire, wind) creating larger patches. Indeed, inherent scales of disturbance in northwest coniferous forests under natural conditions likely represent the range of conditions and landscape mosaics that match evolved life histories of mammal species throughout the region. However, the effects of alterations in the spatial and temporal variability, pattern, and composition of the landscape caused by recent human disturbances, particularly timber harvest, raise questions regarding the continued persistence of diverse mammalian communities in managed forests (Lawlor 2003).

Landscapes are defined as the heterogeneous land area composed of an interacting mosaic of patches, at any scale, relevant to the phenomenon (e.g., species) under consideration (McGarigal and Marks 1995, McGarigal and McComb 1995). Landscapes are the template upon which individuals

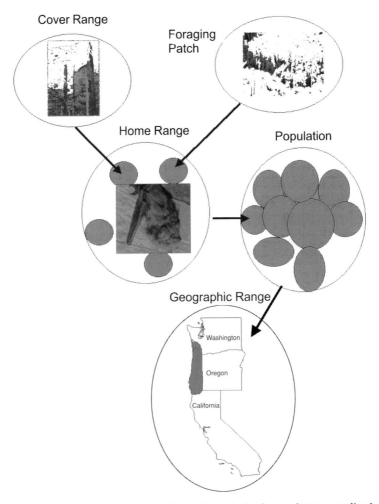

Fig. 16.1. Schematic illustration describing habitat selection by vertebrates as outlined by Johnson's (1980) conceptual framework. Each species scales its environment differently to meet needs at each stage in the habitat selection process.

and populations occur over space and time. Individuals respond to changing local conditions due to the dominant disturbances that collectively result in a population response to changes in the habitat mosaic over space and time. It is this concept that makes consideration of the historic range of variability in landscape conditions intuitively appealing as a basis for formatting management guidelines (Landres et al. 1999). Species respond

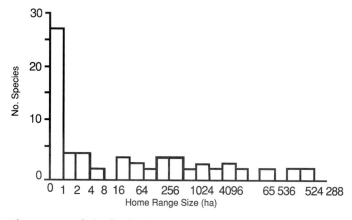

Fig. 16.2. Cumulative distribution of home range sizes for mammals in Pacific Northwest Forests (adapted from Harris 1984).

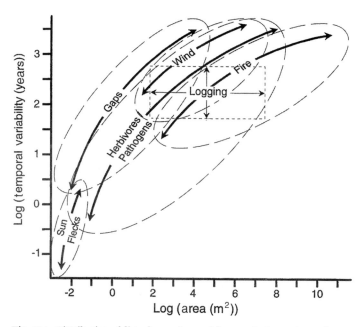

Fig. 16.3. Distribution of disturbance sizes and frequencies in northwest forests (from Spies and Turner 1999).

to these changes differently because they perceive the environment differently, and this is a function of the requirements each species has for survival and persistence. Some species are explicitly associated with specific features of the environment that are not uniformly distributed across forested landscapes. For example, western red-backed voles (*Clethrionomys californicus*) have been associated with coarse woody debris, conifer patch types, and patches that minimize the amount of edge habitat (Tallmon and Mills 1994, Mills 1995, Martin and McComb 2002). Consequently this species tends to be unevenly distributed across landscapes, and may be more likely to form metapopulations on dynamic forested landscapes. At the other end of a habitat specificity spectrum, there are species that can meet their food and cover needs from a wide range of resources and are more panmictic in their distribution, such as deer mice (*Peromyscus maniculatus*). Somewhere between lie other species, such as Pacific marsh shrews (*Sorex bendirii*). This species requires water to meet resource requirements, and is therefore distributed linearly along riparian systems, and may have neither a strict metapopulation nor a panmictic structure.

Spatial structure of mammalian communities and their dynamics are highly variable and are associated with disturbance histories, geomorphology, and life history traits. Each species perceives and utilizes a landscape differently. Often we view landscape configuration and composition through human perceptions, and use pre-defined "habitat types" that represent vegetation composition and seral stages. Many species do not perceive the forested landscape in the same terms as we do, however. Delineation of habitat patches for a Trowbridge's shrew (*Sorex trowbridgii*) would include all coniferous forest habitat types. Pocket gopher (*Thomomys* spp.) patches might be best mapped by considering soil conditions as well as vegetation, but likely cross many vegetation boundaries created by disturbances. Habitat patches for Pacific jumping mice (*Zapus trinotatus*) span many vegetative conditions, but are associated with water (Brown 1985, Doyle 1990, Martin 1998). Structure, composition, changes, and recovery of habitat for any species will differ because habitat is a species-specific concept.

Ensuring the long-term persistence and movement of mammals across forested landscapes will have forest-wide implications. Mammals are an integral part of forested landscapes and contribute to a number of processes and functions that influence the structure and composition of both flora and fauna. Many of the insectivores contribute to energy flow through detrital-based systems (Maser et al. 1988). Other species, such as California red-backed voles and northern flying squirrels (*Glaucomys sabrinus*) are key dispersers of spores of hypogeous mycorrhizal fungi (Z. Maser et al. 1985, C. Maser et al. 1986, Luoma et al. 2003, Mills et al. 2003). Mammals serve as both predators and prey for a number of other vertebrates and invertebrates, thereby contributing to complex trophic structures. In this chapter we provide an overview of landscape composition and configuration of patches within a landscape, consider the relative importance of habitat area and pattern to mammal occurrence or abundance, discuss the implications of changing landscape connectivity across complex landscapes, and provide some basic principles of landscape management that might contribute to maintenance of mammal populations over space and time.

Landscape structure

Landscapes defined for any given species can consist of patches that vary in quality for that species. Probability of occurrence, abundance, or fitness of organisms varies across landscapes in a manner often associated with a subset of physical and vegetative conditions. This distribution may be further modified or refined based on density-dependent factors only secondarily related to the physical, climatic or vegetative condition. Nonetheless, the continuum of habitat quality represented over an area can be defined into discrete patches based on any of the factors related to animal occurrence or fitness. Landscape structure is the composition and spatial configuration of these patches across a landscape. The composition of a landscape is then defined as the area of these patch types over the landscape and spatial configuration is the pattern of these patches across the landscape.

The term "fragmentation" often is confused with loss of area of a particular patch type. Fragmentation is related to but separate from habitat loss (Fig. 16.4). Fragmentation is a process that alters the pattern of a focal patch type, leading to patches that are more complex, isolated, and distributed over the landscape. Perhaps it is more productive to think of landscape configuration as that pattern of patch types that may impact the quality of landscapes for species rather than as fragmentation of forests. Increasing edge density between vegetative conditions, one measure of landscape configuration, will likely increase habitat quality for some species, while decreasing it for others. Characterizing the configuration and composition of a landscape can allow us to understand

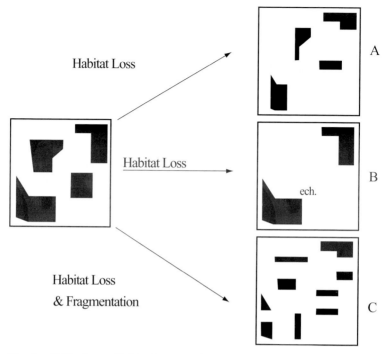

Fig. 16.4. Habitat loss and habitat fragmentation; though often related are different processes (adapted from Fahrig 1997).

implications for a range of species. Pattern of patch types across an area theoretically can influence movements, metapopulation structure, and demographic processes, including genetic representation (Mills 1995, Mills and Tallmon 1999, Mills et al. 2003). Here we consider the existing evidence for associations of mammals with habitat area and habitat patterns to understand the likely consequences of altering forest landscapes through management activities.

Associations between habitat area and animal abundance

Habitat area is probably the most important landscape factor associated with mammal and bird abundance across a range of landscape patterns (McGarigal and McComb 1995, 1999, Trzcinski et al. 1999). It is intuitive that animal abundance will decline with habitat area up to a point where the remaining patches become isolated and/or smaller than the organism's home range. Isolation of organisms results in reduced dispersal,

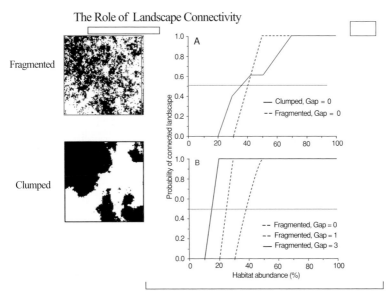

Fig. 16.5. Relationship between loss of habitat and loss of connectivity of habitat (from With 1999). A gap-crossing ability of 0 (i.e., gap = 0) represents a species that is unable to cross any gap; increasing gap numbers (i.e., gap = 1, 2, 3) represent species that have increased gap-crossing abilities.

reduced genetic flow, and increased probability of complete population loss resulting from a catastrophic disturbance. The severity of isolation effects that result from loss of habitat area is dependent on the pattern of habitat loss (With 1999). Two species-specific factors influencing the impact of these changes include demographic potential of the organism and the ability of the organism to survive while traveling through non-habitat areas (Fahrig 1997, 1999). Loss of habitat area will lead to isolation effects when habitat area is reduced below a threshold that is dependent on the pattern of habitat loss (With 1999; Fig. 16.5). As habitat area is further reduced to 10% to 30% of the landscape, species declines will likely accelerate because of factors associated with predation, mortality during dispersal, and potential climatic impacts (Venier and Fahrig 1996, Dooley and Bowers 1998, With 1999). Simulation modeling has shown significant reductions in survival probability once habitat area is reduced below 20% (Fahrig 1999), but this has yet to be demonstrated in natural systems.

In western Oregon, significant associations were detected between habitat area and capture rates of six species of mammals, but no relationship was detected for eight of 14 species (Table 16.1, Martin and

Table 16.1. *Associations between capture rates of 14 small mammal species and the composition and pattern of patch types selected by each species. Small mammals were captured in the central Oregon Coast Range during April and May 1995 and 1996*

Species	Landscape composition[a]				Landscape pattern[b]				
	F	P	R^2	N^c	Variable	F	P	R^2	
Western red-backed vole	6.18	0.02	+0.18	20	Area-weighted mean patch fractal dimension	12.59	0.003	−0.40	
Shrew mole	0.04	0.85	0.001	18	Contrast-weighted edge density	7.10	0.02	−0.32	
Deer mouse	8.50	0.01	+0.23	10	Edge density	113.60	<0.001	+0.32	
White-footed vole	1.81	0.19	0.06	20	Core area density	4.20	0.06	−0.18	
Red tree vole	1.60	0.22	0.05	17	Patch density	4.81	0.05	−0.19	
Coast mole	2.25	0.15	0.07	6^d					
Pocket gopher	1.95	0.17	0.06	4^d					
Oregon vole	19.91	<0.001	+0.42	4^d					
Marsh shrew	2.18	0.15	0.07	17					
Pacific shrew	5.76	0.02	+0.17	26					
Fog shrew	0.26	0.61	0.01	21					
Trowbridge's shrew	0.91	0.35	0.03	16					
Vagrant shrew	11.25	0.002	+0.29	20					
Pacific jumping mouse	7.89	0.01	+0.22	20					

[a] Area of selected patch types across all 30 landscapes sampled.
[b] Pattern variables with a significant relationship ($P \leq 0.05$).
[c] Number of landscapes with > 20% of selected patch types used in pattern analysis.
[d] Sample size is too small to analyze pattern indices.

McComb 2002). Specifically, western red-backed voles, deer mice, Oregon voles (*Microtus oregoni*), Pacific shrews (*S. pacificus*), and Pacific jumping mice were all associated with habitat area at the landscape scale. To our knowledge, this has been the only landscape-based study of habitat area and pattern in northwest coniferous forests. There have been several patch-based analyses which also suggest that habitat amount might be important to mammal abundance or occurrence. Rosenberg and Raphael (1986) reported associations between patch size and abundance in three of 18 species of mammals. In the Olympic Peninsula of Washington, Lomolino and Perault (2000) found no significant differences in species composition or species richness among fragments, corridors, and continuous tracts of old-growth forests. However, capture rates of individual species varied among these configurations of old-growth conditions.

Associations between habitat pattern and animal abundance

Several authors have suggested that characteristics of landscape pattern are associated with the occurrence, abundance, or genetic diversity of species in western coniferous forests. Both positive and negative effects of high-contrast edges have been described, with considerable variability in response to these edges among the species examined (Mills 1995, Martin 1998). Deer mice and Pacific jumping mice in the central Oregon Coast Range were positively associated with edges, but none of the 14 mammalian species captured were negatively associated with edges, including western red-backed voles (Martin 1998). In southwestern Oregon, Mills (1995) reported six times more captures of western red-backed voles at the edges of forest remnants than at traps located in the interior. Consequently, it is not surprising that another measure of adverse edge effects, core area of habitat patches, has also been suggested as an important feature for some species. Few, if any, relationships have been identified between mammal occurrence or fitness and patch isolation; however, Mills (1995) demonstrated the potential for adverse effects on western red-backed voles as a result of patch isolation from commercial clearcutting. Although evidence for associations of mammal abundance or fitness with landscape pattern is often lacking, inconsistent, or at times counterintuitive (Lehmkuhl et al. 1991, Martin and McComb 2002), there is reason to consider pattern effects on mammal populations, especially under conditions of low habitat area (With 1999). Reductions in the habitat area of fragments may result in increases in the population of certain small mammals across a landscape by reducing social costs for individuals (i.e., intra-specific competition) and enhancing food resources for remaining individuals (Dooley and Bowers 1998). Further, With (1999) makes a compelling case for consideration of connectivity across complex landscapes when habitat area drops below 20% to 30% depending on the pattern of habitat loss (Fig. 16.5). There are a number of reasons why investigators may not have detected such effects, perhaps the most important being that no empirical landscape-based study has been specifically designed to test for these effects under conditions of low habitat area. Until data clearly indicate that the dynamic nature of forests mitigates isolation effects, it is prudent to assume that pattern effects can occur when habitat area is reduced below some species-specific threshold. Although there is reason to consider pattern effects at low levels of habitat availability, the theorized effects are highly dependent on gap-crossing ability. We have little to no information on gap-crossing abilities of native mammals, thus a cautious course of action is warranted. Until data prove otherwise, it may be best to assume that this ability is limited for some species.

Species that have evolved in naturally patchy systems tend to occupy patch environments and are more able to disperse across patchy landscapes (Wolff 1999, 2003). In contrast, habitat specialists may have evolved in a relatively homogeneous habitat and find highly fragmented landscapes a barrier to dispersal (Wolff 1999, 2003). An example of a forest habitat specialist would be the marten (*Martes* spp.) (Bissonette and Broekhuizen 1995, Buskirk and Zielinski 2003), a species that rarely travels more than 25 m into open habitat. Preference for arboreal habitats has also been suggested as a potential factor reducing a species' ability to disperse across patchy landscapes (Laurence 1995, Smith et al. 2003). In addition, species that are highly territorial will be limited to the number of breeding territories available within a patch (Wolff 1997, 1999). The interaction of patch size, quality of habitat within a patch, and the home range size of the species are factors in determining the viability of a habitat patch.

Connectivity in landscape mosaics

In theory, some of the negative effects of isolation can be mitigated by identifying and maintaining fixed or dynamic connections across the landscape. Connectivity can be defined as either continuity, the physical connectivity of habitat, or connectedness, the functional connectivity of habitat. Corridors represent one type of connection that can be designed for those species with poor gap-crossing ability. Maintaining a more hospitable matrix condition among patches is an alternative to corridors. With the wide range of species and habitat types comprising western coniferous forests, managers will probably need to incorporate both matrix conditions and corridors into their management plans (Mills et al. 2003). In either case, it may be helpful to think of the demographic basis for developing connections.

Dispersal capabilities vary widely among species, likely being quite limited for some species (e.g., red tree voles (*Arborimus longicaudus*); Hayes 1996) considered dispersal specialists and quite extensive for others considered dispersal generalists (e.g., Pacific jumping mice; Gannon 1988). In either case, the probability of successful dispersal across a landscape can be considered a function of the probability of survival during dispersal. Consider that the daily probability of survival for an organism will vary among patch types. In island biogeography theory, there may be two patch types: habitat (daily probability of survival >0) and non-habitat (daily probability of survival $= \sim 0$). Although island biogeography theory is useful for islands of habitat surrounded by water of non-habitat, its application to forested systems is limited (Bunnell 1999). For most landscapes, patch types may form a mosaic representing a range of survival probabilities. Time-specific probability of survival is equal to the daily probability of survival raised to the power of the number of days spent in the patch type:

$$PS = PSD^d, \text{where}$$

PS $=$ probability of survival in a patch; PSD $=$ daily probability of survival, and $d =$ the number of days spent in the patch. For a dispersing individual, d will be dependent on the rate of movement (distance/day), and the distance traveled in the patch. Species that are more mobile and have higher movement rates will realize higher survival rates in low-quality patches than less mobile species. Large patches of poor-quality habitat suggest that more distance must be covered getting through the patch and hence patch-specific survival probability declines. Clearly, some of the problems associated with low survival probabilities can be overcome by sheer numbers of dispersers. For example, consider that the number of successful dispersers among patches is a function of the cumulative probability of survival per individual among all patches crossed, animal density in the source patch, source patch size, and the percent of the population that are dispersers:

$$NSD = PS \times AD \times AREA \times PROP$$

where, NSD = number of successful dispersers, PS = cumulative probability of survival across the patch mosaic, AD = animal density, AREA = source habitat area, and PROP = the proportion of the population that are dispersers. For species with high reproductive rates, low probability of survival during dispersal may not be significant because of sheer numbers of dispersers. But for species that have low reproductive rates, are rather immobile, occur at low densities for a given body size, and may have daily survival probabilities that vary considerably among patch types, isolation can be a significant problem (Wolff 1999, 2003). In general, both median and maximum dispersal distances are directly correlated with body size (Sutherland et al. 2000). Dietary classification (e.g., omnivore, herbivore, or carnivore) has also been associated with dispersal distances (Sutherland et al. 2000). Use of these characteristics allows us to understand which species are relatively more at risk from isolation effects compared to others. On a continuum of small mammal species, western red-backed voles may represent one extreme and deer mice the other. Maintaining connectivity for western red-backed voles will be a much higher management priority, because it is clearly at the highest risk of adverse effects from isolation (Mills 1995).

Assessing the impacts of connectivity
So what is the best way to increase connectivity within a landscape? Unfortunately, little information is available on the gap-crossing abilities of mammals in northwestern forests and the variability in gap resistance among forest conditions. Gap resistance is a species-specific term and is the inverse of survivability (i.e., low survivability = high gap resistance). Additionally, resistance of habitat probably varies temporally. Cool wet winters and hot dry summers likely represent periods of movement facilitation or reduction for various species. Information is needed on how dispersed patches used as stepping stones during dispersal and corridors are effective in improving connectivity within a landscape. Potential approaches to assess the dispersal capabilities (frequency of dispersal and distance dispersed) of mammals include radiotelemetry, capture-recapture techniques, and use of genetic tools (Mills et al. 2003). These studies should focus on young individuals since they are more likely to disperse than adults, and they should be conducted throughout the year to detect temporal variations in dispersal.

Quality of intervening habitat is a critical component that needs to be quantified for species of concern. Recent analyses of small mammal occurrences across mountaintops in the Great Basin in Western North America found relatively homogenized populations of small mammals (Lawlor 1998). Habitat area and isolation were not correlated with species richness, likely the result of successful dispersers across areas of intervening desert. The author also concluded that area-sensitive species with limited dispersal capabilities may be absent from the mountaintop habitat as a result of earlier extinctions caused by the isolated landscape (Lawlor 2003; but see Lawlor 1998).

Landscape-level conservation strategies for mammals

Based on the work that has been described in this chapter, the following general principles may be used to guide management decisions with regard to reducing risks to species in managed landscapes. Clearly, habitat area is related to animal occurrence and abundance. Obtaining an estimate of the area of habitat for an organism is not the same as obtaining an estimate of the area of a vegetation type. Habitat quality models similar to habitat suitability index models have been developed. They consider not only within-patch conditions (e.g., tree species and sizes, shrub cover, and coarse woody debris biomass), but also matrix conditions in the neighborhood surrounding the patch over an area that is scaled to the home range of the organism (Fig. 16.6). This approach allows patches on a landscape to be scaled in quality for each organism of interest. Ranges of habitat quality can then be defined to allow mapping of habitat quality over the landscape. When linked to a forest dynamics model, these maps allow the manager and planner the opportunity to understand not only the current representation of habitat area across the landscape, but also likely future conditions. By understanding likely outcomes, planners may be able to make management decisions now that can circumvent a future problem that otherwise would not be immediately apparent. It also allows planners the ability to use optimization procedures to find solutions to land-planning problems that must balance species risk with economic interests (Bettinger et al. 2001).

When the area of habitat for a species falls below 20% to 30% of the landscape (With 1999), managers need to identify connections across the landscape to reduce isolation risks. Connections may be in the form of static or dynamic corridors connecting specific patches on the landscape (Sessions et al. 1998, Fig. 16.7). Alternatively, management actions can be taken in

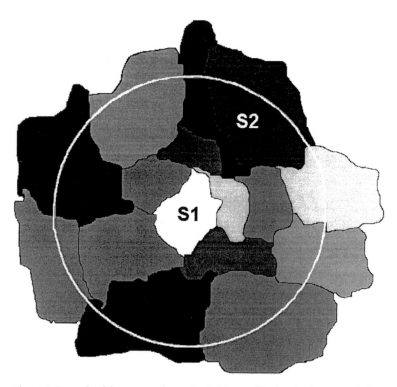

Fig. 16.6. Example of the process of assessing habitat quality for a landscape patch that considers the conditions within the patch and the conditions in the neighborhood surrounding the patch. S1 and S2 represent two patch types that currently provide quality habitat for an organism using this landscape.

the matrix condition among focal patches in an attempt to increase the number of successful dispersers among patches. In addition, maintaining large focal patches to ensure a large number of potential dispersers, keeping focal patches as close together as possible, and considering the risks to focal patches from disturbance may also help mitigate risk of isolation. These guidelines provide a measure of conservatism to management activities that theoretically seems warranted. We will need well-designed studies to determine if such safeguards are necessary to ensure the persistence of mammal populations over these managed landscapes.

Management guidelines

Since species have evolved with natural disturbance regimes, we would expect species to respond in a positive manner to management actions that

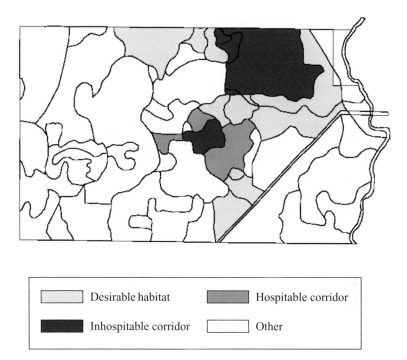

Fig. 16.7. Example of a dynamic corridor connecting patches on a landscape. The polygons represent patches of vegetation on the landscape that change over time through disturbance and re-growth. This figure represents one step in time that shows an acceptable connection among the dark gray polygons. The location of the connecting patches may change over time, while still ensuring connectivity between the focal patches (adapted from Sessions et al. 1998).

mimic those disturbance regimes. As an example, in the Pacific Northwest there is a similar pattern between home range sizes of small mammals and the size of patches created through natural disturbances. With the exception of wilderness areas, land managers do not use natural disturbance techniques to manage landscapes in western forests. Timber harvesting, tree planting, herbicide applications, controlled burning, and thinning are the most commonly used management tools. Using these techniques to mimic natural disturbances is one of the premises behind the field of silviculture. A key component is to manage landscapes with spatial and temporal variation across landscapes observed in natural systems.

Conserving small mammals in the western forests of North America begins with maintaining the landscape composition or habitat area necessary to facilitate long-term survival in these forests. It is also important to consider the pattern of habitat with relation to species-dispersal capabilities and the juxtaposition of habitat types across the landscape. Edges, particularly high-contrast edges, have the potential to impact a variety of species and should be minimized across the landscape. When the area of suitable habitat for a habitat specialist fall below 20% to 30% of the landscape, it may be useful to increase the connectedness of patches across the landscape to ensure long-term survival of these species.

An example of a management guideline for western red-backed voles could be to minimize fragmentation over time and maintain levels of coarse woody debris in both dispersal and suitable habitats. This would allow western red-back voles to persist in available habitat and successfully disperse. Specifically, this would involve managing for large patches within a landscape, with corridors or stepping stones connecting suitable habitats across both spatial and temporal variations. This will often involve long-term modeling of landscapes, integration of spatial and temporal scales through the use of Geographical Information Systems (GIS), and long-term resource planning.

Information needs

Small mammals are abundant and widely distributed, making them an ideal group for landscape-scale investigations. Little is known about the impact of structural change (composition and pattern) of landscapes on small mammals. Important components include habitat associations that are well documented. Areas with relatively little information for species in western forests include dispersal capabilities, impact of unsuitable habitat on dispersal, scales of habitat selection and interaction of scales in the habitat selection process, and the impact of fragmentation on species survival.

Summary

Human activities have changed the variability, composition, and pattern of forested landscapes in the Pacific Northwest. This deviation from natural disturbances has resulted in new landscape mosaics that may have an adverse impact on some mammalian populations while benefiting others. Maintaining habitat area and similar historic patterns necessary to provide resources for indigenous mammals is a management priority in northwest forests if we are to ensure long-term persistence of these species. However, habitat selection and responses to changes in the landscape are species specific, making the role of management challenging. Species that are relatively immobile, have low reproductive rates, and specialized habitat requirements will probably be most vulnerable to changes in the landscape mosaic.

Composition is the most important feature affecting the occurrence and abundance of vertebrates on the landscape. Landscape pattern has also been shown to impact mammalian and avian distributions, but the relative impact of pattern is less studied and varies considerably. Connectivity is a mitigating factor of pattern, providing individuals within a population a habitat in which to disperse among patches. Connectivity provides for exchange of genetic information and the ability to re-populate distant areas as habitat improves.

Species such as the red tree vole and western red-backed vole have been identified as species of concern because of their unique habitat requirements. However, information on how most mammals in the Pacific Northwest coniferous forest react to landscape-scale alterations is unknown. Until further research is conducted to assess the impacts of these alterations, a conservative management approach of maintaining a variety of patch types, patterns, and connectivity within landscape mosaics is warranted.

Acknowledgments

Support for this work was provided in part by the Wisconsin Department of Natural Resources. We thank Cindy Zabel and Robert Anthony for the monumental job of co-ordinating this publication. Richard Schmitz, Kevin McGarigal, Robert Anthony, Karen Martin, and an anonymous reviewer provided valuable comments on earlier drafts of this chapter.

Literature cited

Bettinger, P., K. Boston, J. Sessions, and W.C. McComb. 2001. Integrating wildlife species habitat goals and quantitative land management planning processes. Chapter 23 *in* D.H. Johnson and T.A. O'Neill, managing editors. *Wildlife-Habitat Relationships in Oregon and Washington*. Oregon State University Press, Corvallis, Oregon, USA.

Bissonette, J.A. and S. Broekhuizen. 1995. *Martes* populations as indicators of habitat spatial patterns: the need for a multiscale approach. Pages 95–121 *in* W.Z. Lidicker Jr., editor. *Landscape Approaches in Mammalian Ecology and*

Conservation. University of Minnesota Press, Minneapolis, Minnesota, USA.

Brown, E.R. 1985. Management of wildlife and fish habitats in forests of western Oregon and Washington. Part 2 – Appendices. US Forest Service Publication Number **R6F&WL-192**.

Bunnell, F.L. 1999. What habitat is an island? Pages 1–31 *in* J.P. Rochelle, L.A. Lehman, and J. Wisniewski, editors. *Forest Fragmentation: Wildlife and Management Implications*. Brill Press, Leiden, The Netherlands.

Buskirk, S.W. and W.J. Zielinski. 2003. Small and mid-sized carnivores. Pages 207–249 *in* C.J. Zabel and R.G. Anthony, editors. *Mammal Community Dynamics. Management and Conservation in the Coniferous Forests of Western North America*. Cambridge University Press, Cambridge, UK.

Dooley, J.L., Jr. and M.A. Bowers. 1998. Demographic responses to habitat fragmentation: experimental tests at the landscape and patch scale. *Ecology* **79**:969–980.

Doyle, A.T. 1990. Use of riparian and upland habitats by small mammals. *Journal of Mammalogy* **71**:14–23.

Fahrig, L. 1997. Relative effects of habitat loss and fragmentation on population extinction. *Journal of Wildlife Management* **61**:603–610.

Fahrig, L. 1999. Forest loss and fragmentation: which has the greater effect on persitence of forest-dwelling animals. Pages 87–95 *in* J.P. Rochelle, L.A. Lehman, and J. Wisniewski, editors. *Forest Fragmentation: Wildlife and Management Implications*. Brill Press, Leiden, The Netherlands.

Gannon, W.L. 1988. *Zapus trinotatus*. *Mammalian Species* **315**:1–5.

Johnson, D.H. 1980. The comparison of usage and availability measurements for evaluating resource preference. *Ecology* **61**:65–71.

Harris, L.D. 1984. *The Fragmented Forest: Island Biogeography Theory and the Preservation of Biotic Diversity*. University of Chicago Press, Chicago, Illinois, USA.

Hayes, J.P. 1996. *Arborimus longicaudus*. *Mammalian Species* **532**:1–5.

Landres, P.B., P. Morgan, and F.J. Swanson. 1999. Overview of the use of natural variability concepts in managing ecological systems. *Ecological Applications* **9**:1179–1188.

Laurance, W.F. 1995. Extinction and survival of rainforest mammals in a fragmented tropical landscape. Pages 46–63 *in* W.Z. Lidicker Jr., editor. *Landscape Approaches in Mammalian Ecology and Conservation*. University of Minnesota Press, Minneapolis, Minnesota, USA.

Lawlor, T.E. 1998. Biogeography of Great Basin mammals: paradigm lost? *Journal of Mammalogy* **79**:1111–1130.

Lawlor, T.E. 2003. Faunal composition and distribution of mammals in western coniferous forests. Pages 41–80 *in* C.J. Zabel and R.G. Anthony, editors. *Mammal Community Dynamics. Management and Conservation in the Coniferous Forests of Western North America*. Cambridge University Press, Cambridge, UK.

Lehmkuhl, J.F., L.F. Ruggiero, and P.A. Hall. 1991. Landscape-scale patterns of forest fragmentation and wildlife richness and abundance in the Southern Washington Cascade Range. Pages 425–442 *in* L.F. Ruggiero, K.B. Aubry, A.B. Carey, and M.H. Huff, editors. Wildlife and Vegetation of Unmanaged Douglas-fir Forests. US Forest Service General Technical Report **PNW-285**, Portland, Oregon, USA.

Lomolino, M.V. and D.R. Perault. 2000. Assembly and disassembly of mammal communities in a fragmented temperate rain forest. *Ecology* **81**:1517–1532.

Luoma, D.L., J.M. Trappe, A.W. Claridge, K. Jacobs, and E. Cazares. 2003. Relationships among fungi and small mammals in forested ecosystems. Pages 343–373 *in* C.J. Zabel and R.G. Anthony, editors. *Mammal Community Dynamics. Management and Conservation in the Coniferous Forests of Western North America*. Cambridge University Press, Cambridge, UK.

Martin, K.J. 1998. Habitat associations of small mammals and amphibians in the Central Oregon Coast Range. Dissertation. Oregon State University, Corvallis, Oregon, USA.

Martin, K.J. and W.C. McComb. 2002. Small mammal habitat associations at patch and landscape scales in Oregon. *Forest Science* **48**:255–264.

Maser, C., Z. Maser, J. Witt, and G. Hunt. 1986. The northern flying squirrel: a mycophagist in southwestern Oregon. *Canadian Journal of Zoology* **64**:2086–2089.

Maser, C., R.F. Tarrant, J.M. Trappe, and J.F. Franklin. 1988. From the forest to the sea: a story of fallen trees. US Forest Service General Technical Report **PNW-133**, Portland, Oregon, USA.

Maser, Z., C. Maser, and J.M. Trappe. 1985. Food habits of the northern flying squirrel (*Glaucomys sabrinus*) in Oregon. *Canadian Journal of Zoology* **63**:1084–1088.

McGarigal, K. and B.J. Marks. 1995. FRAGSTATS: spatial pattern analysis program for quantifying landscape structure. US Forest Service General Technical Report **PNW-351**, Portland, Oregon, USA.

McGarigal, K. and W.C. McComb. 1995. Relationships between landscape structure and breeding birds in the Oregon Coast Range. *Ecological Monographs* **65**: 236–260.

McGarigal, K. and W.C. McComb. 1999. Forest fragmentation effects on breeding bird communities in the Oregon Coast Range. Pages 223–246 *in* J.P. Rochelle, L.A. Lehman, and J. Wisniewski, editors. *Forest Fragmentation: Wildlife and Management Implications*. Brill Press, Leiden, The Netherlands.

Mills, L.S. 1995. Edge effects and isolation: red-backed voles on forest remnants. *Conservation Biology* **2**:395–403.

Mills, L.S. and D.A. Tallmon. 1999. The role of genetics in understanding forest fragmentation. Pages 171–186 *in* J.P. Rochelle, L.A. Lehman, and J. Wisniewski, editors. *Forest Fragmentation: Wildlife and Management Implications*. Brill Press, Leiden, The Netherlands.

Mills, L.S., M.K. Schwartz, D.A. Tallmon, and K.P. Lair. 2003. Measuring and interpreting connectivity for mammals in coniferous forests. Pages 587–613 *in* C.J. Zabel and R.G. Anthony, editors. *Mammal Community Dynamics. Management and Conservation in the Coniferous Forests of Western North America*. Cambridge University Press, Cambridge, UK.

Rosenberg, K.V. and M.G. Raphael. 1986. Effects of forest fragmentation on vertebrates in Douglas-fir forests. Pages 263–272 *in* J. Verner, M.L. Morrison, and C.J. Ralph, editors. *Wildlife 2000: Modeling Habitat Relationships of Terrestrial Vertebrates*. University of Wisconsin Press, Madison, Wisconsin, USA.

Sessions, J., G. Reeves, K.N. Johnson, and K. Burnett. 1998. Implementing spatial planning in watersheds. Chapter 18 *in* K.A. Kohm and J.F. Franklin. *Creating a Forestry for the 21st century: the Science of Ecosystem Management*. Island Press, Washington DC, USA.

Smith, W.P., R.G. Anthony, J.R. Waters, N.L. Dodd, and C.J. Zabel. 2003. Ecology and conservation of arboreal rodents of western coniferous forests. Pages 157–206 *in* C.J. Zabel and R.G. Anthony, editors. *Mammal Community Dynamics. Management and Conservation in the Coniferous Forests of Western North America*. Cambridge University Press, Cambridge, UK.

Spies, T.A. and M.G. Turner. 1999. Dynamic forest mosaics. Pages 95–160 *in* M.L. Hunter Jr., editor. *Maintaining Biodiversity in Forested Ecosystems*. Cambridge University Press, Cambridge, UK.

Sutherland, G.D., A.S. Harestad, K. Price, and K.P. Lertzman. 2000. Scaling of natal dispersal distances in terrestrial birds and mammals. *Conservation Ecology* **4(1)**:16. [online] URL: http://www.consecol.org/vol4/iss1/art16.

Tallmon, D. and L.S. Mills. 1994. Use of logs within home ranges of California red-backed voles on a remnant of forest. *Journal of Mammalogy* **75**:97–101.

Trzcinski, M.K., L. Fahrig, and G. Merriam. 1999. Independent effects of forest cover and fragmentation on the distribution of forest breeding brids. *Ecological Applications* **9**:586–593.

Venier, L.A. and L. Fahrig. 1996. Habitat availability causes the species abundance-distribution relationship. *Oikos* **76**:564–570.

With, K.A. 1999. Is landscape connectivity necessary and sufficient for wildlife management? Pages 97–115 *in* J.P. Rochelle, L.A. Lehman, and J. Wisniewski, editors. *Forest Fragmentation: Wildlife and Management Implications*. Brill Press, Leiden, The Netherlands.

Wolff, J.O. 1997. Population regulation in mammals: an evolutionary perspective. *Journal of Animal Ecology* **66**:1–13.

Wolff, J.O. 1999. Behavioral model systems. Pages 11–40 *in* B.W. Barrett and J.D. Peles, editors. *Landscape Ecology of Small Mammals*. Springer-Verlag, New York, New York, USA.

Wolff, J.O. 2003. An evolutionary and behavioral perspective on dispersal and colonization of mammals in fragmented landscapes. Pages 614–630 *in* C.J. Zabel and R.G. Anthony, editors. *Mammal Community Dynamics. Management and Conservation in the Coniferous Forests of Western North America*. Cambridge University Press, Cambridge, UK.

L. SCOTT MILLS, MICHAEL K. SCHWARTZ,
DAVID A. TALLMON AND KEVIN P. LAIR

17

Measuring and interpreting connectivity for mammals in coniferous forests

Introduction

Western coniferous forests have a history of natural disturbance due to fire, disease, and other factors (Agee 1993), but during the past century late-seral forests have been increasingly fragmented due to logging and development. For example, in the Pacific Northwest, less than half of pre-settlement, old-growth Douglas-fir (*Pseudotsuga menziesii*) forest remains, often in relatively small remnants of 100 ha or less in a matrix of clear-cuts and regenerating forest (Booth 1991, Garmon et al. 1999, Jules et al. 1999). Road building has also impacted wildlife habitat, with an average of 3.4 miles of road per square mile on United States Forest Service roaded-lands and approximately twice that on private lands (Federal Budget Consulting Group and Price-Waterhouse LLP 1997, Coghlan and Sowa 1998, Federal Register 2001, USDA 2001).

For certain species associated with late-seral forests, fragmentation due to human perturbations has two consequences: loss of habitat and changes in connectivity among remnants. Habitat loss, and the concomitant decrease in population size for some wildlife species, has garnered the most attention because such loss is painfully obvious both in its occurrence and its effects. The second consequence, the change in connectivity among populations, is more subtle and harder to measure.

Nevertheless, the importance of populations being connected versus isolated has been underscored in the scientific literature for at least 70 years. For example, biologists have long recognized that the interplay between population size, local adaptation, and gene flow (connectivity) will create a unique genetic structure across a landscape (Wright 1931). If populations on large forest tracts become small and isolated via fragmentation, genetic variation can be decreased with subsequent effects on population persistence (*sensu* Mills and Tallmon 1999). Similarly, the demographic consequences of isolation versus connectivity have been emphasized at least since Nicholson and Bailey (1935) and Andrewartha and Birch (1954; see Fahrig and Merriam 1994, Hastings and Harrison 1994). In many cases, the stability of a collection of populations ("metapopulation"; Hanski and Gilpin 1997) may be very sensitive to connectivity. For example, Beier (1993) modeled connectivity for cougars in California and found that overall extinction probability is greatly reduced when connectivity is higher.

Connectivity is also demonstrably important at larger scales. The level of connectivity can affect the spread of geographic ranges and allow response to changing environmental conditions (Pease et al. 1989). Likewise, successful re-introduction can be facilitated by connectivity (Singer et al. 2000). Connectivity can also affect the synchrony across space and the amplitude of population size changes; the classic case is the "traveling-wave" dynamics of Canadian lynx (*Lynx canadensis*) populations emanating from the center of the taiga outward toward the periphery (Blasius et al. 1999, McKelvey et al. 2000, Mowat et al. 2000, Schwartz et al. 2002). Finally, the degree of connectivity has been proposed as a central criterion for determining taxonomy and population distinctiveness, thereby fundamentally affecting whether populations should be treated as distinct management units (Waples 1995, Crandall et al. 2000).

Clearly, the importance of connectivity for understanding population dynamics has implications for forest managers, ranging from evaluating population uniqueness to maintaining population persistence. In addition, re-introduction programs will be heavily weighted by whether or not a species is likely to re-colonize naturally (as in the case of wolf re-introduction and proposed grizzly bear re-introduction in Idaho and Montana). Similarly, decisions of whether supplementation is necessary for extant populations will hinge on the current degree of isolation.

In short, knowledge of connectivity across the landscape is essential to understand wildlife populations (Martin and McComb 2003), and yet connectivity has been enormously challenging to measure, with most of the successful examples of measuring connectivity being experimental studies of small spatial and short temporal scales (reviewed in Debinski and Holt 2000). We will provide an overview of new approaches to measuring connectivity, with examples from some of our own research on mammals in western coniferous forests. We will also consider "how much" connectivity is desirable and some ways of achieving it in a fragmented landscape.

Some definitions

Connectivity is a broad and vague term that implies movement among populations, analogous to what has been described as "transfer" (Ims and Yoccoz 1997). Connectivity can be described more precisely with other terms, such as dispersal, emigration, immigration, colonization, and migration rates. The term "dispersal" addresses behavioral aspects, "constituting movements of individuals (or propagules) away from their home areas, excluding short-term exploratory movements" (Lidicker and Patton 1987:144). "Emigration" and "immigration" refer to dispersal out of and into a target population, respectively. These are demographically meaningful because abundance and population trend are a function of emigration plus deaths leaving the population versus immigration and births adding to it. In addition to dispersal among extant populations, dispersal can occur to areas currently unoccupied by the species ("colonization" if the species has never occupied the site and "re-colonization" if it has).

In population genetics, "migration rate" (m) is the proportion of individuals that move between populations, establish residence, and breed. Thus, migration is equal to "gene flow" in classic population genetics models. Although we will use "migration" in this sense, we recognize that it is different from the traditional ecological use of "migration" as a descriptor of seasonal movements across elevation or latitude (see Webster et al. 2002 for an excellent overview of measuring migratory connectivity). For example, although polar bears (*Ursus maritimus*) are known to have extremely large seasonal movements, gene flow between populations is restricted (Paetkau et al. 1995, 1999).

Measures of connectivity differ in both the timing of movements and whether or not reproduction (gene flow) occurred. Research or management questions may revolve around current movement rates, or around average movement during the recent past, or even in the historical past before widespread human-caused fragmentation. Reproduction is important because dispersal of individuals could have immediate demographic effects, but unless new arrivals breed (i.e., gene flow) they will not affect genetic structure (Ehrlich and Raven 1969) or directly increase long-term population size.

Approaches to measuring connectivity

Historically, connectivity among mammal populations has been evaluated through heroic efforts using radiotelemetry and capture-recapture techniques. Both have the advantages of directly measuring connectivity while also providing insights into natural history, habitat use, individual health, survival, age structure, and population size. In general, trapping grids and telemetry have underestimated dispersal rates and distances because the further an animal goes the less likely it is to be detected, and rare but important dispersal events tend to be missed (Koenig et al. 1996, Peacock 1997). Recent advances in telemetry have been assets for these direct measures of connectivity, and more species will be followed with satellite transmitters (e.g., Ferguson and Messier 2000) as transmitter size, weight, and cost decrease. However, for mammals in western forests, canopy cover may limit the use of satellite technology.

Like the technological advances, development of new analytical methods for capture-recapture (and telemetry) data are also important because they facilitate estimates of both survival and movement among populations (Spendelow et al. 1995, Burnham and Anderson 1998, Powell et al. 2000, Bennetts et al. 2001). However, capture-recapture and telemetry methods cannot readily quantify whether reproduction occurred. On the other hand, recent developments in molecular biology techniques and analyses of genetic data provide a new suite of tools with tremendous potential to monitor rates of gene flow, but the temporal scale of gene flow is typically less clearly defined.

To further explore the utility and limits of these new approaches, we next provide overviews of emerging tools to quantify connectivity based on capture-recapture and genetic analysis. We will focus on approaches most relevant to mammals.

Estimating connectivity with capture-mark-recapture (CMR) approaches

Capture-mark-recapture (CMR) models provide a statistical basis for separating the probability of capture from biologically interesting parameters such as abundance, mortality, and dispersal. Here, we discuss a subset of CMR methods referred to as multi-state or multi-strata models that provide the theoretical framework for estimating transition probabilities among geographic "states" (see Ims and Yoccoz 1997, Nichols and Coffman 1999, Hanski et al. 2000). Multi-state models are extensions of Cormack–Jolly–Seber models that have been used for decades to estimate abundance and survival in open populations (not closed to births, deaths, emigration, and immigration). Following an information theoretic approach to analyzing multi-state CMR data (Lebreton et al. 1992, Burnham and Anderson 1998), a researcher first develops a candidate model set that identifies biologically realistic sources of variation (e.g., age, gender, time of year, population location, etc.) in the parameters of interest. This is accomplished by drawing upon knowledge of the system, previous studies, and biological intuition. Once the data have been obtained, an information criterion, such as Akaike's information criterion (AIC), is used to select the model(s) most consistent with the data.

This model-selection process accomplishes two important tasks. First, model selection tests biological hypotheses by determining which of the models in the candidate set best approximate the data. Second, by determining the candidate models most appropriate for the data, information theoretic approaches provide parameter estimates with minimized bias (given the candidate models and the quality of the data) and estimates of their precision (Lebreton et al. 1992, Burnham and Anderson 1998). Because the inferences that can be made about a system are limited to the validity of the candidate set of models, previous knowledge of a biological system and natural history is important in the development of realistic and valid candidate models (Burnham and Anderson 1998).

One reason why multi-state models have not been used more to estimate connectivity is because capture probabilities (and number of captures) must be high in order to obtain reasonably precise survival and movement estimates (Ims and Yoccoz 1997). High capture probabilities usually require a great deal of effort, and the effort increases as the number of populations (and parameters in the multi-state model) increases. Consequently, there is an implicit trade-off between the number of populations sampled and the precision of parameter estimates.

Estimating connectivity with genetic tools

Genetic tools for measuring connectivity have undergone astonishing developments in the last decade, with advances in both molecular techniques and analytical tools. The greatest leap in molecular techniques has been in applications of the polymerase chain reaction (PCR). Amplification of deoxyribonucleic acid (DNA) via PCR has led to analysis of bits of tissue collected in creative ways from hair, scats, saliva, and blood (Kohn and Wayne 1997, Schwartz et al. 1998, Taberlet et al. 1999). Thus, it is possible to sample non-invasively (Taberlet and Luikart 1999), so that no capture or restraint is necessary. PCR also allows the use of ancient

DNA to examine connectivity among historic populations. For example, a lack of connectivity currently and historically among pocket gophers was inferred from the fact that in Lamar Cave (Yellowstone National Park) gophers had unique *cytochrome b* sequences in samples spanning from the present to 2400 years ago, and these sequences were absent from adjacent localities (Hadley et al. 1998).

On the heels of breakthroughs in non-invasive sampling, there is now an impressive array of approaches for individual and species identification, phylogenetic analysis, and estimation of abundance (Haig 1998, Waits and Leberg 1999, 2000, Woods et al. 1999, Mills et al. 2000*a*, 2000*b*). For measuring connectivity, both nuclear and mitochondrial (mtDNA) markers may be used for complementary insights. Mitochondrial DNA is maternally inherited, so it will not detect connectivity by males alone (male-biased dispersal). In one sense this is a disadvantage of mtDNA. However, if female movement is of primary interest, then the signal movement from mtDNA will be preferred to that of nuclear DNA (Taylor et al. 2000). Furthermore, comparing the signals from mtDNA to nuclear DNA may elucidate sex-biased dispersal patterns of the species of interest.

The analytical and statistical tools to analyze DNA have developed as dramatically as the molecular advances (Rousset and Raymond 1997, Luikart and England 1999, Balding et al. 2001). Next we describe both equilibrium and non-equilibrium measures for estimating connectivity using genetic tools.

Equilibrium genetic measures

Although a number of different metrics have been developed to estimate gene flow from nuclear genetic data (e.g., Slatkin 1995), most derive from Sewall Wright's F_{st} (for history and derivation see Slatkin 1985) so that the mean number of migrants entering each population each generation is inversely related to the variance in gene frequencies among different populations. Specifically, under the assumptions of the island model (Wright 1931; see also Slatkin and Barton 1989, Mills and Allendorf 1996):

$$Nm \approx [1/(4F_{st})] - (1/4) \tag{1}$$

where F_{st} is the proportion of total gene diversity due to divergence among subpopulations, N is the effective population size of each subpopulation, and m is the proportion of migrants entering subpopulations. The product Nm is the number of migrants entering a subpopulation each generation.

These are considered "equilibrium" measures of gene flow because it is assumed that the variance in gene frequencies among populations (F_{st}) represents an equilibrium between genetic drift increasing divergence and gene flow decreasing it. Importantly, the time that it takes for F_{st}, and therefore Nm, to reach equilibrium can be extremely long (hundreds to thousands of generations) when the migration rate is small and/or the population size is large (Varvio et al. 1986, Steinberg and Jordan 1997, Whitlock and McCauley 1999). Therefore, F_{st}-derived estimates of migration rates typically reflect a mix of current and historical gene flow. In certain cases, this actually might be an advantage. For example, if the human-caused perturbation of interest is recent (on the order of tens of generations into the past), equilibrium gene flow measures may "look back" into the period before the perturbation. Thus, F_{st}-type approaches could give at least a qualitative insight into historical connectivity based on the genetic "signature" arising from past differentiation in allele frequencies among populations.

The use of equilibrium-based measures to quantify levels of gene flow for any time period depends on simplifying assumptions that are unlikely to hold true in many cases (for summaries of critiques see Neigel 1996, Bossart and Prowell 1998, Whitlock and McCauley 1999). For example, all populations are assumed to equally contribute and receive migrants; if the number of migrants (Nm) varies among populations (due to population sizes, distance, barriers, and so on), the estimated gene flow is likely to be negatively biased. Variation in gene flow across time, including extinction and colonization of populations, will also affect F_{st} and Nm estimates and inflate the standard errors of gene flow estimates. While new analytical approaches provide migration rate estimates that are robust to violations of some of the fundamental assumptions of F_{st}-based measures (e.g., Beerli and Felsenstein 2001, Vitalis and Couvet 2001), these methods are so new that they have not been thoroughly tested with field data and so it is not known how they perform for samples collected from natural populations. Likewise, a clever approach for estimating distance moved by the dispersing sex based on analysis of isolation by distance for the philopatric sex (applied to lions, Panthera leo, by Spong and Creel 2001) may have potential for application to species with strong sex-biased dispersal (see also Goudet et al. 2002).

In short, we believe that equilibrium gene flow measures can be useful for qualitative comparisons of historical gene flow, and can give some insights into current gene flow if coupled with demographic information (see "Three case studies evaluating connectivity for mammals" below). In general, however, we advocate the use of equilibrium gene flow measures in categorical as opposed to quantitative applications. For example, an $Nm < 1.0$ might be considered "low" due to loss of alleles via genetic drift (Mills and Allendorf 1996), $Nm > 10.0$ might be considered "high" because local adaptations would begin to be swamped (Mills and Allendorf 1996, Vucetich and Waite 2000), and $1.0 < Nm < 10.0$ considered "medium".

Non-equilibrium genetic measures (assignment tests)

Newly developed "assignment tests" represent a major step forward in genetic analyses of connectivity because these approaches do not assume genetic equilibrium and therefore have potential for quantifying *current* gene flow (e.g., Paetkau et al. 1995, Rannala and Mountain 1997, Waser and Strobeck 1998, Cornuet et al. 1999, Davies et al. 1999, Luikart and England 1999, Pritchard et al. 2000, Manel et al. 2003). In essence, assignment tests take advantage of the genetic differences at many loci among populations to assign individuals to the population from which they originated. Connectivity is thereby quantified by estimating the number of first-generation immigrants, individuals born in a population other than the one from which they were sampled (Rannala and Mountain 1997).

In general, assignment tests are more likely to fail when the populations are not distinct (F_{st} less than approximately 0.1), although statistical power can be increased with more polymorphic loci and individuals sampled (Olsen et al. 2000, Manel et al. 2002). Another difficulty is that if the true source population is not included in the analyses, it may be harder to detect an immigrant individual because it cannot be properly categorized with its "home" population (Cornuet et al. 1999). Lastly, the assignment test relies on populations maintaining Hardy–Weinberg proportions, although the test may not be overly sensitive to these assumptions (Cornuet et al. 1999). There is a pressing need for simulations (see Cornuet et al. 1999) to evaluate the power and accuracy of assignment tests under a variety of levels of genetic differentiation, number of samples, and number of loci. Ultimately, the robustness of assignment tests under real-world conditions remains unexplored (Sunnucks 2000) and will determine how generally applicable these tests are under the non-equilibrium conditions that will be the rule in conservation applications. The greatest advances may be in assessing assignment tests in settings where connectivity is known from non-genetic-based measures (e.g., Paetkau et al. 1999).

Three case studies evaluating connectivity for mammals

Much of our research has focused on conservation questions centered on connectivity, isolation, and persistence of forest mammals in western North America, using genetic and demographic approaches. Next we provide three concrete examples of the techniques described above.

Columbian mice on edges in Olympic National Park

We employed capture-mark-recapture (CMR) and information theoretic approaches to examine the effects of forest fragmentation and forest edges on the population dynamics of the Columbian mouse (*Peromyscus keeni oreas*). The study is described completely elsewhere (Lair 2001); here we distill only the aspects directly related to measuring connectivity. Trapping grids were established on the west side of the Olympic Peninsula, Washington at sites where 5- to 10-year-old clear-cuts were adjacent to undisturbed old-growth stands. In 1997 two of the sites (Willoughby and Queets) were trapped and in 1998 two additional sites were added (Hoh and Tacoma). At each of the four sites, trapping grids 16 traps long by three traps wide (48 traps total) ran parallel to the forest–clear-cut boundary in each of three "edge classes": (1) 150 m into the forest from the forest/clear-cut edge; (2) along the edge; (3) 150 m into the clear-cut from the edge. The summer trapping schedule followed the "robust design" (Pollock 1982, Pollock et al. 1990), such that multiple "secondary" sampling sessions (six to seven consecutive nights) made up each "primary" sampling period.

We developed multi-state models to obtain maximum-likelihood estimates of state-specific survival rates (S) and transition probabilities (Ψ) between primary periods, and state-specific capture probability (p) within primary periods. The states in our study corresponded to edge classes within a site, so that transition probabilities were estimates of movement rates among edge classes. We specified a set of candidate models that incorporated biologically realistic sources of variation in S, Ψ, and p based on edge classes, sites, gender, and primary periods (temporal variation).

Model selection and parameter estimation were accomplished using the multi-state routine in program MARK (White and Burnham 1999). The best approximating models for 1997 Columbian mouse movement rate (Ψ) identified only variation among sites as important for transitions (movements) among edge classes (Fig. 17.1). Movement rates among edge classes at the Willoughby site were much greater (0.22; SE = 0.066) than that at the Queets site (0.07; SE = 0.022). These rates can be interpreted

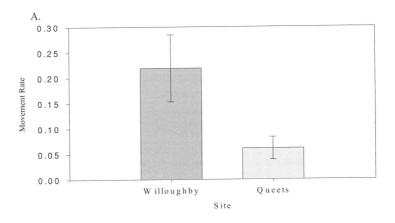

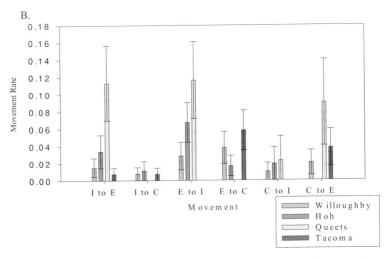

Fig. 17.1. Columbian mouse movement rates between edge classes (forest interior, edge, clear-cut) in 1997 (A) and 1998 (B) under the best approximating multi-state capture-recapture models. In 1997, movement rates were dependent on the site (Queets or Willoughby). In 1998, movement varied among sites and edge classes. Error bars represent one standard error. Letters represent habitat types: I = forest interior, E = forest edge, and C = clear-cut.

as 22% of the mice at the Willoughby site and 7% at the Queets site moved from one edge class to another, on average, every two weeks during summer 1997, with no detectable differences among sexes, time, or specific edge class rates. In 1998, there were again no detectable differences in movement rates among the sexes or over time, but both site and edge class were identified as important sources of variation (Fig. 17.1). Thus, in 1998

Columbian mouse movement rates varied depending upon which habitat the animal was moving to and from. For example, in the Tacoma site movement was greatest from the edge to the clear-cut and the clear-cut to edge, while in the Hoh sites movements were greatest from edge to interior and interior to edge (Fig. 17.1B). In short, quantified movement rates to and from clear-cuts were not radically different than those to and from the other habitats. In addition, our analysis of survival indicated no consistent patterns across sites, primary periods, or years, and movement out of the habitats with higher survival in any year was not greater than movement into these habitats. Together, these results indicate that forest fragmentation is not creating "demographic sinks" for this species on the west side of the Olympic Peninsula (Lair 2001).

Western red-backed voles in Oregon forest fragments

We combined genetic and demographic approaches to evaluate the effects of habitat fragmentation on the connectivity of western red-backed vole (*Clethrionomys californicus*) populations on forest fragments in southwest Oregon. Previous studies suggest this species is closely associated with forests with high litter depth (Rosenberg et al. 1994) and virtually absent from recently clear-cut and burned areas (Tevis 1956, Gashwiler 1970, Mills 1995, 1996), and that voles on forest fragments can show a negative edge effect (Mills 1995, 1996).

To evaluate whether connectivity was affected by fragmentation, we trapped voles in 12 forest fragments, surrounding clear-cuts, and in unharvested control areas in 1997, 1998, and 1999 (a subset of the same sites trapped by Mills 1995). We used capture data and genetic analysis to measure and interpret changes in connectivity (see Tallmon et al. 2002 for details that are summarized here). We were able to consider temporal patterns in genetic variation from populations sampled in 1990–1991, and again in 1998, on two of these fragments and two nearby contiguous control sites. We also used intensive trapping sessions in the summers of 1998 and 1999 to estimate the sizes of these same vole populations.

Our trapping data suggested movement across clear-cuts during summer was extremely limited. In the clear-cuts surrounding the 12 fragments trapped in 1990, 1991, 1997, 1998, and 1999, we detected only 13 voles; only one of these subsequently was trapped in a fragment population. The rate of capture of different voles per trap-night (an index comprised of both number of animals and probability of capture) was much lower on clearcuts compared to fragments or controls: clear-cuts (13 different voles/4728

Table 17.1. *Mitochondrial DNA number of alleles (A) and allelic diversity (ĥ) in vole samples from forest fragments (F1, F2) and control sites (C1, C2, C3)*

1990 and 1991 samples				1998 samples			
Site	n	A	$\hat{h}$ (S.E.)	Site	n	A	$\hat{h}$ (S.E.)
F1	11	1	0.00 (0.00)	F1	22	3	0.17 (0.07)
F2	9	1	0.00 (0.00)	F2	21	1	0.00 (0.00)
C3	20	3	0.34 (0.09)	C1	23	4	0.59 (0.06)
				C2	22	3	0.58 (0.08)

Table 17.2. *The mean number of alleles (Â) and observed heterozygosities (H_o) at 5 microsatellite loci in samples (n) from vole populations on forest fragments (F1, F2) and control sites (C1, C2, C3)*

1990 and 1991 samples			1998 samples		
		Mean (S.E.)			Mean (S.E.)
Site	n	$(\hat{A})$ and (H_o)	Site	n	$(\hat{A})$ and (H_o)
F1	16	10.0 (2.6) 0.66 (0.22)	F1	34	11.2 (2.4) 0.73 (0.20)
F2	9	8.0 (1.0) 0.64 (0.15)	F2	35	10.8 (1.5) 0.77 (0.18)
C3	24	11.0 (2.6) 0.78 (0.12)	C1	36	12.4 (1.8) 0.72 (0.18)
			C2	34	11.6 (2.0) 0.78 (0.13)

trap nights = 0.0027); forest fragments (387 voles/24 505 trap nights = 0.016); and controls (346 voles/12 657 trap nights = 0.027).

Mitochondrial DNA (mtDNA) data from two fragments and two control sites further indicated limited immigration into fragment populations. mtDNA allelic diversity and the number of alleles were lower on the fragments than in the controls in both 1990–1991 and 1998 samples (Table 17.1). This pattern would usually be interpreted as evidence for isolation and possible inbreeding effects in fragment populations. However, in contrast to the trapping and mtDNA data, nuclear DNA heterozygosity and numbers of alleles estimated with five nuclear microsatellite markers did not imply isolation of vole populations on forest fragments (Table 17.2). Instead, we found roughly equivalent levels of genetic variation across fragments and controls – especially in 1998 when sample sizes were large and roughly equal among sites.

The lower variation detected with mtDNA on forest fragments, but not with microsatellite markers, may be explained by the stronger effects of genetic drift on mtDNA (a single haplotype is passed down only by females, in contrast to the biparental inheritance of nuclear genes). An alternative explanation is that these voles exhibit male-biased dispersal among fragments. That is, there may be a bias toward male dispersal that keeps nuclear variation equal while mtDNA variation is reduced within fragments because females do not bring novel haplotypes. We favor this alternative, as male-biased dispersal is common among closely related vole species and many other mammal taxa (see Wolff 2003). In addition, of the 13 voles that we captured in clear-cuts over the five summers of sampling these sites, nine were males; the binomial probability of detecting nine or more males in a sample of 13 voles with an expected 50:50 sex ratio is 0.13.

Estimates of gene flow based on the microsatellite data and F_{st}-based measures are "medium" (on the order of $Nm = 7.7$ with 95% bootstrapped confidence intervals of 6.2 to 10.4). However, as stated above (equilibrial genetic measures), F_{st}-type measures have the potential to confound recent events with historical population structure. A low F_{st} value (and resultant medium or high gene flow estimates) could come from current high levels of gene flow or it could be a genetic "signal" from a large, panmictic population prior to fragmentation (a historical effect). In our case, we teased apart these possibilities by examining demographic data from the fragments. Based on our intensive CMR data, vole populations fluctuated below 50 individuals on the two fragments throughout the summers of 1998 and 1999. We know the genetic effective size of a population is usually between 30% and 50% of the total population size (Frankham 1995, Kalinowski and Waples 2002) and that these fragments have been surrounded by clear-cuts for at least 20 vole generations. Based on population genetics theory (Wright 1931), we would expect the fragments to show at least 12% to 85% reduction in heterozygosity relative to controls if fragments are truly isolated (Tallmon et al. 2002). Instead, fragment populations had heterozygosity levels intermediate to the controls. Consequently, these small fragment populations must be linked by current gene flow (several migrants per generation) to maintain genetic variation equivalent to that in contiguous forest populations. In addition, the reduction in mtDNA variation on fragments relative to controls in both

1990–1991 and 1998 samples suggests current gene flow may be largely male biased.

Determining movement in a fragmented forest landscape for voles has important implications for forest community dynamics. The western red-backed vole is thought to be one of the primary dispersers of hypogeous ectomycorrhizal fungi (as are several other small mammal species; see Luoma et al. 2003). Conifers in this region depend upon mycorrhizal fungi for soil nutrients and water, and the re-establishment success of seedlings in clear-cuts is increased by mycorrhizae formation (e.g., Perry et al. 1989). Because clear-cuts can be depauperate of ectomycorrhizal fungi (Durall et al. 1999, Hagerman et al. 1999), the movement of voles through clear-cuts may be important for ectomycorrhizae dispersal into recently clear-cut areas. Therefore, the detection of movement through clear-cuts using a combination of genetic and demographic data is an important finding for both vole population biology and forest community dynamics.

Canadian lynx (*Lynx canadensis*) in western North America

Canadian lynx are among the most difficult mammals to study in North American coniferous forests because they are elusive, inhabit rugged terrain, and are found at low densities. Although lynx were listed as "Threatened" under the U.S. Endangered Species Act in 2000 (3/24/2000 Federal Register), little is known about intra-population dynamics and even less about inter-population dynamics (Ruggiero et al. 2000). Anecdotal evidence has shown that lynx can travel long distances, with recorded movements of up to 1100 km (Slough and Mowat 1996). However, we do not know if these long distance movements are exploratory trips, dispersal events, or if gene flow occurs.

We used molecular genetic techniques to test two opposing ideas about lynx population sub-structure (see Schwartz et al. 2002, 2003 for details): (1) lynx at the periphery of their geographic range exist in small isolated populations, and (2) movement is ubiquitous among populations regardless of their position on the landscape. We tested these hypotheses using nine DNA microsatellite loci on 599 samples, from 17 populations in Alaska, Western Canada and Montana (Fig. 17.2). All the markers were highly variable, showing between 1 and 20 alleles per population and having an average heterozygosity across populations of 0.66 (SE = 0.074).

The F_{st} estimate was 0.033 (SE = 0.002) from which we estimated approximately six migrants moving among populations on average each generation (a "medium" level of gene flow by our qualitative categories). We believe the F_{st} estimate is produced by current gene flow and is not a

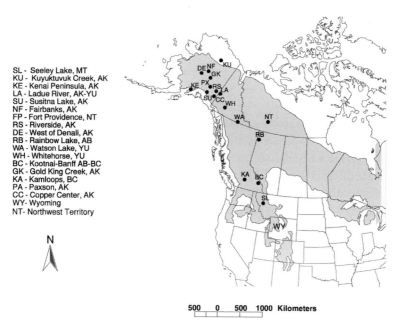

SL - Seeley Lake, MT
KU - Kuyuktuvuk Creek, AK
KE - Kenai Peninsula, AK
LA - Ladue River, AK-YU
SU - Susitna Lake, AK
NF - Fairbanks, AK
FP - Fort Providence, NT
RS - Riverside, AK
DE - West of Denali, AK
RB - Rainbow Lake, AB
WA - Watson Lake, YU
WH - Whitehorse, YU
BC - Kootnai-Banff AB-BC
GK - Gold King Creek, AK
KA - Kamloops, BC
PA - Paxson, AK
CC- Copper Center, AK
WY- Wyoming
NT- Northwest Territory

N

Fig. 17.2. Lynx range map and population identifiers. The shaded area of the map denotes the approximate geographic range of canadian lynx. Each point and two-letter code is a lynx population (see legend). We only considered a group of lynx a population for sampling purposes if at least 5 samples of different lynx were separated by 100 km from other groups or had a human perceived physical barrier between them.

historical artifact among populations because several peripheral lynx populations (e.g., Kenai Peninsula Lake, Montana) have low abundances that would lead to strong genetic drift, rapidly inflating F_{st} unless gene flow occurred (Tallmon et al. 2002). Decomposing the global F_{st} estimate into pairwise F_{st} estimates (range of pairwise F_{st} values = 0.0–0.07) indicated that gene flow was universal among all populations, with the highest level between Fairbanks (NF) and the Ladue River (LA) and the lowest levels between the Kenai Peninsula (KE) and Seeley Lake Montana (SL) and the Kenai Peninsula and Watson Lake (WA). Overall, based on F_{st} it appears that lynx gene flow is high, even compared to other mobile carnivores.

Lastly, we evaluated the usefulness of assignment tests for studying dispersal using GeneClass, the partially Bayesian assignment test of Cornuet et al. (1999). GeneClass assigned only 40.8% of the lynx to the populations from which they were captured, indicative of relatively high gene flow consistent with our F_{st}-based estimates (see Manel et al. 2002 for comparisons with similar numbers and types of markers in different applications). We next used GeneClass to assign two samples from the southern extreme of the lynx's geographic range (Wyoming) that were not included in the previous analysis due to small sample size. We expected both of these samples to be assigned to the nearest large lynx population, Seeley Lake, Montana. However, neither lynx assigned to Seeley Lake, but rather assigned with highest probabilities to the Watson Lake, Yukon (WA) and Northwest Territory (NT) populations (6% and 3% assignment probabilities, respectively). Importantly, GeneClass always classifies an individual to a population, because there is always one population that is more likely than another. However, GeneClass's probabilities of assignment, unlike other assignment tests, do not necessarily add up to one, making it explicit that the true population may not have been sampled (Cornuet et al. 1999). For example, the samples from the southern periphery were assigned to the Northwest Territory with only a 3% probability, in contrast to other assignment programs that would likely assign a much higher (but false) probability because the samples fit the Northwest Territory better than any other population. In our case, the assignment test indicates not that these lynx necessarily migrated the thousands of kilometers from the Northwest Territories or the Yukon (although they may have made such movements), but rather that they are from populations more genetically similar to the Northwest Territories and Yukon lynx than to populations closest to Wyoming.

In this case, genetic tools facilitated insight into connectivity that would not have been possible using conventional mark-recapture or telemetry techniques. We now know that lynx not only move long distances, but do it regularly and transfer genes in the process. Furthermore, our data suggest that populations of lynx in Wyoming did not colonize in a stepwise manner (e.g., from Canada, to Seeley Lake, then to Wyoming). Finally, if lynx persistence in the contiguous U.S. depends upon migration from larger populations, then joint international efforts must be initiated to ensure that connectivity between Southern Canada and the U.S. is maintained. Re-introduction efforts alone, without concomitant maintenance of connectivity among southern and northern lynx populations, are unlikely to prevent local extinctions at the southern periphery (Zager et al. 1995).

How much connectivity is desirable, and how to achieve it?

Although we have given a number of reasons why connectivity is important, and some ways of measuring it, it is not always true that "more is necessarily better." Thus, it is important to consider how much is optimal or desirable, and what is the best way to achieve a desired level of connectivity.

The ability of a population to locally adapt can be overwhelmed when gene flow levels are high, unless selection is very strong (see Allendorf 1983, Slatkin 1985, Lacy 1987, Barton 2001). At the extreme, inappropriately high gene flow can lead to outbreeding depression driven by the breakup of co-adapted gene complexes and the destruction of local adaptation (Leberg 1990, Burton et al. 1999). In addition, connectivity can facilitate ecological problems including subsidizing the movement of exotic species and diseases (Simberloff et al. 1992, Hess 1994).

On the other hand, we have described a number of reasons why connectivity is important and even vital. So how much connectivity is desirable for mammals? The only general "rule of thumb" developed to date is based on purely genetic factors: "one migrant per generation" (OMPG) is sufficient to minimize the loss of heterozygosity within subpopulations while allowing genetic divergence (and local adaptation) among subpopulations (Wright 1931). In an experimental study of the effects of migration on plant fitness, Newman and Tallmon (2001) found that fitness measured in several different traits, and overall, was increased over the 0 migrant control treatment with one migrant per generation, and was approximately equal to that in a 2.5 migrant treatment. At the same time, phenotypic divergence was greater in the 1 migrant treatment than in the 2.5 migrant treatment, providing empirical support for Wright's theory that the trade-off between local adaptation and inbreeding may be optimized by this level of connectivity (see Mills and Allendorf 1996).

If the "rule" is generalized to 1–10 migrants per generation, it appears to be surprisingly robust for genetic considerations (Hedrick 1995, Mills and Allendorf 1996). Of course, ecological and demographic needs can often mandate higher or lower levels of connectivity (see also Vucetich and Waite 2000).

As an alternative to a universal "rule of thumb" such as OMPG, it may be possible to derive an appropriate level of connectivity by estimating historical rates using genetic or ecological data. As mentioned above, emerging genetic tools can provide insight into historical connectivity before human-induced fragmentation. This can be done directly by comparing the genetic structure of extant individuals to that of samples collected prior to fragmentation (e.g., Bouzat et al. 1998), or indirectly by comparing equilibrium estimates of gene flow to assignment test (current) estimates. Also, it may be possible to use radiotelemetry or CMR studies (Ims and Yoccoz 1997) to estimate connectivity across an unfragmented landscape and compare that to a fragmented landscape. This approach can be limited by the difficulty in finding unfragmented control areas and by the spatial/temporal heterogeneity of "natural" fragmentation (fire/disease/etc.) in control areas. Two examples illuminate the potential utility of this approach: Blundell et al. (2002) quantified sex-biased dispersal and gene flow for river otters (*Lontra canadensis*) in wilderness areas of Alaska as a baseline for understanding background levels of connectivity, and Proctor et al. (2003) used assignment tests and equilibrium approaches in areas with and without major highway development to infer that the Highway 3 corridor in British Columbia had led to a fracture in grizzly bear (*Ursus arctos*) movement.

We believe a combination of genetic and demographic tools, coupled with judicious application of the OMPG rule, can provide insights into the appropriate levels of connectivity for mammals. But how to achieve such connectivity? Throughout this chapter, we have avoided equating "connectivity" with "corridors." We agree that the corridor concept has popularized the importance of connectivity (Beier and Noss 1998), and that corridors may be particularly useful for denoting "large, regional connections that are meant to facilitate animal movements and other essential flows between different sections of the landscape" (Dobson et al. 1999:132). But we also think that the term has been overused. Corridors defined as linear strips can be demographic sinks that increase mortality, can "fix" connectivity in places that may be inappropriate, and could take away funding from the acquisition of larger reserves (Soule and Gilpin 1991, Rosenberg et al. 1997, Dobson et al. 1999).

Without a doubt, some animals will use linear patches as either movement corridors or as additional habitat (see Rosenberg et al. 1997), and even low levels of connectivity in low-quality corridors may be better than none at all (Beier 1993, Beier and Noss 1998). But we believe that, to the extent possible, connectivity for a particular species is best facilitated by managing the intervening matrix. This would involve matching the biophysical nature of the routes between patches with the biology and behavior of the dispersing species (Simberloff et al. 1992, Taylor et al. 1993, Doak and Mills 1994, Mills 1996). For example, Roach et al. (2001) determined likely movement routes using equilibrium and non-equilibrium genetic approaches, demonstrating that black-tailed prairie dogs (*Cynomys ludovicianus*) in Colorado have undergone historical extinctions, with re-colonizations occurring along low-lying dry creek drainages.

Conclusions

We are strong advocates of merging demographic and genetic approaches to understand connectivity and population structure. Combining approaches can give insights across temporal and conceptual scales that are not possible using only telemetry, mark-recapture, or genetic measures (Peacock 1997, Lindenmayer and Peakall 2000). The new developments in analysis and techniques using mark-recapture and genetic approaches are enormous and have great potential for facilitating insights into movement of mammals across human-modified landscapes. These scientific advances come at exactly the time that forest managers most need to know which populations should be supplemented, which are likely to re-colonize on their own, and which are most likely to have unique evolutionary trajectories by virtue of being isolated.

Nevertheless, we have much to learn about which tools are best used for particular research or management questions. For example, to obtain precise estimates of connectivity using information theoretic approaches and maximum likelihood parameter estimation methods with CMR data sets, many individuals must be captured often, necessitating the sampling of relatively large areas and frequent trapping sessions. To measure temporal variation in movement, the large-scale trapping needs to occur at various times of year. In short, CMR approaches to estimate movement rates require colossal field efforts. To obtain the necessary data to apply these CMR approaches one may need to trap fewer areas more intensively, thereby decreasing replication and potentially limiting the scope of inference, yet obtaining a rich data set for the populations sampled. In some instances (e.g., clear-cuts are too small for large trapping grids, densities of animals are inherently low), these approaches will simply not be possible, pointing again to merging insights from trapping with those from genetic analysis (see western red-backed vole example).

Similar promise and pitfalls arise in the use of genetic tools. In many cases genetic data are the only means to determine whether populations have historically been connected or are currently connected. Genetic data will become increasingly used as the costs and limitations to their application are reduced. However, it is critical that researchers include standard error estimates and carefully consider the assumptions of the molecular methods and models that must be used to translate genetic data into movement rates (e.g., Steinberg and Jordan 1997, Hedrick 1999). The limitations of equilibrium gene flow estimates lead us to advocate careful qualitative interpretations instead of quantitative point estimates. Genetic information will often have its highest utility in establishing an index of historical movement rates (e.g., whether current movement levels are different than they were during pre-European settlement times), or if movement of organisms is known but breeding is unknown (as in our lynx example). Again, the strongest approach is to use demographic and natural history data to help eliminate confounding interpretations of genetic data or to corroborate conclusions.

We have provided three examples of ways that these techniques can increase our understanding and inform management of mammals in western coniferous forests. By quantifying movement on both sides of forest/clear-cut edges for Columbian mice, we have shown that clear-cuts neither strongly attract nor repel this species from intact forests. Movements of red-backed voles across clear-cuts to forest fragments were relatively common (but still difficult to detect using mark-recapture), with males probably responsible for most movements. This finding has implications for vole population dynamics in a fragmented landscape and potentially for regeneration of forests on the clear-cuts. Finally, despite a fragmented landscape, lynx tend to move across long distances. This may affect both re-introduction strategies and conservation planning across the U.S./Canada border.

Acknowledgments

We appreciate the suggestions on various drafts of the manuscript from Bob Anthony, Paul Beier, Gordon Luikart, Kevin McKelvey, Rod Peakall, Dan Pletscher, and Cindy Zabel. For funding L.S.M. thanks the National Science Foundation (DEB-9870654), D.T. was supported by the NSF Training WEB Program (DGE9553611), NSF MONTS (#291835), USDA McIntire-Stennis, and USDA NRI Competitive Grant Program (97-35101-4355) and K.L. received a Sigma Xi graduate award. We appreciate the dozens of assistants and collaborators on our research projects, and the land management agencies that gave us permission and logistical support.

Literature cited

Agee, J.K. 1993. *Fire Ecology of Pacific Northwest Forests*. Island Press, Washington DC, USA. 493 pages.

Allendorf, F. W. 1983. Isolation, gene flow, and genetic differentiation among populations. Pages 51–65 *in* C. M. Schonewald-Cox, S. M. Chambers, B. MacBryde, and W. L. Thomas, editors. *Genetics and Conservation: a Reference for Managing Wild Animal and Plant Populations*. Benjamin/Cummings, Menlo Park, California, USA.

Andrewartha, H.G. and L.C. Birch. 1954. *The Distribution and Abundance of Animals*. University of Chicago Press, Chicago, Illinois, USA.

Balding, D.J., M.J. Bishop, and C. Cannings. 2001. *Handbook of Statistical Genetics*. John Wiley & Sons, Chichester, UK.

Barton, N.H. 2001. The evolutionary consequences of gene flow and local adaptation: future approaches. *In* J. Clobert, E. Danchin, A.A. Dhondt, and J.D. Nichols. *Dispersal*. Oxford University Press, Oxford, UK.

Beerli, P. and J. Felsenstein. 2001. Maximum likelihood estimation of a migration matrix and effective population sizes in *n* subpopulations by using a coalescent approach. *Proceedings of the National Academy of Sciences, USA* **98**:4563–4568.

Beier, P. 1993. Determining minimum habitat areas and habitat corridors for cougars. *Conservation Biology* **7**:94–108.

Beier, P. and R.F. Noss. 1998. Do habitat corridors provide connectivity? *Conservation Biology* **12**:1241–1252.

Bennetts, R.E., J.D. Nichols, J-O. Lebreton, R. Pradel, J.K. Hines, and W.M. Kitchens. 2001. Methods for estimating dispersal probabilities and related parameters using marked animals. Pages 3–17 *in* J. Clobert, E. Danchin, A.A. Dhondt, and J.D. Nichols. *Dispersal*. Oxford University Press, Oxford, UK.

Blasius, B., A. Huppert, and L. Stone. 1999. Complex dynamics and phase synchronization in spatially extended ecological systems. *Nature* **399**:354–359.

Blundell, G.M., M. Ben-David, P. Groves, R.T. Bowyer, and E. Geffen. 2002. Characteristics of sex-biased dispersal and gene flow in coastal river otters: implications for natural recolonization of extirpated populations. *Molecular Ecology* **11**:289–303.

Booth, D.E. 1991. Estimating prelogging old-growth in the Pacific Northwest. *Journal of Forestry* **89**:25–29.

Bossart, J.L. and D.P. Prowell. 1998. Genetic estimates of population structure and gene flow: limitations, lessons and new directions. *Trends in Ecology and Evolution* **13**:202–205.

Bouzat, J.L., H.A. Lewin, and K.N. Paige. 1998. The ghost of genetic diversity past: historical DNA analysis of the greater prairie chicken. *American Naturalist* **152**:1–6.

Burnham, K.P. and D.R. Anderson. 1998. *Model Selection and Inference: a Practical Information-Theoretic Approach*. Springer-Verlag, New York, New York, USA.

Burton, R.S., P.D. Rawson, and S. Edmands. 1999. Genetic architecture of physiological phenotypes: empirical evidence for coadapted gene complexes. *American Zoologist* **39**:451–462.

Coghlan, G. and R. Sowa. 1998. *National Forest Road System and Use*. USDA, Forest Service, Engineering Staff, Washington Office, USA.

Cornuet, J., S. Piry, G. Luikart, A. Estoup and M. Solignac. 1999. New methods employing multilocus genotypes to select or exclude populations as origins of individuals. *Genetics* **153**:1989–2000.

Crandall, K.A., O.R.P. Bininda-Emonds, G.M. Mace, and R.K. Wayne. 2000. Considering evolutionary processes in conservation biology. *Trends in Ecology and Evolution* **15**:290–295.

Davies, N., F.X. Villablanca, and G.K. Roderick. 1999. Determining the source of individuals: multilocus genotyping in nonequilibrium population genetics. *Trends in Ecology and Evolution* **14**:17–21.

Debinski, D.M. and R.D. Holt. 2000. A survey and overview of habitat fragmentation experiments. *Conservation Biology* **14**:342–354.

Doak, D. and L. S. Mills. 1994. A useful role for theory in conservation. *Ecology* **75**:615–626.

Dobson, A., K. Ralls, M. Foster, M.E. Soule, D. Simberloff, D. Doak, J.A. Estes, L.S. Mills, D. Mattson, R. Dirzo, H. Arita, S. Ryan, E.A. Norse, R.F. Noss, and D. Johns. 1999. Connectivity: maintaining flows in fragmented landscapes. Pages 129–171 *in* M.E. Soulé and J. Terborgh, editors. *Continental Conservation: Scientific Foundations of Regional Reserve Networks*. Island Press, Washington DC, USA.

Durall, D.M., M.D. Jones, E.F. Wright, P. Kroeger, and K.D. Coates. 1999. Species richness of ectomycorrhizal fungi in cutblocks of different sizes in the interior cedar-hemlock forests of northwestern British Columbia: sporocarps and ectomycorrhizae. *Canadian Journal of Forestry Research* **29**:1322–1332.

Ehrlich, P.R. and P.H. Raven. 1969. Differentiation of populations. *Science* **165**:1228–1231.

Fahrig, L. and G. Merriam. 1994. Conservation of fragmented populations. *Conservation Biology* **8**:50–59.

Federal Budget Consulting Group and Price-Waterhouse LLP. 1997. *Financing Roads on the National Forests*. Washington DC, USA.

Federal Register. 2001. Special areas; roadless area conservation 12 January 2001. USDA, Forest Service **36 CFR Part 294**.

Ferguson, M.A.D. and F. Messier 2000. Mass emigration of Arctic tundra caribou from a traditional winter range: population dynamics and physical condition. *Journal of Wildlife Management* **64**:168–178.

Frankham, R. 1995. Effective population size/adult population size ratios in wildlife: a review. *Genetical Research* **66**:95–107.

Garmon, S.L., F.J. Swanson, and T.A. Spies. 1999. Past, present, and future landscape patterns in the Douglas-fir region of the Pacific Northwest. Pages 61–86 *in* J.A. Rochelle, L.A. Lehmann, and J. Wisniewski, editors. *Forest Fragmentation: Wildlife and Management Implications*. Brill, Boston, Massachusetts, USA.

Gashwiler, J.S. 1970. Plant and mammal changes on a clear-cut in West-Central Oregon. *Ecology* **51**:1018–1926.

Goudet, J., N. Perrin, and P. Wasser. 2002. Tests for sex-biased dispersal using bi-parentally inherited genetic markers. *Molecular Ecology* **11**:1103–1114.

Hadley, E.A., M.H. Hohn, J.A. Leonard, and R.K. Wayne. 1998. A genetic record of population isolation in pocket gophers during Holocene climatic change. *Proceedings of the National Academy of Sciences* **95**:6893–6896.

Hagerman, S.M., M.D. Jones, G.E. Bradfield, M. Gillespie, and D.M. Durall. 1999. Effects of clear-cut logging on the diversity and persistence of ectomycorrhizae at a subalpine forest. *Canadian Journal of Forestry Research* **29**:124–134.

Haig, S. M. 1998. Molecular contributions to conservation. *Ecology* **79**:413–425.

Hanski, I. and M. Gilpin. 1997. *Metapopulation Biology: Ecology, Genetics, and Evolution*. Academic Press, San Diego, California, USA.

Hanski, I., J. Alho, and A. Moilanen. 2000. Estimating the parameters of survival and migration of individuals in metapopulations. *Ecology* **81**:239–251.

Hastings, A. and S. Harrison. 1994. Metapopulation dynamics and genetics. *Annual Review of Ecological Systematics* **25**:167–188.

Hedrick, P.W. 1995. Gene flow and genetic restoration: the Florida panther as a case study. *Conservation Biology* **9**:996–1007.

Hedrick, P.W. 1999. Perspective: Highly variable loci and their interpretation in evolution and conservation. *Evolution* **53**:313–318.

Hess, G.R. 1994. Conservation corridors and contagious disease: a cautionary note. *Conservation Biology* **8**:256–262.

Ims, R.A. and N.G. Yoccoz. 1997. Studying transfer processes in metapopulations. Pages 247–265 *in* I.A. Hanski, and M.E. Gilpin, editors. *Metapopulation Biology: Ecology, Genetics, and Evolution*. Academic Press, San Diego, California, USA.

Jules, E., E. Frost, D. Tallmon, and L.S. Mills. 1999. Ecological consequences of forest fragmentation in the Klamath region. *Natural Areas Journal* **19**:368–378.

Kalinowski, S.T. and R.S. Waples. 2002. Relationship of effective to census size in fluctuating populations. *Conservation Biology* **16**:129–136.

Koenig, W.D., D.V. Duren, and P.N. Hooge. 1996. Detectability, philopatry, and the distribution of dispersal distances in vertebrates. *Trends in Ecology and Evolution* **11**:514–517.

Kohn, M.H. and R.K. Wayne. 1997. Facts from feces revisited. *Trends in Ecology and Evolution* **12**:223–227.

Lacy, R.C. 1987. Loss of genetic diversity from managed populations: interacting effects of drift, mutation, immigration, selection, and population subdivision. *Conservation Biology* **1**:143–158.

Lair, K.P. 2001. The effects of forest fragmentation and forest edge on Columbian mouse and southern red-backed vole demography. MS Thesis, University of Montana.

Leberg, P.L. 1990. Genetic considerations in the design of introduction programs. *Transactions of the North American Wildlife and Natural Resource Conference* **55**: 609–619.

Lebreton, J.-D., K.P. Burnham, J. Clobert, and D. R. Anderson. 1992. Modeling survival and testing biological hypotheses using marked animals: a unified approach with case studies. *Ecological Monographs* **62**:67–118.

Lidicker, W.Z. Jr. and J.L. Patton. 1987. Patterns of dispersal and genetic structure in populations of small rodents. Pages 144–161 *in* D. Chepko-Sade and Z. T. Halpin, editors. *Mammalian Dispersal Patterns*. Chicago University Press, Chicago, Illinois, USA.

Lindenmayer, D. and R. Peakall. 2000. The Tumut experiment – integrating demographic and genetic studies to unravel fragmentation effects: a case study of the native bush rat. Pages 172–201 *in* A.G. Young and G.M. Clarke, editors. *Genetics, Demography and Viability of Fragmented Populations*. Cambridge University Press, Cambridge, UK.

Luikart, G. and P. England. 1999. Statistical analysis of microsatellite DNA data. *Trends in Ecology and Evolution* **14**:253–256.

Luoma, D.L., J.M. Trappe, A.W. Claridge, K.M. Jacobs, and E. Cazares, 2003. Relationships among fungi and small mammals in forested ecosystems. Pages 343–373 *in* C.J. Zabel and R.G. Anthony, editors. *Mammal Community Dynamics. Management and Conservation in the Coniferous Forests of Western North America*. Cambridge University Press, Cambridge, UK.

Manel, S., P. Berthier, and G. Luikart. 2002. Detecting wildlife poaching: identifying the origin of individuals with Bayesian assignment tests and multi-locus genotypes. *Conservation Biology* **16**:650–659.

Manel, S., M.K. Schwartz, G. Luikart, and P. Taberlet. 2003. Landscape genetics: combining landscape ecology and population genetics. *Trends in Ecology and Evolution*, in press.

Martin, K.J. and W.C. McComb. 2003. Small mammals in a landscape mosaic: implications for conservation. Pages 567–586 *in* C.J. Zabel and R.G. Anthony, editors. *Mammal Community Dynamics. Management and Conservation in the Coniferous Forests of Western North America*. Cambridge University Press, Cambridge, UK.

McKelvey, K., K.B. Aubry, and Y.K. Ortega. 2000. History and distribution of lynx in the contiguous United States. Pages 207–264 *in* L.F. Ruggiero, K.B. Aubry, S.W. Buskirk, G.M. Koehler, C.J. Krebs, K.S. McKelvey, and J.R. Squires editors. *Ecology and Conservation of Lynx in the United States*. University of Colorado Press, Denver, Colorado, USA.

Mills, L.S. 1995. Edge effects and isolation: red-backed voles on forest remnants. *Conservation Biology* **9**:395–403.

Mills, L.S. 1996. Fragmentation of a natural area: dynamics of isolation for small mammals on forest remnants. Pages 199–219 *in* G. Wright, editor. *National Parks and Protected Areas: their Role in Environmental Protection*. Blackwell Press, Oxford, UK. 470 pages.

Mills, L.S. and F.W. Allendorf. 1996. The one-migrant-per-generation rule in conservation and management. *Conservation Biology* **10**:1509–1518.

Mills, L.S. and D.A. Tallmon. 1999. The role of genetics in understanding forest fragmentation. Pages 171–186 *in* J.A. Rochelle, L.A. Lehmann, and J. Wisniewski, editors. *Forest Fragmentation: Wildlife and Management Implications*. Brill, Boston, Massachusetts, USA.

Mills, L.S., J.J. Citta, K. Lair, M. Schwartz, D. and Tallmon. 2000*a*. Estimating animal abundance using non-invasive DNA sampling: promise and pitfalls. *Ecological Applications* **10**:283–294.

Mills, L.S., K.L. Pilgrim, M.K. Schwartz, and K. McKelvey. 2000*b*. Identifying lynx and other North American felids based on mtDNA analysis. *Conservation Genetics* **1**:285–288.

Mowat, G., K.G. Poole, and M. O'Donoghue. 2000. Ecology of lynx in northern Canada and Alaska. Pages 285–307 *in* L.F. Ruggiero, K.B. Aubry, S.W. Buskirk, G.M. Koehler, C.J. Krebs, K.S. McKelvey, and J.R. Squires editors. *Ecology and Conservation of Lynx in the United States*. University of Colorado Press, Denver, Colorado, USA.

Neigel, J.L. 1996. Estimation of effective population size and migration parameters from genetic data. Pages 329–346 *in* T. Smith, and R. Wayne, editors. *Molecular Conservation Genetics*. Oxford University Press, Oxford, UK.

Newman, D. and D.A Tallmon. 2001. Experimental evidence for beneficial fitness effects of gene flow in recently isolated populations. *Conservation Biology* **15**:1054–1063.

Nichols, J.D. and C.J. Coffman. 1999. Demographic parameter estimation for experimental landscape studies on small mammal populations. Pages 287–309 *in* G.W. Barrett and J.D. Peles, editors. *Landscape Ecology of Small Mammals*. Springer-Verlag, New York, New York, USA.

Nicholson, A.J. and V.A. Bailey. 1935. The balance of animal populations. Part I. *Proceedings of the Zoological Society of London* **3**:551–598.

Olsen, J.B., P. Bentzen, M.A. Banks, J.B. Shaklee, and S. Young. 2000. Microsatellites reveal population identity of individual pink salmon to allow supportive breeding of a population at risk of extinction. *Transactions of the American Fisheries Society* **129**:232–242.

Paetkau, D., W. Calvert, I. Stirling, and C. Strobeck. 1995. Microsatellite analysis of population structure in Canadian polar bears. *Molecular Ecology* **4**:347–354.

Paetkau, D., S.C. Amstrup, E.W. Born, W. Calvert, A.E. Derocher, G.W. Garner, F. Messier, I. Stirling, M.K. Taylor, O Wiig, and C. Strobeck. 1999. Genetic structure of the world's polar bear populations. *Molecular Ecology* **8**:1571–1584.

Peacock, M.M. 1997. Determining natal dispersal patterns in a population of North American pikas (*Ochotona princeps*) using direct mark-resight and indirect genetic methods. *Behavioral Ecology* **8**:340–350.

Pease, C.M., R. Lande, and J.J. Bull. 1989. A model of population growth, dispersal, and evolution in a changing environment. *Ecology* **70**:1657–1664.

Perry, D.A., M.P. Amaranthus, J.G. Borchers, S.L. Borchers, and R.E. Brainerd. 1989. Bootstrapping in ecosystems. *Bioscience* **39**:230–237.

Pollock, K.H. 1982. A capture-recapture design robust to unequal probability of capture. *Journal of Wildlife Management* **46**:757–760.

Pollock, K.H., J.D. Nichols, C. Brownie, and J.E. Hines. 1990. Statistical inference for capture-recapture experiments. *Wildlife Monographs* **107**:1–107.

Powell, L.A., M.J. Conroy, J.E. Hines, J.D. Nichols, and D.G. Krementz. 2000. Simultaneous use of mark-recapture and radiotelemetry to estimate survival, movement, and capture rates. *Journal of Wildlife Management* **64**:302–313.

Pritchard, J.K., M. Stephens, and P. Donnelly. 2000. Inference of population structure using multilocus genotype data. *Genetics* **155**:945–959.

Proctor, M.F., McLellan, B.N., and C. Strobeck. 2003. Population fragmentation of grizzly bears in southeastern British Columbia, Canada. *Ursus* **13**:153–160.

Rannala, B. and J.L. Mountain. 1997. Detecting immigration by using multilocus genotypes. *Proceedings of the National Academy of Sciences* **94**:9197–9201.

Roach, J.L., P. Stapp, B. Van Horne, and M.F. Antolin. 2001. Genetic structure of a metapopulation of black-tailed prairie dogs. *Journal of Mammalogy* **82**:946–959.

Rosenberg, D.K., K.A. Swindle, and R.G. Anthony. 1994. Habitat associations of California red-backed voles in young and old-growth forests in Western Oregon. *Northwest Science* **68**:266–272.

Rosenberg, D.K., B.R. Noon, and C. Meslow. 1997. Biological corridors: form, function and efficacy. *BioScience* **47**:677–687.

Rousset, F. and M. Raymond. 1997. Statistical analyses of population genetic data: new tools, old concepts. *Trends in Ecology and Evolution* **12**:313–317.

Ruggiero, L.F., K.B. Aubry, S.W. Buskirk, G.M. Koehler, C.J. Krebs, K.S. McKelvey, K. and J.R. Squires. 2000. *Ecology and Conservation of Lynx in the United States*. University of Colorado Press, Denver, Colorado, USA. 480 pages.

Schwartz, M.K., D.A. Tallmon, and G.H. Luikart. 1998. Review of DNA-based census and effective population size estimators. *Animal Conservation* **1**:293–299.

Schwartz, M.K., L.S. Mills, K.S. McKelvey, L.F. Ruggiero, and F.W. Allendorf. 2002. DNA reveals high dispersal synchronizing the population dynamics of Canada lynx. *Nature* **415**:520–522.

Schwartz, M.K., L.S. Mills, Y. Ortega, L. Ruggiero, and F.W. Allendorf. 2003. Landscape location affects genetic variation of Canada Lynx (*Lynx canadensis*). *Molecular Ecology*, in press.

Simberloff, D., J.A. Farr, J. Cox, and D.W. Mehlman. 1992. Movement corridors: conservation bargains or poor investments? *Conservation Biology* **6**:493–504.

Singer, F.J., M.E. Moses, S. Bellew, and W. Sloan. 2000. Correlates to colonizations of new patches by translocated populations of bighorn sheep. *Restoration Ecology* **8**:66–74.

Slatkin, M. 1985. Gene flow in natural populations. *Annual Review of Ecology and Systematics* **16**:393–430.

Slatkin, M. 1995. A measure of population subdivision based on microsatellite allele frequencies. *Genetics* **139**:457–462.

Slatkin, M. and N.H. Barton. 1989. A comparison of three indirect methods for estimating average levels of gene flow. *Evolution* **43**:1349–1369.

Slough, B.G. and G. Mowat. 1996. Population dynamics of lynx in a refuge and interactions between harvested and unharvested populations. *Journal of Wildlife Management* **60**:946–961.

Soulé, M.E. and M.E. Gilpin. 1991. The theory of wildlife corridor capability. Pages 3–8 *in* D.A. Saunders and R.J. Hobbs, editors. *Nature Conservation 2: the Role of Corridors*. Surrey Beatty and Sons: Chipping Norton, New South Wales, Australia.

Spendelow, J.A., J.D. Nichols, I.C.T. Nisbet, H. Hays, G.D. Cormons, J. Burger, C. Safina, J.E. Hines, and M. Gochfeld. 1995. Estimating annual survival and movement rates of adults within a metapopulation of roseate terns. *Ecology* **76**:2415–2428.

Spong, G. and S. Creel. 2001. Deriving dispersal distances from genetic data. *Proceedings of the Royal Society of London B* **268**:2571–2574.

Steinberg, E.K. and C.E. Jordan. 1997. Using molecular genetics to learn about the ecology of threatened species: the allure and the illusion of measuring genetic structure in natural populations Pages 440–460 *in* P. Fiedler and P. Kareiva, editors. *Conservation Biology for the Coming Decade*. Chapman and Hall, New York, New York, USA.

Sunnucks, P. 2000. Efficient genetic markers for population biology. *Trends in Ecology and Evolution* **15**:199–203.

Taberlet, P. and G. Luikart. 1999. Non-invasive genetic sampling and individual identification. *Biological Journal of the Linnean Society* **68**:41–55.

Taberlet, P., L.P. Waits, and G. Luikart. 1999. Non-invasive genetic sampling: look before you leap. *Trends in Ecology and Evolution* **14**:323–327.

Tallmon, D.A., H. Draheim, L.S. Mills, and F.W. Allendorf. 2002. Insights into recently fragmented vole populations from combined genetic and demographic data. *Molecular Ecology* **11**:699–707.

Taylor, B. L., S.J. Chivers, S. Sexton, and A.E. Dizon. 2000. Evaluating dispersal estimates using mtDNA data: comparing analytical and simulation approaches. *Conservation Biology* **14**:1287–1297.

Taylor, P.D., L. Fahrig, K. Henein, and G. Merriam. 1993. Connectivity is a vital element of landscape structure. *Oikos* **68**:571–573.

Tevis, L. 1956. Responses of small mammal populations to logging of Douglas-fir. *Journal of Mammalogy* **37**:189–196.

USDA. 2001. *National Forest System Road Management Strategy, Environmental Assessment and Civil Rights Impact Analysis*. January 2001 Forest Service, Washington Office.

Varvio, S., R. Chakraborty, and M. Nei. 1986. Genetic variation in subdivided populations and conservation genetics. *Heredity* **57**:189–198.

Vitalis, R. and D. Couvet. 2001. Estimation of effective population size and migration rate from one- and two-locus identity measures. *Genetics* **157**:911–925.

Vucetich, J.A. and T.A. Waite. 2000. Is one migrant per generation sufficient for the genetic management of fluctuating populations? *Animal Conservation* **3**:261–266.

Waits, J.L. and P.L. Leberg. 1999. Advances in the use of molecular markers for studies of population size and movement. *Transactions of the 64ᵗʰ North American Wildlife and Natural Resources Conference* **1**:191–201.

Waits, J.L. and P.L. Leberg. 2000. Biases associated with population estimation using molecular tagging. *Animal Conservation* **3**:191–199.

Waples, R.S. 1995. Evolutionary significant units and conservation of biological diversity under the endangered species act. Pages 8–27 *in* J.L. Nielsen, editor. *Evolution of the Aquatic Ecosystem: Defining Unique Units in Population Conservation*. American Fisheries Society, Bethesda Maryland, USA.

Waser, P.M. and C. Strobeck. 1998. Genetic signatures of interpopulation dispersal. *Trends in Ecology and Evolution* **13**:43–44.

Webster, M.S., P.M. Marra, S.M. Haig, S. Bensch, and R.T. Holmes. 2002. Links between worlds: unraveling migratory connectivity. *Trends in Research in Ecology and Evolution* **17**:76–83.

White, G.C. and K.P. Burnham. 1999. Program MARK: survival estimation from populations of marked animals. *Bird Study* **46** [Supplement]120–138.

Whitlock, M.C. and D.E. McCauley. 1999. Indirect measures of gene flow and migration: Fst ≠ 1/(4Nm + 1). *Heredity* **82**:117–125.

Wolff, J. O. An evolutionary and behavioral perspective on dispersal and colonization of mammals in fragmented landscapes. Pages 614–630 *in* C.J. Zabel and R.G. Anthony, editors. *Mammal Community Dynamics. Management and Conservation in the Coniferous Forests of Western North America*. Cambridge University Press, Cambridge, UK.

Woods, J.G., D. Paetkau, D. Lewis, B.N. McLellan, M. Proctor, and C. Strobeck. 1999. Genetic tagging of free-ranging black and brown bears. *Wildlife Society Bulletin* **27**:616–627.

Wright, S. 1931. Evolution in Mendelian populations. *Genetics* **16**:97–259.

Zager, P., L.S. Mills, W. Wakkinen, and D.A. Tallmon. 1995. Woodland caribou: a conservation dilemma. *ESA Update* **12**:1–4.

18

An evolutionary and behavioral perspective on dispersal and colonization of mammals in fragmented landscapes

Introduction

Animals are not distributed randomly in time or space; rather their dispersion, movement, and distribution are strongly dictated by their evolutionary history and social organization. Some animals are relatively solitary or territorial, occupy specific microhabitats, and exhibit limited movement, whereas others are gregarious and not territorial, form herds, and migrate over large expanses of heterogeneous habitat. Similarly, not all animals respond the same way to habitat mosaics, habitat fragmentation, movement barriers, and human disturbance. Many of our predictions of how habitat loss and fragmentation affect native populations is through retrospection, speculation, or modeling, rather than by *a priori* predictions of how particular populations or species will respond to habitat perturbations. Our lack of understanding of how species respond to habitat alterations is such that species often are treated as mathematical entities (i.e., all individuals are "average") and individual-, sex-, and species-specific differences in response to fragmentation are not considered (Andrén 1994, Lima and Zollner 1996). Thus, animals are not always distributed as predicted by spatially explicit models, GAP analysis, habitat suitability indices, and so forth (Boone and Krohn 2000).

Some of the observed versus predicted differences in distribution of species across landscapes or in response to fragmentation can be explained by differences in their evolutionary history, dispersal ability, and social organization (Wolff 1999). In this chapter I use evolutionary and behavioral theory to predict some of the demographic consequences of habitat alteration for populations of mammals distributed in continuous and fragmented forest landscapes. Specifically, I focus on those factors that affect the potential for a given species to disperse across matrix habitat and colonize fragmented landscapes. I use a variety of mammalian species to present a theoretical argument to demonstrate how each parameter might affect colonization potential and use examples from mammals of western forests whenever possible.

Dispersal and colonization

An important component of mammalian behavioral systems is dispersal (Stenseth and Lidicker 1992). From an ecological perspective, dispersal has demographic consequences for a population in that it can stabilize densities and provide gene flow and maintain genetic panmixia. From a management or conservation standpoint, dispersal provides propagules for populations to colonize new habitats. From a behavioral perspective, dispersal separates opposite-sex relatives and reduces the chances of inbreeding (Pusey 1987, Brandt 1992, Wolff 1993, 1994). On the other hand, delayed dispersal can restrict gene flow, prevent colonization, and result in delayed sexual maturation and reproductive suppression (e.g., Creel and Waser 1991, Wolff 1992, 1997 and references cited therein), or possible inbreeding (Smith and Ivins 1983). In large continuous populations, animals should be relatively free to move throughout the habitat without consideration of ecological, physical, or behavioral barriers (e.g., McCullough 1985). In fragmented landscapes, however, dispersal can be deterred or prevented depending on the type of barrier (Fahrig and Merriam 1985). Several behavioral parameters can also influence a species potential for dispersal and subsequent colonization of disjunct habitats.

Dispersal distance

One of the main factors that determines whether a patch will be occupied or individuals will move to a particular patch is distance. Dispersal distances obviously differ for various species of mammals and correlate allometrically with body mass using the basic formula of dispersal distance $= ab^x$ where $a =$ constant, $b =$ body mass, and $x =$ slope of the regression line. Dispersal distance varies with sex and trophic level but for all mammals the log (dispersal distance in km) $=$ log (body mass in grams) $\times$ $0.67 - 1.42$ (Wolff 1999). Dispersal distances are greater for carnivores than herbivores and omnivores, and for males than females (Wolff 1999; see also Van Vuren 1998). Some species that show shorter dispersal distances than predicted by the model are highly social, such as ground squirrels and prairie dogs (*Cynomys* spp.), or those that are confined to very patchy habitats such as pikas, *Ochotona* spp. and pond-dwelling muskrats (*Ondatra zibethicus*). Maximum dispersal distances for females are often comparable to those of males, however the proportion of females dispersing and median dispersal distances are generally short (see below). The regression models (Van Vuren 1998, Wolff 1999) predict dispersal distances for all mammals based on their body mass; however, deviations from expected dispersal distances are expected to occur for the reasons discussed throughout this chapter.

Evolutionary history

The rate and (or) probability of a species colonizing distant patches likely is a function of its evolutionary history. A species that evolved in stable continuous habitat may respond very differently to fragmented habitats than a species that evolved in a patchy or frequently disturbed environment (Merriam 1995, Lima and Zollner 1996). For example, in western North America, elk (*Cervus elaphus*) frequently are associated with mature forests or edge habitat, whereas they apparently spent much of their evolutionary history in North America as an open-steppe habitat species (Geist 1971, Guthrie 1982, McCullough 1985). Black bears (*Ursus americanus*) of eastern United States are primarily forest-dwellers, whereas in western and northern North America they frequently are associated with partially open habitats (Powell 1997). Columbian ground squirrels (*Spermophilus columbianus*) probably never evolved dispersal strategies suited to colonization of isolated pockets of habitat because steppe vegetation is stable relative to the lifetime of a ground squirrel (Weddell 1991). Black-tailed prairie dogs (*Cynomys ludovicianus*) likewise do not migrate to unoccupied natural patches (Garrett and Franklin 1988). On the other hand, alpine marmots (*Marmota* spp.), which occupy isolated rock outcrops interspersed in alpine mountains, appear to be adapted to dispersal and colonization of that patchy resource (Van Vuren and Armitage 1994, Van Vuren 1998). White-footed mice (*Peromyscus leucopus*) also readily colonize isolated woodlots and persist as a metapopulation (Middleton and Merriam 1981). Wolves (*Canis lupus*) often follow prey such as caribou (*Rangifer tarandus*) and deer (*Odocoileus* spp.; Weaver et al. 1996), and lynx (*Felis lynx*) disperse over large distances in search of food during declines of snowshoe hares (*Lepus americanus*; Murray et al. 1994). Snowshoe hares (*Lepus americanus*), moose (*Alces alces*), and grassland voles (*Microtus* spp.), which exploit early successional or frequently disturbed habitats, should also be good colonizers (Wolff 1982, Hik 1995), whereas brush rabbits (*Sylvilagus bachmani*), which are confined to a specific brush habitat and avoid open habitat and forest-dwelling voles (e.g., *Clethrionomys*), would not be. Species such as black-tailed jackrabbits (*Lepus californicus*) that have evolved in open plains habitats and avoid forested areas probably would not be good colonists if they had to disperse through wooded habitats. Elk, caribou, and bison (*Bison bison*) have an evolutionary history of annual migrations (McCullough 1985) and would be good colonists. Thus, various aspects of the evolutionary history of a species pre-adapt it to disperse great distances and to move across a habitat mosaic.

Generalists and specialists in habitat mosaics

Habitat suitability indices (HSI) are based on habitat preferences and frequently are used in population viability models (Morrison et al. 1992). Unfortunately, most species do not visualize or utilize habitat based on its description, or on an aerial photo or landsat image. Rather, many species have habitat requirements that may be fine- or coarse-grained and include a mosaic of habitats, each component being necessary, but not sufficient, for successful colonization. For instance, bats typically require a covered roosting site, often with a narrow access passageway, such as caves, tree hollows, or human-made dwellings (Bradbury 1977). Preferred and suitable foraging areas are not necessarily coincident with roosting areas. Bats may feed on nectar, fruit, blood, fish, or flying insects, all of which may or may not be in the immediate vicinity. Opossums (*Didelphis virginiana*) and raccoons (*Procyon lotor*) require hollow trees for nesting, but frequently forage in open habitats, along streams, or in urban settings. Bears (*Ursus* spp.) may shift home range use from mature forest or grazing areas in spring to spawning salmon streams during summer, and berry patches in autumn (Powell 1997, and others), all of which may fall into different vegetation classifications. Marten (*Martes americana*) typically spend 95% of their time in forest habitats but forage extensively for voles in adjacent grassland habitats (Zielinski 1981). Male and female ungulates typically segregate and use different habitats for much of the year (Main et al. 1996, Bleich et al. 1997). Therefore, specific habitat requirements that include all the requisites for life must be considered for species that have different feeding and nesting areas, seasonally available resources, and sex-specific requirements.

In contrast, some species that are habitat specialists avoid mosaics and may perceive them as a barrier to dispersal (e.g., Diffendorfer et al. 1995).

North American red-backed voles (*Clethrionomys* spp.) are forest habitat specialists and would not be good colonists compared to *Peromyscus*, which are habitat generalists and readily cross habitat mosaics (Wegner and Henein 1991). Although few data are available on the response of mammals to fragmented landscapes, predictions can be made regarding probable responses to habitat mosaics by specialists and generalists. For instance, marten are forest specialists and seldom travel greater than 25 m into open habitat (Bissonette and Broekhuizen 1995), which probably restricts their ability to colonize new patches inter-dispersed among an open-habitat matrix (see also Potvin et al. 2000). Arboreality also decreases the probability that a species will colonize patchy habitats. For example red squirrels (*Tamiasciurus hudsonicus*), Douglas squirrels (*T. douglasii*), and tree voles (*Arborimus longicaudus* and *A. pomo*) do not readily cross unforested habitats (Smith et al. 2003). Forest-dwelling spotted skunks (*Spilogale gracilis*) should have a more difficult time dispersing across open fields (Howard and Marsh 1982) than would striped skunks (*Mephitis mephitis*), which are adapted to fragmented landscapes (Godin 1982, Saunders et al. 1982). Similarly, raccoons and opossums, which adapt readily to urbanized landscapes, should be able to cross human-occupied areas more readily than would wolverines (*Gulo gulo*) and fishers (*Martes pennanti*), which tend to avoid human contact.

Species that live on habitat islands – such as muskrats, which are confined to ponds (Messier et al. 1990), and pikas, which occupy isolated talus slopes (Smith and Ivins 1983) – apparently are reluctant to leave their island habitats. Thus, species that evolved within habitat mosaics should be better colonists than habitat specialists, which have evolved within a specific habitat type and are probably reluctant to cross habitat matrices.

Sociality and conspecific attraction

Vacant space within the dispersal distance for a species is not always sufficient for colonization. Some species of mammals exhibit conspecific attraction, such that individuals only disperse among sites already occupied. The tendency to settle near conspecifics rather than colonize new habitat has been demonstrated in pikas (Smith and Peacock 1990), ground squirrels (Weddell 1991), and prairie dogs (Garrett and Franklin 1988). New coteries or populations of prairie dogs and ground squirrels are formed by fission of established colonies (Michener 1983, Halpin 1987) and not by colonization of individuals into vacant patches. Gregarious species such as elk also are more likely to colonize by fission of a large population such that conspecifics leave as a group (McCullough et al. 1996). In contrast, the tendency to disperse and colonize distant patches should be less affected by conspecifics in asocial species, or those that are not attracted to conspecifics *per se*, such as hares, mink (*Mustela vison*), opossums, and moose.

Sex-biased dispersal

For a species to be a good colonist, females must disperse. In most mammals, females are relatively philopatric, and often remain in or near their natal site, and dispersal is male-biased (Greenwood 1980, Boonstra et al. 1987, Pusey 1987, Wolff 1993). Thus, the probability of colonizing and establishing a breeding population at new sites or in distant patches often is less than would be predicted based on an estimated dispersal distance for the species. Females of some species occasionally disperse but, in general, females remain relatively close to their natal home range and often form female kin groups (Greenwood 1980, Holekamp 1984, Kevles 1986, Wolff 1994, 1997). Among large carnivores, female bears normally remain near their natal site and are unlikely to colonize new habitat, whereas female wolves, lynx, and mountain lion (*Felis concolor*) frequently move long distances (Weaver et al. 1996, Sweanor et al. 2000). Most gene flow among populations of grizzlies (*Ursus arctos*) on the islands of the Alaskan Peninsula is through male dispersal, though females occasionally disperse among some closer islands (reviewed in Craighead and Vyse 1996). Genetically distinct populations of bighorn sheep (*Ovis canadensis*) throughout western North America similarly result from lack of female dispersal (Bleich et al. 1996). Females may be more apt to disperse and colonize new habitats if they leave as a group, such as Tule elk (*Cervus nannodes*) of California (McCullough 1985, McCullough et al. 1996). Male mammals are more apt to cross matrix habitat than are females; however, corridors may increase the tendency for females to disperse (Danielson and Hubbard 2000, Davis-Born and Wolff 2000).

Even though both males and females disperse to some extent, males usually disperse farther and at a much higher frequency than do females. Exceptions occur in red squirrels, snowshoe hares, beaver (*Castor canadensis*), porcupines (*Erethizon dorsatum*), red foxes (*Vulpes vulpes*), lynx, coyotes (*Canis latrans*), wolves, and opossums, in which females disperse as far and often as do males (Wolff 1999). General characteristics of species in which both sexes disperse at comparable rates and distances include a monogamous mating system (such as exhibited by many species of canids), species in which both sexes individually defend burrow systems and food caches (such as Douglas squirrels and red squirrels), or species that are not territorial such as hares, porcupines, and opossums; summarized in Wolff (1997) and Wolff and Peterson (1998).

Dispersal and infanticide

Infanticide, the killing of infants by conspecifics, is a major mortality factor affecting juvenile recruitment in many mammalian species. Strange males that have not sired offspring in a given area will frequently kill infants as a means of removing competitive offspring and providing a reproductive opportunity by mating with the victim's mother following termination of lactation (Hausfater and Hrdy 1984 and references cited therein). Infanticide by males commonly occurs in species with altricial young in which males wander over relatively large areas and encounter unrelated offspring (reviewed in Agrell et al. 1998, Ebensperger 1998). Males that have copulated in a given area or with a given female do not commit infanticide, but strangers, immigrants, or those males that have not mated will kill offspring (Wolff and Cicirello 1991, Pusey and Packer 1994, Wielgus and Bunnell 1995). This behavior has important implications for habitat fragmentation where animals are forced to move into new patches or for re-introductions where potentially infanticidal individuals immigrate into new habitats and social groups and encounter unrelated offspring. In fact, low juvenile recruitment of grizzly bears was attributed to infanticide following immigration of strange males after removal of resident dominant males (Wielgus and Bunnell 1995). Infanticide following immigration of males has also been reported in other mammals such as mountain lions (Hornocker 1970), *Peromyscus* spp. (Wolff and Cicirello 1991), lions (*Panthera leo*; Pusey and Packer 1994), and marmots (Coulon et al. 1995).

Ideal free distribution and ideal despotic distribution (female territoriality)

The theory for the relationship between the distribution of animals in space and their fitness is grounded in the ideal free distribution (IFD) and ideal despotic (or dominance) distribution (IDD) models of Fretwell and Lucas (1970). The differences between these models is whether or not a species is territorial; that is, whether members defend exclusive space. Dispersion, dispersal, juvenile recruitment, reproductive suppression, use of source and sink habitats, and potential for colonization in mammalian species are all dependent to a great extent on how females use space. In species in which females are territorial, females compete for offspring-rearing space and breeding is relegated to the territory holders that can successfully defend that space (Wolff and Peterson 1998). In species in which females are not territorial, however, no behavioral mechanism exists to prevent females from breeding (Wolff 1997). Consequently, the size of the breeding population, that in turn affects the rate of population growth and rate of immigration and emigration (see below), should be dependent to some extent on whether or not females are territorial. Thus, territorial species follow an IDD, whereas non-territorial species follow the IFD.

The primary factor that determines whether or not females are territorial is if they have altricial or precocial young (Wolff and Peterson 1998). Female territoriality occurs in species that have altricial young that are deposited in a den or protected nest or natal site. Mammalian species with precocial young or altricial young that are carried with the mother (such as marsupials and primates) are not territorial. Therefore, female territoriality commonly occurs among the insectivores, sciurognath rodents (squirrels, mice, and voles), rabbits, and carnivores, but does not occur among ungulates, hystricognath (e.g., cavies and agoutis) rodents, hares, or marsupials. In territorial species such as red squirrels, wolves, and rabbits, females require an individual territory to breed, whereas in non-territorial species, exclusive space is not a requisite for reproduction (reviewed in Wolff 1997). For instance, in ungulates such as bighorn sheep, elk, or bison, all females have the opportunity to breed irrespective of individual space. Social pressures do not prevent any female from breeding in non-territorial species. The important point here is that in species in which females are territorial, the size of the breeding population should be limited by the number of breeding sites (territories) available in a habitat (Wolff 1997). This relationship should not hold for non-territorial species.

Source-sink habitats and reproduction

Source and sink habitats probably occur for all species (Pulliam 1988), but have a greater influence on territorial than non-territorial species. In territorial species, as soon as all suitable breeding territories are occupied, any additional adults are relegated to suboptimal or sink habitat, whereas in species that do not defend space, individuals can pack much more densely into source habitat. In that females do not require individual space for breeding in non-territorial species, variance in fitness should be greater in territorial than in non-territorial species. Territorial defense of source habitat by dominant individuals, which forces subordinate individuals into sink habitat and affects their fitness, should follow the IDD model for tree squirrels, ground squirrels, and martens, whereas fitness of non-territorial species such as elk, deer, and snowshoe hares would follow an IFD. Thus, knowledge of the social system and some assessment of habitat saturation for a species will help in applying the IDD and IFD models to explain source-sink dynamics and individual variance in reproduction and fitness.

Territoriality, density dependence, and dispersal

The rate of dispersal of individuals away from their natal site is, in part, a function of the ease with which individuals can immigrate into new space. Territoriality can impede movement of animals if all the suitable space is occupied and individuals are thus not able to cross undefended space. This type of barrier to movement is referred to as a social fence (Hestbeck 1982) and results in an inverse density-dependent dispersal pattern in territorial species (Wolff 1997). In contrast, in non-territorial species in which habitat is not actively defended, individuals can move without social impediment at any density. Thus in species such as voles, mice, squirrels, and wolves, dispersal should decrease as density increases whereas in non-territorial species such as deer, elk, porcupines, and opossums, dispersal should not be inhibited by a social fence.

Territoriality also can affect the ways animals use corridors for movement. For instance, if an individual establishes a territory that encompasses the width of the corridor, other individuals will be less able to move along the corridor to adjacent patches than if the corridor were not occupied (Andreassen et al. 1996). The same result should occur if an individual establishes a territory at an entrance to a corridor. For non-territorial species, movement should not be deterred along such stretches of habitat. Corridors that are narrower than the home range diameter of a given species should facilitate greater movement than a wider one occupied and defended against transgression.

Dispersal, philopatry, and kin groups

Usually, only one female breeds on a territory in territorial species; but exceptions do occur. In most species of mammals, young males disperse from the social unit and daughters are philopatric and remain on or near their natal site (Greenwood 1980, Pusey 1987, Brandt 1992, Wolff 1993, also see above). Female philopatry often results in the formation of kin groups or female alliances that share the same space, such that, if space is limited, daughters breed on their mother's territories. This pattern of shared space commonly occurs among prairie dogs (Hoogland 1995), marmots (Armitage 1981) and many species of mice and voles (Jannett 1978, Wolff 1985, McGuire and Getz 1991, Lambin 1994, Wolff 1994, Salvioni and Lidicker 1995). In contrast, only one female breeds on a territory in red foxes (Allen and Sargeant 1993), wolves (Mech 1970), and red squirrels (Price and Boutin 1993). Thus, an understanding of the social relationships among related females and their tolerance of shared breeding space allows more accurate predictions regarding the reproductive potential for a given area of habitat.

Reproductive suppression

A common question asked by wildlife managers and conservation biologists is what factors inhibit sexual maturation and reproduction in a portion of wild animal populations. This is a crucial question in that the rate of population growth is a direct function of the proportion of females breeding. Reproduction can be affected by nutrition, but also by social behavior. Reproductive suppression of young females commonly occurs in high densities of territorial species in which young females delay emigration and remain in their family group or in the presence of other adults. The ultimate causation of reproductive suppression appears to be the prevention of inbreeding with close relatives (Wolff 1992, 1997), or a response to the threat of infanticide from adult females (Wasser and Barash 1983, Abbott 1984, Digby 1995, Wolff 1997). In that inbreeding is deleterious to the inclusive fitness of a given female, it may be in her best interests to delay breeding if the only males available to her are her father or brothers. Similarly, a young female may be more successful in raising young if she waits until she attains higher status or is able to obtain a territory of her own before breeding. In those situations, reproductive inhibition should be alleviated by separating young females from their fathers or brothers to avoid inbreeding or from potentially infanticidal adult females (Wolff 1997). In all of these situations, reproductive suppression is a response to immediate behavioral situations that are created because normal dispersal patterns are prevented. Behavioral reproductive suppression does not appear to occur in non-territorial species. Thus, the behavioral and demographic responses to delayed dispersal appear to be a function of the species' behavioral system.

A behavioral approach to intrinsic population regulation

Animal populations are said to be regulated intrinsically or extrinsically. Extrinsic factors include those such as weather, predation, and food limitation whereas intrinsic factors are those intrinsic to the population, such as behavioral attributes. Below, I summarize the models presented in Wolff (1997) in which I propose an evolutionary pathway that leads to intrinsic and extrinsic population regulation in mammals.

The evolutionary pacemaker for intrinsic population regulation starts with non-mobile altricial young that require a burrow or protected den site. These offspring-rearing sites are defended as territories by females, which, in turn, determine breeding space. Females that do not have breeding space delay sexual maturation, thereby limiting the size of the breeding population. At high densities, dispersal of offspring from the natal site is delayed because of a social fence of territorial neighbors preventing immigration. As a result of delayed dispersal, young females delay sexual maturation if they remain in the vicinity of infanticidal adult females or in the presence of male relatives. Thus, delayed sexual maturation is a mechanism to conserve reproductive effort until chances improve for successful rearing of offspring (as discussed above). At low densities, individuals can disperse from the natal site such that young females can find breeding territories and males can find unrelated females for mates. Consequently, intrinsic population regulation should occur only in species that have non-mobile altricial young.

In non-territorial species, i.e., those that have precocial young or mobile altricial young (such as opossums and other marsupials), infanticide does not occur. Thus females do not defend territories, and no mechanism exists to deter dispersal or suppress reproduction. Consequently, there is no mechanism by which intrinsic population regulation can occur in these species. Population regulation of mammal species with precocial young is limited to extrinsic factors. In that regulation of animal numbers is such an important component of wildlife management and conservation, it is essential to understand the underlying behavioral mechanisms that can affect population dynamics.

Implications for management

Not all species of mammals can be treated as mathematical entities or model species; variance in their dispersion, dispersal patterns, and potential for colonization is dependent to some extent on their evolutionary history and behavioral system. Good colonists are species that evolved in habitat mosaics, are ecological generalists, migratory, asocial, or have female-biased dispersal. Poorer colonists are those species that evolved in homogeneous habitats, are ecological specialists, philopatric or non-migratory, social, and have female philopatry. Fragmentation of habitat, division of populations, and movement of strangers among habitat patches have important implications for infanticide and loss of juvenile recruitment. In that infanticide occurs in some species but not others, knowledge of these differences can be used when assessing the potential costs of removing resident males or planning species re-introductions. A major paradigm relating fitness to the distribution of individuals within habitats rests on the IFD and IDD models. These models differ in their predictions for reproductive success and juvenile recruitment. In that territorial species follow an IDD and non-territorial species follow an IFD, knowledge of behavioral systems will aid in predicting demographic consequences for creating source and sink habitats. The predictions for reproductive success, dispersal, and reproductive inhibition are dependent on whether or not females are distributed freely or despotically. Utilization of source-sink habitat and associated fitness also are functions of the IDD and IFD. Finally, behavioral theory can be used to explain how populations are regulated by intrinsic or extrinsic factors. A thorough knowledge of mammalian behavior systems should contribute substantially to predicting how species will respond to altered habitats.

Acknowledgments

This work was supported in part by NSF Grant 99-96016. I thank Cindy Zabel and Robert Anthony for inviting me to participate in TWS symposium in Nashville, September 2000 and to write this chapter. Dale McCullough and one anonymous reviewer provided helpful comments on an earlier draft of the manuscript.

Literature cited

Abbott, D.H. 1984. Behavioral and physiological suppression of fertility in subordinate marmoset monkeys. *American Journal of Primatology* **6**:169–186.
Agrell, J., J.O. Wolff, and H.Ylönen. 1998. Infanticide in mammals: strategies and counter-strategies. *Oikos* **84**:507–517.
Allen, S.H. and A.B. Sargeant. 1993. Dispersal patterns of red foxes relative to population density. *Journal of Wildlife Management* **57**:526–533.

Andreassen, H.P., S. Halle, and R.A. Ims. 1996. Optimal width of movement corridors for root voles: not too narrow and not too wide. *Journal of Applied Ecology* **33**: 63–70.

Andrén, H. 1994. Effects of habitat fragmentation on birds and mammals in landscapes with different proportions of suitable habitat: a review. *Oikos* **71**:355–366.

Armitage, K.B. 1981. Sociality as a life-history tactic of ground squirrels. *Oecologia* **48**:36–49.

Bissonette, J.A. and S. Broekhuizen. 1995. Martes populations as indicators of habitat spatial patterns: the need for a multiscale approach. Pages 95–121 *in* W.Z. Lidicker, Jr., editor. *Landscape Approaches in Mammalian Ecology and Conservation.* University of Minnesota Press, Minneapolis, Minnesota, USA.

Bleich, V.C., J.D. Wehausen, R.R. Ramey II, and J.L. Rechel. 1996. Metapopulation theory and mountain sheep: implications and conservation. Pages 353–373 *in* D.R. McCullough, editor. *Metapopulations and Wildlife Conservation.* Island Press, Washington DC, USA.

Bleich, V.C., R.T. Bowyer, and J.D. Wehausen. 1997. Sexual segregation in mountain sheep: resources or predation? *Wildlife Monographs* **134**:1–50.

Boone, R.B. and W.B. Krohn. 2000. Predicting broad-scale occurrences of vertebrates in patchy landscapes. *Landscape Ecology* **15**:63–74.

Boonstra, R., C.J. Krebs, M.S. Gaines, M.L. Johnson, and I.T.M. Craine. 1987. Natal philopatry and breeding systems in voles (*Microtus* spp.). *Journal of Animal Ecology* **56**:655–673.

Bradbury, J.W. 1977. Social organization and communication. Pages 1–72 *in* W. Wimsatt, editor. *Biology of Bats.* Academic Press, New York, New York, USA.

Brandt, C.A. 1992. Social factors in immigration and emigration. Pages 96–141 *in* N.C. Stenseth and W.Z. Lidicker, Jr. editors. *Animal Dispersal: Small Mammals as Models.* Chapman and Hall, New York, New York, USA.

Coulon, J., L. Graziani, D. Allainé, M. C. Bel, and S. Puderoux. 1995. Infanticide in the alpine marmot (*Marmota marmota*). *Ethology, Ecology and Evolution* **7**:191–194.

Craighead, F.L. and E.R. Vyse. 1996. Brown/grizzly bear metapopulations. Pages 325–351 *in* D.R. McCullough, editor. *Metapopulations and Wildlife Conservation.* Island Press, Washington DC, USA.

Creel, S.R. and P.M. Waser. 1991. Failure of reproductive suppression in dwarf mongooses (*Helogale parvula*): accident or adaptation? *Behavioral Ecology* **2**:7–15.

Danielson, B.J. and M.W. Hubbard. 2000. The influence of corridors on the movement behavior of individual *Peromyscus polionotus* in experimental landscapes. *Landscape Ecology* **15**:323–331.

Davis-Born, R. and J.O. Wolff. 2000. Age- and sex-specific response of the gray-tailed vole, *Microtus canicaudus*, to connected and unconnected habitat patches. *Canadian Journal of Zoology* **78**:864–870.

Diffendorfer, J.E., M.S. Gaines, and R.D. Holt. 1995. Habitat fragmentation and movements of three small mammals (*Sigmodon*, *Microtus*, and *Peromyscus*). *Ecology* **76**:827–839.

Digby, L. 1995. Infant care, infanticide, and female reproductive strategies in polygynous groups of common marmosets (*Callithrix jacchus*). *Behavioral Ecology and Sociobiology* **37**:51–61.

Ebensperger, L.A. 1998. Strategies and counterstrategies to infanticide in mammals. *Biological Review* **73**:321–346.

Fahrig, L. and G. Merriam. 1985. Habitat patch connectivity and population survival. *Ecology* **66**:1762–1768.

Fretwell, S.D. and H.L. Lucas. 1970. On territorial behaviour and other factors influencing habitat distribution in birds. I. Theoretical development. *Acta Biotheoretica* **19**:16–36.

Garrett, M.G. and W.L. Franklin. 1988. Behavioral ecology of dispersal in the black-tailed prairie dog. *Journal of Mammalogy* **69**:236–250.

Geist, V. 1971. The relation of social evolution and dispersal in ungulates during the Pleistocene with emphasis on the Old World deer and the genus *Bison*. *Quarternary Research* **1**:283–315.

Godin, A.J. 1982. Striped and hood skunks. Pages 674–687 *in* J.A. Chapman and G. A. Feldhammer, editors. *Wild Mammals of North America*. Johns Hopkins University Press, Baltimore, Maryland, USA.

Greenwood, P.J. 1980. Mating systems, philopatry and dispersal in birds and mammals. *Animal Behaviour* **28**:1140–1162.

Guthrie, R.D. 1982. Mammals of the mammoth steppe as paleoenvironmental indicators. Pages 307–329 *in* D. M. Hopkins, editor. *Paleoecology of Beringia*. Academic Press, New York, New York, USA.

Halpin, Z.T. 1987. Natal dispersal and the formation of new social groups in a newly established town of black-tailed prairie dogs (*Cynomys ludovicianus*). Pages 1–4–118 *in* N.D. Chepko-Sade and Z.T. Halpin, editors. *Mammalian Dispersal Patterns*. University of Chicago, Chicago, Illinois, USA.

Hausfater G. and S.B. Hrdy. 1984. *Infanticide: Comparative and Evolutionary Perspectives*. Aldine, New York, New York, USA.

Hestbeck, J.B. 1982. Population regulation of cyclic mammals: the social fence hypothesis. *Oikos* **39**:147–163.

Hik, D.S. 1995. Does risk of predation influence population dynamics? Evidence from the cyclic decline of snowshoe hares. *Wildlife Research* **22**:115–129.

Holekamp, K.E. 1984. Dispersal in ground-dwelling sciurids. Pages 297–320 *in* J.O. Murie and G.R. Michener, editors. *The Biology of Ground-dwelling Sciurids: Annual Cycles, Behavior, Ecology, and Sociality*. University of Nebraska, Lincoln, Nebraska, USA.

Hoogland, J.L. 1995. *The Black-tailed Prairie Dog: Social Life of a Burrowing Mammal*. University of Chicago, Chicago, Illinois, USA.

Hornocker, M.G. 1970. An analysis of mountain lion predation upon mule deer and elk in the Idaho Primitive Area. *Wildlife Monograph* **21**:1–39.

Howard, W.E. and R.E. Marsh. 1982. Spotted and hog-nose skunks. Pages 664–673 *in* J.A. Chapman and G.A. Feldhammer, editors. *Wild Mammals of North America*. Johns Hopkins University Press, Baltimore, Maryland, USA.

Jannett, F.G., Jr. 1978. The density-dependent formation of extended maternal families of the montane vole, *Microtus montanus nanus*. *Behavioral Ecology and Sociobiology* **3**:245–263.

Kevles, B. 1986. *Females of the Species*. Harvard University, Cambridge, Massachusetts, USA.

Lambin, X. 1994. Natal philopatry, competition for resources, and inbreeding avoidance in Townsend's voles (*Microtus townsendii*). *Ecology* **75**:224–235.

Lima, S.L. and P.A. Zollner. 1996. Towards a behavioral ecology of ecological landscapes. *Trends in Ecology and Evolution* **11**:131–135.

Main, M.B., F.W. Weckerly, and V.C. Bleich. 1996. Sexual segregation in ungulates: new directions for research. *Journal of Mammalogy* **77**:449–461.

McCullough, D.R. 1985. Long range movements of large terrestrial mammals. Pages 444–465 *in* M.A. Rankin, editor. *Migration: Mechanisms and Adaptive Significance. Contributions in Marine Science*. Volume **27**, Supplement. University of Texas, Aransas, Texas, USA.

McCullough, D.R., J.K. Fischer, and J.D. Ballou. 1996. From bottleneck to metapopulation: recovery of the tule elk in California. Pages 375–403 *in* D.R. McCullough, editor. *Metapopulations and Wildlife Conservation*. Island Press, Washington DC, USA.

McGuire, B. and L.L. Getz. 1991. Response of young female prairie voles (*Microtus ochrogaster*) to nonresident males: implications for population regulation. *Canadian Journal of Zoology* **69**:1348–1355.

Mech, L.D. 1970. *The Wolf*. Natural History Press, Garden City, New York, USA.

Merriam, G. 1995. Movement in spatially divided populations: responses to landscape structure. Pages 64–77 *in* W.Z. Lidicker, Jr., editor. *Landscape Approaches in Mammalian Ecology and Conservation*. University of Minnesota, Minneapolis, Minnesota, USA.

Messier, F., J.A. Virgl, and L. Marinelli. 1990. Density-dependent habitat selection in muskrats: a test of the ideal free distribution model. *Oecologia* **84**:380–385.

Michener, G.R. 1983. Kin identification, matriarchies, and the evolution of sociality in ground-dwelling sciurids. Pages 528–572 *in* J.F. Eisenberg and D.G. Kleiman, editors. *Advances in the Study of Mammalian Behavior*. Special Publication of the American Society of Mammalogists, Volume **7**. Allen Press, Manhattan, Kansas, USA.

Middleton, J. and G. Merriam. 1981. Woodland mice in a farmland mosaic. *Journal of Applied Ecology* **18**:703–710.

Morrison, M.L., B.G. Marcot, and R.W. Mannan. 1992. *Wildlife-habitat Relationships*. University of Wisconsin, Madison, Wisconsin, USA.

Murray, D.L., S. Boutin, and M. O'Donoghue. 1994. Winter habitat selection by lynx and coyotes in relation to snowshoe hare abundance. *Canadian Journal of Zoology* **72**:1444–1451.

Powell, R. A. 1997. *Ecology and Behavior of North American Black Bears*. Chapman and Hall, New York, New York, USA.

Potvin, F., L. Belanger, and K. Lowell. 2000. Marten habitat selection in a clear-cut boreal landscape. *Conservation Biology* **14**:844–957.

Price, K. and S. Boutin. 1993. Territorial bequeathal by red squirrel mothers. *Behavioral Ecology* **4**:144–150.

Pulliam, H.R. 1988. Sources, sinks, and population regulation. *American Naturalist* **132**:652–661.

Pusey, A.E. 1987. Sex-biased dispersal and inbreeding avoidance in birds and mammals. *Trends in Ecology and Evolution* **2**:295–299.

Pusey A.E. and C. Packer. 1994. Infanticide in lions: consequences and counterstrategies. Pages 277–299 *in* S. Parmigiani and F. vom Saal, editors. *Infanticide and Parental Care*. Harwood, London, UK.

Salvioni, M. and W.Z. Lidicker, Jr. 1995. Social organization and space use in the California vole: seasonal, sexual, and age-specific strategies. *Oecologia* **101**:426–438.

Saunders, A.B., J.R. Greenwood, J.L. Piehl, and W.B. Bicknell. 1982. Recurrence, mortality, and dispersal of prairie striped skunks, *Mephitis mephitis*, and implications to rabies episootiology. *Canadian Field-Naturalist* **96**:312–316.

Smith, A.T. and B.L. Ivins. 1983. Colonization in a pika population: dispersal vs philopatry. *Behavioral Ecology and Sociobiology* **13**:37–47.

Smith, A.T. and M.M. Peacock. 1990. Conspecific attraction and the determination of metapopulation colonization rates. *Conservation Biology* **4**:320–323.

Smith, W., R.G. Anthony, J.R. Waters, N.L. Dodd, and C.J. Zabel. 2003. Ecology and conservation of arboreal rodents of western coniferous forests. Pages 157–206 *in* C.J. Zabel and R.G. Anthony, editors. *Mammal Community Dynamics. Management and Conservation in the Coniferous Forests of Western North America*. Cambridge University Press, Cambridge, UK.

Stenseth, N.C. and W.Z. Lidicker, Jr. 1992. *Animal Dispersal: Small Mammals as a Model*. Chapman and Hall, London, UK.

Sweanor, L.L., K.A. Logan, and M.G. Hornocker. 2000. Cougar dispersal patterns, metapopulation dynamics and conservation. *Conservation Biology* **14**:798–808.

Van Vuren, D. 1998. Mammalian dispersal and reserve design. Pages 363–393 *in* T.M. Caro, editor. *Behavioral Ecology and Conservation Biology*. Oxford University Press, Oxford, UK.

Van Vuren, D. and K.B. Armitage. 1994. Survival of dispersing and philopatric yellow-bellied marmots: what is the cost of dispersal? *Oikos* **69**:179–181.

Wasser, S.K. and D.P. Barash. 1983. Reproductive suppression among female mammals: implications for biomedicine and sexual selection theory. *Quarterly Review of Biology* **58**:513–538.

Weaver, J.L., P.C. Pacquet, and L.F. Ruggiero. 1996. Resilience and conservation of large carnivores in the Rocky Mountains. *Conservation Biology* **10**:964–976.

Weddell, B.J. 1991. Distribution and movements of Columbian ground squirrels (*Spermophilus columbianus* (Ord)): are habitat patches like islands? *Journal of Biogeography* **18**:385–394.

Wegner, J. and K. Henein. 1991. Strategies for survival: white-footed mice and eastern chipmunks in an agricultural landscape. Page 90 *in Proceedings of the World Congress of Landscape Ecology, International Association for Landscape Ecology*. Ottawa, Ontario, Canada.

Wielgus R.B. and F.L. Bunnell. 1995. Tests of hypotheses for sexual segregation in grizzly bears. *Journal of Wildlife Management* **59**:552–560.

Wolff, J.O. 1982. Refugia, dispersal, predation, and geographic variation in snowshoe hare cycles. Pages 441–449 *in* K. Myers and C.D. MacInnes, editors. *Proceedings of the World Lagomorph Conference*. University of Guelph, Guelph, Ontario, Canada.

Wolff, J.O. 1985. Behavior. Pages 340–372 *in* R.H. Tamarin, editor. *Biology of New World Microtus*. American Society of Mammalogists Special Publication Number **8**. Allen Press, Manhattan, Kansas, USA.

Wolff, J.O. 1992. Parents suppress reproduction and stimulate dispersal in opposite-sex juvenile white-footed mice. *Nature* **359**:409–410.

Wolff, J.O. 1993. What is the role of adults in mammalian juvenile dispersal. *Oikos* **68**:173–176.

Wolff, J.O. 1994. More on juvenile dispersal in mammals. *Oikos* **71**:349–352.

Wolff, J.O. 1997. Population regulation in mammals: an evolutionary perspective. *Journal of Animal Ecology* **66**:1–13.

Wolff, J.O. 1999. Behavioral model systems. Pages 11–40 *in* G. Barrett and J. Peles, editors. *Landscape Ecology of Small Mammals*. Springer-Verlag, New York, New York, USA.

Wolff, J.O. and D.M. Cicirello. 1991. Comparative paternal and infanticidal behavior of sympatric white-footed mice (*Peromyscus leucopus noveboracensis*) and deermice (*P. maniculatus nubiterrae*). *Behavioral Ecology* **2**:38–45.

Wolff, J.O. and J.A. Peterson. 1998. An offspring-defense hypothesis for territoriality in female mammals. *Ethology Ecology and Evolution* **10**:227–239.

Zielinski, W.J. 1981. Food habits, activity patterns, and ectoparasites of the pine marten at Sagehen Creek, California. MS Thesis, University of California, Berkeley, California, USA.

19

The functional diversity of mammals in coniferous forests of western North America

The ecological knowledge needed to achieve the goals of ecosystem management will not be limited to understanding the influence of habitat manipulations on desired mammal populations; it will also include an understanding of how those mammals contribute to the functioning of the ecosystems they occupy. Examples of the significant influence that mammals may have on the structure and function of ecosystems include the effects of sea otters (*Enhydra lutris*) on the community structure of coastal marine ecosystems (Estes and Palmisano 1974), the effects of American beavers (*Castor canadensis*) on the hydrology and ecology of temperate riparian ecosystems (Naiman et al. 1986, Anthony et al. 2003), the effects of burrowing mammals on soil fertility and stability (Meadows and Meadows 1991, Ayarbe and Kieft 2000), and the effects of large ungulates on successional processes and the structure of plant communities in a variety of ecosystems (Hobbs 1996). Each of these species or species groups has been described as a potential keystone species (Mills et al. 1993) in the ecosystems they occupy. Because most species of mammals may not influence ecosystem processes to the extent that keystone species do, their ecological contributions are often overlooked. We propose, however, that the collective importance of terrestrial mammals to ecosystem structure and function is substantial and that the decline or loss of forest mammal species could have detrimental effects on ecosystem diversity, productivity, or sustainability.

To determine how management actions may influence ecological conditions, managers must be able to characterize and quantify the contributions of resident organisms to ecosystem function. Here, we refer to the roles played by an organism that directly affect other species or strongly influence environmental conditions in a given ecosystem, as *key ecological functions* (KEFs). A classification system and database of KEFs was first developed for plant, invertebrate, and vertebrate species of the interior West (Marcot et al. 1997; also see Morrison et al. 1998) and later for vertebrates of Washington and Oregon (Marcot and Vander Heyden 2001; see Appendix at the end of this chapter). In these projects, the ecological roles of species were identified by expert panels, organized into hierarchical classifications, and coded into relational databases (primarily as categorical data). By querying the database, one can determine the array of KEFs associated with a given species or species group, the array of species sharing a given KEF category, information about the species' habitat requirements and life history patterns, the potential influence of management activities on key habitat elements and KEFs, and other environmental relations.

In this chapter, we evaluate the contributions of mammals to an array of ecological processes in coniferous forests of western North America using the wildlife–habitat relations and KEF databases from the Species-Habitat Project in Washington and Oregon (Johnson and O'Neil 2001). These databases relate wildlife species to forest types, structural conditions, key environmental correlates (KECs), and KEFs; and KECs to

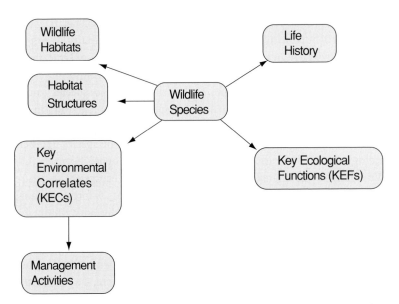

Fig. 19.1. Components of the wildlife–habitat relations database from the Species-Habitat Project in Washington and Oregon (Johnson and O'Neil 2001) used in our functional analyses.

management activities (Fig. 19.1). We present a process to: (1) describe the ecological roles of forest-dwelling mammals; (2) depict the "web of ecological functions" and other parameters associated with various functional groups; (3) link KEFs for each mammal species to forest types, vegetation structure, and KECs; and (4) quantify the potential effects of management actions on forest habitats, the mammals associated with those habitats, and the ecological functions they perform. The manager can use this approach to determine how forest mammals contribute to the functioning of a given ecosystem and estimate the extent to which that ecosystem will retain its functional integrity in response to management actions. The results of such assessments can be treated as repeatable and testable hypotheses of the effects of management activities on forested ecosystems.

Methods

We compared the list of mammals in the Species-Habitat Project database to the list of mammals occurring in forested habitats of western North America and identified the subset of forest-dwelling mammals that occur in Washington and Oregon (hereafter referred to as "forest mammals"). We then queried the databases to evaluate functional roles for various forest mammal assemblages. Because the KEF database consists mostly of categorical data, we used species counts as a unit of measure. We used the taxonomy of functional patterns in KEFs presented by Marcot and Vander Heyden (2001) to structure our investigation. These included community patterns (i.e., functional richness, redundancy, profiles, webs, and homologies), geographic patterns, ecological roles of species (i.e., critical links and functions, and functional breadth and specialization), and the functional responses of species assemblages (i.e., functional resilience and resistance). Definitions of each component of this taxonomy are presented in Results (also see Marcot and Vander Heyden 2001). We also used literature on the ecological roles of mammals to interpret the results of database queries. The information presented here should be viewed as working hypotheses because the specific rates and details of many of the functional relationships we discuss in this chapter have been poorly studied for most forest mammals.

Results

Ecological roles of forest mammals

Some of the ecological functions performed by forest mammals are unique. Mammals are the only vertebrates in western North American forests that feed on bark and cambium (seven species) or that create snags from live trees (three species). These activities add to the structural complexity of forests and provide habitat for a wide array of microorganisms, invertebrates, and cavity-using birds and mammals. Forest mammals are also key players in the dispersal of mushrooms and truffles, including the ectomycorrhizal fungi that play a critical role in the uptake of nutrients by conifer trees (Li et al. 1986, Maser and Maser 1988, Loeb et al. 2000; see Aubry et al. 2003 and Luoma et al. 2003); 11 species of forest mammals (Roosevelt elk (*Cervus elaphus roosevelti*), American pika (*Ochotona princeps*), two of voles, two of mice, and five of squirrels) and only one non-forest mammal (feral pig (*Sus scrofa*)) perform this fungi-dispersal function. In addition, recent work on the food habits of fishers suggests that forest carnivores may also serve as long-distance dispersal agents for fungal spores (see Aubry et al. 2003).

There are several other ecological functions for which forest mammals play a significant role. These include several categories of primary consumption, such as browsing on leaves or stems, eating mushrooms and truffles, eating feces and other excreta, dispersing lichens, excavating rabbit-sized or larger burrows, creating runways or trails, using runways created by other species, impounding water by creating diversions or dams, and altering vegetation structure and composition by browsing on trees or shrubs.

There are also a number of ecological roles, however, for which forest mammals participate the least; these include eating aquatic plants, eating aquatic macroinvertebrates, eating fish, eating fruits, dispersing insects and other invertebrates, dispersing propagules of vascular plants, and excavating cavities. There are an additional 12 KEFs that are not performed at all by forest mammals, including eating sap, creating sapwells in trees, eating freshwater zooplankton, pollinating plants, and serving as nest parasites or hosts. Thus, the array of ecological functions of forest mammals, as an assemblage, is unique and complementary to that of other taxonomic groups.

Community functional patterns
Functional richness and mean functional redundancy
The Species-Habitat Project database listed 733 species of amphibians, reptiles, birds, and mammals that occur in Washington and Oregon, of which 116 (16%) are mammals that inhabit forests, and 58 (8%) are forest mammals closely associated with coniferous forests. Forest mammals

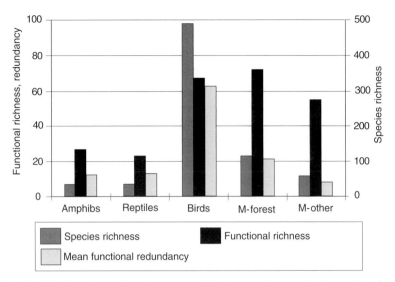

Fig. 19.2. Species richness, functional richness (number of categories of key ecological functions among all species), and mean functional redundancy (average number of categories of key ecological functions/species) of amphibians, reptiles, birds, forest mammals (M-forest), and non-forest mammals (M-other) in Washington and Oregon. Note that forest mammals have relative low species richness but high functional richness.

comprise 67% of mammal species in this region. There are 72 KEFs pertaining to at least one of the 116 forest mammals; this value represents the *total functional richness* (the number of KEF categories) of the forest mammal species assemblage (there is some redundancy in this value, because it includes both categories and subcategories of KEFs; see Appendix). This value is higher than the total functional richness of amphibians, reptiles, birds, and non-forest mammals, even though the total *species* richness of forest mammals is far less than that of birds and only slightly higher than those of each of the other taxonomic groups (Fig. 19.2). Thus, forest mammals fill a disproportionately broad array of ecological roles compared to other terrestrial vertebrate species groups.

On a per-KEF basis, however, the functional diversity represented by forest mammals is relatively low. For each species group (taxonomic group or assemblage), the average number of species performing each KEF is the *mean functional redundancy* of that group, and *functional diversity* is functional richness weighted by mean functional redundancy, analogous to species diversity, species richness, and species abundance (Brown 1995). Forest mammals average a functional redundancy of 21 species per KEF, which exceeds values for non-forest mammals, amphibians, and reptiles, but is much lower than the average value for birds (63 species per KEF; Fig. 19.2). In other words, forest mammals perform a more diverse array of ecological functions than amphibians, reptiles, or non-forest mammals, but contribute less to the functional diversity of forest ecosystems than do birds. Understanding patterns of functional diversity is important, because forest ecosystems with high levels of functional redundancy probably have a higher resilience to perturbations, stresses, and environmental changes (Peterson et al. 1998, Fonseca and Ganade 2001).

Functional profiles

A histogram that compares the functional redundancy among a set of habitats is a *functional profile*. These graphs can be useful for identifying habitats that are particularly rich or poor in specific functions (Marcot and Vander Heyden 2001). Nine forest habitats are described in the Species-Habitat Project database for Washington and Oregon (Chappell et al. 2001). Among these, Eastside Mixed Conifer, Montane Mixed Conifer, and Westside Lowlands Conifer-Hardwood Forests have the highest number of forest mammal species, whereas Western Juniper and Mountain Mahogany Woodlands, and Upland Aspen Forest have the lowest.

The highest functional redundancy in burrow excavation is found in Montane and Eastside Mixed Conifer Forests, whereas the lowest occurs in Western Juniper Forests (Fig. 19.3). However, digging species, which contribute to soil aeration and turnover of soil organic matter (Meadows and Meadows 1991, Butler 1995, Jones et al. 1996), are most numerous in Ponderosa Pine Forests and least numerous in Lodgepole Pine Forests. For dispersers of plant propagules, Montane and Eastside Mixed Conifer forests tend to be most species-rich. Western Juniper and Upland Aspen Forests have the fewest dispersers, which may be related to the relatively simple floras of those forest types compared with mixed-conifer forest types. The objective of comparing such functional profiles among habitats is to determine the habitats that support specific functions the most or the least. Knowing when only a few species provide a specific function in a given habitat may help managers design prescriptions that will reduce the likelihood that these species' ecological functions will be lost from the system.

Functional webs

The array of KEFs performed by an assemblage of species associated with a particular habitat element or structure is a *functional web*. For example, Marcot (2002) and Rose et al. (2001) identified functional webs of

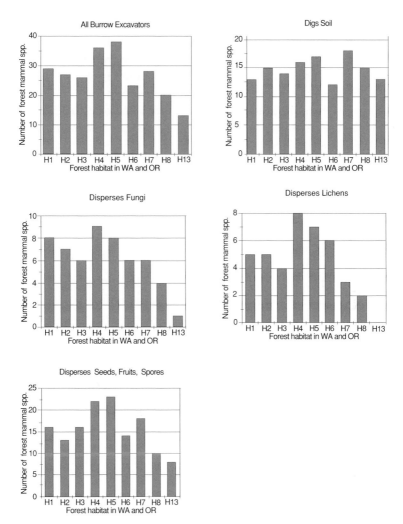

Fig. 19.3. Functional profiles for five selected categories of key ecological functions of forest mammals in Washington (WA) and Oregon (OR). H1 = Westside Lowlands Conifer-Hardwood Forest, H2 = Westside Oak (*Quercus garryana*) and Dry Douglas-fir (*Pseudotsuga menziesii*) Forest and Woodlands, H3 = Southwest Oregon Mixed Conifer-Hardwood Forest, H4 = Montane Mixed Conifer Forest, H5 = Eastside Mixed Conifer Forest, H6 = Lodgepole Pine (*Pinus contorta*) Forest and Woodlands, H7 = Ponderosa Pine (*Pinus ponderosa*) and Eastside White Oak Forests and Woodlands, H8 = Upland Aspen (*Populus tremuloides*) Forest, H13 = Western Juniper (*Juniperus occidentalis*) and Mountain Mahogany (*Cercocarpus montanus*) Woodlands.

forest species (including mammals) that are associated with down wood and snags in forests of Washington and Oregon. Results suggested that down wood was a habitat component for 86 wildlife species, 51 of which are forest mammals, and snags provided habitat for 95 wildlife species including 24 forest mammals. Collectively, these down wood- and snag-using species perform a rather surprisingly broad array of KEFs that could be maintained in the ecosystem by providing adequate amounts of snags and coarse woody debris for the wildlife species that are associated with such structures. Forest mammals play a key role especially in the down wood functional web; all nine wildlife species that are associated with down wood and that disperse fungal spores are forest mammals. Down wood also supports other species of small mammals that are prey for carnivores, disperse plant propagules, and provide an array of other ecological functions (McComb, 2003).

Forest mammals associated with down wood in Westside Lowlands Conifer-Hardwood Forest perform an array of at least 26 ecological functions (Fig. 19.4). Such functional webs can be described for any forest condition or habitat element by querying the databases to determine the array of associated species and the KEFs they perform. By doing so, managers can assess the ecological "value" of providing for a specific habitat element, and gain an understanding of how such structures help support the complex web of ecological functions that characterize coniferous forested ecosystems.

Functional homologies

When different habitats have a similar number of species performing the same KEFs, the habitats can be said to be *functionally homologous*. That is, although species composition may differ, the habitats have similar levels of functional redundancy for the same KEFs (Marcot and Vander Heyden 2001). To what extent are the nine forested habitats in Washington and Oregon functionally homologous? This can be answered by inspecting the functional profile graphs and comparing the number of forest mammal species that perform each KEF across the nine forest types. Results (Fig. 19.5A) indicate that these habitats are not highly homologous for all functions. That is, for some KEFs, the number of forest mammal species (the functional redundancy) varies considerably among forest types. In particular, they vary the most for forest-mammalian functions pertaining to primary excavation of burrows, secondary use of excavated burrows, and dispersal of seeds, fruits, and fungal spores. This means that, at least

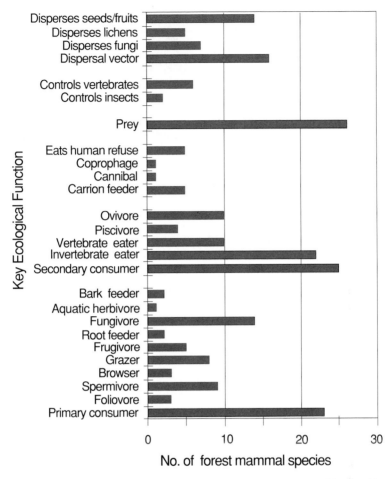

Fig. 19.4. Functional web of forest mammals associated with down wood in Westside Lowlands Conifer-Hardwood Forest. For example, ten species eat eggs (ovivore) as part of their functions within the forested ecosystem.

for forest mammals, the forest types are not strictly the same in terms of their arrays of ecological functions.

Functional homology can also be evaluated by comparing the similarity in the number (functional redundancy) of species per KEF in a cluster classification. Such a comparison (Fig. 19.5B) suggests that Westside Oak and Dry Douglas-fir Forest and Woodlands, and Southwest Oregon Mixed Conifer-Hardwood Forest are quite similar, as are Montane Mixed Conifer Forest, and Eastside Mixed Conifer Forest. The forest type Western

A

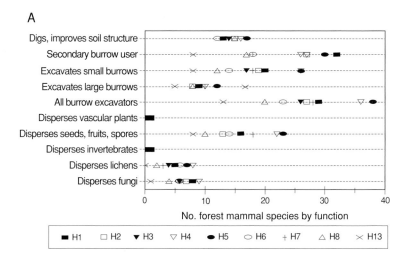

B

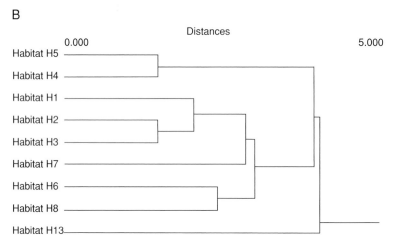

Fig. 19.5. (A) Functional homologies for selected key ecological functions, among nine forest wildlife habitats in Washington and Oregon (see Fig. 19.3 for forest habitat codes). Functional homology is a comparison of numbers of forest mammal species (or functional redundancy) with various key ecological functions (vertical axis) among habitats. (B) Hierarchical cluster classification of the nine forested wildlife habitats in Washington and Oregon, based on the number (functional redundancy) of forest-dwelling mammal species among the ten categories of key ecological functions (KEFs) listed in (A). Clustering was based on single linkage and Euclidean distance metrics. Note that wildlife habitats H2 and H3, and H4 and H5 are the most functionally homologous in terms of number of species performing these KEFs. The most dissimilar (least homologous) habitat functionally is H13.

Juniper and Mountain Mahogany Woodlands is most dissimilar, that is, least functionally homologous to the other forested habitats in Washington and Oregon. (The specific composition of forest mammal species also differs among the nine forested habitats analyzed here, somewhat but not fully paralleling the similarity in functional redundancy, with the least similar species composition occurring in Western Juniper and Mountain Mahogany Woodlands.) Thus, if managers wish to provide for the full array of ecological functions provided by forest mammals in all forested habitats of Washington and Oregon, they would need to provide for conditions that support such functions across the variety of forested habitats.

Geographic functional patterns

Once species' distributions and habitat conditions are accurately mapped, geographic patterns of functional redundancy for any given KEF or set of KEFs can be displayed and analyzed spatially (as has been done for some abiotic functions; see Noronha and Goodchild 1992). For example (Fig. 19.6), the functional redundancy of species that dig soil (most of whom are mammals) can be related to important contributions to soil structure and aeration within the Columbia River Basin in the U.S., and this can be mapped. One type of functional map can depict changes in functional redundancy for this KEF by comparing historic (early 1800's) to current (2000) conditions, showing that there has been a significant decline in functional redundancy of this KEF in many inland valley and basin systems including the Willamette Valley of Oregon, the Columbia Basin of Washington, and the Snake River Basin of Idaho. These are geographic areas where native grasslands and shrublands have been largely converted to agriculture, creating mostly inhospitable conditions for many native soil-digging mammals. Consequently, the soil-digging function currently is not well represented in these locations as compared to historic conditions.

Further, there do not seem to be any corridors of increased redundancy linking these areas. These areas are therefore *functional bottlenecks* that restrict the degree to which this function could operate across the landscape. Managers may wish to know where the geographic weakening or severing of functions might set the stage for further degradation of interacting functional ecosystems (Marcot and Vander Heyden 2001). This may be of particular interest because restoring or maintaining interacting functional ecosystems has been stated as a potential objective for ecosystem management (Strange et al. 1999). In one non-mammal example,

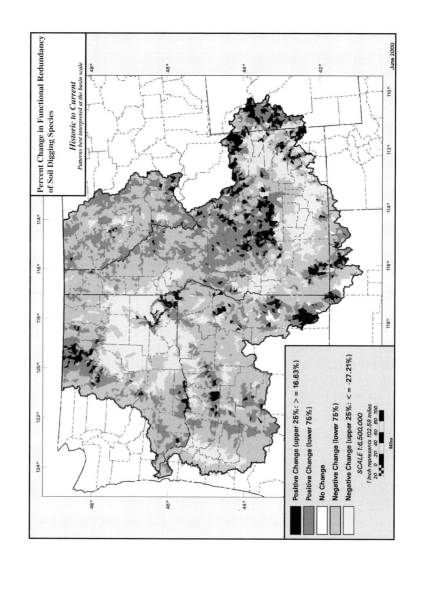

Percent Change in Functional Redundancy of Soil Digging Species

Historic to Current
Patterns best interpreted at the basin scale

Positive Change (upper 25%: > = 16.63%)

Positive Change (lower 75%)

No Change

Negative Change (lower 75%)

Negative Change (upper 25%: < = -27.21%)

SCALE 1:6,500,000
1 inch represents 102.59 miles
20 0 20 40 60 80 100
Miles

June 2000

Nabhan (2001) identified important habitat corridors for 300 species of nectar-feeding pollinators that migrate between Mexico and North America.

In contrast, some of the mountain areas show a significant increase in functional redundancy of the soil-digging and burrowing KEF, including parts of the North Cascades and Blue Mountains in Washington, the Rocky Mountains of southern Idaho, and some of the mountains in the Greater Yellowstone ecosystem of southeastern Idaho and western Wyoming. In these regions, many forested areas have undergone extensive timber harvesting since historic times, reverting them to earlier successional stages that support more wildlife species that dig soil, especially forest mammals. As described previously, communities with higher levels of functional redundancy may be more resilient to stressors and disturbances. However, as viewed throughout the entire Columbia River Basin, the total area of decrease in redundancy of this function (Fig. 19.6) is far greater than the total area of increase, indicating that the Columbia River Basin in the U.S. has suffered an overall decline in redundancy for the soil-digging function.

Such a map could be produced for any KEF, for use by managers to locate specific areas of lowered or lost redundancy or of functional bottlenecks, to help prioritize areas for restoration or maintenance of conditions for specific functions. For example, Dale et al. (2000) noted that "Particular species and networks of interacting species have key, broad-scale ecosystem-level effects." In the Columbia River Basin, patterns of change in other KEFs also vary geographically (Fig. 19.6). Change also can be compared between current and predicted future conditions under various land-planning alternatives (Marcot et al. 2002). In this way, managers can project future geographic effects on ecological functions and, by mapping KEFs, identify land areas needing special attention to avoid significant declines in one or more KEFs. Such an approach can help managers

Fig. 19.6. Map of functional redundancies of soil-digging animals in the Columbia River Basin, showing high and low areas of change in functional redundancy from historic conditions. Although the map depicts terrestrial wildlife of all taxonomic classes, most of the species shown here are mammals (of both forested and non-forested habitats). The map was produced based on wildlife habitats and associated species by sub-watershed, and the different shading denotes quartiles of change categories. For example, the category of highest positive change is shown as the top 25% of sub-watersheds having the highest value in change in functional redundancy, and this represents a change of 16.6% increase in functional redundancy values. (Source: Tom O'Neil, Northwest Habitat Institute, Corvallis, Oregon; used by permission.)

select planning alternatives that best match stated goals for maintaining intact forest mammal communities in specific ecosystems.

Species' functional roles

Critical functions and critical functional link species

When only one or a few species perform a particular ecological function and their removal would signal a serious decline or loss of that *critical function*, such species are designated as *critical functional link species* (Marcot and Vander Heyden 2001). For example, among forest mammals in the Westside Lowlands Conifer-Hardwood Forests of coastal Washington and Oregon, the KEF of creating snags (girdling or killing live trees) is provided by the American beaver, black bear (*Ursus americanus*), and common porcupine (*Erethizon dorsatum*). Insect species and fungal pathogens, along with fire, also provide this function. These forest mammals are the only three vertebrate wildlife species that provide this function in western forests, making this KEF a moderately "critical function" and these species critical functional link species for this particular function.

Although relatively few forest mammals can be considered critical functional link species, several species do provide critical functions in the ecosystems they occupy. In Westside Lowlands Conifer-Hardwood Forests of Washington and Oregon, critical functions include creating ground structures used by other species (bushy-tailed woodrat (*Neotoma cinerea*), Douglas squirrel (*Tamiasciurus douglasii*)), creating aquatic structures and impounding water (American beaver), secondary use of aquatic structures created by other species (fisher (*Martes pennanti*), mink (*Mustela vison*)), and creating wet swales and small ponds by wallowing (Roosevelt elk). No other species than these few forest mammals perform these functions in this particular forest type.

These patterns contrast to those in another forest type, Western Juniper and Mountain Mahogany Woodlands, an arid environment of eastern Washington and Oregon. In this forest type, at least eight KEFs are performed by only a few vertebrates, all of which are forest mammals. These critical functions include eating bark and cambium and creating snags (American beaver, common porcupine), browsing (American beaver, common porcupine, a few ungulates), eating feces (Nuttall's or mountain cottontail (*Sylvilagus nuttallii*)), creating aquatic structures and impounding water (American beaver), secondary use of aquatic structures (mink), and changing vegetation structure or successional stage through intensive herbivory (common porcupine, golden-mantled ground squirrel

(*Spermophilus lateralis*)). However, none of the functions associated with forest mammals in this forest type is an *imperiled function*, that is, the forest mammals listed here are not scarce, greatly declining, or extirpated, so these functions can be reasonably expected to continue. However, in some areas where American beavers have been trapped out, their critical functions listed above may have suffered. Managers could use such information to identify and prioritize functions that may depend on only one or a few forest mammal species. Habitat conditions for these species could then be provided to help maintain high-priority functions that may be in danger of being lost from the system.

Functional breadth and functional specialization of species

The array of functions performed by a species is its *functional breadth*. Species performing very few functions (e.g., fewer than eight associated KEFs) are *functional specialists*, whereas those performing many (e.g., >20 KEFs) are *functional generalists* (Marcot and Vander Heyden 2001). For example, among forest mammals in Westside Lowlands Conifer-Hardwood Forest of Washington and Oregon, functional specialists include the fog shrew (*Sorex sonomae*; five KEFs), Baird's shrew (*Sorex bairdi*; six KEFs), and masked shrew (*Sorex cinereus*), ermine (*Mustela erminea*), long-eared myotis (*Myotis volans*), and Pacific shrew (*Sorex pacificus*) (seven KEFs each). Functional specialists in this forest type tend to be insectivores and some are secondary predators. Functional generalists tend to be omnivorous or herbivorous, and include the black bear (33 KEFs), raccoon (*Procyon lotor*; 27 KEFs), deer mouse (*Peromyscus maniculatus*; 26 KEFs), American beaver (24 KEFs), Douglas squirrel (*Tamiasciurus douglasii*; 23 KEFs), striped skunk (*Mephitis mephitis*; 22 KEFs), and Roosevelt elk (21 KEFs). In contrast, there are only two forest mammals that are functional specialists in Western Juniper and Mountain Mahogany Woodlands of eastern Washington and Oregon: the long-eared myotis and western pipistrelle (*Pipistrellus hesperus*), both of which are insectivorous bats with only seven KEFs each. Functional generalists are limited to the deer mouse (26 KEFs), golden-mantled ground squirrel, and American beaver (24 KEFs each).

Functional specialists in other forest types of Washington and Oregon include wolverine (*Gulo gulo*; five KEFs), lynx (*Lynx canadensis*), northern bog lemming (*Synaptomys borealis*) (six KEFs each), mountain goat (*Oreamnos americanus*), Preble's shrew (*Sorex preblei*), and spotted bat (*Euderma maculatum*) (seven KEFs each); and functional generalists include red squirrel (*Tamiasciurus hudsonicus*; 23 KEFs), black-tailed jackrabbit

(*Lepus californicus*), and grizzly bear (*Ursus arctos*) (22 KEFs each). Although not necessarily a forest-dwelling species, humans (*Homo sapiens*) are the greatest functional generalists of all; our impressive array of 35 KEFs (Appendix) exceeds that of any other vertebrate species. This may explain, in part, why humans have had such an overwhelming influence on so many habitats and wildlife communities (Marcot and Vander Heyden 2001).

Functional responses of species assemblages

Functional resilience and resistance among forest structural classes
Maintaining the biodiversity and productivity of communities or ecosystems may require that they remain resilient or resistant to disturbances (Walker 1992, 1995). The capacity of an ecosystem to rebound to its initial functional pattern following a change from disturbance is its *functional resilience* (Reice et al. 1990, Carpenter and Cottingham 1997, Ludwig et al. 1997, Gunderson 2000, Marcot and Vander Heyden 2001), and *functional resistance* is the capacity of a community to maintain its functional patterns in response to a disturbance (Halpern 1988, Brang 2001, Marcot and Vander Heyden 2001).

Few studies have been conducted on these parameters for individual species or assemblages of forest mammals, although Weaver et al. (1996) discussed the importance of functional resilience to the conservation of large carnivores. These concepts may be useful considerations for managers who want to maintain the functional roles of forest mammals in the presence of disturbance events, especially forest-management activities. Here, we explore the potential changes in functional redundancy of forest mammals among structural and successional stages of Westside Lowlands Conifer-Hardwood Forest as an example of how these concepts can be applied to management. The degree to which forest mammal communities would be able to respond to changes in these forest stages remains to be studied in the field. We intend for this analysis to generate hypotheses regarding the influence of forest management on patterns of functional redundancies in mammals that could be tested empirically.

We compared patterns of functional redundancies among selected KEFs among successional stages and canopy structure conditions in Westside Lowlands Conifer-Hardwood Forest using the Species-Habitat Project database for Washington and Oregon (O'Neil et al. 2001). Successional stages included grass/forb, shrub/seedling, sapling/pole, small tree, medium tree, large tree, and giant tree stages of single-story, closed-canopy forests. Mammal species composition varied among these successional stages. Overall trends in species richness suggested that most forest mammals and most secondary consumers occur in the medium to giant tree stages, with fewest in the sapling/pole and small tree stages (Fig. 19.7A). Functional redundancy of grazing is highest in the grass/forb-closed stage, whereas that of spermivory (seed-eating) is highest in the medium and large tree-single story-closed stages (Fig. 19.7B). Thus, to ensure that the full set of all ecological functions is present, with their highest redundancies, the forest manager may wish to provide for the full array of successional stages. Note that mammal species composition typically varies among successional stages, even for the same ecological function.

The number of forest mammals that are primary consumers tends to be more evenly distributed among the seven stages than are secondary consumers (Fig. 19.7A). KEFs with higher functional redundancies in grass/forb and shrub/seedling stages than in later stages included bark and cambium eaters, grazers, and diggers of small burrows. KEFs with higher functional redundancies in medium, large, and giant tree stages than in earlier stages included seed eaters, fungi eaters, egg eaters, carrion eaters, secondary burrow users, and dispersers of lichens, fungi, seeds, and fruits. KEFs with nearly equal functional redundancies among all seven stages included browsers, root eaters, fruit eaters, cannibals, fish eaters, and diggers of large burrows (Fig. 19.7B–F).

We ran one-way analysis of variance (ANOVAs) with post-hoc Bonferroni multiple comparison tests to determine if levels of functional redundancies among selected KEF categories varied significantly among successional stages, canopy-closure classes (open, moderate, or closed canopy), or number of canopy layers (single or multiple canopies) in Westside Lowlands Conifer-Hardwood Forest (Table 19.1). Results of ANOVA tests suggested that there were no significant effects of successional stage or the number of canopies on the total number of forest mammal species in this forest type (although there was a trend in the means for successional stage as noted above). There was a significant effect of canopy closure, however, with the highest number of species occurring in open-canopy conditions.

Among specific KEF categories, the functional redundancies of carrion eaters were significantly influenced by successional stage and number of canopies, and marginally influenced by canopy closure. The number of large-burrow excavators was influenced significantly by canopy closure but not by successional stage or the number of canopies. The number of small-burrow excavators showed reverse trends, being significantly

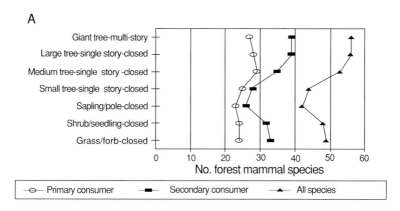

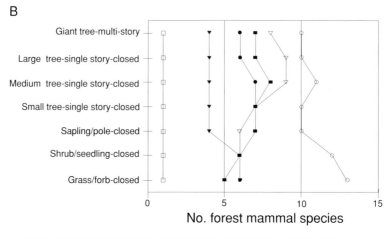

Fig. 19.7A–F. Number of forest mammal species in Westside Lowlands Conifer-Hardwood Forest of Washington and Oregon by seven successional stages of forest growth and selected categories of key ecological functions. Some functions reach their highest number of associated wildlife species (functional redundancy) in early successional stages, whereas other functions reach their highest functional redundancy in late successional stages.

C

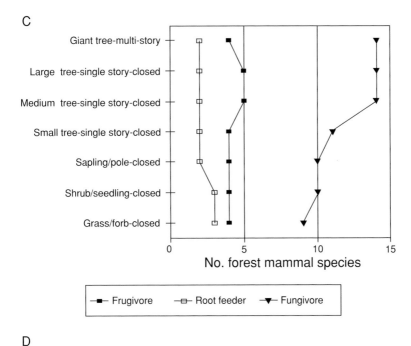

D

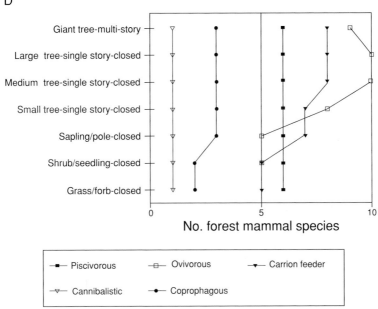

Fig. 19.7A–F. (*cont.*)

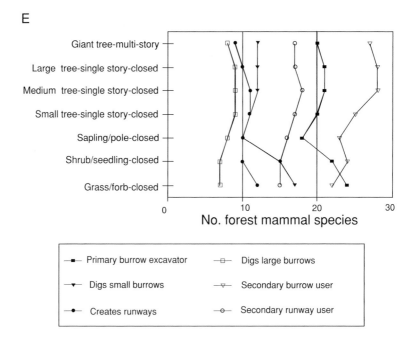

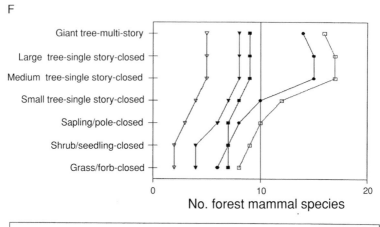

Fig. 19.7A–F. (cont.)

Table 19.1. *One-way analysis of variance tests with post-hoc Bonferroni multiple comparisons, on numbers of species (functional redundancies) of forest mammals in Westside Lowland Conifer-Hardwood Forests, for selected categories of key ecological functions (KEF) across successional stage (tree size) and canopy-closure classes, and number of canopies*

KEF category	F value	df	P value
By successional stage (tree size) class[a]			
All species	1.945	5	0.132
Feeds on carrion	22.034	5	<0.001**
Excavates large (>rabbit-sized) burrows	1.446	5	0.251
Excavates small (rabbit-sized) burrows	18.920	5	<0.001**
Disperses fungi	28.384	5	<0.001**
Digs soil	2.085	5	0.110
By canopy closure class (open, moderate, closed canopy)			
All species	11.471	2	<0.001**
Feeds on carrion	2.924	2	0.074+
Excavates large (>rabbit-sized) burrows	10.866	2	<0.001**
Excavates small (rabbit-sized) burrows	0.526	2	0.598
Disperses fungi	2.085	2	0.147
Digs soil	13.881	2	<0.001**
By number of canopies (single, multiple canopies)[b]			
All species	0.048	1	0.829
Feeds on carrion	6.317	1	0.019*
Excavates large (>rabbit-sized) burrows	2.665	1	0.116
Excavates small (rabbit-sized) burrows	3.546	1	0.072+
Disperses fungi	8.848	1	0.007**
Digs soil	0.476	1	0.497

[a] Successional stage (tree size) classes: grass/forb, shrub/seedling, sapling/pole, small tree, medium tree, large tree, and giant tree. Large tree and giant tree classes were combined in the ANOVAs to reduce number of classes. Tests focused on single-story, closed-canopy conditions of these stages.
[b] ANOVA tests reduce to unpaired Student t-tests.
+ $0.05 \leq P \leq 0.10$.
* $P < 0.05$.
** $P < 0.01$.

influenced by successional stage but not canopy-closure class, and only marginally by the number of canopies. The number of fungi dispersers was influenced significantly by successional stage and number of canopies, but not by canopy closure, and the number of soil diggers was influenced by canopy closure but not successional stage or the number of canopies (Table 19.1).

It is clear that the functional redundancies of different KEFs are influenced by different forest structural attributes. Also, at least for Westside

Lowlands Conifer-Hardwood Forest, no single forest condition (e.g., open, multi-canopy, large-tree forest) provides maximum redundancy of all associated forest mammal KEFs; nor is there a single condition that accounts or provides for all maximum KEF levels of all forest mammals. Thus, a mix of successional stages, canopy-closure classes, and canopy densities would be required to provide for the highest number of forest mammals for the KEFs included in these analyses.

Influence of forest management on ecological functions of forest mammals

The functional redundancy of forest mammals is not only influenced by successional stage, canopy closure, and the number of canopies, but also by the presence of microhabitat elements or substrates. This was discussed previously in the section on functional webs of species associated with down wood and snags in coniferous forests.

Additionally, the size of live trees and snags can have varying influences on different KEFs. For example, some fungi-eating (and therefore potential spore-dispersing) forest mammals in Westside Lowland Conifer-Hardwood Forest are also associated with large trees or large snags (Fig. 19.8), especially trees or snags >36 cm (14 in.) in diameter at breast height (dbh). Thus, if the forest manager wishes to provide for

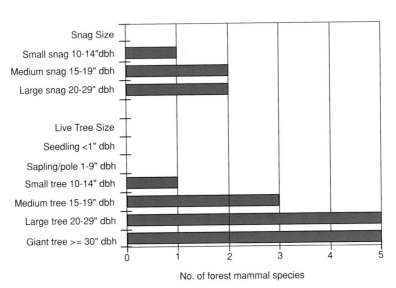

Fig. 19.8. Number (functional redundancy) of forest mammal fungivores that are associated with live tree and snag size classes. dbh stands for diameter at breast height.

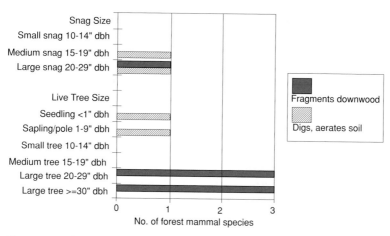

Fig. 19.9. Number of forest mammals that fragment down wood and dig soil and are associated with live tree and snag size classes. Fragmenting down wood and digging soil are natural ecological roles that provide habitat for a variety of fungi, invertebrates, and other organisms, likely speed the uptake of organic matter into soil, and maintain soil productivity. More forest mammals that fragment down wood are associated with large snags and with large and giant live trees, than with smaller snags or live trees, because forest mammals that fragment down wood are largely associated with late successional stages. More forest mammals that turn over soil are associated with medium and large snags than with small snags, and with seedling and sapling/pole size trees than with larger trees, because soil-digging forest mammals are associated with early successional stages. Thus, to maintain the full set of forest mammal species with these two ecological functions pertinent to soil productivity, the manager may provide for large snags in both early- and late-successional stages.

maximum functional redundancy of fungivory, which may be important for dispersal of spores of beneficial fungi, management guidelines could include specifically providing large snags and large live trees.

In another example, only a few forest mammal species provide the function of fragmenting down wood, and these species tend to be associated with large trees and snags >51 cm (20 in.) dbh (Fig. 19.9). On the other hand, forest mammals that dig and aerate soil and are associated with a particular tree or snag size tend to be associated with small, live trees <25 cm (10 in.) dbh, but are also associated with large snags >36 cm dbh.

Managers may want to know the array of microhabitat and substrate elements used by forest mammal species having desired KEFs and ensure that they are provided for in forest-management plans. Most of these habitat elements could be provided relatively easily by ensuring adequate retention of large live trees, snags, down wood, truffle patches, and other elements of older forests (Franklin et al. 2000).

Discussion

Caveats and assumptions of the functional approach

The analyses presented above should not be viewed in isolation from empirical data on the autecology and demography of individual species. Simply because a particular category of KEF is present or maintained in a faunal community does not mean that all native species associated with that habitat are equally well conserved or even present. We intend the kinds of functional assessments we present here to complement, not replace, species-specific conservation.

Marcot and Vander Heyden (2001) listed a number of caveats pertaining to the types of functional assessments presented in this chapter, including the following:

1. In the Species-Habitat Project and Interior Columbia River Basin databases, assignment of KEFs to each species was determined at least as much by the collective judgment of expert panels, as by results of empirical studies. Results should be viewed as testable working hypotheses and should be validated and refined through new field research.

2. Some KEFs are incompletely represented in these databases, especially those relating to nutrient cycling and disease transmission.

3. Results of functional assessments are best interpreted at the level of broad geographic areas, such as ecoprovinces or sub-basins, rather than at the scale of project areas or forest stands. Applying these findings to small geographic areas is likely to lead to overestimations in the number of ecological functions present, unless finer resolution information and local knowledge are applied.

4. Existing databases do not consider how KEFs for a given species might vary in different habitats or with the presence or absence of specific environmental conditions or elements, such as particular prey items. Empirical data on the KEFs for most species, including forest mammals in western coniferous forests and their variation, are generally lacking (Marcot 1997).

For some functions, such as the creation of snags or various influences on soil structure and productivity, other taxonomic groups may have a far greater influence on ecosystem conditions than forest mammals. Thus, the user should have a basic understanding of the relative importance of forest mammals compared to other faunal groups (including invertebrates) for performing various ecological functions.

For example, infestations of bark-feeding cerambicid beetles and some foliage-feeding larval lepidoptera (e.g., spruce budworms) can have far greater influences on the structure of coniferous forests of western North America than the few forest mammals that girdle trees or fragment standing wood (van Hees and Holsten 1994, Williams and Liebhold 2000). Earthworms and other burrowing invertebrates can process soil, enhance uptake of soil organic matter, and engage in soil nutrient cycling at far greater rates than burrowing forest mammals (Hendrix 1995).

One of the basic tenets of the functional assessment approach presented here is that of functional redundancy. However, by definition, each species defines its own niche. Simply because two species share a general category of ecological function does not mean they are completely interchangeable in the ecosystem, that is, define the exact same niche. Each will perform its function in different ways and will interact with different species, use different substrates, and exert its influence at varying intensities or rates. In a sense, the unique attributes of individual species can be depicted by including specific key environmental correlates, habitats, vegetation structural conditions, and even species' life history attributes in queries of wildlife-habitat databases. Ultimately, the notion of one species = one niche becomes that of one species = one or more KEFs. The extent to which sets of forest mammals (or any species assemblage) can provide redundant functions that influence community or ecosystem diversity and stability in equivalent ways is poorly known and needs further study.

Some KEFs may be dependent on the involvement of other species, such as relations involving predators and prey, pollination, and the dispersal of propagules. An example we discussed previously is that of forest rodents that feed on and disperse mycorrhizal fungi, a KEF that in turn aids nitrogen cycling in forests and uptake by trees having obligate symbiotic relations with the fungi (Li et al. 1986). Another example was the relation of beaver to willow and aspen. We speculate that there are probably other co-evolved relations that are mediated by forest mammals but remain undiscovered.

Implications for management

We have described a number of ways that managers might use functional assessments of forest mammals (or other taxa) to guide ecosystem management. Managers may wish to know the influence of historic, current, or potential management actions on ecological functions of organisms, and they may wish to set explicit objectives for conserving or restoring ecological functions to meet the goal of managing for fully functional ecosystems. They could use historic conditions or reference landscapes to assess the extent to which altered landscapes have maintained their functionality. However, managing for ecological functions and functional groups is unlikely to provide for species- or other issue-specific conservation needs.

Because these concepts are new, it may be difficult for some managers to consider a functional assessment approach to forest management and species conservation. However, the impetus of such change is in some existing state and federal mandates for land and wildlife management. For example, a primary purpose of the Endangered Species Act is "... to provide a means whereby the ecosystems upon which endangered species and threatened species depend may be conserved ..." (US Endangered Species Act, Sec. 2b). Ensuring that ecosystems remain fully functional may be integral to meeting these objectives.

Implications for research

As others have argued in defense of the keystone species concept (Simberloff 1998), one of the primary advantages of a functional approach to management is that it involves explicit consideration of the mechanisms that underlie ecosystem structure and function. We believe that an important byproduct of the functional approach to forest mammal conservation is the generation of new understandings about ecosystem linkages, new insights about the ecology of individual species, and new research questions. Many of the "findings" described in this chapter can be re-stated as testable hypotheses for future research.

Although studying and conserving forest mammals from the perspective of ecological functionality is a relatively new concept, several recent studies have employed this approach. For example, Kunkel and Pletscher (1999) found that predation was the primary factor limiting deer and elk populations in Montana, and their research findings suggested that managers may therefore be able to enhance other prey populations by changing ungulate densities. Henke and Bryant (1999) found that coyotes may function as keystone predators in some ecosystems because removal of coyotes changed faunal community structure. McShea and Rappole (2000) suggested that breeding bird populations can be managed by controlling the influence of herbivory by deer populations on plant cover. Sirotnak and Huntly (2000) reported that herbivory by voles in riparian areas influenced nitrogen cycling.

Smallwood et al. (1998) found that burrowing by mammals is not only important to soil formation and intermixing of geologic materials and organic matter, it can also influence the environmental exposure of buried hazardous wastes. Among the key burrowing parameters that might influence such exposures are: the catalog of resident burrowing species and their abundances, typical burrow volumes (void space created by soil displacement), burrow depth profiles, maximum depth of excavation, constituents and structural qualities of excavated soil mounds, and proportion of the ground covered by excavated soil. Other important parameters included the rate of mound construction, depth of den chambers, and volume of burrow backfills.

The new insights these and other studies have generated demonstrate the heuristic value of quantitative autecological research on the functional roles of mammals. Research is also needed on how ecological functions of forest mammals vary among individuals, populations, geographic locations, and habitats, and in different successional stages. Through such studies, functional assessments can begin to be framed in quantitative terms that are based on empirical data, not just informed judgments. Quantitative models can provide a scientific basis for the implementation of ecosystem management. Categories of KEFs explored in our functional assessment can be quantified with rates (e.g., numbers of fungal spores dispersed per unit area per time period, or volume of soil dug per unit area per time period). Process models could be devised to initially hypothesize and, through empirical validation, ultimately explain how functional roles of mammals (and other taxonomic groups) quantitatively influence biodiversity, productivity, and the sustainability of ecosystems.

Summary and conclusions

Understanding and quantifying the ecological roles of mammals in forested ecosystems remain important management and research needs. Much recent literature has addressed this topic. We have built upon this work and offered a practical framework and a set of example assessments that can be done with existing databases. Explicitly considering the functional roles of mammals in ecosystems can complement species-specific conservation of forest mammals and generate new understandings of the contributions that forest mammals make to ecosystem function.

Woodward (1994) asked "How many species are required for a functional ecosystem?" We can now begin to answer this in ways of use to land managers. Analyzing the functional richness (number of KEFs), functional diversity (number of KEFs weighted by number of species per KEF), functional web (interactions among species and KEFs), and other KEF patterns of undisturbed or native forests can essentially define a "fully-functional ecosystem." In turn, this can serve as a baseline from which to clearly and repeatably measure the expected influence on ecosystem function from alternative forest-management actions. Such analyses are already being used to characterize the functional patterns of fish and wildlife communities across broad landscapes (Marcot et al. 2002).

We urge forest managers to consider some of the further questions we raise and to pursue a more functional approach to the management and conservation of forest mammals and the habitats they occupy and, in turn, influence.

Acknowledgments
We thank Larry Irwin, Tom O'Neil, and Cindy Zabel for reviews of the manuscript.

Appendix

Categories of key ecological functions as coded for the Wildlife-Habitat Relationships database for Washington and Oregon (Johnson and O'Neil 2001). Not all of these categories pertain to forest mammals.

1. Trophic relationships[*]
    1.1 Heterotrophic consumer[*]
        1.1.1 Primary consumer (herbivore)[*]
            1.1.1.1 Foliovore (leaf eater)[*]
            1.1.1.2 Spermivore (seed eater)[*]
            1.1.1.3 Browser (leaf, stem eater)
            1.1.1.4 Grazer (grass, forb eater)
            1.1.1.5 Frugivore (fruit eater)[*]
            1.1.1.6 Sap feeder
            1.1.1.7 Root feeder[*]
            1.1.1.8 Nectivore (nectar feeder)
            1.1.1.9 Fungivore (fungus feeder)[*]
            1.1.1.10 Flower/bud/catkin feeder
            1.1.1.11 Aquatic herbivore
            1.1.1.12 Feeds in water on decomposing benthic substrate
            1.1.1.13 Bark/cambium/bole feeder

      1.1.2 Secondary consumer (primary predator or primary carnivore)[*]
        1.1.2.1 Invertebrate eater
          1.1.2.1.1 Terrestrial invertebrates
          1.1.2.1.2 Aquatic macroinvertebrates
          1.1.2.1.3 Freshwater or marine zooplankton
        1.1.2.2 Vertebrate eater (consumer or predator of herbivorous vertebrates)[*]
          1.1.2.2.1 Piscivorous (fish eater)[*]
        1.1.2.3 Ovivorous (egg eater)
      1.1.3 Tertiary consumer (secondary predator or secondary carnivore)
      1.1.4 Carrion feeder
      1.1.5 Cannibalistic
      1.1.6 Coprophagous (feeds on fecal material)
      1.1.7 Feeds on human garbage/refuse
        1.1.7.1 Aquatic (e.g., offal and bycatch of fishing boats)
        1.1.7.2 Terrestrial (e.g., landfills)
    1.2 Prey relationships
      1.2.1 Prey for secondary or tertiary consumer (primary or secondary predator)
2.   Aids in physical transfer of substances for nutrient cycling (C,N,P, etc.)[*]
3.   Organismal relationships[*]
    3.1 Controls or depresses insect population peaks[*]
    3.2 Controls terrestrial vertebrate populations (through predation or displacement)[*]
    3.3 Pollination vector
    3.4 Transportation of viable seeds, spores, plants, or animals[*]
      3.4.1 Disperses fungi
      3.4.2 Disperses lichens
      3.4.3 Disperses bryophytes, including mosses
      3.4.4 Disperses insects and other invertebrates
      3.4.5 Disperses seeds/fruits (through ingestion or caching)
      3.4.6 Disperses vascular plants[*]
    3.5 Creates feeding, roosting, denning, or nesting opportunities for other organisms[*]
      3.5.1 Creates feeding opportunities (other than direct prey relations)[*]
        3.5.1.1 Creates sapwells in trees
      3.5.2 Creates roosting, denning, or nesting opportunities[*]
    3.6 Primary creation of structures (possibly used by other organisms)[*]
      3.6.1 Aerial structures[*]
      3.6.2 Ground structures[*]
      3.6.3 Aquatic structures[*]

       3.7 User of structures created by other species
           3.7.1 Aerial structures
           3.7.2 Ground structures
           3.7.3 Aquatic structures
       3.8 Nest parasite
           3.8.1 Inter-species parasite
           3.8.2 Common inter-specific host
       3.9 Primary cavity excavator in snags or live trees
       3.10 Secondary cavity user
       3.11 Primary burrow excavator (fossorial or underground burrows)
           3.11.1 Creates large burrows (rabbit-sized or larger)
           3.11.2 Creates small burrows (less than rabbit-sized)
       3.12 Uses burrows dug by other species (secondary burrow user)
       3.13 Creates runways (possibly used by other species)
       3.14 Uses runways created by other species
       3.15 Pirates food from other species
       3.16 Inter-specific hybridization
4.   Carrier, transmitter, or reservoir of vertebrate diseases
       4.1 Diseases that affect humans[*]
       4.2 Diseases that affect domestic animals
       4.3 Diseases that affect other wildlife species
5.   Soil relationships[*]
       5.1 Physically affects (improves) soil structure, aeration (typically by digging)[*]
       5.2 Physically affects (degrades) soil structure, aeration (typically by trampling)[*]
6.   Wood structure relationships (either living or dead wood)[*]
       6.1 Physically fragments down wood[*]
       6.2 Physically fragments standing wood[*]
7.   Water relationships[*]
       7.1 Impounds water by creating diversions or dams[*]
       7.2 Creates ponds or wetlands through wallowing
8.   Vegetation structure and composition relationships[*]
       8.1 Creates standing dead trees (snags)[*]
       8.2 Herbivory on trees or shrubs that may alter vegetation structure and composition (browsers)
       8.3 Herbivory on grasses or forbs that may alter vegetation structure and composition (grazers)[*]

---

[*] = Key ecological functions of *Homo sapiens*.

Literature cited

Anthony, R.G., M.A. O'Connell, M.M. Pollock, and J.G. Hallett. 2003. Associations of mammals with riparian ecosystems in Pacific Northwest forests. Pages 510–563 *in* C.J. Zabel and R.G. Anthony, editors. *Mammal Community Dynamics. Management and Conservation in the Coniferous Forests of Western North America.* Cambridge University Press, Cambridge, UK.

Aubry, K.B., J.P. Hayes, B.L. Biswell, and B.G. Marcot. 2003. The ecological role of tree-dwelling mammals in western coniferous forests. Pages 405–443 *in* C.J. Zabel and R.G. Anthony, editors. *Mammal Community Dynamics. Management and Conservation in the Coniferous Forests of Western North America.* Cambridge University Press, Cambridge, UK.

Ayarbe, J.P. and T.L. Kieft. 2000. Mammal mounds stimulate microbial activity in a semiarid shrubland. *Ecology* **81**(4):1150–1154.

Brang, P. 2001. Resistance and elasticity: promising concepts for the management of protection forests in the European Alps. *Forest Ecology and Management* **145**:107–119.

Brown, J.H. 1995. *Macroecology.* The University of Chicago Press, Chicago, Illinois, USA. 269 pp.

Butler, D.R. 1995. *Zoogeomorphology: Animals as Geomorphic Agents.* Cambridge University Press, New York, New York, USA.

Carpenter, S.R. and K.L. Cottingham. 1997. Resilience and restoration of lakes. *Conservation Ecology* **1**:2. [Online at http://www.consecol.org/vol1/iss1/art2.]

Chappell, C.B., R.C. Crawford, C. Barrett, J. Kagan, D.H. Johnson, M. O'Mealy, G.A. Green, H.L. Ferguson, W.D. Edge, E.L. Greda, and T.A. O'Neil. 2001. Wildlife habitats: descriptions, status, trends, and system dynamics. Pages 22–114 *in* D.H. Johnson and T.A. O'Neil, managing directors. *Wildlife-habitat Relationships in Oregon and Washington.* Oregon State University Press, Corvallis, Oregon, USA.

Dale, V.H., S. Brown, R.A. Haeuber, N.T. Hobbs, N. Huntly, R.J. Naiman, W.E. Riebsame, M.G. Turner, and T.J. Valone. 2000. Ecological principles and guidelines for managing the use of land. *Ecological Applications* **10**:639–670.

Estes, J.A. and J.F. Palmisano. 1974. Sea otters: their role in structuring nearshore communities. *Science* **185**:1058–1060.

Fonseca, C.R. and G. Ganade. 2001. Species functional redundancy, random extinctions and the stability of ecosystems. *Journal of Ecology* **89**:118–125.

Franklin, J.F., D. Lindenmayer, J.A. MacMahon, A. McKee, J. Magnuson, D.A. Perry, R. Waide, and D. Foster. 2000. Threads of continuity. *Conservation Biology in Practice* **1**:9–16.

Gunderson, L.H. 2000. Ecological resilience in theory and application. *Annual Review of Ecology and Systematics* **31**:425–440.

Halpern, C.B. 1988. Early successional pathways and the resistance and resilience of forest communities. *Ecology* **69**:1703–1715.

Hendrix, P.F., editor. 1995. *Earthworm Ecology and Biogeography in North America.* Lewis Publishers, Boca Raton, Florida, USA.

Henke, S.E. and F.C. Bryant. 1999. Effects of coyote removal on the faunal community in western Texas. *Journal of Wildlife Management* **63**:1066–1081.

Hobbs, N.T. 1996. Modification of ecosystems by ungulates. *Journal of Wildlife Management* **60**:695–713.

Johnson, D. and T. O'Neil, managing directors. 2001. *Wildlife–habitat Relationships in Oregon and Washington*. Oregon State University Press, Corvallis, Oregon, USA.

Jones, C.G., J.H. Lawton, and M. Shachak. 1996. Organisms as ecosystem engineers. Pages 130–147 *in* F.B. Samson and F.L. Knopf, editors. *Ecosystem Management: Selected Readings*. Springer-Verlag, New York, New York, USA.

Kunkel, K. and D.H. Pletscher. 1999. Species-specific population dynamics of cervids in a multipredator ecosystem. *Journal of Wildlife Management* **63**:1082–1093.

Li, C.Y., C. Maser, Z. Maser, and B.A. Caldwell. 1986. Role of three rodents in forest nitrogen fixation in western Oregon: another aspect of mammal-mycorrhizal fungus-tree mutualism. *Great Basin Naturalist* **46**:411–414.

Loeb, S.C., F.H. Tainter, and E. Cazares. 2000. Habitat associations of hypogeous fungi in the southern Appalachians: implications for the endangered northern flying squirrel (*Glaucomys sabrinus coloratus*). *American Midland Naturalist* **144**(2): 286–296.

Ludwig, D., B. Walker, and C.S. Holling. 1997. Sustainability, stability, and resilience. *Conservation Ecology* **1**:7. [Online at: http://www.consecol.org/vol1/iss1/art7.]

Luoma, D.L., J.M. Trappe, A.W. Claridge, K.M. Jacobs, and E. Cazares. 2003. Relationships among fungi and small mammals in forested ecosystems. Pages 343–373 *in* C.J. Zabel and R.G. Anthony, editors. *Mammal Community Dynamics. Management and Conservation in the Coniferous Forests of Western North America*. Cambridge University Press, Cambridge, UK.

Marcot, B.G. 1997. Research information needs on terrestrial vertebrate species of the interior Columbia River Basin and northern portions of the Klamath and Great Basins. Research Note **PNW-RN-522**. USDA, Forest Service, Pacific Northwest Research Station, Portland, Oregon, USA. Abstract and database available online at http://www.fs.fed.us/pnw/marcot.html.

Marcot, B.G. 2002. An ecological functional basis for managing decaying wood for wildlife. Pages 895–910 *in* W.F. Laudenslayer, Jr., P.J. Shea, B.E. Valentine, C.P. Weatherspoon, and T.E. Lisle, technical coordinators. Proceedings of the symposium on the ecology and management of dead wood in western forests, November 2–4, 1999; Reno, Nevada. General Technical Report **PSW-GTR-181**. USDA, Forest Service, Pacific Southwest Research Station, Fresno, California, USA.

Marcot, B.G. and M. Vander Heyden. 2001. Key ecological functions of wildlife species. Pages 168–186 *in Wildlife–habitat Relationships in Oregon and Washington*. Oregon State University Press, Corvallis, Oregon, USA.

Marcot, B.G., M.A. Castellano, J.A. Christy, L.K. Croft, J.F. Lehmkuhl, R.H. Naney, K. Nelson, C.G. Niwa, R.E. Rosentreter, R.E. Sandquist, B.C. Wales, and E. Zieroth. 1997. Terrestrial ecology assessment. Pages 1497–1713 *in* T.M. Quigley and S.J. Arbelbide, editors. An assessment of ecosystem components in the interior Columbia Basin and portions of the Klamath and Great Basins. Volume III. General Technical Report **PNW-GTR-405**. USDA, Forest Service, Pacific Northwest Research Station, Portland, Oregon, USA.

Marcot, B.G., W.E. McConnaha, P.H. Whitney, T.A. O'Neil, P.J. Paquet, L. Mobrand, G.R. Blair, L.C. Lestelle, and K.M. Malone. 2002. *A Multi-species Framework*

Approach for the Columbia River Basin: Integrating Fish, Wildlife, and Ecological Functions. On CD-ROM and at http://www.edthome.org/framework. Northwest Power Planning Council, Portland, Oregon, USA.

Maser, C. and Z. Maser. 1988. Interactions among squirrels, mycorrhizal fungi, and coniferous forests in Oregon. *Great Basin Naturalist* **48**:358–369.

McComb, W.C. 2003. Ecology of coarse woody debris and its role as habitat for mammals. Pages 374–404 *in* C.J. Zabel and R.G. Anthony, editors. *Mammal Community Dynamics. Management and Conservation in the Coniferous Forests of Western North America*. Cambridge University Press, Cambridge, UK.

McShea, W.J. and J.H. Rappole. 2000. Managing the abundance and diversity of breeding bird populations through manipulation of deer populations. *Conservation Biology* **14**:1161–1170.

Meadows, P.S. and A. Meadows, editors. 1991. *The Environmental Impact of Burrowing Animals and Animal Burrows*. [Symposia of the Zoological Society of London 63.] Oxford University Press, UK.

Mills, L.S., M.E. Soulé, and D.F. Doak. 1993. The keystone-species concept in ecology and conservation. *BioScience* **43**:219–224.

Morrison, M.L., B.G. Marcot, and R.W. Mannan. 1998. *Wildlife–habitat Relationships: Concepts and Applications*. Second Edition. University of Wisconsin Press, Madison, Wisconsin, USA.

Nabhan, G.P. 2001. Nectar trails of migratory pollinators: restoring corridors on private lands. *Conservation Biology in Practice* **2**(1):21–27.

Naiman, R.J., J.M. Melillo, and J.E. Hobbie. 1986. Ecosystem alteration of boreal forest streams by beaver (*Castor canadensis*). *Ecology* **67**:1254–1269.

Noronha, V.T. and M.F. Goodchild. 1992. Modeling interregional interaction: implications for defining functional regions. *Annals of the American Geographers* **82**:86–102.

O'Neil, T.A., K.A. Bettinger, M. Vander Heyden, B.G. Marcot, C. Barrett, T.K. Mellen, W.M. Vander Haegen, D.H. Johnson, P.J. Doran, L. Wunder, and K.M. Boula. 2001. Structural conditions and habitat elements of Oregon and Washington. Pages 115–139 *in* D.H. Johnson and T.A. O'Neil, managing directors. *Wildlife–habitat Relationships in Oregon and Washington*. Oregon State University Press, Corvallis, Oregon, USA.

Peterson, G., C.R. Allen, and C.S. Holling. 1998. Ecological resilience, biodiversity, and scale. *Ecosystems* **1**:6–18.

Reice, S.R., R.C. Wissmar, and R.J. Naiman. 1990. Disturbance regimes, resilience, and recovery of animal communities and habitats in lotic ecosystems. *Environmental Management* **14**:647–659.

Rose, C.L., B.G. Marcot, T.K. Mellen, J.L. Ohmann, K.L. Waddell, D.L. Lindley, and B. Schreiber. 2001. Decaying wood in Pacific Northwest forests: concepts and tools for habitat management. Pages 580–623 *in* D.H. Johnson and T.A. O'Neil, managing directors. *Wildlife–habitat Relationships in Oregon and Washington*. Oregon State University Press, Corvallis, Oregon, USA.

Simberloff, D. 1998. Flagships, umbrellas, and keystones: is single-species management passe in the landscape era? *Biological Conservation* **83**:247–257.

Sirotnak, J.M. and N.J. Huntly. 2000. Direct and indirect effects of herbivores on nitrogen dynamics: voles in riparian areas. *Ecology* **81**:78–87.

Smallwood, K.S., M.L. Morrison, and J. Beyea. 1998. Animal burrowing attributes affecting hazardous waste management. *Environmental Management* 22(6):831–847.

Strange, E.M., K.D. Fausch, and A.P. Covich. 1999. Sustaining ecosystem services in human-dominated watersheds: biohydrology and ecosystem processes in the South Platte River Basin. *Environmental Management* 24:39–54.

van Hees, W.W.S. and E.H. Holsten. 1994. An evaluation of selected spruce bark beetle infestation dynamics using point in time extensive forest inventory data, Kenai Peninsula, Alaska. *Canadian Journal of Forest Research* 24:246–251.

Walker, B.H. 1992. Biodiversity and ecological redundancy. *Conservation Biology* 6:18–23.

Walker, B. 1995. Conserving biological diversity through ecosystem resilience. *Conservation Biology* 9:747–752.

Weaver, J.L., P.C. Paquet, and L.F. Ruggiero. 1996. Resilience and conservation of large carnivores in the Rocky Mountains. *Conservation Biology* 10:964–976.

Williams, D.W. and A.W. Liebhold. 2000. Spatial synchrony of spruce budworm outbreaks in eastern North America. *Ecology* 81:2753–2766.

Woodward, F.I. 1994. How many species are required for a functional ecosystem? Pages 271–291 *in* E.D. Schulze and H. A. Mooney, editors. *Biodiversity and Ecosystem Function*. Springer-Verlag, Berlin, Germany.

20
_____

Synopsis and future perspective

Introduction

Landscapes of western North America have been highly dissected by glaciation, vulcanization, physiographic heterogeneity, and the formation of major river systems. These physical features have served as barriers to movement and gene flow among mammal populations resulting in speciation throughout evolutionary history. As a result of these many influences and resulting speciation, the mammal fauna of this area is highly rich and diverse. About half of the mammalian species in North America occur in western coniferous forests (see Lawlor 2003). Although fewer species are found specifically within these forests, most are widespread in many vegetative communities.

Mammalian faunas of coniferous forests of western North America represent many taxonomic groups, and they perform many functional roles. Shrews are little known but are often the most abundant species in these forests, and they are primarily insectivorous, as are bats. In contrast, ungulates are highly visible and represent large amounts of standing biomass in forested ecosystems. Their browsing and herbivorous diets can have a major influence on plant communities and other species of mammals.

Direct associations between coarse woody debris and mammals vary among species and studies, but most information on this association is for the terrestrial small mammals. For many species, coarse wood debris serves as an important form of cover or as feeding sites (McComb 2003). For example, martens use downed logs as important hunting sites during the winter when snow is deep. The indirect functions of coarse woody debris include retention of soil moisture during the dry seasons and contributions to nutrient cycling throughout the forest. The benefits to mammals of these indirect functions are not well known and will need further investigation.

Multi-layered canopies have been identified as important structures for avian communities, but the benefit to mammals is either not well known or is indirect. Indirect benefits might include increased avian prey for predatory mammals such as martens and raccoons (*Procyon lotor*). To the best of our knowledge, mammalian ecologists have not investigated this aspect to the extent that it has been by avian ecologists. This is likely because mammals are much more difficult to observe, hear, or trap in forest canopies than birds.

Influence of riparian systems

The structure and function of riparian zones are determined by topography, size of the water body, soils, microclimate, vegetation, and the disturbance regimes to which they are subjected. As a result of the diversity of biotic and abiotic features found in riparian zones, mammalian species richness is high and use of these areas by some mammal species is disproportionately higher than in upland areas. Several species of mammals in the Pacific Northwest are considered *riparian obligates*, and they represent a variety of taxonomic groups from small insectivores to large carnivores (see Anthony et al. 2003). Prominent among these species is the beaver (*Castor canadensis*) which functions as a keystone species by its ability to alter riparian vegetation and water flows which, in turn, influences vegetative characteristics and associated animal communities. Many other species of mammals are significantly more abundant, their movements are highly associated with, or they have higher reproductive fitness in, riparian areas. This group of *riparian associates* is represented by most of the taxonomic groups of mammals, including shrews, moles, bats, rodents, carnivores, and ungulates. The degree of association of some mammals to riparian zones varies spatially among different forest types and between the east and west side of the Cascade Mountains. Riparian zones typically have vegetation that is adapted to high disturbance produced by frequent flooding, scouring, and deposition of sediment, which promotes early-successional vegetation and small mammals that are associated with herbaceous vegetation. In addition, many species of mammals are not dependent on riparian areas nor are they more abundant there; however, they may be major components of riparian faunas (e.g., the common deer mouse (*Peromyscus maniculatus*)).

Human disturbances, particularly timber harvest, cattle grazing, and road construction can have major influences on riparian systems and associated mammal communities. With respect to timber harvest, the retention of buffer strips where some but not all of the trees are harvested has been used to manage these areas. The amount of benefit provided to mammals and other animals depends on the width of the buffer, density of trees retained, and the specific habitat associations of the species. More information is needed on the potential benefits of these buffer strips, because we know very little about the use of these areas as refugia, travel corridors, or their role in gene flow among subpopulations. Because riparian zones may serve as travel corridors and provide connectivity among fragmented late-successional forests (Harris 1984), these areas must be managed with these functions in mind.

Important food webs and interactions

Mammals occupy many different food webs and have important interactions with both the abiotic and biotic components of coniferous forests. These food webs have connections among many species of plants and animals, and linkages between the physical and biological components of ecosystems as well. These linkages are not simply one-way pathways of energy flow from the abiotic to the biotic components (direct pathways), but comprise indirect interactions (trophic cascades) among different components of the ecosystem. The potential number of indirect linkages is considerably higher than the number of direct linkages, and the indirect linkages can be very important in ecosystem functions. For example, the linkage among acorn production, gypsy moth outbreaks, and the risk of lyme disease in mammals is an important indirect linkage among components of a terrestrial ecosystem (Jones et al. 1998).

Among the different taxa of mammals there are herbivores, folivores, fungivores, spermivores, omnivores, insectivores, and carnivores that have direct and indirect interactions with other components of the ecosystem. No mammals are known to be detritus feeders, but shrews (*Sorex* spp.) and moles (*Scapanus* spp.) feed on detritivores that inhabit the forest litter and dead wood. Rodents and ungulates represent most of the herbivores, the latter being low in number but high in biomass and are important pathways of energy in forested ecosystems. Rodents on the other hand are numerous but represent much lower biomass on a per-unit area basis. The Microtine rodents are particularly known for their herbivorous foraging behavior. In addition, deer (*Odocoileus* spp.), elk (*Cervus elaphus*), and moose (*Alces alces*) are known to have considerable impacts on forest and riparian vegetation where their numbers reach carrying capacity (Kie et al. 2003, Singer et al. 2003).

The folivores are represented by three species of the genus *Arborimus* that are restricted to coniferous forests west of the crest of the Cascade Mountains in western Oregon and Northwestern California. The tree voles (*Arborimus longicaudus, A. pomo*) are the most highly specialized Microtine rodents in the world and are the most arboreal mammals in North America. They spend most of their lives in the canopy of Douglas-fir (*Pseudotsuga menziesii*) forests and their diet is almost entirely restricted to Douglas-fir needles. They are important species in the food web as they are prey for northern spotted owls (*Strix occidentalis*) (Forsman et al. 1984) and other mammalian and avian species. The third species is the white-footed vole (*A. albipes*), which apparently is not as restricted to an arboreal life style as the tree voles. Recent studies indicate that this species is one of the rarest mammals in the Pacific Northwest, and it is found primarily in riparian forests (Gomez and Anthony 1998) where its diet is comprised mainly of foliage of deciduous trees (Voth et al. 1983). These three species are some of the more unique mammals of this region.

Several species of fungivores inhabit coniferous forests of western North America including voles (*Clethrionomys* spp., *Microtus* spp.), Critcetid rodents (*Peromyscus* spp., *Tamias* spp.), and northern flying squirrels (*Glaucomys sabrinus*). These rodent species consume sporocarps (fruiting bodies) of below-ground ectomycorrhizal fungi and are responsible for dispersal of spores of several species of fungi. The ectomycorrhizal fungi have a symbiotic relationship with the roots of coniferous trees and influence the overall health of these forests (see Luoma et al. 2003 and Smith et al. 2003). The northern flying squirrel also depends on epiphytic lichens in addition to fungi perhaps because they prefer them to cone seeds (Zabel and Waters 1997) and because fungi are not readily available during all parts of the year. All of these species of small mammals are important prey of a number of mammalian and avian predators, so this is one of the important food webs of coniferous forests of western North America.

The fruits and seeds produced in western coniferous forests are not large and conspicuous, which may explain why the concept of a "mast crop" is not as frequently mentioned for western coniferous forests as it is for eastern deciduous forests. Nonetheless, seed production is important to a number of mammalian species, and most species appear to select Douglas-fir or ponderosa pine (*Pinus ponderosa*) seeds. The principal seed eaters are Murid rodents of the genera *Peromyscus* and *Neotoma*; Sciurid rodents of the genera *Tamias*, *Sciurus* and *Tamiasciurus*; and several species of shrews. Population densities of small mammals have been shown to increase in years following abundant seed production (Gashwiler 1965). Many of the omnivores are fruit eaters, at least on a seasonal basis, and include raccoons, black and grizzly bears (*Ursus arctos*), and coyotes (*Canis latrans*).

Shrews, moles, bats, and many rodent species represent insectivores, and their consumption of insect biomass can be most impressive (see Hayes 2003, Hallett et al. 2003). Members of the Insectivora are generally higher in number of species and individuals than their counterparts in eastern deciduous forest of North America, so they occupy important functional roles in these forests. A comparison of the number of species of shrews in western ($n \approx 19$) versus eastern ($n \approx 8$) North America exemplifies this difference in species richness.

Last, but not least, the mammalian carnivores are a diverse group in terms of number of species, size, prey species, and ecological roles (Buskirk and Zielinski 2003, Kunkel 2003). They range in size from the ermine (*Mustela erminea*) to the grizzly bear and some have very restricted diets (e.g., mountain lions (*Felis concolor*)) while others consume a variety of prey and are omnivorous (e.g., raccoons and black bears). The explanation for the large array of carnivores in coniferous forest of western North America likely is the diversity and magnitude of energy pathways, both direct and indirect, that are mentioned above coupled with the aquatic production of anadromous fish on an annual basis (Harris 1984:51). Many of the mammalian carnivores feed on other species of small or medium-sized mammals for most of their food resources. In addition, many avian predators occupy these forests and depend on mammals as their primary prey, including the Accipiter hawks and six species of owls (Harris 1984:51). One threatened species, the northern spotted owl, is dependent almost exclusively on mammals for food, and three species of mammals (northern flying squirrels and two species of woodrats (*Neotoma* spp.)) comprise the majority of its diet (Forsman et al. 1984). Consequently, mammals are important linkages for energy flow in coniferous forest ecosystems, and without them many of the key ecological functions of these forests would be missing or depauperate (Marcot and Aubrey 2003).

Keystone species and important function roles

Mammals, small to large, are important components of food webs in western North America and are "strong interactors" in these ecosystems. Strong interactors play important roles in food webs as consumers, prey, or predators and by cycling nutrients and physically altering some part of the ecosystem. For example, small mammals have considerable influence on tree recruitment in Eastern deciduous forests (Ostfeld et al. 1997), and similar roles have been described for small mammals in coniferous forests (Hallett et al. 2003 and Smith et al. 2003). Predator–prey interactions are the fundamental linkages among species in every ecosystem, and ecologists have characterized these consumer–prey interactions into "bottom-up" and "top-down" control of ecosystem processes (Hunter and Price 1992). The bottom-up view contends that basic resources (e.g., space, nutrients) influence the higher trophic levels of the ecosystem, whereas the top-down view supports the importance of interactions between apex predators and their prey on lower trophic forms. The former was the more common view throughout much of the 1900's, but there is increasing evidence that herbivorous and carnivorous mammals can have important effects on ecosystem structure and function and are "strong interactors" in these systems (Buskirk and Zielinksi 2003, Kie et al. 2003, Kunkel 2003, Singer et al. 2003). Because mammals play important roles in the ecological functions of coniferous forest ecosystems, some have been portrayed as "keystone species." Keystone species are those whose abundance is relatively low but their effect on the community or ecosystem is relatively large (Power et al. 1996). Most mammals are not keystone species but most play important functional roles and are therefore strong interactors in forest ecosystems.

There is no doubt that the beaver functions as a keystone species in riparian ecosystems throughout their geographic range. They have profound effects on water flow regimes, biogeochemical characteristics, streamside vegetation, and faunal characteristics and are particularly important for rich and healthy waterbird and fish populations in riparian ecosystems throughout western North American (Naiman et al. 1994, Pollock et al. 1995). Many of these ecosystems have ceased to function fully since the removal of beaver for fur by European settlers in the 1880's

(Lichatowich 1999:54–57). The importance of beaver for healthy riparian ecosystems is particularly well known by fisheries biologists, who have documented their importance to salmonid populations on the Pacific Northwest (Anthony et al. 2003).

Are there other mammalian species or groups of species that play important functional roles in terrestrial ecosystems? Certainly, the influence of ungulates on nitrogen cycles, plant associations, and faunal characteristics has been documented (see Kie et al. 2003, Singer et al. 2003). However, most of the studies on the effects of ungulates on ecosystem structure and function have been conducted in areas that do not have the full complement of predatory mammals with which ungulates evolved. Consequently, the pursuit of answers to this question has not been easy because many of the apex predators (gray wolves (*Canis lupus*), grizzly bears, mountain lions) in North America have been eliminated or significantly reduced by human persecution; therefore, there are few areas where all of these predators are fully represented in forested ecosystems. However, there is growing evidence that some predatory mammals also play important functional roles in a number of terrestrial ecosystems (Terborgh et al. 2001). A good example is the gray wolf-moose-balsam fir (*Abies balsamea*) system on Isle Royale (McLaren and Peterson 1994), where the number of wolves and thus the intensity of wolf predation influence the moose population. Growth rings on young fir trees show depressed growth rates for periods when wolves were rare and moose abundant, from which they inferred that wolves had effects on the structure and function of the ecosystems. An analogous scenario has been shown for lynx – hare (*Lynx canadensis* and *Lepus americanus*, respectively) cycles in the boreal forest of northern North America (Boutin et al. 2003) where oscillations of lynx and hare numbers have effects on other species in the community. In addition, there is evidence that the elimination or reduction of large carnivores has resulted in competitive release of medium-sized predators such as coyotes, foxes (*Vulpes* spp.) and skunks (*Mephitis mephitis*). Gray wolves at one time may have limited coyotes (Sargeant et al. 1993), and the reduction of coyotes thereby led to increased intensity of predation by the mesopredators and reduction of local extinction of their prey (Estes 1996). Mesopredator release has been proposed or supported for a number of systems including chaparral (Soule et al. 1988), grasslands (Vickery et al. 1992), prairie wetlands (Ball et al. 1995), tropical forests (Terborgh et al. 2001), and coniferous forests (Buskirk and Zielinski 2003). The keystone roles of large apex predators such as wolves and grizzly bears is intriguing but hard to substantiate; but this role is becoming more clear with the recovery and translocation of both species in North America. For example, the establishment of wolves in the north-central U.S. has led to a restriction in the distance from aquatic habitats that beavers forage, which has limited, in turn, the effects of beavers on upland plant associations (Naiman et al. 1994, Pollock et al. 1995). Similarly, the establishment of wolves in other areas has been followed by declines in caribou ((*Rangifer caribou*) Bergerud 1988), moose (Messier and Crete 1985), elk (Singer et al. 2003), and black-tailed deer ((*Odocoileus hemionus*) Hatter and Janz 1994). Re-introduction of wolves into Yellowstone National Park and the wilderness areas of north-central Idaho has provided a natural experiment from which the potential keystone role of wolves may be better understood. While some of the above relations may be speculative and provide case examples, collectively they suggest that mammalian predators have important functional roles, are strong interactors, and may initiate trophic cascades (Estes 1996, Terborgh et al. 2001). This will be an area of important research upon which future management and conservation of large mammals will depend.

Conservation issues

Many of the mammal species of western North America have been persecuted or overexploited during the last two centuries by humans. Beaver were exploited for their fur in the 1800's, and their numbers and influence on riparian systems have not recovered completely (Lichatowich 1999). Similarly, the bison (*Bison bison*) was overexploited for its meat and hides, and it now only exists as remnant populations in a few areas. Wolves and grizzly bears were trapped or shot on sight for many decades because they were perceived to be a threat to humans and their livestock. Collectively, the reduction in number and distribution of these species and the ecological functions they likely carried out have altered mammalian community structure and function in most of the areas that they originally inhabited. Fortunately, wolves and grizzly bears have always been abundant in Canada, which has acted as a source of individuals to supplement populations in the U.S. In addition, the Endangered Species Act of the U.S. government has provided protection for these species and other vertebrates, so that populations have increased naturally and re-introduction of the species into parts of their former range has been possible.

One of the fundamental concerns for mammal populations in western North America besides human persecution and overexploitation is the harvest of late-successional (mature and old-growth) forests. These harvest activities have resulted in loss and fragmentation of the remaining old forests, which has potential influences on the numbers and distribution of some mammal species. Fortunately, there appear to be very few species of mammals that are highly associated with late-successional forests, and most of the medium- to large-sized species are capable of dispersing across large areas of suboptimal habitat. Unfortunately, there have been few studies on the effect of forest fragmentation on the abundance and persistence of mammal populations throughout this area (Martin and McComb 2003). In addition, there is little information on the ability of mammals to disperse across suboptimal habitats and what kinds of habitats provide connectivity between optimal habitats. All mammal species cannot be treated equally in these studies because variance in their dispersion, dispersal patterns, and potential for colonization are dependent on their evolutionary history and behavioral systems (Wolff 2003). Good colonists are species that evolved in habitat mosaics, are ecological generalists, migratory, asocial, and/or have female-based dispersal, whereas poorer colonists evolved in homogeneous habitats, are ecological specialists, philopatric, social, and have female philopatry (Wolff 2003). The above topics will be areas of important research, and a combination of demographic and genetic approaches will be helpful in the pursuit of answers to these questions (Mills et al. 2003).

In the meantime, conservation plans such as the Northwest Forest Plan (FEMAT 1993), the Columbia Basin Ecosystem Management Plan, the Wildlands Project, and Sierra-Nevada Conservation Plan (Johnson et al. 1999) will provide important conservation strategies that will contribute to the persistence of mammal species in the western U.S. In addition, there is need to encourage the re-establishment of beaver, bison, wolves, grizzly bears, and lynx into parts of their former ranges in order to restore mammal community structure and function. These are conservation and management challenges for the future.

The state of our knowledge

There is considerable information available on the biology, ecology, population dynamics, and habitat associations of many species of mammals in forested landscapes of western North America. This is exemplified by the voluminous list of references for many of the chapters in this book. However, there is much less information on community relations and species associations. An exception to this pattern is the extent of our knowledge on lynx–hare cycles of boreal forests, which has been extensively researched over the last five decades. In addition, little is known about the competitive relations of mammalian species in coniferous forests, although the competitive relations of white-tailed deer (*Odocoileus virginianus*) and mule deer (*Odocoileus hemionus*) have been investigated in Alberta, Canada (Kramer 1973), Montana (Martinka 1968), Oregon (Whitney 2001), and Arizona (Anthony and Smith 1977). Community relations and interspecies competition will be areas of fruitful research in the future.

The functional role of mammals in forested ecosystems has not been studied for many species of mammals, primarily because such research is not easily designed or carried out. Use of stable isotopes will be helpful in such investigations in the future. In addition, satellite telemetry will be very important in understanding the seasonal movements of medium- to large-sized mammals as they migrate seasonally or disperse to new habitats. Also, gene flow among populations is not well understood, because sophisticated genetic laboratories have not been available until recently. DNA fingerprinting will be useful in understanding the relations among individuals in populations and among subpopulations.

Much of the research on mammals in the last half-century has been on their individual ecologies and habitat associations. As such, most studies have failed to identify the ecological roles of mammals, especially ungulates and large carnivores, in terrestrial ecosystems, and the omission of mammals as influential grazers has distorted our perspective on the functional roles of mammals in terrestrial ecosystems (Paine 2000). According to Paine (2000), "Experimental studies published since 1960 identify mammals as strong interactors, that their grazing can dominate the rate of primary production, alter floristic composition, and change successional trajectories. When strong interactions have been documented in other ecosystems, they are often associated with a significant increase in indirect effects, alternative assemblage states, and even trophic cascades. Although these features will be more difficult to identify in terrestrial communities because of slow rates of assemblage response and experimental intractability of the interactors, they should be sought." The literature also suggests that the same can be said for large carnivores and the influence of their predation on large ungulates. Because most of our knowledge of mammalian communities has been derived in the absence of the native large carnivores (e.g., wolves, grizzly bears, and cougars) which have been eliminated from most of their historic range, their functional role in these ecosystems is not well understood or appreciated. As some of these species increase in numbers and geographic ranges (e.g., recent introduction of wolves into Yellowstone National Park), it will be important to understand how these apex predators influence the structure and function of ecosystems. As future research addresses these topics, we predict that mammalian ecologists will document even more significant functional roles of mammals in forested ecosystems of North America.

Literature cited

Anthony, R.G. and N.S. Smith. 1977. Ecological relationships between mule deer and white-tailed deer in southeastern Arizona. *Ecological Monographs* **47**:255–277.

Anthony, R.G., M.A. O'Connell, M.M. Pollock, and J.G. Hallett. 2003. Associations of mammals with riparian ecosystems in Pacific Northwest forests. Pages 510–563 *in* C.J. Zabel and R.G. Anthony, editors. *Mammal Community Dynamics. Management and Conservation in the Coniferous Forests of Western North America.* Cambridge University Press, Cambridge, UK.

Ball, I.J., R.L. Eng, and S.K. Ball. 1995. Population density and productivity of ducks on large grassland tracts in northcentral Montana. *Wildlife Society Bulletin* **23**:767–773.

Bergerud, A.T. 1988. Caribou, wolves, and man. *Trends in Ecology and Evolution* **3**:68–72.

Boutin, S., C.J. Krebs, R. Boonstra, and A.R.E. Sinclair, 2003. The role of the lynx–hare cycle in boreal forest community dynamics. Pages 487–509 *in* C.J. Zabel and R.G. Anthony, editors. *Mammal Community Dynamics. Management and Conservation in the Coniferous Forests of Western North America.* Cambridge University Press, Cambridge, UK.

Buskirk, S.W. and W.J. Zielinski. 2003. Small and mid-sized carnivores. Pages 207–249 *in* C.J. Zabel and R.G. Anthony, editors. *Mammal Community Dynamics. Management and Conservation in the Coniferous Forests of Western North America.* Cambridge University Press, Cambridge, UK.

Estes, J.A. 1996. Predators and ecosystem management. *Wildlife Society Bulletin* **24**:390–396.

Forest Ecosystem Management Assessment Team (FEMAT). 1993. *Forest Ecosystem Management: An Ecological, Economic, and Social Assessment.* Departments of Agriculture, Commerce, and Interior and Environmental Protection Agency, Portland, Oregon, USA. 1000+ pp.

Forsman, E.D., E.D. Meslow, and H.M. Wight. 1984. Distribution and biology of the spotted owl in Oregon. *Wildlife Monographs* **87**:1–64.

Gashwiler, J.S. 1965. Tree seed abundance vs. deer mouse populations in Douglas-fir clear-cuts. *Proceedings of Annual Conference Society of American Foresters* **219**:222.

Gomez, D.M. and R. G. Anthony. 1998. Small mammal abundance in riparian and upland areas of five seral stages in western Oregon. *Northwest Science* **72**: 293–302.

Hallett, J.G., M.A. O'Connell, and C.C. Maguire. 2003. Ecological relationships of terrestrial small mammals in western coniferous forests. Pages 120–156 *in* C.J. Zabel and R.G. Anthony, editors. *Mammal Community Dynamics. Management and Conservation in the Coniferous Forests of Western North America*. Cambridge University Press, Cambridge, UK.

Harris, L.D. 1984. *The Fragmented Forest: Island Biogeography Theory and the Preservation of Biotic Diversity*. University of Chicago Press, Chicago, Illinois, USA.

Hatter, I. and D.W. Janz. 1994. The apparent demographic changes in black-tailed deer associated with wolf control in northern Vancouver Islands, Canada. *Canadian Journal of Zoology* **72**:878–884.

Hayes, J.P. 2003. Habitat ecology and conservation of bats in western coniferous forests. Pages 81–119 *in* C.J. Zabel and R.G. Anthony, editors. *Mammal Community Dynamics. Management and Conservation in the Coniferous Forests of Western North America*. Cambridge University Press, Cambridge, UK.

Hunter, M.D. and P.W. Price. 1992. Playing chutes and ladders: heterogeneity and the relative roles of bottom-up and top-down forces in natural communities. *Ecology* **73**:724–732.

Johnson, K.N., F. Swanson, M. Herring, and S. Greene. 1999. *Bioregional Assessments: Science at the Crossroads of Management and Policy*. Island Press, Covelo, California, USA. 398 pp.

Jones, C.G., R.S. Ostfeld, M.P. Richard, E.M. Schauber, and J.O. Wolff. 1998. Chain reactions linking acorns to gypsy moth outbreaks and lyme disease risk. *Science* **279**:1023–1026.

Kie, J.G., R.T. Bowyer, and K.M. Stewart. 2003. Ungulates in western coniferous forests: habitat relationships, population dynamics, and ecosystem processes. Pages 296–341 *in* C.J. Zabel and R.G. Anthony, editors. *Mammal Community Dynamics. Management and Conservation in the Coniferous Forests of Western North America*. Cambridge University Press, Cambridge, UK.

Kramer, A. 1973. Interspecific behavior and dispersion of two sympatric species of deer. *Journal of Wildlife Management* **37**:288–300.

Kunkel, K.E. 2003. Ecology, conservation, and restoration of large carnivores in western North America. Pages 250–295 *in* C.J. Zabel and R.G. Anthony, editors. *Mammal Community Dynamics. Management and Conservation in the Coniferous Forests of Western North America*. Cambridge University Press, Cambridge, UK.

Lawlor, T.E. 2003. Faunal composition and distribution of mammals in western coniferous forests. Pages 41–80 *in* C.J. Zabel and R.G. Anthony, editors. *Mammal Community Dynamics. Management and Conservation in the Coniferous Forests of Western North America*. Cambridge University Press, Cambridge, UK.

Lichatowich, J.A. 1999. *Salmon Without Rivers*. Island Press, Washington DC, USA.

Luoma, D.L., J.M. Trappe, A.W. Claridge, K. Jacobs, and E. Cazares. 2003. Relationships among fungi and small mammals in forested ecosystems. Pages 343–373 *in* C.J. Zabel and R.G. Anthony, editors. *Mammal Community Dynamics. Management and Conservation in the Coniferous Forests of Western North America*. Cambridge University Press, Cambridge, UK.

Marcot, B.G. and K.B. Aubry. 2003. The functional diversity of mammals in coniferous forests of western North America. Pages 631–664 *in* C.J. Zabel and R.G. Anthony, editors. *Mammal Community Dynamics. Management and Conservation in the Coniferous Forests of Western North America.* Cambridge University Press, Cambridge, UK.

Martin, K.J. and W.C. McComb. 2003. Small mammals in a landscape mosaic: implications for conservation. Pages 567–586 *in* C.J. Zabel and R.G. Anthony, editors. *Mammal Community Dynamics. Management and Conservation in the Coniferous Forests of Western North America.* Cambridge University Press, Cambridge, UK.

Martinka, C.J. 1968. Habitat relationships of white-tailed and mule deer in northern Montana. *Journal of Wildlife Management* **32**:558–565.

McComb, W.C. 2003. Ecology of coarse woody debris and its role as habitat for mammals. Pages 374–404 *in* C.J. Zabel and R.G. Anthony, editors. *Mammal Community Dynamics. Management and Conservation in the Coniferous Forests of Western North America.* Cambridge University Press, Cambridge, UK.

McLaren, B.E. and R.O. Peterson. 1994. Wolves, moose, and tree rings on Isle Royale. *Science* **266**:1555–1558.

Messier, R. and M. Crete. 1985. Moose-wolf dynamics and the natural regulation of moose populations. *Oecologia* **65**:503–512.

Mills, L.S., M.K. Schwartz, D.A. Tallmon, and K.P. Lair. 2003. Measuring and interpreting connectivity for mammals in coniferous forests. Pages 587–613 *in* C.J. Zabel and R.G. Anthony, editors. *Mammal Community Dynamics. Management and Conservation in the Coniferous Forests of Western North America.* Cambridge University Press, Cambridge, UK.

Naiman, R.J., G. Pinay, C.A. Johnston, and J. Pastor. 1994. Beaver influences on the long term biogeochemical characteristics of boreal forest drainage networks. *Ecology* **75**:905–921.

Ostfeld, R.S., R.H. Manson, and C.D. Canham. 1997. Effects of rodents on survival of tree seeds and seedlings invading old fields. *Ecology* **78**:1531–1542.

Paine, R.T. 2000. Phycology for mammalogists. *Journal of Mammalogy* **81**:637–648.

Pollock, M.M., R.J. Naiman, H.E. Erickson, C.A. Johnston, J. Pastor, and G. Pinay. 1995. Beaver as engineers: influences on biotic and abiotic characteristics of drainage basins. Pages 117–126 *in* C.G. Jones and J.H. Lawton, editors. *Linking Species and Ecosystems.* Chapman and Hall, New York, New York, USA.

Power, M.E., D. Tilman, J.A. Estes, B.A. Menge, W.J. Bond, L.S. Mills, G. Daily, J.C. Castilla, J. Lubchenco, and R.T. Paine. 1996. Challenges in the quest for keystones. *Bioscience* **46**:609–620.

Sargeant, A.B., R.J. Greenwood, M.A. Sovada, and T.L. Schaffer. 1993. *Distribution and Abundance of Predators that Affect Duck Production: Prairie Pothole Region.* US Fish and Wildlife Service, Resources Publication 194.

Singer, F.J., G. Wang, and N.T. Hobbs. 2003. The role of ungulates and large predators on plant communities and ecosystem processes in national parks. Pages 444–486 *in* C.J. Zabel and R.G. Anthony, editors. *Mammal Community Dynamics. Management and Conservation in the Coniferous Forests of Western North America.* Cambridge University Press, Cambridge, UK.

Smith, W.P., R.G. Anthony, J.R. Waters, N.L. Dodd, and C.J. Zabel. 2003. Ecology and conservation of arboreal rodents of western coniferous forests. Pages 157–206 *in* C.J. Zabel and R.G. Anthony, editors. *Mammal Community Dynamics. Management and Conservation in the Coniferous Forests of Western North America.* Cambridge University Press, Cambridge, UK.

Soule, M.E., D.T. Bolger, A.C. Alberts, J. Wright, M. Sorice, and S. Hill. 1988. Reconstructed dynamics of rapid extinctions of chaparral-requiring birds in urban habitat islands. *Conservation Biology* **2**:75–92.

Terborgh, J., L. Lopez, P. Nunez, M. Rao, G. Shahabuddin, G. Orihuela, M. Riveros, R. Ascanio, G.H. Adler, T.D. Lambert, and L. Balbas. 2001. Ecological meltdown in predator-free forest fragments. *Science* **294**:1923–1926.

Vickery, P.D., M.L. Hunter, Jr., and J.V. Wells. 1992. Evidence of incidental nest predation and its effects on nests of threatened grassland birds. *Oikos* **63**:281–288.

Voth, E., C. Maser, and M. Johnson. 1983. Food habits of *Arborimus albipes*, the white-footed vole in Oregon. *Northwest Science* **57**:1–7.

Whitney, L.W. 2001. Ecological relationships between Columbia white-tailed and black-tailed deer in southwest Oregon. MS Thesis, Oregon State University, Corvallis, Oregon, USA.

Wolff, J.O. 2003. An evolutionary and behavioral perspective on dispersal and colonization of mammals in fragmented landscapes. Pages 614–630 *in* C.J. Zabel and R.G. Anthony, editors. *Mammal Community Dynamics. Management and Conservation in the Coniferous Forests of Western North America*. Cambridge University Press, Cambridge, UK.

Zabel, C.J. and J.R. Waters. 1997. Food preference of captive flying squirrels from the Lassen National Forest in northeastern California. *Northwest Science* **71**:103–107.

Index

Italics are used for *figures* and **bold** for **tables**.
The word *passim* indicates "here and there throughout".